Geophysical Monograph Series

Including
IUGG Volumes
Maurice Ewing Volumes
Mineral Physics Volumes

Geophysical Monograph Series

130 **Atmospheres in the Solar System: Comparative Aeronomy** *Michael Mendillo, Andrew Nagy, and J. H. Waite (Eds.)*

131 **The Ostracoda: Applications in Quaternary Research** *Jonathan A. Holmes and Allan R. Chivas (Eds.)*

132 **Mountain Building in the Uralides Pangea to the Present** *Dennis Brown, Christopher Juhlin, and Victor Puchkov (Eds.)*

133 **Earth's Low-Latitude Boundary Layer** *Patrick T. Newell and Terry Onsage (Eds.)*

134 **The North Atlantic Oscillation: Climatic Significance and Environmental Impact** *James W. Hurrell, Yochanan Kushnir, Geir Ottersen, and Martin Visbeck (Eds.)*

135 **Prediction in Geomorphology** *Peter R. Wilcock and Richard M. Iverson (Eds.)*

136 **The Central Atlantic Magmatic Province: Insights from Fragments of Pangea** *W. Hames, J. G. McHone, P. Renne, and C. Ruppel (Eds.)*

137 **Earth's Climate and Orbital Eccentricity: The Marine Isotope Stage 11 Question** *André W. Droxler, Richard Z. Poore, and Lloyd H. Burckle (Eds.)*

138 **Inside the Subduction Factory** *John Eiler (Ed.)*

139 **Volcanism and the Earth's Atmosphere** *Alan Robock and Clive Oppenheimer (Eds.)*

140 **Explosive Subaqueous Volcanism** *James D. L. White, John L. Smellie, and David A. Clague (Eds.)*

141 **Solar Variability and Its Effects on Climate** *Judit M. Pap and Peter Fox (Eds.)*

142 **Disturbances in Geospace: The Storm-Substorm Relationship** *A. Surjalal Sharma, Yohsuke Kamide, and Gurbax S. Lakhima (Eds.)*

143 **Mt. Etna: Volcano Laboratory** *Alessandro Bonaccorso, Sonia Calvari, Mauro Coltelli, Ciro Del Negro, and Susanna Falsaperla (Eds.)*

144 **The Subseafloor Biosphere at Mid-Ocean Ridges** *William S. D. Wilcock, Edward F. DeLong, Deborah S. Kelley, John A. Baross, and S. Craig Cary (Eds.)*

145 **Timescales of the Paleomagnetic Field** *James E. T. Channell, Dennis V. Kent, William Lowrie, and Joseph G. Meert (Eds.)*

146 **The Extreme Proterozoic: Geology, Geochemistry, and Climate** *Gregory S. Jenkins, Mark A. S. McMenamin, Christopher P. McKay, and Linda Sohl (Eds.)*

147 **Earth's Climate: The Ocean–Atmosphere Interaction** *Chunzai Wang, Shang-Ping Xie, and James A. Carton (Eds.)*

148 **Mid-Ocean Ridges: Hydrothermal Interactions Between the Lithosphere and Oceans** *Christopher R. German, Jian Lin, and Lindsay M. Parson (Eds.)*

149 **Continent-Ocean Interactions Within East Asian Marginal Seas** *Peter Clift, Wolfgang Kuhnt, Pinxian Wang, and Dennis Hayes (Eds.)*

150 **The State of the Planet: Frontiers and Challenges in Geophysics** *Robert Stephen John Sparks and Christopher John Hawkesworth (Eds.)*

151 **The Cenozoic Southern Ocean: Tectonics, Sedimentation, and Climate Change Between Australia and Antarctica** *Neville Exon, James P. Kennett, and Mitchell Malone (Eds.)*

152 **Sea Salt Aerosol Production: Mechanisms, Methods, Measurements, and Models** *Ernie R. Lewis and Stephen E. Schwartz*

153 **Ecosystems and Land Use Change** *Ruth S. DeFries, Gregory P. Anser, and Richard A. Houghton (Eds.)*

154 **The Rocky Mountain Region—An Evolving Lithosphere: Tectonics, Geochemistry, and Geophysics** *Karl E. Karlstrom and G. Randy Keller (Eds.)*

155 **The Inner Magnetosphere: Physics and Modeling** *Tuija I. Pulkkinen, Nikolai A. Tsyganenko, and Reiner H. W. Friedel (Eds.)*

156 **Particle Acceleration in Astrophysical Plasmas: Geospace and Beyond** *Dennis Gallagher, James Horwitz, Joseph Perez, Robert Preece, and John Quenby (Eds.)*

157 **Seismic Earth: Array Analysis of Broadband Seismograms** *Alan Levander and Guust Nolet (Eds.)*

158 **The Nordic Seas: An Integrated Perspective** *Helge Drange, Trond Dokken, Tore Furevik, Rüdiger Gerdes, and Wolfgang Berger (Eds.)*

159 **Inner Magnetosphere Interactions: New Perspectives From Imaging** *James Burch, Michael Schulz, and Harlan Spence (Eds.)*

160 **Earth's Deep Mantle: Structure, Composition, and Evolution** *Robert D. van der Hilst, Jay D. Bass, Jan Matas, and Jeannot Trampert (Eds.)*

161 **Circulation in the Gulf of Mexico: Observations and Models** *Wilton Sturges and Alexis Lugo-Fernandez (Eds.)*

162 **Dynamics of Fluids and Transport Through Fractured Rock** *Boris Faybishenko, Paul A. Witherspoon, and John Gale (Eds.)*

163 **Remote Sensing of Northern Hydrology: Measuring Environmental Change** *Claude R. Duguay and Alain Pietroniro (Eds.)*

164 **Archean Geodynamics and Environments** *Keith Benn, Jean-Claude Mareschal, and Kent C. Condie (Eds.)*

165 **Solar Eruptions and Energetic Particles** *Natchimuthukonar Gopalswamy, Richard Mewaldt, and Jarmo Torsti (Eds.)*

166 **Back-Arc Spreading Systems: Geological, Biological, Chemical, and Physical Interactions** *David M. Christie, Charles Fisher, Sang-Mook Lee, and Sharon Givens (Eds.)*

Geophysical Monograph 167

Recurrent Magnetic Storms: Corotating Solar Wind Streams

Bruce Tsurutani
Robert McPherron
Walter Gonzalez
Gang Lu
José H. A. Sobral
Natchimuthukonar Gopalswamy
Editors

American Geophysical Union
Washington, DC

Published under the aegis of the AGU Books Board

Jean-Louis Bougeret, Chair; Gray E. Bebout, Cassandra G. Fesen, Carl T. Friedrichs, Ralf R. Haese, W. Berry Lyons, Kenneth R. Minschwaner, Andrew Nyblade, Darrell Strobel, and Chunzai Wang, members.

Library of Congress Cataloging-in-Publication Data

Recurrent magnetic storms : corotating solar wind streams / Bruce
    Tsurutani ... [et al.], editors.
       p. cm. – (Geophysical monograph series, ISSN 0065-8448 ; 166)
  Includes bibliographical references.
  ISBN-13: 978-0-87590-432-0
  ISBN-10: 0-87590-432-7
    1. Magnetic storms.   2. Corotating interaction regions.   3. Magneto-
sphere.   4. Atmosphere, Upper.   I. Tsurutani, Bruce T.
  QC835.R37 2006
  538'.744--dc22

2006029672

  ISBN-13: 978-0-87590-432-0
  ISBN-10: 0-87590-432-7

ISSN 0065-8448

Copyright 2006 by the American Geophysical Union
2000 Florida Avenue, N.W.
Washington, DC 20009

Figures, tables and short excerpts may be reprinted in scientific books and journals if the source is properly cited.

Authorization to photocopy items for internal or personal use, or the internal or personal use of specific clients, is granted by the American Geophyscial Union for libraries and other users registered with the Copyright Clearance Center (CCC) Transactional Reporting Service, provided that the base fee of $1.50 per copy plus $0.35 per page is paid directly to CCC, 222 Rosewood Dr., Danvers, MA 01923. 0065-8448/06/$01.50+0.35.

This consent does not extend to other kinds of copying, such as copying for creating new collective works or for resale. The reproduction of multiple copies and the use of full articles or the use of extracts, including figures and tables, for commercial purposes requires permission from the American Geophysical Union.

Printed in the United States of America.

CONTENTS

Preface
Bruce T. Tsurutani, Robert L. McPherron, Walter D. Gonzalez, Gang Lu,
José Humberto A. Sobral, and Nat Gopalswamy i

Foreword
Sir Arthur C. Clarke ix

Magnetic Storms Caused by Corotating Solar Wind Streams
Bruce T. Tsurutani, Robert L. McPherron, Walter D. Gonzalez, Gang Lu, Nat Gopalswamy,
and Fernando L. Guarnieri 1

The Solar Wind: Then and Now
Joseph V. Hollweg 19

**The Role of Comet Tails in the Discovery of the Solar Wind and Its Spatial
and Temporal Variations**
D. A. Mendis 31

The Formation of CIRs at Stream-Stream Interfaces and Resultant Geomagnetic Activity
I. G. Richardson 45

**The Freestream Turbulence Effect in Solar-Wind/Magnetosphere Coupling: Analysis Through
the Solar Cycle and for Various Types of Solar Wind**
Joseph E. Borovsky and John T. Steinberg 59

**Modeling the Behavior of Corotating Interaction Region Driven Storms in Comparison
With Coronal Mass Ejection Driven Storms**
Vania K. Jordanova 77

Ring Current Behavior Inferred From Ground Magnetic and Space Observations
F. Søraas, M. Sørbø, K. Aarsnes, and D. S. Evans 85

**High-Speed Streams, Coronal Mass Ejections, and Interplanetary Shocks: A Comparative Study
of Geoeffectiveness**
G. Lu 97

**Energetics of Magnetic Storms Driven by Corotating Interaction Regions:
A Study of Geoeffectiveness**
Niescja E. Turner, Elizabeth J. Mitchell, Delores J. Knipp, and Barbara A. Emery 113

**The Solar Wind and Geomagnetic Activity as a Function of Time Relative to
Corotating Interaction Regions**
Robert L. McPherron and James Weygand 125

The Role of Radial Transport in Accelerating Radiation Belt Electrons
Xinlin Li 139

Mechanisms for the Acceleration of Radiation Belt Electrons
Richard B. Horne, Nigel P. Meredith, Sarah A. Glauert, Athina Varotsou, Daniel Boscher,
Richard M. Thorne, Yuri Y. Shprits, and Roger R. Anderson 151

Magnetospheric Energetics During HILDCAAs
W. D. Gonzalez, F. L. Guarnieri, A. L. Clua-Gonzalez, E. Echer, M. V. Alves, T. Ogino,
and B. T. Tsurutani 175

Energetic Neutral Atom Observations During Recurrent Magnetic Storms
J.-M. Jahn and H. A. Elliott... 183

Global Auroral Response to Interplanetary Media With Emphasis on Solar Wind Dynamic Pressure Enhancements
Kan Liou... 197

IMF B_y and the Spatio-Temporal Structure of the Dayside Aurora
P. E. Sandholt, C. J. Farrugia, E. J. Lund, and W. F. Denig... 213

The Nature of Auroras During High-Intensity Long-Duration Continuous AE Activity (HILDCAA) Events: 1998 to 2001
F. L. Guarnieri 235

Dayside Ionospheric (GPS) Response to Corotating Solar Wind Streams
B. T. Tsurutani, A. J. Mannucci, B. A. Iijima, A. Komjathy, A. Saito, T. Tsuda, O. P. Verkhoglyadova,
W. D. Gonzalez, and F. L. Guarnieri 245

A Statistical Study of Ionospheric Irregularities Observed With a GPS Network in Japan
Y. Otsuka, T. Aramaki, T. Ogawa, and A. Saito.. 271

Magnetic Storm Associated Disturbance Dynamo Effects in the Low and Equatorial Latitude Ionosphere
M. A. Abdu, J. R. de Souza, J. H. A. Sobral, and I. S. Batista... 283

Selected Upper Atmospheric Storm Effects
Gerd W. Prölss.. 305

Response of the Upper/Middle Atmosphere to Coronal Holes and Powerful High-Speed Solar Wind Streams in 2003
J. U. Kozyra, G. Crowley, B. A. Emery, X. Fang, G. Maris, M. G. Mlynczak, R. J. Niciejewski,
S. E. Palo, L. J. Paxton, C. E. Randall, P.-P. Rong, J. M. Russell III, W. Skinner, S. C. Solomon,
E. R. Talaat, Q. Wu, and J.-H. Yee... 319

PREFACE

This book reviews our current understanding of magnetic storms and geomagnetic activity that occur during the declining and the minimum phases of the solar (sunspot) cycle. The solar, interplanetary, magnetospheric and ionospheric processes that occur during these phases are substantially different from those that occur during the solar maximum phase. Such a collection of papers on this former topic does not presently exist in the literature; thus the pressing need for this book. The various chapters were chosen to cover relevant topics from the sun to the ionosphere/atmosphere, and experts in the field were selected to write them. It is hoped that this book will provide the necessary basics for readers to understand the fundamental physical processes occurring during the declining and minimum activity phases, so that they can launch into their own research.

In the declining and minimum phases of the solar activity cycle, the dominant causes of geomagnetic activity at Earth are related to high speed solar wind streams emanating from coronal holes. Storm main phases are caused by corotating interaction regions (CIRs) impinging on the Earth's magnetosphere. These magnetic storms are only moderate to weak in intensity. The storm "recovery" phases are unique in that they have exceptionally long durations, often lasting from many days to weeks. It is the southward magnetic field component of the embedded large-amplitude Alfvén waves within the high speed streams that lead to frequent and sporadic injections of the plasmasheet plasma into the nightside magnetosphere. This process is responsible for the unusually long magnetic storm "recoveries".

Although geomagnetic activity during the declining and minimum phases of the solar cycle appears to be relatively benign (especially in comparison to the dramatic and very intense magnetic storms caused by interplanetary coronal mass ejections (ICMEs) that predominate during solar maximum), this is misleading. The chapters in this book will demonstrate that the time-averaged, accumulated energy input into the magnetosphere and ionosphere due to high speed streams can be greater during these solar phases than due to ICMEs during solar maximum! This little known feature indicates that the declining and minimum phases may be far more important than previously recognized. Acceleration of relativistic magnetospheric electrons is a predominant feature during this part of the solar cycle, especially during preferred equinoctial intervals. These electrons may be more hazardous to Earth-orbiting satellites than ICME-related magnetic storm particles and solar energetic particles. Reviews on mechanisms to accelerate these electrons are contained in the monograph.

The following chapters give overviews and new information on phenomena that appear throughout the sun-Earth system as high speed streams blow by Earth. These individual phenomena and their underlying physical processes may at first appear to be isolated features, specific to limited spatial regions. However when several are linked together as a chain, explanations of fundamental response of the global system may be possible. One example at the end of such a chain is the destruction of ozone and its possible relation to climate variability. We give the following long line of related phenomena toward this process: solar coronal holes are a source of high speed streams/Alfvén waves which initiate magnetic reconnection at Earth and cause recurrent auroral activations, continuous sporadic plasma injections into the magnetosphere as well as MHD (Kelvin-Helmholz) and plasma wave instabilities (chorus and ion cyclotron waves). MHD and wave-particle interactions accelerate electrons to relativistic energies and also scatter the electrons into the loss cone. Both auroral precipitation and relativistic electron precipitation penetrate deep into the atmosphere, initiating a chain of chemical reactions that result in the production of NOx either directly in the stratosphere or at higher altitudes with subsequent transport down into the stratosphere by winds in the polar night. NO_X causes ozone destruction by the catalytic reactions: $NO + O_3 \rightarrow NO_2 + O_2$ and $NO_2 + O \rightarrow NO + O_2$. This destruction can affect the radiation balance of the atmosphere. Most of the above works/linkages have been performed by independent researchers without consideration that the macrophysical or microphysical processes they were studying were in any way linked to ozone. For a discussion of the end of this chain of related phenomena, we recommend the chapter by *Kozyra et al.* One of the lessons learned is that the study of diverse, seemingly unrelated topics can sometimes be important, even necessary, to understand fundamental, large-scale processes. Microscopic physics may lead to macroscopic chemistry.

We are honored by an introduction to the book given by Sir Arthur C. Clarke of Sri Lanka. Sir Arthur has been and is a visionary for space sciences and space travel. He was the first to suggest placing satellites in geostationary orbit, a feature that we presently use for communications and also for space plasma research (several chapters that follow use data from satellites in these orbits). He has discussed using the solar wind for space travel. Although NASA and ESA now contemplate a more efficient solar source for interplanetary travel (solar photons), Sir Arthur's initial idea led to this later development.

This monograph is a byproduct of an American Geophysical Union Chapman Conference held in Manaus, Amazonas, Brazil, 6 to 12 February, 2005. An accompanying special issue of the *Journal of Geophysical Research* entitled "Corotating Solar Wind Streams and Recurrent Geomagnetic Activity" contains research articles on the same subject. We wish to thank the Chapman Conference Program Committee for the selection of invited speakers and authors of chapters this book. They were: A. Balogh, W. Baumjohann, J. Burkepile, I. Daglis, C.-G. Falthammar, Y. Feldstein, A.J. Foppiano, N.J. Fox, T.J. Fuller-Rowell, A.L.C. Gonzalez, M. Grande, H. Hudson, Y. Kamide, J. Moen, G. Rostoker, P.-E. Sandholt, J.F. Spann, R.M. Thorne, V.M. Vasyliunas, and T. Zurbuchen

Bruce T. Tsurutani
Robert L. McPherron
Walter D. Gonzalez
Gang Lu
José Humberto A. Sobral
Nat Gopalswamy

FOREWORD

I was very interested to hear that you are organizing this conference to discuss new frontiers in research related to the solar wind. I would have attended if it were possible, but it is impossible for me to do so.

I based one of my better-known short stories on this phenomenon. When it was first published in 1963, I gave it the title *The Wind from the Sun*. This book is about a closely related idea whose time has yet to come. What a delightful irony it would be if the real age of sail has yet to dawn – not on the oceans of planet Earth, but in the far wider 'seas' of space…

I am delighted that NASA and ESA are now taking this seriously and studying this possibility. Some of you might still be around to see this dream become a reality in the coming decades.

I realize that your meeting is not about designing solar sails and navigation in the solar wind. It is about understanding the sources of the solar wind and being able to predict its properties. There are also strong consequences for its variability here on Earth. I therefore view your Conference as a first step in being able to predict the solar wind, allowing this new frontier of space travel to blossom fully. I commend your efforts and wish you success.

Sir Arthur C. Clarke
Colombo Sri Lanka
February 2005

(Adapted from the welcome message sent to the Chapman Conference on "Corotating Solar Wind Streams and Recurrent Geomagnetic Activity", Manaus, Amazonas, Brazil, February 2005)

Magnetic Storms Caused by Corotating Solar Wind Streams

Bruce T. Tsurutani[1,2], Robert L. McPherron[3], Walter D. Gonzalez[4],
Gang Lu[5], Nat Gopalswamy[6], and Fernando L. Guarnieri[4]

Geomagnetic activity at Earth due to corotating high speed solar wind streams are reviewed. High density plasma regions in the vicinity of the heliospheric current sheet in the slow solar wind impinge upon the magnetosphere and create magnetic storm intial phases. Dst increases can be higher than those associated with shocks in front of interplanetary coronal mass ejections. High speed streams following the high density interplanetary plasma interact with the upstream slow speed streams and create magnetic field compression regions called corotating interaction regions (CIRs). The southward components of the typically rapid Bz fluctuations within the CIRs, through sporadic magnetic reconnection with the Earth's magnetic fields, lead to weak to moderate intensity magnetic storm main phases, typically Dst >-100 nT. Some CIRs (without southward component Bz fields) cause no perceptible Dst changes at all. The "recovery" phases of CIR-induced magnetic storms can last for a few days up to 27 days. The cause of these particularly long duration storm "recoveries" is near-continuous shallow plasma injections into the magnetosphere. Magnetic reconnection associated with the southward component of the Alfvén waves within the high speed streams with magnetopause fields is the cause of these injections. The auroras during these intervals are continuous and global auroral zone features. The AE/AL maxima are not substorms or convection bays. Relativistic electrons are accelerated/observed during these high speed stream intervals. The electrons first appear in the beginning of the lengthy storm "recovery" phases. Geomagnetic quiet is due to weak interplanetary magnetic fields with a lack of Alfvénic fluctuations. These interplanetary regions generally occur in the decay portion of high speed streams.

[1]Jet Propulsion Laboratory, Pasadena, CA
[2]RISH, Kyoto University, Uji, Japan
[3]University of California at Los Angeles, CA
[4]Instituto Nacional Pesquisas Espaciais (INPE), Sao Jose dos Campos, Sao Paulo, Brazil
[5]High Altitude Observatory, Boulder, CO
[6]Goddard Space Flight Center, Greenbelt, MD

Recurrent Magnetic Storms: Corotating Solar Wind Streams
Geophysical Monograph Series 167
Copyright 2006 by the American Geophysical Union.
10.1029/167GM03

INTRODUCTION

During the declining phase of the solar cycle (and solar minimum), the solar and interplanetary causes of geomagnetic activity are considerably different from those during the solar maximum phase (*Gonzalez et al.*, 1994; *Tsurutani and Gonzalez*, 1997). Corotating, fast solar wind streams emanating from enlarged polar coronal holes (*Krieger et al.*, 1973) sweep past the Earth's magnetosphere every solar rotation, ~27 days (*Sheeley et al.*, 1976; *Tsurutani et al.*, 1995). At the leading edge of the streams, there are regions called corotating interaction regions or CIRs, which are compressed magnetic field and plasma regions created by the fast stream interactions with the upstream slow speed streams (*Smith and Wolf*, 1976; *Pizzo*, 1985; 1991; *Balogh et al.*, 1999; *Richardson*, this issue). The sporadic southward magnetic field components of these CIRs are responsible for weak-to-moderate intensity magnetic storms, presumably through magnetic reconnection between the CIR magnetic fields and the Earth's magnetopause magnetic fields (*Dungey*, 1961).

The high speed solar wind streams that trail the CIRs are also involved in transferring energy to the magnetosphere/magnetotail, but at a lower intensity level than either magnetic storms or substorms. One specific mechanism of solar wind energy transfer is magnetic reconnection associated with the southward components of large amplitude interplanetary Alfvén waves. These Alfvén waves are ever-present in high speed corotating streams and can reach relative amplitudes of $|\mathbf{b}|/|\mathbf{B}| \sim 1$ to 2, where $\mathbf{b}$ is the peak-to-peak wave amplitude and $|\mathbf{B}|$ is the ambient field magnitude (*Tsurutani et al.*, 1994; *Balogh et al.*, 1995). When there is significant southward components in the Alfvén waves, the corresponding geomagnetic activity is typified by near-continuous, intense AE activity, slightly depressed Dst indices and global UV auroras often covering the entire dayside and nightside auroral zones (*Guarnieri*, 2005). This activity has been named High-Intensity Long-Duration Continuous AE Activity (HILDCAA; *Tsurutani and Gonzalez*, 1987) and can last for days to many weeks, essentially as long as the high speed streams and large amplitude Alfvén waves are present.

Clearly other forms of solar wind energy injection into the magnetosphere/magnetotail system are occurring, although quantitative estimates of the magnitudes/relative magnitudes have not been made. The Kelvin Helmholtz instability (*Parker*, 1958; *Chen and Hasegawa*, 1974; *Southwood*, 1974; *Farrugia et al.*, 2001) due to the flow of plasma along the flanks of the magnetosphere is an obvious energy transfer source. Small, sporadic solar wind ram pressure pulses caused by small spatial scale enhanced plasma density increases (magnetic decreases or MDs) present at the edges of interplanetary Alfvén waves (*Tsurutani et al.*, 2002) could generate magnetopause/outer magnetospheric Alfvén waves, which could then propagate to the auroral zones, transferring energy to ionospheric altitudes.

Relativistic electron acceleration (*Paulikas and Blake*, 1979) occurs during the high speed stream proper (*Horne et al.*, this issue). This radiation can be damaging to spacecraft orbiting the Earth (*Wrenn*, 1995). There are several proposed mechanisms for this electron acceleration. The two dominant mechanisms are radial diffusion associated with PC 5 electromagnetic waves (*Hudson et al.*, 1999; *Li and Temerin*, 2001, *O'Brien et al.*, 2001; *Elkington et al.*, 2003; *Li*, this issue) and energy diffusion associated with electromagnetic whistler mode chorus (*Horne et al.*, 1998; 2003; this issue).

It should be noted that the former wave mode is expected to be generated by the Kelvin Helmholtz instability and the latter waves by electron instabilities (*Kennel and Petschek*, 1966; *Tsurutani and Smith*, 1974; 1977) associated with plasma injections into the magnetosphere (due to magnetic reconnection).

We will begin with a discussion of the interplanetary causes of geomagnetic activity associated with upstream slow speed streams and corotating high speed solar wind streams. The basic features of magnetic storms generated by interplanetary coronal mass ejections (ICMEs) will also be reviewed to illustrate the similarities and differences (mostly differences) between the two types of magnetic storms. We will conclude by discussing some of the outstanding space weather problems in the hope of stimulating young researchers to enter this field.

RESULTS

Profiles of a typical magnetic storm caused by an ICME and a typical magnetic storm caused by a CIR-high speed solar wind stream are given in Figure 1. The two different types of magnetic storms qualitatively look the same: both have storm initial phases, both have main phases, and both

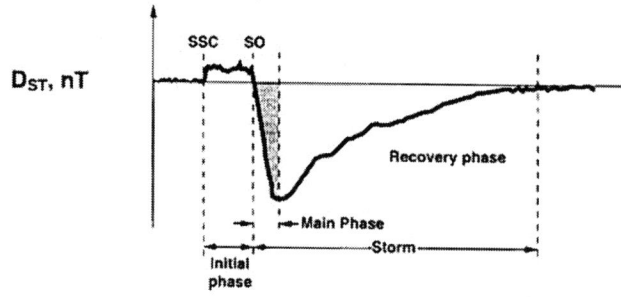

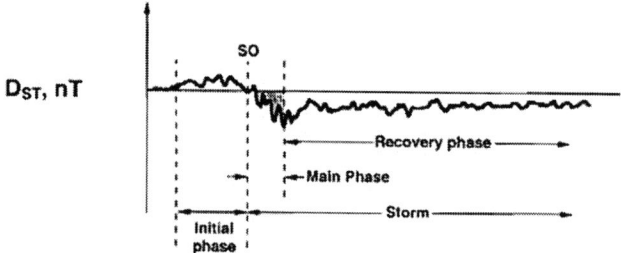

Figure 1. Schematic of a magnetic storm generated by an ICME (top) and by a CIR (bottom). Although the profiles of these two different magnetic storms are qualitatively similar, the physical causes and characteristics of the different storm phases are different. The figure is taken from *Tsurutani* (2000).

have recovery phases. However, there the similarities end. The scale sizes of the different storm phases, both in amplitude and duration, are in fact quite different. Most importantly, the interplanetary causes are quite distinct from each other.

Initial Phase

The magnetic storm initial phase is denoted by a positive increase in the Dst index. The ICME-related magnetic storm initial phases typically start abruptly. The increases in the horizontal component of near-equatorial magnetic fields at Earth can occur on time scales of seconds. These abrupt jumps are caused by sudden ram pressure increases impinging upon the magnetosphere. The ram pressure increases are due to solar wind plasma density and velocity increases at and immediately following fast forward shocks upstream of interplanetary coronal mass ejections (ICMEs). For a discussion of fast shock plasma and magnetic field jump conditions, we refer the reader to *Kennel et al.* (1985). These sudden increases in the Earth's near-equatorial magnetic field have been called Sudden Impulses (SIs) or Storm Sudden Commencements (SSCs). We refer the reader to *Kikuchi and Araki* (1979) for further discussion. The term SSC is used only when the SIs are followed by magnetic storms. However it is known that the physics for both are the same (see discussion in *Joselyn and Tsurutani*, 1990). The increases in the magnetic fields during ICME storm initial phases are often relatively constant. This is due to the relatively constant ram pressures in the interplanetary sheaths following the ICME shocks.

In contrast, the initial phases associated with CIR-related magnetic storms typically have gradual onsets. There are no SIs or SSCs at the beginning of the events. This is caused by the general lack of shocks preceding corotating streams (at 1 AU). The horizontal component of the Earth's magnetic field during the initial phases are often highly variable, indicating variable interplanetary ram pressures. We will show later that this phase of CIR-related magnetic storms are not caused by the high speed streams, but by variable plasma densities in the slow speed solar wind streams ahead of (antisunward of) the high speed streams.

Main Phase

Magnetic storm main phases are caused by the injection of energetic particles into the magnetosphere by large-scale, intense magnetospheric dawn-to-dusk electric fields (*Gonzalez et al.*, 1994). The diamagnetic effect of the ring current particles associated with gradient and curvature drifts of protons, oxygen ions and electrons, cause decreases in the horizontal component of the near-equatorial magnetic field (*Dessler and Parker*, 1959; *Sckopke*, 1966).

The Dst index is constructed from ground-based near-equatorial stations and responds to a number of external (to the Earth) currents: the magnetopause Chapman-Ferraro current, the magnetospheric ring current (see above), field-aligned magnetospheric-ionospheric currents and the magnetotail currents. During magnetic storms, the Dst index is influenced by all of the above currents. Although it is generally agreed that ring current is the dominant influence, other points of view have been expressed in the literature. We refer the reader to a series of articles on this important and highly debated issue (*Baker et al.*, 2001; *Kozyra et al.*, 2002, *Daglis et al.*, 2003 and *Feldstein et al.*, 2005).

The main phases of ICME-related magnetic storms generally develop smoothly, i.e., Dst decreases monotonically with time. The interplanetary causes are well known (see review in *Gonzalez et al.*, 1994). The storm main phases are either caused by step-like decreases in the southward component of the magnetic field across the shocks antisunward of the ICMEs (*Tsurutani et al.*, 1988), or by the southward magnetic fields within magnetic clouds (*Klein and Burlaga*, 1982; *Tsurutani et al.*, 1988; *Farrugia et al.*, 1997). Magnetic clouds are the most geoeffective portions of the ICMEs. In cases where there are southward sheath fields followed by magnetic clouds with southward component magnetic fields, then "two-step" magnetic storm main phases will result (*Kamide et al.*, 1998).

The intensity of magnetic storm main phases caused by ICMEs can be from as little as Dst = -25 nT to as large as minus hundreds of nT. The duration can be as short as a few hours, and in exceptional cases, can be as long as a day. Almost all major (Dst <-100 nT) magnetic storms are caused by magnetic clouds within ICMEs or the upstream sheath fields (see discussion in companion JGR special issue paper by *Tsurutani et al.*, 2006).

Magnetic storm main phases caused by CIRs are highly irregular in profile. They typically last ~a day and they are generally negligible (Dst > -25 nT) to weak-to-moderate (-25 nT $>$ Dst > -75 nT) in intensity. They typically do not reach intensities of Dst <-100 nT. The cause of the storm main phases is the southward component of the interplanetary magnetic field (IMF) Bz within CIRs. Because the magnetic field directionality is highly variable within CIRs, the storm intensities are only weak-to-moderate, and sometimes negligible. Examples will be shown later.

Recovery Phase

The recovery phases of magnetic storms are the decay of the depressed Dst indices back to the ambient quiet time values. After ICME-generated magnetic storm main phases, the losses of the energetic protons, oxygen ions and electrons forming the ring current cause the decrease of the diamagnetic current and thus an increase in the Dst index to quiet-time values.

There are four different ring current particle loss mechanisms: wave-particle interactions, Coulomb collisions, charge exchange, and magnetospheric convection through the dayside magnetopause (*Kozyra et al.*, 2002; *Jordanova*, this issue). Coulomb collision and charge exchange loss processes depend on individual particle pitch angles, kinetic energies and L shell location. Convection losses are, of course, dependent on magnetospheric convection electric fields, magnetospheric shielding, and the temporal profile of the onset and termination of the electric fields. Wave-particle losses depend on resonant wave intensities. If the waves are generated by particles of different energies or species (via plasma instabilities), then the loss process will ultimately depend on the properties of those instabilities and the propagation of the waves from the generation region to the interaction regions. Each of the above processes has different time-scales. Thus from the above discussion, it is clear that the recovery phase of magnetic storms is quite complex, and there is not a single number for the time scales of this phase of magnetic storms However for purposes of modeling magnetic storms, researchers have used a collective loss time of ~7 to 10 hrs.

In sharp contrast, the "recovery" phases of CIR-generated storms are particularly long. They can last for days to many weeks. It will be shown that the term "recovery" in this case is a misnomer, from a physical point of view. There are fresh particle injections occurring continuously during this phase of storms, and the ring current is not simply just "recovering".

Coronal Holes, High Speed Solar Wind Streams, and Nonlinear Alfvén Waves

Coronal Holes are "dark" regions of the corona, when viewed in soft x-rays. This is where their name comes from. Just after solar maximum, coronal holes form at the poles of the sun and expand in size with time. They have their maximum size in the declining phase of the solar cycle and at solar minimum (*Harvey et al.*, 2000).

The source of high speed solar wind streams are these coronal holes (*Krieger et al.*, 1973; *Balogh et al.*, 1999; *Guarnieri*, 2005). The speeds of the streams are a relatively constant ~750 to 800 km/s (except at the edges of the streams where there are gradients). The mechanism for the strong acceleration of coronal plasma up to these high speeds is currently being debated (see *Hollweg and Isenberg*, 2002; *Suzuki and Inutsuka*, 2005; *Hollweg*, this issue). If an asymmetric "finger" of a polar coronal hole comes down to low enough heliographic latitudes, the plasma and magnetic fields of the high speed stream will impact the Earth's magnetosphere. If the coronal hole also has a long duration, then the high speed stream will impact the Earth once every solar rotation period, or once every ~27 days. The solar wind stream will thus appear to "corotate". This is the cause of the "recurrent geomagnetic activity" that was noted by Bartels many years ago.

High speed solar wind streams can also come from isolated near-equatorial coronal holes. However their durations may be considerable shorter and therefore "recurrent" geomagnetic activity may not take place.

High speed streams contain large amplitude magnetic field oscillations. These oscillations are in fact Alfvén waves which are propagating outward from the sun. The solar wind speeds are much faster than the Alfvén wave phase speeds (by a factor of ~5 to 10), so the waves are primarily convected (outward) by the solar wind (*Belcher and Davis*, 1971).

Figure 2 shows the three orthogonal components of the solar wind velocity and magnetic field for a Ulysses pass over the sun's south pole. The plasma and magnetic field are taken by instrumentation from the satellite. The coordinate system is the heliospheric **rtn** system where **r** is the radial direction from the sun towards the spacecraft, $\mathbf{t} = (\hat{\Omega} \times \mathbf{r})/|\hat{\Omega} \times \mathbf{r}|$, where $\hat{\Omega}$ is the north rotation pole of the sun, and $\hat{n}$ completes a right-hand coordinate system. In the sixth panel from the top, it can be noted that the $\mathbf{B_n}$ magnetic field component oscillates between +1.0 and −1.0 nT. The magnetic field magnitude is ~1.2 nT (second from the bottom panel). Thus the Alfvén waves have peak-to-peak amplitudes of ~1 to 2 times the ambient magnetic field. Detailed discussion of Alfvén waves can be found in *Tsurutani et al.* (2005) with evidence that these may be intermediate shocks. During the polar pass, Ulysses was at a heliospheric distance of ~2.0 AU, and therefore the magnetic field was weaker (than at 1 AU) due to the radial expansion of the solar wind to this distance. At 1.0 AU the relative wave amplitude to the magnetic field strength is comparable (this will be shown later).

The next to the bottom panel shows the magnetic field magnitude. There are large field decreases that appear as downward spikes in the plot. These are "magnetic holes" (MHs) or "magnetic decreases" (MDs) (*Turner et al*, 1977; *Tsurutani and Ho*, 1999; *Franz et al.*, 2000; *Neugebauer et al.*, 2001; *Tsurutani et al.*, 2005) that have been noted to occur at the edges of the Alfvén waves (*Tsurutani et al.*, 2002).

Winterhalter et al. (1994) have shown that these MHs/MDs are pressure balance structures. Since the perpendicular temperatures of the protons within MHs/MDs are only 10-20% higher inside the MHs/MDs than outside the structures (*Tsurutani et al.*, 2002), and the magnetic decreases can be as large as 90% of the ambient magnetic field, the plasma densities inside the MHs/MDs may be substantially higher than that of the ambient solar wind. The structures are flowing at the same speed as the ambient solar wind, or are convected structures (*Tsurutani et al.*, 2005). The impact of MHs/MDs onto the Earth's magnetosphere can be significant ram pressure increases (up to a factor of ~2 for an ambient plasma β of ~1).

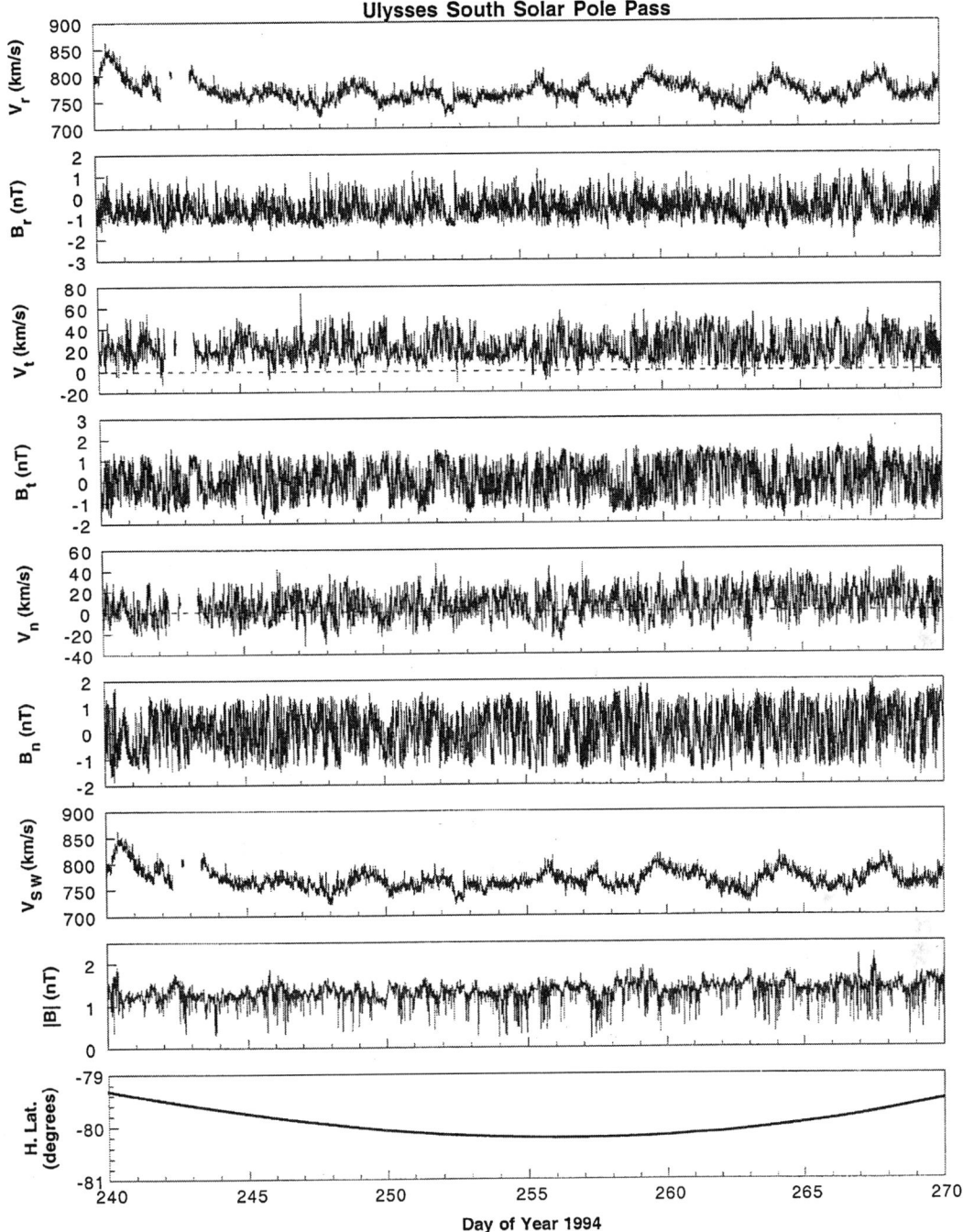

Figure 2. Large amplitude Alfvén waves in the high speed solar wind. Note the large Bn fluctuations. Although these are displayed in the rtn coordinate system, there will be large Bz fluctuations in the GSM coordinate system (not shown). The latter are important for geomagnetic activity. The decreases in the magnetic field magnitude (Magnetic Decreases, or MDs) shown in the next to last panel may also be important for geomagnetic activity.

Stream-Stream Interactions, Corotating Interaction Regions, and Magnetic Storm Main Phases

An example of an interaction between a high speed stream and an upstream slow speed stream is shown in Figure 3. The solar wind speed is given in the top panel. The high speed stream is present on the right, and the slow, ~300 km/s speed stream on the left. A compression in plasma and magnetic field is present from ~00 UT day 177 (June 26) 1974 until ~01 UT day 178 (June 27) 1974. This can be noted in

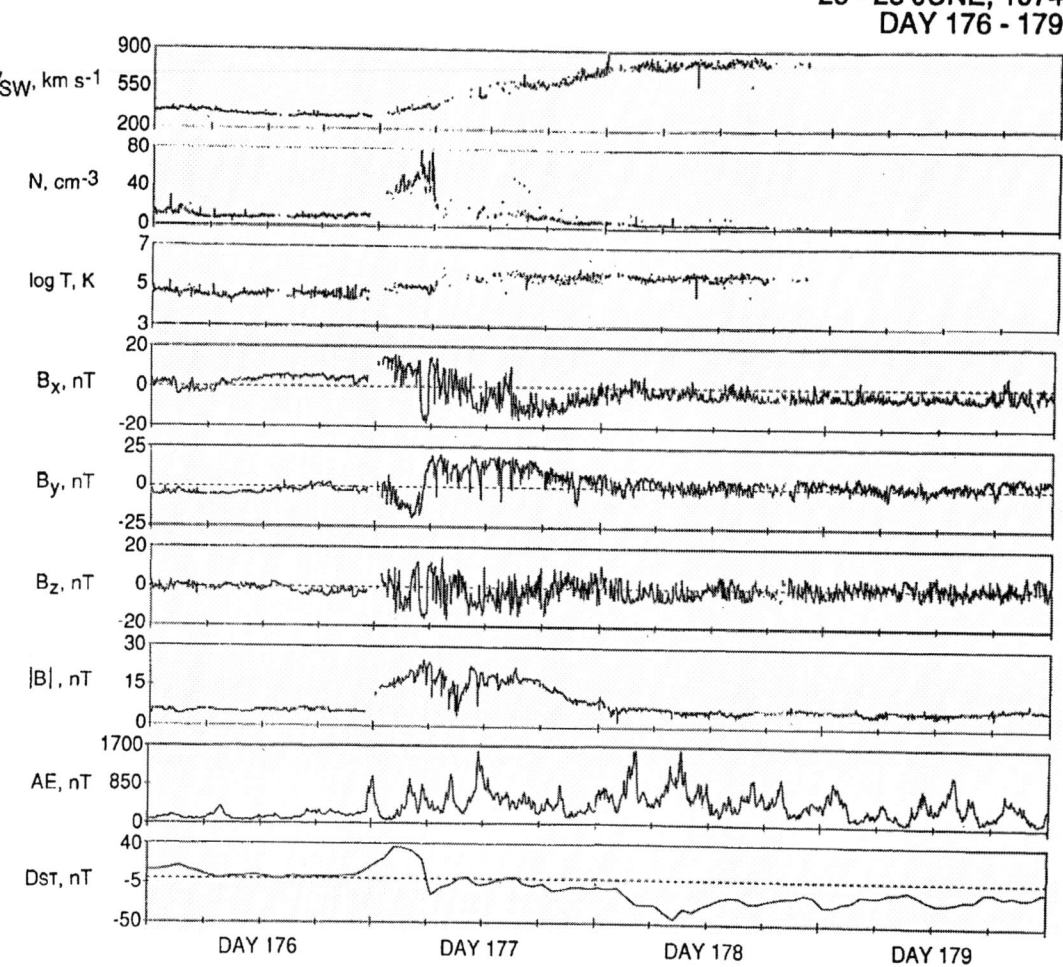

Figure 3. The heliospheric plasmasheet (HPS), a CIR, high speed stream proper in IMP-8 interplanetary data just upstream of the Earth. The above three interplanetary features are responsible for the magnetic storm initial phase, main phase and "recovery" phase, respectively. The figure has been taken from *Tsurutani et al.* (1995).

the plasma densities (second panel) and particularly in the magnetic field magnitudes (third from bottom panel). The compressed magnetic field magnitude region is the CIR. The beginning and end of the CIR are not well defined, however. This is typical of the situation at 1 AU. At larger distances from the sun, say ~2 AU, CIRs are typically bounded by fast forward and fast reverse shocks (*Balogh et al.*, 1999). This boxcar-like configuration of the magnetic field magnitude led to the first identification of these structures (*Smith and Wolf*, 1976). For further discussion of the formation and evolution of CIRs, we refer the reader to *Pizzo* (1985), *Pizzo et al.* (1991) and *Richardson* (this issue).

It is (typically) the southward component of the Bz fluctuations within the CIRs that are responsible for the main phases of the resultant magnetic storms (*Tsurutani et al.*, 1995). In this case the Bz fluctuations created only a weak

Dst <–10 to –25 nT storm main phase (the effects of enhanced ram pressure have not been removed). The maximum negative Dst of –45 nT occurs outside of the CIR, at ~09 UT day 178 where the ram pressure has been significantly reduced.

The plasma densities are given in the second panel of Figure 3 and the GSM Bx and By components in the fourth and fifth panels. In this coordinate system, **x** is the direction from the Earth to the sun, **y** is equal to $(\hat{\Omega}_m \times \mathbf{x})/|\hat{\Omega}_m \times \mathbf{x}|$, where $\hat{\Omega}_m$ is the south magnetic pole of the Earth, and **z** completes a right-hand coordinate system. There is a high density plasma "plug" at the beginning of the CIR, from ~00 UT to 05 UT day 177. The By component is negative prior to ~06 UT day 177 and positive afterward. The Bx component is positive prior to this time and has a temporary large negative decrease at the time that the By component reverses sign.

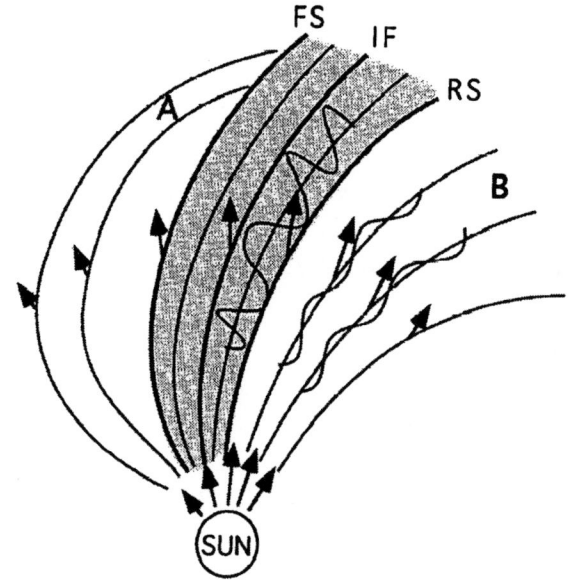

Figure 4. A schematic of the formation of a corotating interaction region due to a fast stream-slow stream interaction. A forward (fast) shock (FS), interface discontinuity (IF), and reverse (fast) shock (RS) are indicated. Alfvén waves in the fast solar wind stream are amplified by reverse shock compression. Shocks typically do not form until ~1.5 to 2.0 AU distance from the sun.

After ~12 UT day 177, Bx remains negative. This "switch" in the By and Bx component polarities is most likely a heliospheric current sheet (HCS) crossing. HCSs occur in the slow solar wind and are characterized by high plasma densities (plasmasheets) surrounding them (*Winterhalter et al.*, 1994). Thus the high density plasma plug at the beginning of the CIR can be interpreted as the heliospheric plasmasheet (HPS) that has been swept up by the high speed stream-slow speed stream interaction.

The bottom two panels are the AE and Dst indices. There is a positive increase in Dst to values approaching ~+40 nT, coincident with impingement of the high density plasma plug onto the magnetosphere. It should be noted that this plug is present in the slow ~300 to 350 km/s solar wind. The positive increase in Dst is the initial phase of the magnetic storm.

A schematic of a CIR is shown in Figure 4. The view is from the north pole of the sun and the cut is in the ecliptic plane. The CIR is indicated by shading. The "boundaries" are the fast forward and fast reverse shocks (from previous discussions, shocks are not typically formed at the boundaries of CIRs by 1 AU). The plasma and fields within CIRs are not from the same solar wind source. The outermost (antisunward) portion of the CIR is compressed, accelerated slow speed solar wind plasma and fields. The innermost (sunward) portion is compressed, decelerated fast solar wind plasma and fields. There is an "interface" separating the two regions. See further discussion on these topics in *Richardson et al.* (this issue).

High Speed Streams, Alfvén Wave Bz Fluctuations and Resultant Geomagnetic Activity

Four days of a high speed solar wind stream are shown in Figure 5. From top to bottom are the solar wind speed, the proton densities, the magnetic field magnitudes, the GSM Bz components (the southward components are shaded), and the AE and Dst indices. At the far left of the Figure, from 00 UT to ~02 UT day 135, is a trailing portion of a CIR. During this four day interval the solar wind speed increased from ~350 km/s to over 600 km/s. This can be interpreted as being the boundary of a high speed solar wind stream.

The magnetic field magnitude is ~8 nT at the beginning of the interval (outside of the CIR) and decreases to ~6 nT by day 139. The Bz fluctuations on days 135 to 137 are large, reaching ~+8 and ~−8 nT. Similar to the Ulysses results shown previously, the Alfvén wave amplitudes are ~1 to 2 times the magnetic field magnitude.

It can be noted in the Figure that the southward components of the IMF Bz fields are extremely well correlated with both AE increases and Dst decreases. For each major negative Bz interval (southward fields), there is an AE increase and also a Dst decrease. One obvious interpretation of this relationship is that the southward IMF Bz components lead to magnetic reconnection, plasma injection into the magnetosphere (Dst decreases) and auroral electrojet intensifications (AE increases).

The above hypothesis was tested using NOAA satellite particle data (*Soraas et al.*, 2004; 2005; this issue). Some results are shown in Figure 6. It is found that the proton injections during these high AE intervals were quite shallow. Particles were essentially injected to a minimum distance of L = 4.0. Deeper injections were not found during these High Intensity Long Duration Continuous AE Activity (HILDCAA) events.

What Is the Nature of the Geomagnetic Activity Associated With the Alfvén Waves During High Speed Streams?

The cause of the AE increases during high speed streams is explored in Plates 1 and 2. Plate 1 shows the nightside sector auroral intensities in the Lyman-Birge-Hopfield (LBH) long (~160 to 180 nm) UV wavelength band. The images were taken by the UVI instrument onboard the Polar satellite. The three panels are representative midnight sector auroras for a magnetic storm, a substorm and a HILDCAA event. The auroral intensities are greater for magnetic storms than for substorms, a feature that is well known. What is interesting from the figure is that the auroral intensities for HILDCAAs

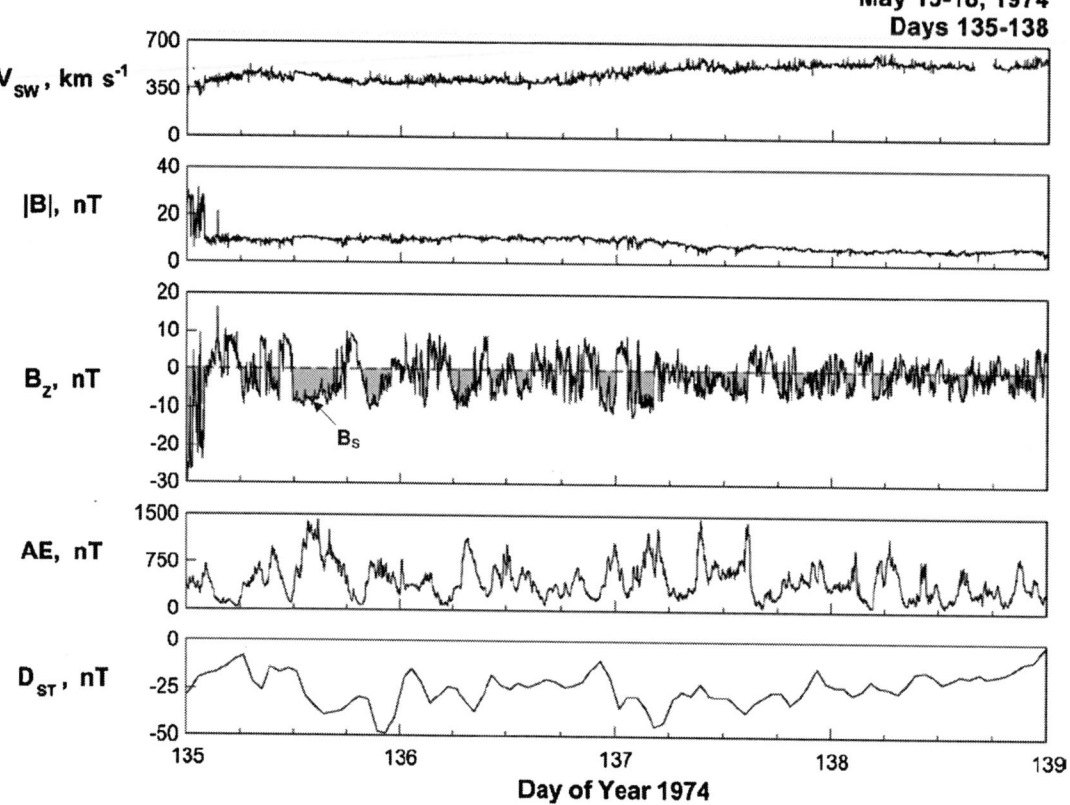

Figure 5. A magnetic storm "recovery" phase following a CIR-induced magnetic storm main phase. There is a one-to-one relationship between negative IMF B_Z intervals (shaded), Dst decreases and AE increases.

are considerably lower than for substorms. This is discussed in detail in *Guarnieri* (2005).

Plate 2 shows another feature of HILDCAA auroras. HILDCAA auroras are noted to be global phenomenon, i.e., auroras are present at all local times along the auroral oval. At times, auroras also cover the whole polar region as well. Further discussion of HILDCAAs possibly being a new type of geomagnetic activity is discussed in *Guarnieri* (2005), *Gonzalez et al.* (this issue), and *Tsurutani et al.* (2006).

The solar wind parameters and the Dst indices are shown for the entire year of 1974 in Figure 7. The interval from 1973 through 1975 was an interval of exceptionally long-lasting corotating solar wind streams, and 1974 was the most geoeffective in terms of annual AE average of the three years (*Tsurutani and Gonzalez*, 1997). There has not been an interval of this type since that time. From top to bottom are: the IMF Bz components, the Dst indices, the solar wind velocities and the magnetic field magnitudes. At the very top of the Figure above the Bz panel are the interplanetary magnetic field polarities. By convention, positive magnetic field sectors are those where the IMF is directed outward from the sun and negative sectors are those where the field points inwards toward the sun. The bars separating the two magnetic polarities are the "sector boundaries". This term has now been replaced by another, more physical term, the "heliospheric current sheet" (HCS) crossings. The original name "sector boundary" was used to describe the two or four (or six) "polarity sectors" that were typically noted during a solar rotation (*Wilcox and Scherrer*, 1972). However the term "heliospheric current sheet" is a better physical description of the three dimensional "ballerina skirt" configuration (*Schulz*, 1973; *H. Alfvén*, personal comm., 1978; *Smith et al.*, 1978) of the interplanetary magnetic fields.

In Figure 7, there are three magnetic storms labeled events A, B and C. The storms are identified as the only "major" magnetic storms (intensities were Dst <-100 nT) that occurred during the year. The A event was the largest, with an intensity of Dst $= -169$ nT. All three events were examined in detail and were found to be caused by fast ICMEs and their upstream shocks/sheaths, and not by corotating solar wind streams/CIRs (*Tsurutani et al.*, 1995). We will not discuss these events further, other than to note the paucity of "major"

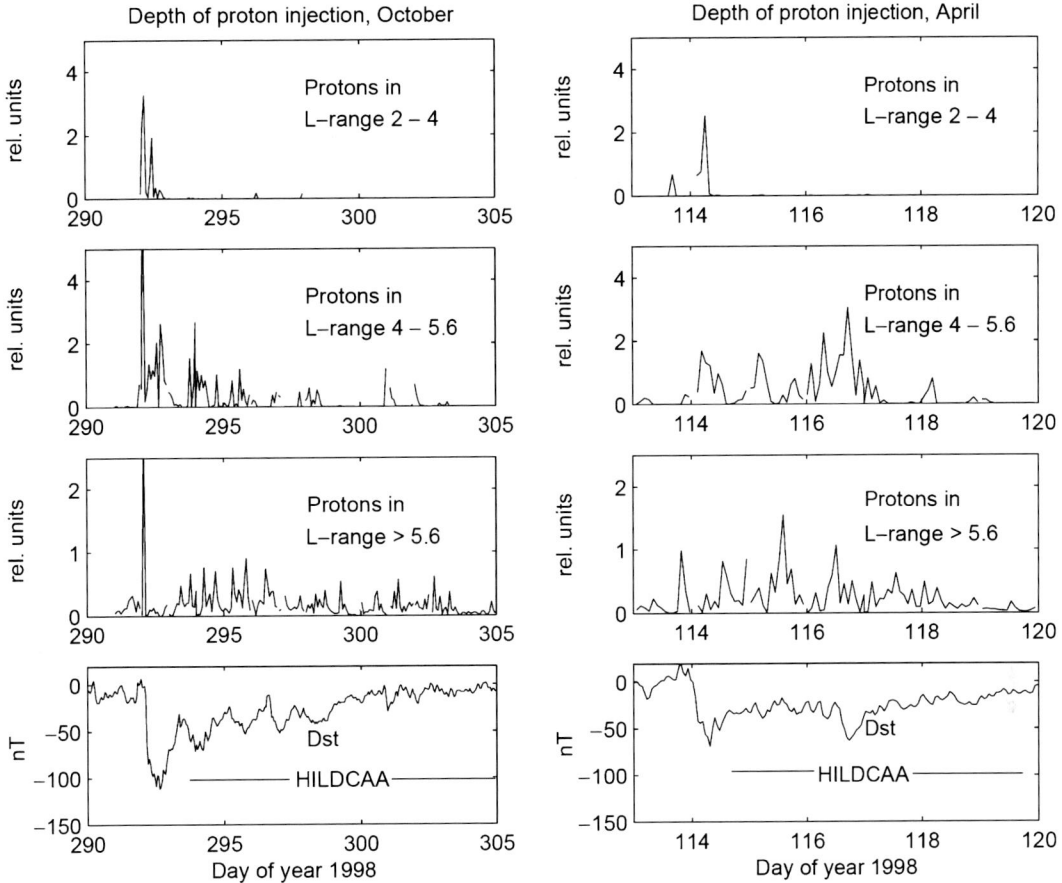

Figure 6. NOAA energetic particle data during two 1998 HILDCAA events. Protons are injected into the nightside magnetosphere to distances L ≥ 4.0 during these events. The Figure was taken from *Soraas et al.* (2004).

magnetic storms during this year. As a comparison, in the solar maximum years 2001 to 2003, there were an average of 15 major magnetic storms per year. This decrease of a factor of ~5 from the solar maximum years to the declining phase is typical (*Tsurutani et al.*, 2006).

In the first half of 1974 there were two clear solar wind corotating streams. These have been labeled streams 1 and 2, denoted in the velocity panel. Each stream recurred with a nearly clock-like repetition of ~27 days. In this case, the two streams were associated with "fingers" of polar coronal holes, one from each hemisphere (see *Sheeley et al.*, 1976; 1977).

In Figure 7, there are many positive magnetic field magnitude spikes, noted in the bottom panel. Two examples are on day 14 and 24. These are CIRs that occur at the leading edges of the sequence 2 and sequence 1 streams, respectively. There are many more events of this type. Just prior to the day 14 magnetic spike, there is a HCS crossing earlier on day 14 and a positive Dst increase on day 13-14. The high plasma densities near the HCS causes the storm initial phase when the structure impinges onto the magnetosphere (see also *Tsurutani and Gonzalez*, 1997; *Tsurutani et al.*, 2006). There are many more of these examples present within this figure.

The numerous small Dst decreases in the Figure indicate the occurrence of many small (minor) magnetic storms. Each small magnetic storm starts with an initial phase. Most of the magnetic storm main phases are associated with CIRs (magnetic field magnitude spikes) (other than the ICMEs discussed earlier).

In the first half of the year, it is clear that there are many more high speed streams than there are magnetic storms. Thus many CIRs have negligible effects on geomagnetic activity.

The storm "recovery" phases sometime last almost a full solar rotation, or through two corotating streams. Some good examples are from day ~80 to ~101 and from ~day 107 to ~day 131. The stream 1 CIRs cause the onsets of the storm main phases. The second stream, stream 2, leads to the continuation of the long storm recovery phase. The CIR of

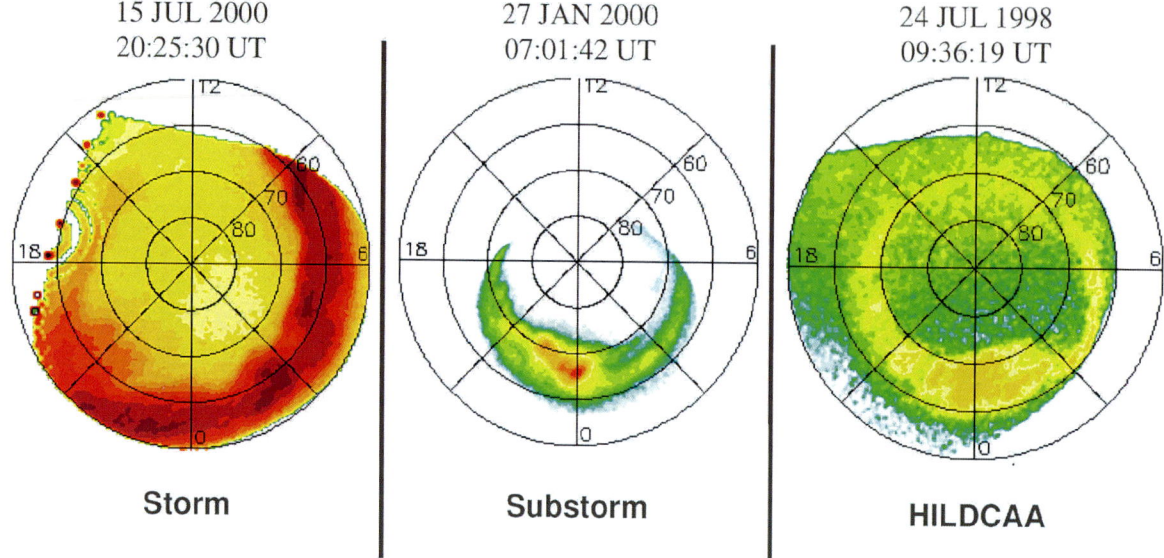

Plate 1. A comparison of typical auroral intensities during a magnetic storm, a substorm and a HILDCAA event. The HILDCAA event has the lowest of the three auroral luminosities. The figure is taken from *Guarnieri* (2005).

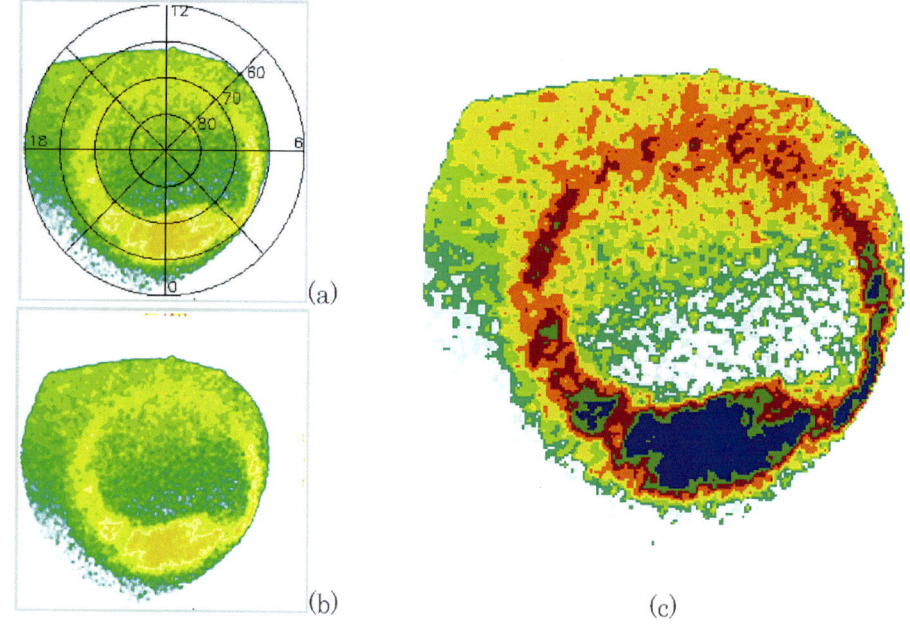

Plate 2. Polar UV images during a HILDCAA event. The aurora covers the entire auroral oval, from night to day.

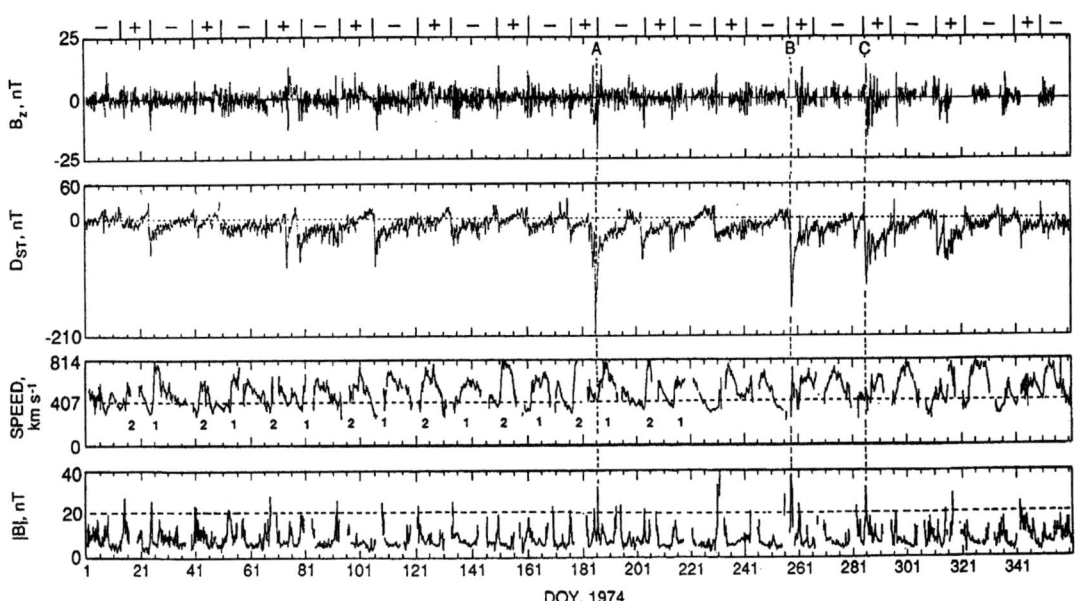

Figure 7. The solar wind and geomagnetic activity for the year 1974. Two ~27 day recurrent high speed stream sequences are indicated (third panel) as sequences 1 and 2. The stream-stream interaction generated CIRs (magnetic field "spikes") are found at the leading edges of the streams. CIRs are noted to cause only minor to moderate intensity magnetic storms. Wherever there is a high speed stream and significant southward components of the IMF, the Dst index remains depressed.

stream 2 in these cases did not cause second storm main phases. There are other examples with similar relationships.

Explanation for this behavior can be found in the magnetic field polarity information. The magnetic field on day ~80 changes from a positive polarity to a negative polarity (the same on ~day 107). The magnetic field polarities on ~day 95 and ~122 have the opposite polarity changes. During this half of the solar cycle, the polar coronal holes had positive polarity fields in the northern solar polar region and negative polarity fields in the south solar polar regions. Thus the region 1 streams are coming from the southern polar coronal hole and the region 2 streams are coming from the northern polar coronal holes.

Why is one CIR more geoeffective than the other? As the solar wind drags the interplanetary magnetic fields from the sun to 1 AU, the magnetic fields obtain a Parker spiral shape. Although the fields are radial (**r**) in orientation near the sun, they have approximately equal **r** and **t** component magnitudes near 1 AU. For positive polarity magnetic fields, the rtn coordinate system **r** component is positive and the **t** component is negative. For geoeffectiveness, it is important to view the magnetic field in the GSM coordinate system (this system was discussed earlier). During spring, negative (rtn coordinate sytem) t component fields will have a positive GSM Bz components, and conversely, during fall, negative (**rtn**) t components will have a negative GSM Bz components. Thus we can now see why one stream (stream 1) is more geoeffective and stream 2 is less geoeffective for the events discussed above. Stream 1 is a negative polarity stream and stream 2 is a positive polarity stream. These events occurred close to the spring equinox. Thus there are stronger negative GSM Bz components in the stream 1 event than in the stream 2 event. This is the reason for these storms having apparent ~27 day "recovery" phases. This effect has been discussed by numerous authors (*McIntosh*, 1959; *Russell and McPherron*, 1973; *Murayama*, 1974; *Berthelier*, 1976; *Crooker and Siscoe*, 1986; *Silverman*, 1986; *Clua de Gonzalez, et al.*, 1993).

Figure 8 shows the geomagnetic indices for the same year, 1974. The profile of the AE indices is especially remarkable. In the above Figure, the AE indices remain high throughout the storm "recoveries". This is consistent with the results shown in Figures 3 and 5, but here it can be noted how often these occur and how they dominate geomagnetic activity for the year.

Geomagnetic Quiet

Figure 8 has a number of intervals where Dst is ~0, AE <100 nT and ap <20 nT. These are typically intervals detected at the ends of high speed streams near the HPS crossings.

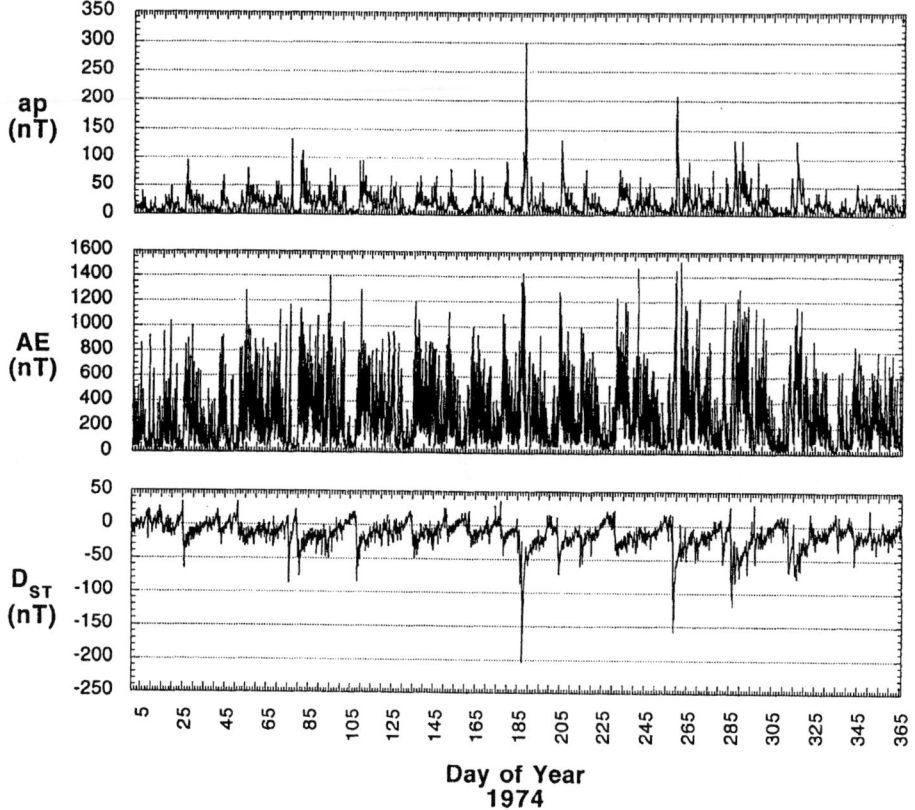

Figure 8. The Ap, AE and Dst indices during 1974. The AE index is exceptionally high throughout the year.

By returning to Figure 3, one can examine one of these intervals in higher time resolution.

In Figure 3, from 00 UT to 23 UT day 176, the solar wind velocity is low (<400 km/s), the magnetic field magnitude is relatively weak (~4 to 5 nT), and there are little or no Bz fluctuations. AE is typically <100 nT and Dst ~0 nT. The HCS occurred on the next day, day 176 and the CIR at the end of day 176 through day 177.

In comparison, the causes of geomagnetic quiet during the solar maximum phase of the solar cycle is different. Magnetic clouds, portions of ICMEs, have large north-south Bz variations or vice versa (*Klein and Burlaga*, 1982; *Lepping et al.*, 1990; 2005; *Farrugia et al.*, 1997; *Tsurutani and Gonzalez*, 1997). The magnetic field strength within magnetic clouds are >10 nT, up to 50 nT, with a general lack of Alfvén waves. The southward Bz component is well recognized as causing magnetic storm main phases (*Gonzalez et al.*, 1994). In sharp contrast, the northward Bz component of magnetic clouds are responsible for extreme geomagnetic quiet (Dst ~0 nT, AE < 100 nT) (*Tsurutani and Gonzalez*, 1995).

Relativistic Electron Acceleration During High Speed Streams

One of the most interesting aspects of the geoeffectiveness of corotating high speed solar wind streams is the acceleration of relativistic electrons during high speed streams. The electrons do not appear in the Earth's magnetosphere or magnetotail during the magnetic storm main phase caused by the CIR southward Bz components, but occur afterwards in the storm "recovery" phase (see recent examples in *Tsurutani et al.*, 2006). As mentioned before, the term "recovery", although correct for the profile of the Dst index, is a misnomer when concerning magnetospheric particle energy. In this phase of the storm, the storm-time radiation belts are not simply decaying away, but in addition to this, fresh particles are being injected into the outer regions of the magnetosphere. Dst is almost in an equilibrium.

There have been numerous works on the mechanisms for electron acceleration since the original discovery of this phenomenon (*Paulikas and Blake*, 1979; *Baker et al.*, 1986). Some of the more pertinent works are: *Horne et al.* (1998),

Hudson et al. (1999), Li and Temerin (2001), O'Brien et al. (2001), Meredith et al. (2003), Elkington et al. (2003) and Trakhtengerts et al. (2003). This issue will contain up-to-date reviews by Jordanova et al., Soraas et al., McPherron and Weygand, Horne et al., and Li. All of these latter review papers will touch on some aspect of the electron acceleration process.

There are two main acceleration mechanisms proposed. Both require resonant wave-particle interactions. Both types of waves are generated by energy transfer from the fast solar wind to the magnetosphere. Large amplitude PC5 waves are observed during magnetic storm main phases and slightly afterward (Tsurutani et al., 2006). One possible generation mechanism is the Kelvin-Helmholtz instability along the flanks of the magnetosphere. Another mechanism is associated with convection events associated with HILDCAAs. These electromagnetic PC5 waves can lead to the breaking of the third adiabatic invariant of electrons drifting around the magnetosphere. This resonant interaction causes radial diffusion of the electrons, a portion of which will gain energy in the process. This radial diffusion electron acceleration mechanism will be discussed in greater detail by Li (this issue). Electromagnetic chorus emissions are also observed during the storm main and recovery phases. These waves are generated by a loss cone plasma instability (Kennel and Petschek, 1966) from electrons injected into the magnetosphere. The electron injections are presumably caused by the magnetospheric electric fields associated with magnetic reconnection within the CIR and Alfvénic southward Bz magnetic fields in the high speed streams, as discussed previously. The cyclotron resonant interaction will be discussed in detail by Horne et al. (this issue).

Summary of Geomagnetic Activity During Corotating Solar Wind Streams

Figure 9 gives a summary for the geomagnetic activity that occurs during corotating solar wind streams. The panels from top to bottom are the solar wind plasma densities, magnetic field magnitudes, solar wind velocities, and the IMF Bz components. The bottom two panels are the geomagnetic AE and Dst indices. The location of the HCS is indicated by a vertical dashed line. This schematic represents the solar wind features and geomagnetic responses discussed in this paper. We will start from the left-hand side of the figure and then move to the right, in the direction of increasing time.

At time 1, during the decay of a previous high speed stream, the magnetic field magnitude is low and the Bz fluctuation amplitudes are low. The result is geomagnetic quiet, with AE <100 nT and Dst ~0 nT.

As the heliospheric plasmasheet is encountered (near the HCS), the high density plasma in the slow solar wind

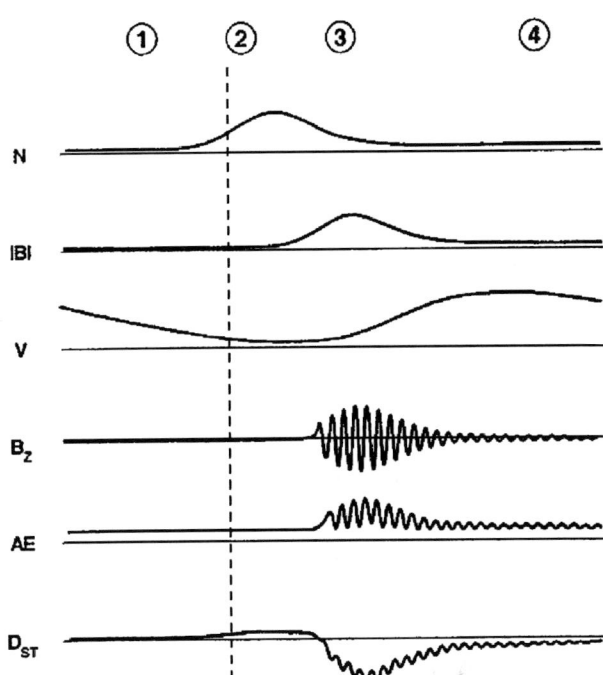

Figure 9. A schematic of the interplanetary causes of CIR-induced geomagnetic storms and geomagnetic quiet.

impinging on the magnetosphere causes positive Dst values. The HPS densities can be as high as 100 particles/cm^3 or higher, thus the Dst increases can be far greater than those which occur from interplanetary shocks. AE is typically <100 nT during these intervals.

At time 3 when the CIR impinges upon the magnetosphere, a magnetic storm main phase may or may not occur. The Bz fields are typically highly variable. If there is substantial southward components in these fields, a weak to moderate magnetic storm main phase will result. If the magnetic fields are northward or not southward (e.g., Bz ~0 nT), then the storm intensity will be weak to negligible.

At time 4, in the high speed stream proper, the IMF Bz fields are highly fluctuating. These are Alfvén waves that are being convected outward by the fast solar wind. The southward components of these fields lead to AE enhancements and Dst depressions. This corresponds to short time-duration plasma injections into the nightside sector of the magnetosphere. The injected electrons will generate electromagnetic chorus emissions by the loss cone instability.

CURRENT OUTSTANDING PROBLEMS

We have attempted to give a brief overview of geomagnetic activity and its interplanetary causes during corotating solar wind streams. The latter occur most often during the solar cycle declining phase or minimum phase when polar coronal holes are the largest (*Harvey et al.*, 2000; *Harvey and Recely*, 2002). We have shown that although magnetic storms appear to be qualitatively similar to those of ICME magnetic storms, the quantitative values as well as the interplanetary causes are different for the initial, main and "recovery" phases of the magnetic storms. Some pertinent questions about storms during this (declining) phase of the solar cycle are:

1. It has been noted that the Bz fields within CIRs are highly fluctuating. The consequential magnetic storm main phases are relatively weak due to this feature. What is not known is the configuration of the radiation belt during these CIR-storms. Are the particle injections patchy, with consequential gaps in the outer belt (before velocity dispersion smoothes them out)? Observations and modeling are needed to answer these questions. See reviews in *Jordanova* (this issue) and *Soraas and Sorbo* (this issue).

   The source of the IMF Bz fluctuations is believed to be due to compressed Alfvén waves that exist in the pre-shocked solar wind plasma. But can wave-wave interactions and plasma instabilities also generate new waves of significant amplitudes? Are the latter sources of CIR Bz fluctuations important, in a geomagnetic sense?

2. The auroras and current systems during HILDCAAs are not well understood. There is a lack of correlation between HILDCAA AE increases and substorms (*Tsurutani et al.*, 2004; *Guarnieri*, 2005). The electrojet AL indices are large in comparison to AU. The auroras are apparently neither substorm-related nor convection bays (*Pytte et al.*, 1978). What are these auroras and the current systems related to them? See review by *Sandholt et al.*, *Liou et al.*, and *Kozyra et al.* (all in this issue).

   What is the nature of ionospheric electric fields/currents during HILDCAA auroras? This topic has not been explored to any depth and needs to be better understood. See reviews by *Otsuka et al.*, *Kozyra et al.*, *Proelss*, and *Abdu et al.* (all this issue).

   HILDCAA auroras are present at all local times and sometimes cover the polar cap (*Guarnieri*, 2005). What are the source(s) of the polar, dayside, dawn and dusk auroras? The Kelvin Helmholtz instability is attractive towards explaining dawn and dusk auroras. This is particularly compelling because the magnetic field is continuously changing directions due to the presence of interplanetary Alfvén waves (randomly oriented magnetic fields) in the high speed streams. Density increases at MDs at the edges of Alfvén waves will cause small scale, short duration pressure pulses on the dayside magnetosphere. These can cause direct particle precipitation due to plasma compression and the consequential loss cone instability (*Zhou and Tsurutani*, 1999).

   Magnetospheric Alfvén waves can be generated by these processes, which can then propagate to lower altitudes and transfer their energy back to plasmas through nonlinear processes (*Haerendel*, 1998; *Tsurutani et al.*, 2001). Both of these mechanisms can result in dayside auroras. Although the above are interesting speculations, solid observational evidence is clearly needed. One can certainly expect surprises.

4. Relativistic electron acceleration processes have been difficult to pin down primarily because electrons have easy access to all parts of the magnetosphere and magnetotail. It is therefore difficult to determine where the acceleration is taking place. Arguments have been led primarily by theoretical modeling and fittings to satellite observations. Significant advances to this problem might be made if one could devise an approach to identify the specific region(s) of particle acceleration.

5. The high AE indices for long periods of time during "recovery" phases of magnetic storms indicates that a lot of energy is injected into the auroral ionosphere. What are the effects of energy deposition into such a confined area? Are the resultant disturbance dynamos (*Proelss*, 1997; *Fuller-Rowell et al.*, 1997; *Buonsanto*, 1999; *Proelss*, this issue) especially significant? What are the consequences for the low latitude atmosphere and ionosphere? See articles by *Proelss*, *Kozyra et al.*, and *Lu et al.* (all this issue).

GLOSSARY

Alfvén Wave (magnetohydrodynamic shear wave)- A transverse wave in magnetized plasma characterized by a change of direction of the magnetic field with no change in either the intensity of the field or the plasma density (note that the interplanetary waves discussed in this paper cannot be fully described by MHD; these waves are "kinetic").

CIR- Corotating Interaction Region, created by the interaction of a high speed stream with a slow speed stream. Although the plasma and fields of the Interaction Region flow more or less radially outward from the sun, because of the rotation of the sun, the interaction region (IR) appears to corotate, thus the name CIR.

CME- Coronal Mass Ejection, a transient outflow of plasma from or through the solar corona.

GSM- Geomagnetic Solar Magnetospheric, a coordinate system used for magnetospheric measurements.

HCS- Heliospheric Current Sheet, a surface dividing the northern and southern magnetic field hemispheres in the solar wind.

HPS- Heliospheric Plasma Sheet, a high density plasma region surrounding the HCS.

HILDCAA- High Intensity, Long-Duration, Continuous AE Activity.

ICME- Interplanetary Coronal Mass Ejection, the interplanetary remains/evolution of a CME.

Magnetic Storm- A worldwide disturbance of the Earth's magnetic field, distinct from regular diurnal variations.

Initial Phase: The interval when there is an increase of the near-equatorial and middle-latitude horizontal magnetic field intensity at the surface of the Earth. However, some storms proceed directly into the main phase without having an initial phase.

Main Phase: Of a magnetic storm, that interval when the horizontal magnetic field at near-equatorial and middle latitudes decreases, owing to the effects of an increasing magnetospheric ring current.

Recovery Phase: Of a magnetic storm, that interval when the depressed horizontal magnetic fields at near-equatorial and middle latitudes return to premain phase levels.

Shock Wave (fast, collisionless)- A shock wave is characterized by a discontinuous change in pressure, density, temperature, and particle streaming velocity, propagating through a compressible fluid or plasma. Collisionless (fast) shock waves occur in the solar wind when fast solar wind overtakes slow solar wind with the difference in speeds being greater than the local magnetosonic speed.

Solar Maximum- The month(s) during the sunspot cycle when the smoothed sunspot number reaches a maximum.

Solar Minimum- The month(s) during the sunspot cycle when the smoothed sunspot number reaches a minimum.

SI- Sudden Impulse, an abrupt increase in the magnetic field at the surface of the Earth; this is caused by ram pressure increase onto the magnetosphere.

SSC- Storm Sudden Commencement, a **SI** which is followed by a magnetic storm main phase.

For further acronyms and definitions, we refer the reader to *Suess and Tsurutani* (1998).

Acknowledgments. Portions of this research were performed at the Jet Propulsion Laboratory, California Institute of Technology under contract with NASA. BTT wishes to thank RISH, Kyoto University, Uji, Japan for hosting him during the portions of the writing of this paper, and A. Clua de Gonzalez for insightful discussions about seasonal variations in geomagnetic activity. FLG wishes to thank CAPES and FAPESP for financial support for portions of this work. The authors thank the participants of the Chapman 2005 Manaus Conference for opportunities to present this material and gain valuable feedback.

REFERENCES

Abdu, M.A., J.R. de Souza, J.H.A. Sobral, and I.S. Batista, Magnetic storm associated disturbance dynamo effects in the low and equatorial latitude ionosphere, in *Recurrent Magnetic Storms: Corotating Solar Wind Streams*, edited by B.T. Tsurutani, R.L. McPherron, W.D. Gonzalez, G. Lu, J.H.A. Sobral, and N. Gopalswamy, *Amer. Geophy. Un. Press*, Wash. D.C., 2006.

Akasofu, S.-I., The development of the auroral substorm, *Planet. Space Sci.*, 12, 273, 1964.

Baker, D.N., J.B. Blake, R.W. Klebesadel, and P.R. Higbie, Highly relativistic electrons in the Earth's outer magnetosphere, 1. Lifetimes and temporal history 1979-1984, *J. Geophys. Res.*, 19, 4265, 1986.

Balogh, A., E.J. Smith, B.T. Tsurutani, D.J. Southwood, R.J. Forsyth, and T.S. Horbury, The heliospheric magnetic field over the south polar region of the sun, *Science*, 268, 1007, 1995.

Balogh, A., J.T. Gosling, J.R. Jokipii, R. Kallenbach, and H. Kunow, editors, Corotating Interaction Regions, *Space Sci. Rev.*, 89, 1999.

Bartels, J., Twenty-seven day recurrences in terrestrial-magnetic and solar activity, 1923-1933, *Terr. Magn.*, 39, 201, 1934.

Belcher, J.W. and L. Davis, Jr., Large amplitude Alfvén waves in the interplanetary medium, 2, *J. Geophys. Res.*, 3534, 1971.

Berthelier, A., Influence of the polarity of the interplanetary magnetic field on the annual and the diurnal variations of magnetic activity, *J. Geophys. Res.*, 81, 4546, 1976.

Buonsanto, M.J., Ionospheric storms- A review, *Space Sci. Rev.*, 88, 563, 1999.

Chen, L. and A. Hasegawa, A theory of long-period magnetic pulsations, 1) Steady state excitation of field-line resonances, *J. Geophys. Res.*, 79, 1024, 1974.

Clua de Gonzalez, A.L., W.D. Gonzalez, S.L.G. Dutra, and B.T. Tsurutani, Periodic variation in the geomagnetic activity: A study based on the Ap index, *J. Geophys. Res.*, 98, 9215, 1993.

Crooker, N.U. and G.L. Siscoe, The effect of solar wind on the terrestrial environment, in *Physics of the Sun and Solar Terr. Rel.*, edited by P.A. Sturrock, T.E. Holzer, D.M. Mihalas, and R.K. Ulrich, 193, D. Reidel, Hingham, MA, 1986.

Daglis, I.A., J.U. Kozyra, Y. Kamide, D. Vassiliadis, A.S. Sharma, M.W. Liemohn, W.D. Gonzalez, B.T. Tsurutani, and G. Lu, Intense space storms: Critical issues and open disputes, *J. Geophys. Res.*, 108, doi:10.1029/2002JA09722, 2003.

Dessler, A.J. and E.N. Parker, Hydromagnetic theory of magnetic storms, *J. Geophys. Res.*, 64, 2239, 1959.

Dungey, J.W., Interplanetary magnetic field and the auroral zone, *Phys. Rev. Lett.*, 6, 47, 1961.

Elkington, S.R., M.K. Hudson, and A.A. Chan, Resonant acceleration and diffusion of outer zone electrons in an asymmetric geomagnetic field, *J. Geophys. Res.*, 108, doi: 10.129/2001JA009202, 2003.

Farrugia, C.J., L.F. Burlaga, and R.P. Lepping, Magnetic clouds and their quiet-storm effects at Earth, in *Magnetic Storms*, edited by B.T. Tsurutani, W.D. Gonzalez, Y. Kamide, and J.K. Arballo, *Amer. Geophys. Un. Press*, 98, 91, 1997.

Farrugia, C.J., F.T. Gratton, and R.B. Torbert, The role of viscous-type processes in solar wind-magnetosphere interactions, in *Challenges to Long-Standing Unsolved Problems in Space Physics in the 20th Century*, 95, 443, 2001.

Feldstein, Y.I., A.E. Levitin, J.U. Kozyra, B.T. Tsurutani, A. Prigancova, L. Alperovich, W.D. Gonzalez, U. Mall, I.I. Alexeev, L.I. Gromova, and L.A. Dremukhina, Self-consistent modeling of the large-scale distortions in the geomagnetic field during the 24-27 September 1998 major magnetic storm, *J. Geophys. Res.*, 110, A11214, doi:10.1029/2004JA010584, 2005.

Franz, M., et al., Magnetic field depressions in the solar wind, *J. Geophys. Res.*, 105, 12,725, 2000.

Fuller-Rowell, T.M., M.V. Codrescu, R.G. Roble, and A.D. Richmond, How does the thermosphere and ionosphere react to a geomagnetic storm?, in *Magnetic Storms*, Geophys. Mon. Series, edited by B.T. Tsurutani, W.D. Gonzalez, Y. Kamide, and J.K. Arballo, *Amer. Geophys. Un. Press*, 98, 203, 1997.

Gonzalez, W.D., J.A. Joselyn, Y. Kamide, H.W. Kroehl, G. Rostoker, B.T. Tsurutani, and V.M. Vasyliunas, What is a geomagnetic storm?, *J. Geophys. Res.*, 99, 5771, 1994.

Gonzalez, W.D., F.L. Guarnieri, A.L.C. Gonzalez, E. Echer, M.V. Alves, T. Ogino, and B.T. Tsurutani, Magnetospheric energetics during HILDCAAs,, in *Recurrent Magnetic Storms: Corotating Solar Wind Streams*, edited by B.T. Tsurutani, R.L. McPherron, W.D. Gonzalez,

G. Lu, J.H.A. Sobral, and N. Gopalswamy, *Amer. Geophy. Un. Press*, Wash. D.C., 2006.

Guarnieri, F.L., *A Study of the Interplanetary and Solar Origin of High Intensity Long Duration and Continuous Auroral Activity Events*, PhD thesis, INPE, Brazil, Feb. 2005.

Harvey, K.L. and F. Recely, Polar coronal holes during cycles 22 and 23, *Solar Phys.*, 211, 31, 2002.

Harvey, K., S. Suess, M. Aschwanden, M. Guhathakurta, J. Harvey, D. Hathaway, B. LaBonte, N. Sheely, and B. Tsurutani, A. NASA workshop on coronal holes near solar maximum and over the solar cycle, *NASA white paper*, December 2000.

Hollweg, J.V. and P.A. Isenberg, Generation of the fast wind: A review with emphasis on the resonant cyclotron interaction, *J. Geophys. Res.*, 107 (A7), SSH 12-1, doi:10.1029/2001JA000270, 2002.

Hollweg, J.V., The solar wind: Then and now, in *Recurrent Magnetic Storms: Corotating Solar Wind Streams*, edited by B.T. Tsurutani, R.L. McPherron, W.D. Gonzalez, G. Lu, J.H.A. Sobral, and N. Gopalswamy, *Amer. Geophy. Un. Press*, Wash. D.C., 2006.

Horne, R.B., The contribution of wave-particle interactions to electron loss and acceleration in the Earth's radiation belts during geomagnetic storms, in *Rev. Radio Sci. 1999-2002*, edited by W.R. Stone, Wiley, N.Y., 801, 1998.

Horne, R.B., R.M. Thorne, N.P. Meredith, and R.R. Anderson, Diffuse auroral electron scattering by electron cyclotron harmonic and whistler mode waves during an isolated substorm, *J. Geophys. Res.*, 108, doi:10.1029/2002JA009736, 2003.

Horne, R.B., N.P. Meredith, S.A. Glauert A. Varotsou, R.M. Thorne, Y.Y. Shprits, and R.R. Anderson, Mechanisms for the acceleration of relativistic electrons, in *Recurrent Magnetic Storms: Corotating Solar Wind Streams*, edited by B.T. Tsurutani, R.L. McPherron, W.D. Gonzalez, G. Lu, J.H.A. Sobral, and N. Gopalswamy, *Amer. Geophy. Un. Press*, Wash. D.C., 2006.

Hudson, M.K., S.R. Elkington, J.G. Lyon, C.C. Goodrich, and T.J. Rosenberg, Simulations of radiation belt dynamics driven by solar wind variations, in *Sun-Earth Plasma Connections*, edited by J. Burch, R.L Carovillano, and S.K. Antiochos, *Amer. Geophys. Un.*, Wash. D.C., 171, 1999.

Jahn, J.-M. and H.A. Elliott, Energetic neutral atom observations during recurrent magnetic storms, in *Recurrent Magnetic Storms: Corotating Solar Wind Streams*, edited by B.T. Tsurutani, R.L. McPherron, W.D. Gonzalez, G. Lu, J.H.A. Sobral, and N. Gopalswamy, *Amer. Geophy. Un. Press*, Wash. D.C., 2006.

Joselyn, J.A. and B.T. Tsurutani, Geomagnetic sudden impulses and storm sudden commencements- A note on terminology, *EOS*, 71, Nov. 20, 1808, 1990.

Kamide, Y., N. Yokoyama, W. Gonzalez, B.T. Tsurutani, I.A. Daglis, A. Brekke, and S. Masuda, Two-step development of geomagnetic storms, *J. Geophys. Res.*, 103, A4, 6917, 1998.

Kennel, C.F. and H.E. Petschek, Limit on stably trapped particle fluxes, *J. Geophys. Res.*, 71, 1, 1966.

Kikuchi T. and T. Araki, Horizontal transmission of the polar electric field, *J. Atmos. Terr. Phys.*, 41, 927, 1979.

Klein, L.W. and L.F. Burlaga, Interplanetary magnetic clouds at 1 AU, *J. Geophys. Res.*, 87, 613, 1982.

Kozyra, J.U., V.K. Jordanova, R.B. Horne, and R.M. Thorne, Modeling of the contribution of electromagnetic ion cyclotron (EMIC) waves to stormtime ring current erosion, in *Magnetic Storms*, edited by B.T. Tsurutani, W.D. Gonzalez, Y. Kamide, and J.K. Arballo, *Amer. Geophys. Un. Press*, Wash. D.C., 98, 187, 1997.

Kozyra, J.U., G. Crowley, G. Lu, *et al.*, Response of the upper/middle atmosphere to high-speed streams in the solar wind: First comprehensive TIMED/CEDAR observations, in *Recurrent Magnetic Storms: Corotating Solar Wind Streams*, edited by B.T. Tsurutani, R.L. McPherron, W.D. Gonzalez, G. Lu, J.H.A. Sobral, and N. Gopalswamy, *Amer. Geophy. Un. Press*, Wash. D.C., 2006.

Krieger, A.S., A.F. Timothy, and E.C. Roelof, A coronal hole and its identification as the source of a high velocity solar wind stream, *Solar Physics*, 23, 123, 1973.

Jordanova, V.K., Modeling the behavior of corotating interaction region driven storms in comparison with coronal mass ejection driven storms, in *Recurrent Magnetic Storms: Corotating Solar Wind Streams*, edited by B.T. Tsurutani, R.L. McPherron, W.D. Gonzalez, G. Lu, J.H.A. Sobral, and N. Gopalswamy, *Amer. Geophy. Un. Press*, Wash. D.C., 2006.

Lepping, R.P., J.A. Jones, and L.F. Burlaga, Magnetic field structure of interplanetary magnetic clouds at 1 AU, *J. Geophys. Res.*, 95, 11,957, 1990.

Lepping, R.P., D.B. Berdichevsky, C.C. Wu, A. Szabo, T. Narock, F. Mariani, A.J. Lazarus, and A.J. Quivers, A summary of WIND magnetic clouds for the years 1995-2003: Model-fitted parameters, associated errors, and classifications, *Annales Geophys.*, submitted, 2005.

Li, X.-L., Radial transport in energizing radiation belt electrons in the magnetosphere, in *Recurrent Magnetic Storms: Corotating Solar Wind Streams*, edited by B.T. Tsurutani, R.L. McPherron, W.D. Gonzalez, G. Lu, J.H.A. Sobral, and N. Gopalswamy, *Amer. Geophy. Un. Press*, Wash. D.C., 2006.

Liou, K., Global auroral response to interplanetary media with emphasis on solar wind dynamic pressure enhancements, in *Recurrent Magnetic Storms: Corotating Solar Wind Streams*, edited by B.T. Tsurutani, R.L. McPherron, W.D. Gonzalez, G. Lu, J.H.A. Sobral, and N. Gopalswamy, *Amer. Geophy. Un. Press*, Wash. D.C., 2006.

McIntosh, D.H., On the annual variation of magnetic disturbances, *Philos. Trans. R. Soc.*, London ser. A, 251, 525, 1959.

McPherron, R.P. and J. Weygand, The solar wind and geomagnetic activity as a function of time relative to corotating interaction regions, in *Recurrent Magnetic Storms: Corotating Solar Wind Streams*, edited by B.T. Tsurutani, R.L. McPherron, W.D. Gonzalez, G. Lu, J.H.A. Sobral, and N. Gopalswamy, *Amer. Geophy. Un. Press*, Wash. D.C., 2006.

Murayama, T., Origin of the semiannual variation of geomagnetic Kp indices, *J. Geophys. Res.*, 79, 297, 1974.

Neugebauer, M., *et al.*, Ion distributions in large magnetic holes in the fast solar wind, *J. Geophys. Res.*, 106, 5635, 2001.

O'Brien, T.P., R.L. McPherron, D. Sornette, G.D. Reeves, R. Friedel, and H.J. Singer, Which magnetic storms produce relativistic electrons at geosynchronous orbit?, *J. Geophys. Res.*, 106, 15533, 2001.

Otsuka, Y.-I., T. Aramaki, T. Ogawa, and A. Saito, A statistical study of ionospheric irregularities observed with a GPS network in Japan, in *Recurrent Magnetic Storms: Corotating Solar Wind Streams*, edited by B.T. Tsurutani, R.L. McPherron, W.D. Gonzalez, G. Lu, J.H.A. Sobral, and N. Gopalswamy, *Amer. Geophy. Un. Press*, Wash. D.C., 2006.

Parker, E.N., Interaction of solar wind with the geomagnetic field, *Phys. Fluids*, 1, 171, 1958.

Paulikas, G. and J.B. Blake, Effects of the solar wind on magnetospheric dynamics: Energetic electrons at the synchronous orbit, in *Quantitative Modeling of Magnetospheric Processes*, edited by W. Olsen, *Amer. Geophys. Un.*, Wash. D.C., 21, 180, 1979.

Pizzo, V.J., Interplanetary shocks on the large scale: A retrospective on the last decade's theoretical efforts, in *Collisionless shocks in the heliosphere: Reviews of Current Research*, edited by B.T. Tsurutani and R.G. Stone, *Amer. Geophys. Un.*, Wash. D.C., 35, 51, 1985.

Pizzo, V.J., The evolution of corotating stream fronts near the ecliptic plane in the inner solar system. II Three-dimensional tilted-dipole fronts, *J. Geophys. Res.*, 96, 5405, 1991.

Proelss, G.W., Magnetic storm associated perturbations of the upper atmosphere, in *Magnetic Storms*, Geophys. Mon. Series, edited by B.T. Tsurutani, W.D. Gonzalez, Y. Kamide, and J.K. Arballo, *Amer. Geophys. Un. Press*, 98, 203, 1997.

Proelss, G.W., Selected upper atmospheric storm effects, in *Recurrent Magnetic Storms: Corotating Solar Wind Streams*, edited by B.T. Tsurutani, R.L. McPherron, W.D. Gonzalez, G. Lu, J.H.A. Sobral, and N. Gopalswamy, *Amer. Geophy. Un. Press*, Wash. D.C., 2006.

Pytte T., R.L. McPherron, E.W. Hones Jr., and H.I. West, Multiple satellite studies of magnetospheric substorms: Distinction between polar magnetic substorm and convection driven negative bay, *Planet. Space Sci.*, 83, 663, 1978.

Richardson, I.G., The formation of CIRs at stream-stream interfaces and resultant geomagnetic activity, in *Recurrent Magnetic Storms: Corotating Solar Wind Streams*, edited by B.T. Tsurutani, R.L. McPherron, W.D. Gonzalez, G. Lu, J.H.A. Sobral, and N. Gopalswamy, *Amer. Geophy. Un. Press*, Wash. D.C., 2006.

Russell, C.T. and R.L. McPherron, Semiannual variation of geomagnetic activity, *J. Geophys. Res.*, 78, 92, 1973.

Sandholt, P.E., C.J. Farrugia, E.J. Lund, and W.F. Denig, IMF By and spatio-temporal structure of the dayside aurora, in *Recurrent Magnetic Storms: Corotating Solar Wind Streams*, edited by B.T. Tsurutani, R.L. McPherron, W.D. Gonzalez, G. Lu, J.H.A. Sobral, and N. Gopalswamy, *Amer. Geophy. Un.* Press, Wash. D.C., 2006.

Schulz, M., Interplanetary sector structure and the heliomagnetic equator, *Astrophys. Space Sci.*, 24, 371, 1973.

Sckopke, N., A general relation between the energy of trapped particles and the disturbance field near the Earth, *J. Geophys. Res.*, 71, 3125, 1966.

Sheeley, N.R. Jr., J.W. Harvey, and W.C. Feldman, Coronal holes, solar wind streams and recurrent geomagnetic disturbances: 1973-1976, *Solar Phys.*, 49, 271, 1976.

Sheeley, N.R. Jr., J.R. Asbridge, S.J. Bame, and J.W. Harvey, A pictoral comparison of interplanetary magnetic field polarity, solar wind speed, and geomagnetic disturbance index during the sunspot cycle, *Solar Phys*, 52, 485, 1977.

Silverman, A.M., Annual variation of aurora and solar wind coupling, in *Solar Wind-Magnetospheric Coupling*, ed. By Y. Kamide and J.A. Slavin, 643, 1986.

Smith, E.J. and J.H. Wolfe, Observations of interaction regions and corotating shocks between one and five AU: Pioneers 10 and 11, *Geophys. Res. Lett.*, 3, 137, 1976.

Smith, E.J., B.T. Tsurutani, and R.L. Rosenberg, Observations of the interplanetary sector structure up to heliographic latitudes of 16°: Pioneer 11, *J. Geophys. Res.*, 83, 717, 1978.

Soraas, F., K. Aarsnes, K. Oksavik, M.I. Sandanger, D.S. Evans, and M.S. Greer, Evidence for particle injection as the cause of Dst reduction during HILDCAA events, *J. Atmos. Solar-Terr. Phys.*, 66, 177, 2004.

Soraas, F., K. Aarsnes, D.V. Carlsen, K. Oksavik, and D.S. Evans Ring current behavior as revealed by energetic proton precipitation, in *The Inner Magnetosphere: Physics and Modeling*, edited by T.I. Pulkkinen, N.A. Tsygenenko, and R.H.W. Friedel, *Amer. Geophys. Un.* Press, Wash. D.C., 155, 237, 2005.

Soraas, F., M. Sorbo, K. Aarsnes, and D.S. Evans, Ring Current behavior as revealed by low altitude satellite and ground observations, in *Recurrent Magnetic Storms: Corotating Solar Wind Streams*, edited by B.T. Tsurutani, R.L. McPherron, W.D. Gonzalez, G. Lu, J.H.A. Sobral, and N. Gopalswamy, *Amer. Geophy. Un.* Press, Wash. D.C., 2006.

Southwood, D.J., Some features of field-line resonance in the magnetosphere, *Planet. Space Sci.*, 22, 483, 1974.

Suess, S.T. and B.T. Tsurutani, editors, *From the Sun: Auroras, Magnetic Storms, Solar Flares, Cosmic Rays*, *Amer. Geophys. Un.* Press, Wash. D.C., 1998.

Suzuki, T.K. and S.-I. Inutsuka, Making the coronal and the fast solar wind: A self-consistent simulation for the low-frequency Alfvén waves from photosphere to 0.3 AU, *Astrophys. J. Lett.*, 632, L49, 2005.

Trakhtengerts, V.Y., M.J. Rycroft, D. Nunn, and A.G. Demekhov, Cyclotron acceleration of radiation belt electrons by whistlers, *J. Geophys. Res.*, 108, doi:10.1029/2002JA009559, 2003.

Trakhtengerts, V.Y., A.G. Demekhov, E.E. Titova, B.V. Kozelov, O. Santolik, D. Gurnett, and M. Parrot, Interpretation of Cluster data on chorus emissions using the backward wave oscillator model, *Phys. Plasmas*, 11, 1345, 2004.

Tsurutani, B.T., Solar/interplanetary plasma phenomena causing geomagnetic activity at Earth, in *Proc. International School of Physics "Enrico Fermi" Course CXLII*, edited by B. Coppi, A. Ferrari, and E. Sindoni, IOS Press, Amsterdam, 273, 2000.

Tsurutani, B.T. and E.J. Smith, Postmidnight chorus: A substorm phenomenon, *J. Geophys. Res.*, 79, 118, 1974.

Tsurutani, B.T. and E.J. Smith, Two types of magnetospheric ELF chorus and their substorm dependences, *J. Geophys. Res*, 82, 5112, 1977.

Tsurutani, B.T. and W.D. Gonzalez, The cause of high intensity long-duration continuous AE activity (HILDCAAs): Interplanetary Alfvén wave trains, *Planet. Space Sci.*, 35, 405, 1987.

Tsurutani, B.T. and W.D. Gonzalez, The efficiency of "viscous interaction" between the solar wind and the magnetosphere during intense northward IMF events, *Geophys. Res. Lett.*, 22, 663, 1995.

Tsurutani, B.T. and W.D. Gonzalez, The interplanetary causes of magnetic storms: A review, in *Magnetic Storms*, edited by B.T. Tsurutani, W.D. Gonzalez, Y. Kamide, and J.K. Arballo, *Amer. Geophys. Un.*, Wash. D.C., 98, 77, 1997.

Tsurutani, B.T. and C.M. Ho, A review of discontinuities and Alfvén waves in interplanetary space: Ulysses results, *Rev. Geophys.*, 37, 517, 1999.

Tsurutani, B.T., W.D. Gonzalez, F. Tang, S.-I. Akasofu, and E.J. Smith, Origin of interplanetary southward magnetic fields responsible for major magnetic storms near solar maximum (1978-1979), *J. Geophys. Res.*, 93, 8519, 1988.

Tsurutani, B.T., C.M. Ho, E.J. Smith, M. Neugebauer, B.E. Goldstein, J.S. Mok, J.K. Arballo, A. Balogh, D.J. Southwood, and W.C. Feldman, The relationship between interplanetary discontinuities and Alfvén waves: Ulysses observations, *Geophys. Res. Lett.*, 21, 2267, 1994.

Tsurutani, B.T., W.D. Gonzalez, A.L.C. Gonzalez, F. Tang, J.K. Arballo and M. Okada, Interplanetary origin of geomagnetic activity in the declining phase of the solar cycle, *J. Geophys. Res.*, 100, 21,717, 1995.

Tsurutani, B.T., X.-Y. Zhou, J.K. Arballo, W.D. Gonzalez, G.S. Lakhina, V. Vasyliunas, J.S. Pickett, T. Araki, H. Yang, G. Rostoker, T.J. Hughes, R.P. Lepping, and D. Berdichevsky, Auroral zone dayside precipitation during magnetic storm initial phases, *J. Atmos. Sol.-Terr. Phys.*, 63, 513, 2001.

Tsurutani, B.T., B. Dasgupta, C. Galvan, M. Neugebauer, G.S. Lakhina, J.K. Arballo, D. Winterhalter, B.E. Goldstein, and B. Buti, Phase-steepened Alfvén waves, proton perpendicular energization and the creation of magnetic holes and magnetic decreases: The ponderomotive force, *Geophys. Res. Lett.*, 29, 86-1, doi:10.1029/2002GL015652, 2002.

Tsurutani, B.T., W.D. Gonzalez, F. Guarnieri, Y. Kamide, X. Zhou, and J.K. Arballo, Are high-intensity long-duration continuous AE activity (HILDCAA) events substorm expansion events?, *J. Atmos. Sol.-Terr. Phys.*, 66, 167, 2004.

Tsurutani, B.T., G.S. Lakhina, J.S. Pickett, F.L. Guarnieri, N. Lin, and B.E. Goldstein, Nonlinear Alfvén waves, discontinuities, proton perpendicular acceleration and magnetic holes/decreases in interplanetary space and the magnetosphere: intermediate shocks?, *Nonl. Proc. Geophys.*, 12, 321, 2005.

Tsurutani, B.T., A.J. Mannucci, B.A. Iijima, A. Komjathy, A. Saito, T. Tsuda, O.P. Verkhoglyadova, W.D. Gonzalez, and F.L. Guarnieri, Dayside ionospheric (GPS) response to corotating solar wind streams,, in *Recurrent Magnetic Storms: Corotating Solar Wind Streams*, edited by B.T. Tsurutani, R.L. McPherron, W.D. Gonzalez, G. Lu, J.H.A. Sobral, and N. Gopalswamy, *Amer. Geophy. Un.* Press, Wash. D.C., 2006.

Tsurutani, B.T., W.D. Gonzalez, A.L.C. Gonzalez, F.L. Guarnieri, N. Gopalswamy, M. Grande, Y. Kamide, Y. Kasahara, G. Lu, I. Mann, R.L. McPherron, F. Soraas, and V.M. Vasyliunas, Corotating solar wind streams and recurrent geomagnetic activity: A review, in press, *J. Geophys. Res.*, 2006.

Turner, J.M., L.F. Burlaga, N.F. Ness, and J.F. Lemaire, Magnetic holes in the solar wind, *J. Geophys. Res.*, 82, 1921, 1977.

Wilcox, J.M. and P.H. Scherrer, An annual and solar magnetic cycle variations in the interplanetary magnetic field, 1926-1971, *J. Geophys. Res.*, 77, 5385, 1972.

Winterhalter, D., E.J. Smith, M.E. Burton, N. Murphy, and D.J. McComas, The heliospheric plasma sheet, *J. Geophys. Res.*, 99, 6667, 1994a.

Winterhalter, D.M. Neugebauer, B.E. Goldstein, E.J. Smith, S.J. Bame, and A. Balogh, Ulysses field and plasma observations of magnetic holes in the solar wind and their relation to mirror-mode structures, *J. Geophys. Res.*, 23,372, 1994b.

Wrenn, G.L., Conclusive evidence for internal dielectric charging anomalies on gesynchronous communications spacecraft, *J. Spacecraft and Rockets*, 32, 514, 1995.

Zhou, X.-Y., and B.T. Tsurutani, Rapid intensification and propagation of the dayside aurora: Large scale interplanetary pressure pulses (fast shocks), *Geophys. Res. Lett.*, 26, 1097, 1999.

The Solar Wind: Then and Now

Joseph V. Hollweg

University of New Hampshire, Durham, New Hampshire, USA

In the original formulation of the solar wind, the electron pressure gradient was the principal accelerating force. This was soon recognized to be insufficient to drive the observed flow speeds, especially of the high-speed streams. The discovery of Alfvén waves in the solar wind led to a long series of models in which wave pressure provided additional acceleration, but these wave-driven models ultimately failed to explain the rapid acceleration of the fast wind close to the Sun. An alternate view was that the pressure of hot protons close to the Sun could explain the rapid acceleration, with the proton (and ion) heating coming from the cyclotron resonance. SOHO has provided remarkable data which have verified some of the predictions of this view, and given impetus to ongoing studies of the ion cyclotron resonance in the fast wind. After a historical review, we discuss the basic ideas behind current research, emphasizing the particle kinetics.

1. INTRODUCTION

The Solar and Heliospheric Observatory (SOHO) has yielded remarkable data concerning the origin of the solar wind and the heating of the corona from which it originates. SOHO has forced us to discard some long-held ideas, and has also stimulated a number of entirely new ideas. The purpose of this review is to survey our current understanding of the fast solar wind. Since we may be at a watershed, we will also review the historical context.

We will for the most part concentrate on the fast solar wind, for four reasons: First, we have a pretty good idea where it comes from, especially the large polar coronal holes; see *Miralles et al.* [2002, 2004] for discussions of solar wind flows out of smaller non-polar holes. Second, the fast wind is much steadier than the slow wind, and seems to be less structured. It has always been hoped that this apparently simpler wind would be more amenable to yielding its secrets. Third, and most importantly, the fast wind has long been known to be less influenced by Coulomb collisions than the slow wind [e.g., *Neugebauer*, 1981]. This means that signatures of other kinetic processes are more readily seen in the data. Indeed, one of our conclusions will be that kinetic processes are an essential ingredient of the physics of the fast wind. Fourth, the physics of the slow wind is widely believed to be very different from that of the fast wind. In particular, magnetic reconnection between open and closed magnetic field lines seems to be the key physical process driving the slow wind, whereas the fast wind exists on quasi-steady open field lines which are not interacting with closed field lines. Thus the slow wind really requires an extensive review of its own; see for example *Einaudi et al.* [1999], *Fisk and Schwadron* [2001], and *Lapenta and Knoll* [2005] and references therein.

We do not have space for a complete literature survey. We do try, however, to reference key papers which themselves provide references to the bulk of the literature. See especially recent reviews by *Hollweg and Isenberg* [2002, hereafter 'HI'], *Cranmer* [2002, 2005], and *Erdös* [2003]. We also recommend the review of the physics of wave-particle interactions by *Tsurutani and Lakhina* [1997]; though applied to the magnetosphere, the concepts discussed are very relevant to the solar wind.

2. THE ELECTRON-DRIVEN WIND

Parker [1958] noted that thermal conductivity leads to a slow decline of T_e in the hot corona (T is temperature and subscript e denotes electrons). For a corona in static equilibrium, Parker showed that the slow decline of T_e leads to a plasma pressure at $r \to \infty$ which is many orders-of-magnitude larger than the interstellar pressure (r is heliocentric distance). He concluded that with nothing to contain the asymptotic pressure, the corona must expand. In essence, the expansion is driven by the electron pressure gradient because T_e remains high.

However, in a later review, *Parker* [1965] compared detailed theoretical predictions with the by then known properties of the solar wind at 1 AU. He concluded "Thus the model for the hypothetical conduction corona leads to a temperature falling too rapidly with radial distance from the Sun, indicating that the actual solar corona is probably actively heated for some considerable distance by the dissipation of waves." We are not yet sure whether waves are responsible, but we now know that he was right about extended coronal heating.

Hartle and Sturrock [1968] presented the first two-fluid model of the solar wind, with separate energy equations for electrons and protons. They obtained a very slow wind: 250 km s^{-1} at 1 AU compared to a typical fast wind speed of 750 km s^{-1} (see *Feldman et al.* [1976] for an early review of the properties of the fast wind). They also found that the model protons were much too cold. They concluded that "departures of the solar wind characteristics near Earth from those of this model are also to be attributed to heating by a flux of non-thermal energy."

3. THE WAVE-DRIVEN WIND

The next major advance was the discovery [*Belcher and Davis*, 1971] of the ubiquitous presence of Alfvén waves in the solar wind. Most of the wave power resides at long periods, of the order of hours. The waves predominantly propagate away from the Sun, especially in the fast wind. The outward propagation strongly suggests that the Sun is the source of these waves.

It was immediately realized that the Alfvén waves might be Parker's "waves", and Hartle and Sturrock's "flux of non-thermal energy." *Belcher* [1971] and *Alazraki and Couturier* [1971] inaugurated the concept of the wave-driven wind by noting that the waves exert a 'wave pressure' $-\nabla <\delta \mathbf{B}^2>/8\pi$ on the wind (in cgs units, which we shall use throughout); $\mathbf{B}$ is magnetic field, the prefix δ denotes a fluctuation, and the angle brackets denote a time-average. *Hollweg* [1973] showed how the wave energy equation could be extended to include dissipation and plasma heating. With heating and wave pressure, the wave-driven models were able to explain the high-speeds and hot protons observed in the fast wind in interplanetary space [e.g., *Hollweg*, 1978]. (*Bretherton and Garrett* [1969] appear to have been the first to mention the radiation pressure of Alfvén waves. Their work was expanded upon by *Dewar* [1970] and *Jacques* [1977].)

These models generally succeeded in explaining solar wind data far from the Sun, but they failed close to the Sun. In the early 1990s, spacecraft gave us new coronal hole density data, which verified previous evidence that the density declines very rapidly with increasing r [*Guhathakurta and Holzer*, 1994; *Fisher and Guhathakurta*, 1995; *Guhathakurta and Fisher*, 1998]. That requires the flow speed to increase very rapidly with r. The wave-driven models could not achieve such rapid accelerations. The reason is simply that, close to the Sun, the wave pressure is small compared to other terms in the momentum balance.

Heavy ions (the best observed being He^{++}) presented other dilemmas. Ions flow faster than the protons, roughly by the Alfvén speed V_A, and they are hotter than the protons, roughly in proportion to their masses; see *Neugebauer* [1992] for a review. These properties are most noticeable in the fast wind, because Coulomb collisions are weaker there. Efforts (mainly in the early 1980s) to explain these observations generally invoked the ion-cyclotron resonance to heat and accelerate the ions; see reviews by *Isenberg* [1983], *Cranmer* [2002, 2005], and *HI*. These models were only partially successful. One difficulty was that these early studies considered resonant effects only far from the Sun, well beyond the acceleration region. In contrast, the SOHO data show extensive heavy ion heating in the acceleration region.

Other data pointed to the cyclotron resonance. *In situ* measurements of proton distribution functions often show that the protons in the vicinity of the peak of the distribution have more thermal energy perpendicular to $\mathbf{B}$ than along $\mathbf{B}$; this is most noticeable in high-speed wind. Moreover, the average magnetic moment, $T_{p\perp}/B$, increases with distance from the Sun ($T_{p\perp}$ is the proton temperature perpendicular to $\mathbf{B}$). (See *Marsch* [1991] for a review.) Both of these observations suggest that perpendicular heating is occurring in interplanetary space, and that in turn suggests the cyclotron resonance.

4. THE PROTON-DRIVEN WIND

Application of these ideas to the solar wind close to the Sun was first made by *Hollweg* [1986] and *Hollweg and Johnson* [1988, hereafter '*HJ*']. They assumed that the Sun launches low-frequency Alfvén waves which undergo a turbulent cascade to high frequencies, where they are dissipated via the cyclotron resonance; *HJ* assumed that the resonant dissipation would heat only the protons. The heating rate was dictated by the rate at which energy cascades to high frequencies; *HJ* took the Kolmogorov rate. These models also included acceleration by the wave pressure. The waves were taken to be outward-propagating in the short-wavelength WKB limit. (We shall later see that this is formally inconsistent with turbulence.) These models succeeded in reproducing the observed high-speed wind far from the Sun, as well as the rapid acceleration close to the Sun. *HJ* found that the protons close to the Sun, in $r > 3r_S$, were considerably hotter than the electrons. The rapid flow acceleration was due mainly to the pressure of the hot protons. The available data at the time indicated that the protons were not hot close to the Sun, and so these models were discarded. *Isenberg* [1990] extended *HJ* by including He^{++}. This model too had hot protons close to the Sun, and was discarded.

As it happened, *HJ* and *Isenberg* [1990] actually predicted the results which SOHO would soon obtain for coronal holes: hot coronal protons, and heavy ions which flow faster and are more than mass-proportionally hotter than the protons close to the Sun. Moreover, the SOHO results, especially from the Ultraviolet Coronagraph Spectrometer (UVCS), indicate that positive particles in coronal holes close to the Sun are heated mainly perpendicular to the magnetic field; this result is firm for O^{+5} [*Kohl et al.*, 1998; *Dodero et al.*, 1998; *Antonucci et al.*, 2000], but less certain for protons. Thus the SOHO/UVCS results suggest: 1. The solar wind is mainly proton-driven, via the proton pressure. 2. Consistent with the perpendicular heating of O^{+5}, the cyclotron resonance is at work. 3. At least in the fast wind, coronal heating is proton and ion-dominated. It is not Joule heating. 4. The peculiar properties of heavy ions originate close to the Sun in the wind's acceleration region. 5. Coronal heating extends through the acceleration region into the supersonic wind. Most remarkable is O^{+5}, which has a temperature 3×10^8 K at $r = 3.5r_S$; the temperature is still an increasing function of *r*, in spite of strong adiabatic cooling [*Kohl et al.*, 1998]. Unlike in early studies which assumed that all coronal heating took place in a thin layer at the coronal base, coronal heating and solar wind acceleration must be treated together. *Parker* [1965] was right about extended heating. 6. Other instruments on SOHO, especially SUMER (Solar Ultraviolet Measurements of Emitted Radiation), also show indications of ion heating low in the corona [e.g., *Tu et al.*, 1998; *Peter and Vocks*, 2003; *Moran*, 2003].

There is, however, one note of caution. The proton and oxygen temperatures are derived from spectral line widths, assuming that non-thermal broadening due to waves or turbulence is negligible. This is almost certainly true for O^{+5}, which shows considerably higher line widths than the protons [*Esser et al.*, 1999; *Cranmer*, 2005]. Moreover, models [e.g., *HI*] suggest that waves or turbulence do not contribute substantially to any of the observed line widths.

Finally, we note that other authors investigated the effects of hot coronal protons [*Esser and Habbal*, 1995; 1996; *Esser et al.*, 1997; *Hansteen and Leer*, 1995; *Axford and McKenzie*, 1996; *McKenzie et al.*, 1995, 1997; *Lie-Svendsen et al.*, 2001], but only *ad hoc* heating functions were used. *Li* [1999, 2002], *Li and Habbal* [1999], and *Li et al.* [1999, 2004] have extended the original models of *HJ* and *Isenberg* [1990] to include thermal anisotropy, wave dispersion, and other effects.

5. THE ION-CYCLOTRON RESONANCE

The cyclotron resonance occurs when the wave frequency seen by a particle matches the particle's cyclotron frequency. Formally the resonance condition is $\omega - k_\| V_\| = \pm \Omega$, where ω is the wave angular frequency, $k_\|$ is the wavenumber along **B**, $V_\|$ is the particle's drift speed along **B**, and Ω is the gyro-frequency. For ions resonating with left-hand polarized waves the + sign is appropriate, while the − sign is used for right-hand waves. When the resonance condition is satisfied, the particle's energy changes secularly. The secular energy change can be a gain or a loss, depending on the phase of the particle's gyro-motion relative to the wave. In a random field, that phase will be random, the particle will gain or lose energy randomly, and consequently the particle will undergo a random walk in velocity space, which in turn leads to velocity space diffusion (frequently referred to as pitch-angle diffusion).

Not all particles can resonate. The heavy curve in Figure 1 shows the dispersion relation for the electromagnetic ion-cyclotron wave propagating along **B** in a cold electron-proton plasma: $k^2 V_{Ap}^2 = \Omega_p \omega^2 / (\Omega_p - \omega)$, where V_{Ap} is the Alfvén speed based on the proton density (we have taken $m_e = 0$ for these low-frequency waves), and Ω_p is proton gyro-frequency; Figure 1 is in the frame moving with the bulk proton flow. In Figure 1, the resonance condition is a straight line with slope $V_\|/V_{Ap}$ and an ordinate intercept at Ω/Ω_p. Near the top of the figure we show the resonance conditions for two test protons, one with $V_\| > 0$ and moving with the wave, and the other with $V_\| < 0$ and moving against the wave. Resonance can only occur if the resonance condition intersects the dispersion relation. For a proton with $V_\| > 0$ there is no resonance. Only protons moving against the wave can resonate. Also shown in Figure 1 are resonance conditions for three O^{+5} test particles, one with $V_\| < 0$ and two with $V_\| > 0$. The two lower lines show that resonance is possible for oxygen ions moving both with and against the wave. In resonant heating and acceleration, this gives the ions a major advantage over the protons. However, Figure 1 also shows that O^{+5} can drop out of resonance if $V_\|$ becomes too large; as will be seen below, this presents a difficulty. (Actually, He^{++} is sufficiently abundant to modify the dispersion relation. As discussed by *Hollweg* [2000a] and *HI*, He^{++} suffers a disadvantage similar to that of the protons.)

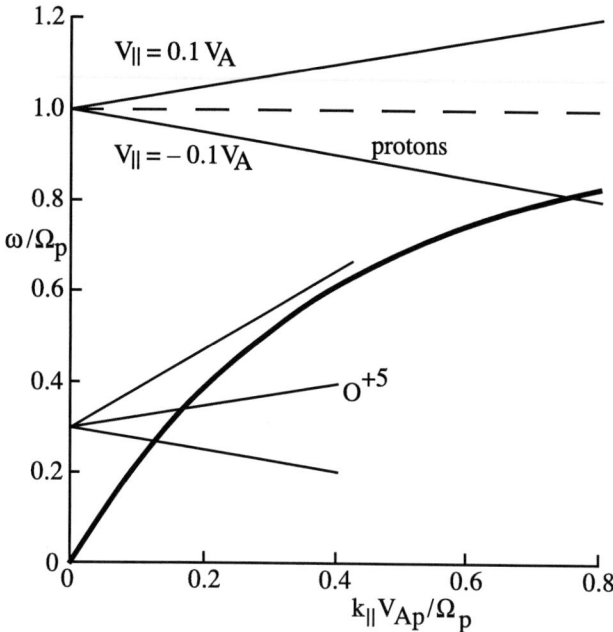

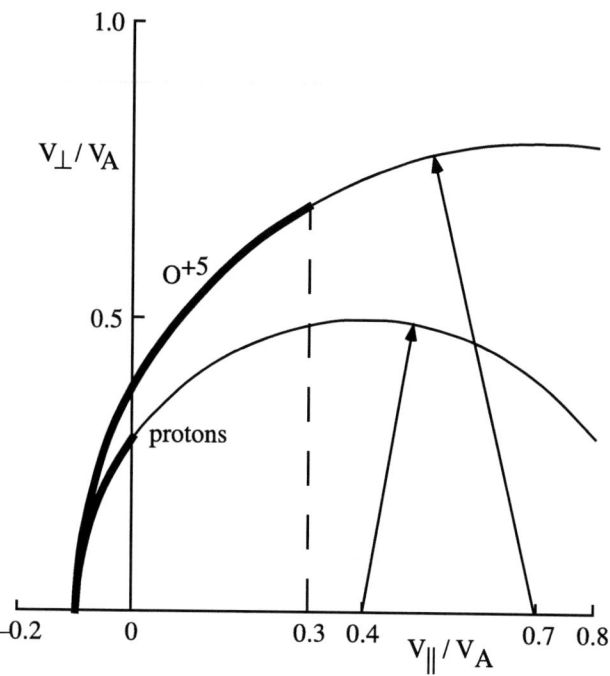

Figure 1. Dispersion relation (heavy curve) for the parallel-propagating ion-cyclotron mode in a cold electron-proton plasma, as viewed in the bulk proton frame. The resonance condition is shown as the thin straight lines, with positive (negative) slopes for test particles moving with (against) the wave. The top two lines are the resonance conditions for protons; only protons moving against the wave can be in resonance. The lower three lines are for O^{+5}; ions moving with and against the wave can be in resonance, but they can drop out of resonance if they move too rapidly.

Figure 2 illustrates several effects associated with velocity space diffusion. To keep things simple, we will here ignore dispersion, so that we can define a wave frame. Since there is no wave electric field in that frame, particles will conserve energy and diffuse along circular arcs $V_\parallel^2 + V_\perp^2 =$ constant ($V_\perp$ is the velocity perpendicular to **B**.) In Figure 2 the circular arcs are centered on the phase speed in the proton frame. In reality protons will resonate with slower-moving waves than O^{+5}, so we have drawn separate circles for those particles. *Marsch and Tu* [2001] have actually found evidence that solar wind protons diffuse along circular arcs.

Consider what happens to a group of particles which start out on the abscissa with $V_\perp = 0$. Those particles will diffuse upward along their arcs, thereby acquiring a perpendicular temperature; this is the most important part of the resonant heating since particles with large $V_\perp$ can subsequently be accelerated via the magnetic mirror force. Particles which start with the same $V_\parallel$ will acquire a spread in $V_\parallel$, and thus an increased parallel temperature, as they diffuse. (It is also possible for $T_\parallel$ to decrease [*Dusenbery and Hollweg*, 1981; *Li et al.*, 1999]. $T_\parallel$ heating or cooling is not a major effect, and

Figure 2. Diffusion of particles with nondispersive waves, as viewed in the bulk proton frame. In the wave frame particle energy is conserved, so particles diffuse along circular arcs centered on the wave phase speed. The arcs for protons and O^{+5} are centered on different values of $V_\parallel$, because protons tend to resonate with slower-moving waves than does O^{+5}. The thicker portions of the arcs indicate the values of $V_\parallel$ for which the particles can be in resonance with the ion-cyclotron mode. O^{+5} (and other heavier ions as well) can diffuse to higher values of $V_\perp$ than can the protons. Thus if the diffusion is fast enough, ions can attain $T_{\perp i} > (m_i/m_p)\, T_{\perp p}$.

we will not consider it further.) As particles diffuse upward along their arcs, they also move to the right. This represents a bulk resonant acceleration of the particles, but it is generally not as important as the magnetic mirroring. Now note the thick portions of the arcs in Figure 2. They indicate the values of $V_\parallel$ for which particles can be in resonance. Diffusion will tend to fill up the dark arcs, but not beyond. This fact, along with the two circles having different radii, clearly allow O^{+5} to attain larger values of $V_\perp$ than the protons. Recalling that temperature is proportional to mass times (thermal speed)^2, we see that diffusion tends to give $T_{\text{oxygen},\perp}/T_{p\perp} > m_{\text{oxygen}}/m_p$, in qualitative agreement with the UVCS/SOHO results (m denotes particle mass). Finally, the oxygen advantage is furthered by the fact that O^{+5} resonates with lower frequency waves which are observed to have more power.

These considerations are qualitatively in accord with the UVCS/SOHO data, but there are quantitative difficulties. *Kohl et al.* [1998] show that the O^{+5} temperature is an increasing function of r in $2 < r/r_S < 3.5$ (the outer limit of the data). But models [*Hollweg*, 2000a; *HI*] involving resonances with outward-propagating waves have been unable to reproduce this result. The resonant heating is effective in producing very hot O^{+5} rather close to the Sun. But the large values of $V_\perp$ then lead to a rapid acceleration via the mirror force, causing the oxygen to drop out of resonance, and to experience strong adiabatic cooling.

Resonances with sunward-propagating waves can save the day. (Sunward-propagating waves can arise from reflections or instabilities.) An outward-moving ion resonating with a sunward-propagating wave will have a resonance condition with negative slope in Figure 1, e.g., the lowest line in the figure. As the particle accelerates outward, the slope of the straight line will become even more negative. Not only will the particle never drop out of resonance, but it will resonate at ever smaller wavenumbers, where there is more power.

Another effect associated with sunward-propagating waves has been suggested by *Isenberg* [2001a], who pointed out that O^{+5} can simultaneously resonate with sunward and antisunward waves, while the protons cannot. In terms of diffusion, protons diffuse only along the arcs, while O^{+5} can diffuse across the arcs. This cross-arc transport is more commonly called second-order Fermi acceleration [e.g., *Terasawa*, 1989]. Since it is available to O^{+5} but not to the protons, it represents an energization mechanism which is inherently preferential to heavy ions. (He^{++} would tend to behave like protons in this scenario.) Quantitative evaluation of the effects of sunward-propagating waves remains to be carried through.

Another puzzle concerns Mg^{+9}, which has also been studied by SOHO/UVCS. Whereas O^{+5} is heated more than mass-proportionally relative to the protons, Mg^{+9} attains temperatures which are roughly mass-proportional. Why is there such different behavior for ions with similar values of q/m? The answer is not known; see discussions in *HI* and *Cranmer* [2002].

6. WHENCE THE ION-CYCLOTRON WAVES?

The UVCS/SOHO data seem to be plangent evidence for the cyclotron resonance. But there is no agreement on where the ion-cyclotron waves come from.

Based on seminal work by *Coleman* [1968] and *Barnes* [1979], one school of thought follows *Hollweg* [1986] and *HJ*: the Sun launches low-frequency waves which undergo a turbulent cascade that produces the resonant waves. This viewpoint is well-motivated observationally. *In situ* observations of magnetic field and velocity fluctuations show most power at low frequencies, but with power law power spectra extending to high frequencies [e.g., *Marsch and Tu*, 1990]. The power law indices have a preference for the −5/3 value expected for a Kolmogorov turbulent cascade. *In situ* data for the corona are not available. Radio studies do give some information about density fluctuations, though their connection with ion-cyclotron waves is unclear. The data are suggestive nonetheless. Most power resides at low frequencies, and the power spectra have the Kolmogorov index at higher frequencies, with some flattening just below the turbulence dissipation range [*Coles and Harmon*, 1989]. In both radio and *in situ* data, the power spectra steepen at the spatial scales expected if the ion-cyclotron resonance is coming into play [e.g., *Yamauchi et al.*, 1998; *Leamon et al.*, 1998a, b, 1999, 2000; *Bale et al.*, 2005; *Harmon and Coles*, 2005]; the spectral steepening is identified with the dissipation range.

It is often suggested that the turbulence scenario has a serious flaw: MHD turbulence tends to produce large cross-field wavenumbers, $k_\perp$, rather than the large values of $k_\parallel$ needed for cyclotron resonance [e.g., *Shebalin et al.*, 1983; *Ng and Bhattacharjee*, 1996; *Milano et al.*, 2001; *Oughton et al.*, 2004]. But the observational fact is that there is substantial power at large $k_\parallel$'s. This is especially true in the high-speed wind at low frequencies [*Dasso et al.*, 2005], but also near the dissipation range [*Leamon et al.*, 1998a, b, 1999, 2000]. *Harmon and Coles* [2005], from radio studies close to the sun, have concluded "radio scattering indicates there must be a substantial parallel component to the wave power which, in the turbulence picture, suggests a substantial non-perpendicular cascade". *Vasquez et al.* [2004] have proposed that large $k_\parallel$'s can be produced by turbulence if the background is spatially structured; the waves advect the structures, which in turn refract the waves, and so on, quickly leading to large $k_\parallel$'s. Perhaps non-MHD processes are at work, particularly near the dissipation range where kinetic effects can come into play. (See also *Cranmer and van Ballegooijen* [2003] for a discussion of the $k_\parallel$ problem.)

If the turbulence scenario is correct, the Sun must launch sufficient power at long periods to drive the fast wind. This issue has been addressed by looking at Faraday rotation fluctuations impressed on a radio signal as it traverses the corona. *Hollweg et al.* [1982] used the linearly polarized signal from the Helios spacecraft. They found that the polarization direction varied with timescales of the order of hours, suggesting a connection with the long-period Alfvén waves observed *in situ*. They showed that the observed Faraday rotation fluctuations in $2r_S < r < 15r_S$ closely matched what would be expected if there were indeed long-period Alfvén waves in the corona with enough power to drive the fast solar wind. Similar results were obtained by *Andreev et al.* [1997]. However, *Mancuso and Spangler* [1999] and *Spangler* [2002] looked at Faraday rotation fluctuations from natural radio sources and concluded that there was not enough long-period wave power to drive the fast wind. But it should be pointed out that there are two basic problems with the Faraday rotation studies. The first is that the observed rms fluctuations depend on the line-of-sight correlation length for the turbulence, which has to be guessed. The second difficulty is that the studies so far have extrapolated the data back to the coronal base without allowing for dissipation.

Harmon and Coles [2005] have used other radio data, sensitive to the density fluctuations, to investigate the energetics of the turbulence cascade close to the Sun. Their conclusion is "the cascade energy dissipated in proton cyclotron damping and electron Landau damping is large enough to be an important factor in solar wind heating and dynamics".

Like the *in situ* data, the Faraday rotation and density fluctuations in the corona have timescales of the order of hours. What on the Sun is responsible for these timescales? We know of only one solar phenomenon which occurs on timescales of hours: the flux cancellation events [*Livi et al.*, 1985; *Martin et al.*, 1985; *Ryutova et al.*, 2003]. It is not unreasonable to suppose that the drastic alteration of magnetic field in a flux cancellation event could launch long-period Alfvén waves with substantial energy fluxes [*Hollweg*, 1990]. This conclusion is probably closely related to a recent study by *Close et al.* [2004]. They found that "the timescale for magnetic flux to be remapped in the quiet-Sun corona is, surprisingly, only 1.4 hr ... implying that the quiet-Sun corona is far more dynamic than previously thought." In view of the long-known presence of hour periods in the low corona and solar wind, their result is not so surprising. (Twisting or shaking of intense magnetic flux tubes by the convective motions should also launch Alfvén waves. The trouble is that the timescales aren't right. The solar granulation has timescales of minutes while the supergranulation has timescales of tens of hours. The photosphere also contains weaker fields, called inter-network or intra-network fields [e.g., *Lites and Socas-Navarro*, 2004, and references therein]; the possible role of these fields in launching waves has to our knowledge not been explored.)

Ulrich [1996] looked at velocity and magnetic field fluctuations in the chromosphere, and found that they were correlated consistent with outgoing Alfvén waves. The time-averaged upward Poynting flux was about the amount required to power the chromosphere and corona in strong field active regions, where the data were taken. The study has not been done, but it is not unreasonable to suppose that the Alfvénic energy flux in regions of weaker field, e.g., in coronal holes, might be sufficient to drive the high-speed solar wind. However, unlike the *in situ* data and the radio data, Ulrich's fluctuations were mainly in the 5 minute band. Might the putative turbulent cascade originate from waves with periods of the order of 5 minutes, rather than hours?

An alternate proposal for the origin of the ion-cyclotron waves is that magnetic reconnection events in the photosphere or chromosphere launch waves with the kilohertz frequencies that are resonant with protons and ions in the corona [e.g., *Schwartz et al.*, 1981; *Axford and McKenzie*, 1992, 1996; *Czechowski et al.*, 1998; *McKenzie et al.*, 1995, 1997; *Ruzmaikin and Berger*, 1998]. To generate kilohertz frequencies, the reconnecting elements would have to be extremely small. The waves originate well below the local ion cyclotron frequencies in the photosphere and chromosphere, but they become resonant in the corona where the magnetic field is much weaker. *Tu and Marsch* [1997] and *Marsch and Tu* [1997] presented detailed solar wind models based on this scenario. They assumed that the waves propagate along **B** with a prescribed power spectrum. *Hollweg* [2000b] suggested that oblique propagation is far more likely, in which case the waves would be weakly compressive. Using a power spectrum specified by *Tu and Marsch*, he calculated the expected power spectrum of density fluctuations, if the waves propagate obliquely. At high wavenumbers, the predicted density spectrum at $r = 5r_S$ is 2-3 orders of magnitude larger than the observations [*Coles and Harmon*, 1989]. Unless the waves really are nearly parallel-propagating, which is difficult to imagine, we conclude that the direct-launching scenario is not viable. (See *HI* for a discussion of other objections to direct launching.)

A third proposal for the origin of the ion-cyclotron waves has emerged in recent years, viz. the waves are locally generated by plasma microinstabilities in the corona or transition region [e.g., *Markovskii*, 2001; *Markovskii and Hollweg*, 2002a; *Voitenko and Goossens*, 2002a, 2003, 2005; *Viñas et al.*, 2000; *Chen and Zhou*, 2003]. In some studies the instabilities are driven by the currents or gradients associated with low-frequency waves which contain most of the power. But the unstable waves are cyclotron resonant with protons and ions. The net result is a direct transfer of wave energy from low to high frequencies, without a 'cascade' through intermediate frequencies.

In other studies the instabilities are driven by electron or proton beams. An example of this type of study (one which we believe to be particularly promising) is the work of *Markovskii and Hollweg* [2002b, 2004, 2005]. They assume that microflares near the coronal base intermittently launch bursts of large electron heat flux outward into the corona. Via a Landau resonance with the distorted thermal electron distribution function, the heat flux drives electrostatic or electromagnetic ion-cyclotron waves unstable. These waves are highly oblique to the magnetic field, but with sunward $\omega/k_\parallel$ (in the local plasma frame). The cyclotron waves give perpendicular proton and ion heating. (Since the generated waves are highly dispersive, the particles diffuse approximately along hyperbolae in velocity space, not the circles used in our previous examples.)

7. A FEW PROMISING RECENT IDEAS

In Section 4 we mentioned that some models were internally inconsistent in that they assumed purely outward-propagating

Alfvén waves in the WKB limit. In that case the nonlinear terms which lead to turbulence sum to zero. For turbulence to develop, there must be a mix of inward- and outward-propagating waves, or the waves must be non-WKB. *Dmitruk et al.* [2002] proposed a simple model based on equations which describe the non-WKB propagation of linear Alfvén waves, but with an additional nonlinear term giving a dissipation rate dimensionally similar to the Kolmogorov rate, but with the important difference that there is no dissipation unless both inward- and outward-propagating waves are present. *Dmitruk et al.* [2002] offer some numerical calculations of the volumetric heating rates in the corona; the derived heating rates are comparable to what is required to drive the fast wind. We believe that this approach contains a significant amount of the physics needed for describing how the Alfvén wave amplitudes evolve as they propagate away from the Sun, and for describing the overall heating in the corona and interplanetary space; at the same this approach offers the virtue of simplicity. (See also *Matthaeus et al.* [1999], *Dmitruk et al.* [2001], and *Dmitruk and Matthaeus* [2003]. *Cranmer and van Ballegooijen* [2005] have shown that the approach of *Dmitruk et al.* [2002] accounts well for what is known about the evolution of wave amplitudes throughout the corona and interplanetary space. The reader might also wish to consult *Isenberg et al.* [2003], *Matthaeus et al.* [2004], and *Breech et al.* [2005] for examples of calculations of the evolution of turbulent fluctuations in the solar wind far from the Sun, where V_A can be neglected compared to the solar wind flow speed. Unlike *Dmitruk et al.* [2002], these models include the effects of velocity shear and pickup ions. Velocity shear may be an important aspect of the evolution, perhaps accounting for some of the differences between the evolution of waves/turbulence in fast and slow solar wind flows [*Breech et al.*, 2005].)

A wave-turbulence phenomenolgy such as we have just discussed can describe many aspects of the wave evolution and plasma heating in the fast wind. But there are other aspects of the wave evolution which turbulence phenomenology does not capture. It should first be noted that if waves supply the energy for the wind, then they must be very linear ($|\delta\mathbf{B}|/B \ll 1$) in the vicinity of the acceleration region; otherwise there would be more energy supplied than can be accounted for. However, even allowing for turbulent dissipation, the tendency to conserve wave action implies that the waves become nonlinear ($|\delta\mathbf{B}|/B \approx 1$) in interplanetary space, as is observed. But even though the magnetic field direction fluctuates strongly, the magnetic field strength is observed to be nearly constant ($\delta|\mathbf{B}|/B \ll 1$); in other words, the tip of the magnetic field vector moves nearly on a sphere. How this comes about has been explained via a nonlinear wave analysis combined with hybrid simulations [*Vasquez and Hollweg*, 1996a,b]. Nonlinear wave theory [*Vasquez and Hollweg*, 1998, 1999a] has also succeeded in explaining why there are many apparently stable rotational discontinuities (essentially sharp-crested Alfvén waves) imbedded within the waves in interplanetary space [*Tsurutani et al.*, 1994], even though an isolated rotational discontinuity would be unstable and very short-lived. Moreover, even in high-speed wind, where the Alfvén waves are most 'pure', there is a weak admixture of fast waves propagating across the background magnetic field and pressure-balanced structures (or perhaps transversely-propagating slow modes) [*Tu and Marsch*, 1994]; this result has also been explained as a natural outcome of the nonlinear evolution of Alfvén waves [*Vasquez and Hollweg*, 1999b]. Another result best explained using wave theory concerns the behavior of He^{++}. We have already said that the He^{++} in interplanetary space flows faster than the protons, by about V_A. But V_A is a decreasing function of r in interplanetary space, implying that the He^{++} must slow down. This, too, has best been explained in terms of He^{++} interacting with nonlinear waves [*Kaghashvili et al.*, 2003].

Thus far, turbulence theory has not addressed what happens in the dissipation range, where non-MHD kinetic processes are critical. Similarly, nonlinear wave studies have not simultaneously reproduced a turbulence cascade and the details of the dissipation range. The goal would be to calculate how the proton and ion distribution functions evolve subject to resonant diffusion in velocity space, along with global forces such as gravity, magnetic mirroring, etc. *Isenberg et al.* [2001] and *Isenberg* [2001b, 2004a,b] have suggested an approximate procedure, which is mostly analytical. They call their approach "the kinetic shell model", the shells being the 3-dimensional versions of the diffusion arcs in Figure 2. (See also *Galinsky and Shevchenko* [2000] for independent work which is closely related to the kinetic shell model.) The essence of this model is shown in Figure 3, which is similar to Figure 2. Velocity space is displayed with several diffusion arcs for protons. The kinetic shell model's key assumption is that the velocity space diffusion is faster than any other timescale, so that the distribution function is always very nearly uniform along each arc. (This implicitly assumes that the wave power spectrum in the resonant range is above some minimum level.) However, the number of particles can differ from shell to shell. Individual shells can be thought of as moving in response to the global forces: gravity, mirroring, and the charge-separation electric field. Consider first sunward-moving protons, which we take to have $V_\parallel < 0$. Recall that these protons resonate with outward-propagating waves. The leftmost shells, which reach to large values of $V_\perp$, experience a net mirror force which overwhelms gravity and the electric field, so those shells are pushed to larger values of $V_\parallel$, as indicated by the leftmost arrow. Other shells having $V_\parallel < 0$, but reaching to smaller values of $V_\perp$, are dominated by the sunward combination of gravity and electric field; those

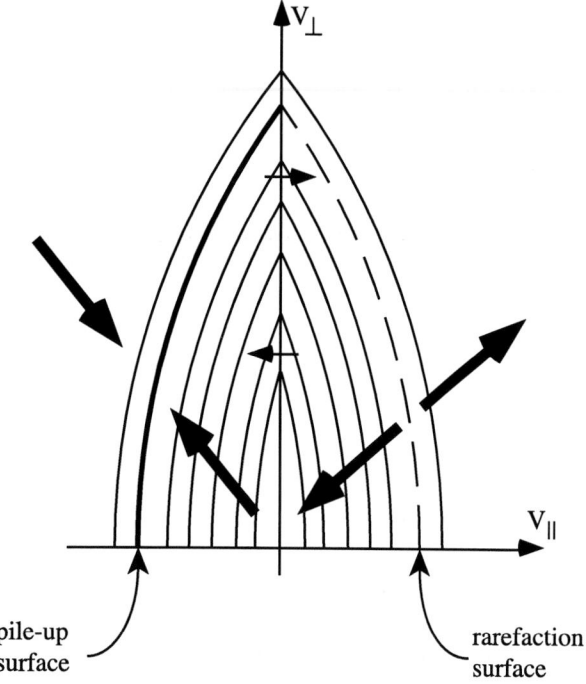

Figure 3. Velocity space diagram illustrating the evolution of the proton distribution in the kinetic shell model, as the plasma moves away from the Sun. Protons interacting with outward (inward) propagating waves diffuse along the arcs in the left (right) half of the diagram. The thick arrows indicate the shell motion, converging on the sunward ($V_\parallel < 0$) side and diverging on the anti-sunward side ($V_\parallel > 0$). The thin arrows indicate the particle transport across the $V_\parallel = 0$ boundary, determined by the nonresonant forces. The pileup and rarefaction surfaces are shown as the thick and dashed curves, respectively.

shells are driven to more negative values of $V_\parallel$, as indicated by the second arrow from the left. The result is that the sunward-moving protons tend to accumulate near a 'pile-up surface' where all forces balance. At $V_\parallel = 0$ the protons are resonant with waves having infinite wavenumber, and zero power. Thus, those protons are unaffected by the waves, and move across the $V_\parallel = 0$ 'boundary' in response to the global forces. If $V_\perp$ is large enough, the particles will be pushed toward $V_\parallel > 0$, as indicated by the right-pointing arrow; conversely for particles with lower $V_\perp$. Particles with $V_\parallel > 0$ resonate with sunward-propagating waves, and follow the diffusion arcs sketched in the figure. The two rightmost arrows indicate the shell motions in response to the global forces; shells which reach to large $V_\perp$ are pushed to larger values of $V_\parallel$, and so on. In this case the opposite of a pileup surface develops: there will be a 'rarefaction surface' on which particles are depleted. Even if there are no sunward-propagating waves initially, in this model they will be generated by those protons at large $V_\perp$ which are pushed into the $V_\parallel > 0$ region by the mirror force. The resulting distribution is highly unstable to the generation of sunward-propagating waves.

With dispersion ignored [*Isenberg*, 2001b], rapidly accelerating fast winds were obtained. But when wave dispersion was included [*Isenberg*, 2004a,b], only slow winds were produced. The reason has to do with the tops of the arcs in Figure 3, which tend to flatten out when the waves are dispersive. Thus the arcs extend to smaller values of $V_\perp$, and the net mirror force is reduced. Nonetheless, in the author's opinion the kinetic shell model has many virtues. Perhaps further developments including the warm plasma dispersion relation and obliquely-propagating waves will save the day.

8. SUMMARY

We have shown how our thinking about the solar wind has progressed from Parker's electron-driven wind, through a wave-driven wind, to our current proton-driven wind, driven mainly the magnetic mirror force if the protons have $T_\perp \gg T_\parallel$. (See *Vásquez et al.* [2003] for a discussion of the importance of thermal anisotropy.). Our current picture also allows for a significant contribution from wave pressure, but this will give a more gradual acceleration acting mainly beyond the sonic point. We have emphasized that coronal heating and solar wind acceleration need to be treated together.

The SOHO fast solar wind data have shown that coronal heating works mainly on the transverse components of protons and ions. It is not Joule heating, as is very commonly presumed to be the case in other parts of the corona. Coronal heating also does not seem to be dominated by viscosity or by heat conduction. Do proton and ion heating dominate other parts of the corona, such as the active region loops, as well?

Heating transverse to the magnetic field strongly implicates the ion-cyclotron resonance. But we still do not know the source of the high-frequency resonant waves. We have given some arguments against the proposal that the Sun directly launches these waves. The Sun could well launch the required energy fluxes at long periods, with the high-frequencies generated by a turbulent cascade. The tendency of MHD turbulence to produce mainly high $k_\perp$ is a difficulty, but we do not believe it is insurmountable. Finally, some workers have begun to explore the promising idea that the high-frequency waves may originate locally in the corona from plasma microinstabilities; this scenario seems worth pursuing.

One thing should be clear from the foregoing discussion: A complete description of coronal heating and fast wind acceleration will require detailed considerations of particle kinetics; MHD does not tell the whole story. The non-Maxwellian particle distribution functions observed by spacecraft [e.g., *Marsch*, 1991] are trying to tell us something about the physics of heating and acceleration closer to the Sun.

Future studies will need to place more emphasis on wave propagation oblique to the magnetic field. The Sun almost certainly launches highly oblique waves which are subsequently refracted. The microinstabilities considered so far tend to produce highly oblique waves. And there is some evidence [*Leamon et al.*, 1998a,b, 1999, 2000; *Bale et al.*, 2005] that the dissipation range of the turbulence contains kinetic Alfvén waves, which are essentially highly oblique Alfvén waves [*Hollweg*, 1999] which are compressive and therefore subject to Landau damping. *Voitenko and Goossens* [2002b, 2004] have emphasized the possible importance of kinetic Alfvén waves. The latter paper considers what happens to test ions interacting with a large-amplitude kinetic Alfvén packet. Surprisingly, under certain circumstances the ions can gain substantial energy even though there is no cyclotron or Landau resonance. This process might be important far from the Sun where the solar wind wave field has large amplitude, but it will not be important close to the Sun where the waves are almost certainly small amplitude with $|\delta \mathbf{B}|/B \ll 1$.

Future studies will also need to expand consideration of inward-, as well as outward-propagating waves, especially since the mix of inward and outward waves is essential for the development of turbulence. Moreover, inward-propagating waves may be responsible for the extended heating of O^{+5}.

Finally, we need to say something about electron heating, which has been completely ignored in the foregoing. It is difficult to say much about the electrons because we do not understand how to model their heat conduction under weakly collisional conditions. It is possible that the electrons need no external heating, and that collisional coupling with the protons suffices to maintain their temperatures. It is also possible that the electrons are strongly heated intermittently by reconnection events at the coronal base. And it may be that electrons are heated throughout the corona via kinetic Alfvén waves, which are compressive and have a parallel electric field; since the electron thermal speed is comparable to V_A in the corona, the electrons can be heated via the Landau resonance [e.g., *Leamon et al.*, 1999, 2000]. From their radio studies close to the Sun, *Harmon and Coles* [2005] conclude "at least half of the total cascade flux goes into electron heating as cascaded power traverses the broad Landau resonance on the way to the cyclotron resonance".

Fisk [2003] and *Gloeckler et al.* [2003] have emphasized observational evidence that the fastest (slowest) solar wind originates from the coronal regions with the coolest (hottest) electrons, a result completely opposite to the model of *Parker* [1958]. They explain this result in terms of 'interchange reconnections' which convert closed loop-like magnetic field lines into open field lines. Alternatively, *Schwadron and McComas* [2003] have suggested that coronal regions with hotter electrons give rise to slower solar wind because those regions conduct more energy down to the transition region where it is lost via radiation, leaving less energy for the kinetic energy of the flow.

Acknowledgments. The author is grateful to S.R. Cranmer, P.A. Isenberg, M.A. Lee, S.A. Markovskii, W.H. Matthaeus, and B.J. Vasquez for many helpful conversations and collaborations. This work was supported by the NASA Sun-Earth Connection Theory Program under grant NAG5-11797 to the University of New Hampshire.

REFERENCES

Alazraki, G., and P. Couturier, Solar wind acceleration caused by the gradient of Alfvén wave pressure, *Astron. Astrophys.*, 13, 380, 1971.

Andreev, V.E., A.I. Efimov, L.N. Samoznaev, I.V. Chashei, and M.K. Bird, Characteristics of coronal Alfvén waves deduced from Helios Faraday rotation measurements, *Solar Phys.*, 176, 387, 1997.

Antonucci, E., M.A. Dodero, and S. Giordano, Fast solar wind velocity in a polar coronal hole during solar minimum, *Solar Phys.*, 197, 115, 2000.

Axford, W.I., and J.F. McKenzie, The origin of high speed solar wind streams, in *Solar Wind Seven*, edited by E. Marsch, and R. Schwenn, p. 1, Pergamon Press, Oxford, 1992.

Axford, W.I., and J.F. McKenzie, The acceleration of the solar wind, in *Solar Wind Eight*, edited by D. Winterhalter, J.T. Gosling, S.R. Habbal, W.S. Kurth, and M. Neugebauer, p. 72, Amer. Inst. Phys., New York, 1996.

Bale, S.D., P.J. Kellogg, F.S. Mozer, T.S. Horbury, and H. Reme, Measurements of the electric fluctuation spectrum of magnetohydrodynamic turbulence, *Phys. Rev. Lett.*, 94, in press, 2005.

Barnes, A., Hydromagnetic waves and turbulence in the solar wind, in *Solar System Plasma Physics, Vol. I*, edited by E.N. Parker, C.F. Kennel, and L.J. Lanzerotti, p. 249, North-Holland, Dordrecht, 1979.

Belcher, J.W., Alfvénic wave pressures and the solar wind, *Astrophys. J.*, 168, 509, 1971.

Belcher, J.W., and L. Davis, Jr., Large-amplitude Alfvén waves in the interplanetary medium,2, *J. Geophys. Res.*, 76, 3534, 1971.

Breech, B., W.H. Matthaeus, J. Minnie, S. Oughton, S. Parhi, J.W. Bieber, and B. Bavassano, Radial evolution of cross helicity in high-latitude solar wind, *Geophys. Res. Lett.*, 32, L06103, doi:10.1029/2004GL022321, 2005.

Bretherton, F.P., and C.J.R. Garrett, Wavetrains in inhomogeneous moving media, *Proc. Roy. Soc. A*, 302, 529, 1969.

Chen, Y.P., and G.C. Zhou, Obliquely propagating shear Alfvén waves exicted by newborn ions in the chromosphere-corona transition region, *Solar Phys.*, 215, 57, 2003.

Close, R.M., C.E. Parnell, D.W. Longcope, and E.R. Priest, Recycling of the solar corona's magnetic field, *Astrophys. J.*, 612, L81, 2004.

Coleman, P.J., Jr., Turbulence, viscosity, and dissipation in the solar wind plasma, *Astrophys. J.*, 153, 371, 1968.

Coles, W.A., and J.K. Harmon, Propagation observations of the solar wind near the sun, *Astrophys. J.*, 337, 1023, 1989.

Cranmer, S.R., Coronal holes and the high-speed solar wind, *Space Sci. Rev.*, 101, 229, 2002.

Cranmer, S.R., Coronal heating versus solar wind acceleration, in *Proceedings SOHO-15: Coronal Heating*, in press, ESA, Noordwijk, 2005.

Cranmer, S.R., and A.A. van Ballegooijen, Alfvénic turbulence in the extended solar corona: kinetic effects and proton heating, *Astrophys. J.*, 594, 573, 2003.

Cranmer, S.R., and A.A. van Ballegooijen, On the generation, propagation, and reflection of Alfvén waves from the solar photosphere to the distant heliosphere, *Astrophys. J. Suppl.*, 156, 265, 2005.

Czechowski, A., R. Ratkiewicz, J.F. McKenzie, and W.I. Axford, Heating and acceleration of minor ions in the solar wind, *Astron. Astrophys.*, 335, 303, 1998.

Dasso, S., L.J. Milano, W.H. Matthaeus, and C.W. Smith, Anisotropy in fast and slow solar wind fluctuations, *Geophys. Res. Lett.*, submitted, 2005.

Dewar, R.L., Interaction between hydromagnetic waves and a time-dependent, inhomogeneous medium, *Phys. Fluids*, 13, 2710, 1970.

Dmitruk, P., L.J. Milano, and W.H. Matthaeus, Wave-driven turbulent coronal heating in open field line regions: nonlinear phenomenological model, *Astrophys. J.*, 548, 482, 2001.

Dmitruk, P., W.H. Matthaeus, L.J. Milano, S. Oughton, G.P. Zank, and D.J. Mullan, Coronal heating distribution due to low-frequency wave-driven turbulence, *Astrophys. J.*, 575, 571, 2002.

Dmitruk, P., and W.H. Matthaeus, Low-frequency waves and turbulence in an open magnetic region: Timescales and heating efficiency, *Astrophys. J.*, 597, 1097, 2003.

Dodero, M.A., E. Antonucci, S. Giordano, and R. Martin, Solar wind velocity and anisotropic coronal kinetic temperature measured with the O VI doublet ratio, *Solar Phys.*, 183, 77, 1998.

Dusenbery, P.B., and J.V. Hollweg, Ion-cyclotron heating and acceleration of solar wind minor ions, *J. Geophys. Res.*, 86, 153, 1981.

Einaudi, G., P. Boncinelli, R.B. Dahlburg, and J.T. Karpen, Formation of the slow solar wind in a coronal streamer, *J. Geophys. Res.*, 104, 521, 1999.

Erdös, G., Waves and turbulence in the solar wind, in *Turbulence, Waves and Instabilities in the Solar Plasma*, edited by R. Erdélyi, K. Petrovay, B. Roberts, and M.J. Aschwanden, p. 367, Kluwer, Dordrecht, 2003.

Esser, R., and S.R. Habbal, Coronal heating and plasma parameters at 1 AU, *Geophys. Res. Lett.*, 22, 2661, 1995.

Esser, R., and S.R. Habbal, Modeling high flow speeds in the inner corona, in *Solar Wind Eight*, edited by D. Winterhalter, J. Gosling, S.R. Habbal, W.S. Kurth, and M. Neugebauer, p. 133, AIP, New York, 1996.

Esser, R., H.R. Habbal, W.A. Coles, and J.V. Hollweg, Hot protons in the inner corona and their effect on the flow properties of the solar wind, *J. Geophys. Res.*, 102, 7063, 1997.

Esser, R., S. Fineschi, D. Dobrzycka, S.R. Habbal, R.J. Edgar, J.C. Raymond, J.L. Kohl, and M. Guhathakurta, Plasma properties in coronal holes derived from measurements of minor ion spectral lines and polarized white light intensity, *Astrophys. J.*, 510, L63, 1999.

Feldman, W.C., J.R. Asbridge, S.J. Bame, and J.T. Gosling, High-speed solar wind flow parameters at 1 AU, *J. Geophys. Res.*, 81, 5054, 1976.

Fisher, R., and M. Guhathakurta, Physical properties of polar coronal rays and holes as observed with the Spartan 201-01 coronagraph, *Astrophys. J.*, 447, L139, 1995.

Fisk, L.A., Acceleration of the solar wind as a result of the reconnection of open magnetic flux with coronal loops, *J. Geophys. Res.*, 108(A4), 1157, doi:10.1029/2002JA009284, 2003.

Fisk, L.A., and N.A. Schwadron, Origin of the solar wind: Theory, *Space Sci. Rev.*, 97, 21, 2001.

Galinsky, V.L., and V.I. Shevchenko, Nonlinear cyclotron resonant wave-particle interaction in a nonuniform magnetic field, *Phys. Rev. Lett.*, 85(1), 90, 2000.

Gloeckler, G., T.H. Zurbuchen, and J. Geiss, Implications of the observed anticorrelation between solar wind speed and coronal electron temperature, *J. Geophys. Res.*, 108(A4), 1158, doi:10.1029/2002JA009286, 2003.

Guhathakurta, M., and T.E. Holzer, Density structure inside a polar coronal hole, *Astrophys. J.*, 426, 782, 1994.

Guhathakurta, M., and R. Fisher, Solar wind consequences of a coronal hole density profile: Spartan 201-203 coronagraph and Ulysses observations from $1.15R_s$ to 4 AU, *Astrophys. J.*, 499, L215, 1998.

Hansteen, V.H., and E. Leer, Coronal heating, densities, and temperatures and solar wind acceleration, *J. Geophys. Res.*, 100, 21577, 1995.

Harmon, J.K., and W.A. Coles, Modeling radio scattering and scintillation observations of the inner solar wind using oblique Alfvén/ion cyclotron waves, *J. Geophys. Res.*, 110, A03101, doi:10.1029/2004JA010834, 2005.

Hartle, R.E., and P.A. Sturrock, Two-fluid model of the solar wind, *Astrophys. J.*, 151, 1155, 1968.

Hollweg, J.V., Alfvén waves in a two-fluid model of the solar wind, *Astrophys. J.*, 181, 547, 1973.

Hollweg, J.V., Some physical processes in the solar wind, *Rev. Geophys.*, 16, 689, 1978.

Hollweg, J.V., Transition region, corona, and solar wind in coronal holes, *J. Geophys. Res.*, 91, 4111, 1986.

Hollweg, J.V., Heating of the solar corona, *Computer Phys. Reports*, 12, 205, 1990.

Hollweg, J.V., The kinetic Alfvén wave revisited, *J. Geophys. Res.*, 104, 14811, 1999.

Hollweg, J.V., The cyclotron resonance in coronal holes. 3. a five-beam turbulence-driven model, *J. Geophys. Res.*, 105, 15699, 2000a.

Hollweg, J.V., Compressibility of ion-cyclotron and whistler waves: can radio measurements detect high-frequency waves of solar origin in the corona?, *J. Geophys. Res.*, 105, 7573, 2000b.

Hollweg, J.V., M. Bird, H. Volland, P. Edenhofer, C. Stelzried, and B. Seidel, Possible evidence for coronal Alfvén waves, *J. Geophys. Res.*, 87, 1, 1982.

Hollweg, J.V., and W. Johnson, Transition region, corona, and solar wind in coronal holes: some two-fluid models, *J. Geophys. Res.*, 93, 9547, 1988.

Hollweg, J.V., and P.A. Isenberg, Generation of the fast solar wind: A review with emphasis on the resonant cyclotron interaction, *J. Geophys. Res.*, 107(A7), 1147, doi:10.1029/2001JA000270, 2002.

Isenberg, P.A., Acceleration of heavy ions in the solar wind, in *Solar Wind Five*, edited by M. Neugebauer, p. 655, NASA, Washington DC, 1983.

Isenberg, P.A., Investigations of a turbulent-driven solar wind model, *J. Geophys. Res.*, 95, 6437, 1990.

Isenberg, P.A., Heating of coronal holes and generation of the solar wind by ion-cyclotron resonance, *Space Sci. Rev.*, 95, 119, 2001a.

Isenberg, P.A., The kinetic shell model of coronal heating and acceleration by ion cyclotron waves. 2. inward and outward propagating waves, *J. Geophys. Res.*, 106, 29249, 2001b.

Isenberg, P.A., The kinetic shell model of coronal heating and acceleration by ion cyclotron waves. 3. the proton halo and dispersive waves, *J. Geophys. Res.*, 109, A03101, doi:10.1029/2002JA009449, 2004a.

Isenberg, P.A., Correction to "The kinetic shell model of coronal heating and acceleration by ion cyclotron waves: 3. The proton halo and dispersive waves", *J. Geophys. Res.*, 109, A06106, doi:10.1029/2004JA010524, 2004b.

Isenberg, P.A., M.A. Lee, and J.V. Hollweg, The kinetic shell model of coronal heating and acceleration by ion-cyclotron waves. 1. outward-propagating waves, *J. Geophys. Res.*, 106, 5649, 2001.

Isenberg, P.A., C.W. Smith, and W.H. Matthaeus, Turbulent heating of the distant solar wind by interstellar pickup ions, *Astrophys. J.*, 592, 564, 2003.

Jacques, S.A., Momentum and energy transport by waves in the solar atmosphere and solar wind, *Astrophys. J.*, 215, 942, 1977.

Kaghashvili, E.K., B.J. Vasquez, and J.V. Hollweg, Deceleration of streaming alpha particles interacting with waves and imbedded rotational discontinuities, *J. Geophys. Res.*, 108(A1), 1036, doi:10.1029/2002JA009623, 2003.

Kohl, J.L., et al., UVCS/SOHO empirical determinations of anisotropic velocity distributions in the solar corona, *Astrophys. J.*, 501, L127, 1998.

Lapenta, G., and D.A. Knoll, Effect of a converging flow at the streamer cusp on the genesis of the slow solar wind, *Astrophys. J.*, 624, 1049, 2005.

Leamon, R.J., W.H. Matthaeus, C.W. Smith, and H.K. Wong, Contribution of cyclotron-resonant damping to kinetic dissipation of interplanetary turbulence, *Astrophys. J.*, 507, L181, 1998a.

Leamon, R.J., C.W. Smith, N.F. Ness, W.H. Matthaeus, and H.K. Wong, Observational constraints on the dynamics of the interplanetary magnetic field dissipation range, *J. Geophys. Res.*, 103, 4775, 1998b.

Leamon, R.J., C.W. Smith, N.F. Ness, and H.K. Wong, Dissipation range dynamics: kinetic Alfvén waves and the importance of β_e, *J. Geophys. Res.*, 104, 22331, 1999.

Leamon, R.J., W.H. Matthaeus, C.W. Smith, G.P. Zank, D.J. Mullan, and S. Oughton, MHD-driven kinetic dissipation in the solar wind and corona, *Astrophys. J.*, 537, 1054, 2000.

Li, B., X. Li, Y.-Q. Hu, and S.R. Habbal, A two-dimensional Alfvén wave-driven solar wind model with proton temperature anisotropy, *J. Geophys. Res.*, 109, A07103, doi:10.1029/2003JA010313, 2004.

Li, X., Proton temperature anisotropy in the fast solar wind: a 16-moment bi-Maxwellian model, *J. Geophys. Res.*, 104, 19773, 1999.

Li, X., Heating in coronal funnels by ion cyclotron waves, *Astrophys. J.*, 571, L67, 2002.

Li, X., and S.R. Habbal, Ion cyclotron waves, instabilities and solar wind heating, *Solar Phys.*, 190, 485, 1999.

Li, X., S.R. Habbal, J.V. Hollweg, and R. Esser, Heating and cooling of protons by turbulence-driven ion cyclotron waves in the fast solar wind, *J. Geophys. Res.*, 104, 2521, 1999.

Lie-Svendsen, Ø., E. Leer, and V.H. Hansteen, A 16-moment solar wind model: from the chromosphere to 1AU, *J. Geophys. Res.*, 106, 8217, 2001.

Lites, B.W., and H. Socas-Navarro, Characterization of magnetic flux in the quiet Sun. II. The internetwork fields at high angular resolution, *Astrophys. J.*, 613, 600, 2004.

Livi, S.H.B., J. Wang, and S.F. Martin, The cancellation of magnetic flux. I. on the quiet sun, *Aust. J. Phys.*, 38, 855, 1985.

Mancuso, S., and S.R. Spangler, Coronal faraday rotation observations: measurements and limits on plasma inhomogeneities, *Astrophys. J.*, 525, 195, 1999.

Markovskii, S.A., Generation of ion-cyclotron waves in coronal holes by a global resonant MHD mode, *Astrophys. J.*, 557, 337, 2001.

Markovskii, S.A., and J.V. Hollweg, Parametric cross-field current instability in solar coronal holes, *J. Geophys. Res.*, 107(A10), 1329, doi:10.1029/2001JA009140, 2002a.

Markovskii, S.A., and J.V. Hollweg, Electron heat flux instabilities in coronal holes: implications for ion heating, *Geophys. Res. Lett.*, 29, 1843, doi:10.1029/2002GL015189, 2002b.

Markovskii, S.A., and J.V. Hollweg, Intermittent heating of the solar corona by heat flux generated ion cyclotron waves, *Astrophys. J.*, 609, 1112, 2004.

Markovskii, S.A., and J.V. Hollweg, Radial evolution of intermittent heat flux in solar coronal holes, *Nonlinear Proc. Geophys.*, in press, 2005.

Marsch, E., Kinetic physics of the solar wind plasma, in *Physics of the Inner Heliosphere, 2, Particles, Waves and Turbulence*, edited by R. Schwenn, and E. Marsch, p. 45, Springer-Verlag, Berlin, 1991.

Marsch, E., and C.-Y. Tu, On the radial evolution of MHD turbulence in the inner heliosphere, *J. Geophys. Res.*, 95, 8211, 1990.

Marsch, E., and C.-Y. Tu, The effects of high-frequency Alfvén waves on coronal heating and solar wind acceleration, *Astron. Astrophys.*, 319, L17, 1997.

Marsch, E., and C.-Y. Tu, Evidence for pitch angle diffusion of solar wind protons in resonance with cyclotron waves, *J. Geophys. Res.*, 106, 8357, 2001.

Martin, S.F., S.H.B. Livi, and J. Wang, The cancellation of magnetic flux. II. in a decaying active region, *Aust. J. Phys.*, 38, 929, 1985.

Matthaeus, W.H., G.P. Zank, S. Oughton, D.J. Mullan, and P. Dmitruk, Coronal heating by magnetohydrodynamic turbulence driven by reflected low-frequency waves, *Astrophys. J.*, 523, L93, 1999.

Matthaeus, W.H., J. Minnie, B. Breech, S. Parhi, J.W. Bieber, and S. Oughton, Transport of cross helicity and radial evolution of Alfvénicity in the solar wind, *Geophys. Res. Lett.*, 31, L12803, doi:10.1029/2004GL019645, 2004.

McKenzie, J.F., M. Banaszkiewicz, and W.I. Axford, Acceleration of the high speed solar wind, *Astron. Astrophys.*, 303, 45, 1995.

McKenzie, J.F., W.I. Axford, and M. Banaszkiewicz, The fast solar wind, *Geophys. Res. Lett.*, 24, 2877, 1997.

Milano, L.J., W.H. Matthaeus, P. Dmitruk, and D.C. Montgomery, Local anisotropy in incompressible magnetohydrodynamic turbulence, *Phys. Plasmas*, 8, 2673, 2001.

Miralles, M.P., S.R. Cranmer, and J.L. Kohl, Cyclical variations in the plasma properties of coronal holes, in *Proc. SOHO 11 Symposium*, edited by A. Wilson, p. 351, ESA SP-508, Noordwijk, 2002.

Miralles, M.P., S.R. Cranmer, and J.L. Kohl, Low-latitude coronal holes during solar maximum, *Adv. Space Res.*, 33, 696, 2004.

Moran, T.G., Test for Alfvén wave signatures in a solar coronal hole, *Astrophys. J.*, 598, 657, 2003.

Neugebauer, M., Observations of solar-wind helium, in *Solar Wind Four*, edited by H. Rosenbauer, p. 425, Max Planck Institut für Aeronomie, Katlenburg-Lindau, 1981.

Neugebauer, M., Knowledge of coronal heating and solar wind acceleration obtained from observations of the solar wind near 1 AU, in *Solar Wind Seven*, edited by E. Marsch, and R. Schwenn, p. 69, Pergamon, Oxford, 1992.

Ng, C.S., and A. Bhattacharjee, Interaction of shear-Alfvén packets: implication for weak magnetohydrodynamic turbulence in astrophysical plasmas, *Astrophys. J.*, 465, 845, 1996.

Oughton, S., P. Dmitruk, and W.H. Matthaeus, Reduced magnetohydrodynamics and parallel spectral transfer, *Phys. Plasmas*, 11, 2214, 2004.

Parker, E.N., Dynamics of the interplanetary gas and magnetic fields, *Astrophys. J.*, 128, 664, 1958.

Parker, E.N., Dynamical theory of the solar wind, *Space Science Rev.*, 4, 666, 1965.

Peter, H., and C. Vocks, Heating the magnetically open ambient background corona of the Sun by Alfvén waves, *Astron. Astrophys.*, 411, L481, 2003.

Ruzmaikin, A., and M.A. Berger, On a source of Alfvén waves heating the solar corona, *Astron. Astrophys.*, 337, L9, 1998.

Ryutova, M., T. Tarbell, and R. Shine, Interaction and dynamics of the photospheric network magnetic elements, *Solar Phys.*, 213, 231, 2003.

Schwadron, N.A., and D.J. McComas, Solar wind scaling law, *Astrophys. J.*, 599, 1395, 2003.

Schwartz, S.J., W.C. Feldman, and S.P. Gary, The source of proton anisotropy in the high-speed solar wind, *J. Geophys. Res.*, 86, 541, 1981.

Shebalin, J.V., W.H. Matthaeus, and D. Montgomery, Anisotropy in MHD turbulence due to a mean magnetic field, *J. Plasma Phys.*, 29, 525, 1983.

Spangler, S.R., The small amplitude of density turbulence in the inner solar wind, *Nonlinear Proc. Geophys.*, 10, 113, 2002.

Terasawa, T., Particle scattering and acceleration in a turbulent plasma around comets, in *Plasma Waves and Instabilities at Comets and in Magnetospheres*, edited by B.T. Tsurutani, and H. Oya, p. 41, Amer. Geophys. Union, Washington, 1989.

Tsurutani, B.T., C.M. Ho, E.J. Smith, M. Neugebauer, B.E. Goldstein, J.S. Mok, J.K. Arballo, A. Balogh, D.J. Southwood, and W.C. Feldman, The relationship between interplanetary discontinuities and Alfvén waves: Ulysses observations, *Geophys. Res. Lett.*, 21, 2267, 1994.

Tsurutani, B.T., and G.S. Lakhina, Some basic concepts of wave-particle interactions in collisionless plasmas, *Rev. Geophys.*, 35, 491, 1997.

Tu, C.-Y., and E. Marsch, On the nature of compressive fluctuations in the solar wind, *J. Geophys. Res.*, 99, 21481, 1994.

Tu, C.-Y., and E. Marsch, Two-fluid model for heating of the solar corona and acceleration of the solar wind by high-frequency Alfvén waves, *Solar Phys.*, 171, 363, 1997.

Tu, C.-Y., E. Marsch, K. Wilhelm, and W. Curdt, Ion temperatures in a solar polar coronal hole observed by SUMER on SOHO, *Astrophys. J.*, 503, 475, 1998.

Ulrich, R.K., Observations of magnetohydrodynamic oscillations in the solar atmosphere with properties of Alfvén waves, *Astrophys. J.*, 465, 436, 1996.

Vasquez, B.J., and J.V. Hollweg, Formation of arc-shaped Alfvén waves and rotational discontinuities from oblique linearly polarized wave trains, *J. Geophys. Res.*, 101, 13527, 1996a.

Vasquez, B.J., and J.V. Hollweg, The making of an Alfvénic fluctuation: resolution of a second-order analysis, in *Solar Wind Eight*, edited by D. Winterhalter, J. Gosling, S.R. Habbal, W.S. Kurth, and M. Neugebauer, p. 331, AIP, New York, 1996b.

Vasquez, B.J., and J.V. Hollweg, Formation of spherically polarized Alfvén waves and imbedded rotational discontinuities from a small number of entirely oblique waves, *J. Geophys. Res.*, 103, 335, 1998.

Vasquez, B.J., and J.V. Hollweg, Nonlinear evolution of Alfvén waves and RDs - hybrid simulations, in *Solar Wind Nine*, edited by S.R. Habbal, R. Esser, J.V. Hollweg, and P.A. Isenberg, p. 167, AIP, Woodbury, 1999a.

Vasquez, B.J., and J.V. Hollweg, Formation of pressure-balanced structures and fast waves from nonlinear Alfvén waves, *J. Geophys. Res.*, 104, 4681, 1999b.

Vasquez, A.M., A.A. van Ballegooijen, and J.C. Raymond, The effect of proton temperature anisotropy on the solar minimum corona and wind, *Astrophys. J.*, 598, 1361, 2003.

Vasquez, B.J., S.A. Markovskii, and J.V. Hollweg, Nonlinear Alfvén waves 2. The influence of wave advection and finite wavelength effects, *J. Geophys. Res.*, 109, A05104, doi:10.1029/2003JA010106, 2004.

Viñas, A., H.K. Wong, and A.J. Klimas, Generation of electron suprathermal tails in the upper solar atmosphere: implications for coronal heating, *Astrophys. J.*, 528, 509, 2000.

Voitenko, Y., and M. Goossens, Excitation of high-frequency Alfvén waves by plasma outflows from coronal reconnection events, *Solar Phys.*, 206, 285, 2002a.

Voitenko, Y., and M. Goossens, Nonlinear excitation of small-scale Alfvén waves by fast waves and plasma heating in the solar atmosphere, *Solar Phys.*, 209, 37, 2002b.

Voitenko, Y., and M. Goossens, Kinetic excitation mechanisms for ion-cyclotron kinetic Alfvén waves in Sun-Earth connection, *Space Sci. Rev.*, 107, 387, 2003.

Voitenko, Y., and M. Goossens, Cross-field heating of coronal ions by low-frequency kinetic Alfvén waves, *Astrophys. J.*, 605, L149, 2004.

Voitenko, Y., and M. Goossens, Cross-scale nonlinear coupling and plasma energization by Alfvén waves, *Phys. Rev. Lett.*, 94, 5003, 2005.

Yamauchi, Y., M. Tokumaru, M. Kojima, P.K. Manoharan, and R. Esser, A study of density fluctuations in the solar wind acceleration region, *J. Geophys. Res.*, 103, 6571, 1998.

Joseph V. Hollweg, Space Science Center, Morse Hall, University of New Hampshire, Durham, NH 03824 USA. (joe.hollweg@unh.edu)

The Role of Comet Tails in the Discovery of the Solar Wind and Its Spatial and Temporal Variations

D.A. Mendis

Department of Electrical and Computer Engineering, University of California, San Diego, La Jolla, California, USA

Ever since Ludwig Biermann used the observed aberration of cometary plasma tails to infer the continuous outflow of plasma from the sun (since referred to as the solar wind) comets have been used to good advantage to delineate the average global flow parameters of the solar wind. Their role as natural probes of the solar wind has not been entirely superceded by the subsequent advent of spacecraft because comets approach the sun at all angles and some go closer to it than any spacecraft has so far. Comets have thus provided useful information about the spatial variations of the solar wind (e.g., the latitudinal variation of its speed). Also sudden and often dramatic changes in the cometary plasma tails (e.g., total disconnections) provide information about discontinuous changes in the solar wind flow. In this brief overview I will discuss these observations and inferences and conclude with some thoughts about the variable nature of the comet-solar wind interaction with heliocentric distance, and the possible role of the forthcoming Rosetta mission in verifying these.

1. INTRODUCTION

The plasma tails of comets are highly useful free natural probes of the solar wind, providing information not only of its mean global properties, but also of its spatial and temporal variations (e.g., high speed streams, magnetic sector boundaries and coronal mass ejections). This role of comets will be the central theme of this paper. I will however start with a brief overview of the early history which is both interesting and instructive (section 2). This will also enable me to briefly discuss the nature of the second class of cometary tail; the dust tail whose dynamics is controlled not by the solar wind but by solar electromagnetic radiation. The "modern" era, beginning with the use of the aberration of the plasma tails of comets to infer the existence of the solar wind will be discussed next in section 3. This will be followed (section 4) by a discussion of our present understanding of the physical nature of the solar wind interaction with a well developed cometary atmosphere. I will next consider the role of the plasma tails of comets as probes of the solar wind including its spatial and temporal variations (section 5). I will conclude (section 6) with a discussion of the variable global nature of the solar wind interaction with a comet approaching the sun and show how this could be used to extend the heliocentric range of comets as probes of the solar wind. In this connection, the important role of the forthcoming Rosetta comet mission will also be commented upon.

2. THE EARLY HISTORY

Since many comets are observed visually just before sunrise or soon after sunset, the fact that their spectacular tail points, more or less, directly away from the sun should have been apparent even to the very early observers. The earliest known written record of this fact is due to Li Chung-feng in 635 AD in The History of the Chin Dynasty. The first Western record of this fact is almost 900 years later when the German mathematician Peter Apian drew attention to it (see Figure 1). There the orientation of the tail of comet P/Halley

Recurrent Magnetic Storms: Corotating Solar Wind Streams
Geophysical Monograph Series 167
Copyright 2006 by the American Geophysical Union.
10.1029/167GM05

Figure 1. Peter Apian's August 1531 observations of a comet (Halley) in the constellation Leo were used to demonstrate the antisolar nature of cometary tails. Woodcut illustration from Apian's Practica auff dz. 1532 Jar.... (Landshut). (Courtesy of the Crawford Library, Edinburgh, Scotland). [From *Yeomans*, 1991.]

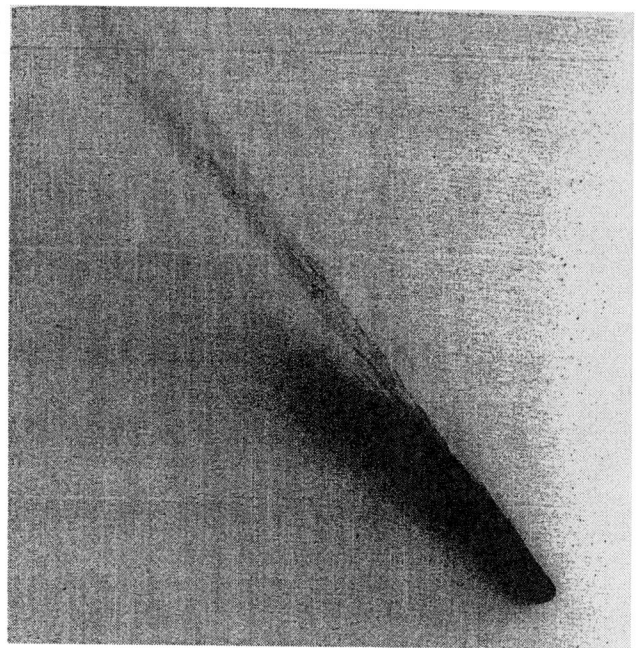

Figure 2. Comet Mrkos (1957d), showing the straight structured type 1 tail and a broad, homogenous type 2 tail, which lags behind the type 1 tail. Cometary orbital motion is toward the right of the figure. [From *Brandt*, 1967b.]

as it moved in the constellation of Leo during its 1531 AD apparition, is clearly shown to be in the anti-sunward direction. Incidentally, it is worth noting that the reason for this long time lag between Chinese and Western records lay in the dominance, in the West, of the Aristotelian dogma that comets were merely "exhalations from the earth," and not celestial objects worthy of astronomical investigation. Interestingly this view prevailed, for a while, even after the Danish astronomer Tycho Brahe used parallax measurements to show that the great comet of 1577 was at a supra-lunar distance, and therefore was a truly celestial (not terrestrial) phenomenon. (For more detailed discussions of this subject, see *Yeomans*, 1991.)

Even after it was generally accepted that the comet tail points in the anti-sunward direction, the reason for this remained unresolved well into the twentieth century. Central to this was the confusion caused by the fact that there are in fact two major types of tails, type 1 and type 2 (see Figure 2). Following the spectroscopic studies, beginning the middle of last century, we now know that narrow (type 1) tail, which points almost directly away from the sun, is composed of plasma (observed from the ground by resonance scattering of solar radiation by various ions; with its bluish color being due to the strong contribution of the violet bands of CO^+) while the broader, more featureless (type 2) tail (which lags behind the type 1 tail; the cometary orbital motion being to the right in this figure) is composed of dust, observed by scattered sunlight, leading to its more yellowish hue. This was of course unknown to the early workers, who mostly thought that both tails were composed of gas. It needs to be pointed out that a few comets, when close to the sun, also exhibit a third type of tail composed of neutral sodium atoms. This is because the resonance scattering of solar photons by the Na atom is much stronger than that by any other cometary chemical species.

Edmund Halley's recognition, based on their orbital elements, that the comets observed in 1531, 1607 and 1682, were one and the same, moving in a highly elliptical orbit, and his successful prediction that this comet (now fittingly named Halley's Comet) would return again in 1758 had firmly established the validity of Isaac Newton's Law of Universal Gravitation postulated in 1689. It soon became clear that while the motion of the center of mass of the comet could be explained by the sunward force of solar gravity, the anti-sunward orientation of comet tails required an outward directed (repulsive) force emanating from the sun. It is remarkable that as early as 1812 the Dutch astronomer, Heinrich Olbers correctly speculated that the cometary tail consisted of "minute particles driven away in the anti-solar direction by a solar repulsive force that is *electrical* in nature." [See *Yeomans*, 1991.]

The first major contribution to the quantitative study of comet tails was due to *Friedrich Bessel* [1836]. He assumed that, whatever the nature of this repulsive force, it would vary inversely as the square of the distance, just like the gravitational force. He was then able to successfully calculate the observed shape of the type 2 (dust) tails assuming that they were moving under an effective (reduced) solar acceleration of magnitude $\frac{\mu G M_\odot}{r^2}$ ($M_\odot$ and r being the solar mass and solar distance respectively, G the universal constant of gravitation and μ the reduction factor) which implies that the ratio of this outward radial force, F_r, to the gravitational force F_g is given by $\left|\frac{F_r}{F_g}\right| = 1 - \mu$. Bessel's mechanical theory was extensively used and analyzed by *Fedor Bredikhin* [1903] who showed that all the type 2 tails he analyzed could be explained by assuming that $(1 - \mu) \sim (0.7 - 2.2)$. He also recognized that if the same model was applied to type 1 tails, typically $(1 - \mu) \geq 100$.

The physical nature of this outward force was first recognized by *Svante Arrhenius* [1900] to be the solar radiation (pressure) force. Incidentally this was the first proposed application of this force in an astronomical context although its existence was discovered theoretically by *James Clarke Maxwell* in 1873. This idea was further developed by *Karl Schwarzschild* [1901] who showed that with

$$1 - \mu = \frac{C}{a \rho_d}, \text{ where } C \sim 6 \times 10^{-5} \, Q_{pr} \, \text{gcm}^{-2}$$

(Q_{pr} being the scattering efficiency for radiation pressure, a the grain radius and ρ_d the grain bulk density), $(1 - \mu) \sim 1$ (typically) and cannot exceed ~ 20 for any grain size or reasonable density, for known compositions.

Following the spectroscopic discovery that type 1 tails were not composed of dust, but of various ions (e.g., CO^+, CO_2^+, N_2^+, etc.), in the early twentieth century, it was established [*Wurm*, 1943] that radiation pressure on these ions typically gave $(1 - \mu) \sim 0.1$. So clearly some other force was required to explain the high accelerations observed in these tails. Even before the nature of the type 1 tails was known, there were speculations that the agency responsible for the observed acceleration were "solar coronal particles" [*Schaeberle*, 1893].

3. THE DISCOVERY OF THE SOLAR WIND

The "modern" era of our understanding of the nature of type 1 tails, which also led to the discovery of the solar wind, began with the central observation of *Hoffmeister* [1943] who noticed that the axis of type 1 tails always lagged slightly behind the sun-comet axis and also that the tangent of this lag (or aberration) angle, ε was proportional to the transverse (i.e., normal to the radius vector in the orbital plane) component of the orbital velocity of the comet. *Ludwig Biermann* [1951] drew the obvious conclusion from this that there was continuous outflow of plasma from the sun

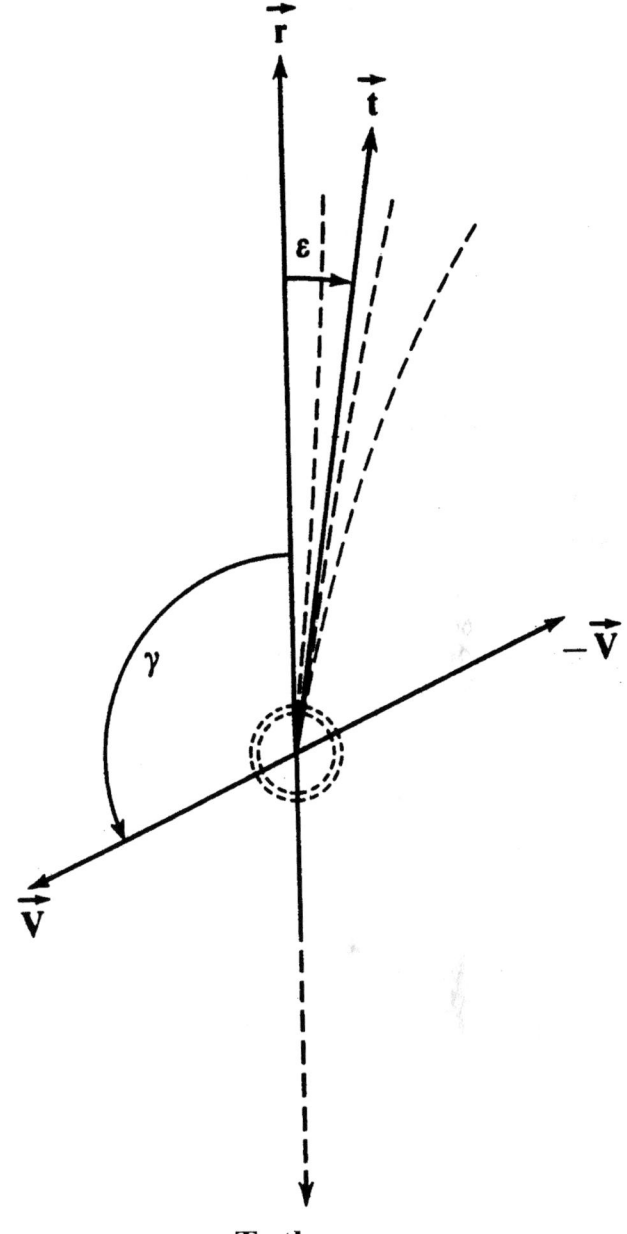

Figure 3. Geometry of the solar wind interaction with the cometary plasma tail. The cometary orbital velocity vector $\vec{V}$ and the radial vector $\vec{r}$ (from the sun to the cometary nucleus) define the orbital plan. The angle between $\vec{r}$ and $\vec{V}$ is γ and the angle between $\vec{r}$ and the plasma tail axis (denoted by the vector $\vec{t}$) is the aberration angle ε. The angle i (used in equation (1)) is the angle between this orbital plane and the plane of the solar equator. [From *Belton and Brandt*, 1966.]

and that due to its interaction with outflowing solar plasma, the cometary plasma tail pointed in the direction of this flow as observed by the moving comet, which explains the aberration (see Figure 3). That there was plasma outflow from the sun was inferred earlier by scientists studying geomagnetic variations [e.g., see *Bartels*, 1932], but they believed that it was intermittent. Plasma tails of comets are of course visible continuously (at least when their heliocentric distance ≤2 AU). So *Biermann's* [1951] paper is credited with first convincing detection of *continuous* outflow of plasma from the sun, referred to as the Solar Wind, following the classic theoretical paper by *Eugene Parker* [1958]. As is clear from Figure 3:

$$\tan\varepsilon = \frac{V\sin\gamma - w_\varphi \cos i}{w_r - V\cos\gamma} \quad (1)$$

where V is the speed of the comet, w_r and w_φ are the radial and longitudinal components of the solar wind velocity, γ is the angle between the cometary velocity vector and the radius vector, ε is the aberration angle (i.e., the angle between the tail axis and the radial direction), and i is the inclination of the comet's orbital plane to the solar equator [e.g., *Belton and Brandt*, 1966].

Biermann [1951] assumed that the solar wind flow was strictly radial and also that the cometary plasma tail lay in the orbital plane of the comet. Then $\tan\varepsilon = \frac{V\sin\gamma}{w_r - V\cos\gamma} \approx \frac{V\sin\gamma}{w_r}$ (assuming that the solar wind speed, $w_r \gg |V\cos\gamma|$, the radial component of the comet's orbital speed). With typical values for the transverse component of the comet's orbital speed $V\sin\gamma \approx 20$ km/s and $\varepsilon \sim 3°$, one obtains $w_r \sim 400$ km/s for the "typical" solar wind speed.

4. THE NATURE OF THE COMET-SOLAR WIND INTERACTION

Following the discovery of the solar wind, *Biermann* directed his attention to the important question of the coupling between the solar wind and the cometary plasma tail by studying the acceleration of several inhomogeneities (e.g., "knots" and condensations) observed down the plasma tails of comets.

There is more than sufficient momentum flux in the solar wind to explain the observed acceleration in the cometary plasma tail, provided there is an efficient mode of coupling between the two plasmas. *Biermann* [1953] suggested that the coupling was due to long-range Coulomb collisions between the two groups of ions, leading to a cometary ion acceleration of $a \sim \frac{e^2 nw}{\sigma m_c}$ where m_c is cometary ion mass, σ is the electrical conductivity (esu), and n and w are the solar wind number density and speed, respectively. Taking $w = 1000$ km/s, $n = 10^3$ cm^{-3}, $\sigma = 5 \times 10^{12}$ esu, and $m_c = 28$ amu, *Biermann* obtained $a \sim 100$ cm/s^2. While this value of a is consistent with the accelerations inferred also from the kinematics of inhomogeneities ("knots" and "condensations") observed down plasma tails, the high values assumed for w and n were not considered too high at the time.

In an important paper *Alfvén* [1957] criticized *Biermann's* mechanism for the production of large accelerations in cometary plasma tails, in particular noting that the high solar wind densities were inconsistent with inferences from coronal white light measurements. *Biermann* [1951] had already noted that the solar wind plasma would probably carry a magnetic field. *Alfvén* [1957] developed this idea qualitatively to produce his "magneto-hydrodynamic model" for the interaction of the solar wind with the cometary plasma (see Figure 4). Briefly the idea is that as the solar wind, with its "frozen-in" magnetic field flows into (4a) and past the comet, this magnetic field gets "hung-up" (4b) in the cometary plasma (ionosphere) and is dragged into the tail (4c, d), as shown. While *Alfvén* noted that this picture is strikingly similar to the "folding umbrella" morphology of plasma rays and streamers in cometary tails, he also noted that observed wavy patterns moving at high velocities may be due to the propagation of hydromagnetic waves down the tail. While *Alfvén* did not develop this phenomenological model quantitatively he made the important point that the plasma tail of the comet must be regarded as an integral part of the comet, fastened to the head by the magnetic field which channel the tail plasma. In other words the cometary plasma tail is a true "windsock" as opposed to *Biermann's* [1951] view which may essentially be described as a "smoke-trail."

While *Alfvén* [1957] identified a central intermediary of the cometary solar wind interaction the second crucial one was identified ten years later by *Biermann et al.* [1967]. This is mass loading of the inflowing solar wind with heavy cometary ions produced either by photoionization or charge exchange, which makes the solar wind interaction with comets so different from its interaction with either strongly magnetized planets (e.g., Earth, Jupiter, Saturn) or with essentially unmagnetized planets with dense atmospheres (e.g., Venus). Noting that the solar wind flow was both supersonic and super-alfvenic, and solving the steady-state 1-D hydrodynamic equations along the sun-comet axis, with further approximations, *Biermann et al.*, recognized that the solution led to an unrealistic self-reversal of the flow unless $\hat{x} = \frac{\rho w}{\rho_\infty w_\infty} \leq \frac{\gamma^2}{(\gamma^2 - 1)}$, where ρ and w are the contaminated solar wind mass density and speed, while γ is the ratio of the specific heats. Taking $\gamma = 2$ as an indirect concession to the existence of the magnetic field they notice that this critical value = 4/3, implying that this corresponded to only a few percent contamination of the solar wind with the heavy cometary ions (e.g., CO^+, CO_2^+, N_2^+, H_2O^+, etc). They also

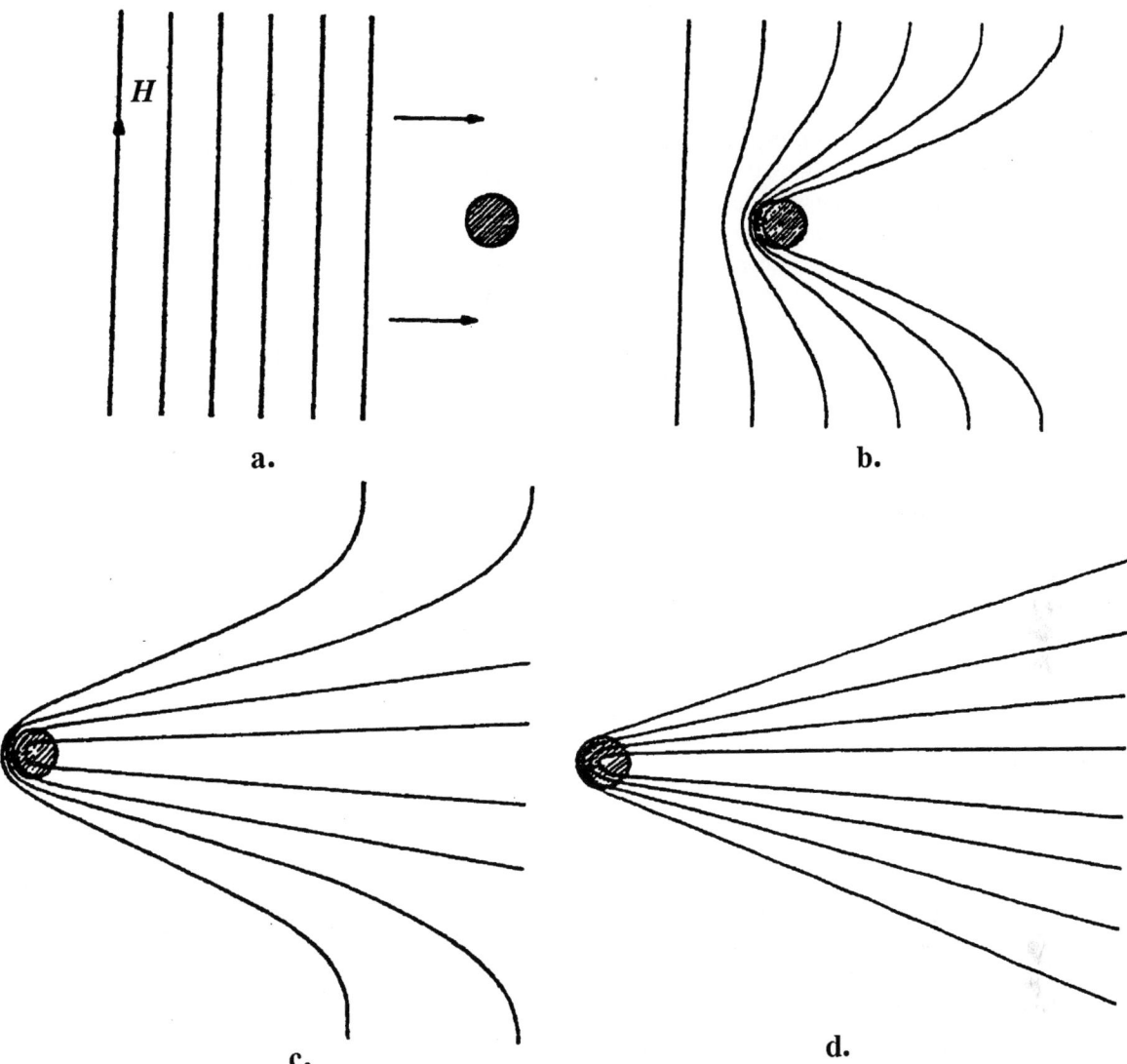

Figure 4. Schematic representation of the "piling-up of the interplanetary magnetic field convected by the solar wind against the cometary ionosphere. [From *Alfvén*, 1957.]

recognized that the implications of this was that a shock will form upstream of the comet to divert the solar wind around the comet before this critical value was reached. Already *Axford* [1964] had pointed out that, since the expanding cometary ionosphere would act as an obstacle to the solar wind, a bow shock should form typically at a distance of $\sim(10^4 - 10^5)\ R_n$ upstream of the cometary nucleus (R_n being the cometary radius). *Biermann et al.* [1967] arbitrarily assumed that the Mach number, M, of this shock would be ≈ 10 (as in the case of the earth), but it was subsequently shown by *Wallis* [1973] that this shock would be much weaker ($M \approx 2$). This because, not only is the inflowing solar wind gradually slowed down by the mass loading, more importantly, it is also heated up because the newly assimilated cometary ions have a thermal speed comparable to the local solar wind speed. Also *Wallis* showed that this shock would occur considerably closer to the cometary nucleus than assumed by *Biermann et al.* [1967]. While subsequent 2-D and 3-D MHD models by several authors have validated *Wallis'* 1973 contention based on a simple 1-D hydrodynamic model [e.g., see *Mendis et al.*, 1985], a semi-kinetic two-component (solar wind protons and cometary ions) model developed by *Wallis and Ong* [1975] and subsequently extended by others provides to an explicit kinetic description of cometary ion assimilation process. For a concise review of this model see *Ip and Axford* [1990].

The kinetics of the pick-up process also leads to an understanding of the momentum coupling between the solar wind and cometary plasma as was first shown by *Wu and Davidson* [1972]. The nature of the process depends on the orientation of the interplanetary magnetic field (IMF) to the solar wind flow direction. In the special case when IMF is normal to the flow the pick-up of cometary ions are entirely due to macroscopic fields: the IMF and the associated motional electric field. The newly created cometary ions gyrate around the local magnetic field with a gyro-speed of w_{sw}, while their guiding centers move with the magnetic field while conserving their magnetic moment at the point of origin. When the solar wind flows obliquely to the IMF the coupling between the solar wind and cometary ions is dominated by microscopic electric and magnetic fields generated by various plasma instabilities. The newly formed ions move along the local magnetic field line with a speed of $w_\parallel = w_{sw} \cos \theta$, while gyrating around this field line with a speed of $w_\perp = w_{sw} \sin \theta$, (where θ = angle between the IMF and the solar wind direction). The beaming of this gyrotropic ring distribution along the magnetic field leads to various plasma instabilities, and the resulting waves cause both pitch angle scattering and energy diffusion of the ions. It can be shown [e.g., see *Flammer*, 1991] that the position of the cometary bow shock depends on, among other parameters, its Mach number and on the ratio of specific heats, γ. Giotto observations at comet Halley showed that while its Mach number, $M \approx 2$, its position was between those corresponding to $\gamma = 2$ (gyrotropic ring) and $\gamma = \frac{5}{3}$ (isotropic shell), which was also consistent with the observation of the ion velocity distribution corresponding to an incomplete shell.

Regarding the structure of the cometary bow-shock what is definitely known from in-situ observations at comet Halley is that it is much thicker than the terrestrial one. While the thickness of the latter is of the order of a proton gyro-radius (≈ 100 km) the thickness of the former is of the order of a pick-up ion gyro-radius ($\geq 10^4$ km). For more details about the structure of the box-shock, see *Ip and Axford* [1990].

As the sub-sonic mass accreting solar wind flows towards the nucleus it continues to slow down while its magnetic field continues to increase, and is eventually brought to stagnation (along the sun-comet axis). The existence of a tangential discontinuity surface (loosely called the cometary ionopause) was already anticipated in the early work of *Biermann et al.* [1967], and the basic mechanism responsible for its formation, which is the balance of the electromagnetic $\mathbf{j} \times \mathbf{B}$ force and the drag of the out-flowing cometary neutrals on the plasma just outside it, was first proposed by *Ip and Axford* [1982], who went on to make an estimate of its linear dimension along the sun-comet axis. In order to do so, they needed to know the strength of the magnetic field just outside the ionopause, which they estimated by assuming that the entire ram pressure of the solar wind was converted to magnetic pressure at the stagnation point.

Subsequently several authors [e.g., *Cravens*, 1989; *Ip and Axford*, 1987; *Haerendel*, 1987] went on to calculate the radial profile (along the sun-comet axis) of the magnetic field in the "magnetic barrier" region just outside the ionopause, while *Wu* [1987] subsequently calculated the 2-D shape of the ionopause, showing it to have a "tear drop" shape (see *Ip and Axford* [1990], for a detailed review).

One of the clearest and most dramatic discoveries of the Giotto mission to Halley's comet was the detection of this ionopause by the magnetometer onboard Giotto [*Neubauer et al.*, 1986]. An essentially magnetic field-free cavity, containing purely cometary ions, separated from the inflowing contaminated solar wind ions by a sharp boundary was observed at a distance ~4700 km inbound and about 3800 km outbound. These encounter distances (by the spacecraft moving at an angle of about 107° to the sun-comet axis) was consistent with the theoretical expectations.

A schematic of the global morphology of the overall comet-solar wind interaction, which summarizes our present knowledge, is shown in Figure 5. The possible existence of an "inner-shock" where the supersonically out-flowing cometary ions are decelerated and diverted into the flanks, was first proposed by *Wallis and Dryer* [1976], and the structure of the shocked layer between the ionopause and this inner shock has been discussed since then by several authors [e.g., see *Damas and Mendis*, 1992].

As is obvious from Figure 5, there is another global feature called the "cometopause" between the bow shock and the ionopause. This transition region where the solar wind proton density drops relatively fast while the cometary ion density increases towards the nucleus, was observed by the VEGA spacecraft as a rather thick region (~10^4 km) at a cometocentric distance ~10^5 km. While a similar transition was observed during the Giotto encounter, it was much more diffuse. Its nature has been discussed by several authors [see *Ip and Axford*, 1990] but it seems fair to say that we do not have a good understanding of its nature at the present time.

A basic consequence of the comet-solar wind interaction, also shown in Figure 5, is the draping of the IMF around the comet. This magnetotail model of *Alfvén* [1957] was spectacularly confirmed by the magnetometer onboard the NASA/ICE spacecraft as it flew through the tail of comet Giacobini-Zinner [*Smith et al.*, 1986]. Two magnetic lobes of opposite polarity separated by a current-carrying neutral sheet was clearly observed (see Figure 6); note in particular the dramatic flipping of the component of the magnetic field, B_x, which is along the sun-comet line, as the spacecraft transits the tail axis).

So far the plasma environments of three comets have been observed in-situ by particle and field experiments carried

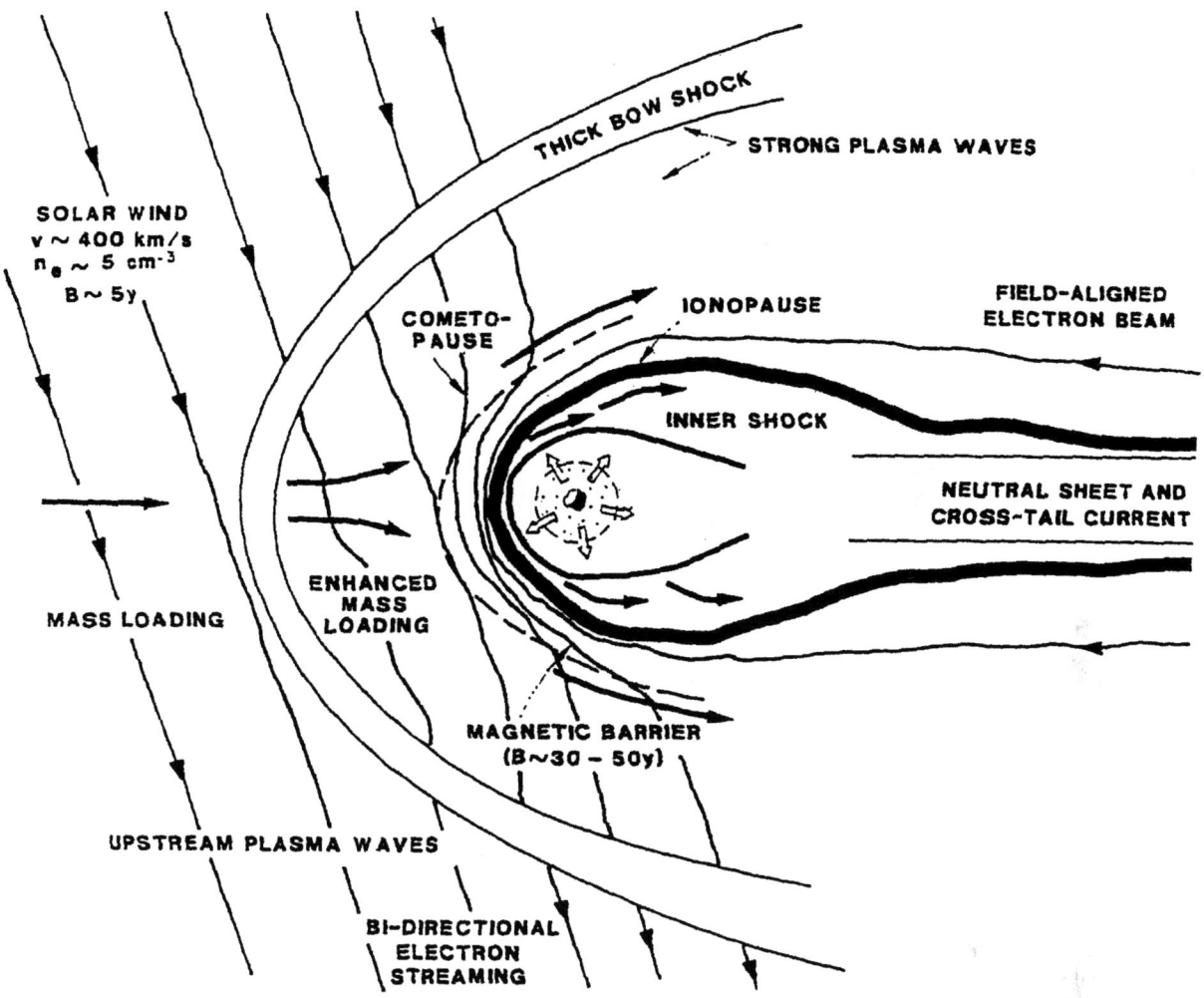

Figure 5. Schematic representation of the global morphology of the solar wind interaction with a well developed cometary atmosphere. [From *Mendis*, 1988.]

several onboard spacecraft. Comet Giacobini-Zinner, by the ICE spacecraft in 1985, comet Halley by ICE, Sakigake, Suisei, VEGA1 and 2 and Giotto, in 1986 and comet Grigg-Skjellerup by Giotto in 1992. What was seen, in all cases, was that these environments were far from quiescent, being characterized by high level of plasma wave activity and turbulence. These observations also led to the detection of a plethora of wave modes. The nature of the waves, e.g., their growth, level of non-linearity and non-coherence (turbulence) varied from comet to comet as well as from place to place within a given comet. For a comprehensive review see *Tsurutani* [1991]. This was not surprising considering the fact that not only were the three encounter geometries very different, but also both the cometary parameters (e.g., the production rate of neutrals) and the solar wind parameters (e.g., the magnitude and orientation of the IMF) were vastly different during these three encounters. Consequently not only were different wave modes observed during the three encounters but even the characteristics of the same mode varied from comet to comet. Theoretical studies have focused on the roles of the variable solar wind and cometary conditions in determining the nature and development of the waves. The most important solar wind parameter appears to be the angle between the IMF and the solar wind flow direction [*Tsurutani and Smith*, 1986]. The most important cometary parameter appears to be the production rate of neutrals, which controls the extent of the region where mass loading of the solar wind with heavy cometary ions is significant.

Plasma waves and turbulence in the mass loaded cometary environment became an important area of investigation following the in-situ observations at comets, leading to sophisticated theoretical models. This is not only because of

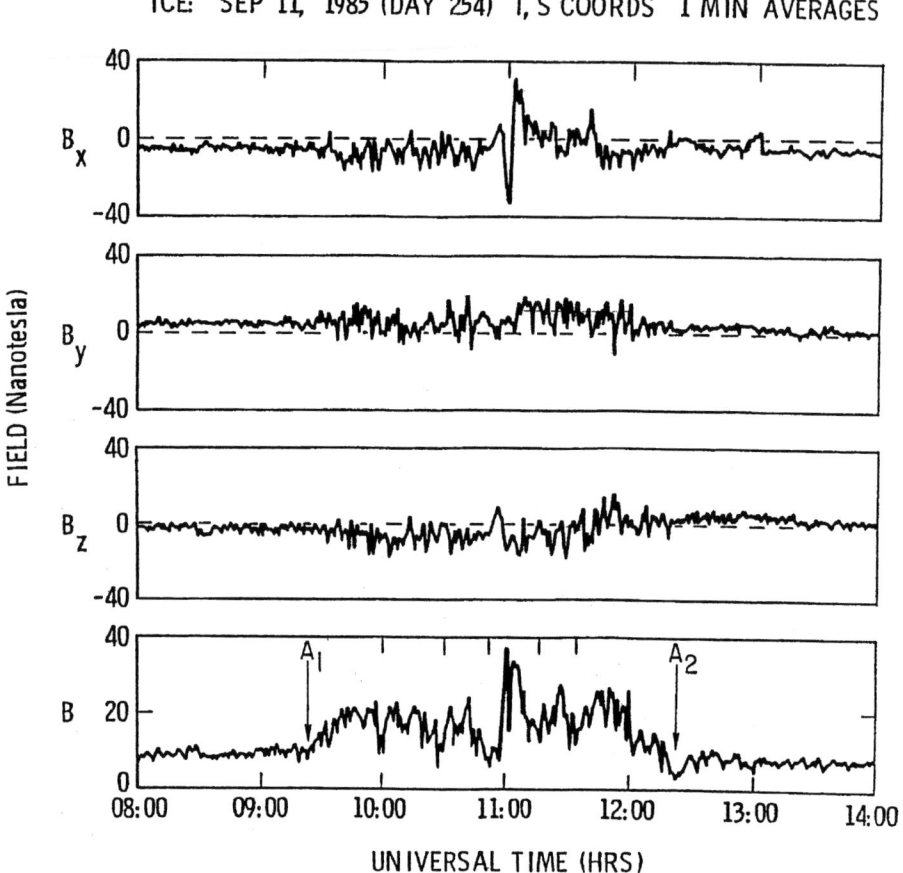

Figure 6. An overview of the magnetic field morphology observed during the NASA/ICE encounter with the tail of Comet Giacobini-Zinner (September 11, 1985). One-minute averages of the three orthogonal components of the magnetic field in *I,S* coordinates, as well as the total magnitude, are shown. Here B_x is along the Sun-comet line. [From *Smith et al.*, 1986.]

the central role that wave particle interactions play in the momentum transfer form the solar to the cometary plasma as we discussed earlier, but also because mass loading takes place in many other solar system situations including the interaction of the solar wind with the atmospheres of unmagnetized planets and the interstellar medium, and comets provide an excellent accessible natural laboratory for the study of mass loaded plasmas under a variety of conditions. The future Rosetta mission which will study the temporal variations of the comet-solar wind interaction over a long period of time, as a comet approaches the sun and gradually develops an atmosphere, will undoubtedly increase our understanding of the subject. I have already mentioned the excellent comprehensive review of the in-situ observations as well as the theoretical studies of plasma waves at comets by *Tsurutani* [1991]. There have also been several good reviews of the subject subsequently, e.g., *Tsurutani et al.* [1997] and *Szegö et al.* [2000].

5. PROBING THE SOLAR WIND WITH PLASMA TAILS

Beginning in the mid nineteen-sixties *John C. Brandt*, together with several co-authors embarked on a systematic study of the global properties of the solar wind, using plasma tail orientation. The earlier studies, which were statistical in nature, and which aimed at deriving "average" values and their spatial variations used a large compilation of tail (both type 1 and type 2) orientations numbering over 1600 [*Belton and Brandt*, 1966]. Examination of equation (1), where it is assumed that the plasma tail and the cometary orbit are coplanar, shows that if w_φ is not zero, it will show up as a systematic difference in the mean aberration angles for direct (prograde) and retrograde orbits (because cos *i* changes sign) with the value corresponding to the retrograde sample, larger. *Brandt* [1967a] obtained $<\varepsilon>_D \approx 3.7°$ and $<\varepsilon>_R \approx 5.5°$ where the suffixes *D* and *R* refer to direct and retrograde, with the mean value for the entire sample $<\varepsilon> \approx 4.7°$. This gave $w_\varphi \approx 9$ km/s and $w_r \approx 450$ km/s for the entire sample. In a subsequent analysis of the same data [*Brandt et al.*, 1972] where the conditions of coplanarity, as well as the assignment of weights to allow for geometrical circumstances was dropped yielded $w_r \approx 415$ km/s and $w_\varphi \approx 6$ km/s. Here, while the uncertainty in w_r is small (≤5%) the uncertainty in w_φ is larger (~25%). Later analyses also obtained the meridional flow speed, w_θ, assuming a theoretical model for its variation with the polar angle θ.

The average values correspond to heliocentric distances around (0.5 – 1.5) AU, since this is the distance range where plasma tails were typically observed. In this distance range the change in the solar wind velocity is small, which justifies the average values. For the same reason, the aberration of comet tails could not be used to estimate the variation of the solar wind velocity with heliocentric distance. On the other hand there could be another type of spatial variation of the solar wind speed, which is with heliographic latitude.

Brandt et al. [1975] next used the aberration of plasma tails in an effort to calculate this possible variation of the radial solar wind speed with heliographic latitude, b, assuming the linear variation:

$$w_r = w_{r^0} + \frac{dw_r}{d|b|}|b| \qquad (2)$$

and obtained $\frac{dw_r}{d|b|} = -0.9 \pm 0.7 \text{ km s}^{-1} \text{ deg}^{-1}$. This result was at odds with variation of the solar wind with heliographic latitude inferred by *Coles et al.* [1980] using radio scintillation observations which gave:

$$\frac{dw_r}{d|b|} \approx (2-3) \text{ km s}^{-1} \text{ deg}^{-1}. \text{ (during the declining phase of the solar cycle)}$$
$$\approx 0 \qquad \text{(at solar maximum)}$$

This latter inference has subsequently been supported by in-situ measurements of the solar wind speed by the Ulysses spacecraft [*McComas et al.*, 2003]. Near solar minimum the solar wind speed shows a characteristic U-shaped profile with speeds increasing monotonically from ≤450 km/s near the equator to ≥750 km/s near the poles. Also while the speed fluctuates greatly near the equator it becomes quite uniform at high latitudes. The profile is much flatter during solar maximum, with a very high degree of fluctuation at all latitudes.

The reason for the apparent discordance between comet observations and the other observations discussed above are apparent. On the one hand the comet calculations used a plasma tail aberration sample which spanned 75 years, i.e., almost four solar cycles. On the other hand this sample was highly weighted toward equatorial with only 58 of a total of 700 observations being polar [e.g., see *Brandt and Snow*, 2000].

Clearly a more direct use of comet tails in this connection is to follow individual, high-inclination comets, measuring their plasma tail aberration at varying heliographic latitudes. This has been done more recently [*Brandt and Snow*, 2000] for three comets: de Vico ($i = 85.4°$) in 1995, Hyakutake ($i = 124.9°$) in 1996 and Hale-Bopp ($i = 89.4°$) in 1997. Their tail aberrations, which are systematically smaller at higher latitudes implied typical solar wind velocities ~750 km/s at high latitudes, and ~450 km/s in the equatorial region. Also while the plasma tails, in the polar regions, were sharp (as would be expected of a steady solar wind) they appear disturbed in the equatorial region (as would be expected from a highly varying solar wind). Since all these observations correspond to a phase in the solar cycle, near solar minimum (8/17/96), they are entirely consistent with the observations of the Ulysses spacecraft. *Brandt and Snow* [2000] speculate that the boundary, between the polar region where the comet tail appears undisturbed and the equatorial region where the tail appears highly disturbed, is the maximum poleward extension of warped heliospheric current sheet (HCS). If correct, this is a useful extension of the use of comet plasma tails as natural probes of the interplanetary medium. Interestingly, *Brandt and Snow* [2000] have reexamined the original data set of *Belton and Brandt* [1966] and find only three comets there are usable, in this connection; these being the ones with sufficient observations both in the equatorial and polar solar wind regions, with only two of these, Mrkos (1957d) and Brooks (1911c) providing reliable data. Both these comets exhibit "transregional" behavior like the three more recent ones. Comet Mrkos goes from having a highly disturbed plasma tail to a relatively undisturbed one around 65° N heliographic latitude, in August 1957, whereas comet Brooks does so around 28° N in October, 1911. *Brandt and Snow* [2000] argue that these two observations are compatible with the fact that the Mrkos transition took place near solar maximum, when the HCS has high poleward extensions, whereas the Brooks transition took place around solar minimum when the HCS is confined to lower latitudes.

Besides the global properties of the solar wind and its spatial (i.e., latitude) variations, the cometary plasma tail can also provide information about its temporal variations. The long wavelength waves which are occasionally observed, propagating down the plasma tails of comets and which were already attributed to MHD waves [*Alfvén*, 1957] have since been discussed by several authors. *Ershkovich and Chernikov* [1973] attributed them to the Kelvin-Helmholtz instability excited by the velocity shear between the solar wind and cometary flows in the tail, when this exceeds a critical value (see *Erchkovich* [1980] for a detailed review). A more direct inference of the sudden variation of the solar wind speed is the appearance of a "kink" or bend down the tail, where the aberration angle changes from a larger value in the more distant part to a smaller value in the part closer to the nucleus. An obvious conclusion reached by many observers is that the comet has encountered a fast solar wind stream; the inner region, with the smaller aberration angle, is already immersed in this fast stream, whereas further out the influence of this stream has not yet been felt. These two regions are separated by an intermediate region which has not yet reached equilibrium with the ambient medium and shows up as a kink in the tail. This phenomenon as well as others associated with the time varying fine structure observed in the plasma tails of comets have been discussed by numerous authors. For a detailed review see e.g., *Mendis et al.* [1985]. Here I will limit myself to a discussion of the most dramatic temporal phenomenon observed in the plasma tails of several comets, namely their occasional total separation from the head of the comet. This is also the subject that has received, by far, the most attention in more recent times.

While this phenomenon was already described by *Barnard* [1920] in connection with what he called the "rejection" of the tail of comet 1919b, its real study of began with a pioneering paper by *Niedner and Brandt* [1978] wherein they attributed this phenomenon, clearly observed in the tail of comet Kohoutek (1973XII), in January 1974, and which they called a "disconnection event" (DE) to the crossing of an interplanetary magnetic sector boundary by the comet. Their basic mechanism is shown in Figure 7. Here magnetic reconnection occurs at the comet's head as a magnetic field of the opposite polarity is pushed against the old field during the passage of the sector boundary. As the tail, gradually peeled off by this process, drifts away from the comet, a new one containing a magnetic field of the opposite polarity is generated. Despite the unavoidable uncertainty involved in the timing (e.g., the extrapolation of solar wind conditions measured at a different time and place by spacecraft to the location of the comet) these authors, in a continuing series of papers (which now covers over 100 DE's), have made a strong case for the connection between DE's and magnetic sector boundary transversals [e.g., *Niedner*, 1982, see also *Brandt*, 1990 and *Brandt and Snow*, 2000]. While it is fair to say that the above model of *Niedner and Brandt* [1978] is the leading one at present, it is not the only one. While several authors have proposed that the DE's could be a consequence of increased pressure-induced effects (e.g., the flute instability; the Rayleigh-Taylor instability) during the encounter of high speed solar wind streams (which are often, though not always, associated with magnetic sector boundaries) and shocks [e.g., *Ip and Mendis*, 1978; *Jockers*, 1981; *Ip and Axford*, 1982], others have also proposed that the responsible mechanism is magnetic reconnection in the tail itself rather than in the head [*Ip*, 1985; *Russell et al.*, 1986].

The coordination of extensive ground-based observations of large scale phenomena in the plasma tail of comet Halley with in-situ spacecraft observations during its fly-by in 1986 provided an excellent opportunity to discriminate between the competing models, but the results could not provide an unambiguous answer. *Niedner and Schwingenschuh* [1987] have argued that ground-based observations of a DE between 8 and 10 March, 1986 was associated with a reversal of the IMF detected by the VEGA 1 and VEGA 2 spacecraft. *Saito et al.* [1986], on the other hand compared data from the Sakigake spacecraft with observation from the ground and noted that there were no apparent signs of DE's between 11 and 14 March, 1986, although the spacecraft crossed the heliospheric current sheet at least four times during this period. These latter authors, therefore note that more than a mere crossing of such a sheet is necessary for the production of a DE, with the encounter geometry presumably playing a essential role. It is also possible that more than one process could trigger DE's. Clearly this is an area that needs further investigation. *Tsurutani et al.* [2006] have pointed out that the high and variable plasma densities observed near the heliospheric current sheet (HCS), referred to as the heliosopheric current sheet plasma sheet or HCSPS [*Winterhalter et al.*, 1994] could have important geomagnetic consequences. What role the HCSPS may play in the initiation of the DE's in comets crossing the HCS has not yet received any attention. This is another area that needs investigation.

There have also been several numerical simulations of this tail disconnection process with different authors reaching different conclusions. Most recently *Konz et al.* [2004] have critiqued these earlier simulations. They point out that ideal MHD simulations do not allow for field line reconnection and that it could arise only as an artifact of the chosen algorithm where some numerical diffusion may lead to it. What is needed is a localized violation of the ideal Ohm's law, which is provided in the above simulation by a current driven production of anomalous resistivity. The same conclusion was reached earlier by *Morrison and Mendis* [1978] who showed that the generation of anomalous resistivity by current-carrying electrons, which may reach the Alfvén energy, $\frac{B_t^2}{4\pi n_e}$ (where B_t and n_e are respectively the magnetic field and the electron density in the tail) could lead to the tearing of the cometary cross-tail current sheet along current flow lines. While the simulation of *Konz et al.* [2004] leads to both day-side and tail reconnection, it like the earlier simulations uses normal geometry for the interaction. In view of the suggestion by *Saito et al.* [1986] noted earlier, future numerical simulations should also attempt to model oblique interaction geometries.

Very recent observations have also shown the effects of the encounter of a fast-moving Interplanetary Coronal Mass Ejection (ICME) by a comet; in this case comet

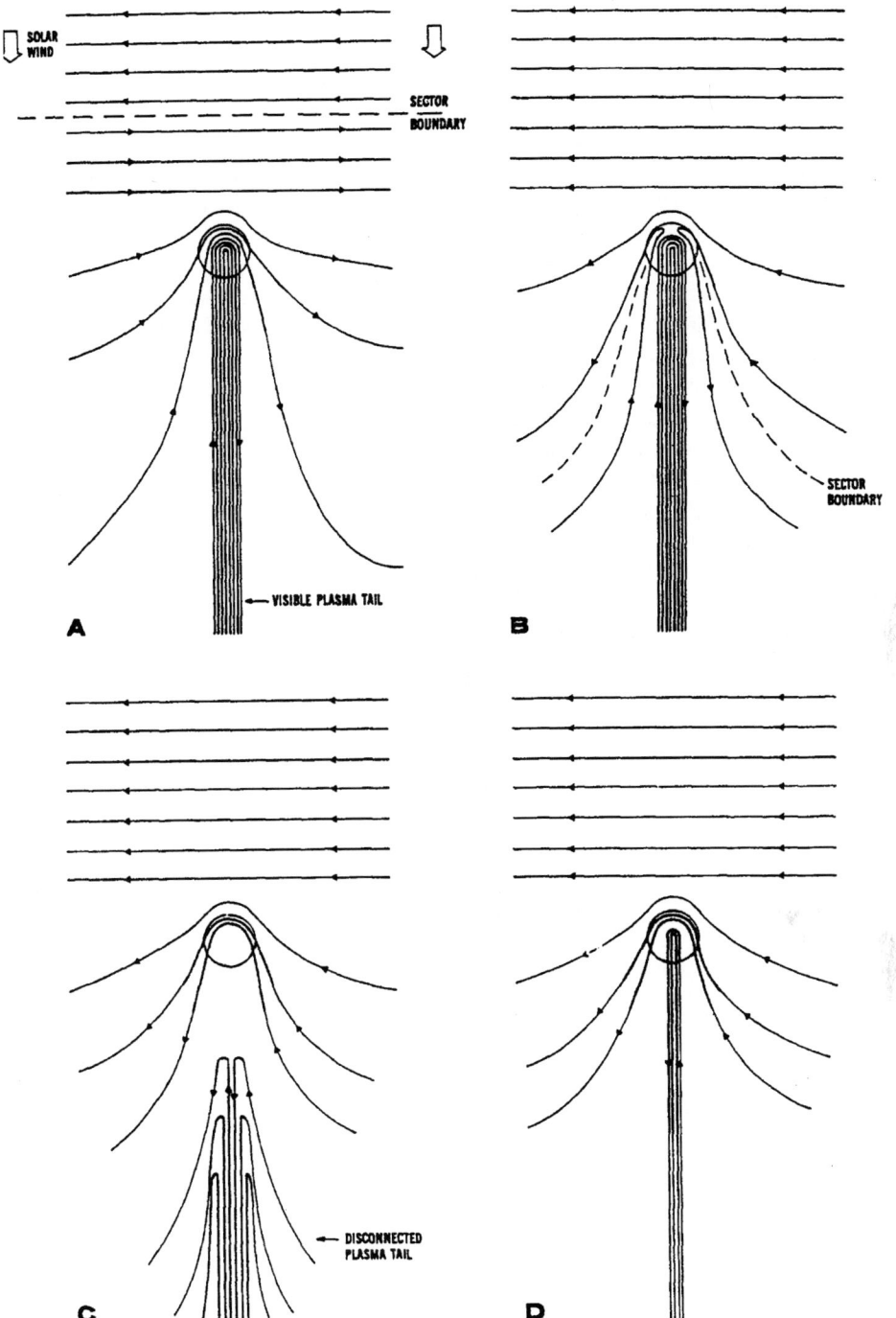

Figure 7. Magnetic sector boundary model of plasma tail disconnection event. [From *Niedner and Brandt*, 1978.]

153P/Ikeya-Zhang [*Jones and Brandt*, 2004]. The impact produces a highly unusual large scale disturbance down the tail, which is characterized by highly "scalloped" appearance. The reason for this unusual appearance is not clear although the above authors speculate that it may be due to the ICME magnetic field wrapping around preexisting tail density enhancements. Nonlinear development of waves exited by Kelvin-Helmholtz instability is also a possibility. Whatever its cause, if more observations in the future could establish that such a scalloped appearance of the tail is characteristic of ICME-comet interactions, then comets could also be used as natural probes of ICMEs.

6. MULTIPLE MODES OF INTERACTION OF THE SOLAR WIND WITH A COMET APPROACHING THE SUN

My discussion, so far, has been limited to the case when a comet has a well developed plasma tail (typically at heliocentric distances ≤3 AU) which arises from the interaction of the solar wind with a well developed cometary atmosphere; the case discussed at length in section 4. I will now consider the possible use of comets as probes of the solar wind at large heliocentric distances. It was shown by *Mendis and Flammer* [1984] [see also *Flammer*, 1991] that the interaction of the solar wind with a comet approaching the sun had not one mode but multiple modes of interaction. At large distances ($r \geq 5$ AU, for a H_2O-controlled comet) the comet has no significant atmosphere and the solar wind flows directly onto the surface. As the comet moves closer an atmosphere begins to develop. Mass loading of the inflowing solar wind can cause it to develop a weak collisionless bow shock upstream of the comet (for comet Halley this would happen when $r \approx 2.5$ AU). However at this time there is not sufficient momentum in the outflowing cometary atmosphere to stagnate the inflowing solar wind ahead of the nucleus. So the subsonic solar wind flows all the way to the nucleus. Closer to the sun (i.e., when $r \approx 2.2$ AU for comet Halley) the contaminated subsonic solar wind is brought to stagnation ahead of the nucleus by a well developed cometary ionopause. This is probably the time when the beginning of a plasma tail is first observed [*Brandt*, 1990]. Indeed the "turn-on time" for the plasma tail of comet Halley (inbound) was when $r \approx 1.8$ AU, while corresponding "turn-off time" (outbound) was when $r \approx 2.3$ AU, which are in reasonable agreement with the predictions of *Mendis and Flammer* [1984] [see also *Brandt*, 1990].

It had been argued by *Mendis et al.* [1981] that the unimpeded flow of the solar wind to the cometary nucleus can cause differential electrostatic charging of the nucleus and thereby lead to the electrostatic levitation and blow-off of fine loose dust on it. They show that this happens mainly on the unlit (night) side which can achieve a large negative electrostatic potential $\propto V_{SW}^2$ due to the buildup of space charge there. The large sporadic brightness fluctuations of comet Halley at heliocentric distances between 8 and 11 AU (inbound), which appear to be entirely associated with dust outbursts, has been attributed to this cause by *Flammer et al.* [1986], who showed a one-to-one correspondence between these outbursts and the possible interaction of the comet with a corotating high-speed solar wind stream emanating from a coronal hole. Incidentally the prevailing view that the coronal holes are the source of corotating high-speed solar wind streams has a long history going all the way back to *Krieger et al.* [1973] [see also *Rickett et al.*, 1976]. The Rosetta mission, which will first intercept its target comet at a distance where the outgassing is expected to be very small, could be used to check the validity of this model. If it turns out to be correct, then the use of comets as natural probes of the solar wind could be extended to much larger distances.

Also the ongoing observations of the Solar Heliospheric Observatory (SOHO) launched in late 1995 may provide an opportunity to use cometary plasma tails to probe the solar wind very close to the solar surface. Its coronagraph cameras, which use occulting discs to produce an artificial solar eclipse, thereby revealing the solar corona and bodies as close to the sun as $2R_\odot$ ($R_\odot$ being the solar radius) has led to the detection of a large number (approaching 1000) of very small "sun grazing" comets with perihelion distances as small as $2R_\odot$ [*Hoffman and Marsden*, 2005]. While many of these mini-comets, believed to be the disruption fragments of a larger parent sun grazer, are observed as uncomet-like specks at the limit of photographic resolution, others exhibit distinct straight narrow tails. While this morphology is characteristic of type 1 (plasma) tails, one cannot jump to that conclusion. The plasma tails of more distant comets are identified not only by their morphology but also by their ionic emissions, with the generally dominant emission for CO^+ providing its characteristic bluish color. There are no identifications of likely cometary ions in the SOHO observations. One does not expect to see ions such as CO^+ so close to the sun, since its "parent" molecule (CO or H_2CO) would have sublimated away long before the comet comes so close to the sun. The likely cometary ions at these solar distances are those that originate from more refractory cometary material (e.g., singly ionized metallic atoms, Mg, Al, Fe, Si, etc.) These cannot be detected by the (ultraviolet and extreme ultraviolet) spectrographs on SOHO [e.g., see *Domingo et al.*, 1995]. If however the electron density in the cometary plasma tail is sufficiently high it may be possible that their Thomson scattering of solar radiation could be detected by the Large Angle Spectroscopic Coronagraph (LASCO) on board SOHO. Also the different polarization characteristics of the Thomson scattered solar radiation by electrons and the Mie scattered solar radiation by dust could presumably provide a way to discriminate between the plasma tail and the dust tail. If a cometary plasma tail can be definitively identified by the SOHO observations, evaluation of its aberration would provide a unique opportunity of estimating the speed of the solar wind below the critical point around $4R_\odot$. This is clearly an area that needs careful investigation.

I will conclude by pointing out that, while comets do not possess intrinsic magnetic fields, the interaction of the solar wind with the cometary ionosphere produces a large, purely induced cometary magnetosphere. Consequently there could be cometary analogs of several terrestrial magnetospheric processes. There has been some discussion of this area: e.g., cometary substorms [*Ip and Mendis*, 1976], and associated cometary "auroral" phenomena including the generation of cometary X rays as the energetic auroral electrons precipitate into the cometary atmosphere [*Hudson et al.*, 1981]. While X-ray emissions from comets have since been detected [e.g., *Dennerl*, 1997] competing processes have been proposed for their generation [e.g., *Cravens*, 1997; *Bingham et al.*, 1997; *Shapiro et al.*, 2005]. This too is clearly an area that needs further study. For instance if one could establish a clear correlation between solar wind activity and cometary X-ray activity, the role of comets as probes of the solar wind would be further extended.

Acknowledgments. The author thanks the two anonymous referees for their careful reading of the original manuscript and their useful suggestions for its improvement. He also thanks his colleague Dr. William A. Coles for useful discussions. Support from the DOE grant: DE-FG02-04ER54804 is acknowledged.

REFERENCES

Alfvén, H., On the theory of comet tails, *Tellus*, 9, 92, 1957.
Arrhenius, S., Über die ursache der nordlichter, *Phys. Zeitschr.*, 2, 81, 1900.
Axford, W.I., The interaction of the solar wind with comets, *Planet. Space Sci.*, 12, 719, 1964.
Barnard, E.E., On comet 1919b and on the rejection of a comet's tail, *Astrophys. J.*, 51, 102, 1920.
Bartels, J., Terrestrial magnetic activity and its relationship to solar phenomena, *Terr. Mag. Atm. Elect.*, 37, 1, 1932.
Belton, M.J.S. and J.C. Brandt, Interplanetary gas. XIII. A category of comet tail orientations, *Astrophys. J. Suppl.*, 13, 125, 1966.
Bessel, F.W., Beobachtungen Über die physische des Halley'schen kometen und dadurch veranlasste bemerkungen, *Astron. Nachr.* 13, 185, 1836.
Biermann, L., Kometenschweife und Solare Korpuskularstrahlung, *Zeit. Astrophys.*, 29, 274, 1951.
Biermann, L., Physical processes in comet tails and their relation to solar activity, La Physique des comètes (Université de Liège), p251, 1953.
Biermann, L., B. Brosowski, and H.U. Schmidt, The interaction of the solar wind with a comet, *Solar Phys.*, 1, 254, 1967.
Bingham, R., J.M. Dawson, V.D. Shapiro, *et al.*, Generation of X rays from comet c/Hayakutake, 1996B2, *Science*, 275, 49, 1997.
Brandt, J.C., Interplanetary gas. XIII. Gross plasma tail velocities from the orientation of ionic comet tails, *Astrophys. J.*, 147, 201, 1967a.
Brandt, J.C., Introduction to the solar wind, W.H. Freeman and Co., San Francisco, p107, 1967b.
Brandt, J.C., The large-scale plasma structure of Halley's comet, 1985-1986, in *Comet Halley: Investigations, Results, Interpretation*, vol. 1, edited by J. Mason, Ellis Harwood, NY, p33, 1990.
Brandt, J.C., R.S. Harrington, and R.G. Roosen, Interplanetary gas XX. Does the radial solar wind speed increase with latitude? *Astrophys. J.*, 196, 877, 1975.
Brandt, J.C., R.J. Roosen and R.S. Harrington, Interplanetary gas XVII. An astrometric determination of solar wind velocities from orientation of ionic comet tails, *Astrophys. J.*, 177, 277, 1972.
Brandt, J.C. and M. Snow, Heliospheric latitude variation of properties of cometary plasma tails: A test of the Ulysses comet watch paradigm, *Icarus*, 52, 2000.
Bredichin, T.H., Mechanische untersuchungen Über kometenformen, in *Syst. Darstellung*, compiled by R. Jaegermann, St. Petersburg, 1903.
Coles, W.A., B.J. Rickett, V.H. Rumsey, *et al.*, Radio scintillation observations, *Nature*, 286, 239, 1980.
Cravens, T.E., A magnetohydrodynamical model of the inner coma of comet Halley, *J. Geophys. Res.*, 94, 15025, 1989.
Cravens, T.E., Comet Hyakutake X-ray source: Charge transfer of solar wind heavy ions, *Geophys. Res. Lett.*, 24, 105, 1997.
Damas, M.C., and D.A. Mendis, A three dimensional axisymmetric photochemical flow model of the cometary "inner" shock layer, *Astrophys. J.*, 396, 704, 1992.
Dennerl, K., J. Englahauser, and J. Trumper, X-ray emission from comets detected in the Röntgen X-ray satellite all sky survey, *Science*, 277, 1625, 1997.
Ershkovich, A., Kelvin-Helmholtz instability in type 1 comet tails and associated phenomena, *Space Sci. Rev.*, 25, 3, 1980.
Ershkovich, A., and A.A. Chernikov, Non-linear waves in type 1 comet tails, *Planet. Space Sci.*, 21, 663, 1973.
Flammer, K.R., The global interaction of comets with the solar wind, in *Comets in the Post-Halley Era*, vol. 2, edited by R.L. Newburn *et al.*, Kluwer Acad. Pub., Dordrecht, Netherlands, 1125, 1991.
Flammer, K.R., B. Jackson, and D.A. Mendis, On the brightness variations of comet Halley at large heliocentric distances, *Earth Moon and Planets*, 35, 203, 1986.
Haerendel, G., Plasma transport near the magnetic cavity surrounding comet Halley, *Geophys. Res. Lett.*, 13, 255, 1987.
Hoffman, T., and B.J. Marsden, The booming science of sungrazing comets, *Sky and Telescope*, 110, 33, 2005.
Hoffmeister, C., Physikalische untersuchungen an kometen. 1. Die beziehungen des primären schweifstrahls zum radusvektor, *Zs. F. Astrophysik.*, 22, 265, 1943.
Hudson, H.S., W.H. Ip, and D.A. Mendis, An Einstein search for X-ray emission from comet Bradfield 1979l, *Planet. Space Sci.*, 29, 1373, 1981.
Ip, W.H., Solar wind interaction with neutral atmospheres, ESA SP-235, 65, 1985.
Ip, W.H., and W.I. Axford, The formation of the magnetic field cavity at comet Halley, *Nature*, 325, 418, 1987.
Ip, W.H., and W.I. Axford, The plasma, in *Physics and Chemistry of Comets*, edited by W.F. Huebner, Springer-Verlag, Berlin, p177, 1990.
Ip, W.H, and W.I. Axford, Theories of physical processes in cometary comas and ion tails, in *Comets*, edited by L.L. Wilkening, Univ. of Ariz. Press, Tucson, Arizona, p588, 1982.
Ip, W.H, and D.A. Mendis, Generation of magnetic fields and electric currents in the cometary plasma tail, *Icarus*, 29, 147, 1976.
Ip, W.H., and D.A. Mendis, The flute instability as the trigger mechanism for the disruption of the cometary plasma tails, *Astrophys. J.*, 223, 671, 1978.
Jockers, K., Plasma dynamics in the tail of comet Kohoutek 1973XII, *Icarus*, 47, 397, 1981.
Jones, G.H., and J.C. Brandt, The interaction of comet 153P/Ikeya-Zhang with interplanetary coronal mass ejections: Identification of fast ICME signatures, *Geophys. Res. Lett.*, 31, L20505, 2004.
Konz, C., G.T. Burk, and H. Lesch, Plasma—neutral gas simulation of reconnection events in cometary tails, *Astron. Astrophys.*, 415, 791, 2004.
Krieger, A.S., A.F. Timothy, and E.C. Roelof, A coronal hole and its identification as a source of a high velocity solar wind stream, *J. Geophys. Res.*, 29, 505, 1973.
McComas, D.J., H.A. Elliott, Schwadron, The three-dimensional solar wind around solar maximum, *Geophys. Res. Lett.*, 30, 1537, 2003.
Mendis, D.A., A postencounter view of comets, *Ann. Rev. Astron. Astrophys.*, 26, 11, 1988.
Mendis, D.A. and K.R. Flammer, The multiple modes of interaction of the solar wind with a comet as it approaches the sun, *Earth, Moon and Planets*, 31, 301, 1984.
Mendis, D.A., J.R. Hill, H.L.F. Houpis, and E.C. Whipple Jr., On the electrostatic charging of the cometary nucleus, *Astrophys. J.*, 249, 787, 1981.
Mendis, D.A., H.L.F. Houpis, and M.L. Marconi, The physics of comets, *Fund. Cosmic Phys.*, 10, 1, 1985.
Morrison, P.J., and D.A. Mendis, On the fine structure of cometary plasma tails, *Astrophys. J.*, 226, 350, 1978.
Neubauer, F.M., K.H. Glassmeir, M. Pohl, *et al.*, First results from the Giotto magnetometer experiment at comet Halley, *Nature*, 321, 352, 1986.
Niedner, M.B. Jr., Interplanetary gas XXVIII. A study of the three-dimensional properties of interplanetary sector boundaries using disconnection events in cometary plasma tails, *Astrophys. J. Suppl.*, 48, 1, 1982.

Niedner, M.B. Jr., and J.C. Brandt, Interplanetary gas XXIII. Plasma tail disconnection events in comets: evidence for magnetic field line reconnection at interplanetary sector boundaries? *Astrophys. J.*, 223, 655, 1978.

Niedner, M.B. Jr., and K. Schwingenschuh, Plasma tail activity at the time of VEGA encounters, *Astron. Astrophys.*, 187, 103, 1987.

Parker, E.N., Dynamics of the interplanetary gas and magnetic fields, *Astrophys. J.*, 128, 644, 1958.

Rickett, B.J., D.G. Sime, N.R. Sheeley, W.R. Crockett, and R. Tousey, High latitude observations of solar wind streams and coronal holes, *J. Geophys. Res.*, 81, 3845, 1976.

Russell, C.T., M.A. Saunders, and J.L. Phillips, Near tail reconnection as a cause of cometary tail disconnections, *J. Geophys. Res.*, 91, 1417, 1986.

Saito, K., T. Saito, T. Aoki, and K. Yumoto, Possible models on disturbances of the ion tails of comet Halley during the 1985–1986 apparition, ESA SP-250(3), 155, 1986.

Schaeberle, J.M., Preliminary note on a mechanical theory of comets, *Astron. J.*, 13, 151, 1893.

Schwarzschild, K., Der druck des lichts auf kleine kugeln und die Arrhenius'sche theorie des kometenschweife, *Sitz. Bayer. Acad. Wiss. München.*, 31, 293, 1901.

Shapiro, V.D., R. Bingham, B.J. Kellett, *et al.*, X-ray emission from comets and nonmagnetic planets: Theory and comparison with CHANDRA observations, *Physica Scripta*, T116, 83, 2005.

Smith, E.J., B.T. Tsurutani, J.A. Slavin, *et al.*, International Cometary Explorer encounter with comet Giacobini-Zinner: Magnetic field observations, *Science*, 232, 382, 1986.

Szegö, K., K.H Glassmier, R. Bingham, *et al.*, Physics of mass loaded plasmas, *Space Sci. Rev.*, 94, 429, 2000.

Tsurutani, B.T., Cometary plasma waves and instabilities, in *Comets in the Post-Halley Era*, vol. 2, edited by R.L. Newburn Jr. *et al.*, Kluwer Acad. Pub., Dordrecht, Netherlands, 1171, 1991.

Tsurutani, B.T., N. Gopalswamy, R.L. McPherron, *et al.*, Magnetic storms caused by corotating solar wind streams, 2006 (This volume).

Tsurutani, B.T., K.H. Glassmeir, and F.M. Neubauer, A review of nonlinear low frequency (LF) wave observations in space plasmas: On the development of plasma turbulence, in *Non Linear Waves and Chaos in Space Plasmas*, edited by T. Hada and H. Matsumoto, Terra Sci. Publ. Co., Tokyo, p1, 1997.

Tsurutani, B.T., and E.J. Smith, Hydromagnetic waves and instabilities associated with cometary ion pickup: ICE Observations, *Geophys. Res. Lett.*, 13, 263, 1986.

Wallis, M.K., Weakly shocked flows of the solar wind plasma through atmospheres of planets and comets, *Planet Space Sci.*, 21, 1647, 1973.

Wallis, M.K., and M. Dryer, The sun and comets as sources in an external flow, *Astrophys. J.*, 205, 895, 1976.

Wallis, M.K., and R.S.B. Ong, Strongly cooled ionized plasma flows with applications to Venus, *Planet Space Sci.*, 23, 713, 1975.

Winterhalter, D., E.J. Smith, M.E. Burton, *et al.*, The heliospheric plasma sheet, *J. Geophys. Res.*, 99, 6667, 1994.

Wu, Z-J, Calculation of the shape of the contact surface at comet Halley, in *Symposium on the Diversity and Similarity of Comets*, edited by E.J. Rolfe and B. Battrick, ESA SP-278, 69, 1987.

Wu, C.S., and R.C. Davidson, Electromagnetic instabilities produced by neutral particle ionization in interplanetary space, *J. Geophys. Res.*, 77, 5399, 1972.

Wurm, K., Die nature der kometen, Mitt. Hamberger Sternwatz, 8, Nr. 51, 1943.

Yeomans, D.K., *Comets: A Chronological History of Observation, Science, Myth and Folklore*, John Wiley, NY, 19, 1991.

The Formation of CIRs at Stream-Stream Interfaces and Resultant Geomagnetic Activity

I.G. Richardson

NASA Goddard Space Flight Center, Greenbelt, Maryland, and the Department of Astronomy, University of Maryland, College Park, Maryland, USA

Corotating interaction regions (CIRs) are regions of compressed plasma formed at the leading edges of corotating high-speed solar wind streams originating in coronal holes as they interact with the preceding slow solar wind. Although particularly prominent features of the solar wind during the declining and minimum phases of the 11-year solar cycle, they may also be present at times of higher solar activity. We describe how CIRs are formed, and their geomagnetic effects, which principally result from brief, southward interplanetary magnetic field excursions associated with Alfvén waves. Seasonal and long-term variations in these effects are briefly discussed.

THE FORMATION OF COROTATING INTERACTION REGIONS

The first in-situ observations of the solar wind were returned by the Mariner 2 spacecraft en route to Venus in 1962. They revealed that the solar wind at that time, during the late-declining phase of solar cycle 19, was organized into streams of fast (>> 400 km/s) flows separated by intervals of slower solar wind. The fast flows lasted for several days and tended to recur at the solar rotation interval (sidereal period = 25.38 days, or 27.28 days as viewed from Earth), suggesting they were long-lived, spatial features corotating with the Sun [*Neugebauer and Snyder*, 1966]. Similar high-speed streams have subsequently been observed in the inner heliosphere, for example by the Helios 1 and 2 spacecraft at 0.3–1 AU [e.g., *Schwenn*, 1990], in the outer heliosphere by Pioneers 10 and 11 and Voyagers 1 and 2 [e.g., *Smith and Wolfe*, 1976, 1977, 1979; *Burlaga et al.*, 1983; *Gazis and Lazarus*, 1983; *Gazis*, 1984], as well as by Ulysses [e.g., *McComas et al.*, 2000] and numerous near-Earth spacecraft.

Corotating interaction regions (CIRs) [*Belcher and Davis*, 1971; *Smith and Wolfe*, 1979] form at the leading edges of corotating high-speed streams as they collide with the preceding slower solar wind. As predicted by the *Parker* [1958] solar wind theory, solar wind flow streamlines, and the magnetic field lines that are carried out with the solar wind plasma, are twisted into Archimedian spirals of the form

$$r - r_o = -V(\phi - \phi_o)/(\Omega \cos\theta), \quad (1)$$

where r is the heliocentric distance, V is the solar wind speed, Ω is the solar angular velocity, θ and ϕ are the heliolatitude and heliolongitude of the observer, and r_o and ϕ_o are the heliocentric distance and heliolongitude of the initial plasma position at the Sun. At low latitudes, streamlines are inclined at an angle

$$\psi = \arctan(r\Omega/V) \quad (2)$$

to the outward radial direction. Thus, streamlines in *faster* solar wind follow spirals that are *less tightly* wound. The resulting collision of the leading edge of a fast solar wind stream with the slower solar wind ahead of it forms a region of enhanced pressure, the corotating interaction region, that lies approximately along the Archimedian spiral [*Parker*, 1963]. At 1 AU, $r\Omega \approx 400$ km/s ~ V, so flow streamlines,

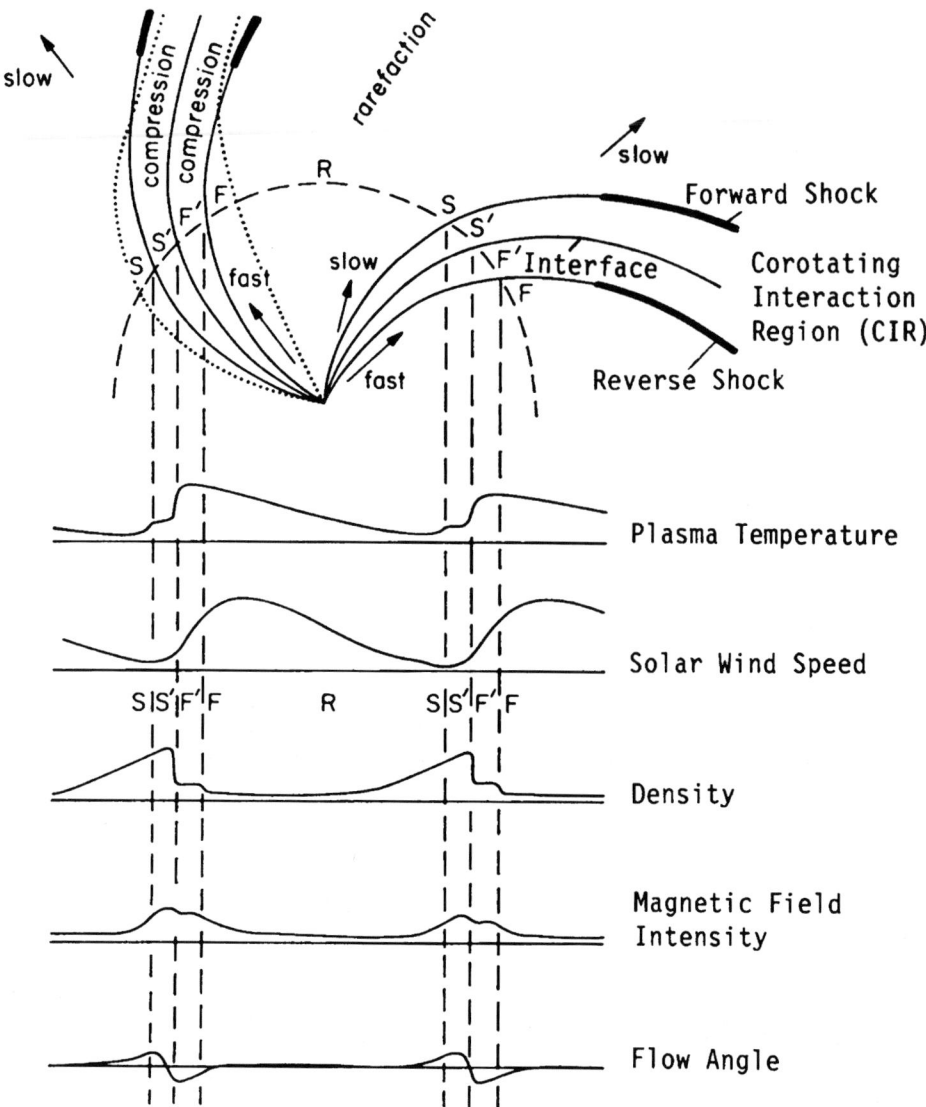

Figure 1. Schematic of two high-speed streams corotating with the Sun [*Richardson et al.*, 1996, after *Belcher and Davis*, 1971]. Variations in plasma parameters at ~1 AU are also shown.

the interplanetary magnetic field, and CIRs are typically inclined at $\psi \sim 45°$ to the radial direction.

Figure 1 [*Richardson et al.*, 1996, after *Belcher and Davis*, 1971] shows the formation of CIRs at the leading edges of two high-speed streams, as viewed from above the north solar pole. Dotted lines indicate representative magnetic field lines/streamlines in the slow and fast solar wind. Variations in plasma parameters at ~1 AU associated with the stream interactions are also shown. It is convenient to identify four regions: the ambient, undisturbed, slow solar wind (S); the compressed, accelerated, slow solar wind plasma (S'); the compressed, decelerated, fast-stream plasma (F'), and the ambient, undisturbed, fast-stream plasma (F). The S' and F' regions form the interaction region, characterized by enhanced plasma densities and magnetic field intensities.

An important feature within a CIR is the boundary between the S' and F' regions, the "stream interface" [*Burlaga*, 1974; *Gosling et al.*, 1978; *Schwenn*, 1990; *Forsyth and Marsch*, 1999; *Crooker et al.*, 1999]. This is typically characterized by a relatively abrupt fall in plasma density (n), and increases in the plasma proton temperature (T_p), solar wind speed, and specific entropy $\propto T_p/n^{\gamma-1}$, where γ is the ratio of specific heats [*Intriligator and Siscoe*, 1994]. The stream-stream interaction tends to deflect solar wind to the west (i.e., in the sense of rotation) ahead of the interface and to the east following the interface. *Belcher and Davis* [1971] suggested that the interface originates as a sharp transition between the two flows near the Sun, ideally a tangential discontinuity. Though challenged by *Burlaga* [1974], who proposed instead that a gradual transition near the Sun becomes steepened by the stream-stream interaction, this interpretation is now generally accepted following the work of *Gosling et al.* [1978]. Thus, for example, the decrease in density across the interface arises because slow solar wind ahead of the interface is intrinsically denser than the fast solar wind. Changes in solar wind composition across the interface [e.g., *Wimmer-Schweingruber et al.*, 1997] are also consistent with the interface separating slow and fast solar wind. Only ~30% of streams at 1 AU include a clear, discontinuous interface, the fraction increasing to ~50% if slightly broader structures are also considered [*Gosling et al.*, 1978; *Schwenn*, 1990]. In some cases, CIRs may include multiple interface-like structures that may be multiple crossings of a wavy boundary separating slow and fast solar wind [*Wimmer-Schweingruber et al.*, 1997].

A close relationship between the stream structure and the recurring sunward- or anti-sunward-directed "sector" structure of the interplanetary magnetic field was first noted by *Wilcox and Ness* [1965]. The sector structure is now known to be associated with crossings of the heliospheric current sheet (HCS), which is embedded in slow, dense solar wind emerging from the "streamer belt" that typically overlays the solar magnetic equator. Thus, the HCS is often crossed in the slow solar wind preceding a CIR, and may be incorporated into the CIR along with the slow solar wind [e.g., *Pizzo and Gosling*, 1994]. Each high-speed stream then has a specific magnetic field polarity that reflects the direction of the field imposed at the Sun.

By tracing spiral streamlines in high-speed flows back to the solar corona, large regions of weak coronal X-ray emission, often persisting for several solar rotations, have been identified as the source regions of corotating high-speed flows [e.g., *Krieger et al.*, 1973; *Zirker*, 1977]. Such "coronal holes" are also evident as regions of reduced ultraviolet emission or weak scattered visible light, and are clearly visible in FeXV images from the Extreme Ultraviolet Imaging Telescope (EIT) on the SOHO spacecraft (Figure 2).

Figure 3 shows CIRs at the leading edges of three corotating high-speed streams observed during one 27-day period by the near-Earth Advanced Composition Explorer (ACE) and IMP 8 spacecraft in December 1999–January 2000, which exhibit the features discussed above. The data illustrated include the magnetic field intensity, polar and azimuthal angles, the plasma proton temperature, density, speed, flow angle, and O^7/O^6 and Mg/O ratios, all from ACE, and the cosmic ray intensity from IMP 8 (specifically,

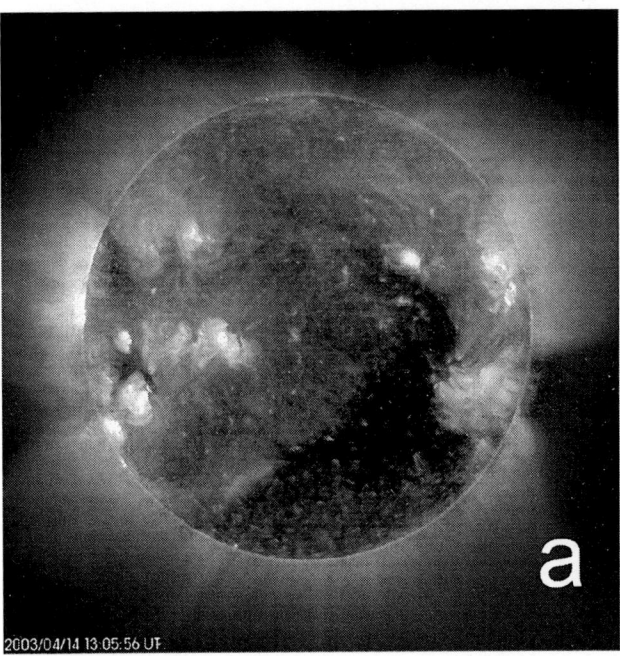

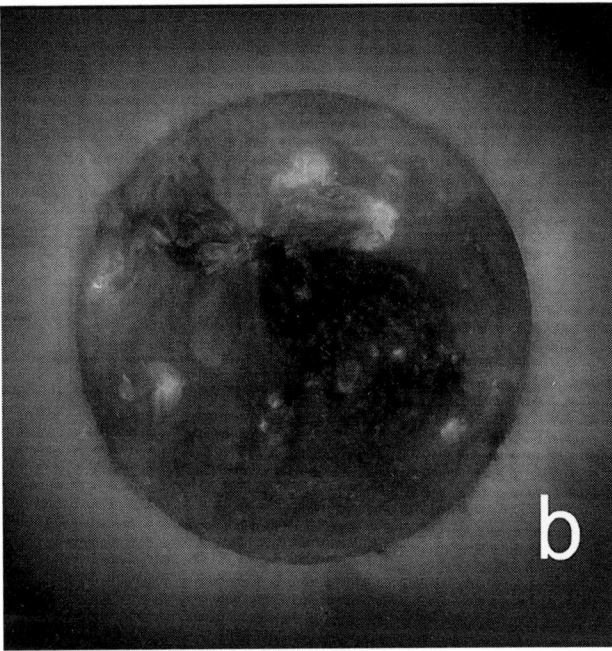

Figure 2. Coronal holes observed by the SOHO EIT instrument in FeXV (284Å) emission (courtesy of the SOHO/EIT consortium). (a) An equatorward extension of a polar coronal hole (observed on April 14, 2003); (b) a near-equatorial coronal hole (December 31,1999) that was the source of the second stream in Figure 3.

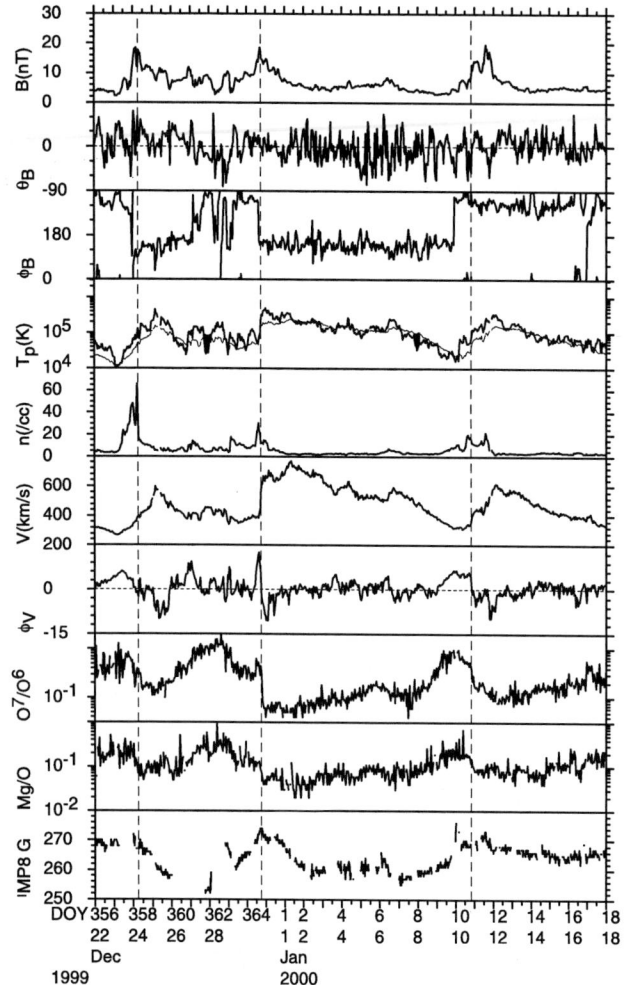

Figure 3. ACE magnetic field and solar wind plasma parameters, and cosmic ray intensity (counting rate of the IMP 8 GME anticoincidence guard) for a 27-day period in December, 1999–January, 2000 when three corotating high-speed streams with CIRs at their leading edges were present. Dashed vertical lines indicate stream interface passages.

hours ahead of the interface, close to the interface in the second CIR, and near the leading edge of the third CIR. Note that these streams were observed near solar maximum–corotating streams were prominent in the near-Earth solar wind during 1998–early 2001 [e.g., *Richardson et al.*, 2002].

As predicted by *Hundhausen* [1973a,b], CIRs can lead to the formation of shocks that corotate with the Sun. The enhanced plasma pressure within a CIR,

$$P = nk(T_e + T_p) + B^2/2\mu_o \quad (3)$$

(where T_e is the plasma electron temperature and B is the magnetic field intensity) causes the CIR to expand into the ambient solar wind. Since the magnetosonic speed in the solar wind,

$$V_f = \sqrt{V_A^2 + V_S^2}, \quad (4)$$

where $V_A = B/\sqrt{\mu_o \rho}$ is the Alfvén speed, and $V_s = \sqrt{\gamma P/\rho}$ is the sound speed (ρ is the plasma mass density), decreases with increasing distance from the Sun, the expanding boundaries of the CIR may eventually steepen into shocks. A "forward" shock propagating into the slow solar wind and away from the Sun may form at the CIR leading edge while a "reverse" shock, propagating into the fast solar wind and toward the Sun, may develop at the trailing edge (see Figure 1). Pioneer 10/11 and Voyager 1/2 observations have demonstrated that CIR shocks tend to form beyond 2 AU [e.g., *Gosling et al.*, 1976; *Hundhausen and Gosling*, 1976; *Smith and Wolfe*, 1976, 1977; *Gazis and Lazarus*, 1983]. However, they may occasionally form by 1 AU [e.g., *Burlaga*, 1970; *Lazarus et al.*, 1970; *Chao et al.*, 1972; *Richardson and Zwickl*, 1984; *Berdichevsky et al.*, 2000] or even closer to the Sun [*Schwenn*, 1990].

Figure 4 [*Crooker et al.*, 1999; *Richardson*, 2004] illustrates the development of a CIR beyond 1 AU. Figure 4(a) shows energetic particle, solar wind plasma, and magnetic field observations at IMP 8 (1 AU) of a CIR in April, 1974. The interface (I) and two possible developing reverse shock-like features (R?) are identified. At Pioneer 11 (3.8 AU; Figure 4(b)), the CIR was bounded by fully developed forward (F) and reverse (R) shocks, while the interface may be identified in the middle of the CIR. Note that the heliospheric current sheet that was located in the slow solar wind ahead of the CIR at 1 AU is now incorporated into the CIR, and crossed in a data gap ahead of the interface. At Pioneer 10 (5.2 AU; Figure 4(c)), again the forward and reverse shocks and stream interface may be identified, and the current sheet is ahead of the interface. *Thomas and Smith* [1981] found that the forward shock had overtaken the current sheet in the majority of CIRs beyond 5 AU.

the count rate of the GME anticoincidence guard [*Richardson*, 2004]). Dashed vertical lines within the magnetic field intensity and plasma density enhancements associated with the CIRs indicate probable stream interface crossings. These are characterized by decreases in density, increases in solar wind speed and proton temperature, the solar wind flow angle moving through the radial direction, changes in the solar wind composition (O^7/O^6 and Mg/O decrease) and the onset of cosmic ray modulations which extend through the high-speed streams. Heliospheric current sheet crossings (abrupt ~180° changes in field azimuth) occur within the first CIR a few

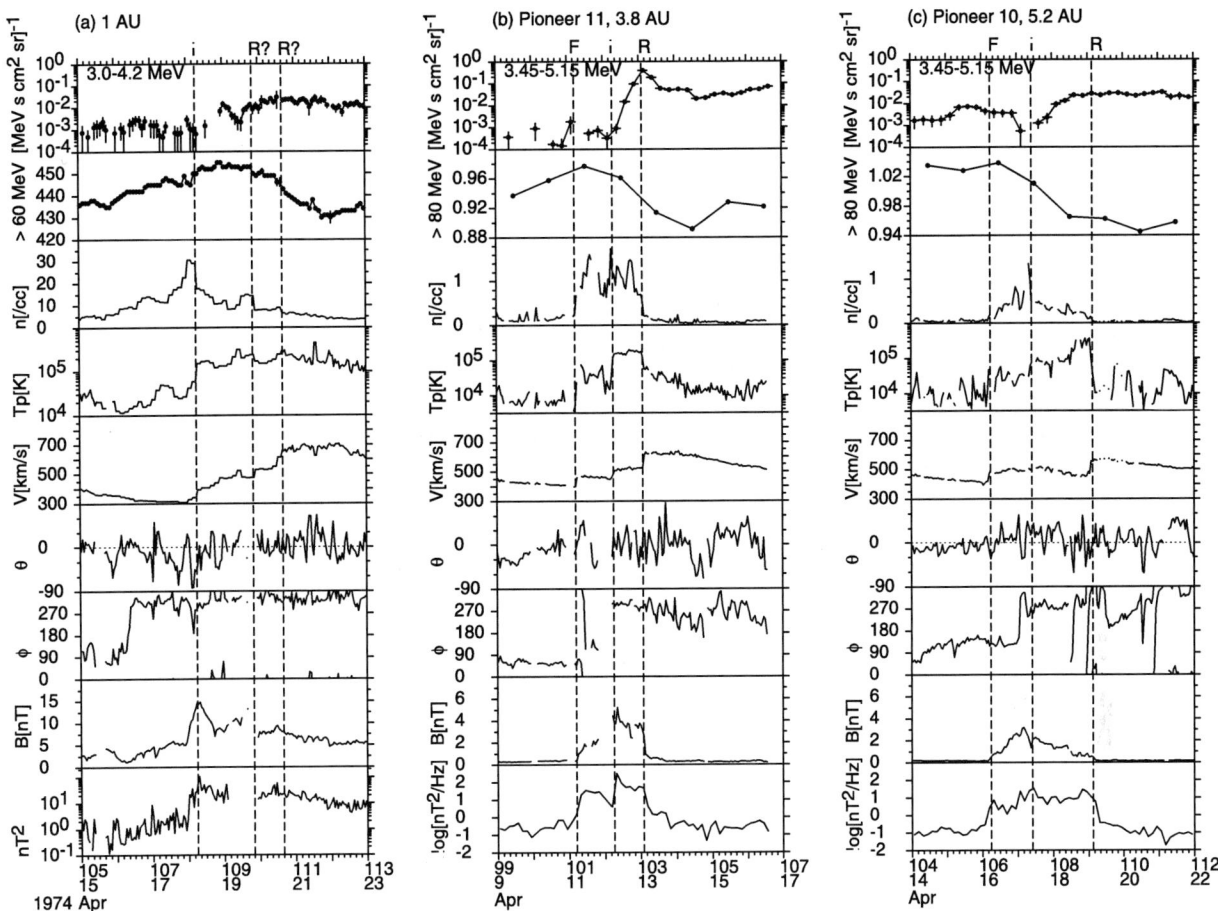

Figure 4. Energetic particle, solar wind plasma and magnetic field observations during the passage of the same CIR at (a) 1 AU, (b) Pioneer 11 at 3.8 AU, and (c) Pioneer 10 at 5.2 AU [*Crooker et al.*, 1999; *Richardson*, 2004]. Forward (F) and reverse (R) shocks and stream interface crossings (I) are indicated. Bottom panels show magnetic field turbulence levels.

The bottom graphs in Figure 4 show how enhanced magnetic field turbulence at 1 AU (here measured by the sum of the squares of the magnetic field component variances) is confined to the region following the stream interface and extends into the high-speed stream. At Pioneer 11 and then Pioneer 10, however, the turbulence (represented by the power at wavenumbers of $1.6 - 3.3 \times 10^{-5}$ km^{-1}; T. Horbury, private communication, 1999) becomes distributed throughout, and is largely confined to, the CIR. The top panels show the intensity of ~4 MeV protons accelerated in the vicinity of CIRs and the modulation of ≥60 MeV galactic cosmic rays. See *Richardson* [2004], and references therein, for further discussion of interplanetary energetic particle effects associated with CIRs.

Although corotating high-speed streams may be present in the near-Earth solar wind at all stages of the solar cycle [e.g., *Richardson et al.*, 2002], nevertheless a profound change in the latitudinal structure of high-speed streams occurs between solar minimum and solar maximum. This was first inferred from interplanetary scintillation measurements [*Rickett and Coles*, 1991] and later confirmed by in-situ observations made by the Ulysses mission, launched in October 1990 to explore the heliosphere to high heliographic latitudes [e.g., *Balogh et al.*, 2001, and references therein]. At solar minimum, the most prominent coronal holes form around the poles of the Sun. Thus, Ulysses observed persistent high-speed coronal hole flows above ~30° latitude [*McComas et al.*, 2000]. At lower latitudes, CIRs form by interactions between high-speed streams originating from equatorward extensions of the polar coronal holes (e.g., Figure 2(a)) and slow solar wind from the streamer belt. Consistent with this configuration, CIR forward (reverse) shocks observed by Ulysses were found to be propagating to lower (higher) latitudes in the slow (fast) solar wind [*Gosling et al.*, 1993; *Gosling*, 1996; *Riley et al.*, 1996].

At solar maximum, the polar coronal holes shrink, and Ulysses observed variable solar wind speeds at all latitudes [*McComas et al.*, 2001, 2003]. High-speed flows near the Earth typically emerge from small, lower latitude, coronal holes [e.g., *Bravo et al.*, 1998; *Luhmann et al.*, 2002; *Neugebauer et al.*, 2002; *Wang and Sheeley*, 2003]. Figure 2(b) shows the low-latitude coronal hole that gave rise to the second high-speed stream in Figure 3.

For further information on the formation and characteristics of CIRs, see *Schwenn and Marsch* [1990], chapters 7 and 8 of *Burlaga* [1995], Volume 89, Nos. 1–2 of *Space Science Reviews* [1999], and *Balogh et al.* [2001].

GEOMAGNETIC ACTIVITY ASSOCIATED WITH CIRS AND COROTATING HIGH-SPEED STREAMS

Enhanced geomagnetic activity is a consequence of an increase in the rate of energy transfer from the solar wind into the Earth's magnetosphere. This is largely determined by the strength and orientation of the interplanetary magnetic field, and the solar wind speed and density. One formulation (not including density) is the ε function of *Perreault and Akasofu* [1978],

$$\varepsilon = l_o^2 V B^2 \sin^4(\theta/2) \quad (5)$$

where l_o^2 is the area of the magnetopause through which the energy enters, and θ is the "clock angle" of the IMF relative to the Sun-Earth line. Although increased energy transfer is expected in faster solar wind, the typical factor of ~2–3 variation in solar wind speed is much less than the variation in the magnetic field dependence, which reflects the efficient energy transfer that occurs when the IMF has a southward component, facilitating reconnection between the solar wind and magnetospheric magnetic fields [*Dungey*, 1961].

The strongest geomagnetic "storms" occur sporadically and are almost invariably produced by interplanetary transients (interplanetary coronal mass ejections, ICMEs) resulting from coronal mass ejections at the Sun [*Gosling et al.*, 1991; *Richardson et al.*, 2001; see also section 3.2]. *Tsurutani and Gonzalez* [1997] give an overview of the interplanetary conditions that drive geomagnetic storms associated with solar wind transients, including southward fields in the CME-related material or in the "sheath" of compressed plasma ahead of the transient.

Geomagnetic activity associated with CIRs and high-speed streams has a rather different nature. Most characteristically, it tends to recur at the solar rotation period. The existence of recurring geomagnetic activity was first clearly established by *Maunder* [1905] who noted that periods of enhanced geomagnetic disturbances during 1882 to 1903 were ordered by solar longitude. He inferred that some "definite and restricted areas" on the Sun gave rise to these geomagnetic disturbances. Furthermore, sunspots did not have to be present, and he could find no better relationship with faculae or prominences. *Maunder* concluded that the recurrence could only be explained "by supposing that the Earth has encountered, time after time, a definite stream ... which, continually supplied from one and the same area of the Sun's surface, appears to us, at our distance, to be rotating with the same speed as the area from which it arises". A stream diameter of ~20° of solar longitude was suggested by the typical duration of the disturbances. This removed the criticism by Lord Kelvin (among others), on the grounds of the unreasonable energy required, of the then-prevailing concept that geomagnetic activity was caused by "magnetic waves" expanding away from the Sun equally in all directions.

Further insight into the origin of recurrent geomagnetic activity came from the work of *Greaves and Newton* [1929], who demonstrated (see Figure 1 of *Cliver* [1995]) that recurrence is a property of smaller geomagnetic storms, and is not exhibited by major storms even though these are often associated with large sunspot regions which may persist for more than one solar rotation. *Bartels* [1932] extended this work using stacked 27-day plots of the geomagnetic C9 index for 1906 to 1931. A plot of C9 for 1971–1974, during the decline of cycle 20, is shown in Figure 5. Numbers of different "weights" help to visualize the variations in the daily C9 index and sunspot number (R9). At high solar activity levels, geomagnetic activity is dominated by sporadic, often intense storms associated with transients. As the sunspot number declines, moderate, recurrent geomagnetic activity becomes predominant, as Bartels noted (see also *Bartels* [1940]). He argued that sunspots were independent of the sources of recurrent activity, which he termed "M (mystery) regions". The mystery was, of course, eventually solved by the 1970's through the combination of the Parker theory of the solar wind, in-situ solar wind observations and the discovery of coronal holes. *Sheeley et al.* [1976, 1977] illustrate 27-day stack plots of C9, solar wind speed, and coronal holes which convincingly show the intimate relationship between recurrent geomagnetic activity, high-speed streams and coronal holes. For further discussion of the historical aspects of recurrent geomagnetic activity, see *Burlaga and Lepping* [1977], *Cliver* [1994, 1995], and references therein. In addition, *Crooker and Cliver* [1994] review the studies that culminated in the identification of "M-regions", and argue that the slow solar wind and CIRs should be included in this identification, in addition to corotating high-speed streams.

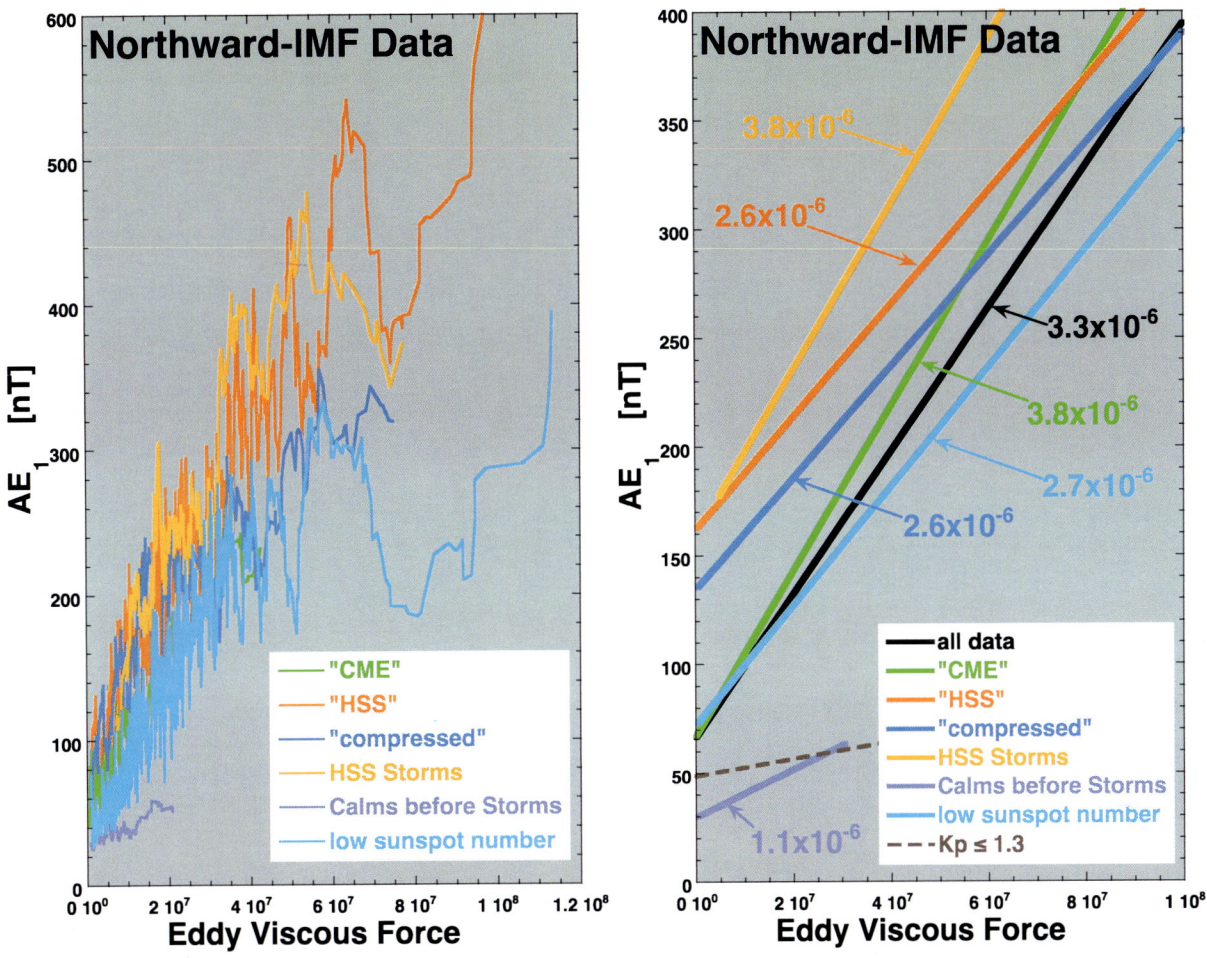

Plate 2. For the various categorizations of solar-wind data, AE (1-hour lagged) is plotted as a function of the solar-wind eddy-viscous force $nv^{5/2}\delta B/B_o$ for northward IMF (1000 nT km/s < vB_z < 3000 nT km/s). In the left panel 20-point running averages of AE_1 are plotted. In the right panel least-square linear-regression fits $AE_1 = \text{slope} \times (nv^{5/2}\delta B/B_o) + \text{intercept}$ are shown, with the various slope values indicated.

effect were seen between the various types of solar wind, with the exception of the calms before the storms in which case data selection produces an anomalous result.

Acknowledgments. The author wishes to thank Herb Funsten and Bob McPherron for helpful conversations, to thank Ruth Skoug and Chuck Smith for ACE spacecraft data, and to thank Mike Henderson for providing modern-era AE index. Supported by NASA RSSW@1AU Program, by the LDRD Program at Los Alamos National Laboratory, and by the U.S. Department of Energy.

REFERENCES

Baumjohann, W., and Y. Kamide, Hemispherical Joule heating and the AE indices, *J. Geophys. Res.*, 89, 383, 1984.

Blair, M.F., Influence of free-stream turbulence on turbulent boundary layer heat transfer and mean profile development part 1: Experimental data, *J. Heat Transf.*, 105, 33, 1983.

Borovsky, J.E., The eddy viscosity and flow properties of the solar wind: CIRs, CME sheaths, and solar-wind/magnetosphere coupling, submitted to *Phys. Plasmas*, 2006.

Borovsky, J.E., and J.T. Gosling, The level of turbulence in the solar wind and the driving of the Earth's magnetosphere, *Eos Trans. Amer. Geophys. Soc.*, 82(20), S368, 2001.

Borovsky, J.E., and H.O. Funsten, The role of solar-wind turbulence in the coupling of the solar wind to the Earth's magnetosphere, *J. Geophys. Res.*, 108, 1246, 2003.

Borovsky, J.E., and J.T. Steinberg, The "calm before the storm" in CIR/magnetosphere interactions: Occurrence statistics, solar-wind statistics, and magnetospheric preconditioning, in press, *J. Geophys. Res.*, 2006.

Chen, H., and D. Montgomery, Turbulent MHD transport coefficients: An attempt at self-consistency, *Plasma Phys. Control. Fusion*, 29, 205, 1987.

Elliott, H.A., D.J. McComas, N.A. Schwadron, J.T. Gosling, R.M. Skoug, G. Gloeckler, and T.H. Zurbuchen, An improved expected temperature formula for identifying ICMEs, *J. Geophys. Res.*, A4, DOI:10.1029/2004JA010794.

Gosling, J.T., A.J. Hundhausen, V.J. Pizzo, and J.R. Asbridge, Compression and rarefactions in the solar wind: Vela 3, *J. Geophys. Res.*, 77, 5442, 1972.

Gosling, J.T., V. Pizzo, and S.J. Bame, Anomalously low proton temperatures in the solar wind following interplanetary shock waves – Evidence for magnetic bottles, *J. Geophys. Res.*, 78, 2001, 1973.

Gosling, J.T., and V.J. Pizzo, Formation and evolution of corotating interaction regions and their three dimensional structure, *Space Sci. Rev.*, 89, 21, 1999.

Ishizawa, A., and Y. Hattori, Large coherent structure formation by magnetic stretching term in two-dimensional MHD turbulence, *J. Phys. Soc. Japan*, 67, 4302, 1998.

King, J.H., and N.E. Papitashvili, Solar wind spatial scales in and comparisons of hourly Wind and ACE plasma and magnetic field data, *J. Geophys. Res.*, 110, 2104, 2005.

Kwok, K.C.S., and W.H. Melbourne, Freestream turbulence effects on galloping, *ASCE J. Engin. Mech. Div.*, 106, 273, 1980.

Lopez, R.E., Solar cycle invariance in solar wind proton temperature relationships, *J. Geophys. Res.*, 92, 11189, 1987.

Mathieu, J., and J. Scott, *An Introduction to Turbulent Flow*, sect. 5.5, Cambridge University Press, New York, 2000.

McComas, D.J., S.J. Bame, P. Barker, W.C. Feldman, J.L. Phillips, P. Riley, and J.W. Griffee, Solar Wind Electron Proton Alpha Monitor (SWEPAM) for the Advanced Composition Explorer, *Space Sci. Rev.*, 86, 563, 1998.

Pal, S., Freestream turbulence effects on wake properties of a flat plate at an incidence, *AIAA J.*, 23, 1868, 1985.

Richardson, I.G., and H.V. Cane, Regions of abnormally low proton temperature in the solar wind (1965-1991) and their association with ejecta, *J. Geophys. Res.*, 100, 23397, 1995.

Schlichting, H., Boundary Layer Theory, Sect. XXI.a, McGraw-Hill, New York, 1979.

Smith, C.W., M.H. Acuna, L.F. Burlaga, J. L'Heureux, N.F. Ness, and J. Scheifele, The ACE Magnetic Fields Experiment, *Space Sci. Rev.*, 86, 611, 1998.

Sullerey, R.K., and M.A. Sayeed Khan, Freestream turbulence effects on compressor cascade wake, *J. Aircraft*, 20, 733, 1983.

Thole, K.A., and D.G. Bogard, High freestream turbulence effects on turbulent boundary layers, *J. Fluids Engin.*, 118, 276, 1996.

Tsurutani, B.T., and W.D. Gonzalez, The cause of high-intensity long-duration continuous AE activity (HILDCAAs): Interplanetary Alfven waves, *Planet. Space Sci.*, 35, 405, 1987.

Tsurutani, B.T., W.D. Gonzalez, A.L.C. Gonzalez, F. Tang, J.K. Arballo, and M. Okada, Interplanetary origin of geomagnetic activity in the declining phase of the solar cycle, *J. Geophys. Res.*, 100, 21717, 1995.

Volino, R.J., M.P. Schultz, and C.M. Pratt, Conditional sampling in a transitional boundary layer under high freestream turbulence conditions, *J. Fluids Engin.*, 125, 28, 2003.

Wu, J.-S., and G.M. Faeth, Sphere wakes at moderate Reynolds numbers in a turbulent environment, *AIAA J.*, 32, 535, 1994.

Yoshizawa, A., and N. Yokoi, Stationary large-scale magnetic fields generated by turbulent motion in a spherical region, *Phys. Plasmas*, 3, 3604, 1996.

Joe Borovsky and John Steinberg, Mail Stop D466, Los Alamos National Laboratory, Los Alamos, NM 87545, USA. (jborovsky@lanl.gov; jsteinberg@lanl.gov)

Modeling the Behavior of Corotating Interaction Region Driven Storms in Comparison With Coronal Mass Ejection Driven Storms

Vania K. Jordanova[1]

Space Science Center, University of New Hampshire, Durham, New Hampshire, USA

There are two types of geoeffective solar disturbances, single events like coronal mass ejections that pass Earth once, or structures like corotating interaction regions which corotate with the Sun and could pass Earth on successive rotations. We use our kinetic ring current-atmosphere interactions model (RAM) to simulate ring current evolution during two geomagnetic storms of similar strength representative of each solar origin, the 15 May 1997 and the 10 March 1998, and compare the mechanisms responsible for trapping particles and for causing their loss. A quiet time ring current distribution inferred from satellite data is used as initial conditions, and ring current intensification due to enhanced plasma inflow from the magnetotail and earthward transport and acceleration is considered. Using an electric potential model driven by interplanetary parameters, we find that the ring current injection calculated with RAM is in good agreement with *Dst* index during May 1997, however, it underestimates *Dst* during March 1998. Additional intensifications by radial diffusion during March 1998 reproduce better its long lasting recovery phase. The dominant ring current ion during both storms is H^+, while O^+ and He^+ are minor constituents. The strongest electromagnetic ion cyclotron waves are generated in the postnoon local time sector and along the plasmapause; these waves cause ~5% decrease of the total ring current energy by proton precipitation during May 1997. Charge exchange is the dominant loss process during the storm recovery phase; the net convective losses are larger than charge exchange losses only for few hours near *Dst* minima.

1. INTRODUCTION

The increased variety and sophistication of present-day technologies placed into space-affected environments requires an improved understanding of this environment, in order to assure the technologies' reliable operation and survivability. Among the physical causes for the disturbances in the space- and ground-based technical systems are the increased electrical current systems in the magnetosphere and ionosphere during geomagnetic storms. Thus for example, the magnetic storm of February 1958 disrupted the cross-Atlantic cable telecommunications and caused a temporary power outage in the Toronto area [*Lanzerotti*, 2001]. Geomagnetic storms have their origin in the structure and dynamics of the solar atmosphere. Two categories of magnetic storms have been identified based upon their solar origin; recurrent, that repeat with the solar rotation period of 27 days, or transient, that occur only once. The recurrent storms are associated with corotating interaction regions (CIR) [*Smith and Wolf*, 1976] that are formed between the high-speed streams from coronal holes and the upstream dense slow-speed solar wind plasma [*Crooker and Cliver*, 1994]. The single non-recurrent events are usually associated with huge eruptions from the Sun of

[1]Now at Los Alamos National Laboratory, Los Alamos, New Mexico, USA

Recurrent Magnetic Storms: Corotating Solar Wind Streams
Geophysical Monograph Series 167
Copyright 2006 by the American Geophysical Union.
10.1029/167GM08

plasma and magnetic flux called coronal mass ejections (CME) and often give rise to the largest geomagnetic storms at Earth [*Tsurutani et al.*, 1992]. For both categories the immediate cause of magnetic storms at Earth is related to periods of strong southward interplanetary magnetic field (IMF) reconnecting with the terrestrial magnetic field and allowing transfer of solar wind energy into the magnetosphere [*Gonzalez et al.*, 1994]. CME are more frequent interplanetary phenomena during solar maximum, while CIR dominate the interplanetary medium during solar minimum when the coronal holes migrate down to lower latitudes. Reviews of the interplanetary origin of intense geomagnetic storms can be found in *Gonzalez et al.* [1999] and *Vieira et al.* [2004].

In this paper we study the geomagnetic activity during two storms representative of each solar origin. The first one is the major storm on 10 March 1998 (with $Dst = -116$ nT and $Kp = 7^+$) triggered by a classic stream-stream interaction region monitored by the Wind spacecraft and identified as a corotating interaction region by *Vieira et al.* [2004]. The interplanetary (IP) observations [*Jordanova et al.*, 2001a] showed a fast stream (speed of ~550 km/s) overtaking a slower stream (speed of ~300 km/s) and forming an interface region between 00 UT and 18 UT on 10 March, where the density, dynamic pressure, and total field were enhanced. The driving of the magnetosphere was mainly through the large negative B_z component of the IMF reaching peak values of -15 nT right at the leading edge of the high-speed stream. In the high-speed flow behind the interaction region occurred a long train of Alfvénic fluctuations with peak-to-peak IMF B_z amplitude of ~8 nT, which caused moderate Dst and Kp activity for several days. The second storm we investigate is the 15 May 1997 major storm (with $Dst = -115$ nT and $Kp = 7^-$) triggered by a coronal mass ejection. Interplanetary data from the instruments on Wind indicated a strong IP shock at ~01 UT causing an increase in the density, bulk speed, and total magnetic field, with large fluctuations in all IMF components for ~10 hours. These were followed by a period of relatively smooth south-to-north IMF B_z excursion characteristic of a magnetic cloud [*Jordanova et al.*, 2001b]. This storm was caused partly by the $B_z < 0$ fields in the sheath region behind the interplanetary shock and partly by the magnetic cloud driving the shock. In the next sections we describe briefly our physics-based model and present numerical simulation results during these two storms, investigating the effect of varying interplanetary conditions on ring current dynamics and why recurrent magnetic storms have long lasting recovery phases.

2. KINETIC RING CURRENT MODEL

We employ the ring current-atmosphere interaction model (RAM) developed by *Jordanova et al.* [2001a,b] and recently extended to relativistic energies including radial diffusion [*Jordanova and Miyoshi*, 2005]. The model includes a region in the equatorial plane from 2 R_E to 6.5 R_E and all magnetic local times (MLT), solving the bounce-averaged kinetic equation for the major ring current particles H^+, O^+, and He^+. Adiabatic drifts of energetic particles are calculated using the ionospheric electric potential model of *Weimer* [2001] mapped to the equatorial plane, a corotation potential, and a dipolar magnetic field of the Earth. To incorporate the effect of magnetic field fluctuations on the injection of high-energy particles we have added a radial diffusion term [*Schulz and Lanzerotti*, 1974], using the Kp-dependent radial diffusion coefficients from the empirical analysis of *Brautigam and Albert* [2000] and considering only magnetic diffusion. The model is coupled with a time-dependent plasmasphere model through the employed electric and magnetic fields. Losses due to charge exchange with neutral exospheric hydrogen, Coulomb collisions with plasmaspheric ions and electrons, atmospheric collisions at low altitudes, and scattering by electromagnetic ion cyclotron (EMIC) waves are included.

The model initial conditions are specified based upon quiet time energetic particle data from the instruments on Polar [*Blake et al.*, 1995] and Equator-S [*Kistler et al.*, 1999] satellites, extrapolated to higher energies after the empirical quiet time radiation belt model AP8 [*Vette*, 1991]. The nightside boundary conditions are updated according to the total ion flux measurements from the MPA and SOPA [*Belian et al.*, 1992] instruments on the Los Alamos National Laboratory (LANL) satellites at geosynchronous orbit [*McComas et al.*, 1993]. These LANL instruments do not provide ion composition, therefore, we divide the measured flux between the three ion species using the ion density ratios parameterized by the $F_{10.7}$ and Kp indices at geosynchronous orbit from *Young et al.* [1982]. The plasma sheet ion densities from the MPA instrument when the LANL satellites were on the nightside (between MLT = 18 and MLT = 6) are shown for the March 1998 and May 1997 storm periods in Figure 1 (left and right, respectively). Enhanced densities are observed during the main phase of both storms; the ion densities decrease during the recovery phase. The nightside data coverage during the storms' main phase is satisfactory except during the ~3 hour gap on March 10 when we linearly interpolate the data. Losses through the dayside magnetopause are included in the simulations allowing free plasma outflow from the dayside boundary.

3. MODEL RESULTS AND DISCUSSION

Data and simulation results from our RAM model during the March 1998 and May 1997 storms are shown in Figure 1. From top to bottom the various panels display the cross polar cap potential drop obtained with *Weimer* [2001] model and the Kp index, the MPA ion density, the calculated ring current

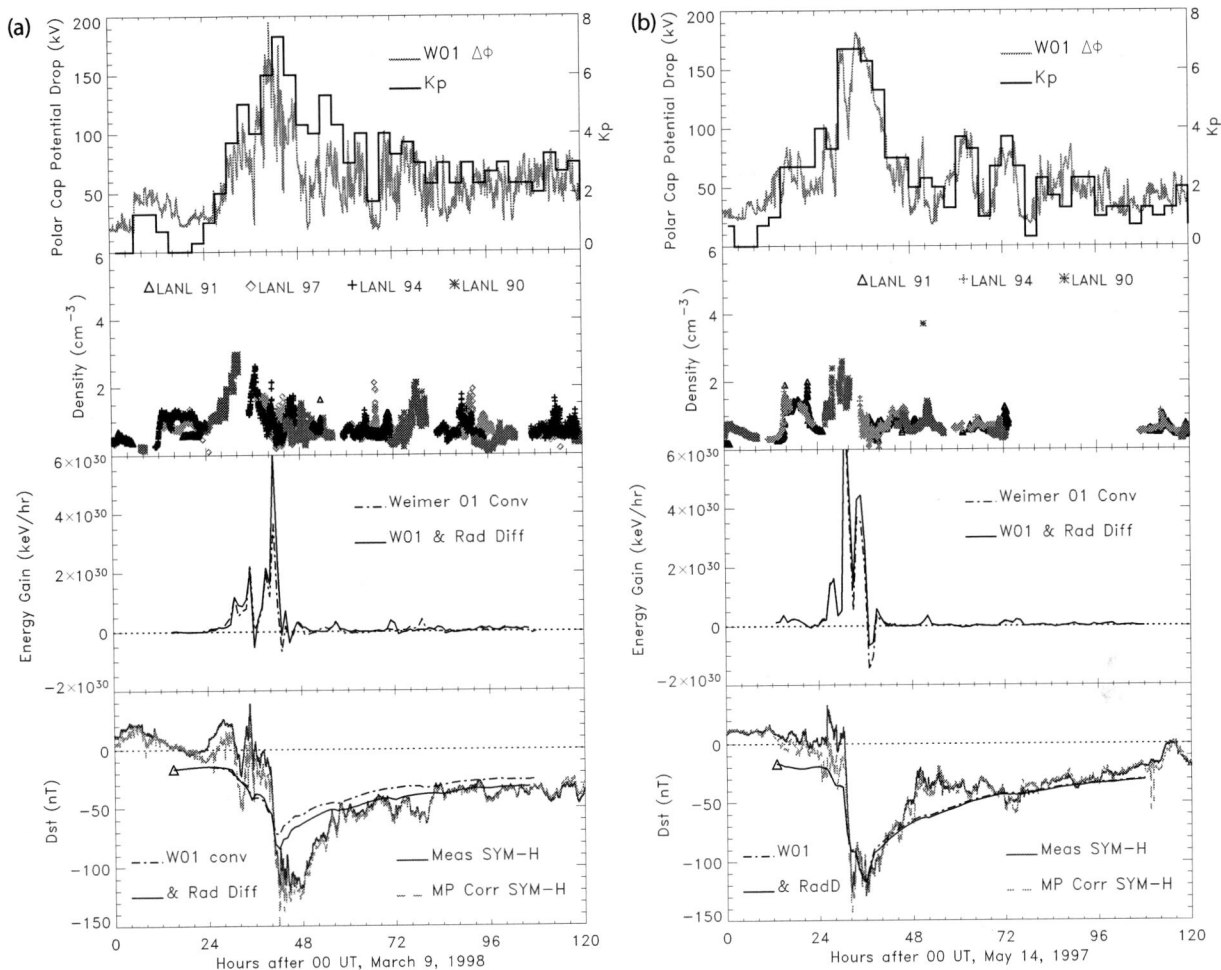

Figure 1. (left) From top to bottom: the polar cap potential drop (solid grey line) obtained with *Weimer* [2001] model and the *Kp* index (solid black line); the nightside plasma sheet ion density at geosynchronous orbit; ring current injection calculated with RAM using Weimer 01 potential without (dashed-dotted line) and with (solid line) radial diffusion; and comparison of model simulations without (dashed-dotted line) and with (solid line) radial diffusion included with measured (solid grey line) and magnetopause currents corrected (dashed line) SYM-H index during the March 1998 storm. (right) Data and simulation results for the May 1997 storm in the same format as Figure 1 left.

injection rate, and the measured and magnetopause currents-corrected SYM-H index compared with model simulations. The ionospheric electric potential model of *Weimer* [2001] is driven by interplanetary parameters and predicts highly variable polar cap potential drop, reaching maximum of ~160 kV at ~16 UT on 10 March, and of ~170 kV at ~10 UT on 15 May, respectively, during the main phase of the storms. The potential pattern exhibits rapid fluctuations of ~30 kV amplitude during the March 1998 storm, while the fluctuations are of smaller amplitude (~10 kV) during the May 1997 storm. Note that the large-scale enhancement in the convection potential during the May 1997 storm is not only stronger but also of longer duration than the one during the March 1998

storm. The potential drop increases abruptly and reaches maximum first at ~6 UT on 15 May during the period of enhanced plasma sheet density and except for a brief interruption at ~8 UT stays enhanced for ~10 hours, while during March 1998 it peaks only for ~6 hours and it is during the period of reduced plasma sheet density; this different behavior results in larger injection rate and stronger ring current buildup during the May 1997 storm. The RAM simulation (Figure 1, bottom) reproduces very well the measured SYM-H minimum during the May 1997 storm, however, it significantly underestimates the SYM-H minimum during the March 1998 storm. One possible reason for this disagreement may be the underestimation of the boundary conditions within the gap in

MPA data on 10 March. The interplay between the magnitudes of magnetospheric convection and plasma sheet density determines ring current evolution. A decrease of either the convection potential or the plasma sheet density could trigger the fast initial ring current decay [*Jordanova et al.*, 2003a]. The recovery phase of the storms thus starts when the ring current injection rate reaches minimum and the plasma sheet density drops to ~1 cm^{-3}.

The RAM simulations (solid line) when diffusive transport due to magnetic field fluctuations was included in addition to convective transport are displayed in the bottom panels of Figure 1. It is evident that radial diffusion did not contribute much to ring current intensification during the May 1997 storm. After the passage of the magnetic cloud on 15 May the geomagnetic activity was rapidly reduced to an average $Kp < 2$ (except for two short-lived enhancements) giving small radial diffusion coefficients as specified by *Brautigam and Albert* [2000] and negligible effect on ring current injection. On the other hand, the Alfvén waves in the high-speed flow behind the interaction region on 10 March produced a long-lasting moderate geomagnetic activity with average $Kp \sim 4$ corresponding to larger radial diffusion coefficients. Diffusive transport increased the injection of high-energy particles at the peak and during the recovery phase of the March 1998 storm, thus enhancing the magnetic field disturbance of the ring current on ground by ~15% and improving the agreement with observations.

The globally averaged O$^+$ energy density percentage obtained from our model (solid line) during both storms is shown in Plate 1 (top panels) together with the O$^+$ density percentage at geosynchronous orbit from *Young et al.* [1982]. As discussed above the *Young et al.* values (dashed-dotted line) are used at the nightside boundary of our model to determine the O$^+$ composition. It is clear that both storms are dominated by H$^+$ ions and the globally averaged contribution of O$^+$ is less than 20% even near Dst minima when the ring current energy maximizes. The contribution of He$^+$ is between 3% and 7% throughout the storms. These results are in agreement with previous statistical studies of ion composition from AMPTE/CHEM observations [*Daglis et al.*, 1993]. Dial plots of the energy density of the three major ion species during (a) prestorm and (b) main phase conditions are shown in the bottom panels of Plate 1. The ring current exhibits pronounced asymmetry during the main phase of both storms, the energy density being larger for all ion species during May 1997 (Plate 1, right). The location of the energy density peak is quite different though, near dusk during the March storm, but midnight to dawn during the May storm. Previous studies [*Fok et al.*, 2003; *Jordanova et al.*, 2003b] have shown that the location of the peak depends strongly on the pattern of the convection electric potential and is usually predicted near dusk using a semi-empirical Volland-Stern electric field model, however, it may be in the midnight-dawn sector when a self-consistently calculated electric field model is used. During the May 1997 storm the energy density peaked in the postmidnight sector even when a Volland-Stern model was used [*Jordanova*, 2003] indicating that this was mostly due to an enhancement in the plasma sheet ion density at geosynchronous orbit in the postmidnight sector during the main phase of this storm.

During magnetic storms the anisotropic ring current distributions are unstable and may excite plasma waves. With our model we calculated the growth rate of He$^+$ band EMIC waves self-consistently with the time-evolving ring current and plasmaspheric populations using the hot plasma dispersion relation; we used quasi-linear theory to include in the model the pitch angle scattering of ring current protons by these waves. The wave gain obtained after integrating the growth rate along field aligned wave paths is shown in Plate 2 (top). The corresponding plasmaspheric electron densities calculated with the coupled model of *Rasmussen et al.* [1993] driven by the *Weimer* [2001] IP-dependent convection model are shown in the bottom panels. As geomagnetic activity increases the plasmasphere erodes on the nightside and high electron densities at larger L shells are confined to the postnoon bulge. Simultaneously, the ring current intensifies and the unstable ion distributions generate EMIC waves. As seen in Plate 2 the maximum wave gain occurs at hours 34 and 48 (respectively for the May 1997 and the March 1998 storms) within the plasmaspheric bulge and along the plasmapause. Data from the magnetic field investigation (MAM) on Equator-S are available during the March 1998 storm [*Mouikis et al.*, 2002]. In Plate 2a we show the position of Equator-S indicating whether EMIC waves are observed (full symbol) or not (open symbol). In agreement with our model predictions, waves were not observed below $L = 6.5$ during prestorm conditions (hour 27), but did occur later in the storm, intensifying near minimum Dst (hour 48) and having reduced amplitude during the recovery phase (hour 90).

The effect of all major loss processes on the time evolution of the total energy of H$^+$ ring current ions during the two storms is shown in Figure 2. Charge exchange losses (solid line) are caused by collisions of ring current ions with neutral hydrogen from the geocorona, producing energetic neutral atoms and low energy protons. Coulomb collisions losses (dashed line) are due to energy transfer to coexisting low-energy plasmaspheric populations, while precipitation losses (dotted line) are due to collisions and removal of particles in the dense atmosphere at low altitudes. Finally, the net energy losses due to convection (triangles) show the energy balance between the convective outflow through the dayside boundary, energy changes inside the simulation domain by convection, and inflow through the nightside boundary. The total energy (dashed-dotted line) of ring current H$^+$ is shown for reference.

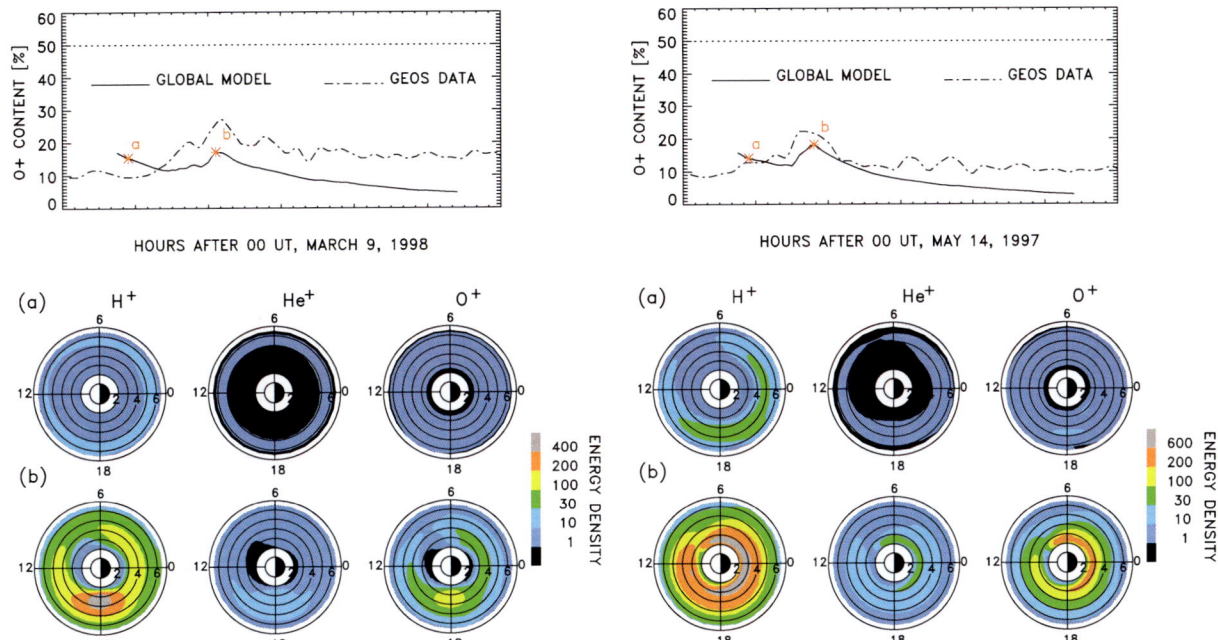

Plate 1. (top) Content of ring current O^+. (bottom) Energy density (keV/cm^3) of H^+, He^+, and O^+ ring current ions as a function of radial distance in the equatorial plane and MLT at selected hours indicated with (a) and (b) in the top panel. The left panels correspond to the March 1998, while the right panels correspond to the May 1997 storm periods.

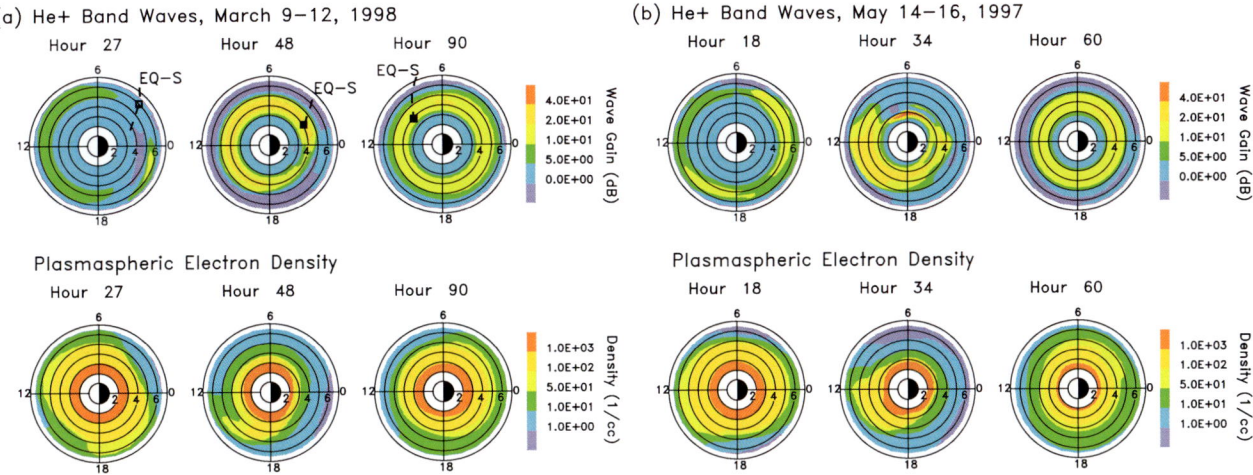

Plate 2. (top) Wave gain of He^+ band EMIC waves and (bottom) plasmaspheric electron density, as a function of radial distance in the equatorial plane and MLT. The dial plots in (a) correspond to selected hours after 00 UT on 9 March 1998, while these in (b) correspond to hours after 00 UT on 14 May 1997. The trajectory (dashed line) of the Equator-S satellite and its position (square symbol) are shown during March 1998.

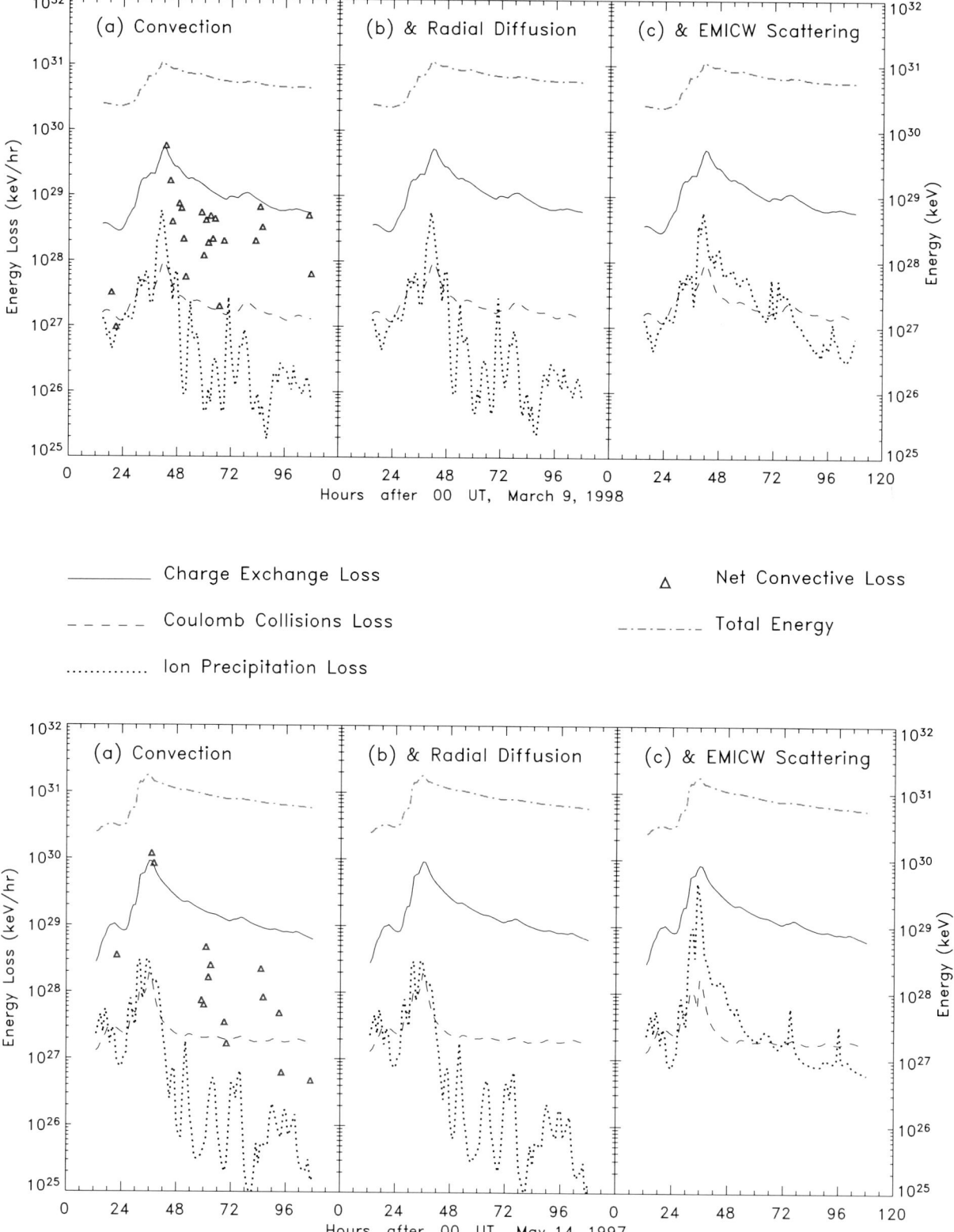

Figure 2. (top) Proton ring current energy (keV) and energy losses (keV/hr) after 00 UT on 9 March 1998 considering (a) only convection, and adding (b) radial diffusion, and (c) plasma waves scattering. (bottom) Model results after 00 UT on 14 May 1997.

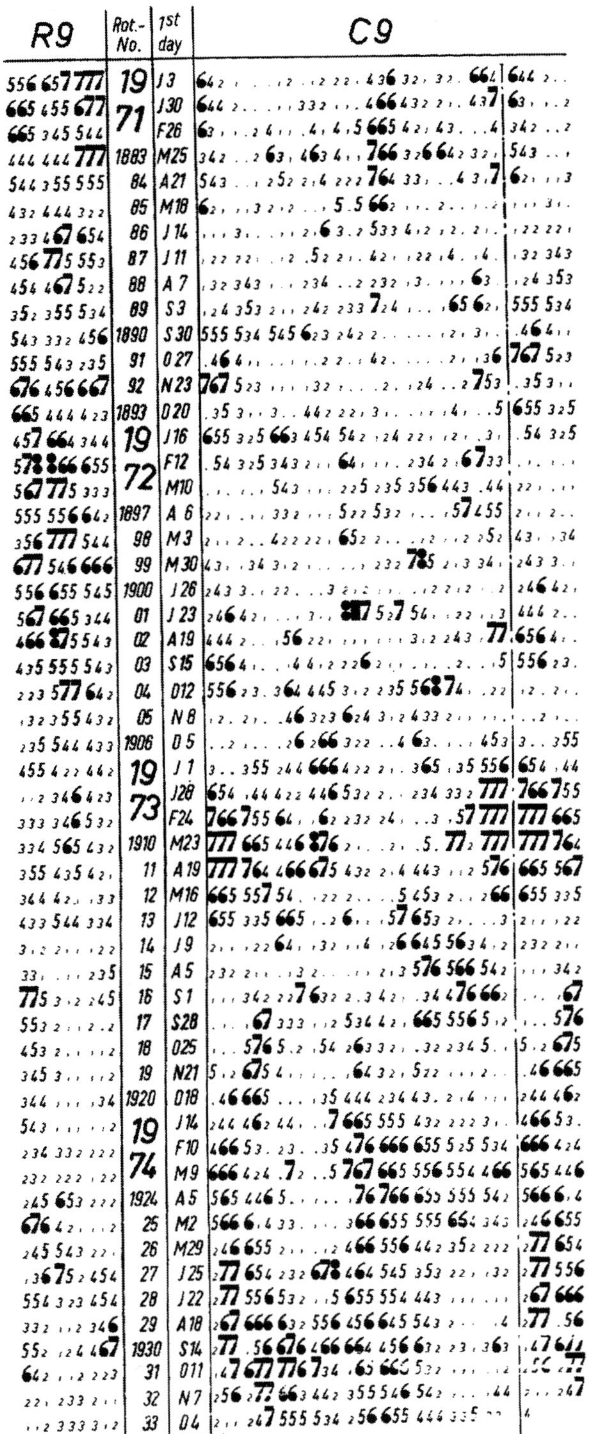

ORIGIN OF CIR/STREAM ASSOCIATED GEOMAGNETIC ACTIVITY

Crooker [2000] notes that a "common misunderstanding about high-speed streams is that the high-speed flow itself causes geomagnetic storms". *Burlaga and Lepping* [1977] were among the first to examine in detail the interplanetary causes of geomagnetic activity associated with CIRs and high-speed streams and demonstrate that the magnetic field plays a prominent role. Figure 6 shows one event from their study. In addition to the solar wind density, speed and magnetic field intensity, which clearly show the stream and CIR, the figure includes the north-south component of the IMF (B_z), the y-component of the interplanetary electric field (B_zV), and the geomagnetic AE index, which measures auroral zone activity that is not necessarily due to storms. *Burlaga and Lepping* [1977] noted a "striking correlation" between the bursts in AE and large southward (negative) values of B_z,

Figure 5. 27-day (Bartels rotation) stackplot of the daily C9 geomagnetic index and 3-day mean sunspot number (R9) for 1971–1974.

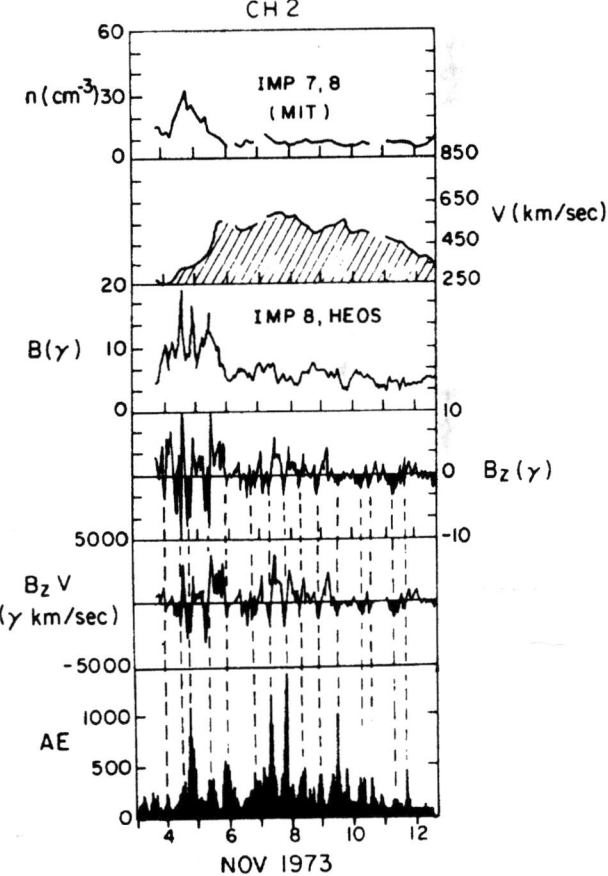

Figure 6. A high-speed stream in November 1973 showing the relationship between bursts of AE activity and brief southward (negative) turnings of the IMF throughout passage of the stream [*Burlaga and Lepping*, 1977].

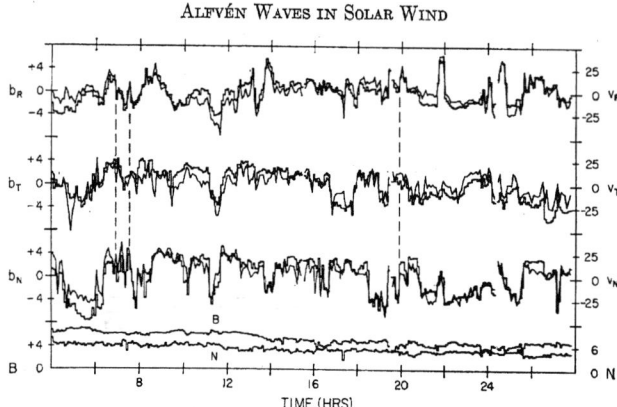

Figure 7. Examples of Alfvén waves showing correlated variations in the three components of the IMF and solar wind speed in RTN coordinates [*Belcher and Davis*, 1971]. The solar wind density (*N*) and total field strength (*B*) are also shown.

concluding that "B_z is an essential factor in causing the geomagnetic activity". The burst-like nature of AE results from the highly variable magnetic field on timescales of a few hours throughout the passage of the high-speed stream. These magnetic field variations in high-speed streams at both low and high heliolatitudes are predominantly large-amplitude Alfvén waves moving outward from the Sun [e.g., *Belcher and Davis*, 1971; *Smith et al.*, 1995; *Tsurutani et al.*, 1995a]. These non-compressive disturbances are perpendicular to, and propagate along, the mean magnetic field direction at the Alfvén speed, and are recognizable from correlated variations in the magnetic field and solar wind velocity components (Figure 7). Transverse fluctuations with $\Delta B/|B| \sim 1\text{--}2$ have been typically reported. *Burlaga and Lepping* [1977] noted that geomagnetic activity tends to be stronger in the vicinity of the CIR-associated magnetic field enhancement, where compression would be expected to enhance any southward fields present. Ulysses results support the idea that Alfvén waves moving out in high-speed streams may strengthen on entering CIRs [*Tsurutani et al.*, 1995a].

Assessment of the geomagnetic impact of CIRs and corotating streams depends on the parameter used to measure this impact since different geomagnetic indices are dominated by activity in particular regions of the magnetosphere. For example, AE measures activity in the auroral zone, indices such as *aa*, *Kp*, and C9 reflect mid-latitude conditions, while *Dst*, the index frequently used to indicate storm amplitudes, is dominated by the strength of the ring current. Figure 8 illustrates schematically the generation of CIR/stream related storms [*Tsurutani et al.*, 1995b, 2006]. The figure shows first the trailing edge of a high-speed stream, where the speed is decreasing, the magnetic field intensity is low, and there are

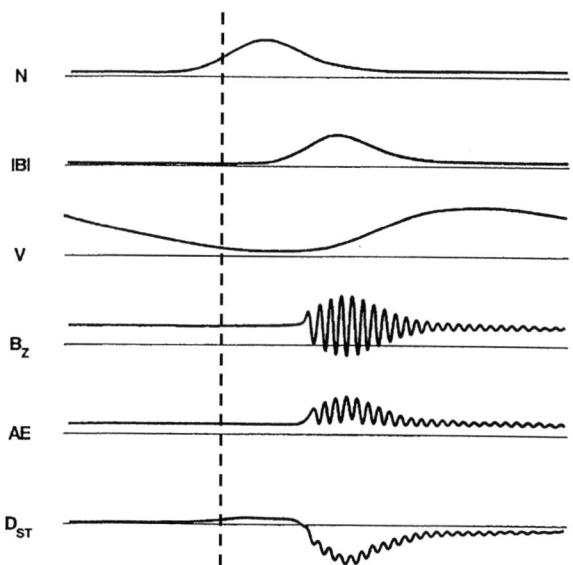

Figure 8. Schematic of the generation of geomagnetic activity by Alfvén waves in high-speed streams, showing the different temporal profiles of the AE and *Dst* geomagnetic indices (*Tsurutani et al.* [2006], updated from *Tsurutani et al.* [1995b]). The vertical dashed line indicates a representative location for the heliospheric current sheet crossing, if present.

few Alfvén waves, so that geomagnetic activity is low. The higher densities associated with the heliospheric plasma sheet and the onset of compressive effects related to the stream-stream interaction lead to a small increase in *Dst* due to inward motion and intensification of the magnetopause currents. Subsequent southward deflections of the IMF result in intervals of geomagnetic activity as measured by AE. In particular, Alfvén waves in the high-speed stream lead to extended intervals of geomagnetic activity that have been termed "High Intensity Long Duration Continuous AE Activity" (HILDCAAs) [*Tsurutani and Gonzalez*, 1987; *Tsurutani et al.*, 1990]. HILDCAA activity is most intense in the enhanced fields associated with the CIR. The *Dst* index shows a decrease to negative values, the storm main phase, associated with the CIR. Because of the fluctuating fields, the main phase may be rather irregular. Recovery occurs where the CIR field enhancement and related southward fields start to decline. Note that when the storm, as measured by *Dst*, has essentially recovered, auroral zone activity continues. A correlation between solar wind dynamic pressure and *Dst* has been reported [*Murayama*, 1982; *Fenrich and Luhman*, 1998]. Nevertheless, this parameter has a relatively weak influence on the strength of geomagnetic activity, and only at times of southward magnetic field [e.g., *Smith et al.*, 1999].

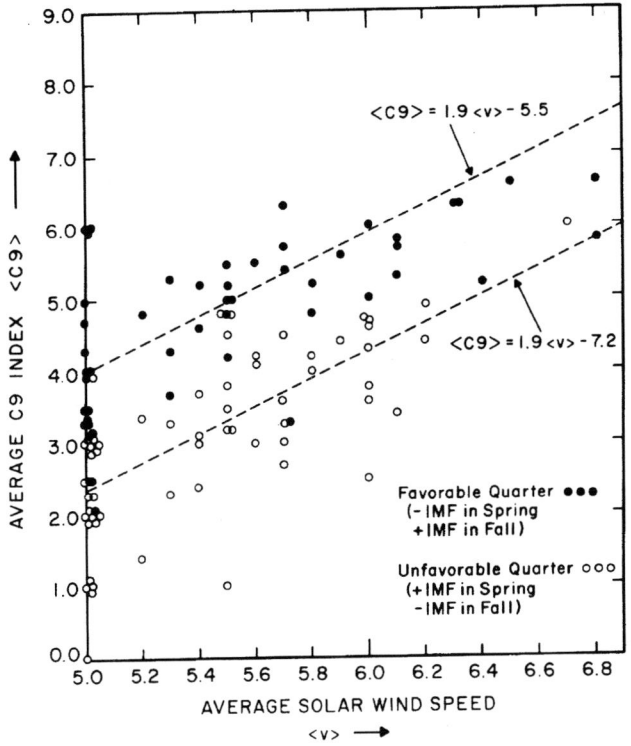

Figure 9. Average C9 index and solar wind speed in a group of high-speed streams observed in spring or fall, showing higher geomagnetic activity levels in streams of similar speeds with favorably-oriented IMFs [*Sheeley et al.*, 1977].

Seasonal Effects

The geoeffectiveness of CIRs and high-speed streams also depends to some extent on the time of year. *Russell and McPherron* [1973] suggested that the changing relative orientations of the axes of the Sun and Earth along the Earth's orbit gives rise to a larger southward component of the Parker spiral field, which favors stronger activity levels, in the (northern hemisphere) autumn at times of outward IMF and in the spring for periods of sunward IMF. For example, Figure 9 [*Sheeley et al.*, 1977] illustrates that the average C9 index during a number of streams in 1962–1975 was indeed slightly higher in those streams in which the seasonal effect was expected to be favorable (solid circles) rather than unfavorable (open circles). A seasonal effect might also be expected to result from the Earth penetrating more deeply into high-speed streams in the southern (northern) hemisphere in the spring (autumn) because of the inclination of the Earth's orbit relative to the solar equator. However, this effect will be complicated by changes associated with temporal development of the source coronal holes [e.g., *Crooker et al.*, 1996]. The relative importance of various seasonal effects has been discussed by *Cliver et al.* [2000].

Long-Term Variations

There are several interesting topics related to "long-term" variations in geomagnetic activity associated with CIRs and high-speed streams. These include variations in the pattern of recurrent activity from cycle to cycle, the frequency of geomagnetic storms associated with these structures as compared with those associated with solar wind transients, variations in average geomagnetic activity levels during the solar cycle, and variations on century timescales.

Figure 10 summarizes the presence of recurrent activity in the C9 index during 1950–2004 using a wavelet analysis [*Torrence and Compo*, 1998] which shows the power at periods of 10–50 days. Each individual 10-year plot *approximately* encompasses a solar cycle, with solar minimum being near the center of each plot. Intermittent ~1 and ~ half-solar rotation components (corresponding to one or two prominent corotating streams present) are evident. Although these components tend to be most prevalent during the declining and minimum phases of the solar cycle, the temporal pattern of power clearly varies from cycle to cycle. For example, ~27-day recurrent activity was prominent at the time of the Mariner 2 observations of high-speed streams in 1962, but was unusually weak during the remainder of the mid-1960's solar minimum, whereas it occurred during much of the next solar minimum. This analysis suggests that the pattern of recurrent activity, which presumably is dictated by the configuration of coronal holes which gives rise to the related high-speed streams, cannot simply be "predicted" from observations in previous cycles. *Cliver et al.* [1996] have concluded from examining the *aa* geomagnetic index [*Mayaud*, 1972] for 1844–1994 that recurrent activity tends to be stronger during the declining phase of even numbered solar cycles, and suggested that a more rapid equatorward expansion of coronal holes than in odd cycles may be responsible. However, this "rule" is not particularly obvious in the presentation in Figure 10, where the early 1950's, 1970's and 1990's are declining phases of even cycles.

Turning to geomagnetic storms, Figure 11 illustrates the variation in the solar wind "drivers" of geomagnetic storms during recent solar cycles [updated from *Richardson et al.*, 2001]. Specifically, variations in the yearly numbers of "small", "medium", and "large" + "major" storms (based on the geomagnetic *Kp* index; for details see *Richardson et al.* [2001]) associated with CME-related structures (ICMEs and related interplanetary shocks) and corotating streams, are shown for 1972–2004. The storm sizes correspond roughly to small: $Dst <~ -50$ nT; medium: $Dst <~ -75$ nT; large: $Dst <~ -150$ nT; and major: $Dst <~ -200$ nT. The sunspot number is shown at the top of the figure. Storms associated with streams are most prevalent during the decay phase of the solar cycle, but continue to be present at all phases of the cycle. These storms are generally small or medium in size.

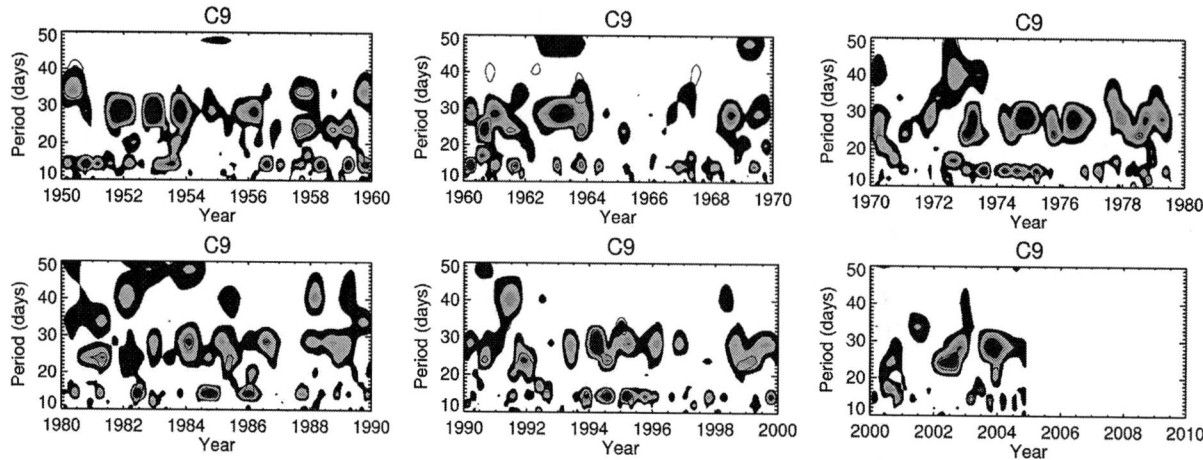

Figure 10. Wavelet analysis of the C9 index for 1950–2004, showing power (arbitrary units) at periods of 10–50 days versus time. Each 10-year panel approximately encompasses a solar cycle, with solar minimum near the center of the panel. Although ~13 and ~27-day (half- and one-solar rotation) components are generally dominant, the temporal patterns do vary from cycle to cycle.

Occasional stronger stream-associated storms do occur, predominately during the declining phase of the solar cycle. Overall, we estimate that ~26% of "large" storms and ~3% of "major" storms in 1972–2004 were driven by streams, while CMEs contribute the majority of these larger storms. The occurrence rate of CME-related storms tends to follow the sunspot cycle. Nevertheless, a temporary decrease in the occurrence rate of ≥ medium CME-related storms is often seen near solar maximum. This is clearly present in 1979–1980 and 1990, while the present cycle is complicated by an additional decrease due to the temporary decrease in the ICME rate [*Cane and Richardson*, 2003] and prevalence of streams in 1999. The temporary reduction in the occurrence rate of strong CME-related storms near solar maximum may be associated with the decrease in the frequency of energetic solar events around the time of the solar magnetic field reversal [e.g., *Feminella et al.*, 1997]. Thus, the typical "double peak" in storm activity during the solar cycle [e.g., *Gonzalez et al.*, 1990] may be the superposition of a double peak in the rate of CME-associated storms and a further peak during the declining phase associated with corotating streams, which may merge with the second CME-associated peak.

Although they are rarely responsible for major geomagnetic storms, high-speed streams do make important contributions to average geomagnetic conditions on longer (>> solar rotation) timescales because of their extended duration. For example, Figure 12 [updated from *Richardson et al.*, 2002] shows the monthly sunspot number and 3-rotation averages of the *aa* geomagnetic index (a mid-latitude index including both auroral zone and equatorial disturbances) averaged over all solar wind, and in CME-related solar wind, corotating high-speed streams, and slow solar wind, in 1972–early 2005, covering ~3 solar cycles. The "all solar wind" results are repeated in the bottom three panels. Note that *aa* is rather poorly correlated with the sunspot number ($cc = 0.306$). Reasons for this include the decrease in geomagnetic activity near sunspot maximum, which appears to reflect principally a temporary decrease in the strength of the

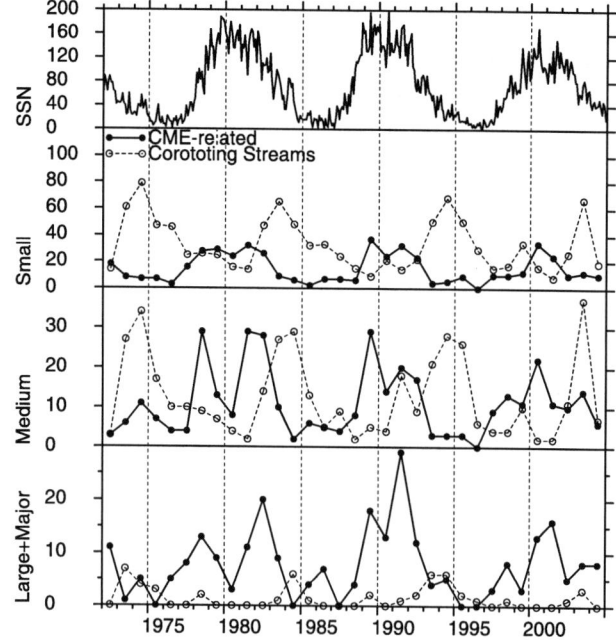

Figure 11. Occurrence rates (/year) of small, medium and large + major storms in 1972–2004 associated with CMEs and corotating streams. The sunspot number is in the top panel (updated from *Richardson et al.* [2001]).

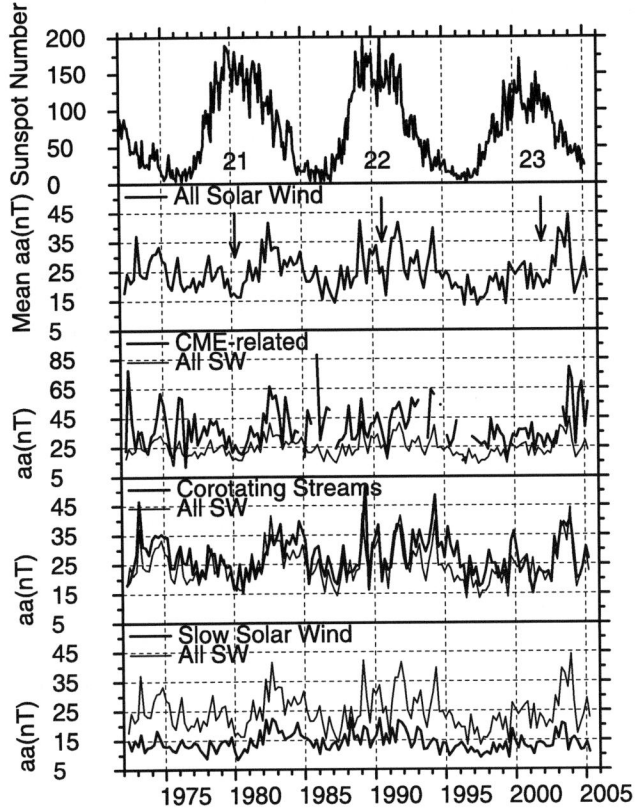

Figure 12. The sunspot number, and 3-rotation averages of the *aa* geomagnetic index associated with all solar wind, "CME-related" solar wind (ICMEs and post-shock flows), high-speed streams, and slow solar wind in 1972–early 2005 (updated from *Richardson et al.* [2002]). The "all solar wind" average, repeated in the bottom three panels, closely tracks activity levels in high-speed streams.

IMF around solar field reversal, and activity during the declining phase of the solar cycle associated with high-speed streams. Overall, mean values of *aa* at all solar activity levels closely track those found in high-speed streams. Thus, long-term averages in *aa* are determined predominantly by activity levels in the background (non-CME-related) solar wind, in particular associated with high-speed streams.

The *aa* index is of particular interest because it extends back to the middle of the 19th century, as mentioned above. An interesting feature of this index is that it shows a clear increase during the first half of the 20th century at all activity levels [*Vennestroem*, 2000]. Although there is some debate about this feature [e.g., *Svalgaard et al.*, 2004], a possible interpretation is that the solar wind magnetic field strength increased during this period, raising the general level of geomagnetic activity [*Lockwood et al.*, 1999]. Figure 13 [*Richardson et al.*, 2002] compares the *aa* index during the solar minimum years of 1977 and 1901 (shaded), and clearly shows the lower geomagnetic activity levels at the beginning of the 20th century. Note that recurrent geomagnetic activity did occur in 1901 (examples are indicated by arrowheads), suggesting that stream interactions and hence variations in solar wind speed, were present at this time.

ICME-CIR INTERACTIONS

We finally briefly mention an interesting situation that occurs occasionally [e.g., *Zhao*, 1992; *Cane and Richardson*, 1997; *Fenrich and Luhmann*, 1998; *Crooker*, 2000] and has been modeled by *Odstrcil and Pizzo* [1999]. This is when a high-speed stream interacts with, and compresses, the trailing

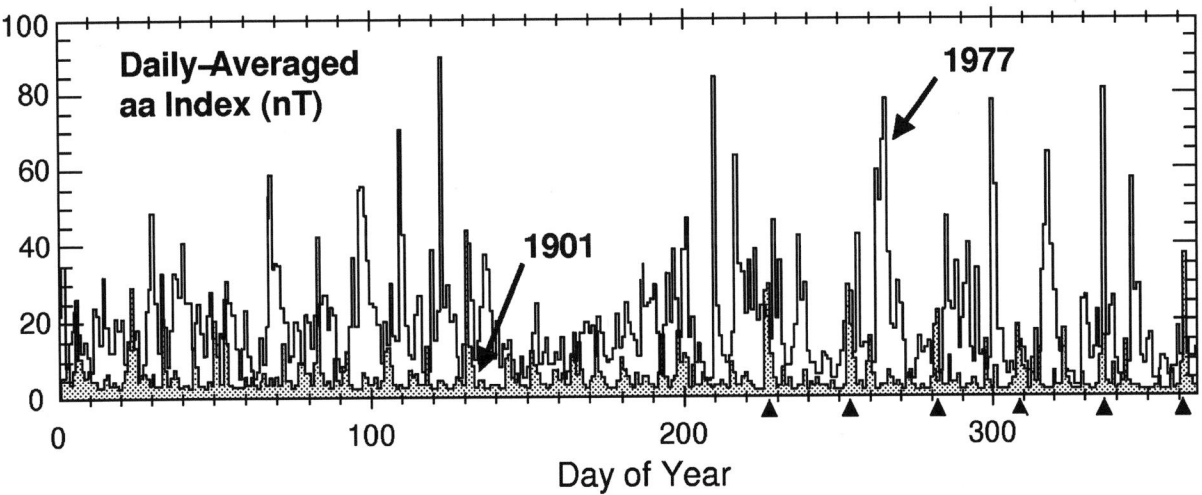

Figure 13. The *aa* index during 1901 (shaded) and 1977, both years of solar minimum, illustrating the lower geomagnetic activity levels at the beginning of the 20th century. Examples of recurrent activity in 1901 are indicated by arrowheads.

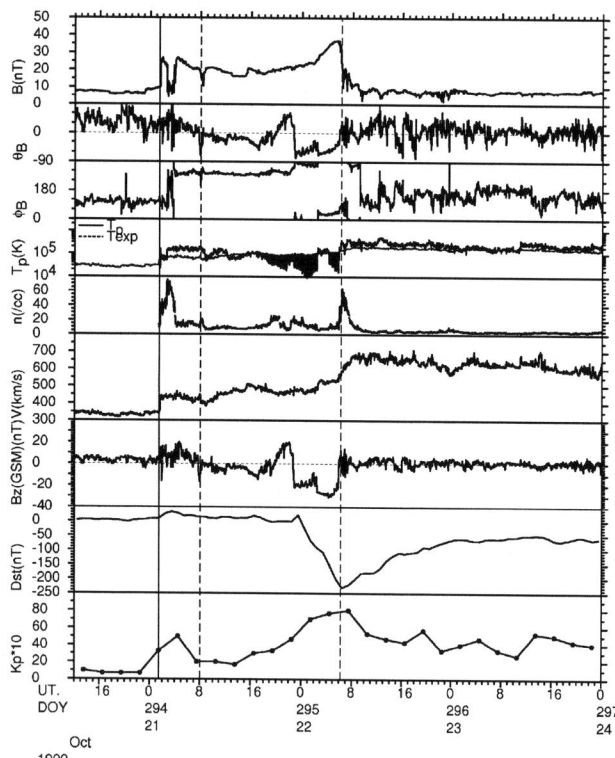

Figure 14. An intense geomagnetic storm in October 1999 resulting from the interaction of a high-speed stream with a CME-related transient (between the dashed lines) that has southward magnetic field inside the trailing edge. The solid line marks the passage of a shock ahead of the transient. In-situ observations are from ACE.

edge of a preceding, slower ICME. If the magnetic field inside the compressed region of the ICME includes a southward component, interaction with the high-speed stream may enhance the geoeffectiveness of the ICME. Some "unexpectedly strong" geomagnetic storms have resulted from this configuration. An example in October, 1999 is shown in Figure 14. The probable ICME [*Cane and Richardson*, 2003] is bounded by dashed vertical lines (the upstream shock is indicated by the solid line). The trailing edge evidently formed the interface with the following high-speed stream. The interaction apparently enhanced the region of southward magnetic field inside the trailing edge, resulting in the major geomagnetic storm indicated by the *Dst* and Kp indices. The storm then decayed within the high-speed flow where the field was weaker and variable in direction.

SUMMARY

Corotating streams, and the CIRs formed at their leading edges by stream-stream interaction, are typically associated with modest levels of geomagnetic activity that recurs at the solar rotation period. The principal causes appear to be intermittent, brief intervals of southward magnetic field resulting from field fluctuations associated with Alfvén waves propagating away from the Sun. These may be amplified (together with their possible geomagnetic effects) when propagating into the CIR. CIRs and streams are typically associated with weak geomagnetic storms, but occasionally with significant storms, most frequently during the declining phase of the solar cycle. Unexpectedly strong storms may also result when a stream interacts with, and compresses, an ICME with an embedded southward field. Longer-term studies suggest that the pattern of recurrent geomagnetic activity varies from cycle-to-cycle, streams are a major contributor to average geomagnetic activity levels, and that recurrent activity, and hence presumably streams of fast and slower solar wind were present at the beginning of the 20th century, when geomagnetic activity levels appear to have been generally lower than during the last ~50 years.

Acknowledgments. The meeting organizers are thanked for the opportunity to prepare this review, and for assistance with travel to Manaus. The use of ACE data, available from the ACE Science Center, and data from the National Space Science Data Center and National Geophysical Data Center, is gratefully acknowledged.

REFERENCES

Balogh, A., R.G. Marsden, and E.J. Smith (eds.), *The Heliosphere Near Solar Minimum - The Ulysses Perspective*, Springer-Verlag, Berlin, Heidelberg, New York, 2001.
Bartels, J., Terrestrial magnetic activity and its relation to solar phenomena, *Terr. Magn. Atmos. Elect.*, 37, 1, 1932.
Bartels, J., Solar activity and geomagnetism, *Terr. Magn. Atmos. Elect.*, 45, 339, 1940.
Belcher, J.W., and L. Davis, Large amplitude Alfvén waves in the interplanetary medium, 2, *J. Geophys. Res.*, 76, 3534, 1971.
Berdichevsky, D.B., A. Szabo, R.P. Lepping, A.F. Viñas, and F. Mariani, Interplanetary fast shocks and associated drivers observed through the 23rd solar minimum by Wind over its first 2.5 years, *J. Geophys. Res.*, 105, 27,289, 2000.
Bravo, S., G.A. Stewart, and X. Blanco-Cano, The varying multipolar structure of the Sun's magnetic field and the evolution of the solar magnetosphere through the solar cycle, *Solar Phys.*, 179, 223, 1998.
Burlaga, L.F., A reverse hydromagnetic shock in the solar wind, *Cosmic Electrodyn.*, 1, 233, 1970.
Burlaga, L.F., Interplanetary stream interfaces, *J. Geophys. Res.*, 79, 3717, 1974.
Burlaga, L.F., *Interplanetary Magnetodynamics*, Oxford University Press, New York and Oxford, 1995.
Burlaga, L.F., and R.P. Lepping, The causes of recurrent geomagnetic storms, *Planet. Space Sci.*, 25, 1151, 1977.
Burlaga, L.F., R. Schwenn, R., and H. Rosenbauer, Dynamical evolution of interplanetary magnetic fields and flows between 0.3 AU and 8.5 AU: Entrainment, *Geophys. Res. Lett.*, 10, 413, 1983.
Cane, H.V., and I.G. Richardson, What caused the large geomagnetic storm of November 1978?, *J. Geophys. Res.*, 102, 17, 445, 1997.
Cane, H.V., and I.G. Richardson, Interplanetary coronal mass ejections in the near-earth solar wind during 1996-2002, *J. Geophys. Res.*, 108(4), 10.1029/2002JA009817, 2003.
Chao, J.K., V. Formisano, and P.C. Hedgecock, Shock pair observation, in C.P. Sonett, P.J. Coleman, Jr., and J.M. Wilcox (eds.), *Solar Wind*, NASA SP-308, p. 435, 1972.

Cliver, E.W., Solar activity and geomagnetic storms: The corpuscular hypothesis, *Eos Trans. AGU*, 75(52), 609, 1994.

Cliver, E.W., Solar activity and geomagnetic storms: From M regions and flares to coronal holes and CMEs, *Eos Trans. AGU*, 76(8), 75, 1995.

Cliver, E.W., V. Boriakoff, and K.H. Bounar, The 22-year cycle of geomagnetic and solar wind activity, *J. Geophys. Res.*, 101, 27,091, 1996.

Cliver, E.W., Y. Kamide, and A.G. Ling, Mountains versus valleys: Semiannual variation of geomagnetic activity, *J. Geophys. Res.*, 105, 2413, 2000.

Crooker, N.U., Solar and heliospheric geomagnetic disturbances, *J. Atmosph. Solar Terr. Phys.*, 62, 1071, 2000.

Crooker, N.U., and E.W. Cliver, Postmodern view of M-regions, *J. Geophys. Res.*, 99, 23,383, 1994.

Crooker, N.U., *et al.*, A two-stream, four-sector, recurrence pattern: Implications from WIND for the 22-year geomagnetic activity cycle, *J. Geophys. Res.*, 23, 1275, 1996.

Crooker, N.U., *et al.*, CIR Morphology, turbulence, discontinuities and energetic particles: Report of Working Group 2, *Space Sci. Rev.* 89, 179, 1999.

Dungey, J.W., Interplanetary magnetic field and the auroral zones, *Phys. Rev. Lett.*, 6, 47, 1961.

Feminella, F., and M. Storini, Large scale dynamical phenomena during solar activity cycles, *Astron. Astrophys.*, 322, 311, 1997.

Fenrich, F.R., and J.G. Luhmann, Geomagnetic response to magnetic clouds of different polarity, *Geophys. Res. Lett.*, 25, 2999, 1998.

Forsyth, R.J., and E. Marsch, Solar origin and interplanetary evolution of stream interfaces, *Space Sci. Rev.*, 89, 7, 1999.

Gazis, P.R.: 1984, Observations of plasma bulk parameters and the energy balance between 1 and 10 AU, *J. Geophys. Res.*, 89, 775, 1984.

Gazis, P.R., and A.J. Lazarus, The radial evolution of the solar wind, 1–10 AU, in M. Neugebauer(ed.), *Solar Wind Five*, NASA Conf. Publ. 2280, Washington D.C., p. 509, 1983.

Gonzalez, W.D., A.L.C. Gonzalez, and B.T. Tsurutani, Dual peak solar cycle distribution of intense geomagnetic storms, *Planet. Space Sci.*, 38, 181, 1990.

Gosling, J.T., Corotating and transient solar wind flows in three dimensions, *Ann. Rev. Astron. Astrophys.*, 34, 35, 1996.

Gosling, J.T., and V.J. Pizzo, Formation and evolution of corotating interaction regions and their three dimensional structure, *Space Sci. Rev.*, 89, 21, 1999.

Gosling, J.T., A.J. Hundhausen, and S.J. Bame, Solar wind evolution at large heliocentric distances: Experimental demonstration and test of a model, *J. Geophys. Res.*, 81, 2111, 1976.

Gosling, J.T., J.R. Asbridge, S.J. Bame, and W.C. Feldman, Solar wind stream interfaces, *J. Geophys. Res.*, 83, 1401, 1978.

Gosling, J.T., D.J. McComas, J.L. Phillips, and S.J. Bame, Geomagnetic activity associated with Earth passage of interplanetary shock disturbances and coronal mass ejections, *J. Geophys. Res.*, 96, 7831, 1991.

Gosling, J.T., *et al.*, Latitudinal variation of solar wind corotating stream interaction regions, *Geophys. Res. Lett.*, 20, 2789, 1993.

Greaves, W.M.H., and H.W. Newton, On the recurrence of magnetic storms, *Mon. Not. R. Astron. Soc.*, 89, 641, 1929.

Hundhausen, A.J., Nonlinear model of high-speed solar wind streams, *J. Geophys. Res.*, 78, 1528, 1973a.

Hundhausen, A.J., Evolution of large-scale solar wind structures beyond 1 AU, *J. Geophys. Res.*, 78, 2035, 1973b.

Hundhausen, A.J., and J.T. Gosling, Solar wind structure at large heliocentric distances: An interpretation of Pioneer 10 observations, *J. Geophys. Res.*, 81, 1436, 1976.

Intriligator, D.S., and G.L. Siscoe, Stream interfaces and energetic ions closer than expected: Analyzes of Pioneers 10 and 11 observations, *Geophys. Res. Lett.*, 21, 1117, 1994.

Krieger, A.S., A.F. Timothy, and E.C. Roelof, A coronal hole and its identification as the source of a high velocity solar wind stream, *Sol. Phys.*, 29, 505, 1973.

Lazarus, A.J., K.W. Ogilvie, and L.F. Burlaga, Interplanetary shock observations made by Mariner 2 and Explorer 34, *Solar Phys.*, 13, 232, 1970.

Lockwood, M., R. Stamper, and N.M. Wild, A doubling of the sun's coronal magnetic field during the past 100 years, *Nature*, 399, 437, 1999.

Luhmann, J.G., Y. Li, C.N. Arge, P. Gazis, and R. Ulrich, Solar cycle changes in coronal holes and space weather cycles, *J. Geophys. Res.*, 107, 10.1029/2001JA007550, 2002.

Maunder, E.W., Magnetic disturbances, 1882 to 1903, as recorded at the Royal Observatory, Greenwich, and their association with sunspots, *Mon. Not. R. Astron. Soc. London*, 65, 2, 1905.

Mayaud, P.N., The aa indices: a 100-year series characterising the geomagnetic activity, *J. Geophys. Res.*, 72, 6870, 1972.

McComas, D.J. et al., Solar wind observations over Ulysses first full polar orbit, *J. Geophys. Res.*, 105, 10,419, 2000.

McComas, D.J., R. Goldstein, J.T. Gosling, and R.M. Skoug, Ulysses second orbit: Remarkably different solar wind, *Space Sci. Rev.*, 97, 99, 2001.

McComas, D.J., H.A. Elliott, N.A. Schwadron, J.T. Gosling, R.M. Skoug, and B.E. Goldstein, The three-dimensional solar wind around solar maximum, *Geophys. Res. Lett.*, 30, 1517, doi:10.1029/2003GL017136, 2003.

Murayama, T., Coupling function between solar wind parameters and geomagnetic indices, *Rev. Geophys. and Space Phys.*, 20, 623, 1982.

Neugebauer, M., and C.W. Snyder, Mariner 2 observations of the solar wind, 1: Average properties, *J. Geophys. Res.*, 71, 4469, 1966.

Neugebauer, M., P.C. Liewer, E.J. Smith, R.M. Skoug, and T.H. Zurbuchen, Sources of the solar wind at solar activity maximum, *J. Geophys. Res.*, 107, 10.1029/2001JA000306, 2002.

Odstrcil, D., and V.J. Pizzo, Three-dimensional propagation of coronal mass ejections (CMEs) in a structured solar wind flow, 1. CME launched within the streamer belt, *J. Geophys. Res.*, 104, 483, 1999.

Parker, E.N., Dynamics of the Interplanetary Gas and Magnetic Fields, *Astrophys. J.*, 128, 664, 1958.

Parker, E.N., Interplanetary Dynamical Processes, John Wiley, New York, 1963.

Perreault, P., and S.-I. Akasofu. A study of geomagnetic storms, *Geophys. J. R. Astr. Soc.*, 54, 547, 1978.

Pizzo, V.J., and J.T. Gosling, Three-dimensional simulation of high-latitude interaction regions: Comparison with Ulysses results, *Geophys. Res. Lett.*, 21, 2063, 1994.

Richardson, I.G., Energetic particles and corotating interaction regions in the solar wind, *Space Sci. Rev.*, 111, 267, 2004.

Richardson, I.G., and R.D. Zwickl, Low energy ions in corotating interaction regions at 1 AU: Observations, *Planet. Space Sci.*, 32, 1179, 1984.

Richardson, I.G., G. Wibberenz, and H.V. Cane, The relationship between recurring cosmic ray depressions and corotating solar wind streams at ≤1 AU: IMP 8 and Helios 1 and 2 anticoincidence guard rate observations, *J. Geophys. Res.*, 101, 13,483, 1996.

Richardson, I.G., H.V. Cane, and E.W. Cliver, Sources of geomagnetic storms for solar minimum and maximum conditions during 1972–2000, *Geophys. Res. Lett.*, 28, 2569, 2001.

Richardson, I.G., H.V. Cane, and E.W. Cliver, Sources of geomagnetic activity during nearly three solar cycles (1972-2000), *J. Geophys. Res.*, 107, 10.1029/2001JA000504, 2002.

Rickett, B.J., and W.A. Coles, Evolution of the solar wind structure over a solar cycle: Interplanetary scintillation velocity measurements compared with coronal observations, *J. Geophys. Res.*, 96, 1717, 1991.

Riley, P., J.T. Gosling, L.A. Weiss, L.A., and V.J. Pizzo, The tilts of corotating interaction regions at midheliographic latitudes, *J. Geophys. Res.*, 101, 24,349, 1996.

Russell, C.T., and R.L. McPherron, Semiannual variation of geomagnetic activity, *J. Geophys. Res.*, 78, 92-108, 1973.

Schwenn, R., Large-scale structure of the interplanetary medium, in R. Schwenn and E. Marsch (eds.), *Physics of the Inner Heliosphere.*, Vol. 1, p. 99, Springer-Verlag, Berlin, Heidelberg, New York, 1990.

Schwenn, R., and E. Marsch (eds.), *Physics of the Inner Heliosphere*, Springer-Verlag, Berlin, Heidelberg, New York, 1990.

Sheeley, N.R., Jr., J.W. Harvey, and W.C. Feldman, Coronal holes, solar wind streams, and recurrent geomagnetic disturbances, 1973–1976, *Solar Phys.*, 49, 271, 1976.

Sheeley, N.R., Jr., J.S. Asbridge, S.J. Bame, and J.W. Harvey, A pictorial comparison of interplanetary magnetic field polarity, solar wind speed, and geomagnetic disturbance index during the sunspot cycle, *Solar Phys.*, 52, 485, 1977.

Smith, E.J., and J.H. Wolfe, Observations of interaction regions and co-rotating shocks between one and five AU: Pioneer 10 and 11, *Geophys. Res. Lett.*, 3, 137, 1976.

Smith, E.J., and J.H. Wolfe, Pioneer 10 and 11 Observations of evolving solar wind streams and shocks beyond 1 AU, in M. Shea, D.F. Smart, and S.T. Wu (eds.), *Study of Traveling Interplanetary Phenomena*, D. Riedel, Hingham, MA, p. 227, 1977.

Smith, E.J., and J.H. Wolfe, Fields and plasmas in the outer solar system, *Space Sci. Rev.*, 23, 217, 1979.

Smith, E.J., A. Balogh, M. Neugebauer, and D. McComas, Ulysses observations of Alfvén waves in the southern and northern solar hemispheres, *Geophys Res. Lett.*, 22, 3381, 1995.

Smith, J.P., M.F. Thomsen, J.E. Borovsky, and M. Collier, Solar wind density as a driver for the ring current in mild storms, *Geophys. Res. Lett.*, 26, 1797, 1999.

Svalgaard, L., E.W. Cliver, and P. Le Sager, IHV: a new long-term geomagnetic index, *Adv. Space Res.*, 34(2), 436, 2004.

Torrence, C., and G.P. Compo, A practical guide to wavelet analysis, *Bull. Am. Meteorol. Soc.*, 79, 61, 1998.

Thomas, B.T. and E.J. Smith, The structure and dynamics of the heliospheric current sheet, *J. Geophys. Res.*, 86, 11,105, 1981.

Tsurutani, B.T., and W.D. Gonzalez, The cause of high intensity long-duration continuous AE activity (HILDCAAs); Interplanetary Alfvén wave trains, *Planet. Space Sci.*, 35, 405, 1987.

Tsurutani, B.T., and W.D. Gonzalez, The interplanetary causes of magnetic storms: A review, in B.T. Tsurutani, W.D. Gonzalez, Y. Kamide, and J.K. Arballo (eds.), *Magnetic Storms*, Geophys. Monogr., 98, AGU, Washington, D.C., p. 77, 1997.

Tsurutani, B.T., T. Gould, B.E. Goldstein, W.D. Gonzalez, and M. Sugiura, Interplanetary Alfvén waves and auroral (substorm) activity: IMP 8, *J. Geophys. Res.*, 95, 2241, 1990.

Tsurutani, B.T., C.M. Ho, J.K. Araballo, B.E. Goldstein, and A. Balogh, Large amplitude IMF fluctuations in corotating interaction regions: Ulysses at midlatitudes, *Geophys. Res. Lett.*, 22, 3397, 1995a.

Tsurutani, B.T., W.D. Gonzalez, A.L.C. Gonzalez, F. Tang, J.K. Araballo, and M. Okada, Interplanetary origin of geomagnetic activity in the declining phase of the solar cycle, *J. Geophys. Res.*, 100, 21,717, 1995b.

Tsurutani, B.T., N. Gopalswamy, R.L. McPherron, W.D. Gonzalez, G. Lu, and F.L. Guarnieri, Magnetic storms caused by corotating solar wind steams, this volume, 2006.

Vennerstroem, S., Long-term rise in geomagnetic activity: A close connection between quiet days and storms, *Geophys. Res. Lett.*, 27, 69, 2000.

Wang, Y.-M., and N.R. Sheeley, Jr., The solar wind and its magnetic sources at sunspot maximum, *Astrophys. J.*, 587, 818, 2003.

Wilcox, J., and N. Ness, Quasi-stationary corotating structure in the interplanetary medium, *J. Geophys. Res.*, 70, 5793, 1965.

Wimmer-Schweingruber, R.F., R. von Steiger, and R. Paerli, Solar wind stream interfaces in corotating interaction regions: SWICS/Ulysses results, *J. Geophys. Res.*, 102, 17,407, 1997.

Zhao, X., Interaction of fast steady flow with slow transient flow: A new cause of shock pair and interplanetary B_z event, *J. Geophys. Res.*, 97, 15,051, 1992.

Zirker, J.B. (ed.), *Coronal Holes and High Speed Wind Streams, Skylab Solar Workshop*, Colorado University Press, Boulder, CO, 1977.

I.G. Richardson, Code 661, NASA Goddard Space Flight Center, Greenbelt, MD, 20771. (ian.richardson@gsfc.nasa.gov)

The Freestream Turbulence Effect in Solar-Wind/Magnetosphere Coupling: Analysis Through the Solar Cycle and for Various Types of Solar Wind

Joseph E. Borovsky and John T. Steinberg

Los Alamos National Laboratory, Los Alamos, New Mexico, USA

The freestream turbulence effect is a viscous interaction between the solar wind and the Earth's magnetosphere wherein the driving of the magnetosphere is stronger when the amplitude of the solar-wind turbulence upstream of the Earth is stronger. The origin of the effect is an eddy viscosity of the solar wind controlled by the amplitude of the ambient MHD turbulence. The present study investigates outstanding issues about the operation of the turbulence effect. For this study, a large solar-wind/magnetosphere data set is assembled: to the 1963-2001 OMNI2 data set, 1-hour-lagged values of the auroral-electrojet index are added. The 1963-2001 OMNI2-AE_1 data set has ~193,000 hours of data that can be used to study magnetospheric driving by the turbulence effect. The freestream turbulence effect is analyzed for southward IMF by holding $-vB_z$ fixed in the data set. Under modest southward IMF, the turbulence effect has the same strength as it does under northward IMF. For large $-vB_z$ where reconnection driving is strong the ability to measure the freestream turbulence effect deteriorates owing to signal-to-noise issues. The freestream turbulence effect was analyzed for 3 solar cycles of data: a weak solar-cycle dependence to the strength of the turbulence effect was found. A categorization of the solar wind was developed from the temperature-versus-speed plot for the solar wind. Using these categorizations, plus data with restrictions on the sunspot number, plus catalogs of solar-wind events, no differences in the freestream turbulence effect were seen between different types of solar wind hitting the magnetosphere.

1. INTRODUCTION

A. Motivation

It is a well known effect in aerodynamics that the presence of upstream turbulence in a flow alters the manner in which a flow couples to an obstacle or boundary. This is known as the "freestream turbulence effect" [*Kwok and Melborne*, 1980; *Sullerey and Sayeed Khan*, 1983; *Pal*, 1985; *Thole and Bogard*, 1996]. This effect is quantified by wind-tunnel experiments wherein coupling parameters are measured as functions of the amplitudes of upstream turbulence. This quantification for one such experiment [*Blair*, 1983] is shown in Figure 1. In the wind tunnel, turbulence is created in by variable grids upstream of the obstacle (an edge-on fin) and the viscous force on the fin is plotted as a function of the amplitude of the turbulence upstream of the fin. The individual data points are plotted, as is a 4-point running average of the data. As can be seen, the higher the amplitude of the turbulence, the greater the viscous force on the fin by the flow. The interpretation of the flow experiments

---

Recurrent Magnetic Storms: Corotating Solar Wind Streams
Geophysical Monograph Series 167
Copyright 2006 by the American Geophysical Union.
10.1029/167GM07

60 THE FREESTREAM TURBULENCE EFFECT

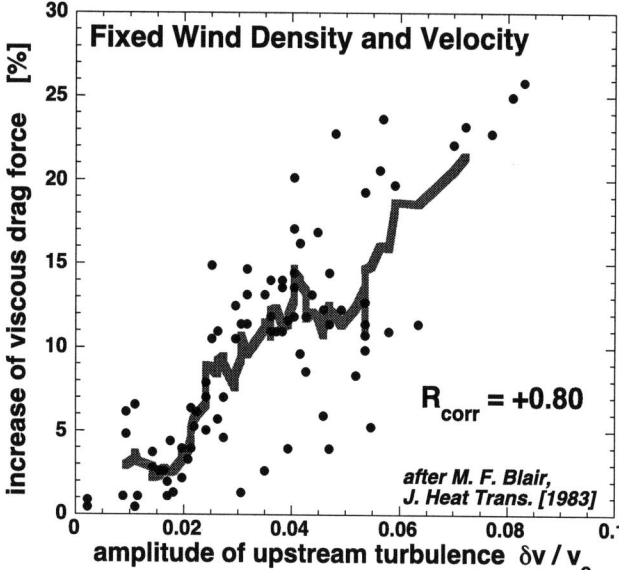

Figure 1. Data points from the wind-tunnel experiment of Blair [1983] are plotted (black points): the viscous force on an edge-on fin versus the amplitude of turbulence in the wind tunnel upstream of the fin. On the vertical axis, the amount of viscous force in the absence of turbulence is subtracted off. The gray curve is a 7-point running average of the data points.

[e.g., *Volino et al.*, 2003] is (1) that the turbulent flow in the wind tunnel has an eddy viscosity (also called a turbulent viscosity) that enhances the viscous coupling of the flow to the obstacle and (2) that the amplitude of the upstream turbulence controls the magnitude of this eddy viscosity.

It must be clarified for the reader that the freestream turbulence effect differs from the more-familiar case of turbulence affecting the coupling wherein the turbulence is generated at the location where a laminar (non-turbulent) flow interacts with the obstacle [e.g., *Schlichting*, 1979; *Mathieu and Scott*, 2000]. In the freestream turbulence effect we are looking at the action of pre-existing turbulence that is not caused by the flow-obstacle interaction. Freestream turbulence, not turbulent boundary layers.

Theoretically, MHD fluids should also exhibit eddy viscosity if they are turbulent [*Chen and Montgomery*, 1987; *Ishizawa and Hattori*, 1998; *Yoshizawa and Yokoi*, 1996]. It was conjectured [*Borovsky and Gosling*, 2001] that since the turbulent solar wind is believed to be described by MHD, then the turbulent solar wind should exhibit this "freestream turbulence effect" in its coupling to the Earth's magnetosphere. That conjecture was followed by the study of *Borovsky and Funsten* [2003].

B. First Study of the Freestream Turbulence Effect

Borovsky and Funsten [2003] set out to test the conjecture that the turbulence effect should act in solar-wind/magnetosphere coupling. The study confirmed that the freestream turbulence effect acts: i.e., that the amplitude of the turbulence in the upstream solar wind statistically affects the strength of the coupling of the solar wind to the magnetosphere as measured by the resulting level of geomagnetic activity.

The summary of the *Borovsky and Funsten* [2003] findings are the following. (1) An increase in the amplitude of the solar-wind turbulence correlates with an increased coupling of the solar wind to the Earth's magnetosphere, as measured by the amplitude of various geomagnetic-activity indices. (2) The freestream turbulence effect acts under both northward and southward IMF. (3) The turbulence effect can be responsible for about 150 nT of the AE index. The freestream turbulence effect is a driver that is small compared with dayside reconnection as controlled by vB_z of the solar wind. For quiet times the turbulence effect is the dominant driver of geomagnetic activity; for storms it is a 10% effect.

The interpretation of the turbulence effect in solar-wind/magnetosphere coupling is: (a) an increase in the amplitude of solar-wind turbulence leads to an increase in the eddy viscosity of the solar wind, (b) this increased eddy viscosity leads to increased momentum transport from the solar wind into the magnetosheath-magnetosphere boundary layer, (c) this increased momentum transport leads to an increased convection in the magnetosphere, (d) the increased convection leads to increased convection currents in the magnetosphere, (e) the closing of these increased currents in the ionosphere leads to increased currents in the ionosphere, (f) the stronger currents in the ionosphere are detected on the ground as stronger geomagnetic indices.

The *Borovsky and Funsten* [2003] study might be the first experimental evidence for eddy viscosity in MHD flows.

It is important to point out to the reader that the freestream turbulence effect differs from the HILDCAA effect [*Tsurutani and Gonzalez*, 1987; *Tsurutani et al.*, 1995]. The freestream turbulence effect is an enhancement of the viscous coupling of the solar wind to the magnetosphere caused by higher-frequency (~30-sec-period) fluctuations giving rise to enhanced momentum transport via an eddy viscosity; the HILDCAA effect is an enhancement of the reconnection coupling of the solar wind to the magnetosphere caused by lower-frequency (≥30-min-period) fluctuations giving rise to intervals of strong southward IMF. Both mechanisms operate as a result of MHD turbulence in the solar wind which contains fluctuations with periods of a few hours and less, but they should nevertheless not be confused. The differences are discussed further in section 6 of *Borovsky and Funsten* [2003].

C. Unfinished Issues

The study of the freestream turbulence effect for solar-wind/magnetosphere coupling is in its infancy. It is important to confirm the effect with more data and to extend the knowledge about how the effect operates in the different types of solar wind that occur (a) sporadically and (b) systematically through the solar cycle.

For solar-wind/magnetosphere coupling, the freestream turbulence effect has some complications. Two major issues are the following. First, the solar-wind flow past the Earth is supersonic and superAlfvenic, and the flow is modified by a shock wave before it interacts with the Earth. The relation between the properties of the turbulence in the solar wind upstream of the bow shock and the properties of the turbulence in the solar wind (magnetosheath) behind the bow shock has not yet been established. Second, a quantification of the viscous force of the solar wind on the Earth has not been made, only a derivation of its proportionality to the solar-wind turbulence amplitude. Hence the numerical value of any coefficient of proportionality in eddy-viscosity expressions has not been evaluated experimentally.

Without solving either of these two major issues, improvement to the *Borovsky and Funsten* [2003] study of the freestream turbulence effect are presented in this report. These improvements deal with: (1) confirming the results with a much larger data set; (2) clarifying the effect by removing the effects of solar-wind speed changes, solar-wind density variations, and solar-wind magnetic-field-strength variations; (3) exploring methods to improve the correlations between solar-wind turbulence and geomagnetic activity such as cleaning the data or using intervals of steady solar wind; (4) investigating the magnitude of the turbulence effect versus the phase of the solar cycle and; and (5) investigating the magnitude of the turbulence effect versus the type of solar wind. Another important improvement is the use of δb ($=\delta B/(4\pi\rho)^{1/2}$) instead of δB as a measure of the amplitude of the solar-wind turbulence.

This manuscript is organized as follows. In Section 2 the large data set used for the present study is described. In Section 3 the checks are made to determine whether there are systematic differences between the various solar-wind spacecraft that contributed to the OMNI data set. In Section 4 the turbulence effect under southward IMF is investigated. Section 5 contains an investigation of the turbulence effect through the solar cycle and for various types of solar wind. The results are summarized in Section 6.

2. THE LARGER DATA SET

The correlations of upstream solar-wind turbulence with geomagnetic activity are studied here with a data set that is much larger than the one used in *Borovsky and Funsten* [2003]. This is done for a number of reasons: (1) to confirm the earlier results with different data, (2) confirming that result that the turbulence effect operates under both northward and southward IMF, (3) to have sufficient data to fix various parameters while performing correlations between other parameters, (4) to have a long enough data set to study the freestream turbulence effect through the solar cycle, and (5) to have a large enough data set to be able to separately study the freestream turbulence effect in various types of solar wind.

The data set used is the 1963-2001 OMNI2 solar-wind data [*King and Papitashvili*, 2005], supplemented with the geomagnetic index AE. For the years 1995-2001, hourly averages of cleaned preliminary AE were provided by Mike Henderson. When cross-correlating solar-wind data with the AE index, a 1-hour time lag is added to AE since the magnetosphere has about a 1 hour time lag to solar-wind driving (see Table 7 of *Borovsky and Funsten* [2003]). In this report, this 1-hour-lagged AE index is denoted as AE_1.

The hourly averaged AE index is chosen for this study because of its availability over several solar cycles and because its temporal resolution matches that of the OMNI data set. For the limited data (1979-1981) used in the *Borovsky and Funsten* [2003] study of the turbulence effect, similar results are obtained for several other geomagnetic indices. As pointed out by a referee, the AE index has the complication that is driven both by electric-field effects and by particle-precipitation effects, and these two effects could be reacting to different processes in solar-wind/magnetosphere coupling. However, for a wide range of values it does linearly represent the power input to the ionosphere from solar-wind driving [e.g., *Baumjohann and Kamide*, 1984].

The 1963-2001 OMNI2-AE_1 data set provides 192,744 hours of data with which the turbulence effect can be studied. In the OMNI2 data set the amplitude of the solar-wind MHD turbulence is measured by the rms amplitude of the magnetic-field variations during each hour of data (see discussion below). To ensure the quality of the turbulence-amplitude measure, hours of data in which there are less than 15 measurements of the magnetic field are removed. This leaves 172,702 hours of data. The OMNI2 data utilized was not cleaned to remove interplanetary shocks as was the OMNI1 data used in the study of *Borovsky and Funsten* [2003].

An important subset of the full data set that will be repeatedly used in this study is a northward-IMF (dayside reconnection off) data set wherein vB_z of the solar wind is in the range 1000 km/sec nT $\leq vB_z \leq$ 3000 km/sec nT. This northward-IMF data set contains 25,088 hours of data.

There are several ways to measure the amplitude of the MHD turbulence in the solar wind. Four will be used in the present studies. Certainly one important measure is δv, the rms amplitude of the solar-wind velocity fluctuations. However, as discussed in *Borovsky and Funsten* [2003], the measures of δv that appear in the OMNI data set can be inaccurate: (a) the various directional components of δv are poorly cross correlated with each other, (b) they are sometimes poorly correlated with the amplitude of the magnetic-field fluctuations, and (c) they are poorly cross-correlated with the amplitude of geomagnetic activity. Hence, δv will not be used in the present studies. Two measures that will be used are δB (the rms amplitude of the magnetic-field-vector fluctuations during one hour of magnetic-field measurements) and $\delta B/B_o$ (the rms amplitude of the fluctuations divided by the hourly averaged magnetic-field strength). The quantity $\delta B/B_o$ is a measure of the angular fluctuation amplitude of the magnetic-field variations. A third measure that will be used is δb, which is defined as $\delta b = \delta B/(4\pi\rho)^{1/2}$, where ρ is the mass density of the solar-wind plasma. The units of δb are km/s; δb is the amplitude of the magnetic-field fluctuations in terms of Alfven units. As demonstrated in Figure 2, in the MHD turbulence of the solar wind δb can be used as a proxy for δv. Here, using 6.3 years of high-time-resolution measurements from the ACE spacecraft in the solar wind, three measures of the magnetic-field-fluctuation amplitude of the solar-wind turbulence are plotted as a function of the measured value of δv. As can be seen δb and δv are highly correlated ($R_{corr} = +0.94$ in the ACE data set) and a statistical fit to the AEC data yields

$$\delta v \approx (0.99 \pm 0.27)\, \delta b \qquad (1)$$

A fourth measure of the amplitude of the MHD turbulence is the viscous force F_{visc} on the magnetosphere owing to the MHD eddy viscosity ν_{eddy} of the solar wind; in *Borovsky and Funsten* [2003] the expression $F_{visc} \propto n v^{5/2}(\delta B/B_o)$ was derived, where n is the number density of the solar wind, and v is the speed of the solar wind. This functional form of the viscous force comes about from a standard viscous-force expression for an obstacle in a fluid with a viscous-drag coefficient chosen for a bullet-shaped obstacle and the *Wu and Faeth* [1994] and *Volino* [1998] prescription of replacing the kinematic viscosity in the expressions with the eddy viscosity (see *Borovsky and Funsten* [2003] or *Borovsky* [2006] for details).

In Figure 3a, for the OMNI2-AE_1 data set the level of geomagnetic activity as measured by the AE index (with a 1 hour time lag) is plotted as a function of vB_z of the solar wind. (Everywhere in this study, B_z is in GSM coordinates.) To clarify the trends in the data a 1500-point running average of AE_1 is also plotted. As can be seen, for vB_z positive (northward IMF, dayside reconnection off) the AE index is nearly independent of the value of vB_z, whereas for vB_z negative (southward IMF, dayside reconnection on) the value of the AE index increases strongly as the magnitude of $-vB_z$

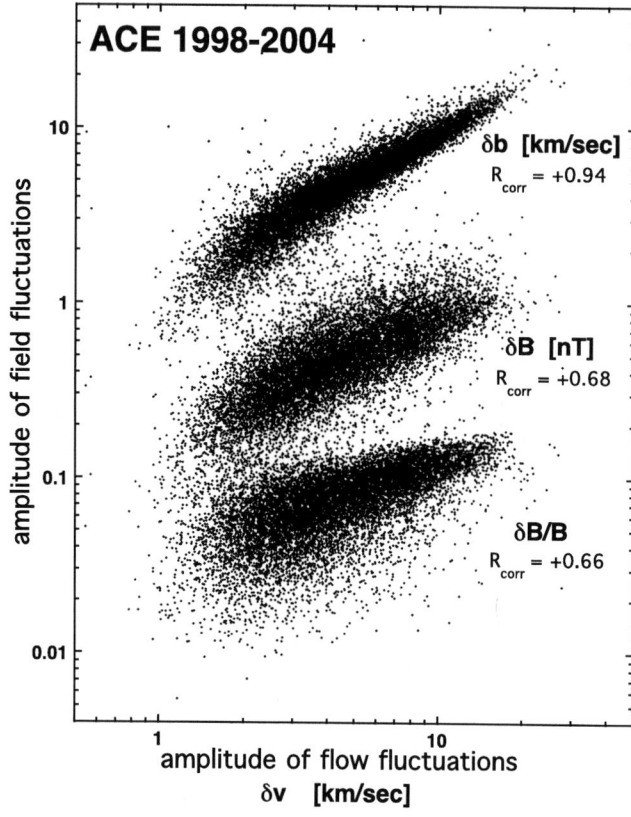

Figure 2. For 12,297 time intervals each 4.5 hours long, the amplitude of the magnetic-field fluctuations of the turbulent solar wind during each interval are plotted as a function of the amplitude of the velocity fluctuations. The fluctuations are measured by the ACE SWEPAM [*McComas et al.*, 1998] and MAG [*Smith et al.*, 1998] instruments at 1 AU. The measurements are high-pass Fourier filtered to periods of 2 minutes.

increases. In Figure 3a, some of the scatter of the data points in the vertical direction about the trend curve is owed to the freestream turbulence effect, wherein variations in AE result from variations in the amplitude of the MHD turbulence in the upstream solar wind.

In Figure 3b, for the northward-IMF (1000 nT km/s $\leq vB_z \leq$ 3000 nT km/s) subset of the OMNI2-AE_1 data set, the level of geomagnetic activity as measured by the AE index (with a 1 hour time lag) is plotted as a function of the eddy-viscous force $nv^{5/2}\delta B/B_o$ of the solar wind. To clarify the trend in the data a 300-point running average of AE_1 is also plotted. As can be seen, for the reconnection-off data the value of the AE index increases as the magnitude of $nv^{5/2}\delta B/B_o$ in the solar wind increases. This is the freestream turbulence effect. The plot in Figure 3b for the solar wind is analogous to the wind-tunnel-experiment plot in Figure 1. The linear correlation coefficient between AE_1 and $nv^{5/2}\delta B/B_o$ in Figure 3b is $R_{corr} = +0.46$.

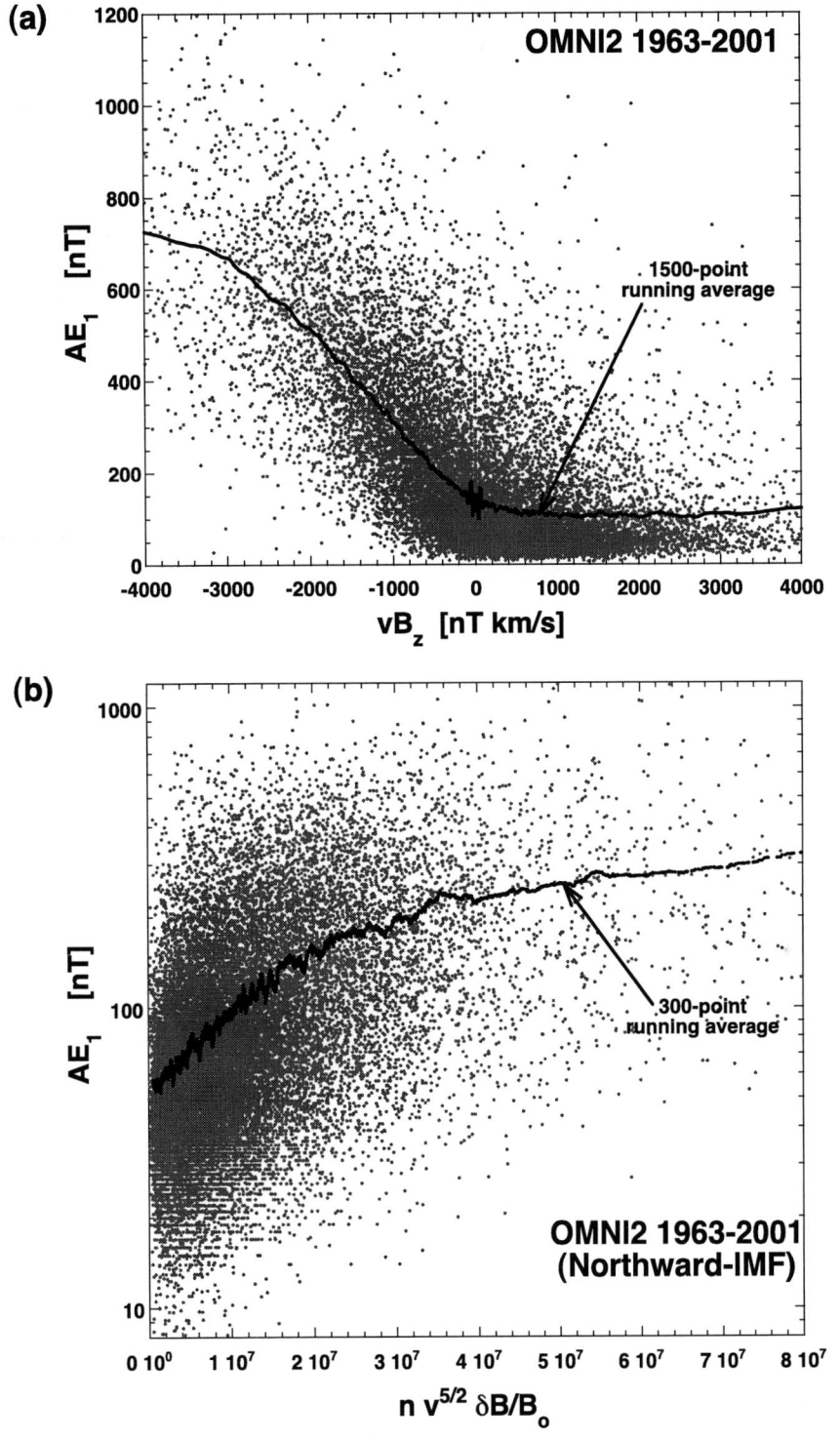

Figure 3. (a) For the full 1963-2001 OMNI2-AE_1 data set the geomagnetic activity AE (with a 1-hour lag) is plotted (gray points) as a function of vB_z of the solar wind. (b) For the northward-IMF (1000 nT km/s $\leq vB_z \leq$ 3000 nT km/s) subset of the 1963-2001 OMNI2-AE_1 data set the level of geomagnetic activity AE (with a 1-hour lag) is plotted (gray points) as a function of the eddy-viscous force of the turbulent solar wind on the magnetosphere. In the top figure, every 10th point is plotted. To visualize the underlying trends to the data points, in panel (a) a 1500-point running average of the AE_1 data is plotted (black curve) and in panel (b) a 300-point running average of the AE_1 data is plotted (black curve).

Table 1. For the northward-IMF data set (1000 nT km/s ≤ vB_z ≤ 3000 nT km/s), linear-regression fits are made for AE (1-hour lagged) as functions of four different measures of the upstream turbulence in the solar wind and as a function of vB_z of the solar wind. The columns of the table are the slope of the linear fit, the offset of the fit, and the linear correlation coefficient

	Slope	Offset	R_{corr}
$AE_1 \leftrightarrow \delta B$	25.7	33.3	+0.43
$AE_1 \leftrightarrow \delta B/B_o$	167.6	43.0	+0.2933
$AE_1 \leftrightarrow \delta b$	2.75	35.3	+0.394
$AE_1 \leftrightarrow nv^{5/2}\delta B/B_o$	3.28×10^{-6}	66.3	+0.461
$AE_1 \leftrightarrow vB_z$	-3.71×10^{-4}	107.7	-0.002

In Table 1, the parameters of linear-regressions fits to AE_1 as functions of the four different measures of the amplitude of the upstream MHD turbulence in the solar wind are collected. The fits utilize the full northward-IMF data set (1000 nT km/s ≤ vB_z ≤ 3000 nT km/s). As can be seen by comparing the δB and $\delta B/B_o$ correlation coefficients in Table 1 to the equivalent numbers in the first row of Table 3 of *Borovsky and Funsten* [2003], substantially improved correlation coefficients are obtained with the present large data set. Improved correlation coefficients translate into improved values for the linear-regression fits. Also shown in the last line of the Table 1 are the parameters of the $AE_1 \leftrightarrow vB_z$ linear-regression fit for the northward-IMF data; as can be seen AE_1 is uncorrelated with vB_z in this northward-IMF (reconnection-off) data set.

3. TESTING FOR DIFFERENCES BETWEEN SOLAR-WIND SPACECRAFT AND EXAMINING LONG-DURATION NORTHWARD-IMF INTERVALS

In this section two examinations are performed to gain a better understanding of the freestream turbulence effect in the large data set. First, the 1963-2001 OMNI2-AE_1 data set is broken up according to the solar-wind spacecraft that contribute measurements to the data set and a comparison is made of the analysis of the turbulence effect for the separate spacecraft. Second, a subset of the 1963-2001 OMNI2-AE_1 data set is created using only time intervals wherein the hourly averaged IMF is northward for 5 or more hours, and then the turbulence effect for these intervals is compared with the effect during all northward-IMF data.

The OMNI2 solar-wind data set is assembled from measurements of the solar wind made by many spacecraft. As a test to ensure that there are no strong trends that are spacecraft dependent in the data set, AE_1 is plotted as a function of the eddy-viscous force $nv^{5/2}\delta B/B_o$ separately for the various spacecraft contributing to the 1963-2001 OMNI2 data set. In Figure 4 linear fits to the separated data are plotted with

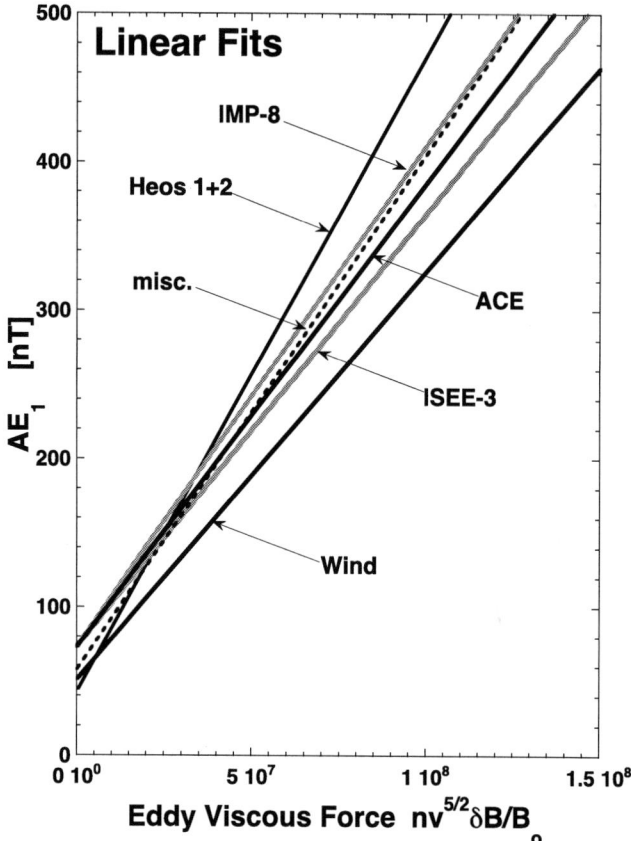

Figure 4. For the various spacecraft contributing solar-wind data to the OMNI2 data set, the turbulence effect under northward IMF (1000 nT km/s ≤ vB_z ≤ 3000 nT km/s) is individually tested. Plotted are the least-squares linear-regression fits to the AE (1-hour-lag) versus eddy-viscous force data for the various spacecraft as labeled. In the plot, misc. stands for the combined data from AIMP-1, AIMP-2, IMP-4, IMP-5, IMP-6, and IMP-7.

labels indicating the spacecraft of origin. The ratio of the maximum slope of a fit (HEOS 1+2) to the minimum slope of a fit (Wind) is 1.56. As can be seen by comparing the fits, there are slight variations from spacecraft to spacecraft, but no overwhelming trends stand out and there are no anomalous spacecraft. (Note that in looking at δv in the OMNI2 data set, there *are* anomalous spacecraft.) Note also that it is possible that there are solar-cycle biases to the various spacecraft data sets.

To determine whether the magnetospheric response to solar-wind turbulence differs if the dayside reconnection is off for a long period of time (3 hours or more), the turbulence effect is examined during long intervals of northward IMF. A data subset is created from the 1963-2001 OMNI2-AE_1 data set wherein the hourly averaged IMF is continuously northward for at least 5 hours: 3 hours before the hour taken, the hour taken, and 1 hour past the hour taken. For the hour

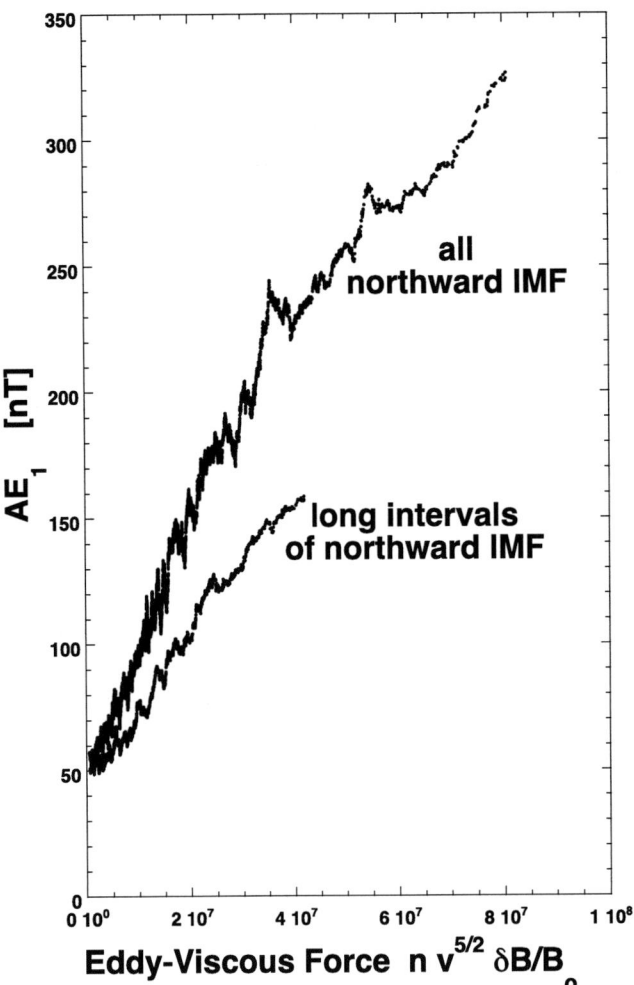

Figure 5. A comparison of the turbulence effect for all northward-IMF data and for extended intervals of northward IMF. Here, to be included in the extended-northward-interval data, the IMF must be northward on average for at least 5 hours: 3 hours prior to the hour used and for 1 hour after the hour used. The turbulence effect is a factor of 1.5 times stronger in the all-northward-IMF data set than it is in the extended-intervals-of-northward-IMF data set.

the full data set where there is no requirement that the IMF be northward for long intervals of time. One can imagine several possible reasons for the elevated AE values in the times following or preceding southward IMF: (1) there is a persistence to AE so northward-IMF hours used in the correlations that are preceded by southward-IMF hours have elevated AE values; (2) since 1-hour-lagged AE is used, northward-IMF hours used in the correlations that are followed by southward-IMF hours have AE_1 values that are partially responding to the southward-IMF driving, (3) there is some high-frequency HILDCAA-type effect wherein even if the hourly average of the IMF is northward there is some southward IMF present and dayside-reconnection driving is occurring, and (4) a persistence of the conductivity of the nightside ionosphere that changes the manner in which the turbulence effect drives the AE index. The first three of these effects would produce an offset to the AE index under northward IMF rather than change the slope of the curves in Figure 5. (The third effect was studied in *Borovsky and Funsten* [2003] (see Figs. 5 and 6 of that paper) and the conclusion of that study was that high-frequency HILDCAA-type effects were not dominating measurements of the turbulence effect.) The fourth effect may produce a change in the slopes, however in Section 5 a sunspot-number effect will be seen that will argue that ionospheric conductivity might have the opposite effect on the slopes.

4. THE FREESTREAM TURBULENCE EFFECT ACTING UNDER SOUTHWARD IMF

It is clear that the freestream turbulence effect acts when the IMF is northward (see, e.g., Figure 3b). And under northward IMF it is straightforward to quantify the turbulence effect (see, e.g., Table 1). When the dayside-reconnection effect which acts when vB_z is negative is shut off, the turbulence effect dominates the driving of the Earth's magnetosphere and the variations in the AE index owed to the turbulence effect are easy to discern in a solar-wind/magnetosphere data set. Multivariate linear-regression fits to the AE index as functions of the solar-wind parameters (eqs. (15) and (16) of *Borovsky and Funsten* [2003]) showed that the turbulence effect also acts when the IMF is southward, but with dayside reconnection ongoing under southward IMF much more of the variance of the AE index is controlled by variations in vB_z and the turbulence effect is more difficult to discern.

One way to investigate the turbulence effect when the dayside-reconnection effect is ongoing under southward IMF is to restrict the variation of vB_z in the data set and look for variations of AE associated with variations of the turbulence amplitude. This requires a large data set. For the investigation, the 1963-2001 OMNI2-AE_1 data set is sorted according to the value of vB_z and the data set is divided up into numerous

taken, it is further required that vB_z is in the range 1000 nT km/s $\leq vB_z \leq$ 3000 nT km/s. In Figure 5 the 1-hour-shifted value of AE is plotted as a function of the eddy viscous force and a 300-point running average of AE_1 is made. This curve is labeled "long intervals of northward IMF". Also plotted is a similar curve for the full northward-IMF data set (1000 nT km/s $\leq vB_z \leq$ 3000 nT km/s), with a 300-point running average of those AE_1 values. As can be seen, the slope of the trend curve for the persistent-northward data is about 2/3 of the slope of the trend curve of the all-northward data. This factor of 2/3 holds also for the slopes of linear-regression fits to the data. The values of AE are elevated in

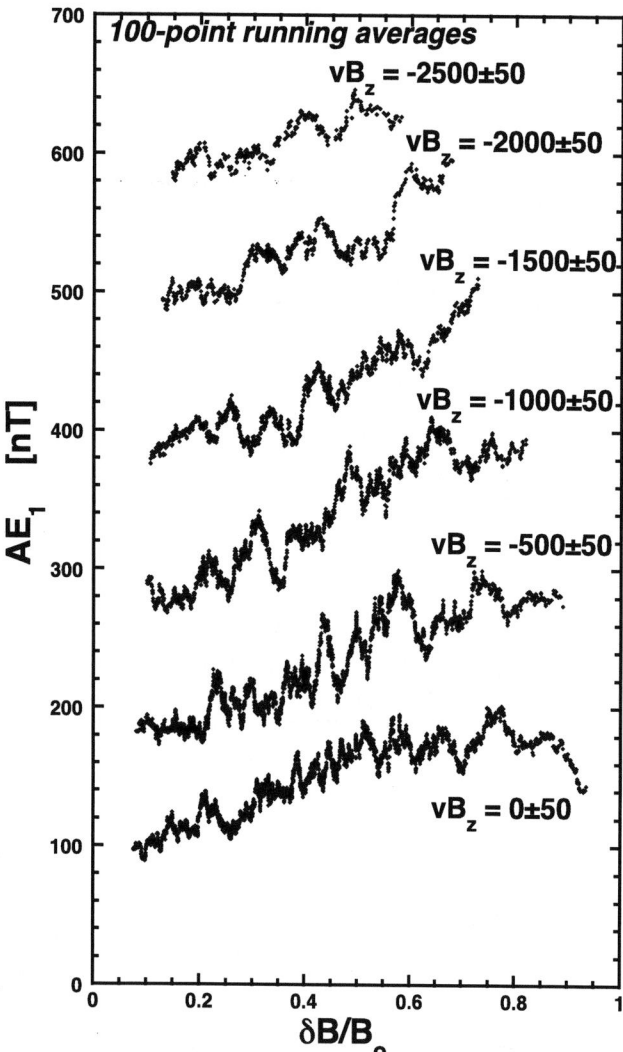

Figure 6. The freestream turbulence effect is investigated with six data subsets, each with vB_z of the solar wind varying by only 100 nT km/s. Shown for the six data subsets are 100-point running averages (trend curves) for AE (1-hour lagged) plotted as a function of the upstream turbulence amplitude $\delta B/B_o$.

data subsets each with vB_z varying by only 100 nT km/s. Figure 6 is produced using 6 of these data subsets (five of which have southward IMF) by plotting AE (1-hour lagged) as a function of the amplitude of the turbulence $\delta B/B_o$. The data points shown are 100-point running averages to indicate the trends. As can be seen, for all 6 curves the AE index increases as the amplitude of the turbulence in the solar wind increases. The positive slopes of the curves indicate that the freestream turbulence effect is operating when the IMF is southward. Note that the 6 curves are vertically offset from each other; this offset is owed to the differing amounts of vB_z reconnection driving in the 6 data subsets.

In Figure 7a, 61 data subsets each with restricted ranges of vB_z are used. Plotted are the slopes of linear regression fits to AE (1-hour lagged) as functions of four different measures of the solar-wind turbulence amplitude, plotted as functions of the value of vB_z of the data subset. As can be seen, for each of the four measurements of the turbulence amplitude, the freestream turbulence effect that is present under northward IMF (positive vB_z) persists into southward IMF with about the same intensity. Note however that for strong negative vB_z, the slopes of the linear-regression fits weaken; this weakening indicates that in this regime a smaller fraction of the variation of AE is described by the turbulence effect (see the discussion in the following two paragraphs). In Figure 7b the intercept of the linear-regression fits of $AE = AE(\delta b)$ is plotted as a function of vB_z of the data subsets. Also plotted as the dashed curve is a 5000-point running average of AE from the entire 1963-2001 OMNI2-AE_1 data set (cf. Figure 3a). In the two panels of Figure 7, driving of the Earth's magnetosphere by the freestream turbulence effect is described by the slopes of the linear-regression fits (top panel) and the driving of the magnetosphere by dayside reconnection is described by the intercept of the fits.

Note that for strong negative vB_z the slopes plotted in Figure 7a deteriorate and the freestream turbulence effect is not being discerned in the data analysis. In this regime, the AE index is large, as shown in Figure 7b. In this paragraph and the next it is argued that when AE is large, the ability to statistically determine AE is poor, and the statistical inaccuracy of AE produces a variance in AE that is larger than the variance in AE owed to the driving of the magnetosphere by the turbulence effect. Hence, when AE is large owed to strong negative vB_z, the signal-to-noise is poor and the turbulence effect cannot be discerned. In Figure 8, for AE (1-hour lagged) plotted as a function of vB_z of the solar wind, several statistical quantities from the 1963-2001 OMNI2-AE_1 data set are shown. The thin curve is the median value of AE_1 as determined from a 2001-point running statistical analysis. The other two curves are two measures of the variance of AE_1. The upper thick curve is the difference between the 90th percentile of AE_1 and the 10th percentile of AE_1 as determined from a running 2001-point statistical analysis. The lower thick curve is the average deviation of AE_1 about the median of AE_1 (defined as $N^{-1}\sum_{i=1,N}|AE_i - AE_m|$, where $N = 2001$ is the number of points, the AE_i are the individual values of AE_1, AE_m is the median value of AE, and the sum Σ goes from $i = 1$ to $i = N$). As can be seen, the median value of AE_1 increases for increasing negative vB_z (see also the mean in Figure 3a). As this happens, the spread in the AE_1 values (as indicated by the two thick curves in Figure 8) also increases. The variance in AE is owed to "noise" in AE plus the turbulence effect, where "noise" here includes uncertainties in the determination of AE plus any other effect besides

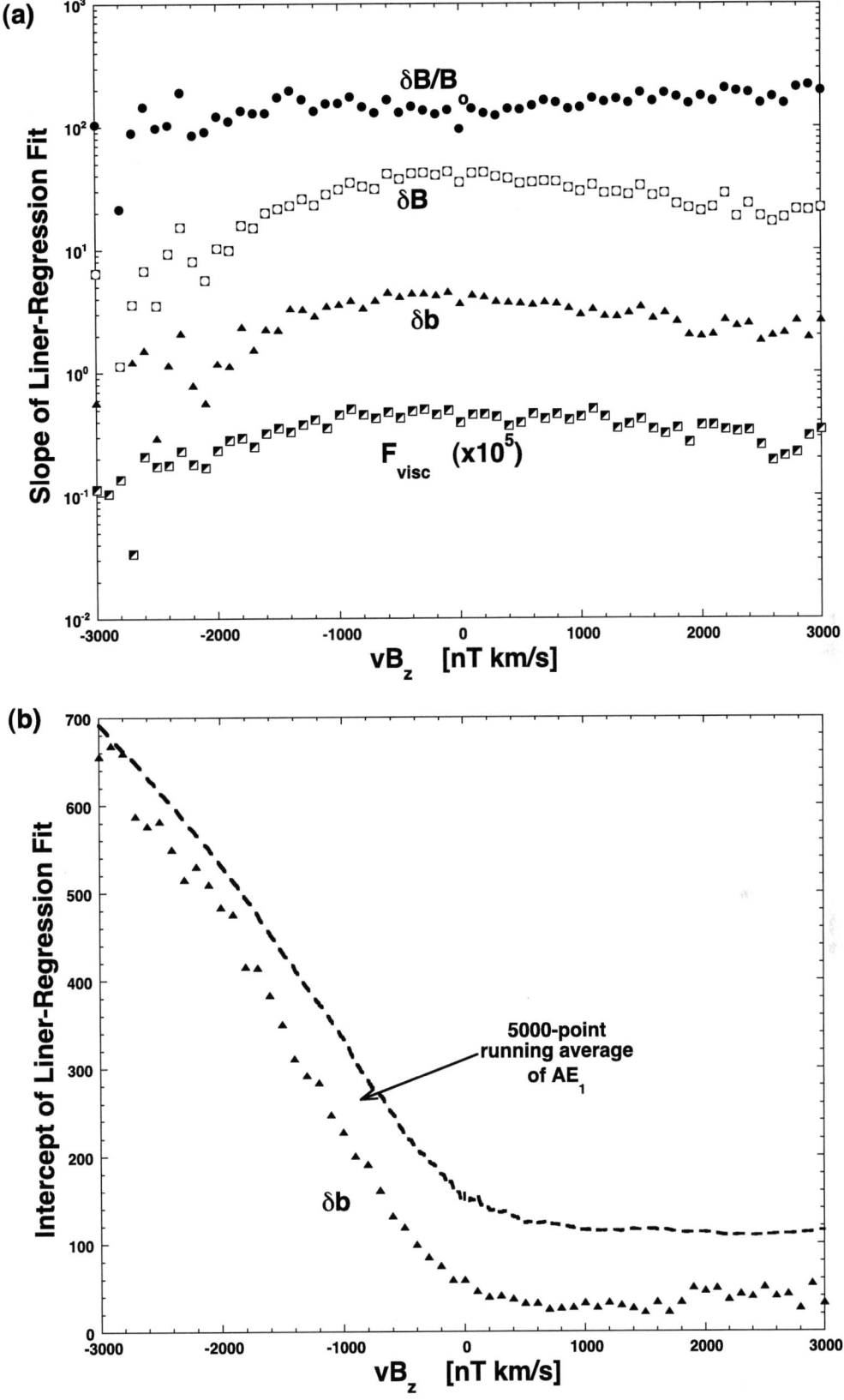

Figure 7. The OMNI2-AE_1 data set is divided into bins that are 100 nT km/sec wide in vB_z and linear-regression fits to $AE = AE(\delta B/B_o)$, $AE = AE(\delta B)$, $AE = AE(\delta b)$, and $AE = AE(F_{visc})$ are made for each bin. In panel (a) the slopes of the linear regression fits are plotted as functions of vB_z. In panel (b) the intercept of the linear-regression fit is plotted as a function of vB_z. For comparison with the intercept (the offset in the fit), a 5000-point running average of AE as a function of vB_z is plotted as the dashed line in panel (b). Note that AE_1 is a 1-hour time lagged AE.

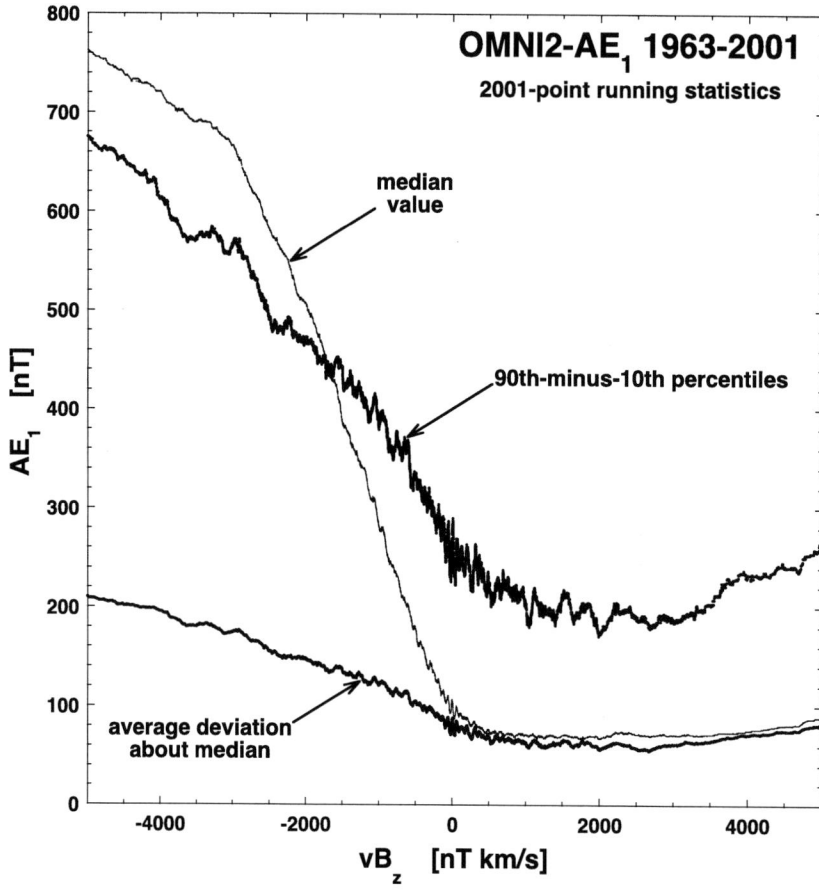

Figure 8. Using the 1963-2001 OMNI2-AE$_1$ data set, AE (1-hour lagged) is plotted as a function of vB$_z$ of the solar wind and various statistical measures are plotted: the 2001-point running median of AE$_1$ versus vB$_z$, the 2001-point average deviation of AE$_1$ about its median value, and the difference between the 2001-point 90th percentile and the 10th percentile of AE$_1$.

the turbulence effect that makes AE vary (such as substorm energy release, etc.). If the amount of variation of AE caused by the turbulence effect is invariant to the value of vB$_z$, then (a) the increasing variance of AE in Figure 8 is owed to increases in the "noise" and (b) the best "signal-to-noise" for determining the turbulence effect occurs where the variance is lowest, which is for modest positive values of vB$_z$. Two indications that the "noise" in AE increases as the value of AE increases are presented in the next paragraph.

In the two panels of Figure 9 the "noise", "uncertainty", or "error" in AE is investigated. The 1-minute-resolution AE-index dataset for the year 1995 is used and hourly averages (denoted in this paragraph as $\langle AE \rangle$) are constructed from the 1-minute AE values. In the left-hand panel of Figure 9 the standard deviation σAE of the 60 1-minute values of AE that go into each hour of data is plotted as a function of the hourly average $\langle AE \rangle$ for that hour. As can be seen, the standard deviation (square root of the variance) of the AE index increases as the value of AE increases. This left-hand panel indicates that there will be an uncertainty or statistical error to the determination of $\langle AE \rangle$ that increases as the value of the $\langle AE \rangle$ increases. Hence, higher-AE values are noisier than lower-AE values. To produce the plot in the right-hand panel of Figure 9 the 1-minute values of AE are used to construct one set of hourly averaged values $\langle AE \rangle$, then the time window is shifted by 30 minutes and the 1-minute values are used to construct a second set of hourly averaged values $\langle AE \rangle_{30}$ shifted by 30 minutes. In the right-hand panel of Figure 9, the difference $|\langle AE \rangle - \langle AE \rangle_{30}|$ between the hourly average and the hourly average shifted by 30 minutes is plotted as a function of the hourly average. As can be seen in the plot, the change in the value of the hourly average of AE that results from shifting the time window is proportional to the value of the hourly average $\langle AE \rangle$. This indicates that the variability of $\langle AE \rangle$ increases and the value of $\langle AE \rangle$ increases, and that in a sense the arbitrariness of the AE hourly averaged value increases as AE increases. Note that none of the results from Figure 9 are surprising.

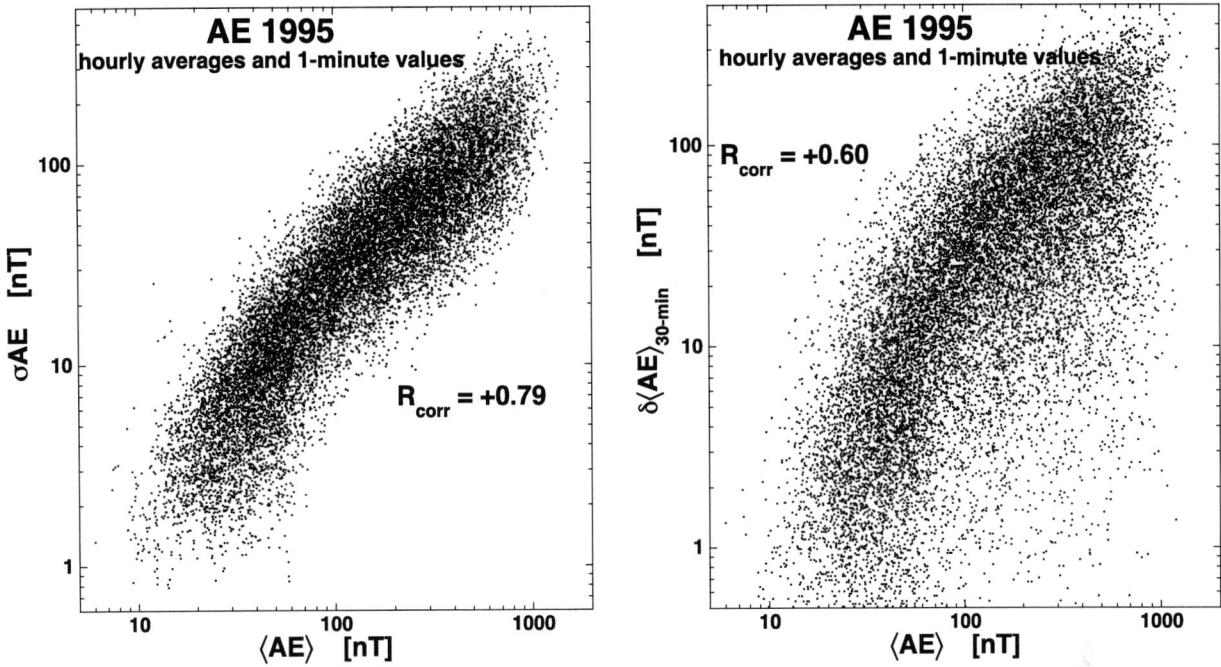

Figure 9. Using the 1-minute values of the AE index for 1995, hourly averages of AE are constructed and the standard deviation of the 1-minute values in each hour are calculated. Here, the standard deviation is plotted as a function of the hourly average.

From Figures 8 and 9 it is concluded that the deterioration of the slopes in Figure 7a for strong negative vB_z probably does not indicate that the turbulence effect is not acting for strong negative vB_z, rather the deterioration is caused by increased variance in AE that is owed to effects other than the turbulence effect and that this increased variance leads to lower correlations between the variations of AE and variations in the amplitude of the turbulence in the solar wind. When correlation coefficients deteriorate, least-squares-fit linear-regression slopes decrease in magnitude.

5. THE TURBULENCE EFFECT THROUGH THE SOLAR CYCLE AND IN VARIOUS TYPES OF SOLAR WIND

To determine whether the freestream turbulence effect operates to the same degree in different types of solar wind, the effect is investigated as a function of time through the three solar cycles. For northward-IMF (1000 nT km/s < vB_z < 3000 nT km/s) the 1963-2001 OMNI2-AE_1 data set is utilized. The northward-IMF data is arranged chronologically and groupings of 1500 data points are extracted. For each grouping least-squares linear-regression fits of AE_1 to three measures of the turbulence are made. The slopes of the three fits are plotted in the first two panels of Figure 10 as functions of the median time of the 1500-point groupings. Also plotted in the bottom panel is the sunspot number. As can be seen by comparing the slopes in the top panel with the sunspot number in the bottom panel, there may be a solar-cycle dependence with the turbulence effect being stronger during solar minima. In particular, for the solar maxima at ~1970, ~1981, and ~1990 there are dramatic decreases in the slopes for $AE_1 = AE_1(\delta B/B)$ and $AE_1 = AE_1(\delta b)$, but such a decrease is not seen for the ~2001 solar maximum. The lack of decrease for the ~2001 solar minima might be owed to a change in the AE index after 1995, wherein the AE ground stations in the Russian sector became unavailable. No clear solar-cycle dependence to the turbulence effect is seen when the eddy-viscous force F_{visc} curve in the second panel of Figure 10 is examined, and the dramatic decreases in the slopes seen in the top panel are greatly reduced in the $AE_1 = AE_1(F_{visc})$ curve of the second panel.

For completeness, the dayside-reconnection effect is examined through the solar cycle with 1500-point chronological groupings of data points are taken for southern-IMF (−3000 nT km/s < vB_z < −1000 nT km/s) in the 1963-2001 OMNI2-AE_1 data set. In the third panel of Figure 10 the slopes of the least-square linear-regression fits to $AE_1 = AE_1(-vB_z)$ are plotted as functions of the median time of the 1500-point groupings. Notice two things about the plot in the third panel: (1) a slight solar-cycle dependence to the driving of AE by $-vB_z$ of the solar wind can be seen and (2) the level

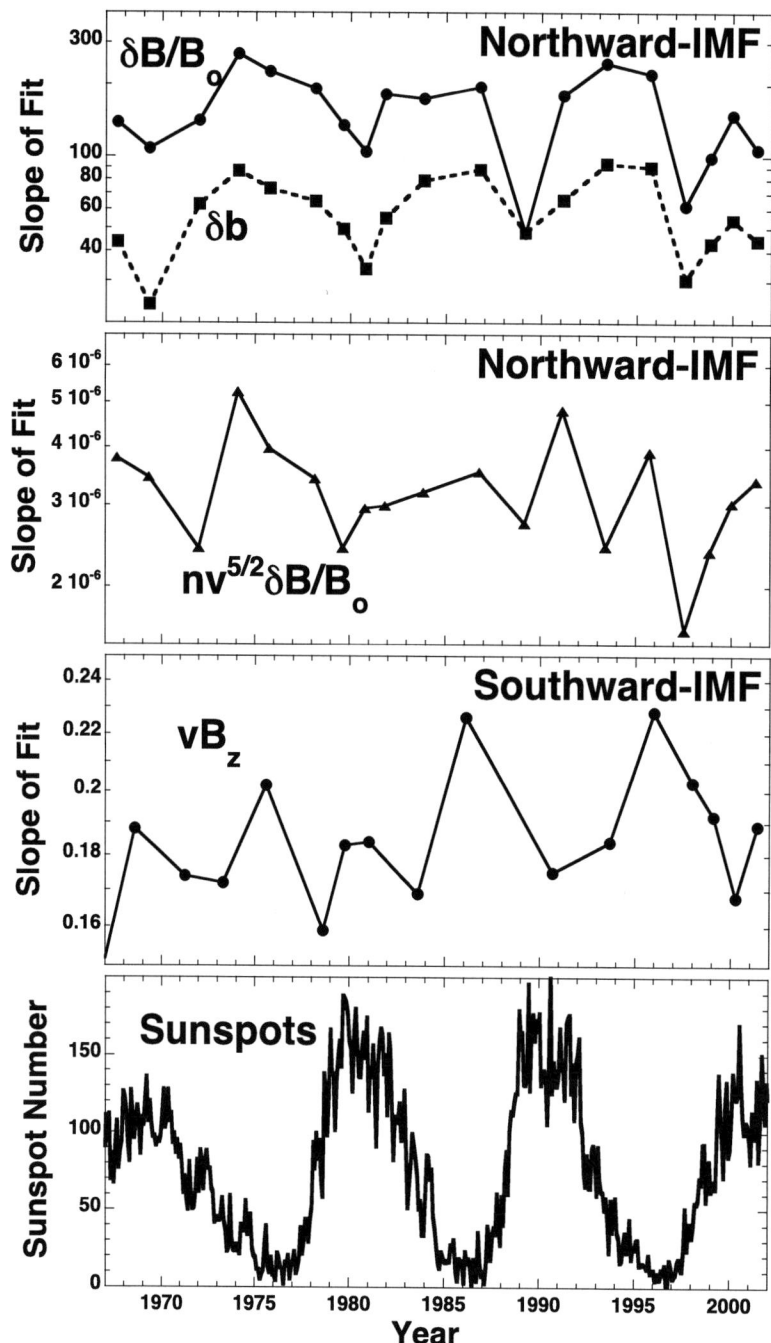

Figure 10. For three solar cycles, the linear-regression-fit coefficients (slopes) between three measures of the turbulence effect and the one-hour-shifted AE index are plotted. The three measures are the fractional amplitude of the magnetic-field fluctuations $\delta B/B$, the amplitude of the magnetic-field fluctuations measured in Alfven units δb, and the eddy-viscous force $nv^{5/2}(\delta B/B)$. In the plot the eddy-viscous-force coefficient is multiplied by 1×10^7. Also plotted (right axis) is the sunspot number.

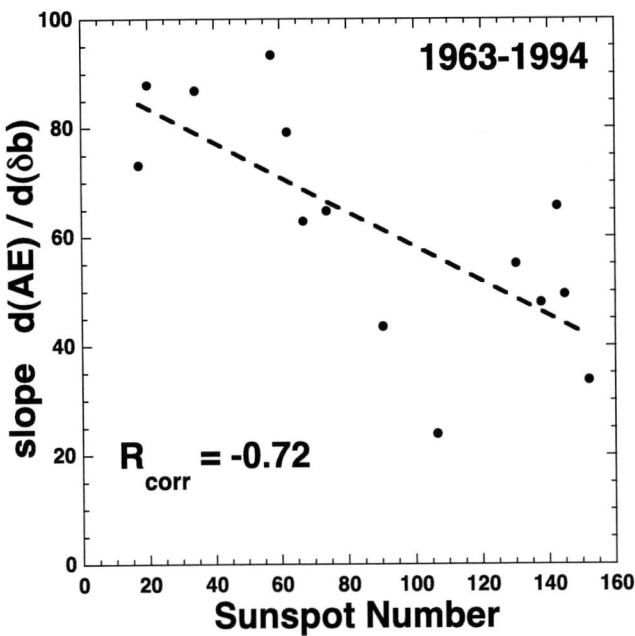

Figure 11. For the years 1963-1994, the slope of the linear-regression fit between δb and AE for northward IMF is plotted as a function of the sunspot number(solid points). A linear-regression fit to these data is shown as the dashed line.

Table 2. For 1963-1994, the linear correlation coefficients R_{corr} between the sunspot number and the slope of the linear fits between AE and various solar-wind drivers are listed. The R_{corr} values are all negative, indicating that all drivers drive the AE index more efficiently when the sunspot number is lower.

Slope of Fit	R_{corr} with Sunspot Number	Data Set
$AE_1 \leftrightarrow \delta B$	−0.426	Northward-IMF
$AE_1 \leftrightarrow \delta B/B_o$	−0.691	Northward-IMF
$AE_1 \leftrightarrow \delta b$	−0.723	Northward-IMF
$AE_1 \leftrightarrow nv^{5/2}\delta B/B_o$	−0.286	Northward-IMF
$AE_1 \leftrightarrow -vB_z$	−0.369	Southward-IMF

of fluctuations of the slope from grouping to grouping is less for $-vB_z$ than it is for the turbulence effect (top two panels). The tendency of the solar-cycle dependence is that the slope of the $-vB_z$-versus-AE coupling is stronger during solar minima. The smaller fluctuation level is probably owed to the fact that $-vB_z$ variations yield larger variations in AE than turbulence-amplitude variations do, so the correlation coefficients are higher and the linear-regression fits are more accurate (less noisy) from grouping to grouping.

The solar-cycle dependences of the solar-wind/magnetospheric coupling are further explored via Figure 11 and Table 2. In Figure 11 the slopes of the δB-versus-AE fits (which are the square points in the top panel of Figure 10) are plotted as a function of the sunspot number for the years 1963-1996 where the AE index was of high quality. As can be seen in Figure 11 there is a distinct anticorrelation between this slope (which is the derivative d(AE)/d(δb)) and the sunspot number. A linear fit to the data is shown as the dashed line and the linear-correlation coefficient R_{corr} = −0.72 is indicated. This correlation coefficient is collected into Table 2, along with other linear-correlation coefficients between the AE-versus-driver slopes and the sunspot number. As can be seen in Table 2, all slopes are anticorrelated with the sunspot number which indicates that the driving of the AE index by all of the solar-wind drivers is weaker when the sunspot number is higher. This could be caused by a number of effects: for examples (1) differences in the nature of the solar wind in different parts of the solar cycle, (2) the way the AE index responds to magnetospheric driver when the solar-irradiation-driven ionospheric conductivity differs, or (3) a nonlinear effect in the solar-wind magnetosphere coupling where the coupling efficiency is weaker when the driving (at solar maxima) is stronger.

To specifically determine whether the turbulence effect operates in the same manner in different types of solar wind, the 1963-2001 OMNI2-AE_1 data set is separated according to the type of solar wind. This is done by three methods: (1) producing a categorization from the temperature-versus-speed solar-wind plot, (2) producing a categorization according to the value of the sunspot number, and (3) using catalogs of solar-wind events. First the categorization from the T_i-versus-v plot is discussed.

In Figure 12 the ion temperature T_i of the solar wind is plotted as a function of the velocity v of the solar wind for the 1963-2003 OMNI2 data set (gray points). Using a running 2001-point statistical analysis of the data points, three quantities are extracted and plotted (irregular black curves) in the figure: the median value of T_i, the 90th percentile of T_i, and the 10th percentile of T_i. Shown for comparison are the *Richardson and Cane* [1995] expected value of T_i (from the first equation pair in that paper, obtained by *Lopez* [1987]). The three curves are very well fit (smooth black curves in the figure) as follows. The median of T_i is fit by

$$T_i = 1.28 \times 10^{-8} v^{3.324} \quad \text{(for } v \leq 372 \text{ km/s)} \quad (2a)$$

$$T_i = 0.0572 \, v - 16.79 \quad \text{(for } v \geq 372 \text{ km/s)} \quad (2b)$$

the 90th percentile of Ti is fit by

$$T_i = 6.32 \quad \text{(for } v \leq 327 \text{ km/s)} \quad (3b)$$

$$T_i = 0.1079 \, v - 28.94 \quad \text{(for } v \geq 327 \text{ km/s)} \quad (3b)$$

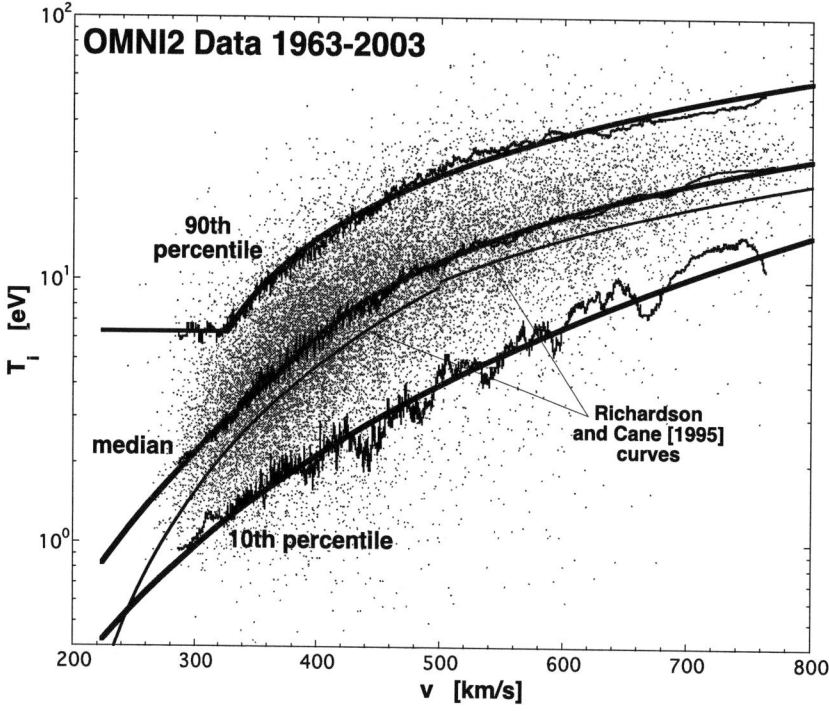

Figure 12. For T_i plotted as a function of v with the OMNI2 data set (gray points, every 10th point plotted), a 2001-point running median is produced, as are 2001-point running 90th percentiles and 10th percentiles (black points). These values are fit with curves (smooth black curves) whose expressions are given in the text. For comparison the *Richardson and Cane* [1995] (*Lopez* [1987]) curves are plotted.

and the 10th percentile of Ti is fit by

$$T_i = 1.169 \times 10^{-7} v^{2.791} \quad \text{(all v)} \qquad (4)$$

where T_i is in eV and v is in km/s. Using the formulas of expressions (2)-(4), an approximate categorization of the solar wind is made as follows (see Plate 1). Since coronal mass ejections (CME) tend to be cooler than normal solar wind [*Gosling et al.*, 1973; *Richardson and Cane*, 1995], times when the solar-wind ion temperature T_i is below the 10th percentile curve as given by expression (4) are categorized as "CME". These are the green data points in Plate 1. High speed streams (HSS) tend to contain solar-wind plasma of ordinary temperature [*Richardson and Cane*, 1995], so to avoid contamination with high-velocity CME wind the requirement that the ion temperature T_i be above the median value as given by expressions (2) is imposed. Wind with v > 550 km/s with T_i above the median is designated as "HSS". These are the red points in Plate 1. The compressed wind of corotating interaction regions (CIR) has a temperature elevated above normal [cf. *Gosling et al.*, 1972; *Gosling and Pizzo*, 1999; *Elliott et al.*, 2005]. Times when T_i is above the 90th percentile as given by expressions (3) and v is in the range 425 km/s < v < 600 km/s are referred to as "compressed" (not as "CIR" as will be noted below). These are the blue data points in Plate 1. As a check on these three categorizations, the occurrence frequency versus solar cycle of the three types of wind categories was examined. It was found that "HSS" wind tends to occur during the declining phase of the cycle, as high speed streams are known to. It was found that "CME" wind tends to occur during solar maximum, as coronal mass ejections are known to. However, "compressed" wind was found to occur during solar maxima and during the declining phase. If this were CIRs, an occurrence primarily during declining phase would be expected. The fact that there is also solar-maximum occurrence probably means that "compressed" wind is CIR wind plus the compressed sheaths of CMEs.

A categorization of solar-minimum solar wind is obtained from the 1963-2001 OMNI2-AE1 data set by collected data taken at times when the sunspot number was 20 or below. This solar-minimum collection contains solar wind that is relatively free of transients such as CMEs, CIRs, and interplanetary shocks.

Two further categorizations of the solar wind come from catalogs of recurring high-speed-stream-driven storms (HSS

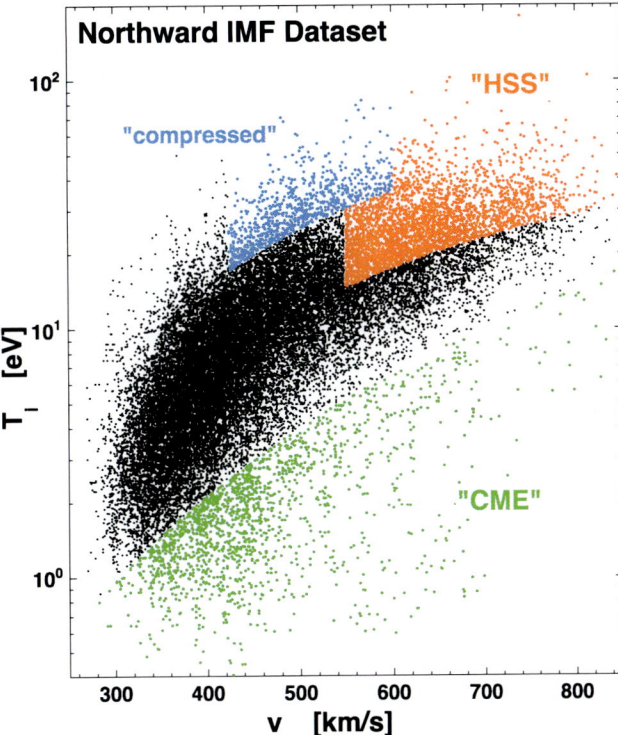

Plate 1. For all northward-IMF data in the OMNI2 data set, the ion temperature T_i of the solar wind is plotted as a function of the solar-wind speed v. Using the curves from Figure 11, the solar wind is approximately categorized into "high-speed streams" (HSS), "coronal mass ejections" (CME), and "compressed" regions. The rules for categorization are given in the text. Note that there is overlap between "HSS" and "compressed".

Storms) and of calms before recurring high-speed-stream-driven storms (Calms before Storms). These storms and calms were collected from three declining phases of the solar cycle (1973-1975, 1983-1984, and 1993-1995) by examining solar-wind and magnetospheric-activity measurements [*Borovsky and Steinberg*, 2006]. Each high-speed-stream-driven storm in the catalog is associated with a 27-day recurrent stream. The storm interval during a given high-speed stream is defined to commence when Kp first exceeds 4.5 and is defined to end when Kp drops below 4.5 for the last time during the stream. Each calm before the storm is an interval of extremely quiet geomagnetic activity that terminates less than 24 hours before a high-speed-stream-driven storm commences. Calms before the storm are presently being extensively analyzed [cf. *Borovsky and Steinberg*, 2006].

The freestream turbulence effect for these various categories of solar wind is analyzed in the two panels of Plate 2. Here only northward-IMF data with 1000 nT km/s < vB_z < 3000 nT km/s is used. AE (1-hour lagged) is plotted as a function of the eddy viscous force $nv^{5/2}\delta B/B_o$. In the left-hand panel 20-point running averages of AE_1 are plotted to show the trends in the data. In the right-hand panel least-squares linear-regression fits to the data are shown with the slopes of the lines indicated. As can be seen in both panels, the freestream turbulence effect operates to similar degrees in the various types of solar wind, with an apparent exception being the solar wind that drives calms before the storms. However, the anomalous result for calms before the storms is probably caused by a data-selection effect, which is explained as follows. Calms before the storm are by definition intervals that are restricted to Kp ≤ 1.3, among other conditions. That is, the calm-before-the-storm intervals are selected by restricting geomagnetic activity (the vertical direction in the plot), but not by restricting the value of $nv^{5/2}(\delta B/B_o)$ (the horizontal direction in the plot). This will result in a fit to the data that is lowered and has a lower slope. This fact is demonstrated in the right-hand panel of Plate 2 as the dashed line, where a gedankenexperiment is made taking all of the northward data and removing everything with Kp > 1.3. As can be seen by comparing the brown dashed line with the black line, the fit is lowered and has a lower slope.

6. SUMMARY

In this study, a large solar-wind/magnetosphere data set was assembled and used to improve and extend the *Borovsky and Funsten* [2003] cross-correlation study of the freestream turbulence effect in solar-wind/magnetosphere interactions. The 1963-2001 OMNI2-AE_1 data set contains almost 193,000 hours of measurements that can be cross correlated.

The larger data set was used to improve the statistics of the turbulence effect providing better fits of AE_1 to four measures of the solar-wind MHD turbulence amplitude: the magnetic-field fluctuation amplitude δB, the fractional amplitude of the magnetic-field fluctuations $\delta B/B_o$, the magnetic-field fluctuation amplitude in Alfven units δb, and the eddy viscous force of the solar wind on the magnetosphere $F_{visc} \propto nv^{5/2}\delta B/B_o$.

The large data set was separated according to the spacecraft contributing solar-wind measurements and the freestream turbulence effect was studied separately for each spacecraft. No significant difference was found between the various spacecraft data sets.

Long intervals (≥5 hours) of northward-IMF were separately examined. It was found that the turbulence effect in this long-interval data set is 33% weaker than it is in the full data set. Speculation as to why was given.

The freestream turbulence effect under southward IMF was made evident by removing variations of $-vB_z$ in the large data set and then cross correlating solar-wind turbulence parameters with geomagnetic activity. The strength of the turbulence effect under modest southward-IMF driving of the magnetosphere is the same as the turbulence-effect strength under northward IMF.

The ability to detect the freestream turbulence effect under large negative values of vB_z breaks down. It was suspected that this break down is a signal-to-noise issue. Noise in the AE index was investigated and it was found that the variability and uncertainty of AE was proportional to the magnitude of AE. The reason for the break down of detection was argued to as follows: (a) strong negative vB_z reconnection driving makes large AE, (b) the noise in AE being proportional to the magnitude of AE makes the noise in AE larger for larger negative vB_z, (c) the noise in AE then dominates the variance in AE owed to the turbulence effect, (d) the correlation coefficients between variations in AE and variations in the turbulence amplitude are then weak, and (e) weak correlations lead to weak slopes in fits to the data.

The freestream turbulence effect was examined through 3 solar cycles. There is some evidence for the turbulence effect being stronger during solar minima and weaker during solar maxima, but all indicators do not agree on this. There is evidence that the turbulence effect is stronger when the sunspot number is lower.

A scheme was developed to categorize three types of solar wind from the solar-wind temperature-versus-speed curve. The three types were high-speed streams, coronal mass ejections, and compressed solar wind.

The freestream turbulence effect was studied separately for the 3 types of solar wind from the categorizations, for solar-minimum solar wind, and for 2 types of solar wind from catalogs of events (high speed streams that drive storms and wind that drives geomagnetic calms before high-speed-stream-driven storms). No differences in the turbulence

Figure 2 shows comparisons of model results from RAM simulations a) including convective transport using *Weimer* [2001] electric potential model, b) adding radial diffusion from magnetic field fluctuations, and c) adding scattering by EMIC waves. The ring current energy increased due to magnetospheric convection during the storm main phase, maximized near Dst minimum, and decreased during the recovery phase. Radial diffusion enhanced the earthward transport of particles and increased the total H^+ energy only during the March 1998 storm; the continuous injection through radial diffusion during the long train of Alfvén waves thus contributed to its longer recovery phase. For both storms the dominant loss process for the ring current protons was charge exchange, except near minimum Dst when convective losses through the dayside magnetopause exceeded it for few hours. Losses due to Coulomb collisions were about two orders of magnitude smaller. Scattering of ring current protons by EMIC waves into the loss cone enhanced significantly the ion precipitation losses only during the May 1997 storm. The ion precipitation losses were of the order of charge exchange losses near minimum Dst and reduced by ~5% the globally averaged energy of ring current protons.

4. SUMMARY AND CONCLUSIONS

Previous modeling work has focused primarily on studying ring current dynamics during geomagnetic storms triggered by interplanetary manifestations of CME and magnetic clouds [*Jordanova et al.*, 1998; *Liemohn et al.*, 2002; *Chen et al.*, 2003]. These interplanetary configurations are the major drivers of storm activity during solar maximum and the elicited storms are usually large to intense [*Burlaga et al.*, 1987; *Tsurutani et al.*, 1988, 1992]. At solar minimum, however, the geomagnetic storms are mostly caused by the Earth passage of corotating high-speed solar wind streams and associated CIR and they are usually of moderate intensity [*Tsurutani et al.*, 1995; *Richardson et al.*, 2000]. The primary causes of magnetic storms are strong and persistent IMF oriented southward, though the shape, intensity, and duration of the southward IMF component vary significantly for different IP structures. Consequently, the interplanetary duskward electric field and the energy injection function vary [*Vieira et al.*, 2004].

In this paper we studied the CIR driven magnetic storm of 10 March 1998 and compared its behavior with the CME driven magnetic storm of 15 May 1997 of similar strength. Although both storms were caused by the strong negative B_z component of the IMF, there were significant differences regarding the coupling of the solar wind energy to the magnetosphere in terms of its timing and duration. The May 1997 storm started during Earth passage of the sheath region behind an interplanetary shock and continued during the passage of the magnetic cloud, while the March 1998 storm was caused by a B_z profile which fluctuated about zero on various time scales in Alfvén waves carried on the faster flows. The IP-parameterized potential model of *Weimer* [2001] thus predicted a strong enhancement and of longer duration during the May 1997 storm, overlapping with the period of enhanced plasma sheet density measured at geosynchronous orbit. The enhancement of the polar cap potential drop during March 1998 was smaller and occurred during the period of already decreased plasma sheet ion density.

We simulated the evolution of H^+, O^+, and He^+ ring current distributions with our coupled ring current-plasmasphere model (RAM) and found:

1. A large ring current injection due to magnetospheric convection was obtained, and in good agreement with Dst index during May 1997; however, the model underestimated the Dst minimum during the March 1998 storm.
2. An additional injection by radial diffusion near storm peak and during the recovery phase gave better agreement with Dst (although still underestimating its peak values) during March 1998; radial diffusion did not contribute during May 1997.
3. Both storms were dominated by H^+ ions. Ring current O^+ contributed less than 20% even near Dst minima, while He^+ had a minor contribution of several percents throughout the storms.
4. The simulated ion energy density showed large ring current asymmetry during the main phase of both storms and exhibited a typical maximum near dusk during March 1998, while the maximum was located in the postmidnight sector during May 1997 mainly due to a local enhancement of the plasma sheet ion density.
5. Electromagnetic ion cyclotron waves were excited by the anisotropic ring current distributions, with temporal and spatial evolution in reasonable agreement with EMIC wave satellite observations. These waves caused enhanced particle precipitation into the atmosphere and ~5% decrease in total energy during May 1997; their contribution to the energy loss during March 1998 was smaller.

Comparing our simulation results during both storms we conclude that a) ring current injection maximizes when increased magnetospheric convection coincides with periods of enhanced plasma sheet ion density; b) radial diffusion contributes significantly to ring current intensification during CIR driven storms; and c) losses due to EMIC waves scattering are larger during CME driven storms. These results suggest that the increased injection during the recovery phase combined with the smaller losses may cause the slower ring current decay during recurrent storms. More comparative simulation studies of magnetic storms of both solar origin will be performed to verify these conclusions. Another subject that will be investigated in future work is the electron contribution, which is expected to be larger during CIR driven storms and may result in better agreement with the *Dst* index.

Acknowledgments. This work was supported in part by NSF under grant ATM-0309585 and NASA under grant NAG5-13512. Special thanks are due to M. Thomsen and G. Reeves for help with LANL data processing. The *Dst* and *Kp* indices are provided by the World Data Center in Kyoto, Japan.

REFERENCES

Belian, R.D., G.R. Gisler, T. Cayton, and R. Christensen, High-Z energetic particles at geosynchronous orbit during the great solar proton event series of October 1989, *J. Geophys. Res.*, 97, 16897, 1992.

Blake, J.B., J.F. Fennell, L.M. Friesen, B.M. Johnson, W.A. Kolasinski, et al., CEPPAD: Comprehensive Energetic Particle and Pitch Angle Distribution Experiment on POLAR, *Space Sci. Rev.*, 71, 531, 1995.

Brautigam, D.H., and J.M. Albert, Radial diffusion analysis of outer radiation belt electrons during the October 9, 1990, magnetic storm, *J. Geophys. Res.*, 105, 291, 2000.

Burlaga, L.F., K.W. Behannon, and L.W. Klein, Compound streams, magnetic clouds, and major geomagnetic storms, *J. Geophys. Res.*, 92, 5725, 1987.

Chen M.W., M. Schulz, G. Lu, L.R. Lyons, Quasi-steady drift paths in a model magnetosphere with AMIE electric field: Implications for ring current formation, *J. Geophys. Res.*, 108 (A5), 1180, doi:10.1029/2002JA009584, 2003.

Crooker, N.U., and E.W. Cliver, Postmodern view of M-regions, *J. Geophys. Res.*, 99, 23383, 1994.

Daglis, I.A., E.T. Sarris, and B. Wilken, AMPTE/CCE CHEM observations of the energetic ion population at geosynchronous altitudes, *Ann. Geophys.*, 11, 685, 1993.

Fok, M.-C., T.E. Moore, G.R. Wilson, J.D. Perez, X.X. Zhang, P.C. Son Brandt, D.G. Mitchell, E.C. Roelof, J.-M. Jahn, C.J. Pollock, and R.A. Wolf, Global ENA image simulations, *Space Sci. Rev.*, 109, 77, 2003.

Gonzalez, W.D., J.A. Joselyn, Y. Kamide, H.W. Kroehl, G. Rostoker, B.T. Tsurutani, and V.M. Vasyliunas, What is a geomagnetic storm?, *J. Geophys. Res.*, 99, 5771, 1994.

Gonzalez, W.D., B.T. Tsurutani, and A.L.C. De Gonzalez, Interplanetary origin of geomagnetic storms, *Space Sci. Rev.*, 88, 529, 1999.

Jordanova, V.K., C.J. Farrugia, L. Janoo, J.M. Quinn, R.B. Torbert, K.W. Ogilvie, R.P. Lepping, J.T. Steinberg, D.J. McComas, R.D. Belian, October 1995 magnetic cloud and accompanying storm activity: Ring current evolution, *J. Geophys. Res.*, 103(A1), 79-92, 10.1029/97JA02367, 1998.

Jordanova, V.K., L.M. Kistler, C.J. Farrugia, and R.B. Torbert, Effects of inner magnetospheric convection on ring current dynamics: March 10-12, 1998, *J. Geophys. Res.*, 106, 29705, 2001a.

Jordanova, V.K., C.J. Farrugia, R.M. Thorne, G.V. Khazanov, G.D. Reeves, and M.F. Thomsen, Modeling ring current proton precipitation by electromagnetic ion cyclotron waves during the May 14-16, 1997, storm, *J. Geophys. Res.*, 106, 7, 2001b.

Jordanova, V.K., New insights on geomagnetic storms from model simulations using multi-spacecraft data, *Space Sci. Rev.*, 107(1-2), 157, 2003.

Jordanova, V.K., L.M. Kistler, M.F. Thomsen, and C.G. Mouikis, Effects of plasma sheet variability on the fast initial ring current decay, *Geophys. Res. Lett.*, 30, 1311, doi:10.1029/2002GL016576, 2003a.

Jordanova, V.K., A. Boonsiriseth, R.M. Thorne, and Y. Dotan, Ring current asymmetry from global simulations using a high-resolution electric field model, *J. Geophys. Res.*, 108(A12), 1443, doi:10.1029/2003JA009993, 2003b.

Jordanova, V.K., and Y. Miyoshi, Relativistic model of ring current and radiation belt ions and electrons: Initial results, *Geophys. Res. Lett.*, 32, L14104, doi:10.1029/2005GL023020, 2005.

Kistler, L.M., B. Klecker, V.K. Jordanova, E. Mobius, M.A. Popecki, et al., Testing electric field models using ring current ion energy spectra from the Equator-S ion composition (ESIC) instrument, *Ann. Geophys.*, 17, 1611, 1999.

Lanzerotti, L.J., Space weather effects on technologies, in *Space Weather, Geophys. Monogr. Ser.*, vol. 125, edited by P. Song et al., p. 11, AGU, Washington, DC, 2001.

Liemohn M.W., J.U. Kozyra, C.R. Clauer, G.V. Khazanov, and M.F. Thomsen, Adiabatic energization in the ring current and its relation to other source and loss terms, *J. Geophys. Res.*, 107 (A4), doi:10.1029/2001JA000243, 2002.

McComas, D.J., S.J. Bame, B.L. Barraclough, J.R. Donart, R.C. Elphic, J.T. Gosling, M.B. Moldwin, K.R. Moore, and M.F. Thomsen, Magnetospheric plasma analyzer: Initial three-spacecraft observations from geosynchronous orbit, *J. Geophys. Res.*, 98, 13453, 1993.

Mouikis C.G., L.M. Kistler, W. Baumjohann, E.J. Lund, A. Korth, et al., Equator-S observations of He^+ energization by EMIC waves in the dawnside equatorial magnetosphere, *Geophys. Res. Lett.*, 29, doi:10.1029/2001GL013899, 2002.

Rasmussen, C.E., S.M. Guiter, and S.G. Thomas, Two-dimensional model of the plasmasphere: refilling time constants, *Planet. Space Sci.*, 41, 35, 1993.

Richardson, I.G., E.W. Cliver, and H.V. Cane, Sources of geomagnetic activity over the solar cycle: Relative importance of coronal mass ejections, high-speed streams, and slow solar wind, *J. Geophys. Res.*, 105, 18203, 2000.

Schulz M., and L.J. Lanzerotti, Particle diffusion in the radiation belts, Springer-Verlag, Berlin, 1974.

Smith, E., J., and J.H. Wolf, Observations of interaction regions and corotating shocks between one and five AU: Pioneers 10 and 11, *Geophys. Res. Lett.*, 3, 137, 1976.

Tsurutani, B.T., W.D. Gonzalez, F. Tang, S.-I. Akasofu, and E.J. Smith, Origin of interplanetary southward magnetic fields responsible for major magnetic storms near solar maximum (1978-1979), *J. Geophys. Res.*, 93, 8519, 1988.

Tsurutani, B.T., W.D. Gonzalez, F. Tang, and Y.T. Lee, Great magnetic storms, *Geophys. Res. Lett.*, 19, 73, 1992.

Tsurutani, B.T., W.D. Gonzalez, A.L.C. Gonzalez, F. Tang, J.K. Arballo, M. Okada, Interplanetary origin of geomagnetic activity in the declining phase of the solar cycle, *J. Geophys. Res.*, 100(A11), 21717, 10.1029/95JA01476, 1995.

Vette, J.I., The NASA/National Space Science Data Center Trapped Radiation Environment Model Program (1964-1991), *Rep. NSSDC/WDC-A-RS 91-29*, Greenbelt, MD, 1991.

Vieira, L.E.A., W.D. Gonzalez, E. Echer, and B.T. Tsurutani, Storm-intensity criteria for several classes of the driving interplanetary structures, *Solar Phys.*, 223, 245, 2004.

Weimer, D.R., An improved model of ionospheric electric potentials including substorm perturbations and application to the Geospace Environment Modeling November 24, 1996, event, *J. Geophys. Res.*, 106, 407, 2001.

Young, D.T., H. Balsiger, and J. Geiss, Correlations of magnetospheric ion composition with geomagnetic and solar activity, *J. Geophys. Res.*, 87, 9077, 1982.

V.K. Jordanova, Space Science Center, 410 Morse Hall, University of New Hampshire, Durham, NH 03824, USA.

Ring Current Behavior Inferred From Ground Magnetic and Space Observations

F. Søraas[1], M. Sørbø[1], K. Aarsnes[1], and D.S. Evans[2]

The existence of an electric current encircling the Earth at a distance of several Earth radii was first predicted from ground based magnetic observations. These large variations in the Earth's magnetic field were called geomagnetic storms. The magnetic field at the Earth's surface exhibit an appreciable Magnetic Local Time (MLT) dependence in the initial and main phase of the storm. The field depression is very asymmetric, with the largest depression in the evening to midnight MLT sector. During such storms a well defined Storm Time Equatorial Belt (STEB) of Energetic Neutral Atoms (ENA) and ions is found to exist at low altitudes around the geomagnetic equator. Most of the particles measured at the equator by the vertical viewing detector on the NOAA satellites at an altitude of 800 km will be ENA. Ring Current (RC) asymmetry and symmetry inferred from the STEB are in accordance with results from ground based magnetic observations. The STEB first appears in the midnight/evening sector and then it appears in the morning sector largely consistent with the expected drift of RC ions. The local magnetic observations, on the other hand, show signatures both in the dusk and morning sectors simultaneously during the storm main phases. One possible explanation is that the large partial RC in the evening sector causes significant magnetic depression in the morning sector during deep injection ($L \sim 2.0$) events. STEB observations show that large convection fields can prevent the RC ions from passing local noon in their drift motion.

1. INTRODUCTION

The build up of the ring current is commonly attributed to electric fields which inject and convect the electrons and ions from the tail plasma sheet towards the Earth. The electrons drift towards the morning sector and the ions towards the evening sector thus creating the RC (ring current). The main physical cause for the ground magnetic perturbations at low latitudes, seen for instance in the Dst-index, is the variability of the RC. The existence of a RC encircling the Earth was first proposed by Carl Størmer [*Størmer*, 1910] in order to overcome a discrepancy in his theory, which predicted the aurora far closer to the magnetic pole than where it was observed. Adolf Schmidt suggested that a ring current was also the cause of the main phase of magnetic storms [*Schmidt*, 1924] and *Chapman* [1932] supported this view, but the existence of the ring current was first confirmed when satellites made in situ observations [*Frank*, 1967]. For a review of the RC see *Tsurutani and Gonzalez* [1997] and *McPherron* [1997].

The dynamics of the RC build up, decay and spatial location are revealed both in particles precipitating in the auroral zone [*Søraas et al.*, 2002] and in the behavior of the STEB (Storm Time Equatorial Belt) observed at low altitudes in

[1]Department of Physics and Technology, University of Bergen, Bergen, Norway
[2]NOAA Space Environment Center, Boulder, Colorado, USA

Recurrent Magnetic Storms: Corotating Solar Wind Streams
Geophysical Monograph Series 167
Copyright 2006 by the American Geophysical Union.
10.1029/167GM09

the equatorial region [Søraas et al., 2003]. The first low altitude observations of energetic charged particles at equatorial latitudes were obtained by the German research satellite Azur in 1969 and 1970 [Moritz, 1972; Hovestadt et al., 1972]. Moritz suggested that the source region of the low altitude particles detected by Azur were reionized ENAs. In the later work of Søraas et al. [2003] it was shown that the STEB can reveal the symmetric and asymmetric phase of the RC. It will therefore be of interest to see if magnetic field observations at the Earth's surface and observations of the STEB give a consistent picture of the RC development. The STEB observations also reveal how strong convection fields prevent the ions from drifting past noon.

Sørbø et al., [2006] have shown that the STEB can have a latitudinal extent of ±40° spanning the geomagnetic equator and that the STEB is broadest in latitudinal extent in the main phase of the storm. Away from the equator the flux enhancements are, however, short lived.

2. INSTRUMENTATION

The present study uses observations from the MEPED instrument on board the NOAA series of polar orbiting satellites. Protons are measured in three energy channels ranging from 30 to 800 keV, and electrons in integral channels above 30, 100 and 300 keV. Observations of both species are made at 10 and 80 degrees to the local vertical. The geometric factor for the proton detectors is $9.5 \cdot 10^{-3} cm^2 sr$. Notice that the vertical detector measures precipitating protons in the auroral zone, while it observes trapped or neutral particles that are moving vertically and thus transverse to the magnetic field in the equatorial region. The MEPED instrument can not distinguish between the different ions. We therefore use the term protons in this study. The detectors are, however equally responsive to both charged and neutral hydrogen.

The orbits of the spacecraft are circular at an altitude of about 800 km. A full description of the satellites and their instrumentation are given by Evans and Greer [2000]. The approximate local time ascending node equatorial crossing (LTAN) for the satellites are: NOAA 15 at 19 MLT, NOAA 16 at 14 MLT and NOAA 17 at 22 MLT. Combined observations from these satellite crossings give a fairly good MLT/ILAT coverage of the particles observed at 800 km altitude.

3. OBSERVATIONS

Two major geomagnetic storms will be considered.

3.1. The March 31, 2001, Storm

Figure 1 shows the solar wind parameters as observed by the ACE satellite during the days of March 30 to April 1 2001. From the top: the magnetic field $|B|$, B_x, B_y and B_z components of the IMF in GSM coordinates, followed by the V_x component of the solar wind. The bottom panel shows the Dst.

The storm started early on March 31 around 03 UT when B_z first became negative. At 08 UT B_z became positive and remained positive for several hours. Around 14 UT there was a second B_z negative phase that lasted to 22 UT. The storm thus exhibited two periods with B_z negative, the first to $-40 nT$ and the second to $-30 nT$. During these two time intervals there were large convection fields, giving rise to intense particle precipitation at high latitudes and injections into the RC.

Figure 2 shows the proton precipitation observed by NOAA-15 at high latitudes during this storm. The three top panels show the night side proton precipitation in the southern hemisphere for three energy channels, while the three bottom panels provide the same information for the day side. The Dst-index is given in the middle panel. Notice that particle enhancements in night sector is simultaneous with the decrease in Dst, but slightly delayed on the day side. Søraas et al. [2002] have shown the isotropic precipitation at high latitudes is well correlated with injections of protons into the RC. The particle precipitation is less intense on the day side as evident from the three bottom panels. The particle injections are seen in all three energy channels. While the largest enhancement is in the 30 to 80 keV channel, typical of RC energies, particles with several hundred of keV energy are also injected into the RC during this storm. On the night side the proton precipitation was observed down to ILAT 49° showing that the inner edge of the RC penetrates to $L = 2.3$.

3.1.1. STEB observations. Figure 3 shows schematically a NOAA satellite transit from north to south across the equator. The vertical detector views radially away from Earth and the horizontal detector views antiparallel to the satellite velocity. This figure also demonstrates how the protons in the RC can be demagnetized through charge exchange. They leave the magnetic field as energetic neutral atoms (ENA), move towards equator and are detected at low altitudes.

Figure 4 shows NOAA 15 proton/ENA data obtained on a pass from pole to pole. The observations encompass from the southern auroral zone across the equator and into the northern auroral zone. Data from the vertical detector in the energy range 30 to 80 keV is shown. The top panel refer to non storm condition. The high proton intensity in the auroral zone is seen, but the intensity in the equatorial region is very low. The bottom panel shows data from the early recovery phase of the March 31, 2001 storm. Here one sees that the auroral zone precipitation has expanded equator-ward and that the intensity at the magnetic equator has increased by a

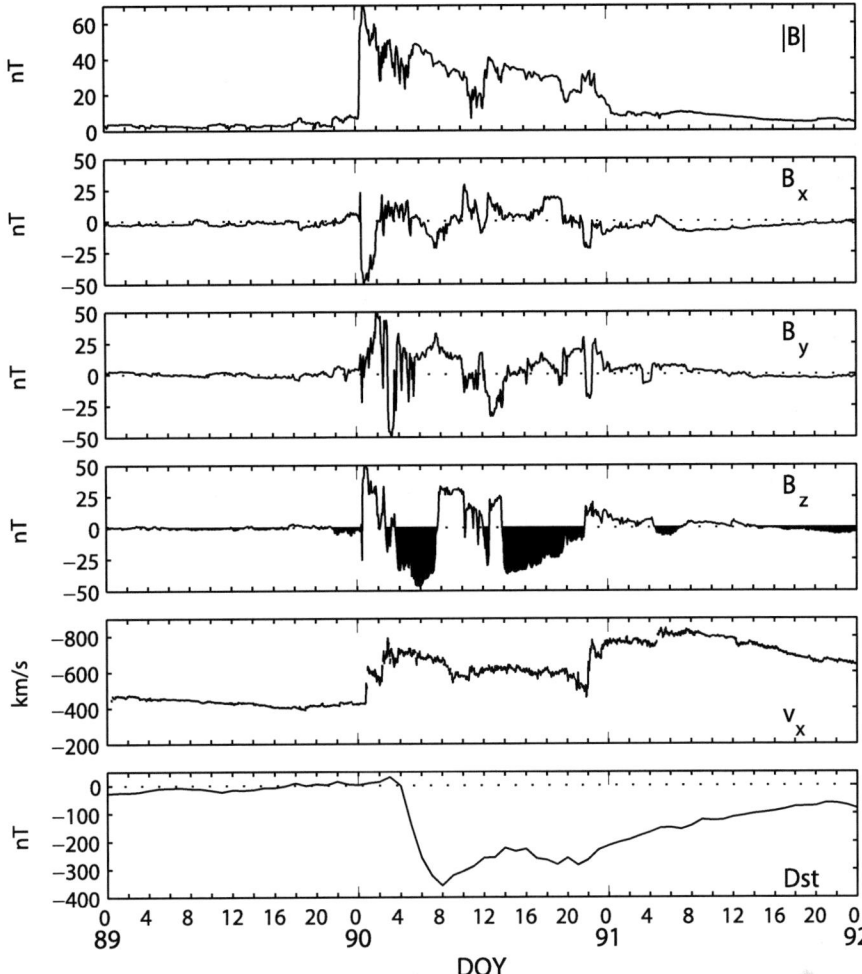

Figure 1. The solar wind parameters as observed by the ACE satellite during the days 30 March to April 1 2001 are shown. From the top the different panels show: the IMF magnetic field $|B|$, B_x, B_y and B_z components of the IMF, the V_x component of the solar wind and the Dst index.

factor of around 100. This region with enhanced particle intensity observed near the equator during geomagnetic storms is the STEB *Søraas et al.* [2003].

3.1.2. RC asymmetry and symmetry inferred from magnetic field and STEB observations. Mid latitude magnetic observations have been used by many investigators as a tool for the study of magnetospheric substorms and RC formation [*Clauer and McPherron*, 1974]. Traditionally the local time dependence of the RC has been determined from observations of the magnetic field at low latitude stations. In the initial phase of the storm the ground magnetic measurements show that the field depression is largest in the midnight to dusk sector. From such measurements the picture has emerged that the RC starts out being asymmetric in the storm main phase and then develops into a symmetric RC independent of MLT.

In order to determine the magnetic field depression at the Earth's surface as a function of MLT, 6 magnetic stations located at near equatorial latitudes, Bangui (long. 18.6), Addis Ababa (long. 38.8), Apia (long. 182.2), Guam (long. 144.9), Kouru (long. 307.3), and Mbour (long. 343.0) have been used. A method similar to the one developed by *Clauer and McPherron* [1974] has been used. The difference between disturbed and quiet days defines the magnetic field depression. Figure 5 show the magnetic field depression for these stations during March 31 through April 1. These data provide the profile of the disturbance field as a function of local time around the world. A vertical line indicates local midnight for each station.

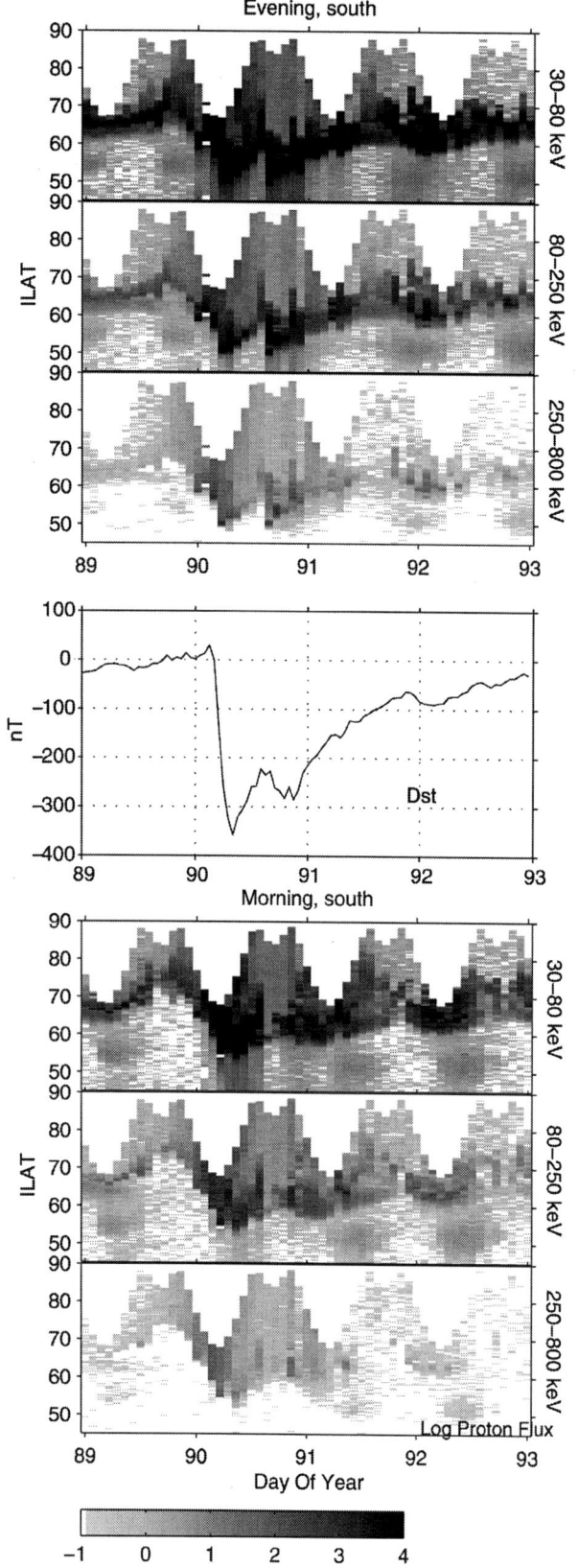

From the magnetic field depression at the different stations it can be seen that the first injection takes place at 08 UT on March 31. This injection gives rise to the largest field depression at Apia (MLT 23), Guam (MLT 21) and Kouru (MLT 03). These stations span local midnight and indicate the maximum magnetic field depression occurred slightly after midnight at around 01 MLT.

The second injection, at 18 UT on March 31, produced the maximum magnetic field depression at Kouru (MLT 15), Mbour (MLT 18) and Bangui (MLT 20) and so indicated a maximum magnetic field depression located towards the evening side.

These surface magnetic field observations have been interpolated in MLT and UT to produce the contour plot of magnetic field depression that is shown in the top panel of Figure 6. In this presentation both the temporal (in UT) and spatial (in MLT) variation of the disturbance field during the March 31 to April 1 storm are shown. While a limited number of stations can provide only a crude indication of the field depression it is enough to expose the main features of the global magnetic field depression.

The STEB observations for four LTs (shown in Figure 7) have been interpolated, in a manner similar used for estimating the magnetic field depression shown in the second panel of Figure 6. This plot can be directly compared with the magnetic disturbance contour plot in the top panel. The comparison shows that during the first injection the STEB is concentrated slightly before midnight (23 MLT) and it extends over a fairly wide MLT region reaching towards dusk. The magnetic disturbance field also exhibited a depression at this MLT region, indicating a strong RC in the evening sector.

The second injection into the RC was observed by the STEB to have its maximum around 18 MLT, which is slightly more towards dusk when compared with the magnetic field depression observed at the ground. Both observations, however, indicate a maximum of the RC in the midnight/dusk sector. Later in the storm both the STEB and the magnetic field are largely independent of MLT indicating a symmetric ring current.

Thus the asymmetric and the symmetric phases of the ring current inferred from both the STEB and the magnetic field at the Earth's surface agree well. The ENA observations at the magnetic equator give a picture of the RC evolution in accordance with the traditional method that identifies the asymmetric and the symmetric phase of the RC by means of magnetic field observations.

Figure 2. The three top panels show the intensity of protons in the evening local time sector from March 31 to April 1 in 2001. Each NOAA 15 pass is plotted vs. ILAT and UT. The observed proton flux is gray scale-coded. The three bottom panels gives similar information related to the day side. The middle panel gives the Dst-index.

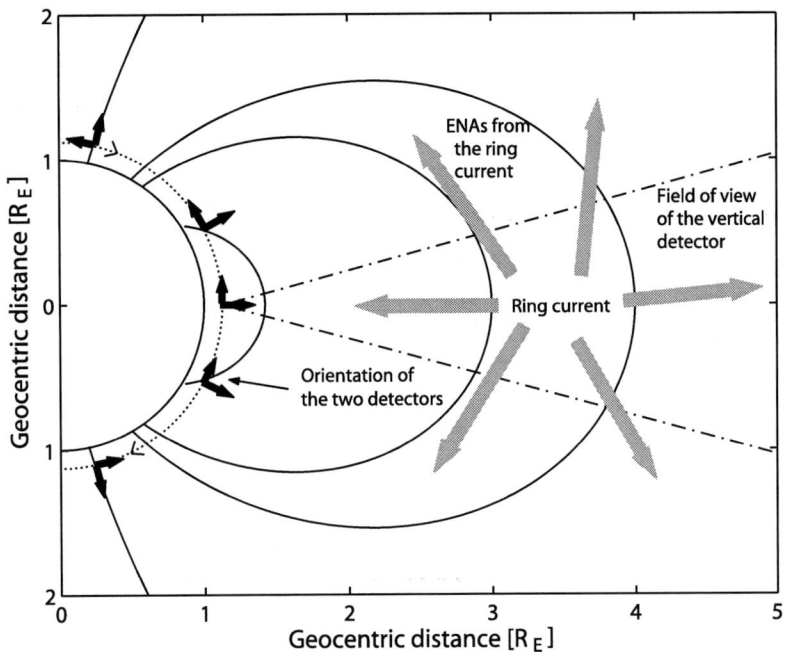

Figure 3. The looking direction for of the horizontal and the vertical detector on a passe from north to south. The horizontal detector looks in the anti-velocity direction. When the RC ions charge exchange they spread in all direction.

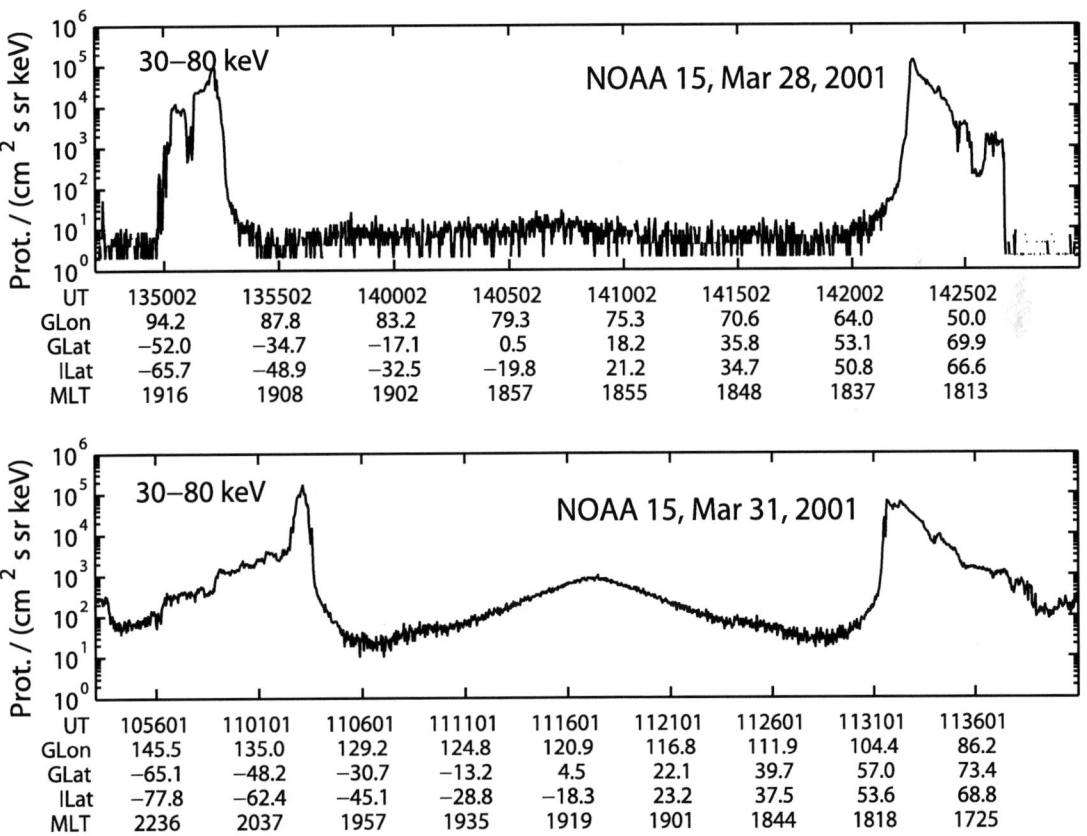

Figure 4. The NOAA 15 proton/ENA data obtained on a pass from pole to pole. Data from the vertical detector in the energy band 30-80 keV are shown. The upper panel refer to non-storm condition, while in the bottom panel data from the March 31 2001 storm are exhibited.

90 RING CURRENT BEHAVIOR INFERRED FROM GROUND MAGNETIC AND SPACE OBSERVATIONS

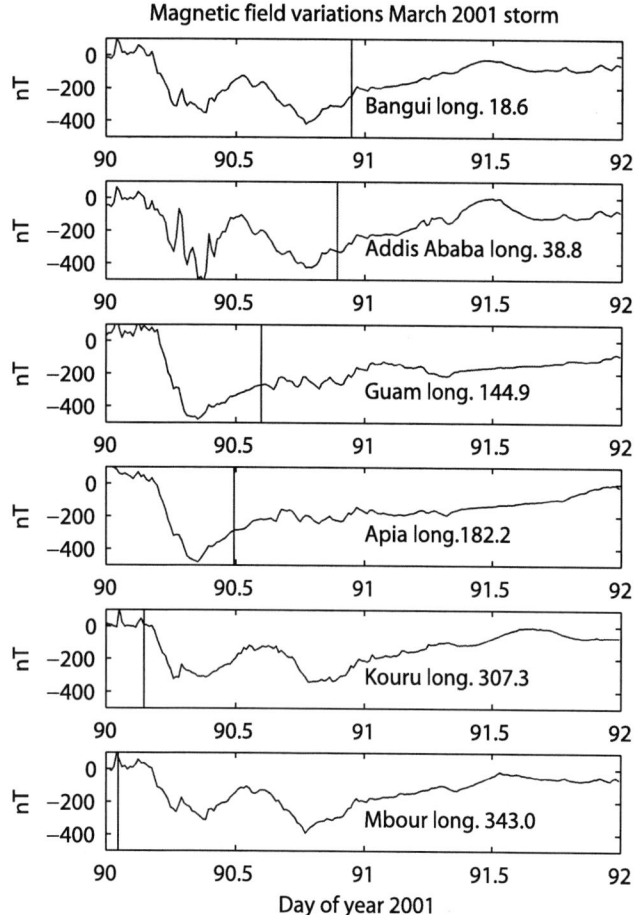

Figure 5. The magnetic field disturbance as observed by six equatorial stations spread in longitude around the world.

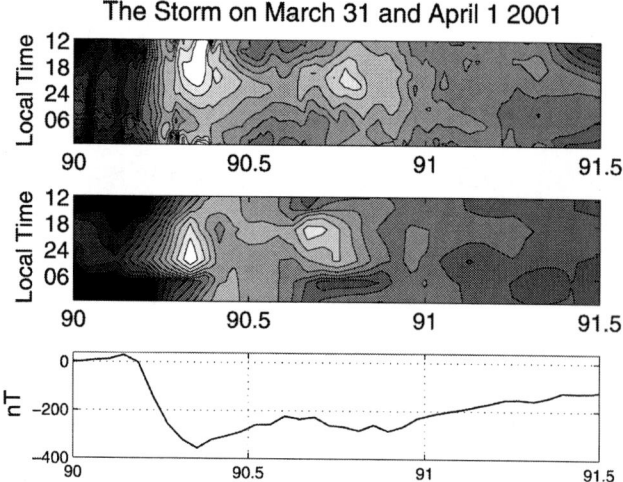

Figure 6. The top panel show a contour plot of the magnetic field variations at the Earths surface in LT/UT. The second panel shows a similar plot for the STEB variations in LT and UT. The bottom panel shows the Dst.

3.1.3. Different views. In order to make a more accurate comparison between the magnetic field observations and the STEB, the STEB observations at the four LTs are compared with the magnetic field observations at the same LTs. The magnetic field at these LTs are obtained from the contour plot shown in the top panel of Figure 6. From this plot the time variations of the magnetic field at the Earth's surface referring to a fixed LT can be obtained. In Figure 7 the STEB observations are shown by circles with a fully drawn line between them. The STEB intensity is given in $ENA/cm^2 ssrkeV$. The magnetic field observations are given by a dotted line and are multiplied by a factor −4 in order to fit the observations into the same diagram as the STEB.

In the figure it is seen that the magnetic field depression starts at the same time independent of LT. The top panel refer to LT = 02 and it is seen that both the magnetic field and the STEB changes simultaneously at the beginning of the storm main phase. This is seen from the Dst in the bottom panel.

At this local time the STEB and the magnetic field exhibits the same time behavior. The two injections are clearly seen in both the magnetic field and in the STEB and they exhibit the same time behavior also in the recovery phase. The same type of behavior is also seen at LT 19 and 14, the two quantities track each other well both in the injection events and in the recovery phase.

At LT = 07 things look different. The peak in the STEB related to the first injection is delayed by 4 to 5 hours compared with the first injection peak at LTs 02 and 19. This show that the magnetic field on the morning side experience a large depression several hours before the ions in the RC have reached the morning side. In the early main phase of the storm the magnetic field depression on the morning side is thus not due to the ions in the RC as they have not yet reached this local time sector.

It is interesting to notice that the time behavior of the magnetic field depression is generally the same at all local times. The depression starts at the same time, the two injection events are seen more or less simultaneous at all local times and in the recovery phase the magnetic field at the stations recover in a similar way back to its quiet level. During the storm main phase the magnetic field depression is noticeable larger, by a factor of 2, in the midnight to evening local time sector than in the 07 local time sector.

3.2. The November 20, 2003, Storm

The SW data and the Dst index for the great storm of 20 November 2003 are shown in Figure 8. The storm was associated with a large Coronal Mass Ejection (CME).

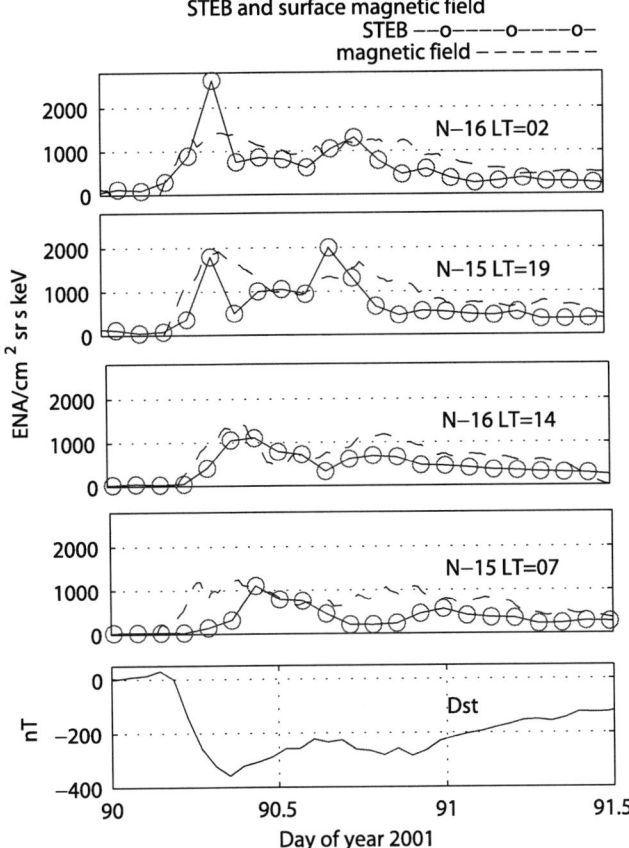

Figure 7. The STEB observed at the magnetic equator at four LTs (02, 19, 14 and 07) during the storm is shown by a fully drawn line with a circle. The dotted line gives the magnetic field variations (multiplied by a factor −4) at the surface of the Earth referring to the same LTs as the STEB observations. The magnetic field is inverted to ease the comparison with the STEB. The bottom panel gives the Dst-index.

The activity started early on November 20 when B_z began gradually to go negative together with an increase in the scalar magnetic field. This led to an energy injection into the RC that is reflected in the gradual decrease in the Dst-index. During this time period the SW velocity stayed constant at around 450 km/s.

Around 10 UT the Interplanetary Coronal Mass Ejection (ICME) reached the ACE-satellite. There was an abrupt increase in the total magnetic field, mainly in the B_y component. Somewhat later B_z began to go strongly negative, reaching −50 nT at 15 UT, while the B_y component exhibited a gradual decline becoming negative at around 17 UT. The absolute value of B maximized around 15 UT at nearly 60 nT. The magnetic field rotates counter clockwise in direction when observed from the Sun. During the event there were only small variations in B_x fluctuating around zero. There is about a 35 minutes delay between the ACE observations and the arrival of the disturbance at Earth. The data presented has not been corrected for this time shift.

The energy injection into the RC, seen from Dst in the bottom panel, started around 11 UT when B_z first turned sharply negative. The Dst reaches its minimum value of −472 nT at 19 UT and the storm enters its recovery phase.

Even though B_z is as negative as −15 nT the RC started to recover around 19 UT, based upon Dst. With such a large negative value for B_z appreciable merging on the magnetopause is expected, but, even so, the recovery in Dst indicates that the loss rate from the RC is larger. This is probably due to the large charge exchange losses occurring for this storm because the RC penetrated deep into the magnetosphere, down to $L = 2.0$ where the geocorona density is high. This would lead to high ENA production and a fast RC decay at its inner edge. Due to the still high convection field (see fourth panel in Figure 9) there will also be convection losses through the dayside magnetopause. The energy injection into the RC would not be large enough to balance the losses.

3.2.1. RC injection and convection losses. The top panel of Figure 9 shows, in the form of a contour plot, the MLT/UT dependence of the surface magnetic field depression during the November 20-21 storm. The plot is constructed by linear interpolation of the magnetic data from the same geomagnetic stations as used in Figure 5. In this presentation both the temporal (in UT) and spatial (in MLT) variation of the disturbance field during the November storm are shown. It is seen that the magnetic field depression is largest around 19:30 UT and MLT 19. The field depression is covering a large region in MLT.

The contour plot of the STEB, based on six equatorial crossings is shown in the second panel of the figure and it is most intense slightly before local midnight (23 MLT) around 19 UT on November 20. This is somewhat before the minimum in Dst (bottom panel) but taking into account the time resolution of the measurements, about 1.5 hours, one can conclude that the maximum in STEB is coincident with the minimum in Dst when the RC has maximum kinetic energy.

The third panel shows an MLT vs UT plot of the total power deposition into the ionosphere from precipitating protons in the energy range 30 to 250 keV integrated over the ILAT range between 42° to 80°. It is seen that the particle injection maximized when the Dst had its most negative slope which took place two hours before the maximum ENA production.

Panel 4 shows the E_y convection field calculated from the solar wind speed v_x and B_z, and it is clear that the proton power into the ionosphere is most intense shortly after the convection field reached its maximum value (35 mV/m). When the convection field is high the RC is not able to pass

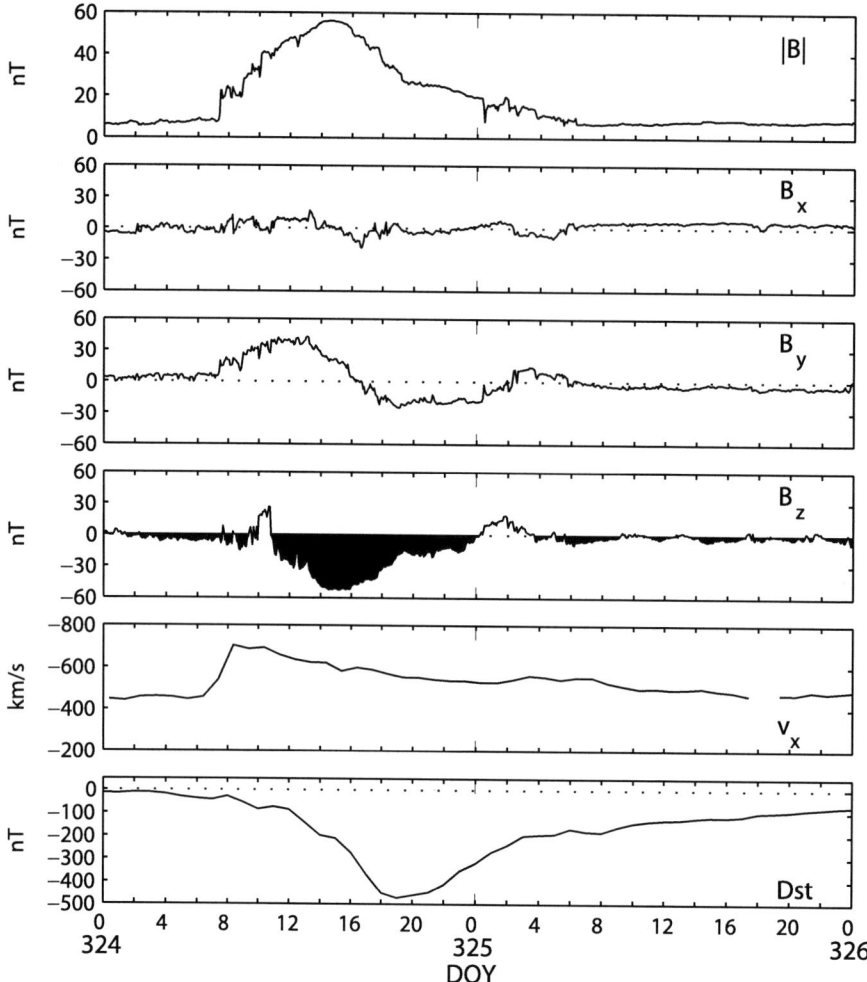

Figure 8. The solar wind parameters as observed by the ACE satellite during the days 20 and 21 November 2003 are shown. At the top: the IMF magnetic field $|B|$, B_x, B_y and B_z components of the IMF, and then the V_x component of the solar wind velocity. At the bottom the Dst-index.

local noon. First when E_y is below $10\,mV/m$ the RC is able to pass noon as seen in the STEB observations around 21 UT (second panel). By the time the convection field decreased to around $10\,mV/m$ the STEB had extended to the noon sector, indicating that the RC ions have drifted past noon. It may be concluded that before 19 UT, when E_y is above $10\,mV/m$, the RC particles were prevented from reaching noon because the large convection field resulted in the particles being lost through the day side magnetopause.

A 50 keV proton at L = 2.5 will take about 50 minutes to drift 3 hours in MLT. The MLT/UT evolution of the STEB shown in the second panel of Figure 9 confirm that the appearance of increased fluxes at 19 MLT and 14 MLT are in agreement with the expected drift times from the midnight sector. However, it required an additional 3 hours after the STEB was observed at MLT 14 for ENA to appear at 11 MLT. This is about 4 times longer than the expected drift time between these locations. This provides additional support for the view that the RC protons are prevented from drifting to noon because of the large convection field.

Figure 10 displays the STEB, as observed at the magnetic equator, during this storm. The STEB observations are shown with a fully drawn line with circles. In this storm, data from three NOAA satellites are available and so the STEB can be observed at six different MLTs. The most intense STEB is observed by the NOAA 17 satellite at 23 MLT. In the 30 to 80 keV channel the maximum intensity reaches $1.2 \cdot 10^4\,particles/cm^2ssrkeV$ which is a factor of 6 larger than observed during the March 2001 storm. The second largest intensity is observed by NOAA 16 located at 03 MLT. The time history at these two MLTs are similar suggesting that the injection took place around local midnight.

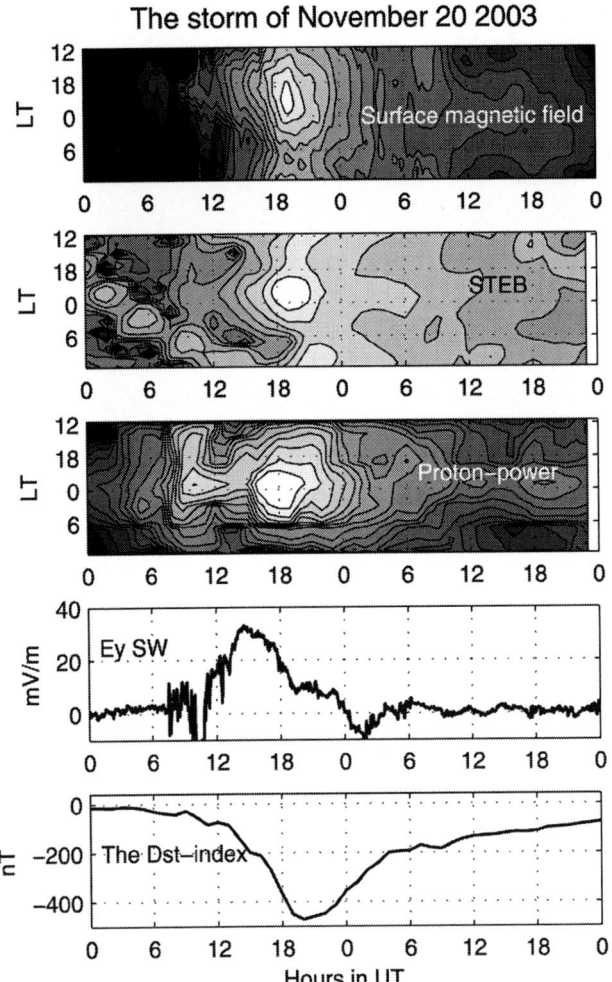

Figure 9. The top panel show a contour plot of the magnetic field variations at the Earths surface in LT/UT. The second and third panel show similar plots for the STEB variations and the proton power of protons into the atmosphere. Data from six equatorial stations and six equatorial satellite crossings are used. The fourth panel gives the convection field calculated from SW parameters, and the bottom panel gives the Dst-index.

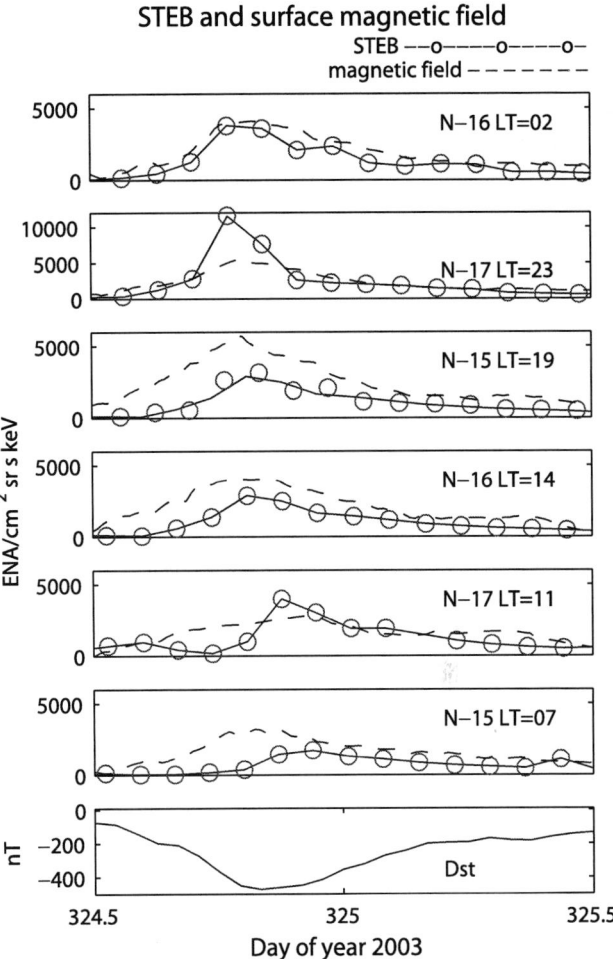

Figure 10. The STEB observed at the magnetic equator at six LTs (03, 23, 19, 14, 11 and 07) during the storm are shown by a fully drawn line with a circle. The dotted line gives the magnetic field variations (multiplied by a factor −8) at the surface of the Earth referring to the same LTs as the STEB observations. The magnetic field is inverted to ease the comparison with the STEB. The bottom panel gives the Dst-index.

In order to make a more accurate comparison between the magnetic field observations and the STEB, the STEB observations at the six LTs are compared with the magnetic field observations at the same LTs. The magnetic field at these LTs are obtained from the contour plot shown in the top panel of Figure 9. From this plot the time variations of the magnetic field at the Earth's surface referring to a fixed LT can be obtained. The magnetic field observations are given by a dotted line.

The top panel refer to LT = 02 and it is seen that both the magnetic field and the STEB changes simultaneously at the beginning of the storm main phase. This is seen from the Dst in the bottom panel. At this local time the STEB and the magnetic field exhibit the same time behavior. The injection is clearly seen in both the magnetic field and in the STEB and in the recovery phase they exhibit the same time behavior.

The same type of behavior is also seen at LT 23, 19 and 14. The two quantities exhibits the same time evolution, both during the injection and in the recovery phase. At LT = 11 and 07 things are different. The peak in the STEB is delayed with respect to the magnetic field. This show that the magnetic field on the morning side experience a large depression several hours before the ions in the RC has reached the morning side.

In the early main phase of the storm the magnetic field depression on the morning side is thus not due to the ions in the RC as they have not yet reached the morning sector. The magnetic field and the STEB are thus related to each other in a similar way during both the big storms taking place in March 2001 and November 2003.

4. SUMMARY AND DISCUSSION

The behavior of the RC during two large geomagnetic storms are studied using ground magnetic observations and particle observations from low altitude polar orbiting satellites. The particles are ENA arriving directly from the RC above the magnetic equator.

1. The ground based magnetic observations support the conventional view of the RC: it is asymmetric during the initial and main phases of the storm and a exhibits more symmetric structure during the recovery phase. This traditional view of the RC is supported by observations of ENA at the equatorial latitudes at several local times.
2. There is, however, a difference between the view supported by the magnetic field observations and the particle observations. The magnetic observations show the storms to be worldwide, displaying essential the same signature all around the equator from the start of the main phase. During the storm main phase the magnetic disturbance was about a factor of two larger in the evening sector than in the morning sector (see Figures 7 and 10).

Contrary to this the STEB appears first in the midnight to evening sector and then it appears delayed in the morning sector. The STEB shows that the injection region of the RC particles is in the midnight to dusk MLT sector and they then drift towards noon.

3. The magnetic field depression in the morning sector is thus not caused by RC ions in that sector, but must be due to other current systems. Possibly the magnetic field depression in the morning sector could partly be due to the large partial RC in the evening sector. An estimate of this follows:

During these two storms the isotropic proton precipitation penetrated to about L = 2. If the asymmetric RC is approximated by a line current flowing at L = 2 this current will give rise to a magnetic field depression on the evening side of the Earth. On the other side of the Earth, $4R_e$, away the magnetic field would be reduced to the half. This is in rough agreement with the observations, neglecting any influence of the Earth on the magnetic field distribution.

On the other hand RC electrons drifting eastward from the night sector will contribute to the magnetic field depression in the morning sector, but due to their low energy density it is unlikely that they can account for the observed field depression. Few RC studies have included the effects of electrons, but *Frank* [1967] has shown that they can contribute to ~25% to the stormtime RC energy content. To reveal the effect of the electrons further investigations are needed.

4. During the 2001 storm there were two particle injections concurrent with the two B_z negative events in the IMF. The first injection was concentrated slightly after midnight while the second one was located earlier in MLT around 20. This is evident both in the magnetic field observations, and the observations of the STEB. The STEB observations allow one to further determine the injection region for the RC and its extension from midnight towards the evening MLT sector. The temporal development of the STEB as a function of MLT is consistent with the expected gradient drift of the RC protons.
5. The minimum value of Dst during the 2001 storm was −387 nT and −472 nT during the 2003 storm. The RC injection reached into a radial distance of L = 2.3 during the March 2001 storm and to L = 2.0 for the November 2003 storm. The intensity of the ENA recorded at the equator was more than 6 times larger during the 2003 storm compared with the 2001 storm (12000 vs 2000) *ENA/cm²ssrkeV*. The proton intensity at the inner edge of the RC was about equal for the two storms, $3 \cdot 10^5$ *protons/cm²ssrkeV* in the energy range 30 to 80 keV.

The larger ENA intensity in the 2003 storm is most likely caused by the deeper RC injection towards the Earth during this storm. A movement of the inner edge of the RC from 2.3 to 2.0 would mean the RC particles would encounter a geocorona density higher by a factor of 2 with consequent increased ENA production. The latitudinal width of the isotropic proton precipitation region, a proxy for radial extent of the RC, was also larger in the 2003 storm than in the 2001 storm. This also would lead to a larger ENA production for the 2003 storm, but detailed modeling of ENA production is necessary to account for the differences in ENA intensities between these two storms.

6. It is of further interest to note that the 2003 storm began to recover even though the B_z component of the SW was −15 nT that would indicate substantial energy injection into the RC, but apparently not enough to balance the charge exchange and convection losses.
7. The maximum proton power injection rate into the auroral zone occurred when the time derivative of the Dst is the largest. That is, the maximum change in Dst is coincident with maximum energy injection into the RC. The STEB, on the other hand, has its maximum intensity at the minimum Dst. This is as expected because ENA production relates to the RC intensity and its spatial extent rather than RC injection rate.
8. The observations of the STEB shows that the RC particles are inhibited from drifting past noon when there was a large convection field transporting the particles out through the dayside magnetopause. This is in accordance with *Liemohn et al.* [1999] who showed that convective drift loss

out the dayside magnetopause is the dominant process in removing RC particles during the initial recovery.

Observations of STEB at several MLTs provided unique opportunities to follow the build up and decay of the RC. The STEB is a good indicator for the RC because it only depends on the flux of ions in the RC and the density of the geocorona. It is a local measurement only depending on the RC. The magnetic field measured at equatorial stations are on the other hand influenced by many current systems in the magnetosphere.

Using several satellites it is possible to get a fair time resolution of the evolution of the STEB and thus of the parent ion RC. A low altitude, polar orbiting satellite is mostly below the radiation belt and in a low background radiation environment. The satellite can thus observe the ENA from the RC with minimum disturbance. Combining these observations with ENA, ion and electron measurements outside and within the RC could provide a unique data set with the potential to gain new information about RC processes.

Acknowledgments. The authors thank the Research Council of Norway, projets 147652/V30 and 165557 (M.Sørbø) for financial support and the referees for useful comments.

REFERENCES

Chapman, S., and V.C.A. Ferraro, A theory of magnetic storms, *Terr. Mag.*, 36, 77-97, 1931.

Clauer, C.R., and R.L. McPherron, Mapping the local time development 1of magnetospheric substorms using mid-latitude magnetic observations, *J. Geophys. Res.*, 79(19), 2811, 1974.

Evans, D.S., and M.S. Greer, Polar orbiting environmental satellite space environment monitor - 2: Instrument descriptions and archive data documentation, *NOAA Technical Memorandum OAR SEC-93*, Boulder, Colorado, 2000.

Frank, L.A., On the extraterrestrial ring current during geomagnetic storms, *J. Geophys. Res.*, 72, 3753, 1967.

Hovestadt, D., B. Hausler, and M. Scholer, Observations of energetic particles at very low altitudes near the geomagnetic equator, *Phys. Rev. Lett.*, 28, 1340, 1972.

Liemohn, M.W., J.U. Kozyra, V.K. Jordanova, G.V. Khazanov, M.F. Thomsen, and T.E. Cayton, Analysis of early phase ring current recovery machanisms during magnetic storms, *J. Geophys. Res.*, 26, 2845, 1999.

McPherron, R.L., The Role of Substorms in the generation of Magnetic Storms, in *Magnetic Storms*, edited by B.T. Tsurutani *et al.*, Geophysical Monograph 98, *Amer. Geophy. Un.*, 1997.

Moritz J., Energetic protons at low equatorial altitudes: A newly discovered radiation belt phenomenon and its explanation, *Z. Geophys.*, 38, 701, 1972.

Schmidt, A., Das erdmagnetische Aussenfeld, *Z. Geophys.*, I, 3, 1924.

Søraas, F., K. Aarsnes, K. Oksavik, and D.S. Evans, Ring current intensity estimated from low-altitude proton observations, *J. Geophys. Res.*, 107, A7, doi:10.1029/2001JA000123, 2002.

Søraas, F., K. Oksavik, K. Aarsnes, D.S. Evans, and M.S. Greer, Storm time equatorial belt - an "image" of RC behavior, *Geophys. Res. Lett.*, 30(2),1052, doi:10.1029/2002GL015636, 2003.

Sørbø, M., F. Søraas, K. Aarsnes, K. Oksavik, and D.S. Evans, Latitude distribution of vertically precipitating energetic neutral atoms observed at low altitudes, *Geophys. Res. Lett.*, 33, L06108, doi:10.1029/2005GL025240, 2006.

Størmer, C., Sur la situation de la zone de frequence maximum des aurores boreales la theorie corpusculaire, *C.R. Acad. Sci.*, 151736, 1910.

Tsurutani, B.T., and W.D. Gonzalez, The interplanetary cause of magnetic storms: A Review, in *Magnetic Storms*, edited by B.T. Tsurutani *et al.*, Geophysical Monograph 98, *Amer. Geophy. Un.*, 1997.

High-Speed Streams, Coronal Mass Ejections, and Interplanetary Shocks: A Comparative Study of Geoeffectiveness

G. Lu

High Altitude Observatory, National Center for Atmospheric Research, Boulder, CO, USA

Corotating high-speed solar wind streams, coronal mass ejections, and interplanetary shocks are common causes of geomagnetic disturbances in Earth's magnetosphere and ionosphere. This paper examines the geoeffectiveness of the different types of interplanetary structures. Several ionospheric electrodynamic parameters as well as geomagnetic indices are used to assess quantitatively the effectiveness of the various solar-driven interplanetary structures on the ionosphere and magnetosphere, including the polar-cap potential drop, Joule and auroral energy dissipation, and the AE and Dst indices. Epoch analysis is performed for three types of events associated with the different interplanetary structures, including 17 events of high-speed streams (HSSs), 18 events of shocks followed by complex ejecta, and 18 events of shocks followed by magnetic clouds. On average, the HSS events resembles a weak geomagnetic storm, with a minimum Dst value of –40 nT. The average behavior of the shock/ejecta and shock/cloud events displays the characteristics of a two-step main phase storm, showing the first Dst dip in the sheath region and the second Dst dip in the following cloud or ejecta. In addition to a positive excursion in Dst, solar wind dynamic pressure enhancement associated with interplanetary shocks also induces a prompt increase in ionospheric electric potentials, Joule heating, auroral precipitation, as well as the AE index. There are 8 superstorms with the minimum $Dst < -250$ nT amongst the events studied here. Four of them are caused by the sheath region and 4 by magnetic clouds and ejecta combined, making the sheath region as "geoeffective" as the cloud and ejecta in producing superstorms. The average ratio of the solar wind electromagnetic power as represented by the ε parameter to the magnetospheric energy dissipation is 1.80 for the HSS events, 1.10 for the shock/ejecta events, and about 0.80 for the shock/cloud events. By comparison, the ratio of solar wind kinetic power to the magnetospheric energy dissipation is only around 0.05 for all types of events.

1. INTRODUCTION

Coronal mass ejections (CMEs) are transient, large-scale eruptions of plasma and magnetic fields from the Sun. They are often referred to as interplanetary coronal mass ejections (ICMEs) after they have propagated away from the Sun into the solar wind and interplanetary media. Magnetic clouds are a subset of ICMEs which possess a number of specific characteristics, such as high magnetic field strength, low ion temperature and plasma beta compared to the ambient solar wind, and a smooth rotation of the magnetic field orientation [e.g., *Burlaga et al.*, 1981]. About 30~55% of all ICMEs are classified as magnetic clouds [*Gosling*, 1990; *Cane et al.*, 1997]. *Cane and Richardson* [2003] further found that the fraction of ICMEs that are magnetic clouds is solar cycle dependent, varying from nearly 100% at solar minimum to about 15% at solar maximum.

Fast ICMEs, which travel supermagnetosonically with respect to the ambient solar wind, often drive interplanetary shocks upstream. Between the shock and the ICME is a region of compressed solar wind plasma and interplanetary magnetic field (IMF) called "sheath" [*Gosling*, 1990]. Inside the sheath region, the magnetic field strength is amplified and the plasmas become denser and hotter due to the compression. It has been reported that an overwhelming (or 97%) portion of transient shocks at 1 AU is associated with ICMEs [*Sheeley et al.*, 1985; *Howard and Tappin*, 2005]. Conversely, not all ICMEs observed at 1 AU are preceded by a shock. *Wu and Lepping* [2002] showed that approximately 1/4 of observed magnetic clouds have no upstream pressure pulse/shock, but all have a density increase.

In addition to ICMEs and interplanetary shocks, high-speed solar wind streams emanating continuously from the Sun's coronal holes form another important type of interplanetary structures. As a consequence of solar rotation, high-speed streams (HSSs) reappear in a roughly 27-day periodicity and they may last for many solar rotations. When the high-speed wind runs into the slower wind ahead, regions of compression develop between the fast and slow winds. Since these compressive interaction regions also corotate with the Sun, they are called corotating interaction regions (CIRs) [*Smith and Wolfe*, 1976]. CIRs are bounded by forward and reverse waves that typically steepen into forward and reverse shocks at heliocentric distances beyond 2 AU [*Balogh et al.*, 1999]. Alfvénic fluctuations in B_z are commonly observed in HSSs due to stream-stream interactions [*Tsurutani et al.*, 1995].

Owing to increased awareness of space weather, considerable attention has been devoted to explore the solar and interplanetary causes of geomagnetic storms. All the above mentioned interplanetary structures are found to potentially (though not always) produce geomagnetic storms. Generally speaking, CIRs and HSSs cause weak (-50 nT $< Dst \leq -25$ nT) to moderate (-100 nT $< Dst \leq -50$ nT) intensity geomagnetic storms [*Tsurutani et al.*, 1995], while intense storms with $Dst < -100$ nT are commonly associated with either ICMEs themselves or shock/sheath disturbances driven by ICMEs [*Tsurutani et al.*, 1988; *Gonzalez et al.*, 1999; *Huttunen et al.*, 2004]. But most (more than 70%) of extremely intense storms with $Dst < -200$ nT are caused by magnetic clouds [*Li and Luhmann*, 2004].

This paper investigates the geoeffectiveness of three types of interplanetary structures, namely HSSs, ICMEs, and interplanetary shocks. Emphasis is placed on the general behaviors of the magnetosphere and ionosphere in response to the different solar and interplanetary drivers. Studies on geoeffectiveness so far are mostly based on the relationship between the interplanetary structures and the standard 1-hour resolution *Dst* index [e.g., *Huttunen et al.*, 2005; *Wu and Lepping*, 2002; *Zhang et al.*, 2004; *Ether and Gonzalez*, 2004; *Li and Luhmann*, 2004]. The purpose of this paper is to expand the scope of geoeffectiveness studies by examining other important ionospheric electrodynamic parameters, such as the cross-polar-cap potential drop, Joule heating and auroral energy dissipation. We also discuss the issue of global energy coupling efficiency associated with the different drivers. We adopt the criteria proposed by *Gonzalez et al.* [1994] to classify geomagnetic storms, that is, with $-30 < Dst_{MIN} \leq -50$ nT for weak storms, -50 nT $< Dst_{MIN} \leq -100$ nT for moderate storms, and $Dst_{MIN} < -100$ nT for intense storms, where Dst_{MIN} denotes the minimum value of the *Dst* index during a given event. But storms with $Dst_{MIN} < -250$ nT are named superstorms in this study.

2. RESULTS

2.1. Selection of Events

Three types of the interplanetary structures are examined in this study. A list of corotating interaction regions (CIRs) from 1994 to 2004 is provided by *Ian Richardson* (private communication, 2005). The HSS events are identified when the speed increases to more than 100 km/s above the background solar wind following the CIRs. The ICME events are largely drawn from the list of *Cane and Richardson* [2003] that covers the period of 1996-2002. Further references are obtained from the list of the Wind magnetic clouds (http://lepmfi.gsfc.nasa.gov/mfi/mag_cloud_publ.html) and the list of ACE interplanetary shocks (http://www.ssg.sr.unh.edu/mag/ace/ACElists/obs_list.html). Some cloud-like events are not marked as well-defined or not included (e.g., events after 2002) in *Cane and Richardson* [2003]. For those events, we refer to several published studies [e.g., *Zhang et al.*, 2004; *Skoug et al.*, 2004; *Huttunen et al.*, 2005] to confirm our classification of events. The event selection is also limited by the availability of events that have been analyzed using the assimilative mapping of ionospheric electrodynamics (AMIE) procedure [*Richmond and Kamide*, 1988]. As a result, only a small fraction of events between 1995 and 2003 are being examined in this study, which includes 17 HSS events, 18 shock/magnetic cloud events, and 18 shock/ejecta events. A complete list of these selected events is shown in Table 1. According to *Cane and Richardson* [2003], all but three of the interplanetary shocks listed in Table 1 are associated with the ICMEs. Also, all but one of the magnetic clouds studied here (e.g., the event of 10-11 July 2001, which had a maximum speed of 395 km/s in the cloud) drive interplanetary shocks.

Table 1. List of events examined in this study. The interplanetary shocks that are not driven by ICMEs are indicated by the asterisk signs

Shock Arrival	Start Time	End Time	max V (km/s)	min Bz (nT)	min Dst (nT)	max AE (nT)
	Magnetic Clouds					
1995/10/18 11:22	1995/10/18 19:42	1995/10/20 00:00	439	−21	−129	1592
1996/05/27 12:20	1996/05/27 15:18	1996/05/29 01:00	419	−8	−36	950
1997/01/10 01:12	1997/01/10 04:57	1997/01/11 02:25	480	−15	−83	1917
1997/04/10 17:50	1997/04/11 06:50	1997/04/11 20:00	495	−14	−79	1579
1997/05/15 01:53	1997/05/15 10:35	1997/05/16 01:00	512	−24	−129	2044
1998/05/01 21:15	1998/05/02 12:03	1998/05/03 12:03	695	−13	−95	2209
1998/06/24 10:15	1998/06/24 13:43	1998/06/25 16:00	555	−7	−41	1127
1998/09/24 23:41	1998/09/25 06:05	1998/09/26 16:00	870	−22	−186	3113
1998/10/18 19:50	1998/10/19 04:50	1998/10/20 07:00	436	−19	−132	2036
2000/02/11 23:54	2000/02/12 12:00	2000/02/13 00:00	603	−20	−195	2834
2000/07/15 14:30	2000/07/15 20:00	2000/07/16 10:00	1190	−59	−387	3329
−	2001/07/10 10:30	2001/07/11 12:00	395	−7	−41	1081
2001/10/31 14:16	2001/10/31 21:56	2001/11/02 13:00	404	−14	−102	1246
2002/04/17 11:11	2002/04/17 21:45	2002/04/19 08:30	631	−28	−135	2158
2002/04/19 22:30	2002/04/20 11:00	2002/04/21 16:00	668	−19	−153	2540
2003/10/29 06:40	2003/10/29 08:40	2003/10/30 16:00	2077	−61	−405	3384
2003/10/30 16:45	2003/10/31 02:40	2003/11/02 18:00	1708	−35	−455	3031
2003/11/20 08:07	2003/11/20 11:37	2003/11/21 01:00	761	−53	−566	3205
	Complex Ejecta					
1998/05/04 02:58*	1998/05/04 08:00	1998/05/06 00:00	901	−34	−252	3714
1998/06/25 16:23	1998/06/26 00:00	1998/06/26 19:30	509	−14	−128	1769
1998/10/23 13:00*	1998/10/23 20:00	1998/10/25 10:00	647	−8	−49	1129
1999/09/22 12:00	1999/09/22 19:46	1999/09/24 18:00	621	−21	−207	2171
1999/10/21 02:42	1999/10/21 08:00	1999/10/22 07:00	578	−31	−230	2202
2000/02/14 07:31	2000/02/15 00:00	2000/02/16 00:00	692	−9	−88	1548
2000/04/06 16:40	2000/04/07 08:00	2000/04/08 18:00	637	−32	−383	2673
2000/11/10 06:29	2000/11/10 10:00	2000/11/11 04:00	947	−13	−111	1765
2000/11/11 04:19*	2000/11/11 08:00	2000/11/12 00:00	968	−5	−59	884
2001/03/31 00:53	2001/03/31 04:00	2001/03/31 23:00	773	−47	−389	2076
2001/08/17 11:01	2001/08/17 20:00	2001/08/20 00:00	602	−20	−131	1773
2001/08/27 19:52	2001/08/28 20:00	2001/08/29 00:00	615	−9	−23	904
2001/10/11 17:01	2001/10/12 02:00	2001/10/12 11:00	606	−20	−94	1683
2001/10/21 16:48	2001/10/22 00:00	2001/10/25 08:00	685	−27	−223	2015
2001/10/26 18:35	2001/10/27 02:00	2001/10/28 03:50	444	−7	−47	958
2001/10/28 03:50	2001/10/29 22:00	2001/10/31 13:00	524	−19	−164	1446
2001/11/06 02:08	2001/11/06 20:00	2001/11/09 05:00	769	−78	−339	3164
2003/10/24 15:42	2003/10/24 22:37	2003/10/25 14:37	63	−17	−91	2518
	HSSs					
	1995/10/20 08:00	1995/10/23 00:00	589	−9	−82	1714
	1996/05/29 23:43	1996/05/31 00:00	478	−4	−10	613
	1996/11/24 14:20	1996/11/26 00:00	503	−8	−29	958
	1997/01/11 03:00	1997/01/14 00:00	596	−8	−46	1416
	1998/03/10 13:00	1998/03/15 00:00	580	−17	−111	1626
	1998/03/21 14:00	1998/03/23 20:00	648	−14	−102	1912
	1998/03/26 10:00	1998/03/28 06:00	533	−9	−46	1037
	1998/03/28 18:00	1998/03/31 14:00	517	−9	−60	1810
	1998/05/08 10:00	1998/05/09 13:00	692	−6	−64	1343
	1998/10/20 18:00	1998/10/23 13:00	751	−9	−72	1542
	1999/09/12 12:00	1999/09/19 12:00	738	−13	−73	1661
	1999/10/22 10:00	1999/10/26 16:00	725	−7	−156	1621
	2000/04/08 06:00	2000/04/10 12:00	603	−6	−75	1181
	2001/08/21 03:00	2001/08/24 12:00	715	−8	−28	1094
	2001/10/08 13:00	2001/10/10 20:00	510	−11	−78	1197
	2001/11/15 14:40	2001/11/17 00:00	414	−15	−38	1141
	2003/10/13 18:00	2003/10/23 16:00	785	−17	−137	1802

High resolution, multi-station derived geomagnetic indices are used in this study. The AE index is obtained from 55 to 80 ground magnetometer stations located between $|55°|$ and $|76°|$ magnetic latitudes north and south, and the Dst index is obtained typically from 20 to 45 stations located below $|40°|$ magnetic latitude. An example of the multi-station derived AE and Dst indices as well as their relationship with the standard AE and Dst indices has been shown by *Lu et al.* [1996]. These AE and Dst indices have a time resolution of 1 or 5 minutes. The solar wind parameters are obtained either from the ACE or Wind spacecraft. The ACE data have a time resolution of 16 second for magnetic fields and 64 seconds for plasmas, and the Wind data have a 92-second resolution for magnetic fields and 97-second for plasmas. The plasma data have been interpolated to the magnetic field data resolution when calculating other related quantities.

Ionospheric electrodynamic parameters such as the polar-cap potential (PCP) drop, the Joule heating rate, and auroral precipitating electron energy flux are derived using the AMIE procedure. AMIE performs an optimally constrained, weighted least squres fit of coefficients to the observed data. Energy fluxes and mean energies of precipitating auroral electrons are either measured in situ by polar-orbiting satellites or inferred from global auroral images, and they can also be inferred from ground magnetometer perturbations using empirical formulas. Electric fields or plasma drifts are measured by high-frequency coherent and incoherent scatter radars, and by polar-orbiting satellites. Ionospheric currents are derived from space and ground based magnetic field measurements (see *Lu et al.* [1996; 1998] for further information). It should be noted that data coverage varies from event to event, depending on the availability of the measurements. Since the southern hemisphere usually has a poorer data coverage, in this study we focus our attention to the northern hemispheric response to the different interplanetary drivers.

2.2. Event Analysis

Figure 1 shows the solar wind measurements from the Wind spacecraft as well as the corresponding geomagnetic and ionospheric parameters derived from AMIE for the period of 18-24 October 1998. Wind was located at (100, 32, 6) R_E in GSE (X, Y, Z) coordinates. Plotted from top to bottom are (a) the solar wind bulk speed; (b) solar wind plasma density (dashed line) and dynamic pressure (solid line); (c) the z-component (dashed) in GSM coordinates and the magnitude (solid) of the interplanetary magnetic field (IMF); (d) the ion temperature; (e) the Dst index (solid) and the pressure-corrected Dst (dashed); (f) the AE (solid), the reversed AL or $-AL$ index (dashed), and the AU index (dotted); (g) the polar-cap potential drop; and (h) the hemispheric integrated Joule heating rate (solid) and auroral precipitation (dashed).

All three types of interplanetary structures were presented during this 7-day period. The magnetic cloud arrived at 0445 UT on 19 October and lasted for 26 hours, and it was preceded by an interplanetary shock that arrived at 2000 UT on 18 October. Immediately behind the interplanetary shock and ahead of the magnetic cloud was the sheath region with elevated solar wind plasma density and temperature and oscillating IMF B_z. The interval between 2000 UT on 18 October and 0700 UT on 20 October is therefore marked as a "shock/cloud" event. Behind the magnetic cloud is a region of high-speed streams (HSSs), which began at about 1800 UT on 20 October when the solar wind bulk speed gradually increased to more than 400 km/s. The HSSs were then taken over by a weak interplanetary shock at 1325 UT on 23 October, which was followed by an ICME beginning at ~2000 UT on 23 October. According to *Cane and Richardson* [2003], the ICME lasted until 1000 UT on 25 October. But the AMIE runs were performed only till the end of 24 October. Since the second ICME did not satisfy the definition of a magnetic cloud, we have designated the interval between 1325 UT on 22 October and 2400 UT on 24 October as a "shock/ejecta" event.

The Dst index shown in Figure 1e was derived from 34 ground magnetometer stations located below $|40°|$ magnetic latitude. It features a positive excursion, which is often referred to as the storm sudden commencement (SSC) caused by solar wind dynamic pressure pulse. The pressure-corrected Dst is calculated from $Dst^* = Dst - a\sqrt{P} + b$, where P is the solar wind dynamic pressure in eV cm^{-3}, $a = 0.2$ nT/(eV cm^{-3})$^{1/2}$ and $b = 20$ nT [*Burton et al.*, 1975]. The Dst^* index, when a proper time shift for the solar wind parameters measured by the spacecraft is applied (in this case the Wind measurements were shifted by 25 minutes), can eliminate the positive excursion in Dst induced by the magnetopause Chapmen-Ferraro currents due to the rapid solar wind pressure enhancement. Using the 1-minute and multi-station derived Dst, along with the 95-second solar wind data from Wind, the accuracy of the time delay thus estimated is within 2 minutes. This method has been used in this study to estimate the time delay of solar wind propagation from the spacecraft (Wind or ACE) location to the ionosphere when the SSC-like positive excursion in Dst is apparent. Otherwise, the time delay is calculated based on the x-distance of the spacecraft from the dayside magnetopause divided by the average solar wind speed.

The compressed sheath materials preceded the magnetic cloud consisted of oscillating IMF B_z. The storm main phase started when B_z turned southward in the sheath region at about 0300 UT on 19 October, nearly 2 hours prior to the arrival of the magnetic cloud. Inside the magnetic cloud the

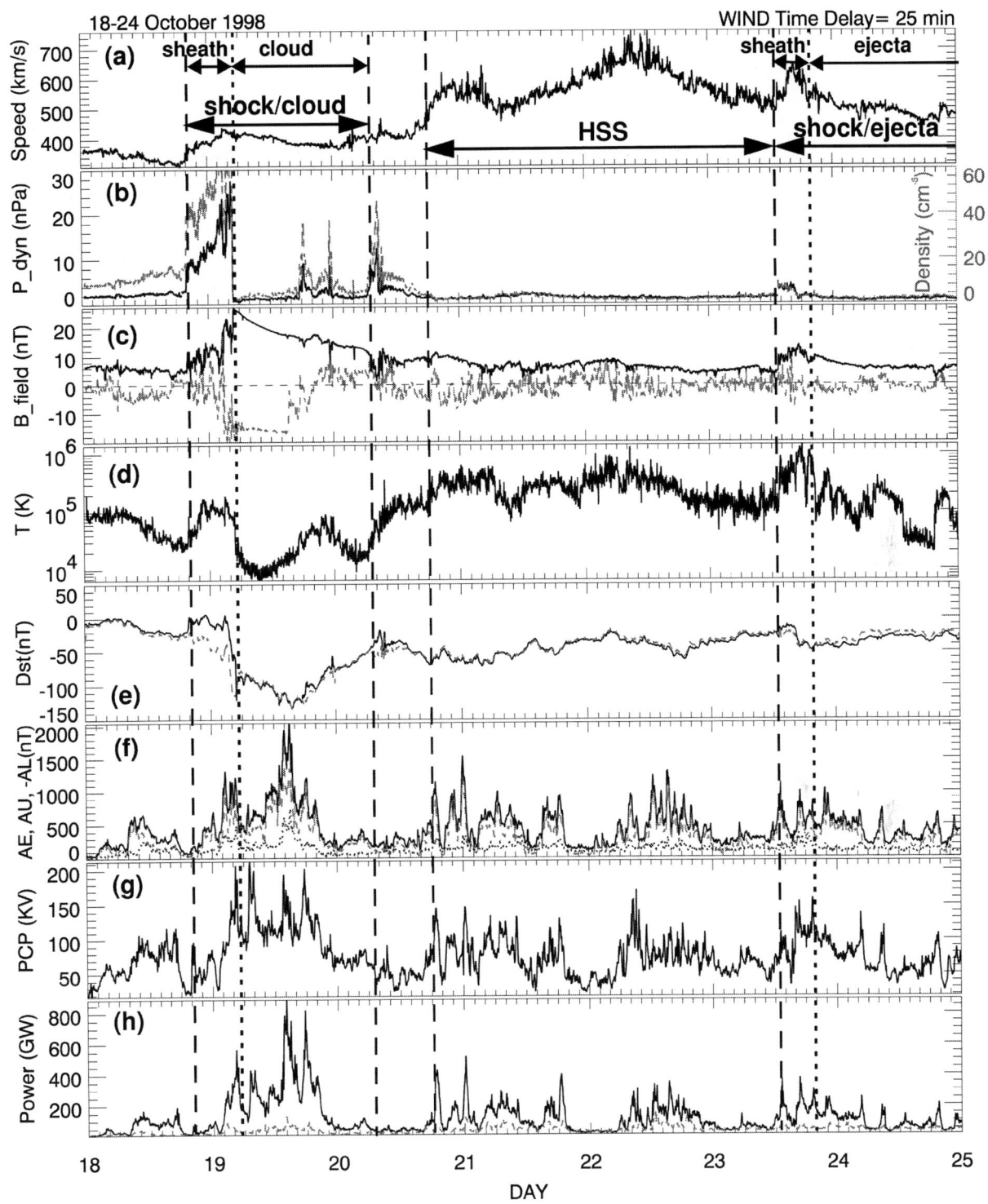

Figure 1. Solar wind, geomagnetic, and ionospheric conditions for 18-24 October 1998: (a) solar wind speed, (b) solar wind dynamic pressure (solid) and plasma density (dashed), (c) the magnitude of the interplanetary magnetic field (solid) and IMF B_z (dashed), (d) ion temperature, (e) Dst (solid) and the pressure-corrected Dst (dashed), (f) AE (solid), $-AL$ (dashed), and AU (dotted), (g) polar-cap potential drop, and (h) Joule heating rate (solid) and auroral electron energy flux (dashed).

B_z component was southward at the leading edge and gradually turned northward around 1600 UT, marking the recovery phase of the storm after Dst had reached its minimum value of −132 nT. During the HSS and shock/ejecta intervals, B_z alternated between northward and southward with a small magnitude ($|B_z| < 5$ nT), resulting in a varying but relatively small Dst value between −10 and −60 nT. The auroral electrojet indices shown in Figure 1f were calculated from the north-south component of the magnetic perturbations measured by 71 magnetometer stations. The maximum AE value of 2047 nT was reached during the sustained southward B_z period inside the magnetic cloud, with 66% contribution from the westward electrojets (or the AL index) and 34% from the eastward electrojets (or the AU index). The polar-cap potential drop varied from 10~30 kV under northward IMF to over 100 kV under southward IMF, and the maximum potential drop reached 210 kV during the passage of the magnetic cloud. Joule heating and auroral precipitation are the two main forms of magnetospheric energy input to the upper atmosphere. During geomagnetically quiet conditions both Joule heating and auroral precipitation were small and their magnitudes were comparable. During geomagnetically active times, however, Joule heat greatly exceeded auroral precipitation. The maximum Joule heating rate was about 890 GW and the maximum auroral energy flux was 150 GW, and both were reached during the magnetic cloud.

During the nearly 3-day period of HSSs between 20 and 23 October, the peak value of AE exceeded 1000 nT and AE only occasionally dropped below 200 nT. Therefore, the HSS interval represents a classic case of the so-called high-intensity long-duration continuous AE activity (HILDCAA) [*Tsurutani and Gonzalez*, 1987]. The PCP varied from 30 kV to 150 kV, and the Joule heating rate was in the range between 50 and 500 GW. The response of the ionosphere to the shock/ejecta interval was remarkably similar to HSSs, but with a slightly smaller magnitude in Dst, AE, PCP, and Joule heating.

2.3. Epoch Analysis

Plates 1, 2, and 3 show the superposed epoch analysis of the HSS events, the shock/ejecta events, and the shock/cloud events, respectively. The HSS events are superposed at the onset time marked by the vertical dotted line. In order to better characterizing the difference and/or similarity between ICMEs and the corresponding shock and sheath regions, the shock/ejecta and shock/cloud events have been rescaled in time so that the different events are overlaid according to the shock arrival time (the first vertical dotted line) and the cloud/ejecta arrival time (the second vertical dotted line) as well as the duration of the cloud/ejecta. In each panel the black solid line shows the median value, and the red dashed line is the mean value. For the HSS events shown in Plate 1, despite the scattering from one event to another, the mean and median values are roughly the same. The minimum Dst value or Dst_{MIN} is greater than −150 nT during HSSs. There are two intense storms with $Dst_{MIN} < -150$ nT (e.g., 6–7 April 2000 and 22 October 1999, respectively); however, both are caused by complex ejecta preceding the HSSs. The superposed shock/ejecta events also show a large event-to-event scattering in all parameters plotted in Plate 2. Again, the mean and median values of all events closely resemble each other. There are four major geomagnetic storms with Dst_{MIN} dropping below −250 nT, amongst them three are associated with the sheath region and one occurs within an ejecta. There are also four superstorms out of the 18 shock/cloud events shown in Plate 3, with three of them taking place during the passage of magnetic clouds and one in the sheath region.

The mean values of the three types of events are replotted in Figure 2, along with the mean values of auroral power and the ε parameter. The solid lines are the mean values for the shock/ejecta events, the dashed lines for the shock/cloud events, and the dotted lines for the HSS events. Although a direct one-to-one comparison is somewhat difficult to make since the different types of events are scaled differently in time, several features as represented by the mean values are worthy of noticing:

- The increase in speed is gradual at the onset of the HSS events, but very rapid at the shock front for the shock/cloud and shock/ejecta events. On average, the solar wind speed jumps from ~450 km/s to over 600 km/s at the shock front, compared to a gradual increase to 500 km/s during HSSs.

- The mean value of the IMF B_z within HSSs is small ($|B_z| < 4$ nT) and highly fluctuating, indicative of the presence of Alfvén waves even in the mean value of the superposed HSS events. B_z also tends to be more oscillating in the sheath region than inside a cloud or ejecta, consistent with the compression that is manifested by the high dynamic pressure in the sheath region (that is, between the two vertical dotted lines). For the 18 magnetic cloud events studied here, the average behavior of the IMF B_z exhibits a southward leading field and a northward trailing field. The average distribution of the eastward solar wind electric field $E_y = -V_x B_z$, where V_x is the x-component of the solar wind velocity, is essentially opposite to that of the IMF B_z.

- The increase of solar wind dynamic pressure is prompt at the shock front; on average, the solar wind dynamic pressure increases from 2~3 nPa to over 8 nPa. For the HSS events, the dynamic pressure rises gradually about 20 hours prior to the onset of HSSs and then falls after the onset. As discussed by *Gosling and Pizzo* [1999], the plasma pressure within a CIR peaks in the vicinity of the stream interface. Thus our identification of the onset time of HSSs appears to be very close to the stream interface on average.

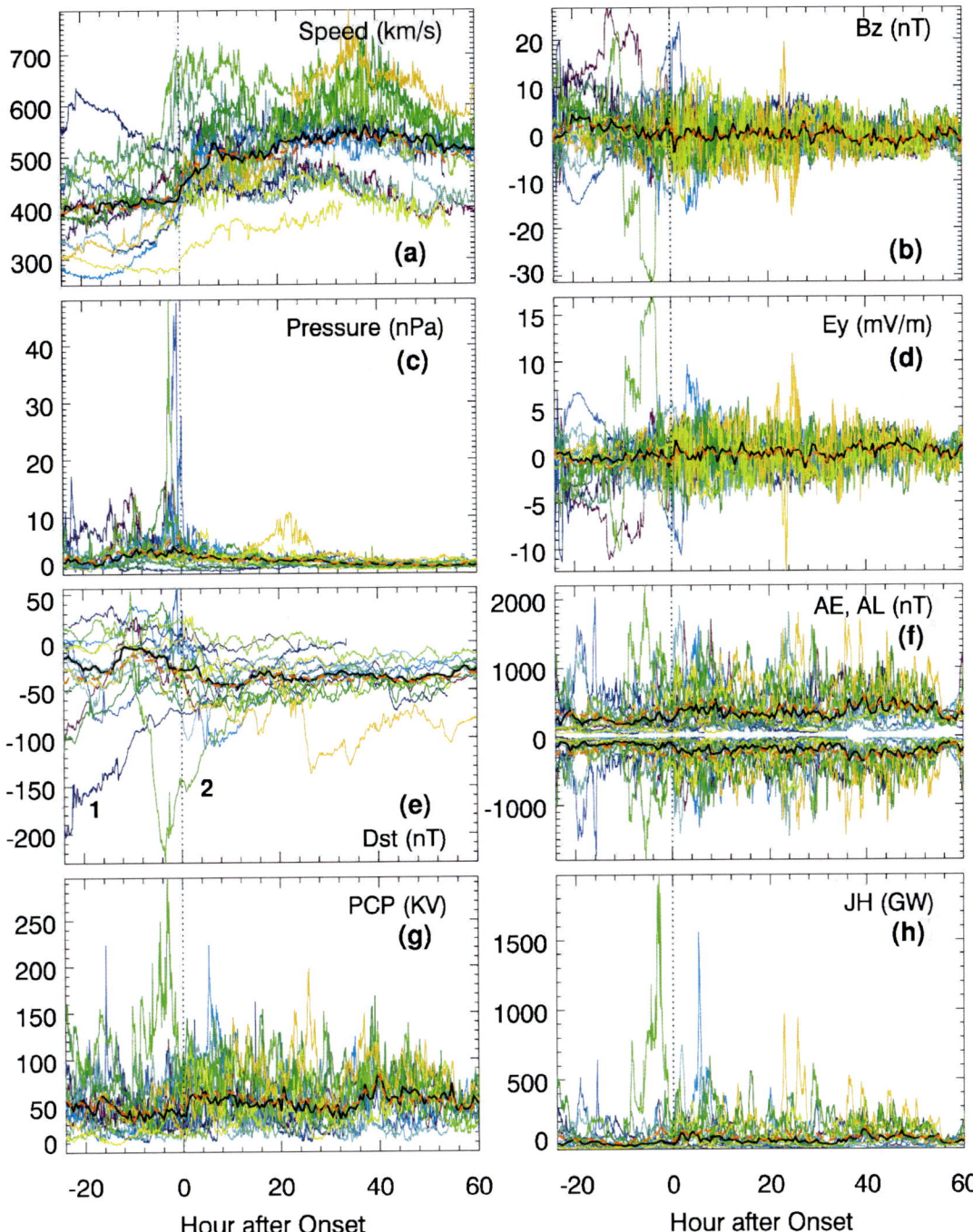

Plate 1. Superposed epoch plot of the HSS events: (a) solar wind speed, (b) IMF B_z, (c) solar wind dynamic pressure, (d) the eastward solar wind electric field, (e) Dst, (f) AE and AL, (g) polar-cap potential drop, and (h) Joule heating rate. The vertical dotted line marks the arrival time of HSSs. In each panel the black solid line indicates the median value and the red dashed line shows the mean value. The two marked intense storm events are (1) 6-7 April 2000 and (2) 22 October 1999.

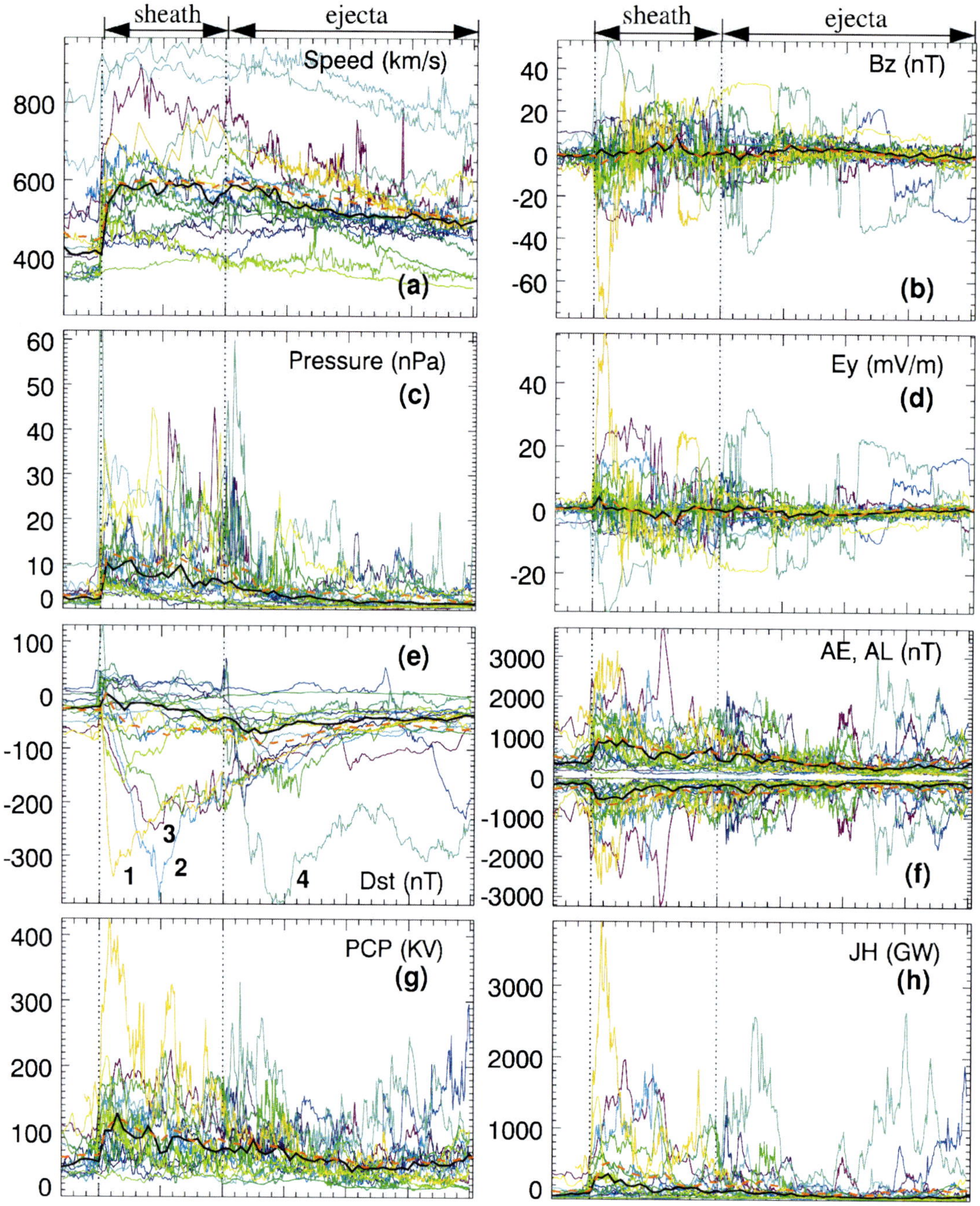

Plate 2. Superposed epoch plot of the shock/ejecta events. The different events are rescaled in time so that the first vertical dotted line marks the arrival of shocks and the second vertical dotted line marks the arrival time of ejecta, and the interval between the second vertical dotted line and the end of the plot corresponds to the duration of ejecta. The black solid line indicates the median value and the red dashed line shows the mean value of each paramter. The four marked superstorms are (1) 6-8 November 2001, (2) 6-8 April 2000, (3) 4-8 May 1998, and (4) 31 March 2001.

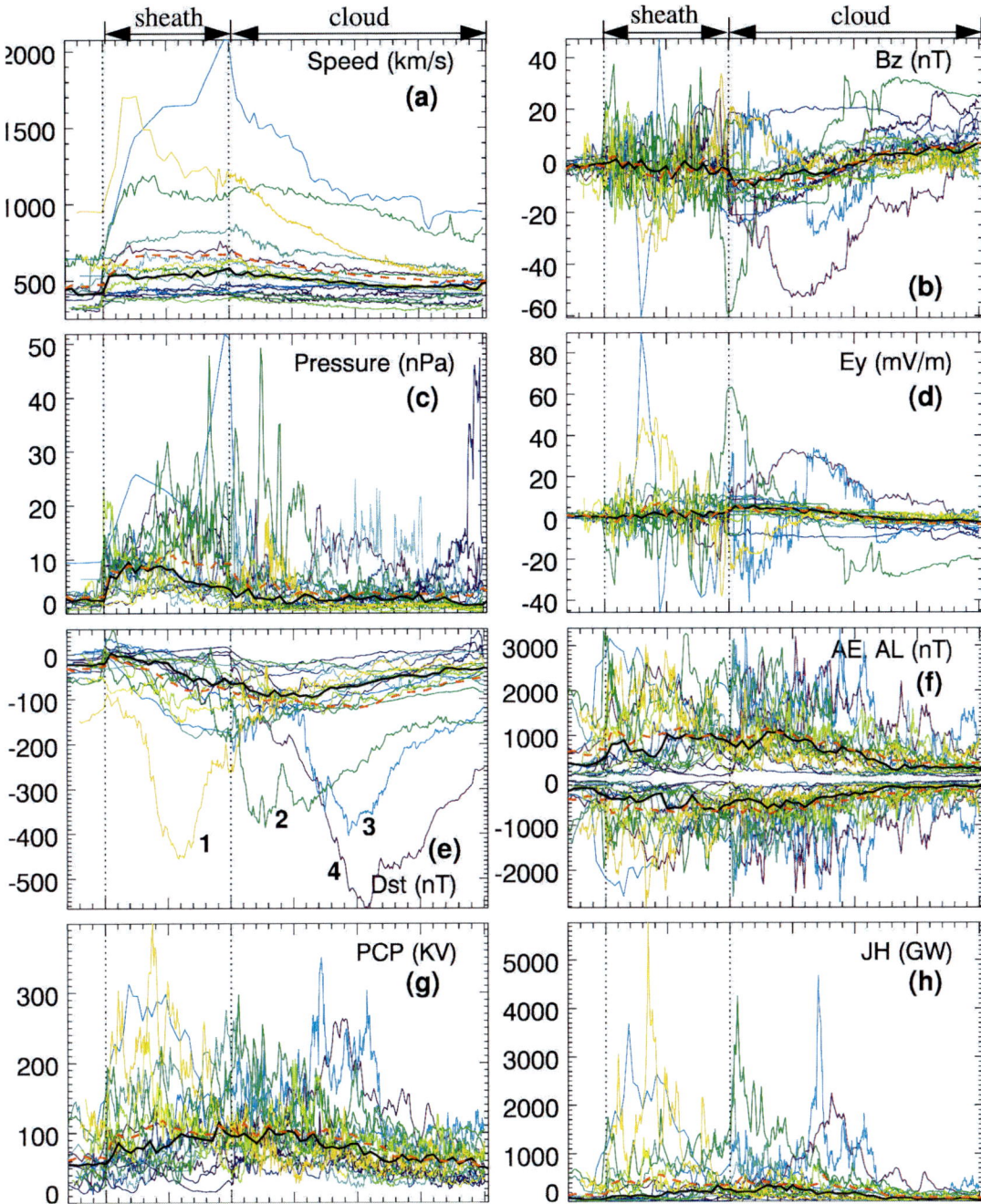

Plate 3. Superposed epoch plot of the shock/cloud events. The vertical dotted lines are the arrival times of the shock front and magnetic cloud, respectively. The black solid line indicates the median value and the red dashed line shows the mean value of each paramter. The four marked superstorms are (1) 29-30 October 2003, (2) 15-16 July 2000, (3) 30-31 October 2003, and (4) 20-21 November 2003.

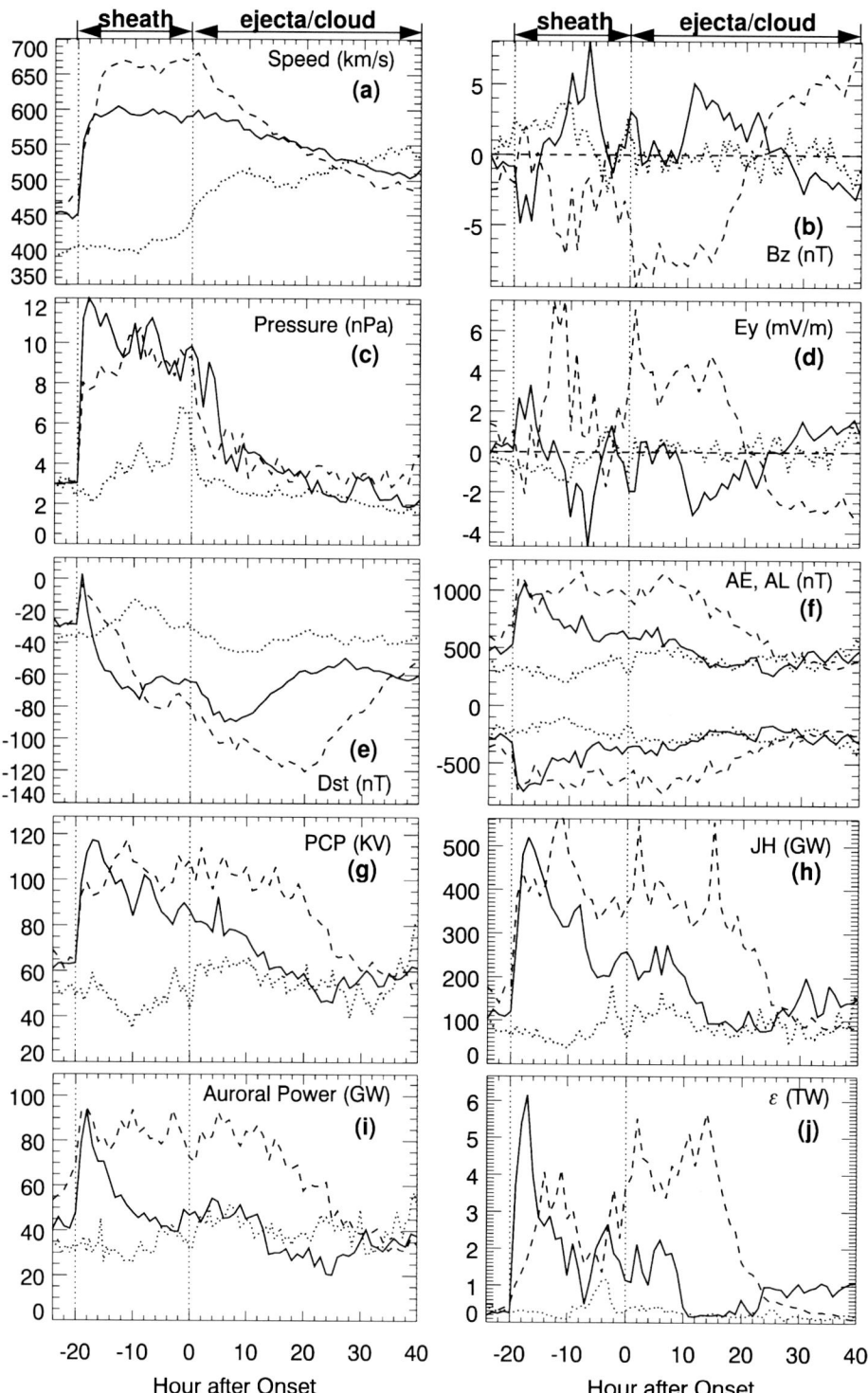

Figure 2. Comparison of the mean value of each parameter for the shock/ejecta events (solid lines), the shock/cloud events (dashed lines), and the HSS events (dotted lines). The bottom time labels apply only the HSS events, whereas the shock/ejecta and shock/cloud events are scaled in the same fashion as in Figures 3 and 4.

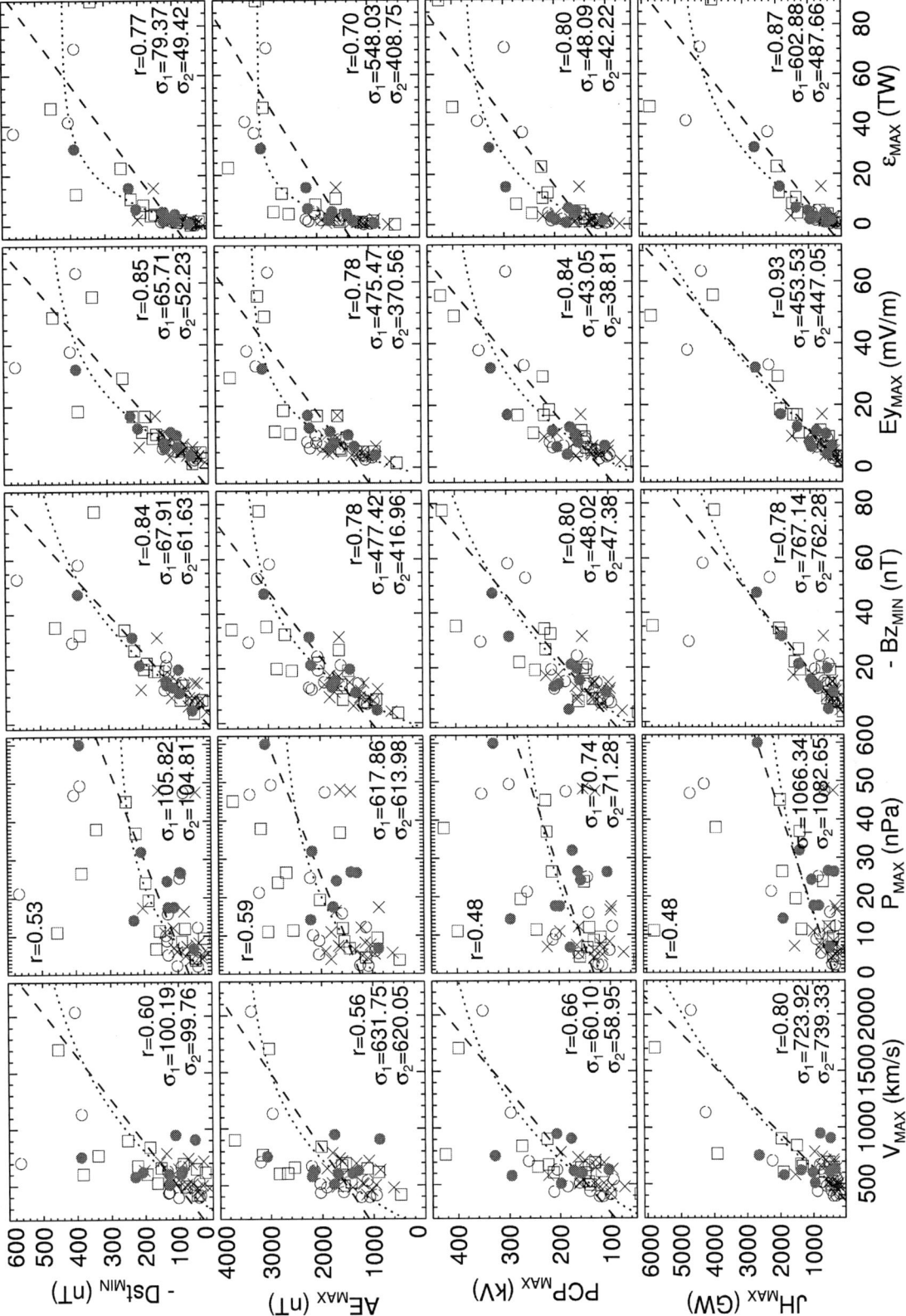

Figure 3. Scatterplots of various solar wind parameters versus the minimum Dst (top row), the maximum AE (second row), the maximum polar-cap potential drop (third row), and the maximum Joule heating rate (bottom row). r is the linear correlation coefficient, σ_1 is the standard deviation of the linear fitting (dashed line), and σ_2 is the standard deviation of the exponential fitting (dotted line). The different symbols represent the interplanetary structures corresponding to the minimum Dst value for a give event. The crosses are for the corresponding HSSs, the squares for the sheath region, the filled circles for ejecta, and the open circles for magnetic clouds.

- The average *Dst* distribution shows a clear SSC-like positive excursion at the shock front. For both shock/cloud and shock/ejecta events, the mean *Dst* value shows the characteristics of a two-step main phase storm, which correspond to the two southward B_z excursions in the sheath region and just ahead of cloud or ejecta, respectively. The average *Dst* distribution for the HSS events resembles a weak storm, with Dst_{MIN} of –40 nT.

- The *AE* and *AL* indices, the PCP, Joule heating, and auroral power all show a prompt increase at the onset of the interplanetary shocks, implying that solar wind dynamic pressure has a direct impact on the ionosphere. The mean values of these parameters during the sheath interval are comparable or even higher than in the following cloud or ejecta. For the HSS events, there is a very modest increase of these parameters, which starts about 6 hours prior to the onset time. This corresponds to the increase of the solar wind dynamic pressure inside CIR while the average IMF B_z is mostly positive there.

- For the shock/cloud events, the mean value of the ε parameter (see next section for further information) is larger in the leading edge of the cloud than in the sheath region due to larger southward B_z. For the shock/ejecta events, the ε parameter peaks near the shock front where the average B_z also points southward. For the HSS events, the mean value of the ε parameter peaks about 2 hour prior to the onset time when the average B_z has a brief southward excursion. Overall, the distribution of the ε parameter is closely related to the magnitude of the southward B_z.

2.4. Global Energy Coupling Efficiency

Interplanetary structures, such as HSSs, ICMEs and shocks, are the ultimate energy sources which are responsible for virtually all electromagnetic disturbances taking place in the magnetosphere and ionosphere. Figure 3 shows the correlation between the various solar wind quantities and geomagnetic/ionospheric parameters. For a given event only the peak values of the solar wind quantities (e.g., the solar wind speed *V*, the dynamic pressure *P*, the southward IMF B_z component or $-B_z$, the eastward interplanetary electric field E_y, and the ε paramter) and the geomagnetic indices (e.g., the reversed *Dst* index or –*Dst*, and the *AE* index) and ionospheric parameters (e.g., the PCP drop and the Joule heating rate) are considered. The crosses represent the events when the peak value of –*Dst* is reached during HSS intervals, the squares correspond to events when the peak value of –*Dst* is reached during the sheath interval, the open circles for events associated with a magnetic cloud, and the filled circles for events associated with a complex ejecta. Amongst the various solar wind parameters, E_y shows the highest overall correlation coefficient with the geomagnetic indices and ionospheric parameters, and $-B_z$ places a close second. The solar wind dynamic pressure has the lowest overall correlation with the various geomagnetic and ionospheric parameters, followed by the solar wind speed. The ε parameter shows an intermediate correlation, with a correlation coefficient *r* ranging from 0.70 to 0.87. Both linear (dashed line) and exponential (dotted line) fittings are performed for each parameter. There is a clear trend of saturation in *Dst*, *AE*, PCP, and Joule heating rate for high values of E_y and ε, and the standard deviation of the exponential fitting σ_2 is about 10~40% smaller than the standard deviation of the corresponding linear fitting σ_1.

Figure 4 illustrates the global energy coupling efficiency for the three types of events. The energy coupling efficiency is calculated as the ratio of total energy deposition into the magnetosphere versus the solar wind energy inputs. The total magnetospheric power U_T is defined as $U_T = U_A + U_{JH} + U_R$, where U_A and U_{JH} are the auroral electron energy flux and Joule heating rate integrated over the northern and southern hemisphere, and U_R is the ring current energy injection rate. However, since not all events studied here have the AMIE analysis done for both hemispheres, the total Joule heating and auroral powers are simply taken as twice of the northern hemispheric integrated powers so all events can be examined in the same fashion. The ring current injection rate U_R is calculated by following the formula of *Akasofu* [1981]:

$$U_R(GW) = -4 \times 10^4 \left(\frac{\delta Dst}{\delta t} + \frac{Dst}{\tau} \right),$$

where τ is the ring current particle lifetime. In this study τ is derived using the

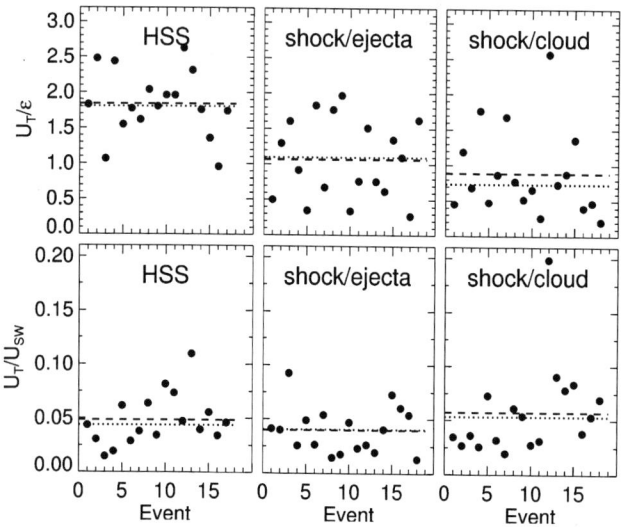

Figure 4. (Top row) ratio of total magnetospheric power versus the ε parameter; (bottom row) ratio of total magnetospheric power versus solar wind kinetic power. In each panel the dashed line is the mean value of all data points, and the dotted line is the median value.

empirical formula of *O'Brien and Mcpherron* [2000]: $\tau(hours) = 2.40e^{9.74/(4.69 + VBs)}$, where VBs in mV/m equals $|VB_z|$ for $B_z < 0$ and zero for $B_z > 0$. The solar wind kinetic power $U_{SW} = 1/2\rho V^3 A$, where ρ is the solar wind mass density, V is the solar wind radial speed, and A is the magnetopause cross section which varies with the solar wind dynamic pressure [*Shue et al.*, 1997]. The solar wind electromagnetic power is represented by the ε parameter of *Perreault and Akasofu* [1978]: $\varepsilon = VB^2 \sin^4(\theta/2)l_0^2$, where B is the IMF strength, θ is the clock angle between the B_z and the IMF projection in the y-z plane in GSM coordinates, and l_0 equals $7R_E$.

The top panels show the ratio of the solar wind electromagnetic power (e.g., the ε parameter), and the bottom panels show the ratio of the solar wind kinetic power. The horizontal dashed and dotted lines are the mean and median values, respectively. The average ratio of the solar wind electromagnetic energy input to the magnetosphere is 1.80 for the HSS events, 1.10 for the shock/ejecta events, and 0.73~0.89 for the shock/cloud events. The ratio of the solar wind kinetic energy input to the magnetosphere is substantially lower for all types of events, with an average value of ~0.05.

Table 2 shows the detailed global energy partitioning for some selected magnetic cloud events. For the moderate storms of 10-11 January 1997 and 2-3 May 1998, the ratio of the total magnetospheric power U_T to the ε parameter is 0.69 and 0.86, respectively. Our estimate of the magnetospheric energy budget did not include the energies that go to the very energetic particles (>1 MeV) forming the radiation belt and to plasma sheet heating, nor that associated with plasmoid ejections in the magnetotail and with lobe field reconfiguration. The ratio for these moderate storms would indicate that about 15-30% of the incident solar wind electromagnetic energy input goes to the unaccounted magnetospheric processes. For the superstorms of 15-16 July 2000 and 29-30 October 2003, however, the ratio of U_T to ε becomes very low (0.26 and 0.39, respectively). Both events featured strongly southward IMF and high-speed solar wind. It is likely that the ε parameter overestimates solar wind electromagnetic power during those superstorms. For weak storms like 10-11 July 2001, this ratio rises to 2.50, implying that the solar wind electromagnetic energy input as represented by the ε parameter becomes insufficient to power the magnetosphere and additional solar wind kinetic power has to tap in.

It is worth noticing that the energy budget for 10-11 January 1997 in Table 2 is somewhat larger than that shown in *Lu et al.* [1998]. The main reason for such a difference is largely due to selecting the interval of concern. The values reported in *Lu et al.* [1998] (e.g., 90 GW for auroral precipitation, 190 GW for Joule heating rate, and 120 GW for the ring current injection) were averaged over the entire 2-day period of 10-11 January, including the early portion of the HSS interval following the magnetic cloud when the IMF was mostly northward and geomagnetic condition was relatively quiet. Whereas the values shown in Table 2 have been averaged over the period between ~0100 UT on 10 January and 0100 UT on 11 January (e.g., the interval between the shock and the passage of the magnetic cloud). Therefore caution should be taken when assessing qualitatively the coupling efficiency between the solar wind energy input and the magnetospheric energy dissipation as both are highly time dependent.

3. SUMMARY AND DISCUSSION

A total of 53 selected events from 1995-2003 have been analyzed using the AMIE procedure based on multi-instrument observations. These events are divided into three groups according to their respective interplanetary structures, with 17 events associated with HSSs, 18 events associated with interplanetary shocks followed by magnetic clouds, and 18 events associated with shocks followed by ejecta. We have demonstrated for the first time the variations of ionospheric parameters, such as the polar-cap potential drop, the Joule heating rate and auroral precipitation, in response to the different solar and interplanetary drivers.

Table 2. Global energy partition for selected shock/cloud events

	15-16 Jul 2000	29-30 Oct 2003	10-11 Jan 1997	2-3 May 1998	10-11 Jul 2001
$U_{SW}(GW)$	76,410	61,530	13,720	24,390	2660
$\varepsilon(GW)$	9220	5940	720	900	206
$U_A(GW)$	171	248	158	134	97
$U_{JH}(GW)$	1593	1522	202	464	374
$U_R(GW)$	624	581	135	179	58
$U_T = U_A + U_{JH} + U_R$	2388	2351	495	777	529
U_T/ε	0.26	0.39	0.69	0.86	2.57
U_T/U_{SW}	0.03	0.04	0.04	0.03	0.20

For both the shock/cloud and shock/ejecta events, the enhancement of solar wind dynamic pressure induces a sharp increase in all ionospheric parameters (e.g., the PCP, Joule heating, auroral power, and the *AE* index). This is consistent with several recent studies that show prompt auroral intensification in response to interplanetary shocks (e.g., *Zhou and Tsurutani*, 1999; 2001; *Liou et al.*, 2004; *Boudouridis et al.*, 2003; 2005). Our epoch analysis also shows that the sheath region ahead of ICMEs has a significant impact on the high-latitude ionosphere. This is evident in Figure 2, which shows that the mean values of these ionospheric parameters during the sheath interval are comparable or even larger than that during the magnetic cloud or complex ejecta. The *Dst* index, which reflects mainly the inner magnetospheric response to storms, shows a double-dip feature. The first, and relatively smaller but rapid, dip is associated with sheath, whereas the second, larger but more gradual, dip associated with cloud or ejecta. Comparable results have been reported by *Vieira et al.* [2004], who found that the storm main phase as represented by the symmetric (SYM) *H* index evolves faster when driven by the sheath region than by the ejectas and HSSs. *Huttnen and Koskinen* [2004] also found that the sheath region is the leading cause of the intense storm during 1997-2002, and our study attests to that. Amongst the 8 superstorms of $Dst_{MIN} < -250$ nT that have been studied here, 4 of them are found to be associated with the sheath region and 4 are associated with magnetic cloud or ejecta. Therefore the sheath region is as geoeffective as magnetic clouds or ejecta in producing superstorms.

A close examination of the IMF B_z (Figure 1c) along with the *Dst* (Figure 1e) and *AE* (Figure 1f) indices shows that each southward turning of B_z is accompanied by an increase in *AE* and a decrease in *Dst*, independent of the interplanetary structures (e.g., sheath, magnetic cloud, HSS, or ejecta) that the IMF is associated with. The intensification of *Dst* and *AE* appears proportional to the amplitude increase of the southward B_z. This fact reiterates the importance of the southward IMF component as a major contributor to the energy and momentum transfer from the solar wind in the magnetosphere [*Tsurutani et al.*, 1988; *Gonzalez et al.*, 1994]. On average, the magnetic clouds examined in this study have a southward leading field and a northward trailing field. This is probably incidental since the events studied here mostly occurred in solar cycle 23. According to *Li and Luhmann* [2004] the polarity of magnetic clouds tends to be predominantly south-to-north in odd solar cycles and reverses in even solar cycles, which is consistent with the earlier finding by *Bothmer and Schwenn* [1998]. On average, the *Dst* distribution of the HSS events resembles a weak geomagnetic storm, with a minimum *Dst* value of −40 nT. The average behavior of the shock/ejecta and shock/cloud events exhibits the characteristics of a two-step main phase storm, and each decrease in *Dst* corresponds to a southward turning of B_z.

As discussed by *Gonzalez et al.* [1994], the prime energy source of geomagnetic storms comes from the transfer of solar wind energy via magnetic reconnection between the IMF and the Earth's magnetic field, which becomes most effective when the IMF is southward. The secondary source is the viscous interaction, and its relative importance increases when the IMF becomes northward. At present, the ε parameter of *Perrault and Akasofu* [1978] is widely used as a proxy for the solar wind electromagnetic energy input to the magnetosphere and ionosphere. However, the justification of the ε function by *Akasofu* [1981] was based on the comparison of ε with the total magnetospheric energy consumption estimated from empirical formulas for storms with $Dst_{MIN} > -200$ nT. It is thus expected that the solar wind energy input as represented by the ε parameter will have some uncertainty, but a quantification of this uncertainty has not been sysmetically done.

Using the solar wind electromagnetic power as represented by the ε parameter, we have found that the average ratio of the solar wind electromagnetic energy input to the magnetosphere is 1.80 for the HSS events, 1.10 for the shock/ejecta events, and 0.73 (median value) to 0.89 (mean value) for the shock/cloud events. The more detailed examination of storms with different intensity as shown in Table 2 clearly calls for the need of a better formulation than the ε parameter to represent the solar wind energy transfer rate to the magnetosphere during both weak and major storms. Indeed, *Koskinen and Tanskanen* [2002] have elucidated several important issues regarding the ε parameter. They also propose a scaling factor of 1.5~2 for the ε parameter in order to eliminate the "energy crisis" problem such as what we have encountered here as well as reported elsewhere [e.g., *Feldstein et al.*, 2003], and to account for the energy loss due to plasmoid release that has not been included in the parameter. This scaling factor seems to be reasonable for the HSS and shock/ejecta events on average; however, it is not a "one size fit all" solution owing to the fact that the scaling factor varies significantly from case to case as evident from the scattering shown in Figure 4. Other effort has also been made to estimate the solar wind energy input using global MHD simulations [e.g., *Palmroth et al.*, 2003]. While global MHD models are a powerful tool to examine the energy flow from the solar wind to the magnetosphere and ionosphere, they have their own limitations with regard to how adequately the physical processes can be described by the MHD approach under various solar wind conditions and how accurately the magnetopause can be determined by the numerical algorithms that the models employ. In summary, fully understanding the energy coupling between the solar wind and the magnetosphere-ionosphere system remains a challenge to the space physics community.

Acknowledgments. This work was supported by NASA Sun-Earth Connection Theory and Guest Investigator programs.

REFERENCES

Akasofu, S.-I., Energy coupling between the solar wind and the magnetosphere, *Space Sci. Rev.*, 28, 121-193, 1981.

Balogh, A., J.T. Gosling, J.R. Jokipii, R. Kallenbach, and H. Know, Introduction, *Space Sci. Rev.*, 89, 1-3, 1999.

Bothmer, V., and R. Schwenn, The structure and origin of magnetic clouds in the solar wind, *Ann. Geophysicae*, 16, 1-24, 1998.

Boudouridis, A., E. Zesta, L.R. Lyons, P.C. Anderson, and D. Lummerzheim, Effect of solar wind pressure pulse on the size and strength of the auroral oval, *J. Geophys. Res.*, 108(A04), 8012, doi:10.1029/2002JA009373, 2003.

Boudouridis, A., E. Zesta, L.R. Lyons, P.C. Anderson, and D. Lummerzheim, Enhanced solar wind geoeffectiveness after a sudden increase in dynamic pressure during southward IMF orientation, *J. Geophys. Res.*, 110, A05214, doi:10.1029/2004JA010704, 2005.

Burlaga, L.E., E. Sittler, F. Miriani, and R. Schwenn, Magnetic loop behind an interplanetary shock: Voyager, Helio, and IMP 8 observations, *J. Geophys. Res.*, 86, 6673-6684, 1981.

Burton, R.K., R.L. McPherron, and C.T. Russell, An empirical relationship between interplanetary conditions and Dst, *J. Geophys. Res.*, 80, 4204-4214, 1975.

Cane, H.V., and I.G. Richardson, Helios 1 and 2 observations of particle decreases, ejecta, and magnetic clouds, *J. Geophys. Res.*, 102(A4), 17075-7086, 1997.

Cane, H.V., and I.G. Richardson, Interplanetary coronal mass ejections in the near-Earth solar wind during 1996-2002, *J. Geophys. Res.*, 108(A4), 1156, doi:10.1029/2002JA009817, 2003.

Echer, E., and W.D. Gonzalez, Geoeffectiveness of interplanetary shocks, magnetic clouds, sector boundary crossing and their combined occurrence, *Geophys. Res. Lett.*, 31, L09808, doi:10.1029/2003GL019199, 2004.

Gonzalez, W.D., J.A. Joselyn, Y. Kamide, H.W. Kroehl, G. Rostoker, B.T. Tsurutani, and V.M. Vasyliunas, What is a geomagnetic storm? *J. Geophys. Res.*, 99(A4), 5771-5792, 1994.

Gonzalez, W.D., B.T. Tsurutani, A.L.C. Gonzalez, Interplanetary origin of geomagnetic storms, *Space Sci. Rev.*, 88, 529-562, 1999.

Gosling, J.T., Coronal mass ejections and magnetic flux ropes in interplanetary space, in Physics of Magnetic Flux Ropes, *Geophysical Monogr.*, vol. 58, edited by C.T. Russell, E.R. Priest, and L.C. Lee, pp. 343-364, AGU, Washington, D.C., 1990.

Gosling, J.T., and V.J. Pizzo, Formation and evolution of corotating interaction regions and their dimensional structure, *Space Sci. Rew.*, 89, 21-52, 1999.

Howard, T.A., and S.J. Tappin, Statistical survey of Earthbound interplanetary shocks, associated coronal mass ejections and their space weather consequences, A&A preprint doi:10.1051/ 0004-6361:20053109, 2005.

Huttunen, K.E.J., and H.E.J. Koskinen, Importance of post-shock streams and sheath region as drivers of intense magnetospheric storms and high-latitude activity, *Ann. Geophysicae*, 22, 1729-1738, 2004.

Huttunen, K.E.J., R. Schwenn, V. Bothmer, and H.E.J. Koskinen, Properties and geoeffectiveness of magnetic clouds in the rising, maximum and early declining phases of solar cycle 23, *Ann. Geophysicae*, 23, 625-641, 2005.

Koskinen, H.E.J., and E.I. Tanskanen, Magnetospheric energy budget and the epsilon parameter, *J. Geophys. Res.*, 107(A11), 1415, doi:10.1029/2002JA009283, 2002.

Li, Y., and J. Luhmann, Solar cycle control of the magnetic polarity and the geoeffectiveness, *J. Atmosph. Solar Terrs. Phys.*, 66, 323-331, 2004.

Liou, K., P.T. Newell, C.-I. Meng, C.-C. Wu, and R.P. Lepping, On the relationship between shock-induced polar magnetic bays and solar wind parameters, *J. Geophys. Res.*, 108(A06), 306, doi:10.1029/2004JA010400, 2004.

Lu, G., B.A. Emery, A.S. Roger, M. Lester, J.R. Taylor, D.S. Evans, J.M. Ruohoniemi, W.F. Denig, O. de la Beaujardiere, R.A. Frahm, J.D. Winninghan, and D.L. Chenette, High-latitude ionospheric electrodynamics as determined by the assimilative mapping of ionospheric electrodyanmics procedure for the conjunctive SUNDIAL/ATLAS 1/GEM period of March 28-29, 1992, *J. Geophys. Res.*, 101, 26,697-26,718, 1996.

Lu, G., D.N. Baker, C.J. Farrugia, D. Lummerzheim, J.M. Ruohoniemi, F.J. Rich, D.S. Evans, R.P. Lepping, M. Brittnacher, X. Li, R. Greenwald, G. Sofko, J. Villain, M. Lester, J. Thayer, T. Moretto, D. Milling, O. Troshichev, A. Zaitzev, G. Makarov, and K. Hayashi, Global energy deposition during the January 1997 magnetic cloud event, *J. Geophys. Res.*, 103, 11,695-11,684, 1998.

O'Brien, T.P., and R.L. McPherron, An empirical phase space analysis of ring current dynamics: Solar wind control of injection and decay, *J. Geophys. res.*, 105, 7707-7719, 2000.

Palmroth, M., T.I. Pulkkinen, P. Janhunen, and C.C. Wu, Stormtime energy transfer in global MHD simulation, *J. Geophys. Res.*, 108(A1), 1048, doi:10.1029/2002JA009446, 2003.

Perrealt, P., and S.-I. Akasofu, A study of geomagnetic storm, *Geophys. J. R. Asrton. Soc.*, 54, 547, 1978.

Richmond, A.D., and Y. Kamide, Mapping electrodynamic features of the high-latitude ionosphere from localized observations: Technique, *J. Geophys. Res.*, 93, 5741-5759, 1988.

Sheeley, N.R., R.A. Howard, M.J. Koomen, D.J. Michels, R. Schwenn, K.-H. Muhlhauser, and H. Rosenbauer, Coronal mass ejections and interplanetary shocks, *J. Geophys. Res.*, 90, 163-175, 1985.

Shue, J.-H., J.K. Chao, H.C. Fu, C.T. Russell, P. Song, K.K. Khurana, and H. Singer, A new functional form to study the solar wind control of the magnetopause size and shape, *J. Geophys. Res.*, 102, 9497, 1997.

Skoug, R.M., J.T. Gosling, J.T. Steinberg, D.J. Mccomas, C.W. Smith, N.F. Ness, Q. Hu, and L.F. Burlaga, Extremely high-speed solar wind: 29-30 October 2003, *J. Geophys. Res.*, 109, doi:10.1029/2004JA010494, 2004.

Smith, E.J., and J.H. Wolfe, Observations of interaction regions and corotating shocks between one and five AU: Pioneers 10 and 11, *J. Geophys. Res.*, 3, 137-140, 1976.

Tsurutani, B.T., and W.D. Gonzalez, The cause of high-intensity long-duration continuous *AE* activity (HILDCAAs): interplanetary Alfvén wave trains, *Plant. Space Sci.*, 35, 405, 1987.

Tsurutani, B.T., W.D. Gonzalez, F. Tang, S.-I. Akasofu, and E.J. Smith, Origin of interplanetary southward magnetic fields responsible for major magnetic storms near solar maximum (1978-1979), *J. Geophys. res.*, 94, 8519-8531, 1988.

Tsurutani, B.T., W.D. Gonzalez, A.L.C. Gonzalez, F. Tang, J.K. Arballo, and M. Okada, Interplanetary origin of geomagnetic activity in the declining phase of the solar cycle, *J. Geophys. res.*, 100, 21,717-21,733, 1995.

Vieira, L.E.A., W.D. Gonzalez, E. Echer, and B.T. Tsurutani, Storm-intensity criteria for several classes of the driving interplanetary structures, *Solar Physics*, 223, 245-258, 2004.

Wu, C.-C., and R.P. Lepping, Efects of magnetic clouds on the geomagnetic storms: The first 4 years of Wind, *J. Geophys. Res.*, 107(A10), 1314, doi:10.1029/2001JA00161, 2002.

Zhang, J., M. Liemohn, J.U. Kozyra, B.J. Linch, and T.H. Zurbuchen, A statistical study of the geoeffectiveness of magnetic clouds during high solar activity years, *J. Geophys. Res.*, 109, doi:10.1029/2004JA010410, 2004.

Zhou, X., and B.T. Tsurutani, Rapid intensification and propagation of the dayside aurora: Large scale interplanetary pressure pulses (fast shocks), *Geophys. Res. lett.*, 26, 1097-1100, 1999.

Zhou, X., and B.T. Tsurutani, Interplanetary shock triggering of nightside geomagnetic activity: Substorms, psuedobreakups, and quiescent events, *J. Geophys. res.*, 106, 18,957-18,967, 2001.

Energetics of Magnetic Storms Driven by Corotating Interaction Regions: A Study of Geoeffectiveness

Niescja E. Turner and Elizabeth J. Mitchell

Florida Institute of Technology, Melbourne, Florida

Delores J. Knipp

United States Air Force Academy, Colorado Springs, Colorado

Barbara A. Emery

High Altitude Observatory, National Center for Atmospheric Research, Boulder, Colorado

We investigate the energetics of magnetic storms associated with corotating interaction regions (CIRs). We analyze 24 storms driven by CIRs and compare to 18 driven by ejecta-related events to determine how they differ in overall properties and in particular in their distribution of energy. To compare these different types of events, we look at events with comparable input parameters such as the epsilon parameter and note the properties of the resulting storms. We estimate the energy output by looking at the ring current energy along with ionospheric Joule heating derived from the PC and *Dst* indices. We also include the energy of auroral precipitation, estimated from NOAA/TIROS and DMSP observations. In general, ejecta-driven storms produce more intense events, as parameterized by *Dst**, but they are usually not as long lasting, and in most cases deposit less energy. This is observed even for events that have similar input quantities, such as epsilon. This may be related to the high speed of the solar wind, in that an increased magnetosonic Mach number may influence the reconnection rate and therefore the coupling. Additionally, we find the efficiency of the coupling varies greatly from CIR-driven to ejecta-driven storms, with the CIR-driven storms coupling substantially more efficiently, particularly in the recovery phase. The efficiency of coupling (output energy divided by input energy) for CIR-driven storms in recovery phase was double that of ejecta-driven storms.

1. INTRODUCTION

1.1. Geoeffectiveness

Geoeffectiveness refers to the efficiency of energy coupling from the solar wind into the magnetosphere. It can be estimated by looking at the solar wind input and the corresponding magnetospheric output for a particular time period. The question becomes, then, how does one quantify the solar wind input and the magnetospheric output? Solar wind input has been parameterized in several key ways over the years, usually in the form of a Poynting flux, and this remains the most widespread estimate. Magnetospheric output, however, is less clear-cut. Some use widely known magnetospheric activity indices such as *Dst* or *Kp* to estimate the response to solar wind drivers. Other researchers may be more interested in the radiation belt response, for example, and have different approaches to quantifying that response. Since the radiation belts are less significant energetically than other sinks, they are not included in this study. For the purposes of this paper, magnetospheric response will be regarded as the total amount of energy (or, in some cases, the instantaneous power) being deposited into the primary magnetospheric energy sinks. These sinks include the ring current, ionospheric Joule heating, and auroral precipitation. As will become clear, this definition of magnetospheric response will often produce different results than other definitions, especially when it comes to defining the types of solar wind drivers that are most geoeffective.

1.2. Magnetospheric Energy Input

Precise measurements of the total amount of energy entering the magnetosphere from the solar wind are simply not possible. Over the years, however, estimates have been made. The epsilon parameter is one such estimate. Epsilon is defined (in SI units) as:

$$\varepsilon = (4\pi/\mu_0)\, vB^2 \sin^4(\theta/2)\, l_0^2$$

where v is the solar wind speed, B is the magnitude of the IMF, l_0 is a characteristic length scale representing the coupling area available for solar wind-magnetosphere interactions, usually approximated as 7 R_E, [*Perreault and Akasofu*, 1978], μ_0 is the permeability of free space, and θ is defined as $\tan^{-1}(|B_Y|/B_Z)$. Epsilon is a measure of the Poynting flux in the solar wind over the magnetospheric collecting area. It uses a "leaky" filter of $\sin^4(\theta/2)$, which means that, while energy coupling is greatly enhanced for southward B_Z, there is still some coupling for northward B_Z as well. The form of epsilon is based on empirical studies of the estimated energy dissipation in the magnetosphere. Thus, while the form of epsilon is shown to replicate the pattern of energy dissipation, its scaling should be considered somewhat arbitrary, as it is based on estimates of energy output rather than quantitative knowledge of energy input. Additionally, the l_0^2 term in this equation does not vary, although the magnetopause area is known to vary with solar wind conditions [see, e.g., *Monreal-MacMahon and Gonzalez*, 1997].

While epsilon is the most commonly used parameterization of solar wind energy input, there are others. One effort to revisit the issue of energy input into the magnetosphere was conducted by *Bargatze et al.* [1985]. They derived a more complex coupling equation which ultimately also used the same $\sin^4(\theta/2)$ term and was similar in form to the epsilon parameter.

Other researchers parameterize solar wind energy input with vB_S, where v is the solar wind velocity and B_S is the southward component of the IMF [e.g., *O'Brien and McPherron*, 2000], or would include the solar wind kinetic energy flux [e.g., *Lu et al.*, 1998]. It is likely that the actual coupling involves some combination of these, and certain parameters may become more or less dominant under different conditions. Since the kinetic energy flux is large in comparison with the epsilon-derived energy input, it would not require a very strong coupling in this regard to have a pronounced effect on the magnetosphere.

2. PREVIOUS WORK

Many researchers have investigated the flow of energy in the magnetosphere [e.g., *Turner et al.*, 2000b; *Baker et al.*, 2001; *Weiss et al.*, 1992; *Vichare et al.*, 2005]. *Lu et al.* [1998] investigated energy budgets in the magnetic storm interval that occurred in January of 1997. They used Assimilative Mapping of Ionospheric Electrodynamics (AMIE) calculations to determine the energy lost to ionospheric processes, and used the standard *Dst* index to estimate the ring current injection rate. Overall, in the January 10 and 11, 1997 case, *Lu et al.*, estimated that the magnetosphere-ionosphere system dissipated an average of about 4.0×10^{11} W. Of this, 1.9×10^{11} W (or 48%) went into Joule heating, 1.2×10^{11} W (or 30%) went into ring current injection, and 0.9×10^{11} W (or 22%) went into auroral precipitation. They did not estimate the energy lost to plasmoids streaming down the magnetotail. *Knipp et al.* [1998] analyzed the November 1993 storm, which was a hybrid event where a high-speed stream followed a CME. They found that high-speed streams could be enormously geoeffective, and for this extreme event the ionospheric heating was $\sim 190 \times 10^{15}$ J, with 30% of that generated within 24 hours of storm onset.

Gonzalez et al. [1989] specifically analyzed energy coupling during intense ($Dst < -100$ nT) storm events and tested the responses of several coupling functions to investigate the *Dst* response. They found that solar wind ram pressure played a role in ring current energization and that during the strong events they studied, there seemed to be a decoupling of auroral response from inner magnetospheric response for the solar wind-magnetosphere coupling functions they analyzed.

Six storms were analyzed by *Turner et al.* [2000b] to determine their energy input and deposition rates. Their calculation incorporated AMIE data to determine the ionospheric loss and a pressure- and tail-corrected form of *Dst* to track the ring current energy, and they also included a term for plasmoid ejection loss. In all cases, epsilon was shown to correlate with the energy output, and in 5 of the 6 events epsilon was estimated to be larger than the output energy. The results of this analysis showed a clear dominance of ionospheric energy deposition over other processes. In fact, Joule heating alone typically accounted for around half or more of the total energy output. The ring current contribution was less than in previous estimates, largely due to a reevaluation of the ring current strength compared to Dst* [*Turner et al.*, 2001, 2000a], and also due to the AMIE analysis suggesting a larger ionospheric loss. This analysis shows the ring current energy to be approximately 10%-15% of the total energy output.

The polar cap (PC) index can be used as a proxy indicator describing the amount of energy deposited into the ionosphere in the form of Joule heating and auroral precipitation. *Chun et al.* [1999], based on comparisons with AMIE data assimilation results, have shown a quadratic relationship between the PC index and the hemispheric integrated Joule heating rate, and recent work shows a linear relationship between PC and electron precipitation. More recent work by *Knipp et al.* [2004] has shown a better fit to the data if both PC and *Dst* are used as inputs.

The study by *Turner et al.* [2000b] of the global energy budget of the magnetosphere analyzed several storms over a fairly small time frame (about 2 years), so it was limited to a small portion of the solar cycle. It is known that solar wind driving conditions vary over the solar cycle, with corotating interaction regions (CIRs) being more common during solar minimum, and coronal mass ejections (CMEs) being more common toward solar maximum. Many researchers have observed differences in the dynamics of storms during times of different types of solar wind driving conditions, such as the existence of High-Intensity Long-Duration Continuous Auroral Activity (HILDCAA) events in the recovery phase of CIR-driven events [e.g., *Tsurutani et al.*, 2004, 2006 (this volume)]. On average, CIRs have less steady B_Z and higher bulk speed than CMEs, and the resulting storms differ in some fundamental properties. Some researchers have studied the ability of different types of solar wind structures to produce storms [see, e.g., *Zhang et al.*, 2004]. *Echer and Gonzalez* [2004] found that compound interplanetary structures were more geoeffective than isolated structures. In another study, *Huttunen et al.* [2002] looked at storms from 1996 to 1999. They found that almost all the intense ($Dst < -100$ nT) storms were associated with CMEs, but for the moderate storms, streams more often generated high Kp storms, while ejecta-related events more often drove stronger *Dst* changes. This could suggest that the relative impacts on the ring current and the ionosphere could vary by type of solar wind driver. *Gonzalez et al.* [1999] found that complex interplanetary structures, including in rare circumstances the influence of subsequent CMEs, could drive particularly intense geomagnetic storms.

3. METHODOLOGY

A list of events sorted by solar wind drivers (Ian Richardson, private communication) was analyzed. These data were classified by time intervals of CMEs and CIRs during 1995-1998. Ionospheric Joule heating power was calculated according to the relations derived by *Knipp et al.* [2004]. The relations for the Joule heating in GW for the northern hemisphere are:

$$JH_{summer} = 29.27\ |PC| + 8.18\ PC^2 - 0.04\ |Dst| + 0.0126\ Dst^2$$
$$JH_{equinox} = 29.14\ |PC| + 2.54\ PC^2 + 0.21\ |Dst| + 0.0023\ Dst^2$$
$$JH_{winter} = 13.36\ |PC| + 5.08\ PC^2 + 0.47\ |Dst| + 0.0011\ Dst^2,$$

where summer is defined as 21 April – 20 August, winter is 21 October – 20 February, and equinox is 21 February – 20 April and 21 August – 20 October. For equinox times, northern hemisphere values were doubled to obtain a global value. For summer and winter dates, a Joule heating estimate for summer was added to a winter estimate to account for the hemispheric seasonal differences.

Global auroral precipitation values were computed from NOAA/TIROS and DMSP satellite measurements of high-latitude precipitating energy flux carried by ions and electrons with energies between 300 eV and 20 keV (NOAA/TIROS) or carried by electrons with energies between 460 eV and 30 keV (DMSP). The energy flux observations made during a single pass over the polar regions are used to estimate the total precipitating power input to a single hemisphere at that time. The power index was devised by Dave Evans for NOAA/TIROS and adapted for DMSP by Frederick Rich and William Denig [*Emery et al.*, 2005, 2006]. Global values were calculated by adding a southern hemisphere estimate to a northern hemisphere estimate.

All solar wind data used in this study were offset to allow propagation time from the satellite to the magnetopause.

The *Dst* index was pressure corrected to *Dst** using the *Burton et al.* [1975] equation, $Dst^* = Dst - b\ P^{1/2} + c$, where P is the solar wind dynamic pressure and the constants b and c are $b = 7.26$ and $c = 11.0$ as derived by *O'Brien and McPherron* [2000].

After pressure correction, the ring current energy was estimated using *Dst**. As many researchers have pointed out, [e.g., *Campbell*, 1996], the *Dst* index measures the effects of many key current systems and cannot single out the ring current. Magnetotail currents are among the primary current systems that can perturb the *Dst* index [see *Turner et al.*, 2000a; *Ohtani et al.*, 2001; *Feldstein et al.*, 2005], as well as induced ground currents. *Dst** was corrected to account for these using the relation described in *Turner et al.* [2001]. This calculation is made by halving the (pressure-corrected) *Dst** to remove the influence of induced ground currents and tail currents and then applying the standard Dessler-Parker-Sckopke [*Dessler and Parker*, 1959; *Sckopke*, 1966] relation. The Dessler-Parker-Sckopke (DPS) relationship between global magnetic field perturbation and particle energy is as follows: $\Delta B = -(\mu_0/2\pi) W_{particles}/(B_0 R_E^3)$, where ΔB is approximated by the corrected *Dst*, $W_{particles}$ is the particle energy, μ_0 is the permeability of free space, B_0 is the surface dipole strength at the equator, and R_E is the radius of the Earth.

So, to calculate the ring current energy from *Dst*, the *Dst* index was first pressure corrected, then halved, and then plugged into the DPS relation to solve for the particle energy. The ring current injection rate was then calculated using the relationship derived by *Akasofu* [1981], $Q = -4 \times 10^{13} (\partial Dst/\partial t + Dst/\tau)$ in SI units, with the ring current decay time set to $\tau = 6$ hours. The epsilon parameter was used to estimate energy input.

Storm intervals were identified where *Dst** was less than –50 nT and a storm recovery phase was observed with at least 80% *Dst** recovery. From these criteria, 24 CIR-driven storms and 18 ejecta-driven storms were identified and analyzed. For a complete list of events in this study, see Tables 1 and 2.

Two example events are shown in Figure 1. A CIR-driven event from April 16-21, 1997 is shown to scale alongside a CME-driven event from April 21-23, 1997. These events are representative of the population of storms in this study with the CIR-driven event lasting approximately 2.75 times longer than the CME-driven storm and containing similar integrated epsilon values. Table 3 contains the energy, minimum *Dst**, and event duration values for these events.

Superposed epoch analyses were conducted for *Dst**, B_Z, solar wind number density, solar wind speed, magnetosonic Mach number, epsilon, ring current injection rate, Joule heating power, and auroral precipitation power. Data were separated into main phase and recovery phase as defined by the minimum *Dst** point in each event. Integrated values of the parameters were calculated for each event.

4. OBSERVATIONS

Figure 2 shows IMF B_Z for CIR and ejecta-driven storms. The CIR storms (Panel A) show rapidly varying B_Z which hovers near zero throughout the events, and continues to oscillate rapidly around $B_Z = 0$ well into the storm recovery. The ejecta storms (Panel B) show a slowly changing B_Z that gradually moves from southward to northward orientation and then largely stays northward. Panel C shows a superposition of B_Z for CIR and ejecta-related events. It is clear from the superposition that for the events in this study, when B_Z in ejecta-related drivers goes northward it stays there, allowing recovery, while CIR-driven events show a B_Z hovering near zero.

Figure 3 shows the superposed solar wind inputs of the entire set of events. Panel A shows the solar wind speed: note the much faster flows during CIR events. The second panel has the epsilon parameter, used as an indicator of input energy. For the storms in this set, the ejecta-related events

Table 1. List of all CIR-driven events in the study

Start Date	Hour	End Date	Hour	Minimum Dst*
29-Jan-95	9	5-Feb-95	5	–74.22
11-Feb-95	6	16-Feb-95	22	–52.13
7-Apr-95	1	12-Apr-95	13	–152.53
22-Apr-95	13	26-Apr-95	5	–54.56
16-May-95	4	21-May-95	12	–89.81
24-Jul-95	4	26-Jul-95	0	–55.06
2-Oct-95	5	11-Oct-95	1	–96.71
20-Oct-95	0	26-Oct-95	6	–66.28
30-Oct-95	11	4-Nov-95	22	–58.68
24-Dec-95	8	25-Dec-95	23	–66.72
10-Mar-96	11	14-Mar-96	14	–60.01
19-Mar-96	7	24-Mar-96	11	–69.72
17-Oct-96	9	26-Oct-96	17	–105.40
27-Feb-97	0	4-Mar-97	15	–92.18
16-Apr-97	18	21-Apr-97	6	–77.31
1-May-97	17	5-May-97	1	–66.29
10-Mar-98	11	14-Mar-98	12	–116.65
20-Mar-98	10	23-Mar-98	7	–87.77
24-Apr-98	1	28-Apr-98	12	–73.19
6-Jun-98	16	9-Jun-98	7	–50.37
16-Jul-98	3	18-Jul-98	5	–58.69
30-Sep-98	15	2-Oct-98	12	–57.69
2-Oct-98	14	5-Oct-98	5	–55.26
7-Oct-98	0	11-Oct-98	12	–67.41

Table 2. List of all ejecta-driven events in the study

Start Date	Hour	End Date	Hour	Minimum Dst*
6-Feb-95	11	9-Feb-95	9	–79.23
4-Mar-95	12	7-Mar-95	4	–91.89
22-Aug-95	17	23-Aug-95	17	–60.67
18-Oct-95	13	19-Oct-95	19	–122.96
10-Jan-97	5	10-Jan-97	23	–76.16
10-Apr-97	22	11-Apr-97	15	–87.15
21-Apr-97	13	23-Apr-97	4	–106.66
15-May-97	6	18-May-97	11	–115.93
26-May-97	16	29-May-97	0	–72.48
5-Nov-97	8	9-Nov-97	8	–114.98
10-Dec-97	11	13-Dec-97	1	–60.14
6-Jan-98	16	8-Jan-98	6	–76.80
29-Jan-98	22	31-Jan-98	17	–55.16
17-Feb-98	14	20-Feb-98	10	–106.35
26-Jun-98	0	26-Jun-98	13	–100.91
6-Aug-98	3	9-Aug-98	1	–148.97
20-Aug-98	8	21-Aug-98	8	–66.47
8-Nov-98	21	11-Nov-98	9	–142.11

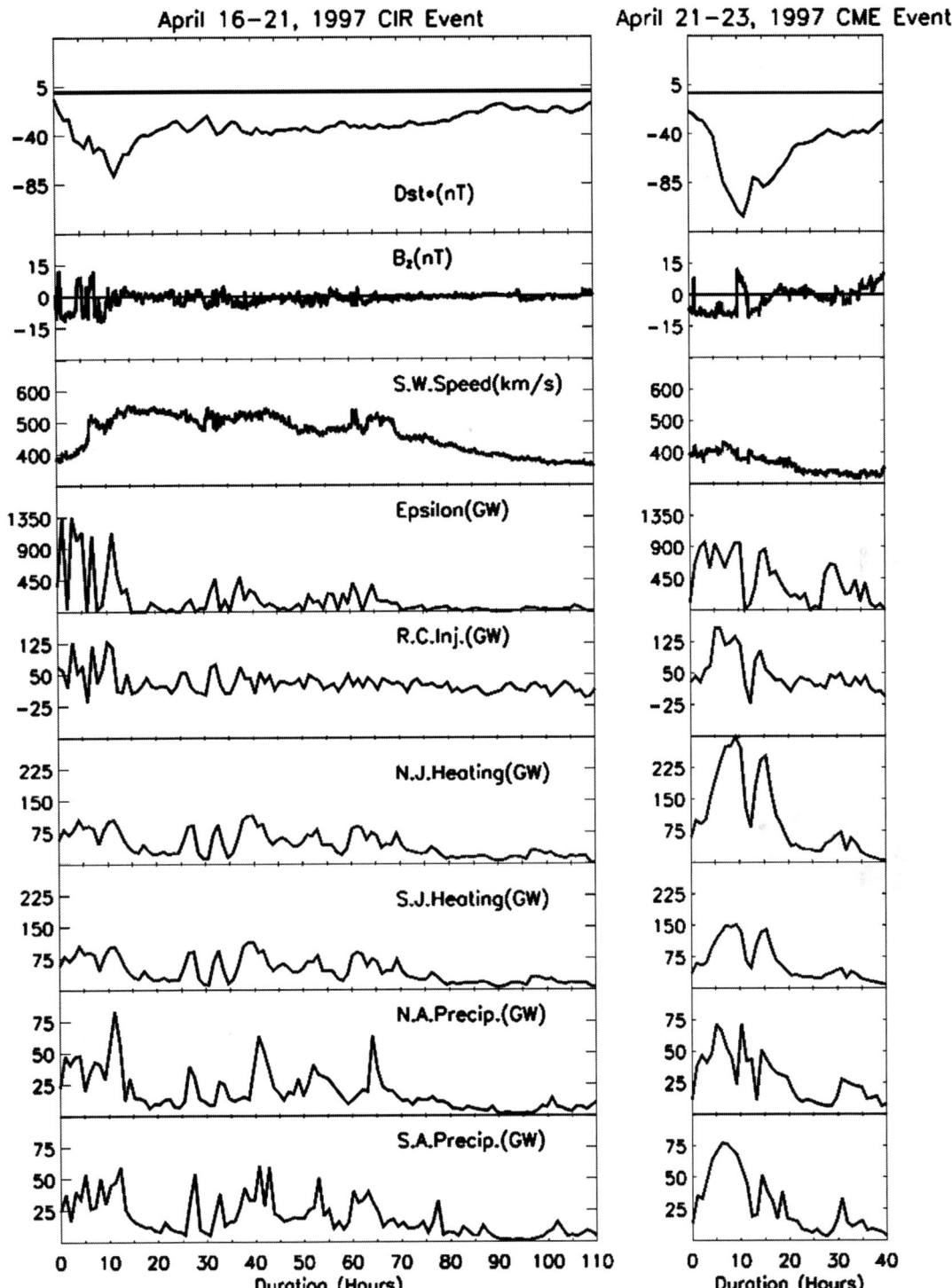

Figure 1. Two example events: a CIR-driven event on April 16-21, 1997 and a CME-driven event on April 21-23, 1997.

Table 3. Table of energies, minimum $Dst*$, and event durations for example events in Figure 1

Type of Event	Date	Duration (Hours)	Min Dst* (nT)	Input (10^{15} J)	Ring Current (10^{15} J)	Joule Heating (10^{15} J)	Auroral Precipitation (10^{15} J)
CIR Event	Apr 16-21, 1997	109	−77.31	62.60	11.24	35.19	14.73
Ejecta Event	Apr 21-23, 1997	40	−106.66	60.02	6.75	23.42	8.04

show a higher epsilon during the main and early recovery phases. Epsilon peaks about 3 hours before the $Dst*$ minimum, and is nearly double for ejecta what is it for CIRs. The magnetosonic Mach number is plotted in Panel C, and again indicates the much faster flows common to CIR events. Panels D and E show the Joule heating and ring current injection rates, respectively, with both showing only a moderate advantage for ejecta-driven events. Panel F shows the superposed $Dst*$, which shows the clear advantage of ejecta-driven events in producing large $Dst*$ excursions.

Figure 4 shows the total energy output for all studied events. While the input energies are similar for the CIR and ejecta events, the output quantities are larger and outside the error bars for every measured quantity. The duration of the CIR storms, it should also be noted, is substantially larger than for ejecta storms, which could play a role in allowing greater energy deposition and may be a result of continuing driving of the system well into the recovery.

In Figure 5 the averaged energy per hour for all storms in the dataset is shown. The input power for ejecta-driven storms is about double that for CIR-driven storms, and the $Dst*$ is somewhat higher for them as well. The output power for ring current, auroral precipitation, and Joule heating are larger for the ejecta events, but not nearly with the same

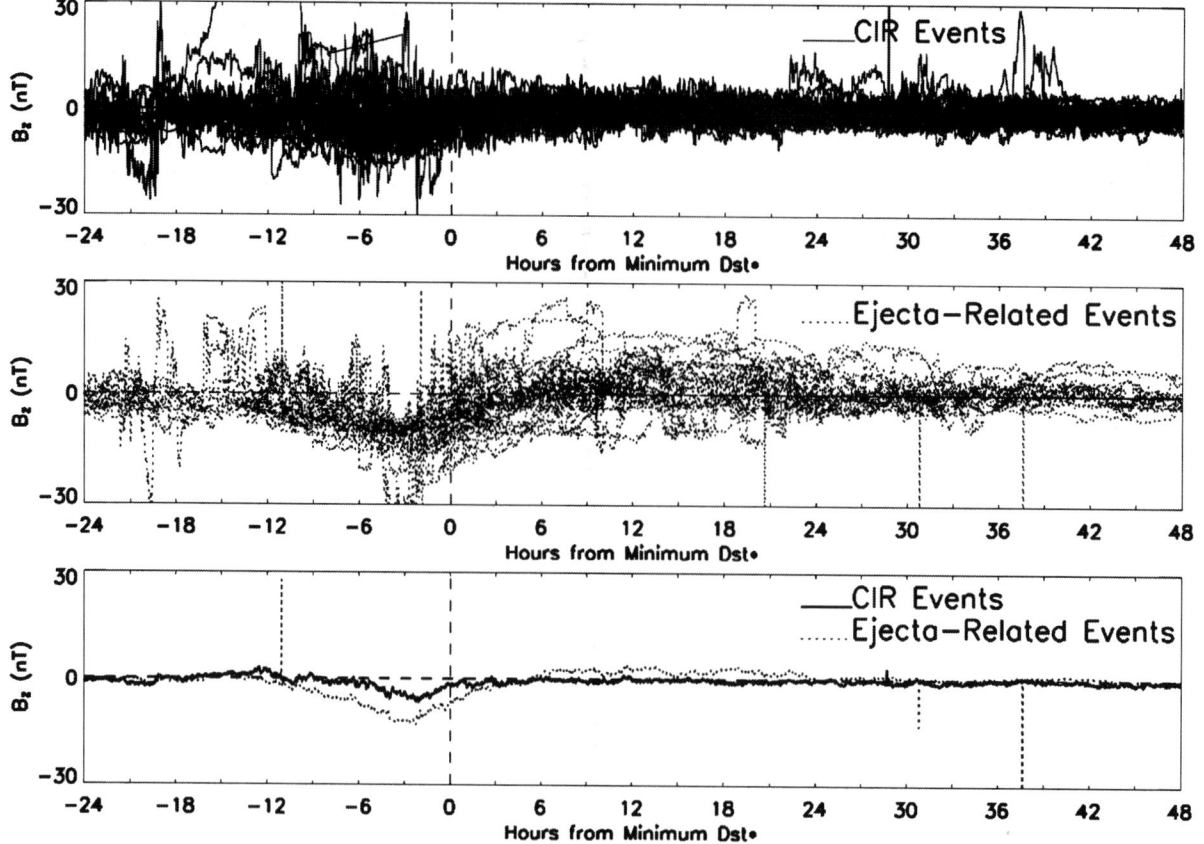

Figure 2. IMF B_Z for all events in study. Panel A shows all CIR events, Panel B shows ejecta-related events, and Panel C shows a superposed epoch analysis for all events in the study. Vertical line shows time of minimum $Dst*$.

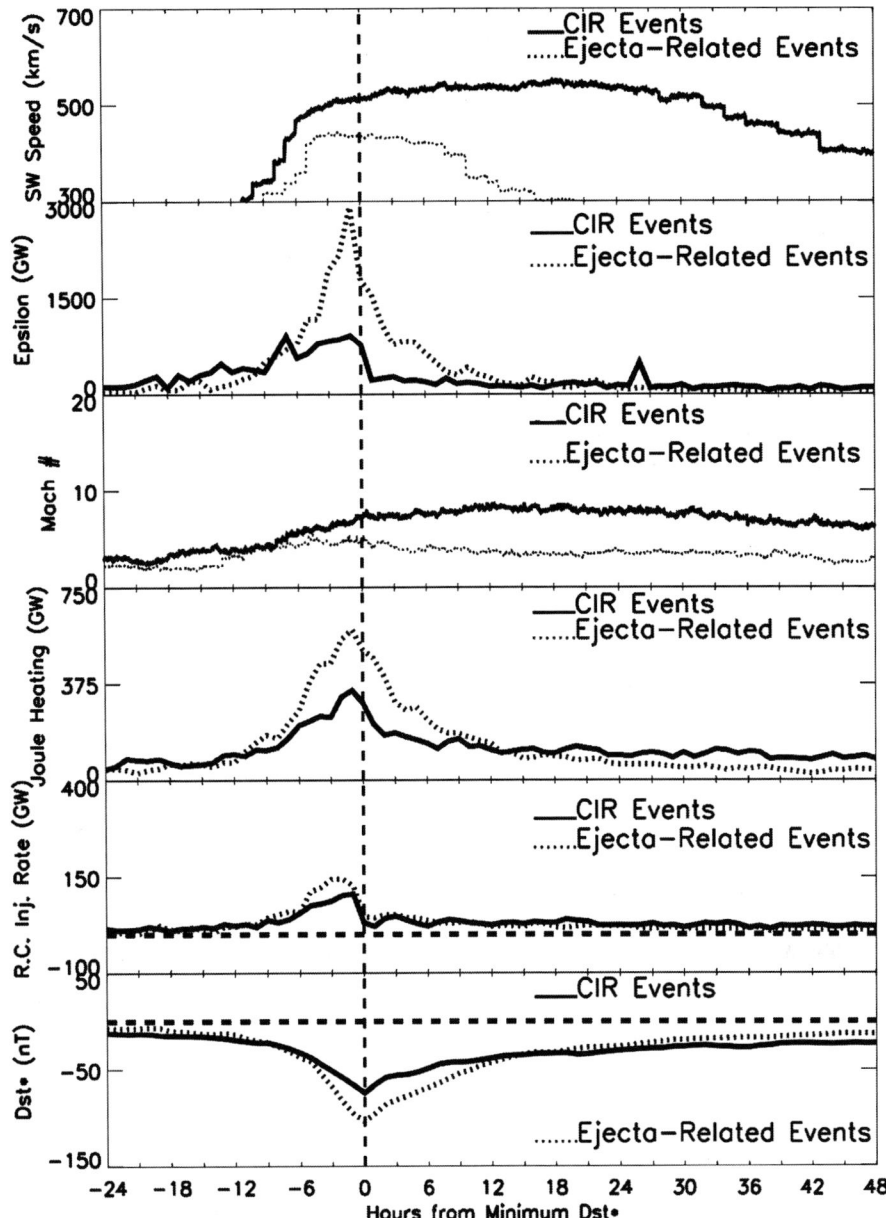

Figure 3. Superposed epoch analysis for solar wind input parameters (Solar wind speed, epsilon, and magnetosonic Mach number) and magnetospheric response (Joule heating, ring current injection rate, and Dst^*) for all events. Vertical line denotes time of minimum Dst^*.

margin as the input power. This difference in coupling efficiency will become more evident in later figures.

Figure 6 shows the input and output power for the main phase. While the input power for ejecta-driven storms is 70% larger than for CIR-driven storms, but the ring current injection rate is only 25% larger, the auroral precipitation rate is almost identical, and the Joule heating rate is 50% larger.

Figure 7 shows the input and output power for the recovery phase. The input power for recovery phase of the storms in the study averages 2.5 times larger for ejecta-driven than CIR-driven storms, and the ring current injection rate is nearly identical, the auroral precipitation rate is nearly identical, and the Joule heating is 33% larger. This gives further evidence of stronger coupling efficiency for CIR-driven storms.

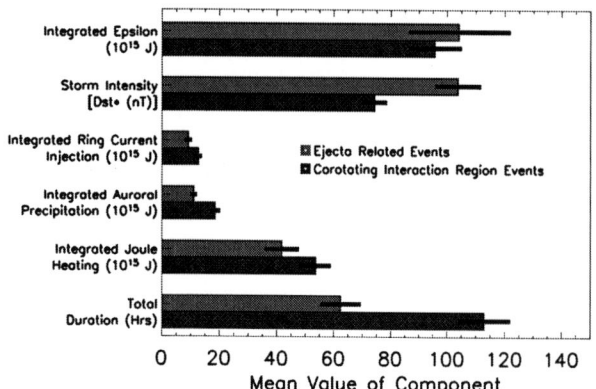

Figure 4. Bar graph depicting storm-integrated solar wind and magnetospheric response parameters for all events.

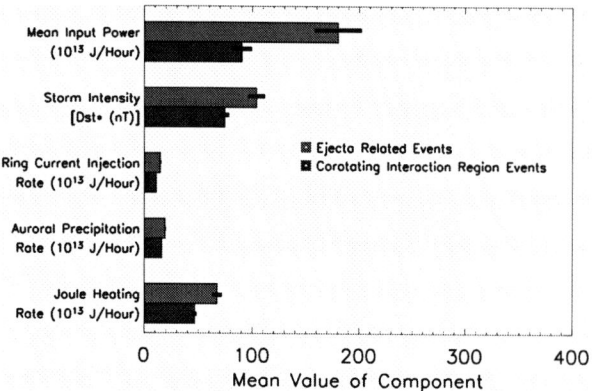

Figure 5. Bar graph depicting values of storm parameters averaged per hour over the storms' duration for all events.

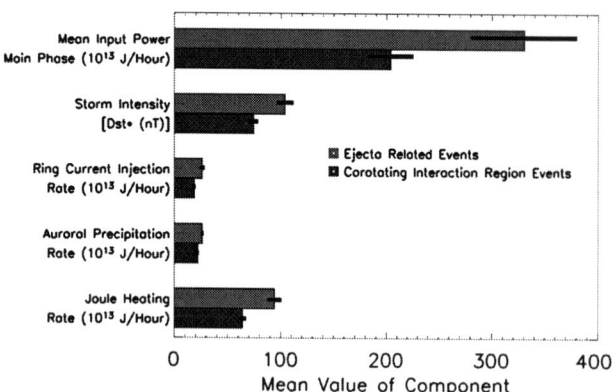

Figure 6. Bar graph depicting values of storm parameters averaged per hour over the storms' main phase for all events.

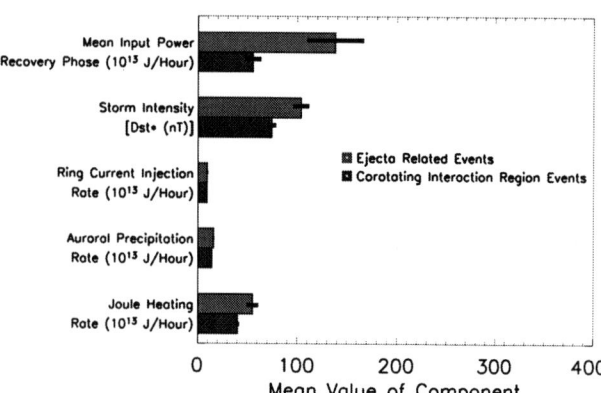

Figure 7. Bar graph depicting values of storm parameters averaged per hour over the storms' recovery phase for all events.

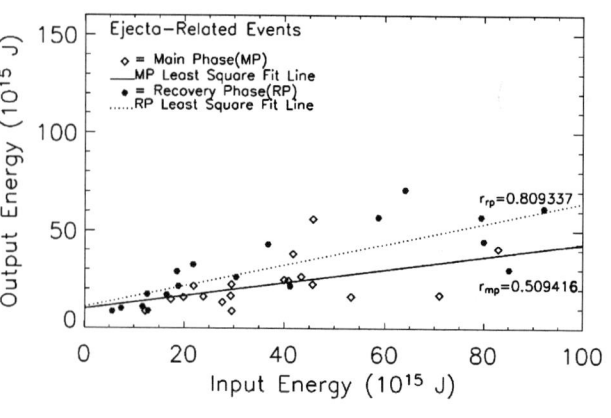

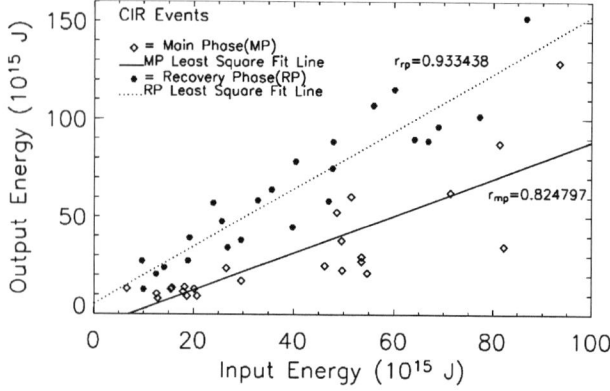

Figure 8. Panel A: Output energy versus input energy (integrated epsilon) for recovery phase and main phase for ejecta-driven storms, shown with linear fit and correlation coefficient for each. Panel B: Same parameters plotted for CIR-driven events.

Figure 8 illustrates in greater detail the differences observed in the preceding plots. In this figure is plotted the output energy on the y-axis and input energy on the x-axis. All storms in the dataset are plotted here. Panel A shows the data for ejecta-driven events. Points corresponding to the main phase are indicated as diamonds, while points corresponding to recovery phase are indicated as filled circles. Two least squares fits are shown. The slopes of these lines approximate

Table 4. Table of energy input, output and geoeffectiveness

Main Phase	Input (10^{15} J)	Ring Current (10^{15} J)	Joule Heating (10^{15} J)	Auroral Precipitation (10^{15} J)	Output/Input
CIR-Driven	40	4.4	17	5.7	0.68
Ejecta-Driven	38	3.3	13	3.5	0.52
Recovery Phase	Input (10^{15} J)	Ring Current (10^{15} J)	Joule Heating (10^{15} J)	Auroral Precipitation (10^{15} J)	Output/Input
CIR-Driven	40	8.2	36	12	1.39
Ejecta-Driven	39	3.3	18	4.8	0.67

the coupling efficiency for main and recovery phases of storms. The slopes of these lines are very different, showing a much greater coupling efficiency for the CIR storms than for the ejecta storms, particularly in the recovery phase.

Table 4 shows the energy input and output for all events. The last column shows the output/input, which is a way to quantify the coupling efficiency. From this measure, the CIR-driven storms show a higher geoeffectiveness than do the ejecta-driven storms.

Figure 9 illustrates the recovery phase energies for storms in different stages of recovery. Panel A shows 20% recovery of the initial Dst^* excursion value, Panel B shows 40%, and Panel C shows 80%. The trend of CIR-driven storms containing greater output energies in the recovery phase is consistent throughout the panels.

5. DISCUSSION

5.1. Solar Wind Driving Conditions

From the B_Z plots in Figure 2 it is clear that the solar wind driving conditions are substantively different in CIRs than ejecta-related events. In the events studied the IMF B_Z moves, on average, towards a more northward configuration over time in CIRs, but keeps oscillating about $B_Z = 0$ for a long duration, typically days. These long-lasting variations could have the effect of driving the magnetosphere long after the main CIR-driven storm has begun to recover. Since it moves repeatedly back into southward B_Z configurations, sporadic reconnection may be driven and energy coupling may be correspondingly enhanced. The ejecta-related events typically show a clearer cutoff to the southward B_Z and the resulting storms appear to recover with less interruption.

During the CIR-driven storms' recovery phases, high-intensity long-duration continuous AE activity (HILDCAA) events are known to occur [e.g., *Tsurutani et al.*, 2004]. These events are a result of the continual driving of the storms during the recovery phase by high-speed solar wind streams and they are evident in the very slow recovery of *Dst* for these events. It is believed that one mechanism of energy transfer during these events involves reconnection associated with the southward components of large amplitude interplanetary Alfvén waves that are present in high-speed streams [*Tsurutani et al.*, 2006 (this volume)].

Another aspect that may affect the coupling efficiency of CIRs is the high solar wind speed itself. As Figure 3 shows, the CIRs have, on average, markedly higher solar wind speeds. As has been pointed out by *Lu et al.* [1998], the kinetic energy flux is much greater than the Poynting flux for typical solar wind conditions. For example, *Lu et al.* [1998] found the kinetic power to be two orders of magnitude higher than the electromagnetic input (epsilon parameter) for the January 1997 storm. So if even a small percentage of kinetic energy flux is important in driving the magnetosphere, it could produce a large effect. Secondly, this also produces an increased magnetosonic Mach number, which may then increase the reconnection rate, thereby coupling more energy into the magnetosphere-ionosphere system. These important differences in the driving conditions may be responsible for enhanced efficiency of energy coupling from CIRs relative to ejecta.

5.2. Geoeffectiveness

Geoeffectiveness was extensively analyzed for the events in this study. It is clear from the bar graphs in Figure 4 that CIRs are more efficient at coupling energy into the magnetosphere than ejecta events. Even with the same input energy available to the system (epsilon), the output was measurably higher in every energy sink evaluated.

Since the CIR-driven events last typically so much longer than the ejecta events, the possibility that this longer duration was the cause of increased output was considered. Since ejecta storms typically produce large Dst^* excursions (see Figures), it could be the case that they couple more energy faster and the CIRs "catch up" only over the long duration of those storms. As seen in Figures 5, 6, 7, and 9, this was not found to be the case. If anything, the power output for CIR and ejecta events looked very similar, so the total energy

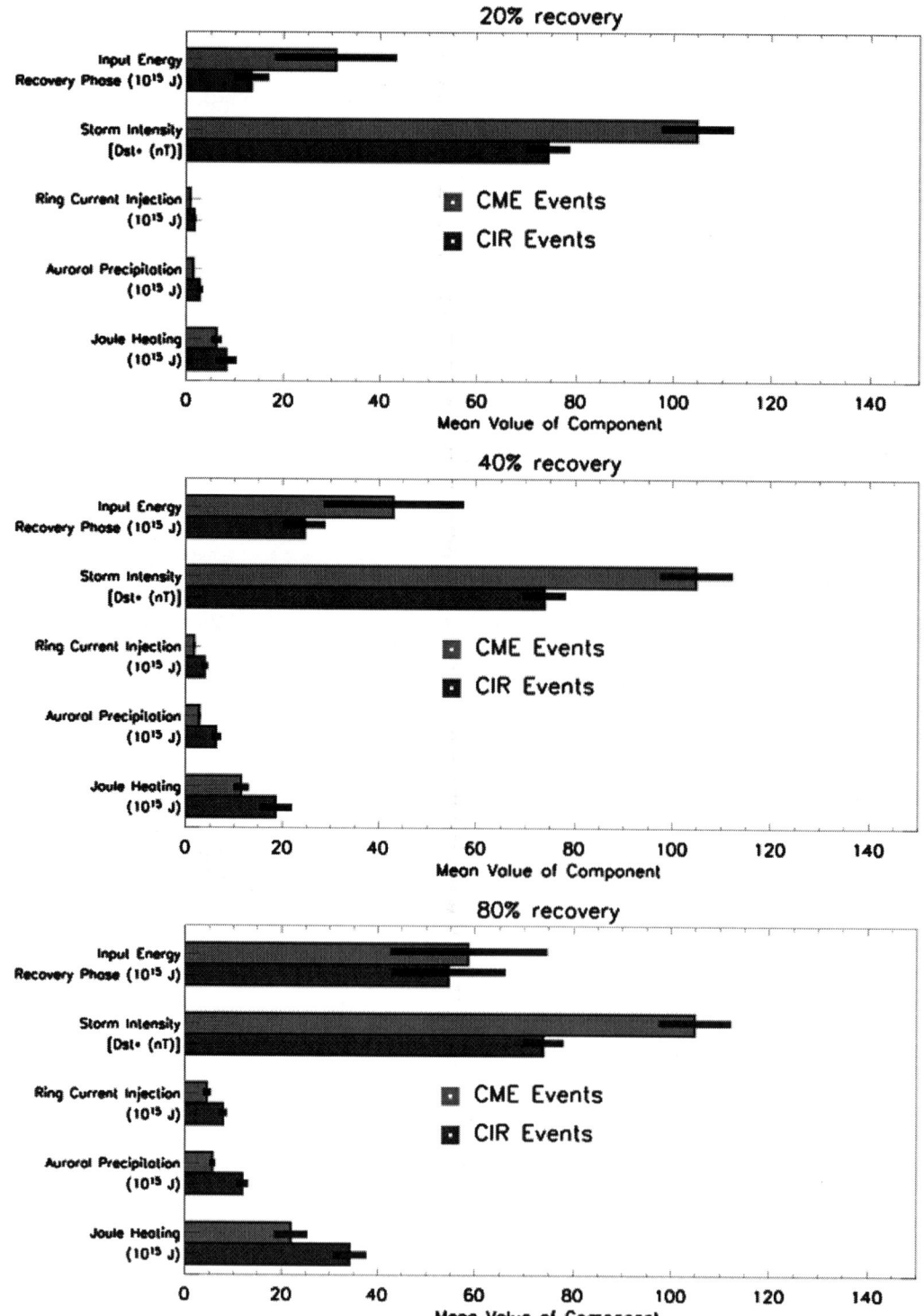

Figure 9. Bar graphs depicting storm parameters for different stages of recovery in all events in the study. Panel A: Values for storms after 20% of Dst^* recovery. Panel B: Values for storms after 40% of Dst^* recovery. Panel C: Values for 80% of Dst^* recovery.

output is higher for CIRs and the per-hour values are similar for both in the main phase. In Figure 9, further confirmation of this is seen as the trends remain for all stages of recovery.

Table 4 shows the output energy divided by the input energy, as an estimate of the coupling efficiency. Clearly, the CIRs are more geoeffective in these events than are the CMEs, and particularly so in the recovery phase.

The fact that the output/input for recovery phase in CIR-driven storms is greater than 1.0 indicates again the limitations of the epsilon parameter, as clearly the output cannot exceed the input. This indicates either that the Poynting flux was underestimated by epsilon or that there is a measurable influence from another type of energy coupling, for example the kinetic energy flux.

5.3. Recovery Phase Energy Coupling

It is in the recovery phase where the dynamics of CIR and ejecta storms diverge the most. Figure 7 shows the power output from both types of storms in recovery phase, where full recovery is considered to be when the Dst^* recovers to 80% of its initial value. Even though the estimated input power for ejecta-driven storms is about 2.5 times that for CIRs, the output power is very similar. This greater coupling efficiency appears as a consistent pattern, as shown clearly in Figures 8 and 9 and Table 4.

From Figure 8 it is clear that the recovery phase behavior of CIR and ejecta-driven storms is very different. Both the clear correlation between input and output energy in the recovery phase and the high geoeffectiveness suggest that the phase of the storm that exists after the Dst^* minimum in a CIR-driven event may not be a pure recovery. This may be due to the repeated excursions into southward B_Z that are characteristic of the solar wind conditions in CIR events. Ejecta-driven events showed weaker responses to recovery-phase driving, which likely accounts for their faster recovery and overall shorter duration.

6. CONCLUSIONS

From this study, it has been shown that the solar wind-magnetosphere energy coupling of CIR-driven and ejecta-driven storms differ in several important properties. In particular, while the ejecta-driven events typically produce greater Dst excursions than the CIR events, the CIRs are more geoeffective in the sense of greater overall energy output than are the ejecta events. When CIR and ejecta events with the same input energies are compared, CIR events have greater energy output. Output power is similar for the two types of events in the storms' main phases, but in recovery phase the CIRs have less input power and the similar output power to ejecta events. More specifically, the measured geoeffectiveness during the recovery phases of CIR-driven storms is double that of ejecta-driven events. These key differences, especially in the recovery phase, are likely due to the rapidly oscillating IMF B_Z that is typical of CIR-driven events well into storm recovery and the energy coupling that produces.

Acknowledgments. The authors acknowledge Ian Richardson for identification of CIR and ejecta events. Work of NT and EM is supported by NSF Career Grant ATM-0454685. EM was also supported by the NIH MARC program. The CEDAR database is sponsored by the NSF. Intersatellite calibration work by BE is supported by National Space Weather Program Grant # 0208145. NT acknowledges fruitful discussions and inspiration from the Chapman Conference on Corotating Solar Wind Streams and Recurrent Geomagnetic Activity. The authors also acknowledge OMNIWeb and the Kyoto Geophysical Data Center for making solar wind and Dst data available and convenient, and gratefully thank the WIND instrument teams for making their data available for use. Finally, the authors thank the reviewers for their valuable comments.

REFERENCES

Baker, D.N., N.E. Turner, and T.I. Pulkkinen, Energy transport and dissipation in the magnetosphere during geomagnetic storms, *J. Atmos. Solar-Terr. Phys.*, 63, 421-429, 2001.

Bargatze, L.F., R.L. McPherron, and D.N. Baker, Solar wind-magnetosphere energy input functions, in *Solar Wind-Magnetosphere Coupling*, Edited by Y. Kamide and J.A. Slavin, 101-109, 1985.

Campbell, W.H., Geomagnetic storms, the Dst ring current myth and lognormal distributions, *J. Atmos. Solar-Terr. Phys.*, 58, 1171-1187, 1996.

Chun, F.K., D.J. Knipp, M.G. McHarg, G. Lu, B.A. Emery, S. Vennerstrom, O.A. Troshichev, Polar cap index as a proxy for hemispheric Joule heating, *Geophys. Res. Lett.*, 26, 1101-1104, 1999.

Dessler, A.J., and E.N. Parker, Hydromagnetic theory of magnetic storms, *J. Geophys. Res.*, 64, 2239-2259, 1959.

Echer, E. and W.D. Gonzalez, Geoeffectiveness of interplanetary shocks, magnetic clouds, sector boundary crossings and their combined occurrence, *Geophys. Res. Lett.*, 31, doi:10.1029/2003GL019199, 2004.

Emery, B.A., D.S. Evans, M.S. Greer, K. Kadinsky-Cade, E. Holeman, F.J. Rich and W. Xu, NOAA and DMSP Intersatellite Adjusted Hemispheric Power Data Sets, *http://cedarweb.hao.ucar.edu* and *http://cedarweb.hao.ucar.edu/instruments/ehp.html, Coupling, Energetics and Dynamics of Atmospheric Regions (CEDAR) Database* at the National Center for Atmospheric Research (NCAR), Boulder, Colorado, USA, 2005.

Emery, B.A., D.S. Evans, M.S. Greer, K. Kadinsky-Cade, E. Holeman, F.J. Rich and W. Xu, The low energy auroral electron and ion hemispheric power after NOAA and DMSP intersatellite adjustments, *NCAR Scientific and Technical Report, TN-470 + STR*, 2006.

Feldstein, Y.I., A.E. Levin, J.U. Kozyra, B.T. Tsurutani, A. Prigancova, L. Alperovich, W.D. Gonzalez, U. Mall, I.I. Alexeev, L.I. Gromova, and L.A. Dremukhina, Self-consistent modeling of large-scale distortions in the geomagnetic field during the 24-27 September 1998 major magnetic storm, *J. Geophys. Res.*, 110, A11214, doi:10.1029/2004JA010584, 2005.

Gonzalez, W.D., B.T. Tsurutani, A.L.C. Gonzalez, E.J. Smith, F. Tang, and S.-I. Akasofu, Solar wind-magnetosphere coupling during intense magnetic storms (1978-1979), *J. Geophys. Res.*, 94, 8835-8851, 1989.

Gonzalez, W.D., B.T. Tsurutani, and A.L. Clúa de Gonzalez, Interplanetary origin of geomagnetic storms, *Space Science Reviews*, 88: 529-562, 1999.

Huttunen, K.E., H.E. Koskinen, R. Schwenn, Variability of magnetospheric storms driven by different solar wind perturbations, *J. Geophys. Res.*, 107, doi:10.1029/2001JA900171, 2002.

Knipp, D.J., B.A. Emery, M. Engebretson, X. Li, A.H. McAllister, T. Mukai, S. Kokubun, G.D. Reeves, D. Evans, T. Obara, X. Pi, T. Rosenberg, A. Weatherwax, M.G. McHarg, F. Chun, K. Mosely, M. Codrescu, L. Lanzerotti, F.J. Rich, J. Shriver, and P. Wilkinson, An overview of the early November 1993 geomagnetic storm, *J. Geophys. Res.*, 103, 26197, 1998.

Knipp, D.J., W.K. Tobiska, and B.A. Emery, Direct and indirect thermospheric heating sources for solar cycles 21-23, *Solar Physics*, 224, 495-505, 2004.

Lu, G., D.N. Baker, R.L. McPherron, C.J. Farrugia, D. Lummerzheim, J.M. Ruohoniemi, F.J. Rich, D.S. Evans, R.P. Lepping, M. Brittnacher, X. Li, R. Greenwald, G. Sofko, J. Villain, M. Lester, J. Thayer, T. Moretto, D. Milling, O. Troshichev, A. Zaitzev, V. Odintzov, G. Makarov, and K. Hayashi, Global energy deposition during the January 1997 magnetic cloud event, *J. Geophys. Res.*, 103, 11,685-11,694, 1998.

Monreal-MacMahon, R. and W.D. Gonzalez, Energetics during the main phase of geomagnetic superstorms, *J. Geophys. Res.*, 102, 14,199-14,207, 1997.

O'Brien, T.P., and R.L. McPherron, An empirical phase space analysis of ring current dynamics: Solar wind control of injection and decay, *J. Geophys. Res.*, 105, 7707-7719, 2000.

O'Brien, T.P., R.L. McPherron, and M.W. Liemohn, Continued convection and the initial recovery of Dst, *Geophys. Res. Lett.*, 29(23) 2143, doi:10.1029/2002GL015556, 2002.

Ohtani, S., M. Nosé, G. Rostoker, A.T.Y. Lui, and M. Nakamura, Storm-substorm relationship: Contribution of the tail current to Dst, *J. Geophys. Res.*, 106, 21999-21209, 2001.

Perreault, P., and S.-I. Akasofu, A study of geomagnetic storms, *Geophys. J. R. Astr. Soc.*, 54, 547, 1978.

Pulkkinen, T.I., N. Ganushkina, D.N. Baker, N.E. Turner, J.F. Fennell, J. Roeder, T.A. Fritz, M. Grande, B. Kellett, G. Kettmann, Ring current ion composition during solar minimum and rising solar activity: Polar CAMMICE/MICS results, *J. Geophys. Res.*, 106, 19131-19147, 2001.

Sckopke, N., A general relation between the energy of trapped particles and the disturbance field near the Earth, *J. Geophys. Res.*, 71, 3125-3130, 1966.

Tsurutani, B.T., W.D. Gonzalez, F. Guarnieri, Y. Kamide, X. Zhou, and J.K. Arballo, Are high-intensity long-duration continuous AE activity (HILDCAA) events substorm expansion events?, *J. Atmos. & Solar-Terr. Phys.*, 66, 167-176, 2004.

Tsurutani, B.T., N. Gopalswamy, R.L. McPherron, W.D. Gonzalez, G. Lu, F.L. Guarnieri, Magnetic storms caused by corotating solar wind streams, *AGU Geophysical Monograph Series, this volume*, 2006.

Turner, N.E., D.N. Baker, T.I. Pulkkinen, and R.L. McPherron, Evaluation of the tail current contribution to Dst, *J. Geophys. Res.*, 105, 5431-5439, 2000a.

Turner, N.E., Solar wind-magnetosphere coupling and global energy budgets in the Earth's magnetosphere, *Doctoral Dissertation*, 2000b.

Turner, N.E., D.N. Baker, T.I. Pulkkinen, J.L. Roeder, J.F. Fennell, and V.K. Jordanova, Energy content in the storm time ring current, *J. Geophys. Res.*, 106, 19149-19156, 2001.

Vichare, G., S. Alex, and G.S. Lakhina, Some characteristics of intense geomagnetic storms and their energy budget, *J. Geophys. Res.*, 110, A03204, doi:10.1029/2004JA010418, 2005.

Weiss, L.A., P.H. Reiff, J.J. Moses, R.A. Heelis, and B.D. Moore, Energy dissipation in substorms, in *Substorms 1*, edited by B. Hultqvist and S. Akasofu, 309-318, 1992.

Zhang, J., M.W. Liemohn, J.U. Kozyra, B.J. Lynch, and T.H. Zurbuchen, A statistical study of the geoeffectiveness of magnetic clouds during high solar activity years, *J. Geophys. Res.*, 109, A09101, doi:10.1029/2004JA010410, 2004.

Niescja E. Turner and Elizabeth J. Mitchell, Physics and Space Sciences, Florida Institute of Technology, Melbourne, Florida 32901

Delores J. Knipp, Physics Department, US Air Force Academy, USAFA, Colorado 80840

Barbara A. Emery, High Altitude Observatory, National Center for Atmospheric Research, Boulder, Colorado 80307

The Solar Wind and Geomagnetic Activity as a Function of Time Relative to Corotating Interaction Regions

Robert L. McPherron

*Institute of Geophysics and Planetary Physics and Department of Earth and Space Science
University of California, Los Angeles, California, USA*

James Weygand

Institute of Geophysics and Planetary Physics at University of California, Los Angeles, California, USA

Corotating interaction regions during the declining phase of the solar cycle are the cause of recurrent geomagnetic storms and are responsible for the generation of high fluxes of relativistic electrons. These regions are produced by the collision of a high-speed stream of solar wind with a slow-speed stream. The interface between the two streams is easily identified with plasma and field data from a solar wind monitor upstream of the Earth. The properties of the solar wind and interplanetary magnetic field are systematic functions of time relative to the stream interface. Consequently the coupling of the solar wind to the Earth's magnetosphere produces a predictable sequence of events. Because the streams persist for many solar rotations it should be possible to use terrestrial observations of past magnetic activity to predict future activity. Also the high-speed streams are produced by large unipolar magnetic regions on the Sun so that empirical models can be used to predict the velocity profile of a stream expected at the Earth. In either case knowledge of the statistical properties of the solar wind and geomagnetic activity as a function of time relative to a stream interface provides the basis for medium term forecasting of geomagnetic activity. In this report we use lists of stream interfaces identified in solar wind data during the years 1995 and 2004 to develop probability distribution functions for a variety of different variables as a function of time relative to the interface. The results are presented as temporal profiles of the quartiles of the cumulative probability distributions of these variables. We demonstrate that the storms produced by these interaction regions are generally very weak. Despite this the fluxes of relativistic electrons produced during these storms are the highest seen in the solar cycle. We attribute this to the specific sequence of events produced by the organization of the solar wind relative to the stream interfaces. We also show that there are large quantitative differences in various parameters between the two cycles.

1. INTRODUCTION

At the Earth the declining phase of the solar cycle is characterized by weak magnetic storms that recur with a period of 27 days. There are no obvious features in the Sun's photosphere to account for these storms so it was originally postulated that they were caused by magnetically effective or "M regions" [*Chapman and Bartels*, 1962]. With the advent of the space age it was found that these storms began with the arrival of a high-speed stream of solar wind [*Hundhausen*, 1972]. When these streams were tracked back to the Sun and compared to maps of the solar magnetic field it was evident that the high-speed streams originated in large regions of unipolar magnetic field. With the launch of Skylab in 1973 it was shown that above these magnetic regions the corona was much darker than elsewhere. These regions were therefore named coronal holes.

It is neither the coronal holes nor the high-speed stream that cause these magnetic storms. Instead it is the interaction region between the high-speed stream and the slow solar wind ahead of it that creates the conditions that drive geomagnetic activity (see *Balogh et al.* [1999] for summary of properties of CIRs). This region is centered on the interface between the two streams [*Gosling et al.*, 1978]. Because of the interplanetary magnetic field (IMF) the two streams can not interpenetrate. Consequently the high-speed plasma is slowed and deflected east of the Sun while the slow-speed plasma is accelerated and deflected west of the Sun. The plasma and magnetic field on either side of the interface is compressed with the total pressure and total magnetic field rising to peak values at the interface.

Since the Sun is rotating the stream interface is a spiral intermediate to that expected in the two streams. The peak in plasma pressure at the interface propagates outward into both streams carrying information about the presence of the interface. Inside of 1 AU these are ordinary pressure waves, but beyond this distance changes in the properties of the solar wind with distance cause these two waves to develop into shocks. The region between the two waves is called a corotating interaction region or CIR.

There are several factors associated with the CIR that cause geomagnetic activity. First is the compression of the magnetic field [*Belcher and Davis*, 1971]. A stronger magnetic field means a larger z-component of the magnetic field in GSM coordinates. When this component is negative the IMF merges with the Earth's magnetic field and drives magnetic activity. Second is the possibility that there is an increase in the fluctuations of the IMF near the interface because of the shear in velocity across the interface [*Belcher and Davis*, 1971]. More and larger fluctuations lead to stronger GSM B_z and more geomagnetic activity. A third and more important factor is that high-speed streams tend to be filled with Alfven waves propagating away from the Sun [*Belcher and Davis*, 1971]. Following the interface the field contains large amplitude fluctuations, which if southward in GSM coordinates, drive activity. Finally, a fourth factor is the high solar wind velocity on the Sunward side of the interface.

The actual driver of geomagnetic activity is the GSM dawn-dusk electric field, VB_z. The combination of large V and frequent strong intervals of southward B_z causes elevated geomagnetic activity.

The association of geomagnetic activity with stream interfaces leads to the possibility of forecasting geomagnetic activity based on predictions of the arrival of the interface. Activity is weak before the interface, very strong at the interface, and then decaying slowly after the interface. This type of forecasting has been referred to as "probabilistic forecasting by air mass climatology" [*McPherron and Siscoe*, 2004]. In this case the air mass is the CIR and the climatology is the average behavior of various indices of magnetospheric activity relative to the interface.

The purpose of this paper is to investigate the average behavior of a number of solar wind and magnetospheric variables relative to a stream interface. We are able to do this well in two solar cycles since full-time monitoring of the solar wind began in January 1995 and continues now 10 years later. The minimum of the previous cycle #22 occurred in June 1996 and the minimum of cycle #23 is expected in December 2006. Thus we have data from the declining phase of an even-numbered and an odd-numbered solar cycle. In this paper we compare the behavior of the solar wind and geomagnetic activity relative to stream interfaces in 1995 and 2004.

Our primary analysis technique is superposed epoch analysis of an ensemble of traces of a given variable relative to the times of the stream interfaces in each year. The results are presented as time series plots of the quartiles of the cumulative probability distributions. The range between the upper and lower quartile at any give epoch time provides a means of probabilistic forecasting provided one can predict the arrival time of the interface.

Some of our results for the solar wind plasma are well known from previous analysis [*Borrini et al.*, 1981; *Gosling et al.*, 1978; *Gosling et al.*, 1981]. However, we have extended the previous work by including the behavior of the interplanetary magnetic field (not previously published), and of several magnetospheric activity indices. A crude analysis of some of the latter has been published [*Schatten and Wilcox*, 1967; *Wilcox and Ness*, 1965; *Wilcox et al.*, 1967]. Also new is our comparison of the behavior of the solar wind and indices in two successive solar cycles.

We will show that all solar wind variables exhibit highly systematic behavior relative to the time of an interface, but that there is a significant quantitative difference between the two solar cycles. We attribute this difference to a combination of three factors. The first is the Russell-McPherron (RM) effect [*Russell and McPherron*, 1973]. The RM effect is most important around the equinoxes when at certain universal times each day the Earth's dipole approaches a tilt of 34° with respect to the ecliptic pole. Since the GSM-z axis is defined as a vector perpendicular to the Sun vector lying in a plane containing the Sun vector and the dipole, it is also tilted towards the ecliptic plane by this amount. An interplanetary magnetic field lying in the ecliptic plane along the Parker spiral then has a large projection of By on the GSM-z axis. If this projection is negative magnetic reconnection and geomagnetic activity are produced. Since the GSM coordinate system is fixed in the Earth with its x-axis pointing to the Sun this system rotates once per year. Thus the orientation of the GSM-y axis reverses from one equinox to the next, reversing the sign of the IMF By projection on GSM-z. A mnemonic rule characterizing which orientations of the IMF are geo-effective is "spring to fall away". Here "to" means the IMF points inward towards the Sun and "away" means it points outward.

The second factor is the Rosenberg-Coleman (RC) effect [*Rosenberg and Coleman*, 1969] which is also most important near the Equinoxes. The rotation axis of the Sun is tilted 7° with respect to the ecliptic plane so that near spring equinox on March 5 the Earth is at maximum southern heliographic latitudes and near fall equinox on September 6 it is at maximum northern latitudes. *Rosenberg* and *Coleman* discovered that the dominant polarity of the IMF at times of most northern and southern latitude is the same as the polarity of the corresponding pole on the Sun. Thus when the northern pole of the Sun is positive the IMF is away from the Sun above the heliographic equator. The Earth is above the equator in fall and according to the RC effect will be dominated by IMF pointing away from the Sun. According to the RM rule this is a geo-effective orientation. Six months later the Earth will be at high southern latitudes where the IMF is toward the Sun. This situation is also geo-effective. Thus throughout an 11-year solar cycle the ordinary IMF is conducive to the production of geomagnetic activity. Observations of the solar magnetic field have established that the phase of the 22-year cycle is such that the Sun's north magnetic pole becomes positive (outward field) just after the maximum of even numbered solar cycles. The last even cycle #22 reached maximum in August 1989. Thus from about 1991 to 2002 the orientation of the IMF was conducive to geomagnetic activity.

About two years after solar maximum the polarity of the Sun's magnetic field reverses. In 2002 the northern pole became negative and the IMF above the equatorial plane was toward the Sun. The Earth is above the equator in fall and needs an IMF pointing away from the Sun to produce geomagnetic activity. Thus this orientation of the IMF is not geo-effective. This situation persists for the entire 11-year cycle until the solar field again reverses. This combination of the RM and RC effects creates a 22-year cycle in geomagnetic activity.

In addition to the two preceding geometrical effects there appears to be an intrinsic 22-year variation of solar activity such that "...the maxima of odd-numbered cycles in even-odd pairs are always larger." [*Cliver et al.*, 1996]. This 22-year cycle of sunspot activity is called the "Hale cycle" [*Hale et al.*, 1919]. *Cliver et al.* argue that this variation in solar activity is the primary cause of the 22-year cycle in geomagnetic activity.

2. DATA BASE AND ANALYSIS METHOD

The data used in this investigation were obtained from the NSSDC as either binary CDFs or ASCII files. For download we selected a subset of the original data corresponding to plasma and magnetic field measurements in GSE coordinates. These data were interpolated to 1-minute resolution using cubic splines. The data were then propagated to the subsolar bow shock (+17 Re, 0, 0) using a modified version of the Weimer minimum variance algorithm [*Bargatze*, 2005; *Weimer et al.*, 2003]. At the chosen point the data were again interpolated to 1-minute samples and the results transformed to GSM coordinates.

An interactive program was developed to display the data at high resolution so that a cross hair could be used to define the time of a stream interface imbedded within a corotating interaction region. Figure 1 shows the parameters used to define an interface in this analysis. They include solar wind velocity and density, IMF field strength, and the azimuthal flow angle in GSE coordinates. A stream interface was

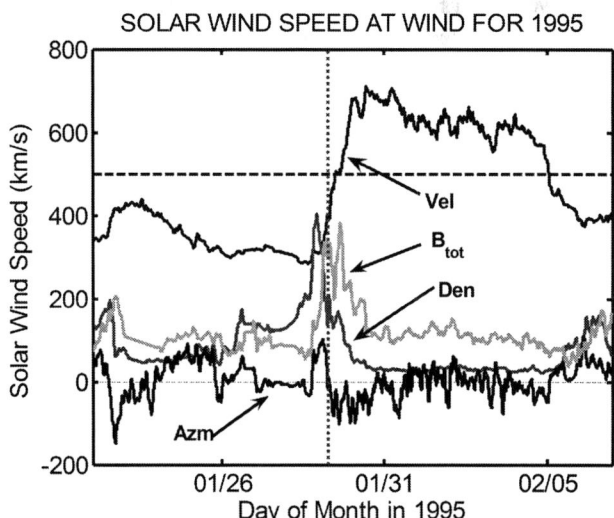

Figure 1. Selection of stream interfaces was performed interactively using a cursor to choose the time of a zero crossing in the azimuthal flow angle. Only crossings associated with an increase in solar wind velocity were selected. The density and magnetic field usually peak before and after the interface respectively.

defined by the following criteria. (1) The solar wind velocity changed rapidly from a value below to a value above 500 km/s. (2) The velocity decreased slowly from its elevated value over a number of days. (3) A peak in density followed by a peak in total field was associated with the rapid rise in velocity. (4) The azimuthal flow changed from positive to negative angles. The time of the zero crossing in flow angle was selected as the time of the stream interface.

These criteria are based on previous results reported by *Gosling* [1978] and *Borrini et al.*, [1981]. Although these studies used different reference times in their superposed epoch analysis than we have they found that solar wind velocity is below 400 km/s before a CIR and near 600 km/s after the CIR. They also found that the velocity increase occurs in about 3-4 days, but the decrease takes 10 days. They showed that the solar wind density and thermal pressure peak near the time of most rapid rise in velocity. In examining plots of solar wind data we found that the time of most rapid rise in velocity is close to the time of a significant zero crossing in the azimuthal flow angle of the solar wind. Taking into consideration the physical explanation for the formation of a CIR it is clear that the zero crossing of the flow angle is the boundary between the slow-speed and high-speed plasma flows. Comparison of the alpha to proton density ratio in regions of positive and negative flow angles demonstrates that the two plasmas are quite different.

Stream interfaces satisfying these criteria are generally observed only in the declining phase of the solar cycle a few years before solar minimum. In the last solar cycle (#22) well developed streams were observed in the years 1994-1996. Since we do not have high resolution solar wind data for the year 1994, and because the stream structure vanished in mid-1996, for this study we used data only from the year 1995. A total of 26 interfaces were found in this year. For the current solar cycle (#23) we used data for the year 2004, the last year for which complete data is available. In 2004 we identified 42 interfaces.

A stack plot showing the recurrent nature of high-speed streams during 2004 is presented in Figure 2. Each trace in the figure displays 30 days of data starting three full days before the start of a 27-day Bartel's rotation interval and ending two days after. The time axis is day in the rotation interval. The vertical dashed line on the left denotes the beginning of day 1 of each 27-day interval. The dashed line at the right signifies the end of day 27 and also the beginning of day 1 of the next interval. The times of stream interfaces are shown by triangles. A persistent stream interface occurring on about the 15th day of each Bartel's rotation period is evident near the middle of each trace.

The lists of stream interface times were used to select 10-day segments of data centered on every stream interface identified in a particular year. These segments were stored as

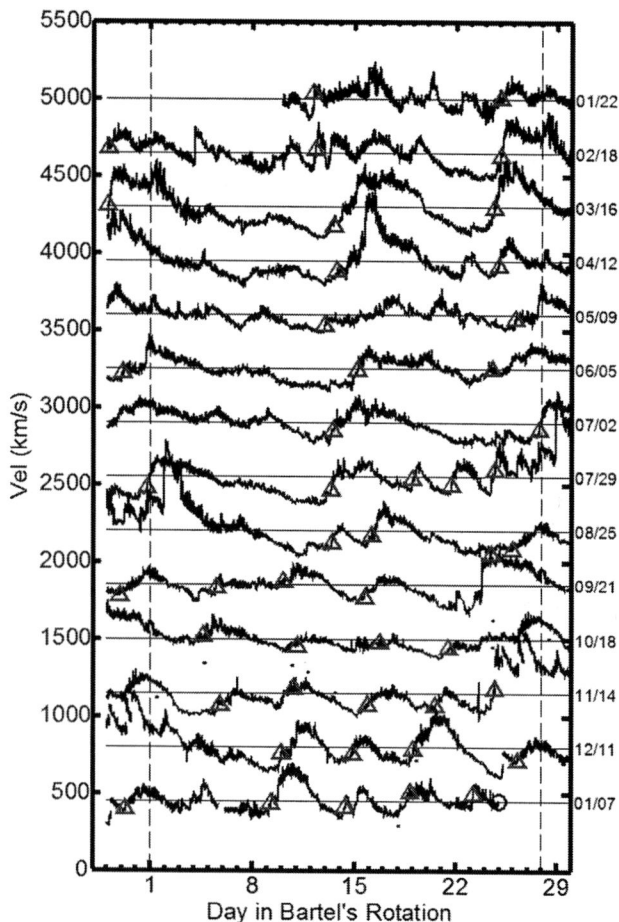

Figure 2. The Solar wind velocity during the year 2004 is displayed as a stack plot with 30 day segments advanced 27 days in successive rows. The date of the last point in each trace is shown along the right side. Vertical dashed lines denote the beginning and end of each 27-day interval. Triangles depict the times selected as a stream interface.

rows of an "ensemble array" for each year. An analysis window of width 2-hours (121 samples) was stepped across the array to determine the cumulative distribution function (CDF) as a function of epoch time. In each step all data points falling within the window (121 * # events) were used to calculate a CDF. The CDF was then sampled at all equally spaced values of the independent variable lying within the range of the lower and upper limits of the graph. If there were no occurrences of a particular value in the analysis window the CDF was set to a flag. The array was then contoured at 10-percentile levels. A plot showing the CDFs of solar wind velocity for the years 1995 and 2004 is presented in Figure 3. Heavy lines in these panels depict the quartiles of the CDF.

The presence of persistent high-speed streams in both years is clearly evident. Before the stream interface the

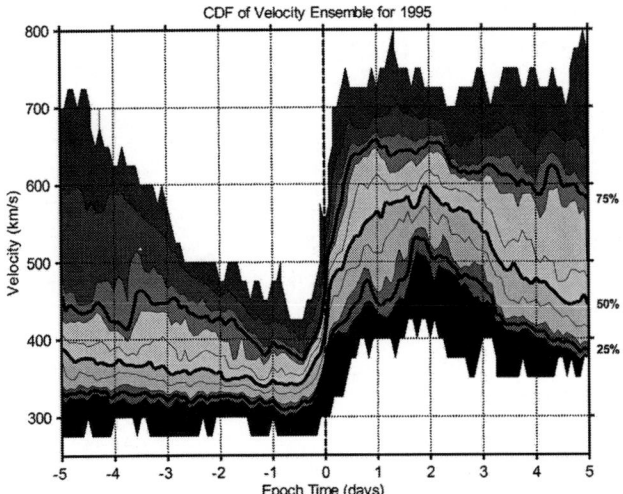

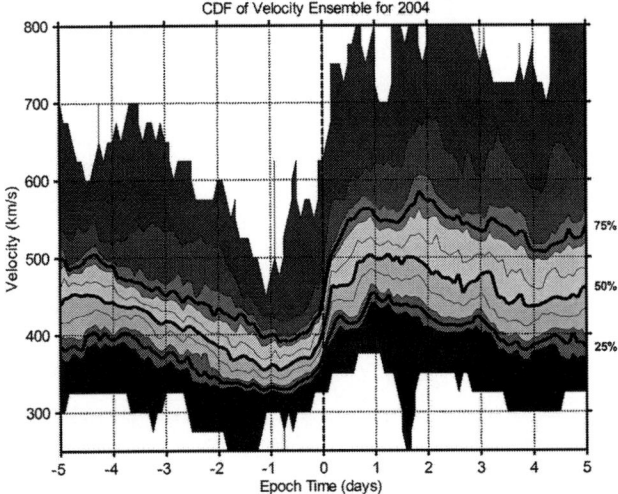

Figure 3. Cumulative distribution functions of the solar wind velocity ensembles in 1995 and 2004 relative to the time of a stream interface. Thin contour lines are drawn for every 10-percentile level Heavy lines show the quartiles of the distribution of values at each epoch time.

median solar wind velocity is about 350 km/s. At the interface the velocity has increased and is rising at its most rapid rate. Two days after the interface the median velocity peaks at >500 km/s and thereafter decays slowly. The streams in 1995 are somewhat better developed than those in 2004 with more contrast between the low velocities before the interface and high velocities after. Possibly this is because the streams are not yet fully developed in 2004.

The same superposed epoch analysis was performed for a number of solar wind and magnetospheric variables using the times of the stream interfaces as epoch zero. In the following presentation we show only the quartiles of each cumulative probability distribution.

3. RESULTS OF ANALYSIS

3.1. Variations of Solar Wind Near Stream Interface

Results of our superposed epoch analysis are summarized in Figures 4-7. Figure 4 presents results for five solar wind variables in the two different years. From the top down these include azimuthal flow angle, solar wind velocity, density, mean ion thermal speed, and total pressure given by the sum of the magnetic and thermal pressure. The vertical dashed line at zero epoch time is the time of the stream interface. The range of variation defined by the upper and lower quartiles of the ensemble of each variable is shown by shaded patchs. Superimposed on each patch is a heavy line depicting the median variation of the variable. Data for 1995 are presented on the left side and data for 2004 are on the right side. We begin by describing the behavior of each variable in the year 1995 and then later contrast the behavior in the two years.

The top panel presents the azimuthal flow angle used to define the time of stream interfaces. The flow direction changes systematically over a two-day interval peaking six hours before and after the interface. The median deflection at the extrema is about five degrees. The sense of the deflection before the interface is westward in the direction of the Earth's motion. This graph verifies the well known result that the flow of the solar wind is deflected westward before the interface between a high-speed and slow-speed stream, and is deflected eastward behind.

The second panel repeats the quartiles of the solar wind velocity discussed above. The velocity begins to increase about 8 hours before the interface, and is increasing most rapidly at the interface. Median velocity peaks about two days after the interface.

The third panel presents the quartiles of the solar wind density. Density begins to slowly increase from a value near 8 particles per cc about 2.5 days before the interface. It reaches a peak of 25 per cc a few hours before the interface and at the interface drops rapidly to a constant value of about 4 per cc. The variation is quite asymmetric relative to the interface with all of the increase and the peak occurring before the interface.

The mean thermal speed of the ions is presented in the fourth panel. In the low velocity solar wind before the stream interface the thermal speed is low with a value near 25 km/s. About 12 hours before the interface the thermal speed begins to increase achieving a peak value of 60 km/s about 4-6 hours after the interface. It subsequently decreases slowly as the bulk velocity also decreases.

The bottom panel presents the total pressure (sum of thermal and magnetic pressure) of the solar wind. Pressure begins to increase 12 hours before the interface. It reaches its

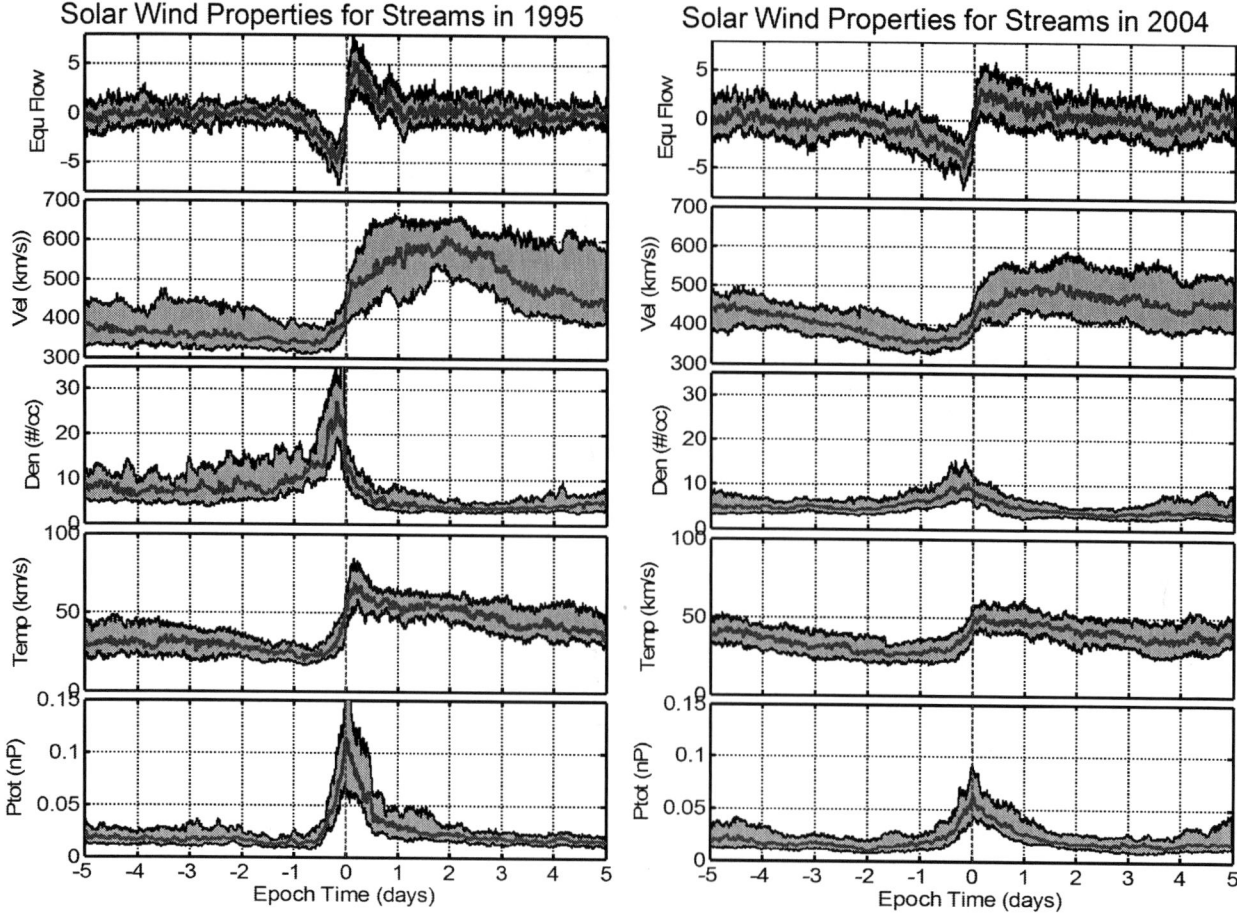

Figure 4. Quartiles of the cumulative distribution of various solar wind variables measured by Wind are plotted versus time relative to stream interfaces. Left panels show data from near solar minimum in 1995 and right panels show a similar period approaching minimum in 2004. From the top down the panels include: azimuthal flow angle, velocity, density, temperature, and total pressure. The 2004 interval is clearly less disturbed.

peak value exactly at the interface, and then decreases to normal values about a day after the interface.

The behavior of these five solar wind plasma parameters during the current solar cycle is shown in the panels on the right side of Figure 4. Qualitatively the behavior is the same as in 1995, but quantitatively the streams appear to be weaker. Difference in peak values range from 25% to 50% lower. This difference could be a result of the streams not being fully developed. However, it is also possible that the current solar cycle is producing weaker streams, and as we will show below, weaker magnetospheric activity.

3.2. Variations of the Interplanetary Magnetic Field

The behavior of the IMF near a stream interface is summarized in Figure 5. The top panel is a repeat of the azimuthal flow angle used to define the time of a stream interface. Panel 2 presents the total magnetic field. The behavior of the total magnetic field is nearly identical to that of the total pressure shown in the previous figure. It begins to increase 12 hours before the interface; reaches a maximum of 15 nT at the interface; and returns to normal about a day after the interface.

The third panel shows IMF B_z in GSE coordinates. B_z is highly variable on the time scale of these plots and the quartiles reflect the magnitude of fluctuations. These are controlled by the behavior of the total field relative to the stream interface. Thus B_z begins to increase 12 hours before the interface; it peaks at the interface with values near 5 nT; and it decays back to near normal values within a day. There is some indication that the level of fluctuations is marginally higher than normal for at least two days after the interface. A careful examination of the median trace reflects the unusual behavior that occurred in the 22nd solar cycle. At stream

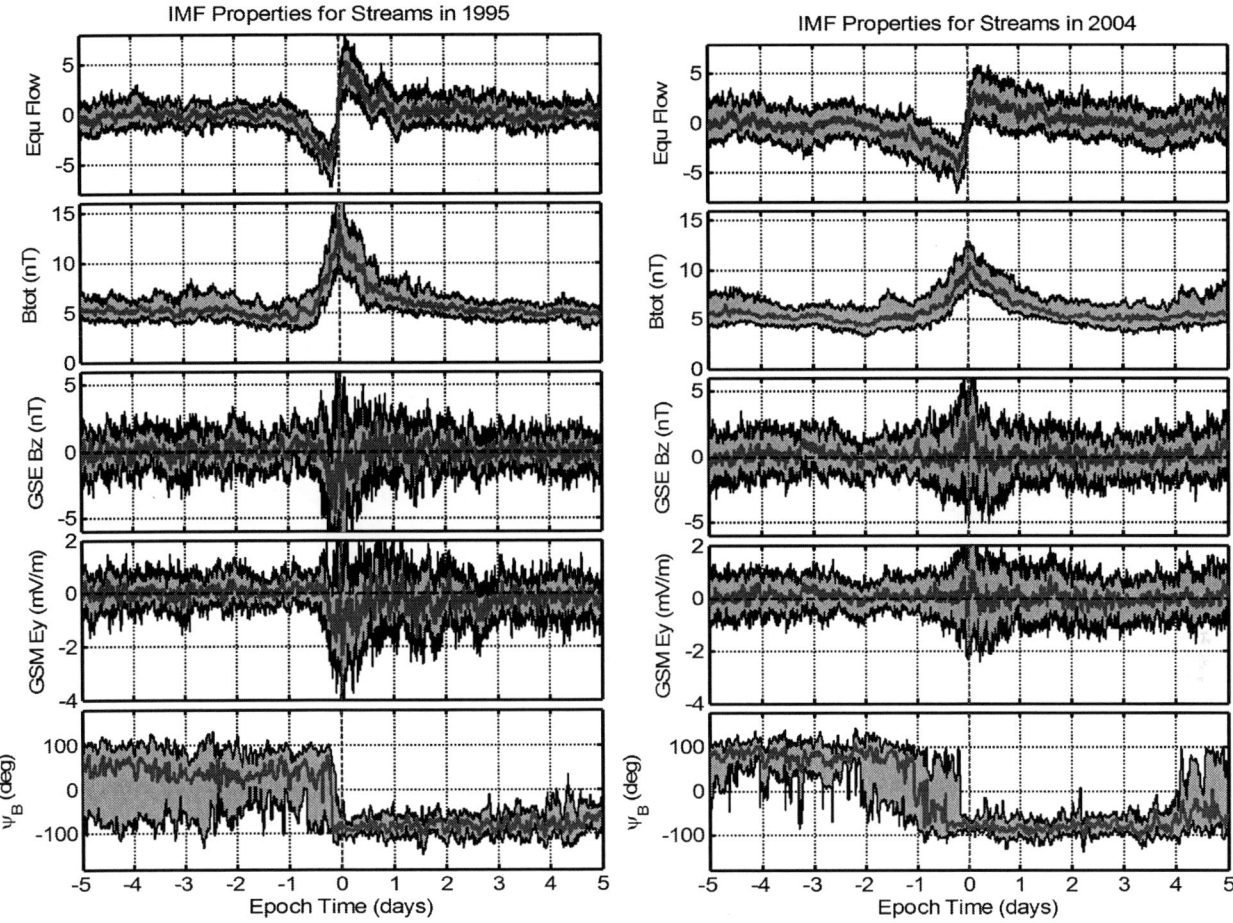

Figure 5. A quartile plot similar to Figure 4 for IMF variables as function of time relative to stream interfaces. From the top down the panels include: azimuthal flow angle, magnetic field magnitude, Bz in GSE, Ey in GSM and the spiral angle of the IMF in the ecliptic plane. The IMF field magnitude and hence Bz and Ey are smaller in 2004 than in 1995.

interfaces the median B_z was biased southward. Both this and the increased magnitude of fluctuations are partially responsible for elevated magnetic activity at the times stream interfaces pass the Earth.

Panel 4 presents the dawn-dusk component of the solar wind electric field (VBs) in GSM coordinates. This quantity is the primary driver for geomagnetic activity. Fluctuations in E_y are very strong at the interface but are elevated for many days afterward. To some extent this reflects the behavior of fluctuations in B_z. However, it is primarily a consequence of the high velocity of the stream following the interface. A second and very important characteristic of E_y in GSM coordinates is a persistent negative bias in the median value that lasts for at least four days. Although not demonstrated here this bias is a consequence of the Russell-McPherron effect discussed in the introduction. During solar cycle #22 virtually all high-speed streams that occurred near equinox were geo-effective according to the Russell-McPherron rule "spring to fall away".

The foregoing result does not imply that the IMF was unidirectional as is demonstrated in the fifth panel. This panel shows the spiral angle of the IMF in a coordinate system rotated 45° counter clockwise about the GSE z-axis. In this system an IMF pointing sunward makes an angle of –90° relative to the rotated x-axis. Note, however, that for fall data we have reversed the sign of the spiral angle so that the spiral angle has the same sign as in spring data. In addition this panel was constructed using only geo-effective events which in this case was 22 our of 26 events.

It is apparent that before the stream interface the selected events were geo-ineffective, i.e. an IMF in the ecliptic plane projects onto the GSM-z axis with a positive projection. (+90° implies an away sector which is ineffective in spring). However, the Earth passed through the heliospheric current sheet before the stream interface reversing the polarity of the IMF so that the IMF was geo-effective during the high-speed stream. Usually this passage occurred about 6 hours before the interface, however, some transitions occured earlier. The preponderance of geo-effective orientation of the IMF during the high-speed stream is the cause of the persistent bias in GSM E_y shown in the fourth panel.

The right hand panels of Figure 5 show the IMF variations during 2004. The qualitative behavior is similar to what has already been presented but all median values are weaker. However, a more important difference is that B_z fluctuations were weaker at the interface and there was no bias in the GSM E_y after the interface. The spiral angle for geo-effective events is plotted in the bottom panel. During 2004 the polarity of the IMF in the high-speed stream was almost equally distributed between geo-effective and ineffective orientations. In this panel we show only the 16 of 42 events that were geo-effective in the high-speed stream.

3.3. Variations of Magnetospheric Activity

The relation of geomagnetic activity to stream interfaces is summarized in Figure 6. Panel 1 repeats the flow angle used to define the interface. Panel 2 shows solar wind dynamic pressure. In 1995 the dynamic pressure variation is slightly asymmetric relative to the interface. The median curve begins to increase about one day before the interface; it peaks at 6 nPa just before the interface; and it returns to normal values of 2 nPa by 12 hours after the interface. Panel 3 shows the GSM E_y component described earlier. Panel 4 illustrates the behavior of the 3-hr ap index. For this analysis we used nearest neighbor interpolation to change the time resolution to one minute. The constant nature of ap for three-hour intervals is not apparent because the time of epoch zero for each data segment is randomly phased relative to the time the ap index changes value. The median ap obtains its lowest value of about 5 nT one day before the interface. Subsequently it

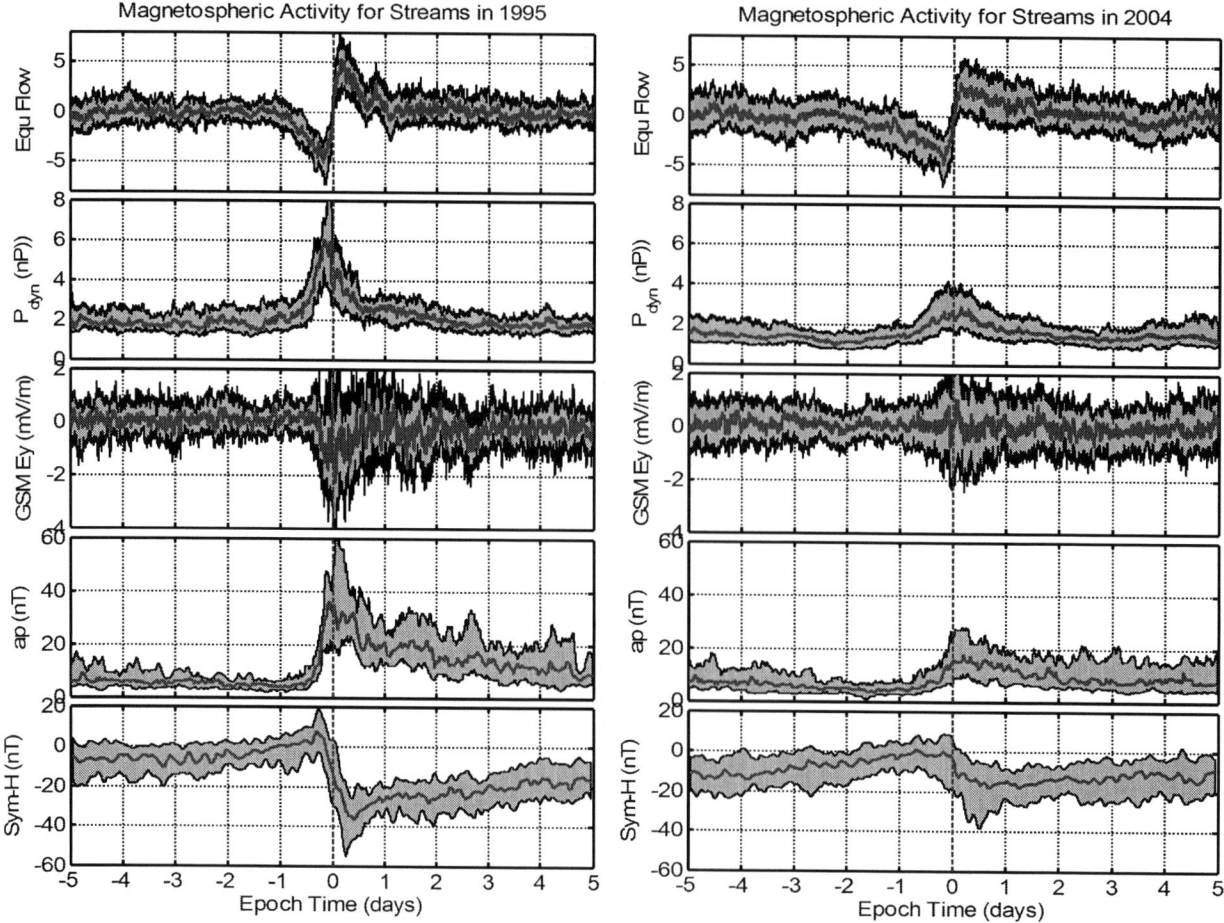

Figure 6. The bottom two panels show quartiles of the 3-hr ap index and 1-min Sym-H index produced by various solar wind drivers in the upper panels. Both auroral activity (ap) and ring current activity (Sym-H) are distinctly weaker in the 2004 solar minimum than they were in 1995.

increases rapidly reaching a peak value of 40 nT at the time of the interface. It then decays exponentially reaching quiet values again in about five days. The bottom panel presents the Sym-H index. Sym-H is a 1-minute approximation of the Dst index and hence monitors the behavior of the ring current. Median Sym-H begins a very gradual increase two days before the stream interface. By the time of the interface it has increased only 5-6 nT. This increase corresponds to the initial phase of a magnetic storm, but in a single event would be undetectable. About 4 hours before the interface Sym-H begins a rapid drop into a storm main phase. The median storm minimum occurs about 6 hours after the interface. Subsequently Sym-H recovers slowly and more or less linearly reaching quiet levels within about a week.

The behavior of the various drivers and response variables during 2004 are summarized in the right hand panels of Figure 6. Qualitatively the behavior is the same as in the year 1995, but quantitatively activity is much weaker. It is clear that this is because both the solar wind velocity and IMF field strength are weaker in this solar cycle than they were in the previous cycle. In 2004 peak values of ap and Sym-H were roughly half of what they were in the year 1995. Clearly an event in which minimum Sym-H is no less than 20 nT would not be classified as a magnetic storm. Despite this the quartiles show that geomagnetic activity is organized by the stream interfaces even if most of these events are too weak to detect in single traces.

Two additional magnetospheric response parameters are available for 1995 and are shown for a 20-day interval in Figure 7. The top four panels present velocity, E_y, ap, and Sym-H and have already been described. The fifth panel shows an index of ULF power in the Pc 5 frequency band (150-500 seconds period). A description of this index can be found in the paper by *O'Brien et al.* [2001]. These waves have been shown to be correlated with the appearance of relativistic electrons at synchronous orbit. Since the period of these waves is comparable to the drift period of relativistic electrons around the Earth it is likely that they are an important cause of inward radial diffusion of lower energy electrons from further out in the magnetosphere.

Prior to the arrival of the CIR magnetic activity is very low and electrons accelerated previously are gradually lost from the radiation belts. About one day before the arrival of the stream interface the azimuthal flow and dynamic pressure begin to increase while at synchronous orbit fluxes of relativistic electron fluxes begin to decrease more rapidly. Twelve hours before the interface the solar wind electric field begins to become more negative, geomagnetic activity picks up, and electron fluxes are dropping most rapidly. Four hours before the interface the main phase of a magnetic storm begins and Pc5 ULF wave activity begins to increase. At the interface the E_y fluctuations are at their largest, and geomagnetic activity

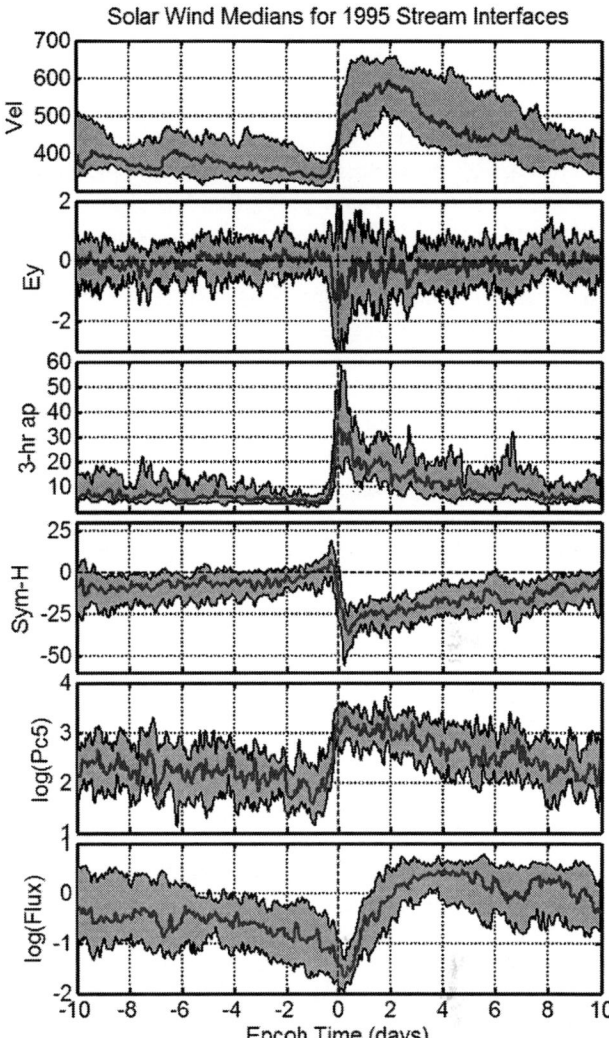

Figure 7. The four bottom panels compare four indices of magnetospheric activity in 1995 as a function of time relative to a stream interface. The bottom two panels show an index of ground Pc 5 power and the noon flux of relativistic electrons.

and ULF wave power reach peaks. Finally, four hours after the interface the main phase ends and electron fluxes reach their lowest value. For the next four days solar wind electric field and geomagnetic activity gradually decrease in strength with ULF power dropping somewhat more slowly. Throughout this time the ring current recovers slowly while the flux of relativistic electrons continues to increase to a peak 100 times higher than they were at the end of the main phase. Subsequently all parameters decay slowly with the electron fluxes still high ten days after the end of the main phase.

4. DISCUSSION

In the Introduction we summarized current understanding of the formation of a corotating interaction region (CIR). This description is based on analysis of solar wind observations similar to those presented here. For example *Gosling et al.* [1978] performed superposed epoch analysis of many different plasma variables obtaining results some of which we have duplicated. However, our work differs from the previous work in several respects. First, *Gosling et al.* [1978] used speed, density, and temperature to identify stream interfaces as compared to our use of speed, density, magnetic field magnitude, and azimuthal flow angle. Second, these authors selected only events for which there was a discontinuous change in the parameters during a 1-3 minute interval whereas we used the zero crossing of the azimuthal flow angle as the fiducial time. Because of these differences *Gosling et al.* [1978] identified only 28 events in three years of data while we found about this many events per year. A more important difference between the two studies is that we have included the components of the magnetic field and several magnetic indices in our analysis. These additional variables make it easier to understand the manner in which the stream interfaces organize magnetospheric activity.

4.1. Recurrent Magnetic Storms

About 1-2 days before the stream interface, fluctuations in GSE B_z reach a minimum as does the solar wind velocity. Transformed to GSM coordinates these produce a weak electric field, which if B_z is southward, will drive weak activity. About 12 hours before the interface the magnetic field magnitude begins to increase due to compression of the slow stream by the fast stream so B_z increases as well. The field magnitude peaks at the interface with a value 2-3 times the value in the normal solar wind. Consequently B_z is largest at this time. In addition the velocity of the solar wind has increased above its value in the slow stream so that the electric field fluctuations reach their maximum amplitude at the interface. We therefore expect geomagnetic activity to be a maximum at this time as well. The profile of the perturbation in magnetic field magnitude is asymmetric with respect to the interface taking only 12 hours to rise to a peak but 48 hours to decay. The velocity profile is also asymmetric being low before the interface and high for more than five days afterward. Because of these asymmetries the amplitude of the electric field fluctuations decays rather slowly after the interface. Consequently we expect magnetic activity to persist for some time.

The behavior of two indices of magnetospheric magnetic activity was summarized in the description of Figure 6. The ap index in the fourth panel is a measure of auroral electrojet activity. The Sym-H index in the fifth panel measures the strength of the ring current. Both indices are clearly organized by the solar wind stream interfaces in the manner expected based on the variations of solar wind variables described above.

Auroral zone activity begins to increase about 12 hours before the interface, peaks at the interface, and then decays slowly for several days after the interface. The strongest auroral zone activity is limited to a ~12 hour interval centered on the interface where the fluctuations in E_y are largest.

The ring current index increases gradually for nearly two days before the interface. This is a consequence of a combination of ring current decay during very quiet times and the increase in dynamic pressure beginning a day before the interface. A few hours before the interface ions begin to be injected into the ring current by the enhanced electric field and their magnetic effects overcome the effect of dynamic pressure so that Sym-H decreases. As long as the electric field fluctuations are large the ring current continues to grow causing Sym-H to decrease further. This growth produces the main phase of a weak magnetic storm that ends 6-12 hours after the interface. Subsequently the ring current recovers very slowly with time. The apparent recovery is much slower than expected from charge exchange. Most likely this is a consequence of a slowly changing equilibrium between frequent injections into the ring current by progressively weakening intervals of southward B_z and charge exchange of ring current ions with atmospheric neutrals.

4.2. Acceleration of Relativistic Electrons

As we showed in the discussion of Figure 7 relativistic electron fluxes decay slowly during the quiet times produced by the slow-speed stream. As the leading edge of a CIR arrives they begin to decrease more rapidly, and during the main phase of the CIR storm they drop to very low values. It is in the first four days of the recovery phase of these storms that electrons are accelerated most efficiently. During this time substorms occur frequently and ULF power is high. Other data not shown here indicate that during the storm recovery phase whistler mode chorus is very strong outside the plasmapause between midnight and dawn [*Meredith et al.*, 2003; *Meredith et al.*, 2001]. These observations support a complex theory explaining how electrons are energized to relativistic energies.

This theory [*Horne et al.*, 2006; *Horne et al.*, 2005] suggests that the first step in the process is an increase in dynamic pressure in the compression region of the CIR which moves the magnetopause closer to the Earth. This converts closed electron drift paths outside of synchronous orbit to open paths allowing electrons to be lost to the magnetosheath. The second step occurs as the main phase develops while dynamic pressure is still elevated. The "Dst effect" [*Kim and Chan*, 1997] of the growing ring current causes a decrease in magnetic flux in the inner magnetosphere. To conserve their third adiabatic invariant relativistic electrons must move outward to maintain constant magnetic flux through their drift shells. As these particles encounter the magnetopause they are also lost. During this time the power in Pc 5 waves is increasing rapidly. These waves may play a role in the rapid loss of relativistic electrons during the main phase. If the peak in phase space density of relativistic electrons is near synchronous orbit ULF waves may actually drive outward diffusion placing relativistic electrons on open drift paths. This process continues until the ring current stops growing at the end of the main phase and the dynamic pressure relaxes to its normal value. At this time the electron fluxes reach their lowest values.

Meanwhile, during the 8-10 hours of the main phase several substorms have taken place each injecting lower energy electrons ~100 keV into the magnetosphere beyond synchronous orbit. During this interval Pc 5 waves are generated at the magnetopause by the Kelvin-Helmholtz instability and inside the magnetosphere by the substorms [*Nose et al.*, 1995; *Vennerstrom*, 1999]. These waves drive inward radial diffusion of the lower energy electrons increasing their energies by conservation of the first two invariants as they drift closer to the Earth [*Elkington et al.*, 2003]. The radial diffusion creates an electron pitch angle distribution peaked at 90° that is unstable to the electron cyclotron instability for low energy electrons near the loss cone. Because of the instability these electrons lose energy and are scattered into the loss cone in the region outside the plasmapause between midnight and dawn. The waves generated in this manner also interact with higher energy electrons over a broad range of pitch angles scattering them to larger pitch angles and higher energy. Each time these electrons drift around the Earth they encounter the chorus and are pumped to even higher energies. Eventually a peak in phase space density of relativistic electrons is produced inside of synchronous orbit. Radial diffusion driven by fluctuations moves these electrons both inward and outward with those electrons that move inward gaining still more energy.

As the stream interface passes the Earth and the main phase ends the Earth is immersed in the high-speed stream. This stream contains large amplitude Alfvén waves of several hour period that quasi periodically turn the IMF southward at the magnetopause. Each time this happens another substorm is driven by magnetic reconnection. More electrons are injected in the outer magnetosphere and are radially diffused inward by the Pc 5 waves created by the K-H instability at the magnetopause. As the electrons move inward radial diffusion becomes less important and the cyclotron instability takes over producing chorus that scatters higher energy electrons over a wide range of pitch angles to larger pitch angles and higher energies.

This process lasts for many days as the velocity of the solar wind slowly decreases, the Alfvén waves become less dominant in the stream, substorms occur less frequently, and fewer electrons are accelerated to relativistic energies. After four days the electron fluxes reach their maximum values and then slowly begin to decay as loss processes begin to dominate over injection and acceleration.

4.3. Probabilistic Forecasting

The systematic behavior of the solar wind relative to stream interfaces provides a possible means for forecasting space weather [*McPherron and Siscoe*, 2004; *McPherron et al.*, 2004a; *McPherron et al.*, 2004b]. If the arrival time of a stream interface can be predicted in advance then one can use the quartiles of various magnetic indices relative to this time to predict activity indices. One forecasting method would specify the range of values within which activity would fall 50% of the time. This range is given by the upper and lower quartiles shown in our plots by the shaded patches. An alternative would be to calculate the cumulative probability distribution at each time and specify the probability that an activity index will exceed some specified threshold.

A serious obstacle to the success of these prediction schemes is change in the probability distributions between cycles. It is apparent from our comparison of the stream structure in 1995 and 2004 that the two cycles differ by nearly a factor of two in the magnitude of changes in different parameters. In 2004 flow deflections, velocity change, density enhancement, temperature increase, and fluctuations in GSE-B_z are all smaller than they were in 1995.

An additional difference between the two solar cycles is apparent in the traces of median GSE-B_z and the spiral angle of the IMF. In 1995 almost all high-speed streams that occurred near the equinoxes were geo-effective according to the Russell-McPherron rule "spring to fall away" [*Russell and McPherron*, 1973]. Thus the heliospheric current sheet crossing that typically occurs before the stream interface converted an ineffective radial IMF orientation to an effective orientation after the interface. This is apparent from the negative bias of the median GSM E_y that lasted for several days after the interface (see Figure 5). In contrast in 2004 the IMF orientations before and after the interfaces were almost equally distributed between effective and ineffective orientations. An even more perplexing fact is that in 1995 GSM E_y was strongly negative for several hours around the interface, but not so in 2004.

One possible explanation for differences between the plots for 1995 and 2004 is the phase of these years in the solar cycle. The year 1995 ends only 6 months before the minimum of cycle #22 while 2004 ends at least two years before the projected minimum of cycle #23. Although not shown here, analysis of the years 1994 and 1996 produced results very similar to those shown for 1995. Thus it seems likely that our results are explained by true differences in the Sun and solar wind between the two cycles and not by slight differences in phase.

A more likely explanation for the differences is the 22-year double solar cycle exhibited by the Sun's magnetic field [*Chernosky*, 1966; *Hale et al.*, 1919]. In each 11-year cycle of sunspot activity the polar magnetic field of the Sun reverses about two years after solar maximum. Consequently it takes 22 years for the Sun's magnetic dipole moment to repeat its orientation relative to the rotation axis. But the rotation axis of the Sun is tilted 7° with respect to the ecliptic pole in a direction such that the Earth is furthest below the solar equator on March 6, and furthest above on September 6. Thus near fall equinox the Earth is likely to be immersed in magnetic field lines connected to the north magnetic pole of the Sun.

Observations of the solar magnetic field have established that the phase of the 22-year cycle is such that the Sun's north magnetic pole becomes positive (outward field) just after the maximum of even numbered solar cycles. The last even cycle was #22 which reached its maximum in August 1989. Thus from about 1991 to 2002 the north pole of the Sun was positive. Consequently the interplanetary magnetic field (IMF) at the Earth was away from the Sun at fall equinox. Six months later near spring equinox the Earth was on the opposite side of the Sun and the IMF was toward the Sun.

This particular orientation of the IMF is geo-effective at the Earth because of the Russell-McPherron effect [*Russell and McPherron*, 1973] described in the Introduction. However, in the second half of the 22-year cycle the solar field reverses creating an away sector in spring and a toward sector in fall. These orientations of the IMF have positive projections on the GSM-z axis that suppress geomagnetic activity.

A second factor affecting the level of activity in a given cycle may be the Rosenberg-Coleman effect [*Rosenberg and Coleman*, 1969] also described in the Introduction. The dipole axis of the Sun is usually tilted at some angle to the rotation axis. Thus as the Sun rotates every 27 days the Sun's magnetic equator and its extension, the heliospheric current sheet, wobbles up and down at the Earth. Usually the Earth crosses the current sheet two or more times per solar rotation. Depending on the tilt of the dipole axis the Earth will spend more time on one side of the sheet than the other. If it spends more time on the side with a geo-effective orientation geomagnetic activity is enhanced.

The foregoing explanation is discussed at some length in a paper by *Cliver et al.* [1996], and found by these authors to be incomplete. The authors argue that "… an intrinsic solar variation (other than polarity reversal) … is the dominant cause of the 22-year cycle in geomagnetic activity." This variation leads to more coronal mass ejections in the first half of odd-numbered cycles and longer-duration 27-day recurrent streams in the second half of even-numbered cycles.

Our results support the *Cliver et al.* [1996] conjecture that there are intrinsic variations on the Sun responsible for strong recurrent streams in even-numbered cycles and weak streams in odd-numbered cycles. Figure 4 demonstrates this behavior. For the year 2004 of the odd-numbered cycle #23 the velocity contrast is smaller; the flow deflections are weaker; there is a smaller density enhancement; and the temperature change is weaker than in the preceding even-numbered cycle. The weaker streams lead to less compression of the IMF at the stream interface and hence to smaller fluctuations in B_z.

While we can not rule out the possibility that stronger high-speed streams will be observed in the next two years it seems unlikely because the strongest streams in the last cycle were seen in 1994 – two years before the end of the cycle, at basically the same phase in the cycle as the year 2004. Results for 1994 (not shown) are similar to those for 1995 and contrast sharply with those for 2004 so that we feel the differences between the two cycles are real. The implication of this difference is that the climatology derived from a single solar cycle can not be applied to the next cycle. For space weather forecasting it will be necessary to develop separate climatology's for odd and even solar cycles.

Acknowledgments. The authors would like to acknowledge support of this work from the NSF through grants NSF ATM 02-1798 and NSF ATM 02-08501. Additional support was provided by NASA through grant NNG-04GA93G. RLM also acknowledges support from LASP of the Univ. of Colorado through CISM, which is funded by the NSF STC Program under Agreement Number ATM-0120950. Spacecraft data were obtained from both the NSSDC and Wind project data centers. We thank the principal investigators of the magnetometer and plasma experiments, R. Lepping and K. Ogilvie for making their data available. GOES particle data were provided by NOAA SEC courtesy of H. Singer. Magnetic indices were obtained from WDC for Geomagnetism in Kyoto, Japan and WDC for Geomagnetism in Copenhagen, Denmark.

REFERENCES

Balogh, A., V. Bothmer, N.U. Crooker, R.J. Forsyth, G. Gloeckler, A. Hewish, M. Hilchenbach, R. Kallenbach, B. Klecker, J.A. Linker, E. Lucek, G. Mann, E. Marsch, A. Posner, I.G. Richardson, J.M. Schmidt, M. Scholer, Y.M. Wang, R.F. Wimmer-Schweingruber, M.R. Aellig, P. Bochsler, S. Hefti, and Z. Mikic, The solar origin of corotating interaction regions and their formation in the inner heliosphere, *Space Sci. Revs.*, 89(1-2), 141-178, 1999.

Bargatze, L.F., R.L. McPherron, J. Minamora, and D. Weimer, A new interpretation of Weimer et al.'s solar wind propagation delay technique, *J. Geophys. Res.*, 110(A7), 1-12, 2005.

Belcher, J.W., and L. Davis, Jr. Large-amplitude Alfven waves in the interplanetary medium. II, *J. Geophys. Res.*, 76(16), 3534-3563, 1971.

Borrini, G., J.M. Wilcox, J.T. Gosling, S.J. Bame, and W.C. Feldman, Solar wind helium and hydrogen structure near the heliospheric current sheet: A signal of coronal streamers at 1 AU, *J. Geophys. Res.*, 86(A6), 4565-4573, 1981.

Chapman, S., and J. Bartels, Geomagnetism, vol. 2, 541-1049 pp., Clarendon Press, Oxford, 1962.

Chernosky, E.J., Double sunspot-cycle Variation in terresterial magnetic activity, 1884-1963, *J. Geophys. Res.*, 71(3), 965-974, 1966.

Cliver, E.W., V. Boriakoff, and K.H. Bounar, The 22-year cycle of geomagnetic and solar wind activity, *J. Geophys. Res.*, 101(A12), 27091-27109, 1996.

Elkington, S.R., M.K. Hudson, and A.A. Chan, Resonant acceleration and diffusion of outer zone electrons in an asymmetric geomagnetic field, *J. Geophys. Res.*, 108(A3), SMP11-11-15, 2003.

Gosling, J.T., J.R. Asbridge, S.J. Bame, and W.C. Feldman, Solar wind stream interfaces, *J. Geophys. Res.*, 83(A4), 1401-1411, 1978.

Gosling, J.T., G. Borrini, J.R. Asbridge, S.J. Bame, W.C. Feldman, and R.T. Hansen, Coronal streamers in the solar wind at 1 AU, *J. Geophys. Res.*, 86(A7), 5438-5448, 1981.

Hale, G.E., V. Boriakoff, and K.H. Bounar, The magnetic polarity of sunspots, *Astrophys. J.*, 49153, 1919.

Horne, R.B., N.P. Meredith, S.A. Glauert, A. Varotsou, R.M. Thorne, Y.Y. Shprits, and R.R. Anderson, Mechanisms for the acceleration of radiation belt electrons, in Corotating Solar Wind Streams and Recurrent Geomagnetic Activity, edited by B. Tsurutani, *et al.*, American Geophysical Union, Washington, D.C., 2006.

Horne, R.B., R.M. Thorne, Y.Y. Shprits, N.P. Meredith, S.A. Glauert, A.J. Smith, S.G. Kanekal, D.N. Baker, M.J. Engebretson, J.L. Posch, C. Spasojevic, U.S. Inan, J.S. Pickett, and P.M.E. Decreau, Wave acceleration of electrons in the Van Allen radiation belts, *Nature*, 437(7056), 227-230, 2005.

Hundhausen, A.J., Coronal Expansion and Solar Wind, p238, Springer-Verlag, New York, 1972.

Kim, H.-J., and A.A. Chan, Fully adiabatic changes in storm time relativistic electron fluxes, *J. Geophys. Res.*, 102(A10), 22,107-122,116, 1997.

Krieger, A.S., A.F. Timothy, and E.C. Roelof, A coronal hole and its identification as the source of a high velocity solar wind stream, *Solar Physics*, 29(2), 505-525, 1973.

McPherron, R.L., and G. Siscoe, Probabilistic forecasting of geomagnetic indices using solar wind air mass analysis Space Weather 2(1), 1-10, 2004.

McPherron, R.L., G. Siscoe, and N. Arge, Probabilistic forecasting of the 3-h ap index, IEEE Transactions on Plasma Science, 32(4), 1425-1438, 2004a.

McPherron, R.L., G.L. Siscoe, N.U. Crooker, and N. Arge, Probabilistic Forecasting of the Dst Index, in The Inner Magnetosphere: Physics and Modeling, edited by T.I. Pulkkinen, *et al.*, pp. 203-210, American Geophysical Union, Helsinki, Finland, 2004b.

Meredith, N.P., M. Cain, R.B. Horne, R.M. Thorne, D. Summers, and R.R. Anderson, Evidence for chorus-driven electron acceleration to relativistic energies from a survey of geomagnetically disturbed periods, *J. Geophys. Res.*, 108(A6), SMP15-11-14, 2003.

Meredith, N.P., R.B. Home, R.M. Thorne, D. Summers, and R.R. Anderson, Substorm dependence of plasmaspheric hiss, *J. Geophys. Res.*, 109(A6), p14, 2004.

Nose, M., T. Iyemori, M. Sugiura, and J.A. Slavin, A strong dawn/dusk asymmetry in Pc 5 pulsation occurrence observed by the DE-1 satellite, *Geophys. Res. Lett.*, 22(15), 2053-2056, 1995.

O'Brien, T.P., R.L. McPherron, D. Sornette, G.D. Reeves, R. Friedel, and H.J. Singer, Which magnetic storms produce relativistic electrons at geosynchronous orbit?, *J. Geophys. Res.*, 106(A8), 15533-15544, 2001.

Rosenberg, R., and P.J. Coleman, Heliographic latitude dependence of the dominant polarity of the interplanetary magnetic field, *J. Geophys. Res.*, 74(24), 5611-5622, 1969.

Russell, C.T., and R.L. McPherron, Semiannual variation of geomagnetic activity, *J. Geophys. Res.*, 78(1), 92-108, 1973.

Schatten, K.H., and J.M. Wilcox, Response of the geomagnetic activity index Kp to the interplanetary magnetic field, *J. Geophys. Res.*, 72(21), 5185-5191, 1967.

Vennerstrom, S., Dayside magnetic ULF power at high latitudes: a possible long-term proxy for the solar wind velocity?, *J. Geophys. Res.*, 104(A5), 10145-10157, 1999.

Weimer, D.R., D.M. Ober, N.C. Maynard, M.R. Collier, D.J. McComas, N.F. Ness, C.W. Smith, and J. Watermann, Predicting interplanetary magnetic field (IMF) propagation delay times using the minimum variance technique, *J. Geophys. Res.*, 108(A1), SMP 16-11 - SMP 16-12, doi:10.1029/2002JA009405, 2003.

Wilcox, J., and N.F. Ness, Quasi-stationary corotating structure in the interplanetary medium, *J. Geophys. Res.*, 70(23), 5793-5805, 1965.

Wilcox, J.M., K.H. Schatten, and N.F. Ness, Influence of interplanetary magnetic field and plasma on geomagnetic activity during quiet-sun conditions, *J. Geophys. Res.*, 72(1), 19-26, 1967.

Robert L. McPherron, Institute of Geophysics and Planetary Physics, University of California Los Angeles, Los Angeles, CA 90095-1567, USA. (rmcpherron@igpp.ucla.edu)

James Weygand, Institute of Geophysics and Planetary Physics, University of California Los Angeles, Los Angeles, CA 90095-1567, USA. (jweygand@igpp.ucla.edu)

The Role of Radial Transport in Accelerating Radiation Belt Electrons

Xinlin Li[1]

Laboratory for Atmospheric and Space Physics and Department of Aerospace Engineering Sciences, University of Colorado, Boulder, Colorado, USA

Several theories of radiation belt electron acceleration have been proposed that involve breaking one or more of the electrons' adiabatic invariants. These theories can be categorized into two classes: inward radial transport and in situ acceleration. Because the actual electron population is ultimately the net result of a delicate balance among acceleration, transport, and loss, it has been very difficult to determine, mainly due to lack of systematic measurements, whether radial transport or in situ heating contributes more to the observed electron enhancements. Observations and models, however, do provide some insight into acceleration mechanisms. It is generally agreed that a strong interplanetary shock impacting the magnetosphere can result in a very fast inward radial transport of pre-existing particles from large L to form a new radiation belt at a smaller L. Test-particle tracings in fields generated by global MHD simulations show that energetic electrons can be transported from the magnetotail into the inner magnetosphere. Simultaneous observations at geosynchronous orbit and L=4.2 show that the phase space density is almost always higher and increases first at geosynchronous orbit, and then enhances at L=4.2, though a possible phase space density peak in between may exist. Long term predictions of MeV electrons at L=4 as well as L=6 based on the radial diffusion equation driven by solar wind parameters suggest that radial diffusion can be a major acceleration mechanism because of the good match between the predicted and measured electrons. This paper will review recent achievements in the understanding of inward radial transport as a means of accelerating relativistic electrons in the Earth's magnetosphere.

INTRODUCTION

The relative importance of acceleration processes responsible for relativistic electrons in the Earth's radiation belt is still an unsolved problem. Earth's outer radiation belt consists of electrons in the energy range from keV to MeV. Compared to the inner radiation belt that usually contains somewhat less energetic electrons and an extremely intense population of protons extending in energy up to several hundred MeV or even a few GeV, the outer belt consists of energetic electrons that show a great deal of variability which is well correlated with magnetospheric storms and high speed solar wind streams [*Williams*, 1966; *Paulikas and Blake*,

---

[1]Also affiliated with Laboratory for Space Weather, Chinese Academy of Sciences, P. O. Box, 8701, Beijing 100080, China

1979]. The region between the inner belt and outer belt, around L=2-3, is referred to as the slot region, the actual location of which is energy dependent. There is usually a low population of energetic electrons in the slot region, which can, however, be filled during active times.

Charged particle motion in magnetic and electric fields is well understood from the Lorentz force law. In a quasi-dipole magnetic field such as that of the Earth, charged particles can be trapped by the magnetic field and follow three distinctive types of motions: gyration, bounce, and drift. Electrons drift eastward as they bounce between the stronger magnetic fields of the northern and southern hemispheres while gyrating around the local magnetic field. More energetic particles bounce and drift faster. These three motions have well separated time scales. For example, a 1 MeV electron with an equatorial pitch angle of 60° at r=6 R_E has gyration, bounce, and drift periods of about 10^{-3}, 10^0, and 10^3 seconds, respectively. An adiabatic invariant is associated with each of these motions. As long as the magnetic and electric fields change slowly over the respective time-scale, the corresponding invariant remains constant [*Northrop*, 1963].

Inward radial transport and in situ acceleration are the two processes important in the acceleration of the radiation belt electrons. Radial transport accelerates electrons by bringing less-energetic electrons inward from larger L-shells, conserving the first and second adiabatic invariants (μ and J). Inward transport requires a larger phase space density of less-energetic electrons at larger L-shells to enhance the flux at smaller L-shells. The process is enhanced by an elevated level of magnetospheric electric and magnetic field fluctuations with frequencies comparable to the electrons' drift frequencies (in mHz range) [*Schulz and Lanzerotti*, 1974; *Rostoker et al.*, 1998; *Baker et al.*, 1998; *Hudson et al.*, 2000; *O'Brien et al.*, 2003; *Mann et al.*, 2004; *Miyoshi et al.*, 2004; *Sarris et al.*, 2006] and by enhanced inductive electric fields due to a strong interplanetary shock impacting the magnetosphere [*Li et al.*, 1993; *Li et al.*, 2003a; *Gannon and Li*, 2005; *Gannon et al.*, 2005]. The former is usually referred to as enhanced radial diffusion, and the latter as interplanetary shock-associated acceleration and transport, which occurs on a time scale less than the electron's drift period.

In situ acceleration occurs by violating the first and second adiabatic invariants of less-energetic electrons. This acceleration can be realized on the same L-shell by electron interaction with whistler mode waves [*Temerin et al.*, 1994; *Horne and Thorne*, 1998; *Summers et al.*, 1998; *Meredith et al.*, 2001; *Albert*, 2002; *Horne et al.*, 2003; *O'Brien et al.*, 2003; *Horne et al.*, 2006].

In this paper, the recent enhancements in the understanding of inward radial transport as a means of accelerating relativistic electrons in the Earth's magnetosphere will be reviewed.

MARCH 24, 1991, EVENT AND ITS CONSEQUENCES

Research into the Earth's radiation belts had been nearly dormant for 10-15 years before the startling measurements from the CRRES satellite of a sudden four-orders-of-magnitude enhancement of >10 MeV electron fluxes near L=2.5 within tens of seconds on March 24, 1991. The initial enhancement and the subsequent peaks, which were drift echoes due to the reappearance of the electrons after a drift around the Earth, were clearly registered in the Cerenkov counter on CRRES [*Blake et al.*, 1992], as illustrated in the top panel on the left column of Figure 1. The measured variations of the main components of the electric and magnetic fields are shown in the lower panels [*Wygant et al.*, 1994]. The time-dependent electric field was modeled by an asymmetric bipolar pulse that is associated with the compression and relaxation of the Earth's magnetic field. A population of relativistic electrons was traced in the modeled fields using a guiding center code. A good agreement was produced between the observed and simulated electron drift echoes [*Li et al.*, 1993], as shown in the right column of Figure 1. In the model, pre-existing ~1-2 MeV electrons at L~8 were brought into L~2.5 in less than one drift period (150 s) of the 15 MeV electrons by a mainly inductive electric field resulting from the shock compression of the magnetosphere. The successful simulation and explanation of the prompt formation of the new radiation belt on March 24, 1991 caused a major addition in our understanding of the possible range of particle transport in the radiation belts.

Later, tracing of test-particles in the fields generated by global MHD simulations associated with the shock impact gave rise to similar results [*Hudson et al.*, 1997; *Elkington et al.*, 2004]. However, no comparable injection of >10 MeV deep into magnetosphere like on March 24, 1991 has been seen since.

Recently, a parametric study of shock-induced transport and acceleration of relativistic electrons in the magnetosphere was conducted by *Gannon et al.* [2006], based on the *Li et al.* [1993] model, in order to understand why shock-induced injections of electrons with energies above 10 MeV to L~2.5 are so rare and to determine whether even larger injections are possible. They varied the electric field amplitude of pulses within the magnetosphere in the *Li et al.* [1993] model from 70 to 400 mV/m. They found that a stronger electric field shifted the peak of the resultant relativistic electron population towards the Earth. Doubling the electric field amplitude from 120 to 240 mV/m moved the peak of the injected electrons with energies above 13 MeV from 2.8 to 2.4. However, as the electric field pulse becomes larger, the response has an asymptotic behavior (Figure 2), such that it is extremely difficult to produce injections of 10 MeV electrons inside L=2. The simulations show that

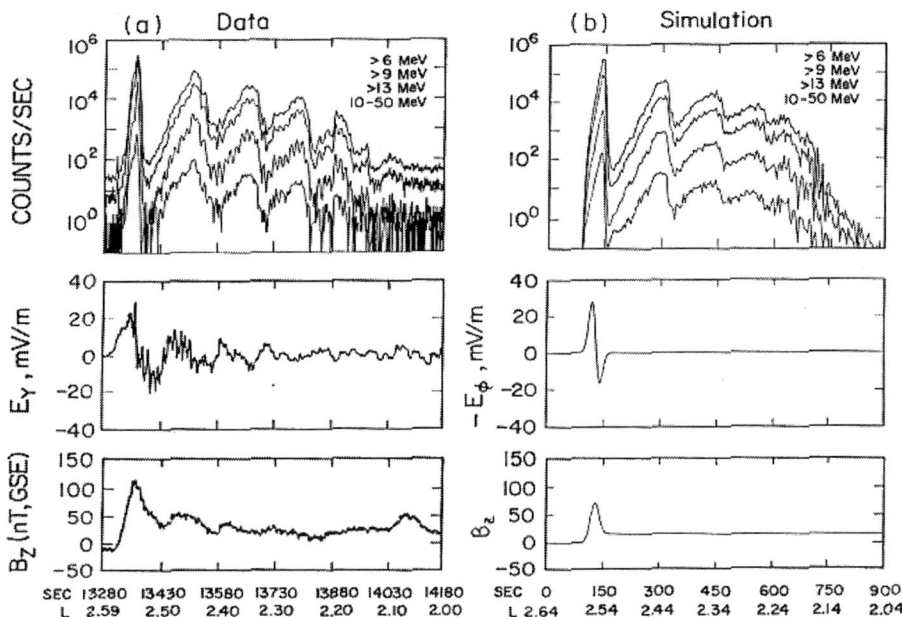

Figure 1. (Figure 1 of [*Li et al.*, 1993]) (a) Data from the CRRES satellite at the time of the March 24, 1991 Storm Sudden Commencement. Top panel shows count rates as a function of time from four energetic electron channels measuring integral counts above a threshold energy indicated, and also between 10-50 MeV. Middle and bottom panels show the measured electric field E_y in a co-rotational frame and the B_z magnetic field component with a model magnetic field subtracted, in GSE coordinates over the same time interval. (b) Simulated results in the same format as (a) measured at a spatial location corresponding to the trajectory of the CRRES satellite.

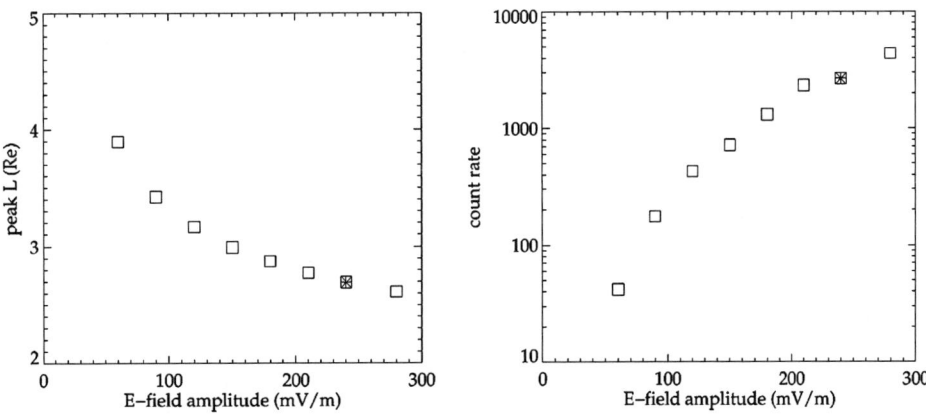

Figure 2. (Figure 2 of [*Gannon et al.*, 2006]) Simulation result summary vs amplitude of modeled electric field for electrons greater than 13 MeV. The left panel shows the location of resultant peak position. This is the L position of the highest point of flux for a set of parameters ($\delta L = 0.1$). The right panel shows the relative countrate level at the above peak position. The symbol * denotes the parameters used to simulate the March 24, 1991 event.

fields on the order of 100 mV/m are required to produce injections inside of L=3. Since it is rare to see even tens of mV/m electric fields near the equatorial plane, injections of electrons with energies above 10 MeV deep into magnetosphere are also very rare.

Nonetheless, the new electron radiation belt, formed in less than a minute on March 24, 1991, lasted for many years, as shown in Plate 1 from subsequent SAMPEX measurements. While the belt slowly decays and diffused inward, the traces of the belt still remain. Plate 1 also shows several injections of electrons with energies above 10 MeV into $L<4$. But these injections were above L=2.5 (except for one in late October of 2003 [*Baker et al.*, 2004; *Looper et al.*, 2005a]) and none lasted as long.

It should be pointed out that injections of electrons with lower energies to higher L associated with interplanetary shocks or solar wind pressure enhancements may occur more often [*Li et al.*, 2003a], but the net effect of these injections is difficult to measure because of the background electron populations. In fact, a more fundamental question is: How are these background electrons (still in the MeV range) accelerated.

LONG TERM OBSERVATIONS OF THE AVERAGED RADIATION ELECTRONS AND THE DST INDEX

Before we further discuss the acceleration mechanisms, let us consider the average picture of radiation belt electrons and their association with solar wind speed and geomagnetic activity, indicated by the Dst index, and have some general discussions about the variations of radiation belt electrons on time scales of solar cycle, semiannual, and solar rotation time scales.

Plate 2 shows SAMPEX's measurements of radiation belt electrons from launch to early 2004, together with the solar wind speed, the sunspot number, and the Dst index. It is evident that radiation belt electron fluxes in the magnetosphere are weakest during sunspot minimum (1996-1997), and more intense during the ascending phase and the maximum of the solar cycle (1997-1999). However, contrary to what one might expect, they are even more intense during the descending phase of the sunspot cycle (1993-1995 and 2003-2004), when recurrent high speed solar wind streams emanating from persistent trans-equatorial coronal holes become dominant, as indicated by the red curve in the upper panel. Energetic electrons are, on average, not as intense during sunspot maximum when the occurrence of coronal mass ejections (CME) is greatest. While fast CME's are very capable of producing magnetic storms and accelerating radiation belt electrons, as indicated by the deep penetrations of the electrons during the ascending phase and the maximum of the solar cycle, CME's do not occur as often or last as long as the recurrent high speed solar wind streams during the descending phase of the solar cycle.

Another distinctive feature in the lower panel of Plate 2 is the seasonal variations. Radiation belt electron fluxes and geomagnetic activity, indicated by the larger negative magnitude of the Dst index, which is a measure of the average change in the magnetic field near the equator and is used as an index to determine the strength of magnetic storms. are most intense near the equinoxes [*Baker et al.*, 1999; *Cliver et al.*, 2000; *Li et al.*, 2001b], marked by the vertical bars along the horizontal axis, and least intense during the solstices. Also energetic electrons penetrate deeper into the magnetosphere near the equinoxes than they do near the solstices [*Li et al.*, 2001b]. Another remarkable feature of Plate 2 is the correlation of the inward extent of MeV electrons with the Dst index.

RADIAL DIFFUSION

The classic explanation of the electron radiation belt is radial diffusion. The tendency of radial diffusion is to equalize the phase space density of the electrons with the same value of μ and J [*Schulz and Lanzerotti*, 1974]. Since the phase space density at a given μ and J usually (but certainly not always) increases with radial distance, the usual effect of radial diffusion is to increase the flux at smaller L at any given energy and position by radially diffusing electrons from larger L. Radial diffusion works best when fluctuations in magnetospheric electric and magnetic fields with frequencies comparable to the electrons' drift frequencies are enhanced. However, in contrast to shock-induced radial transport, radial transport due to diffusion during one drift period is insignificant.

While radial diffusion and shock-induced radial transport can usually explain the increase in the electron flux at almost any L and energy, it cannot explain the origin of the outer radiation belt as a whole. Radial transport as an acceleration mechanism normally requires a source population of electrons that has higher phase space density at larger L.

Phase Space Density Comparisons

Hilmer et al. [2000] compared the phase space density of equatorial electrons of $\mu \sim 2.1 \times 10^3$ MeV/G near L=4.2 and at geosynchronous orbit for 26 well-defined high speed solar wind streams detected by Wind between December 1994 and September 1996. For all 26 events the phase space density is consistently greater at geosynchronous orbit. The critical factor leading to L=4.2 electron flux enhancements was elevated geomagnetic activity levels of Kp ~ 3.0-3.5 and above for extended periods. Their results are consistent with the picture of inward radial diffusion.

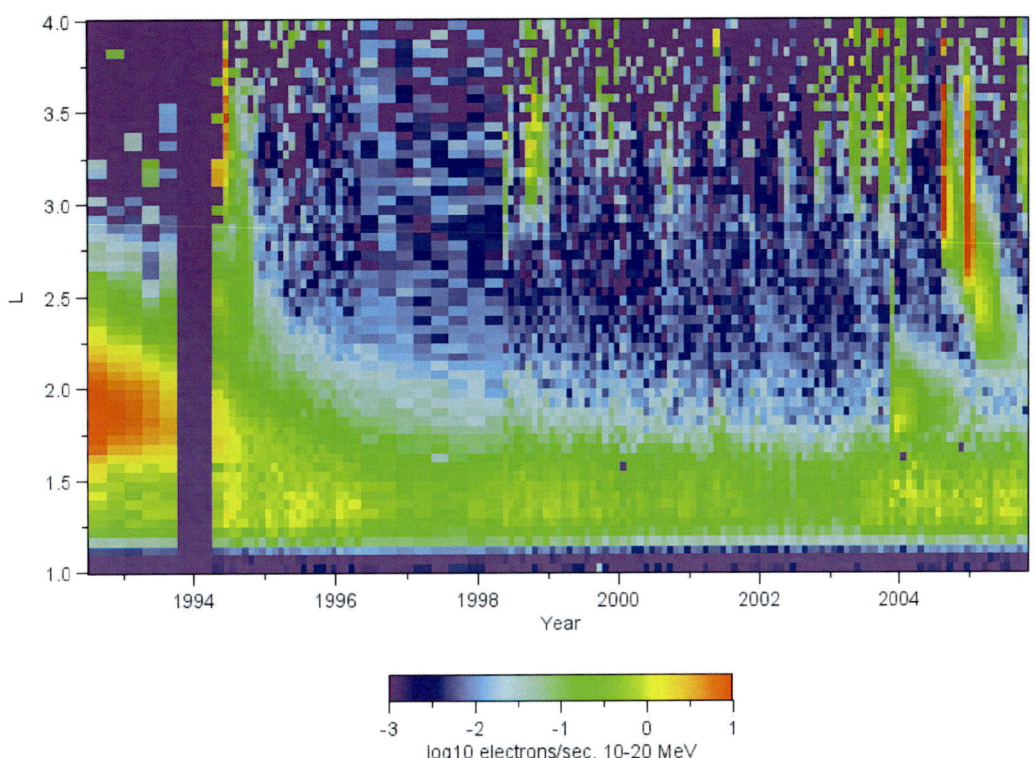

Plate 1. (An embellishment of Figure 5 of [*Looper et al.*, 2005a] and presented in Fall AGU, 2005 [*Looper et al.*, 2005b]) Monthly window-averaged (3-month averaged prior to the middle of 1994) countrate of 10-20 MeV electrons, mirroring near 475 km, throughout SAMPEX mission since its launch (July 3, 1992) into a low-altitude and highly inclined orbit. The "Halloween" (Oct.-Nov. 2003) belt stands out as the deepest injection of 10-20 MeV electrons since March 24, 1991, but two stronger events at larger L can be seen in 2004 (and weaker one from February 1994). All artifacts of instrument mode, attitude, and orbital decay have been taken out [*Looper et al.*, 2005b].

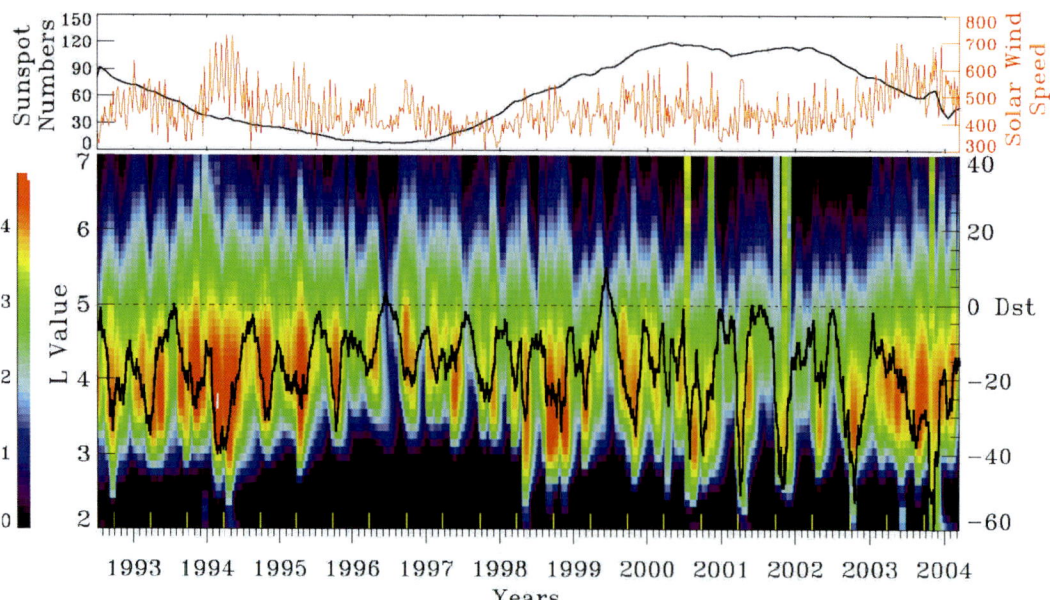

Plate 2. (An embellishment of Figure 1 of [*Li et al.*, 2003b]) Upper panel shows yearly window-averaged sunspot numbers and weekly window-averaged solar wind speed (km/s). The lower panel shows selected SAMPEX measurements of electrons of 2-6 MeV (#/cm²-s-sr in logarithm, in bins of 0.1 L) since launch (July 3, 1992) to early 2004. The Dst index for the same period is superimposed as a black curve on the lower panel. The electron and Dst index is window-averaged over a 30-day period in order to show the overall feature. The yellow vertical bars on the horizontal axis are marks of equinoxes.

However, comparing phase space density at only two spatial points may miss a possible phase-space density peak in between when those two points straddles the peak. *Selesnick and Blake* [2000] used data from the High Sensitivity Telescope (HIST) energetic detector on the Polar spacecraft [*Blake et al.*, 1995] around January and May 1998 to constrain the source location of outer radiation belt electrons and found phase space density peaks at L > 4. Due to the discrepancies in the results obtained with different magnetic field models and at different local times, they treated the conclusion as a tentative one. Later, *Green and Kivelson* [2004] used data principally from Polar/HIST and the *Tsyganenko and Stern* [1996] field model to obtain phase space densities. They found that the phase space density of $\mu \sim 1000$ MeV/G peaks near L = 5 and concluded that the data are best explained by in situ acceleration near L = 5.

However, as a result of the competition between radial diffusion and L-dependent losses there can sometimes be local peaks in phase space density [*Li*, 2004]. Therefore, it is not obvious how to determine the relative importance of different acceleration mechanisms based on the phase space density radial profile only.

Difficulties in Distinguishing the Acceleration Mechanisms

Interplanetary shock induced acceleration is distinguishable from other acceleration mechanisms because of its fast time scale. However, it is difficult to distinguish between radial diffusion and in situ acceleration as discussed above. In both of these latter scenarios, the less-energetic electrons usually have a substantially larger phase space density and thus, less-energetic electrons on either larger L-shells or on the same L-shell are a possible source of electrons to be further accelerated.

Since inward radial diffusion is driven by magnetic and electric field fluctuations on a time scale of the drift period of the radiation belt electrons (~10 minutes) and the observed field fluctuations almost always have more power at lower frequencies, less-energetic electrons diffuse inward faster while more-energetic electrons diffuse inward slower [*Schulz and Lanzerotti*, 1974], leading to a longer delay for the enhancement of the more-energetic electrons, which is a characteristic feature of the enhancement of radiation belt electrons at a given location [*Li et al.*, 2005]. Meanwhile, the longer time delay for the enhancement of more-energetic electrons can also be explained by in-situ heating of electrons by VLF waves on the same L-shell, because it would also take time to accelerate electrons to higher energies on the same L-shell.

Given a particular shape of a phase-space density peak in L, it is not indicative of a certain acceleration process, because either a local acceleration or an inward radial diffusion with a L-dependent loss can create such a peak. In the latter case the peak rises out of the more usual outward gradient whereas in the former case the peak is created by erosion of the density at higher L. Therefore, the detailed temporal evolution of the peak should be revealing the acceleration mechanism responsible [*Selesnick and Blake*, 2000; *Green and Kivelson*, 2004]. However, it is rather difficult to quantify the detailed temporal evolution without systematic multiple spacecraft measurements.

Recently, *Mann et al.* (2004) compared 10 years of magnetometer data from a range of mid-latitude L-shells to the MeV electron response for a range of L (L = 6, and L = 5 and L = 4). The correlation between the ultralow frequency (ULF) power and the MeV flux is clear and strong from geosynchronous orbit to L = 4. Of special interest is the observation that the correlation between solar wind speed, or ULF power, with the MeV electron flux peaks first at geosynchronous orbit, followed by L = 5 and then L = 4 - an evidence (at least for any acceleration processes correlated with solar wind speed, such as radial diffusion) that the flux is transported inwards. Though the time delay with decreasing L may indicate radial diffusion, but the decreasing correlation of solar wind velocity and ULF power with decreasing L is also indicative of another acceleration mechanism contributing at lower L.

Furthermore, there exists a good correlation of low energy electrons with higher energy electrons on the same L-shell [*Li et al.*, 2005], which suggests that these lower energy electrons may be the source of the higher energy electrons through in situ acceleration. A possible explanation is that enhancements of lower energy electrons produce waves that accelerate the high energy electrons to higher energies [e.g., *Summers et al.*, 1998].

Again, due to the limited observations available and the complicated nature of the electron variations, it is extremely difficult to definitely state how much inward radial diffusion and in situ acceleration contribute to the observed electron enhancements. The observed electron enhancement is the net result of a delicate balance among acceleration, transport, and loss. Any quantification work has to resort to a model that takes these factors into account.

MODELING THE ENHANCEMENTS OF RADIATION BELT ELECTRONS THROUGH RADIAL DIFFUSION

Modeling radial diffusion is relatively more straightforward than modeling in situ acceleration in terms of comparing the model results with measurements. A variety of radial diffusion models have been developed and have made quantitative assessments of the contribution of radial diffusion to the observed enhancements of radiation belt electrons.

Using a radial diffusion model, *Brautigam and Albert* [2000] have investigated the electron variations measured by CRRES during a magnetic storm in October 1991. They concluded that the enhancement of lower energy electrons (<1 MeV) at L=4-5 can be explained by radial diffusion, although radial diffusion alone is inadequate to explain the enhancement of higher energy electrons (>1 MeV).

Elkington et al. [2004] traced the guiding centers of electrons in the fields generated from a global MHD simulation for a magnetic storm on March 31, 2001. They found that a significant portion of electrons starting from the magnetotail can be radially transported into the inner magnetosphere and be trapped. This represents significant progress in validating the effects of radial transport because the test-particle tracing was performed in rather realistic fields obtained from a global MHD simulation. The combined effects of induced and convection electric fields lead to the acceleration of plasmasheet particles from 10s of keV in the tail to energies exceeding 1 MeV in the inner magnetosphere.

Radial diffusion models have also been used to make quantitative predictions of radiation belt electron enhancements with good success. Plate 3 shows a comparison of two years of the daily average of 0.7-1.8 MeV electron flux measured at geosynchronous orbit with a prediction based on a radial diffusion equation (with the diffusion coefficient determined by the solar wind speed and the interplanetary magnetic field, IMF). A prediction efficiency (PE = [1-(mean squared residual)/ (variance of data)]) of 0.81 and a linear correlation coefficient of 0.90 were achieved for the two year period, which was during the descending phase toward the minimum of the solar cycle. The prediction results, using the same set of modeling parameters, are not as good for the ascending phase and the maximum of the solar cycle. This radial diffusion model [*Li et al.*, 2001a] has been updated and is making real-time forecasts of daily averaged >2 MeV electron fluxes at geosynchronous orbit [*Li*, 2004]. The real-time forecast results, focusing on one geocentric distance, are currently available on the web (http://lasp.colorado.edu/~lix).

More recently, *Barker et al.* [2005] extended the *Li et al.* [2001a] model to simulate the MeV electron phase space density variations from L=3 to 8, allowing for comparison with measurements at more than one L while retaining a similar form of diffusion coefficient. This presents an important progress since the phase space density is now converted to differential flux of different energy at different location (conserving the μ) in order to compare with measurements. The extended model achieves a PE of 0.61 at L=4 and 0.52 at L=6, when the phase space density is converted to differential flux and compared with orbit-averaged Polar 2 MeV measurements at L=4 and daily-averaged LANL 0.7–1.8 MeV geosynchronous measurements for the year of 1998 (see Plate 4). From the flux measurements (red curves) in the bottom two panels, it is evident that the electron flux varies on very different time scales at the two L-shells. The model's ability to predict well the flux at both L suggests that radial diffusion plays an important role in accelerating the relativistic electrons in the radiation belt. The model is still unable to completely describe the magnitude of the flux variations at both L-shells. This indicates either that the model does not sufficiently simulate radial diffusion or that other processes, such as in situ acceleration, are also contributing. Nonetheless, the results suggest that radial diffusion is a significant factor in radiation belt electron acceleration.

SUMMARY

There is little doubt that energetic electrons at large L can be transported inward and accelerated on a time scale of less than the electron's drift period, if a strong interplanetary shock impacts the magnetosphere. Such shock-induced radial transport and acceleration of radiation belt electrons is not routinely observable. It requires a sufficiently large inductive electric field from a shock impact to transport pre-existing electrons at large L to the slot region. Nonetheless, a big event like March 24, 1991 can have a very long-lasting effect. Injections of energetic electrons associated with normal interplanetary shocks or solar wind pressure enhancements should occur more often, but the radial transport and acceleration are limited. Thus, the net effects of such injections cannot be easily measured because the injected electrons usually add to the background population of radiation belt electrons, which also vary due to other mechanisms.

Given the observed enhancements of radiation belt electrons on the time scale of hours to days, distinguishing whether they are due to enhanced radial transport or due to in situ acceleration is a difficult problem. Contributing to the problem are the abundant source populations for both mechanisms and the consistency of the time scale of the enhancements with both mechanisms. Modeling efforts on both mechanisms have been ongoing. Modeling radial diffusion processes has produced quantitative estimates of the contribution of radial diffusion to the observed enhancements. Models based on radial diffusion equations have also been used to make real-time quantitative forecasts of the variations of MeV electrons at geosynchronous orbit. However, a significant portion of the variation is still not accounted for by radial diffusion, which suggests that either the radial diffusion model does not sufficiently simulate the radial diffusion or, more likely, other processes such as in situ acceleration should be taken into account.

In the near future, we should gain more insight of the acceleration mechanism by improving our modeling by including local acceleration as a possible source term in the radial diffusion model, validating the outer boundary conditions, and

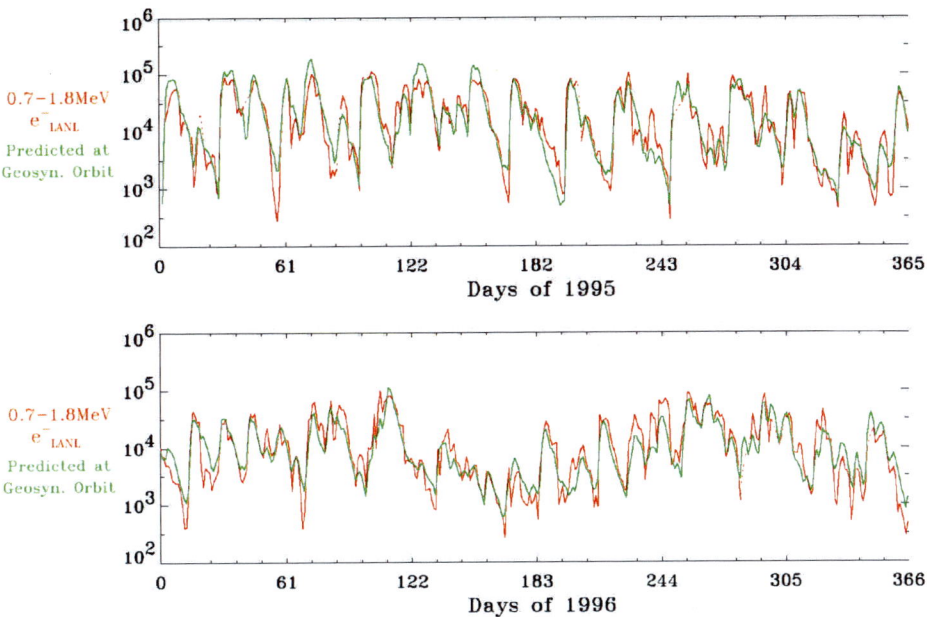

Plate 3. (Figure 2 of [*Li et al.*, 2003b]) A comparison of two years of daily averages of 0.7-1.8 MeV electron flux measured at geosynchronous orbit (averaged from four spacecraft spaced in longitude) with prediction based solely on solar wind measurements.

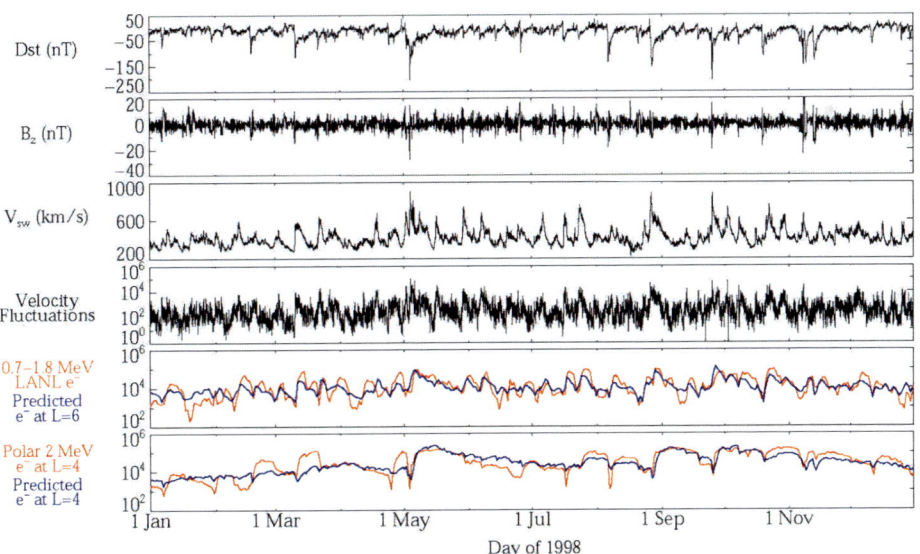

Plate 4. (Figure 4 of [*Barker et al.*, 2005]) (top to bottom) Dst measurement, interplanetary magnetic field strength in the z-direction, solar wind speed, solar wind speed fluctuations, comparison of prediction at L=6 and LANL geosynchronous electron data, comparison of prediction at L=4 and Polar electron data at L=4 for the year of 1998.

using realistic loss lifetimes, to see how significantly better we can reproduce the available observations.

However, a prerequisite to completely understanding the basic acceleration mechanisms is a systematic measurement of the radiation belt environment. This systematic measurement must consist of multiple spacecraft in geotransfer orbits, which would allow the spacecraft to cut across different L and measure the majority of the electron population because of the low inclination, carrying detectors with a high resolution for particle energy and pitch angle and for electric and magnetic fields with wave measurements over a broad spectrum. Such correctly-instrumented satellites in the right orbits have not yet been flown in the 45 years since the discovery of the radiation belts. NASA's LWS/Radiation Belt Storm Probes, consisting of two such satellites, plus a possible third through the mission of opportunity program, scheduled for launch in 2012, will be the best opportunity to completely unravel the acceleration mechanisms.

Acknowledgments. The author likes to thank Dr. M. Looper for providing Plate 1 and also thank both referees for their constructive comments and suggestions. This work was supported in part by NASA grant (NAG-13518) and NSF grants (CISM and ATM-0233302), and also in part by National Natural Science Foundation of China Grant No. 40125012 and the International Collaboration Research Team Program of the Chinese Academy of Sciences.

REFERENCES

Albert, J.M., Nonlinear interaction of outer zone electrons with VLF waves, *Geophys. Res. Lett.*, 29, No. 8, 2002.

Baker, D.N., T.I. Pulkkinen, X. Li, S. Kanekal, K.W. Ogilvie, R.P. Lepping, J.B. Blake, L.B. Callis, G. Rostoker, H.J. Singer, and G.D. Reeves, A strong CME-related magnetic cloud interaction with the Earth's magnetosphere: ISTP observation of rapid relativistic electron acceleration on May 15, 1997, *Geophys. Res. Lett.*, 25, 2975, 1998.

Baker, D.N., S.G. Kanekal, T.I. Pulkkinen, and J.B. Blake, Equinoctial and solstitial averages of magnetospheric relativistic electrons: A strong semiannual modulation, *Geophys. Res. Lett.*, 26, 3193, 1999.

Baker, D.N., S.G. Kanekal, X. Li, S.P. Monk, J. Goldstein, and J.L. Burch, An extreme distortion of the Van Allen belt arising from the 'Halloween' solar storm in 2003, doi:10.1038/nature03116, *Nature*, 2004.

Barker, A.B., X. Li, and R.S. Selesnick, Case Study of Radiation Belt Electrons During Magnetic Storms Based on Solar Wind Measurements, *Space Weather*, Vol. 3, No. 10, S10003 10.1029/2004SW000118, 13 October 2005.

Blake J.B., W.A. Kolasinski, R.W. Fillius, and E.G. Mullen, Injection of electrons and protons with energies of tens of MeV into L<3 on March 24, 1991, *Geophys. Res. Lett.*, 19, 821, 1992.

Blake, J.B., et al., CEPPAD: Comprehensive Energetic Particle and Pitch Angle Distribution Experiment on Polar, *Space Sci. Rev.*, 71, 531, 1995.

Brautigam, D.H., and J.M. Albert, Radial diffusion analysis of outer radiation belt electrons during the 9 October 1990 magnetic storm, jgr 105, 291, 2000.

Cliver, E.W., Y. Kamide, and A.G. Ling, Mountains versus valleys: Semiannual variation of geomagnetic activity, *J. Geophys. Res.*, 105, 2413, 2000.

Elkington, S.R., M. Wiltberger, A.A. Chan, and D.N. Baker, Physical models of the geospace radiation environment, *J. of Atmospheric and Solar-Terrestrial Physics*, Volume 66, Issue 15-16, 1371, Oct-Nov, 2004.

Gannon, J.L., and X. Li, Electron phase space density analysis based on test-particle simulations of magnetospheric compression events, AGU Monograph 155, page 205-214, 10.1029/156GM23, 2005.

Gannon, J.L., X. Li, and M. Temerin, Parametric study of shock-induced transport and energization of relativistic Electrons in the Magnetosphere, *J. Geophys. Res.*, 110, A12206, doi:10.1029/2004JA010679, 2005.

Green, J.C., M.G. Kivelson, Relativistic electrons in the outer radiation belt: Differentiating between acceleration mechanisms, *J. Geophys. Res.*, 109, No. A3, doi:10.1029/2003JA010153, 2004.

Hilmer, R.V., G.P. Ginet, and T.E. Cayton, Enhancement of equatorial energetic electron fluxes near L=4.2 as a result of high speed solar wind streams, *J. Geophys. Res.*, 105, 23,311, 2000.

Horne, R.B., and R.M. Thorne, Potential waves for relativistic electron scattering and stochastic acceleration during magnetic storms, *Geophys. Res. Lett.*, 25, 3011, 1998.

Horne, R.B., N.P. Meredith, R.M. Thorne, D. Heynderickx, R.H.A. Iles, R.R. Anderson, Evolution of energetic electron pitch angle distributions during storm time electron acceleration to megaelectronvolt energies, *J. Geophys. Res.*, 108, 1016, doi:10.1029/2001JA009165, 2003.

Horne, R.B., N.P. Meredith, S.A. Glauert, A. Varotsou, R.M. Thorne, Y.Y. Shprits, and R.R. Anderson, Mechanisms for the acceleration of radiation belt electrons, in *Recurrent Magnetic Storms: Corotating Solar Wind Streams*, edited by B.T. Tsurutani, R.L. McPherron, W.D. Gonzalez, G. Lu, J.H.A. Sobral, and N. Gopalswamy, *Amer. Geophys.* Un. Press, Wash. D.C., 2006.

Hudson, M.K., S.R. Elkington, J.G. Lyon, V.A. Machenko, I. Roth, M. Temerin, J.B. Blake, M.S. Gussenhoven, and J.R. Wygant, Simulation of radiation belt formation during storm sudden commencements, *J. Geophys. Res.*, 102, 14087, 1997.

Hudson, M.K., S.R. Elkington, J.G. Lyon, C.C. Goodrich, Increase in relativistic electron flux in the inner magnetosphere: ULF wave structure, *Adv. Space Res.*, 25, 2327, 2000.

Li, X., I. Roth, M. Temerin, J. Wygant, M.K. Hudson, and J.B. Blake, Simulation of the prompt energization and transport of radiation particles during the March 23, 1991 SSC, *Geophys. Res. Lett.*, 20, 2423, 1993.

Li, X., M. Temerin, D.N. Baker, G.D. Reeves, and D. Larson, Quantitative prediction of radiation belt electrons at geostationary orbit based on solar wind measurements, *Geophys. Res. Lett.*, 28, 1887, 2001a.

Li, X., D.N. Baker, S.G. Kanekal, M. Looper, M. Temerin, SAMPEX long term observations of meV electrons, *Geophys. Res. Lett.*, 28, 3827, 2001b.

Li, X., D.N. Baker, S. Elkington, M. Temerin, G.D. Reeves, R.D. Belian, J.B. Blake, H.J. Singer, W. Peria, G. Parks, Energetic particle injections in the inner magnetosphere as a response to an interplanetary shock, *J. of Atmospheric and Solar-Terrestrial Physics*, Volume 65, Issue 2, 233-244, January 2003a.

Li, X., M. Temerin, D.N. Baker, G.D. Reeves, D. Larson, and S.G. Kanekal, The Predictability of the Magnetosphere and Space Weather, AGU, EOS, Vol. 84, No. 37, 16 Sept. 2003b.

Li, X., Variations of 0.7-6.0 MeV electrons at geosynchronous orbit as a function of solar wind, *Space Weather*, 2, No. 3, S0300610.1029/ 2003SW000017, 2004.

Li, X., D.N. Baker, M. Temerin, G.D. Reeves, R. Friedel, and C. Shen, Energetic electrons, 50 keV–6 MeV, at geosynchronous orbit: their responses to solar wind variations, *Space Weather*, 3, S04001, doi:10.1029/2004SW00015, 2005.

Looper, M.D., J.B. Blake, and R.A. Mewaldt, Response of the inner radiation belt to the violent Sun-Earth connection events of October/November 2003, *Geophys. Res. Lett.*, 32, L03S06, doi:10.1029/2004GL021502, 2005a.

Looper, M.D., J.B. Blake, J.E. Mazur, and R.A. Mewaldt, Response of Inner Zone to Recent Strong Geomagnetic Events: SAMPEX Observations, SM41D-04, Fall AGU, San Francisco, 2005b.

Mann, I.R., T.P. O'Brien, D.K. Milling, Correlations between ULF wave power, solar wind speed, and relativistic electron flux in the magnetosphere: solar cycle dependence, *J. of Atmospheric and Solar-Terrestrial Physics*, Volume 66, 187-198, Oct-Nov, 2004.

Meredith, N.P., R.B. Horne, and R.R. Anderson, Substorm dependence of chorus amplitudes: Implications for the acceleration of electrons relativistic energies, *J. Geophys. Res.*, 106, 13165, 2001.

Miyoshi, Y.S., V.K. Jordanova, A. Morioka, D.S. Evans, Solar cycle variations of the electron radiation belts: Observations and radial diffusion simulation, *Space Weather*, 2, S10S102, doi:10.1029/2004SW000070, 2004.

Northrop, T.G., The Adiabatic Motion of charged Particles. Interscience Publishers, New York, 1963.

O'Brien, T.P., K.R. Lorentzen, I.R. Mann, N.P. Meredith, J.B. Blake, J.F. Fennell, M.D. Looper, D.K. Milling, and R.R. Anderson, Energization of relativistic electrons in the presence of ULF power and MeV microbursts: Evidence for dual ULF and VLF acceleration, *J. Geophys. Res.*, 108, 1329, doi:10.1029/2002JA009784, 2003.

Paulikas, G.A., and J.B. Blake, Effects of the solar wind on magnetospheric dynamics: Energetic electrons at the synchronous orbit, *Quantitative Modeling of Magnetospheric Processes* 21, Geophys. Monograph Series, 1979.

Rostoker, G., S. Skone, and D.N. Baker, On the origin of relativistic electrons in the magnetosphere associated with some geomagnetic storms, *Geophys. Res. Lett.*, 25, 3701, 1998.

Sarris, T.E., X. Li, and M. Temerin, Simulating radial diffusion of energetic (MeV) electrons through a model of fluctuating electric and magnetic fields, *Annales Geophysicae*, submitted, 2006.

Schulz, M. and L. Lanzerotti, Particle Diffusion in the Radiation Belts, Springer, New York, 1974.

Selesnick, R.S., and J.B. Blake, On the source location of radiation belt relativistic electrons, *J. Geophys. Res.*, 105, 2607, 2000.

Summers, D., R.M. Thorne, and F. Xiao, Relativistic theory of wave-particle resonant diffusion with application to electron acceleration in the magnetosphere, *J. Geophys. Res.*, 103, 20487, 1998.

Temerin, M., I. Roth, M.K. Hudson, J.R. Wygant, New paradigm for the transport and energization of radiation belt particles, *AGU, Eos*, Nov. 1, 1994, page 538.

Williams, D.J., A 27-day periodicity in outer zone trapped electron intensities, *J. Geophys. Res.*, 71, 1815, 1966.

Wygant, J., F. Mozer, M. Temerin, J. Blake, N. Maynard, H. Singer, M. Smiddy, Large amplitude electric and magnetic field signatures in the inner magnetosphere during injection of 15 MeV electron drift echoes, *Geophys. Res. Lett.*, 21, 1739-1742, 1994.

X. Li, LASP/CU, 1234 Innovation Drive, Boulder, CO 80303-7814, USA. (lix@lasp.colorado.edu)

Mechanisms for the Acceleration of Radiation Belt Electrons

Richard B. Horne,[1] Nigel P. Meredith,[1] Sarah A. Glauert,[1] Athina Varotsou,[2] Daniel Boscher,[2] Richard M. Thorne,[3] Yuri Y. Shprits,[3] and Roger R. Anderson[4]

During the declining phase of the solar cycle fast solar wind streams produce corotating interaction regions (CIRs) that drive moderate geomagnetic storms. These storms often have an unusually long recovery phase and produce high fluxes of relativistic electrons. Here we investigate the physical mechanisms responsible for accelerating electrons to relativistic energies inside the outer radiation belt. We review the most important electron acceleration and loss mechanisms, and present global simulations that combine radial diffusion with acceleration and loss by whistler mode chorus waves. We show that acceleration by chorus waves alone can increase the ~MeV electron phase space density between $4.5 < L < 6.5$ by up to three orders of magnitude. When radial diffusion and wave acceleration are included accelerated electrons are transported both inwards *and outwards* and increase the phase space density by a factor of 10 between $3.5 < L < 7$. At lower energies of ~0.1 to a few hundred keV, chorus waves cause electron precipitation that enhances inward radial diffusion. We conclude that chorus wave acceleration and loss play a major role in the dynamics of the outer radiation belt. We suggest that during the declining phase of the solar cycle Alfvénic wave activity in the fast solar wind provides continuous inward transport of ~1-100 keV electrons inside the magnetosphere which maintains whistler mode wave power long enough to accelerate electrons up to ~MeV energies, and drives radial diffusion to fill up the entire outer radiation belt.

1. INTRODUCTION

Solar variability in the form of coronal mass ejections, fast solar wind streams, interplanetary shocks, and magnetic clouds can disrupt the Earth's magnetic field on a global scale resulting in major magnetic storms and distortions to the Van Allen radiation belts [e.g., *Baker et al.*, 2004]. The main form of energy input to the magnetosphere is via magnetic reconnection, which has an efficiency of ~10% during intense magnetic storms [*Gonzalez et al.*, 1989], with a much smaller contribution from viscous interactions, which have an efficiency of ~0.2% [*Tsurutani et al.*, 1992]. It is therefore intriguing that of all forms of solar wind variability, increases in the relativistic electron flux (few MeV) in the Earth's outer van Allen radiation belt are best correlated with fast solar wind speed [*Paulikas and Blake*, 1979, *Bühler and Desorgher*, 2002; *Blake et al.*, 1997; *Li et al.*, 1997; *Baker et al.*, 1997; *O'Brien et al.*, 2001] rather than parameters that measure energy input via magnetic reconnection, such as the Akasofu epsilon factor [*Perreault and Akasofu*, 1978]. An increase in the solar wind speed by a factor of 2 can increase the total relativistic electron content in the outer belt by a factor of 100 or more [e.g., *Iles et al.*, 2002], which represents an enormous amplification of solar variability.

---

[1]British Antarctic Survey, Cambridge, England.
[2]ONERA/Department of Space Environment, Toulouse, France.
[3]Department of Atmospheric and Oceanic Sciences, University of California, Los Angeles, USA.
[4]Department of Physics and Astronomy, University of Iowa, Iowa City, Iowa, USA.

Recurrent Magnetic Storms: Corotating Solar Wind Streams
Geophysical Monograph Series 167
Copyright 2006 by the American Geophysical Union.
10.1029/167GM14

At solar minimum at low heliospheric latitudes the solar wind speed is typically 300-400 km s^{-1}, but at high heliospheric latitudes it is much faster, typically 750-800 km s^{-1} [*Phillips et al.*, 1995]. The fast solar wind is associated with coronal holes [*Krieger et al.*, 1973] that rotate with the ~27 day synodic rotation period of the Sun and migrate towards the solar equator during the solar cycle [*Harvey and Recely*, 2002]. The declining phase of the solar cycle is characterized by fast solar wind streams emanating from coronal holes at low heliospheric latitudes and interacting with the Earth's magnetosphere [*Sheeley et al.*, 1976]. However, the rotation of the Sun leads to a more complex solar wind whereby the fast streams 'overtake' the slow solar wind resulting in corotating interaction regions (CIRs) [*Smith and Wolfe*, 1976; see also the review by *Tsurutani et al.*, 2006, this issue]. CIRs are characterized by compressed plasma and magnetic fields (~20 nT), large scale nonlinear magnetic field fluctuations, and forward (reverse) propagating compressional waves on the leading (trailing) edges that usually develop into shocks beyond ~1 AU [e.g., see *Kamide et al.*, 1998]. The compressed magnetic field in a CIR can fluctuate southwards for periods of up to a few hours at a time, sufficient to cause magnetic reconnection and weak to moderate magnetic storms. Typically the minimum *Dst* usually remains above −100 nT in a CIR driven magnetic storm [*Tsurutani et al.*, 2006, this issue]. For weak (minimum *Dst* between −30 and −50 nT) or moderate (−50 to −100 nT) storms superposed epoch analysis (where CIR or interplanetary coronal mass ejections (ICMEs) were not distinguished) shows that the recovery phase is usually one to two days [*Loewe and Prölss*, 1997]. However, large amplitude Alfvén waves are often associated with the fast solar wind streams following the CIR where the ratio of magnetic perturbations to the background field can be significantly larger than $\delta B/B$ ~1. The southward components of the fluctuating magnetic field can then cause additional magnetic reconnection resulting in small plasma injections and a much longer recovery phase to the magnetic storm. These types of events are known as High Intensity Long Duration Continuous Auroral Activity (HILDCAA) events [*Tsurutani and Gonzalez*, 1987]. The response of the radiation belts to this form of solar variability, made more complex through the passage from the Sun to the Earth, is significantly different from other forms of solar wind drivers, and provides important information on particle acceleration and loss processes inside the Earth's radiation belts.

Radiation belt electrons are accelerated up to energies of a few MeV inside the Earth's magnetic field from a source that originates from either the ionosphere at <1 eV, or the solar wind at ~10 eV. In effect, the magnetosphere is a gigantic particle accelerator. The acceleration is thought to occur in a multi-step process [*Baker et al.*, 1986] and has become an important research topic in recent years for a variety of reasons. Scientifically, there has been a step change in our thinking about the processes responsible for the acceleration of energetic electrons. The old established ideas about inward radial diffusion (betatron acceleration) [*Falthamar*, 1965; *Schulz and Landzerotti*, 1974] as the dominant process are now in doubt as acceleration by wave-particle interactions is shown to be increasingly important [e.g., *Horne et al.*, 2005b]. This has opened up an exciting new area for research, and will be a major issue for the International Living with a Star (ILWS) program to be addressed through the Radiation Belt Storm Probes mission (USA), and other missions proposed in Canada, Japan and Europe. Loss of energetic electrons to the atmosphere also creates nitric oxide at ~50-60 km that depletes high altitude ozone [e.g., *Thorne*, 1977; *Randall et al.*, 2005] and may affect wind patterns in the mesosphere, and radiation balance in the atmosphere.

In practical terms, substantial increases in the energetic electron flux are a health hazard to humans in space, and are linked to satellite malfunctions (anomalies) on orbit [*Wrenn*, 1995; *Wrenn et al.*, 2002; *Baker*, 2001; *Webb and Allen*, 2004]. In some cases they may have caused complete satellite loss [*Baker et al.*, 1998]. Since a modern telecommunications spacecraft costs about US$200 M to build, US$100 M to launch into geosynchronous orbit, and about 5% (of the sum insured) to insure each year, and since there are more than 300 spacecraft in geosynchronous orbit alone, there is a huge commercial investment. This investment is growing as we rely more and more on satellite systems. The risk is also becoming higher as more commercial operators choose to insure their satellite fleets themselves by having spare capacity rather than pay high insurance premiums. These drivers provide an impetus to understand the physical processes responsible for electron flux variations so that knowledge can be transferred into applications, for example, to help analyze risk and protect our assets in space.

This paper is a result of the presentation given at the Chapman Conference on Corotating Solar Wind Streams and Recurrent Geomagnetic Activity, held in Manaus, Brazil, in February 2005. We provide some new observations on the effects of corotating solar wind streams on magnetic activity at the Earth, and radiation belt electron flux. We provide a brief review of the most important mechanisms likely to be responsible for electron acceleration during these events, and raise questions over existing ideas about electron acceleration at geosynchronous orbit. We present global simulations to suggest a new concept, that ~MeV flux increases in the outer radiation belt are due to a combination of inward radial diffusion at low energies, local acceleration to high energies by whistler mode waves near $4.5 \leq L \leq 6.5$ combined with inward and outward radial diffusion to fill up the entire outer radiation belt.

2. MAGNETIC ACTIVITY AND ELECTRON FLUX

The electron flux in the outer radiation belt exhibits variations on a variety of different timescales. Over a solar cycle the flux at ~1 MeV in the outer belt is highest during the declining phase of the solar cycle, and the peak of the outer belt tends to be located closer to the Earth [*Li et al.*, 2001; *Miyoshi et al.*, 2004]. When the flux and *Dst* index are averaged over 30 days, there is a remarkable correlation between the penetration of the outer belt to lower L (~3), and negative *Dst* ($\leq$–30 nT) [*Li et al.*, 2001]. These observations suggest that the solar wind is a strong driver for the radiation belts. They also suggest that the increased flux is related to the fast solar wind streams associated with coronal holes observed during the declining phase [*Sheeley et al.*, 1976; *Baker et al.*, 1997].

On timescales of a few days, *Reeves et al.* [2003] found that about 50% of magnetic storms result in a net increase in the MeV electron flux. They defined magnetic storms as having a minimum *Dst* $\leq$–50 nT and did not distinguish between storms driven by ICMEs or CIRs. Typically the flux drops during the main phase and then increases above the pre-storm level during the recovery phase. The flux often remains enhanced above the pre-storm level even when the *Dst* index has returned to near zero indicating that the process is non adiabatic, i.e., there is net acceleration [e.g., *Kim and Chan*, 1997]. Other statistical analysis based on the occurrence of CIRs shows that although the storms produced by CIRs are weak they produce some of the highest relativistic electron fluxes of the solar cycle [*McPherron and Weygand*, 1996, this issue].

One of the distinguishing features of magnetic storms driven by CIRs is that the recovery phase can last for several days and even up to a few weeks for the reasons discussed above [see also *Tsurutani et al.*, 2006, this issue]. In contrast, the recovery phase of most weak or moderate magnetic storms (from a superposed epoch analysis where the ICME or CIR driver were not distinguished) is usually one or two days. In general, the recovery tends to be longer when minimum *Dst* is larger (more negative) [*Loewe and Prölss*, 1997]. Observations show that there is a delay between an increase in solar wind speed and increases in relativistic electron flux of about two days [*Baker et al.*, 1990; *Vassiliadis et al.*, 2002; *Mann et al.*, 2004]. This suggests that the duration of the storm recovery phase, or perhaps prolonged periods of magnetic activity, is an important factor for enhancing the ~MeV electron radiation belt flux and that any non-adiabatic acceleration mechanism operates on a timescale of 1-2 days.

To test the idea that it is the duration of the storm that is the important factor we computed the number of events where the *Kp* index, smoothed over a 15-hour running mean (Kp_{sm}), remains above a given value for different time periods and compared this with the sunspot number for almost 4 solar cycles (1964-2003). Comparison with the sunspot number enables us to determine when these events occur in relation to the phase of the solar cycle. We use the *Kp* index since this is a global measure of the disturbance in the Earth's magnetic field, and is often used as an input to models for the radiation belts. The number of occasions where Kp_{sm} is greater than 2 for less than a day (Plate 1, top left) tends to occur mainly during the rising phase of the solar cycle. This is consistent with substorm disturbances and short period storms. However, the number of occasions that $Kp_{sm} \geq 2$ for three or more days tends to peak during the declining phase (top right), when CIRs and fast solar wind streams are more likely to drive weak magnetic storms. Occurrence during the declining phase is even more pronounced when $Kp_{sm} \geq 3$ for 3 or more days (bottom right).

Observations also show that the electron flux has a seasonal dependence [*Russell and McPherron*, 1973; *Li et al.*, 2001] where during the equinoxes the flux is almost a factor of three higher than at the solstices [*Baker et al.*, 1999]. Three ideas have been suggested for this effect [see *Li et al.*, 2001], but the link to an increase in solar wind speed due to a change (increase) in heliographic latitude [*Rhodes and Smith*, 1976a,b], and the fact that Alfvén wave intensities increase with solar wind speed, may be the most important [*Cliver et al.*, 2000; *Li et al.*, 2001]. By calculating the number of events over a three month period centered on solstice and equinox, Plate 2 shows that prolonged periods of magnetic activity $Kp_{sm} > 2$ (top) and $Kp_{sm} > 3$ (bottom) also peak during spring and autumn. This suggests that electron flux increases at equinox may have a common cause to those during the declining phase. The data here emphasize that it is not only the magnitude of the disturbance, but the length of time a weak or moderate geomagnetic disturbance lasts that is the distinguishing factor.

3. ELECTRON ACCELERATION AND LOSS

Several theories have been proposed for acceleration and loss of relativistic electrons in the outer belt [see the reviews by *Li and Temerin*, 2001; *Friedel et al.*, 2002, and *Horne*, 2002]. Apart from shock acceleration, which is very important for rapid electron acceleration on timescales of minutes [*Hudson et al.*, 1997], the two leading theories for acceleration on timescales of several hours to a few days are inward radial diffusion, and (Doppler shifted) cyclotron resonance with whistler mode chorus waves.

154 MECHANISMS FOR ACCELERATION

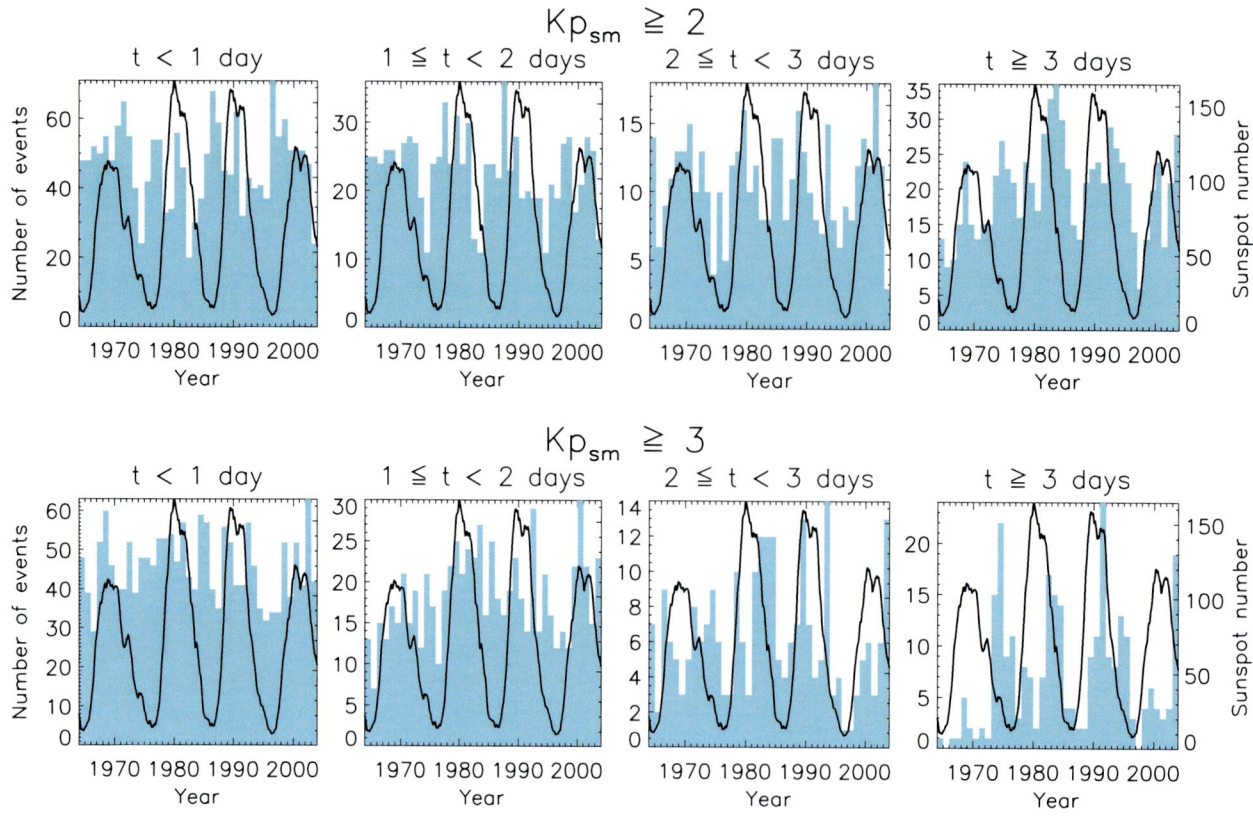

Plate 1. The number of events where the smoothed Kp index (Kp_{sm}) remains above a given level for a specified period of time. (Top) $Kp_{sm} \geq 2$ and (bottom) $Kp_{sm} \geq 3$. The time period increases left to right. The Kp index is smoothed using a 15-hour running mean. The black line shows the sunspot number. The data are for 1964-2003.

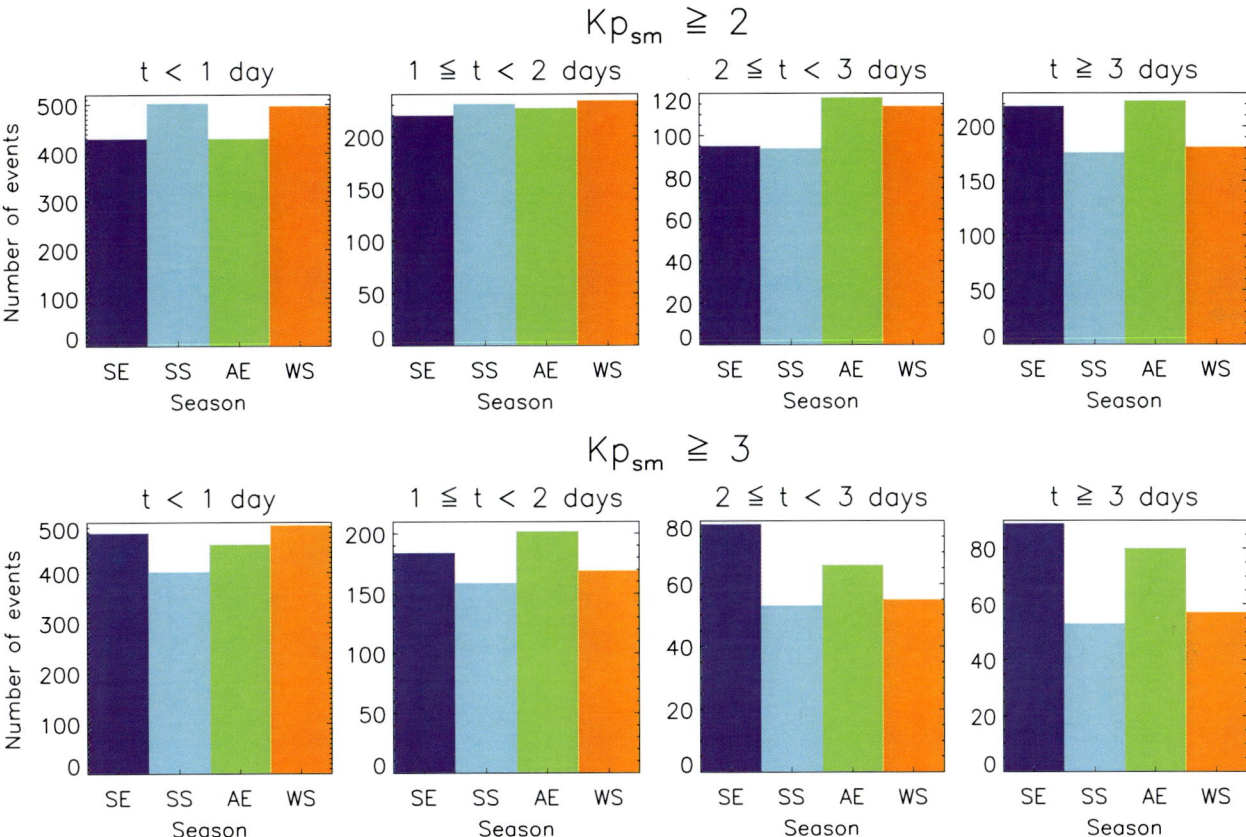

Plate 2. The number of events over a three month period centred on spring equinox (SE), summer solstice (SS), autumn equinox (AE), and winter solstice (WS) where the smoothed *Kp* index (Kp_{sm}) remains above a given level for a specified period of time. (Top) $Kp_{sm} \geq 2$ and (bottom) $Kp_{sm} \geq 3$. The time period increases left to right. The *Kp* index is smoothed using a 15-hour running mean. The data are for 1964-2003.

3.1. Acceleration by Inward Radial Diffusion

Inward radial diffusion is one of the oldest and most important theories for electron acceleration [*Falthamar*, 1965; *Cornwall*, 1968; *Schulz and Eviatar*, 1969; *Schulz and Lanzerotti*, 1974; *Li*, 2006, this issue]. Energetic electrons at ~1 MeV drift around the Earth due to the curvature and gradient of the magnetic field on a timescale of about 10 minutes at $L = 6$ (~75 minutes for 0.1 MeV electrons). As the electrons drift around the Earth, global scale fluctuations in the Earth's magnetic and electrostatic fields at frequencies comparable to the electron drift period (typically 0.1-few mHz) scatter electrons across the magnetic field. The process is considered as stochastic diffusion where the first two adiabatic invariants are conserved, but the third is violated. By conservation of the first adiabatic invariant, electrons are accelerated if they diffuse toward the planet, which requires an outward gradient in phase space density. Energy is effectively transferred from the fluctuating fields into the electrons by betatron acceleration, but this represents a very small energy loss from the fluctuating fields. Radial diffusion tries to reduce the gradients responsible for the diffusion.

The original problem with radial diffusion is that the time for acceleration is slow compared to the variations observed during magnetic storms. However, the process is enhanced when ULF waves are present at frequencies close to the electron drift frequency [*Elkington et al.*, 1999, 2003; *Hudson et al.*, 1999]. To understand the process in more detail consider an electron in a static dipole magnetic field, compressed on the dayside due to the solar wind, and mirroring at the magnetic equator. By conservation of the first adiabatic invariant the electron will drift around the Earth along a surface of constant magnetic field strength and will have a radial velocity component towards the Earth on one half of the drift orbit and away from the Earth on the other half. If a global, monochromatic, torriodal ($m = 2$) field line resonance is present then the electron will experience a positive radial electric field on one half of the drift orbit and a negative electric field on the other half. Over the whole drift orbit there is a net increase, or decrease, in energy depending on the initial phase. Since the first two adiabatic invariants are conserved this results in a radial displacement of the electron; thus ULF waves enhance the rate of radial diffusion. As the electron is diffused to smaller (larger) L the drift period will change. Therefore to continue transporting the electron across the magnetic field ULF waves must be present at the electron drift frequency at each L shell, and over a range of L. Theoretical work shows that when a convection electric field and a non-dipole magnetic field are included drift resonance can be extended over a range of frequencies [*Elkington et al.*, 2003].

Observations show that an increase in solar wind speed is correlated with an increase in ULF wave power in the Pc 5 (2-7 mHz) band near dawn [*Greenstadt et al.*, 1979; *Junginger and Baumjohann*, 1988; *Kokubun et al.*, 1989]. Furthermore, ULF wave power is correlated with an increase in relativistic electron flux but with a delay of about 2 days [*Mathie and Mann*, 2000; *Mann et al.*, 2004]. The source of power for the waves is thought to be either Kelvin-Helmholtz instabilities from the flow of the solar wind past the magnetospheric boundary [e.g., *Southwood*, 1979; *Miura*, 1992; *Cahill and Winkler*, 1992] or variations in the solar wind pressure propagating into the magnetosphere [e.g., *Lysak and Lee*, 1992]. During the initial phase of a storm ULF wave activity is broadband at all local times, consistent with the needs of radial diffusion, but during the recovery phase it is mainly narrow band and most intense in the dawn-noon sector [*Posch et al.*, 2003]. These observations are important evidence in favor of radial diffusion, but the narrow frequency bandwidth of the waves during the recovery phase and the confinement of intense waves in magnetic local time (MLT) still need to be incorporated into models.

Figure 1 shows the variation in electron energy for constant first invariant $M = p^2 \sin^2 \alpha / (2m_0 B_0)$, where p is the electron

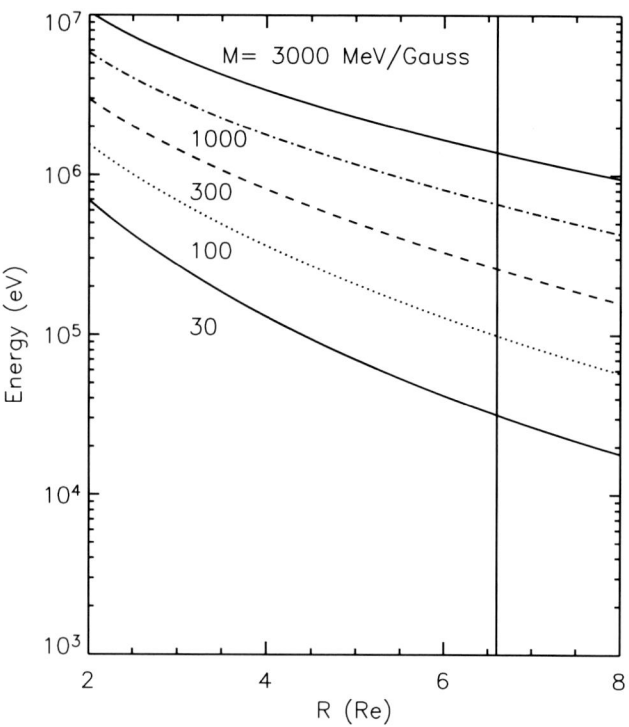

Figure 1. Variation of electron energy for constant magnetic moment M in the Earth's dipole magnetic field in the equatorial plane. The vertical line corresponds to geosynchronous orbit.

momentum, α is pitch angle and m_0 is rest mass and B_0 is the ambient magnetic field. A dipole magnetic field is assumed. To obtain an energy of ~2 MeV at $L \approx 4$ by inward diffusion a seed population with an energy of ~600 keV is required at geosynchronous orbit. Thus particle acceleration is believed to take place as a multi-step process [*Baker et al.*, 1986]. Since substorms generally produce electrons with energies up to about ~100 keV, comparable to the cross polar cap potential, they can supply the seed population for the lower energy portion of the radiation belt, but it is not yet clear how the higher energy part of the seed population is obtained. To obtain ~2 MeV electrons at geosynchronous orbit poses an even more difficult problem. The position of the last closed drift path furthest away from the Earth, or outer trapping boundary, can be no further than the magnetopause and the characteristics of the seed population at that location and the effectiveness of inward radial diffusion in a non-dipole magnetic field are unclear.

Several attempts to reproduce electron flux variations during individual magnetic storms have been made using a form of the radial diffusion equation given by [*Schulz and Lanzerotti*, 1974]

$$\frac{\partial f}{\partial t} = L^2 \frac{\partial}{\partial L}\left(\frac{D_{LL}}{L^2}\frac{\partial f}{\partial L}\right) - \frac{f}{\tau} \quad (1)$$

where the first term on the right is due to radial diffusion and the second term represents losses. Here f is the phase space density, D_{LL} is the radial diffusion coefficient and τ is the loss timescale. Radial diffusion is critically dependent on the form of the diffusion coefficient D_{LL}, which is usually represented as a sum of coefficients for magnetic D_{LL}^M and electrostatic D_{LL}^E field fluctuations. A scaling of D_{LL} for different levels of Kp has been obtained from satellite data and shows that in general $D_{LL}^M > D_{LL}^E$ [*Brautigam and Albert*, 2000; *Brautigam et al.*, 2005]. The magnetic diffusion coefficient is given by

$$D_{LL}^M = 10^{(0.506Kp - 9.325)} L^{10} \quad (2)$$

for $1 \leq Kp \leq 6$ and is shown in Figure 2. Radial diffusion becomes very large at large L, which suggests that radial diffusion should be most important in the outer radiation belt, out to the outer trapping boundary.

Individual case studies generally show that radial diffusion can reproduce electron flux variations below about 500 keV, but is unable to account for observations at a few MeV [*Brautigam and Albert*, 2000; *Miyoshi et al.*, 2003].

To determine how well radial diffusion can account for flux variations over a longer time period, Plate 3 shows the results of radial diffusion using (1) and (2) obtained by *Shprits et al.* [2005]. The model results (top panel) are compared to data (middle panel) obtained from 500 orbits of the Combined Release and Radiation Effects Satellite (CRRES). The input data for the model is a time series of the Kp index (bottom panel) which is used to calculate the diffusion coefficients in (2). The inner and outer boundaries were set at $L = 1$ and 7 and the flux at the outer boundary was found by a fit to CRRES data given by $J = 8222.6 \exp(-7.068K)$ in $cm^{-2} s^{-1} sr^{-1} keV^{-1}$ where K is kinetic energy in MeV. The flux at the outer boundary was kept constant. The timescale for loss τ was parameterized by Kp [*Shprits et al.*, 2005] and was $\tau = 3/Kp$ days, i.e., typically 2-3 days during quiet times and less during active times. When Kp increases radial diffusion is enhanced and electrons are transported towards lower L resulting in a flux increase. As Kp decreases the flux decays according to the loss timescale τ. The model is able to reproduce the location of the peak flux and inner boundary of the outer radiation belt reasonably well on a timescale of a few days, but is not able to reproduce the duration of flux enhancements of many storms, nor the gradual build up of flux during storms. The model underestimates the flux by up to an order of magnitude during many storms, which is attributed to the neglect of local wave acceleration.

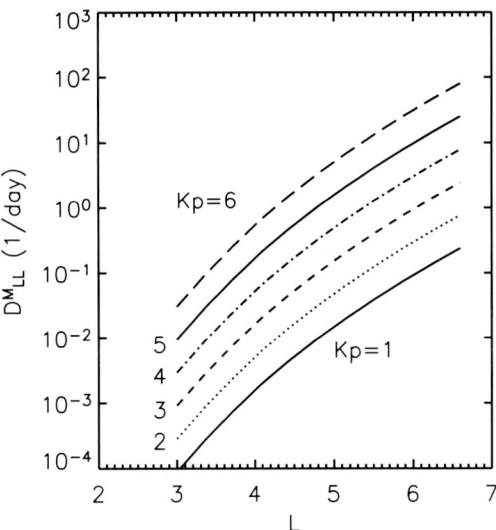

Figure 2. The magnetic diffusion coefficient D_{LL}^M for different levels of the Kp index using the model of *Brautigam and Albert* [2000].

Other evidence also suggests that radial diffusion alone cannot account for flux variations in the outer radiation belt. Observations show that the electron phase space density peaks near $L = 5$, and gradually builds up during the recovery phase of a storm [*Green and Kivelson*, 2004; *Iles et al.*, 2006]. This is inconsistent with an inward radial diffusion model which predicts a monotonic profile in phase space

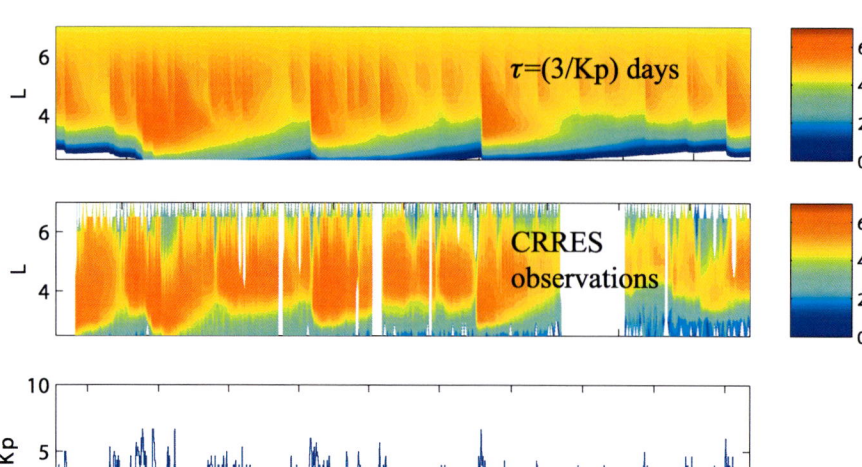

Plate 3. Comparison between the electron flux calculated from the radial diffusion model and CRRES observations. (Top) electron flux at 0.95 MeV obtained from the radial diffusion model with a variabile loss timescale of $\tau = 3/Kp$ days, (middle) electron flux measured by CRRES, and (bottom) the evolution of the Kp index. The units of flux are cm^{-2} sr^{-1} s^{-1} MeV^{-1} on a log scale.

density that is either flat or peaks at the outer boundary. Furthermore, by conservation of the first two adiabatic invariants radial diffusion should result in pitch angle distributions that are peaked near 90° as the electrons diffuse inwards. Observations show that during magnetic storms the pitch angle distribution has a 'flat top' distribution which could be partly due to drift shell splitting [*West et al.*, 1973; *Selesnick and Blake*, 2002], but could also be a result of wave acceleration [*Horne et al.*, 2003a].

3.2. Acceleration by Whistler Mode Chorus Waves

The alternative leading theory is wave acceleration by whistler mode chorus waves [*Horne and Thorne*, 1998; *Summers et al.*, 1998]. The idea is that during the recovery phase of a storm, inward convection, substorms, and inward radial diffusion provide a seed population of ~1 keV to ~100 keV electrons. By conservation of the first two adiabatic invariants the electron distribution becomes anisotropic as the particles are transported towards the planet, with $T_\perp > T_\parallel$ (where $T_\perp$ and $T_\parallel$ are temperatures perpendicular and parallel to the ambient magnetic field, respectively) and excite whistler mode chorus waves at frequencies below the electron gyrofrequency f_{ce} ($= |q_e|B_0/(2\pi m_e)$ where B_0 is the ambient magnetic field). The waves grow by scattering electrons into the loss cone at small pitch angles [*Kennel and Petschek*, 1966], possibly via nonlinear interactions [*Helliwell*, 1967], but they also scatter electrons to higher energies at large pitch angles which remain trapped in the magnetic field. The waves effectively transfer energy from a large number of electrons at low energies to accelerate a smaller number to high energies.

Plate 4 shows an example of whistler mode chorus waves observed by the CRRES spacecraft on 12 September 1990. Chorus waves can be seen near $f_{ce}/2$ (dashed white line) outside the high density plasmasphere region from about 18:30UT onwards, until the spacecraft re-enters the plasmasphere just after 01:30 UT the following day. The top panel shows the auroral electrojet (*AE*) index, and illustrates that chorus wave power enhancements during this event are associated with elevated values of *AE* [*Meredith et al.*, 2000, 2001].

Whistler mode chorus waves tend to propagate along the magnetic field, and their properties are very dependent on MLT. At the magnetic equator, chorus wave intensities peak outside the plasmapause from just before midnight, through dawn to just after midday MLT [*Meredith et al.*, 2001, 2003a]. Chorus wave power is enhanced during substorms [*Tsurutani and Smith*, 1974, 1977], and inward convection [*Lyons et al.*, 2005], and both substorms and inward convection result in high levels of *AE* [*Pytte et al.*, 1978; *Sergeev and Lennartsson*, 1988]. Using *AE* as a proxy for magnetic activity, wave intensities increase significantly with *AE*, for example, by up to two orders of magnitude between *AE* < 100 nT and *AE* > 300 nT (see also Plate 7 below for increases in wave intensities with *Kp*). The main reason why chorus intensities peak at night, through dawn to the dayside is due to the significant increase in the flux of 1-150 keV electrons that drive the loss cone instability. Instability growth rates are proportional to the number of resonant electrons [*Kennel and Petschek*, 1966]. At $L = 5$, the electron flux at 28 keV increases by an order of magnitude or more between low (*AE* < 100 nT) and high (*AE* > 300 nT) levels of *AE* [*Meredith et al.*, 2004]. On the dayside the flux of these electrons becomes smaller with increasing MLT due to precipitation and due to the drift path which may not be closed at these energies. Typically, for *AE* > 300 nT and $L \approx 5$ the average flux of 28 keV electrons after 14:00 MLT is 5 to 10 times smaller than that at dawn [*Meredith et al.*, 2004]. Chorus waves can also propagate to the ground and are observed during the recovery phase of magnetic storms [*Smith et al.*, 2004a,b]. Ground observations also show that chorus wave power peaks after dawn through to the dayside, consistent with the idea that the waves are generated by unstable distributions as a result of inward convection and electron drift.

The October 9, 1990 storm is an event where both chorus waves and electrons were measured by CRRES during a storm with an extended recovery phase (Plate 5). On October 10, the solar wind speed reached 500 km s^{-1} and the *Dst* response shows a strong storm with a minimum *Dst* of −133 nT, and a long recovery phase of 3 days or so. The *Kp* index was greater than 3 for nearly the whole recovery phase, with several periods where *Kp* > 5. During the main phase of the storm on October 10, there was a rapid reduction in the ~1 MeV electron flux near $L = 4$ (top panel) followed by an increase to values greater than the pre-storm level over a period of a few days. Strong chorus wave amplitudes in the frequency range $0.1 < f/f_{ce} < 0.5$ (second panel) were observed outside the plasmapause (solid white line) to beyond geosynchronous orbit. The event illustrates the connection between a storm with a prolonged recovery period where *AE* is enhanced, and inward particle transport is expected, enhanced levels of chorus wave amplitudes, and an increase in the ~1 MeV electron flux in the outer radiation belt [*Meredith et al.*, 2002, 2003b]. The flux of relativistic electrons can also be enhanced during prolonged periods of elevated *AE* that are not associated with storms [*Meredith et al.*, 2003b]. However, no significant flux enhancements are observed during the recovery phase of storms, irrespective of the magnitude of the storm, when *AE* stays small and where chorus amplitudes are low [*Meredith et al.*, 2002, 2003b].

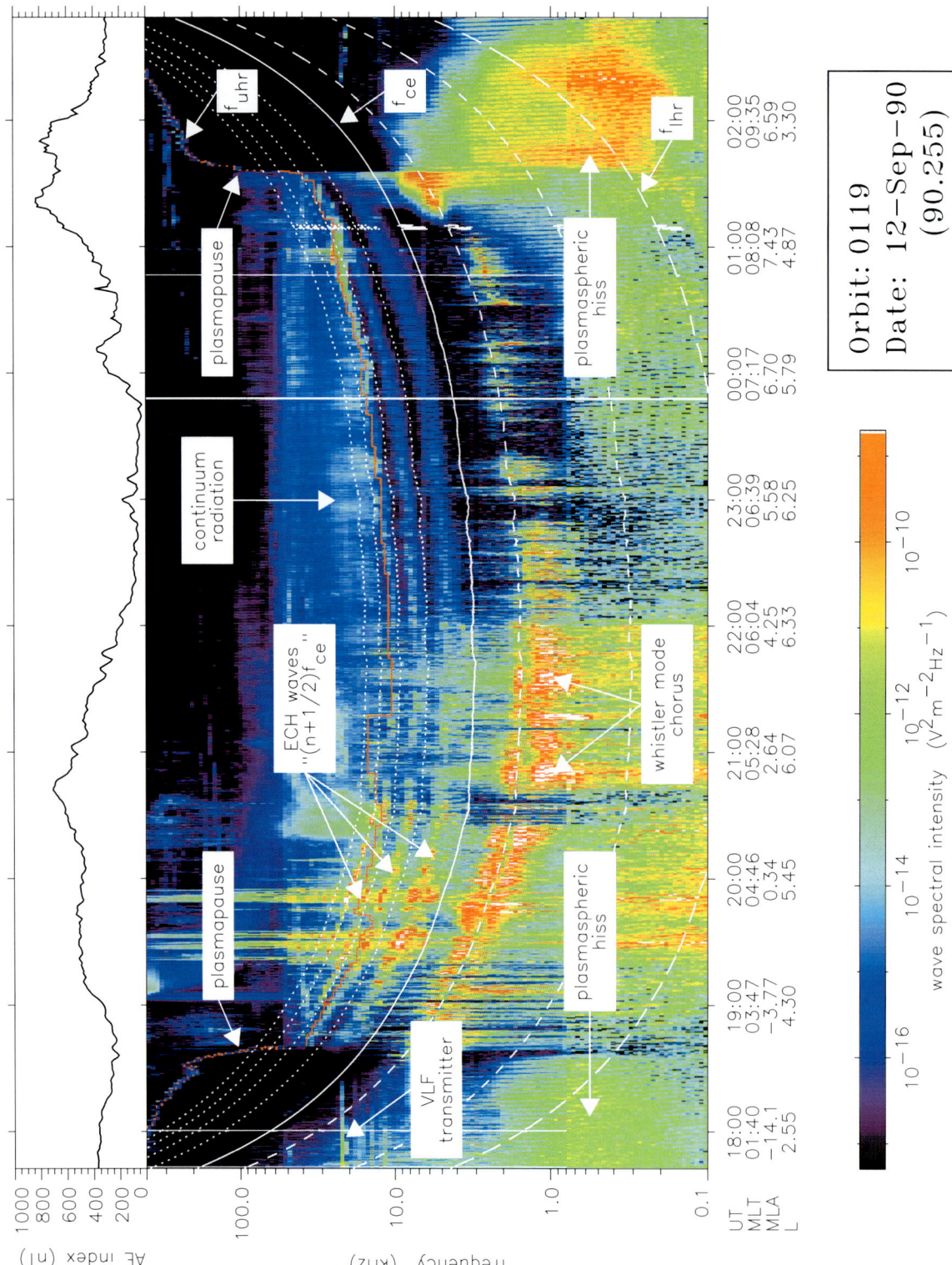

Plate 4. Survey plot of the wave electric field spectral intensity observed by CRRES during orbit 119 together with the auroral electrojet (AE) index (top). Whistler mode chorus is indicated near $0.5f_{ce}$ and occurs mainly outside the plasmapause. The upper hybrid resonance f_{UHR}, f_{ce} and lower hybrid resonance f_{LHR} frequencies are shown by the solid red, white, and long dashed lines, respectively. The dashed lines show $0.1f_{ce}$ and $0.5f_{ce}$, and the dotted lines show harmonics nf_{ce}. After Meredith et al. [2004].

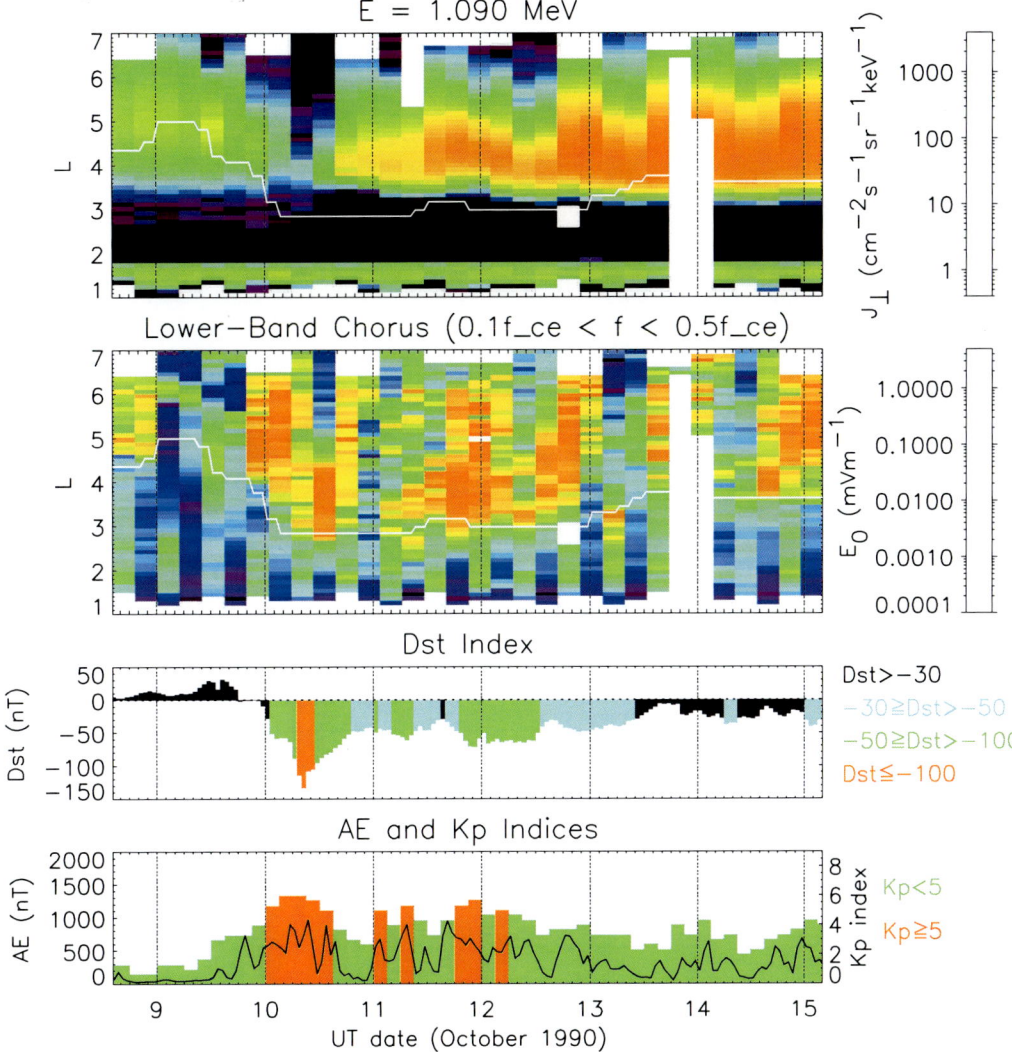

Plate 5. CRRES data and geophysical parameters for the magnetic storm of October 9, 1990. From top to bottom, electron flux at 1.09 MeV, lower band whistler mode chorus amplitudes $0.1 < f/fce < 0.5$, *Dst* index (color coded), and on the same bottom panel AE index (solid line) and *Kp* (color coded).

Acceleration by whistler mode chorus waves is only possible above a minimum energy, typically >1 keV for lower band chorus (<$0.5f_{ce}$) outside the plasmapause in the equatorial region. Electrons >1 keV are easily supplied by inward convection and substorm injection into regions just outside the plasmapause. The energy of the seed population required for wave acceleration near $L \sim 4$ is much lower than that required by radial diffusion, and is easily obtainable at low L. Once excited, chorus can resonate with electrons over a large range of energies, up to several MeV [*Horne and Thorne*, 1998]. However, the acceleration is most effective in regions where the ratio of the electron plasma frequency to the electron gyrofrequency (f_{pe}/f_{ce}) is small, typically less than 4 [*Horne et al.*, 2003b, 2005a]. Low values of f_{pe}/f_{ce} increase the phase velocity of the waves and enables more effective energy diffusion. Since the plasmapause extends to larger L in the afternoon and evening MLT sector [*Carpenter and Park*, 1973; *Carpenter and Anderson*, 1992] the region where f_{pe}/f_{ce} is smallest lies just outside the plasmapause from midnight, through dawn to about midday MLT, and therefore this is the region where acceleration by whistler mode waves should be most effective. This was certainly the case during the strong 2003 Halloween magnetic storms when the plasmapause was compressed inside $L \approx 2$, except near dusk [*Baker et al.*, 2004], and electrons were accelerated by whistler mode chorus waves in the slot region [*Horne et al.*, 2005b].

The timescale for wave acceleration can be calculated using quasi-linear theory. In quasi-linear theory it is assumed that there is a broad band of waves composed of a succession of individual wave packets where each wave packet has a slightly different frequency and wavenumber to the next, and have random phase. Particles are scattered in pitch angle and energy by many wave packets and, since the waves have random phase, perform a random walk in momentum space. The process is treated as a diffusion problem. The total bandwidth of the waves is sufficiently large so that nonlinear particle trapping can be neglected, and also ensures that resonance can occur over a wide range of energies. The process can be formulated as a diffusion equation [*Kennel and Engelmann*, 1966; *Lyons and Williams*, 1984] where the diffusion coefficients depend on wave power. Using typically observed wave powers, and assuming that the waves propagate along the magnetic field, solving the diffusion equation provides a timescale for acceleration, which is typically 1-2 days [*Summers et al.*, 2002, 2004; *Horne et al.*, 2003b]. When wave-particle interactions along the bounce orbit of the particle are taken into account, and the change in spectral properties of the waves with MLT are included the timescale is still found to be of the order of ~1 day [*Horne et al.*, 2005a], providing strong support for wave acceleration as a viable mechanism for producing radiation belt electrons.

Although particle scattering by chorus has been treated as a diffusion problem, in fact chorus is composed of discrete rising frequency elements superimposed on an enhanced broad band background [*Tsurutani and Smith*, 1977]. Scattering by these individual discrete elements has yet to be fully quantified, but nonlinear phase trapping by the waves could contribute to electron acceleration [*Albert*, 2002]. Chorus bursts are also thought to be responsible for very short duration (<1 s) microbursts of relativistic electron precipitation [e.g., *Lorentzen et al.*, 2001] (see also below).

3.3. Electron Loss

The electron flux in the radiation belts is a result of a delicate balance between acceleration, transport and loss. There are several loss processes, but the most important are losses to the atmosphere via pitch angle scattering by wave-particle interactions, and losses onto open drift paths via de-trapping. The relative contribution of each process is unknown.

During the main phase of even weak or moderate storms the ~MeV electron flux often exhibits a rapid reduction, or 'drop out', for example, as shown in Plate 5. Adiabatic effects can lead to a reduction in flux via the '*Dst* effect' [*Kim and Chan*, 1997]. For example, if *Dst* becomes negative on a timescale that is slow compared to the drift orbit, which is ~10 minutes, then to maintain the flux enclosed by the drift orbit (to conserve the third adiabatic invariant) energetic electrons are transported to larger L by an electric field induced by the changing magnetic field. By conservation of the first invariant electrons lose energy as they are transported outwards. This process is reversible and so there should be no net gain or loss of particles. However, if the magnetopause is compressed, or if there are significant changes in the electrostatic fields, some electrons could be transported onto open drift orbits, encounter the magnetopause and be lost from the system. Magnetopause crossing may produce inward phase space density gradients and drive outward radial diffusion.

Several types of waves can contribute to losses via precipitation into the atmosphere. Outside the plasmapause bursts of ~MeV precipitation have been inferred from X ray observations from balloons [*Winckler et al.*, 1962; *Parks*, 1967; *Milan et al.*, 2002] and have been observed directly on satellites [*Lorentzen et al.*, 2001; *O'Brien et al.*, 2004; *Thorne et al.*, 2005a,b]. These bursts are believed to be caused by pitch angle scattering by individual chorus wave elements as they propagate along the field line. The process should be most effective at latitudes $\lambda > 20°$ where the resonant energy increases into the ~MeV range [*Horne and Thorne*, 2003]. Thus chorus can contribute to both loss and acceleration.

Low altitude satellites also observe bands of precipitation which have been associated with electromagnetic ion cyclotron (EMIC) waves [*Lorentzen et al.*, 2000]. In a hydrogen, helium, oxygen plasma, typical of the magnetosphere, EMIC waves usually propagate at frequencies just below the gyrofrequency of each species [*Anderson et al.*, 1992b; *Fraser et al.*, 1996]. The presence of heavy ions enables both left and right hand polarized waves for parallel propagation, but as the angle of propagation increases the polarization becomes more linear (i.e., elliptical). The waves are generated by anisotropic distributions of H^+ and O^+ [*Kozyra et al.*, 1984; *Horne and Thorne*, 1993] and simulations show that they are enhanced during the storm time ring current [*Kozyra et al.*, 1997]. The polarization is often observed to be left handed to linear, with a smaller percentage of right handed to linear polarization. Since the wave frequency is much less than the electron gyrofrequency, cyclotron resonance between right hand polarized waves and electrons occurs when the wave frequency is Doppler shifted upwards by the velocity of the electrons along the magnetic field. This occurs when the waves and electrons travel in opposite directions. Electrons gyrate in the opposite sense to left hand polarized waves but resonance is still possible if the electrons 'overtake' the wave. In this case the sense of wave polarization is effectively reversed in the frame of reference of the electron. As a result, left hand-linear polarized waves resonate with lower energy electrons than right hand polarized waves, above ~0.5 MeV for left hand polarized waves and above ~2 MeV for right hand polarized waves [*Meredith et al.*, 2003c]. EMIC waves are observed in regions of high density near the plasmapause [*Fraser and Nguyen*, 2001; *Meredith et al.*, 2003c] and at larger L, typically beyond $L = 7$ [*Anderson et al.*, 1992a,b]. EMIC waves only contribute to electron loss [*Horne and Thorne*, 1998], and calculations of pitch angle diffusion rates show that they can be very effective in precipitating electrons of ~1 MeV and above in regions of high density [*Summers and Thorne*, 2003; *Albert*, 2003]. However, as electrons drift around the Earth in the heart of the radiation belts near $L \sim 4$ it is thought that they only encounter EMIC waves for a fraction of the electron drift orbit, at the edge of the plasmapause and through the plume region where the ring current overlaps with the high density plasma. As yet the effectiveness of EMIC waves for losses on a global scale has yet to be fully evaluated.

Inside the plasmasphere several types of waves can contribute to electron loss in the energy range ~0.1-1 MeV. They include whistler mode hiss at frequencies between 0.1-2 kHz [e.g., *Lyons et al.*, 1972; *Smith et al.*, 1974; *Meredith et al.*, 2004, 2006], VLF transmitter signals, and lightning generated whistlers [*Abel and Thorne*, 1998a,b].

In summary, to obtain any substantial increase in electron flux the acceleration process must be efficient enough to overcome all losses, and to do this on a timescale of 1-2 days.

4. GLOBAL MODELING

So far there has only been one attempt to incorporate radial diffusion and resonant interactions with whistler mode chorus waves into a global model of the Earth's radiation belts [*Varotsou et al.*, 2005]. The results, which are specific for a model magnetic storm, show that wave acceleration is very important throughout the outer radiation belt, from the plasmapause to beyond geosynchronous orbit. Here we extend these modeling results to investigate the variation in flux during a period of prolonged magnetic activity relevant to fast solar wind streams associated with CIR events. We include the effects of radial diffusion, electron loss, and acceleration due to chorus waves associated with moderate levels of magnetic activity lasting up to 4 days.

We use the Salammbô code [*Beutier and Boscher*, 1995]. This code calculates the change in phase space density from a Fokker Planck equation given by

$$\frac{\partial f}{\partial t} = L^2 \frac{\partial}{\partial L}\left(\frac{D_{LL}}{L^2}\frac{\partial f}{\partial L}\right) + \frac{1}{yT}\frac{\partial}{\partial y}\left(yTD_{yy}\frac{\partial f}{\partial y}\right) + \frac{1}{a}\frac{\partial}{\partial E}\left(aD_{EE}\frac{\partial f}{\partial E}\right) - \frac{1}{a}\frac{\partial}{\partial E}\left(a\frac{dE}{dt}f\right) \quad (3)$$

where the terms on the right hand side are due to radial diffusion, pitch angle diffusion, energy diffusion and energy loss due to Coulomb interactions with electrons and neutrals in the upper atmosphere, respectively. Here $y = \sin \alpha_{eq}$, $a = (E + E_0)(E + 2E_0)^{1/2}E^{1/2}$, E_0 is the electron rest mass energy, and the function $T(y)$ is defined by *Schulz* [1991]. Convection of low energy electrons is not included.

The radial diffusion coefficient D_{LL} is given by (2) described above. The pitch angle diffusion coefficient D_{yy} is a sum of scattering due to Coulomb collisions with atmospheric particles, scattering by whistlers, hiss and VLF transmitters inside the plasmasphere (given by *Beutier and Boscher*, [1995]), and chorus waves outside the plasmasphere (given below). Finally, energy loss due to Coulomb interactions with the atmosphere, and with cold electrons in the plasmasphere, are also included [*Beutier and Boscher*, 1995].

To include loss and acceleration due to chorus waves the bounce averaged pitch-angle $\langle D_{\alpha\alpha}\rangle$ and energy $\langle D_{EE}\rangle$ diffusion rates were calculated using the PADIE code [*Glauert and Horne*, 2005] and converted into D_{yy} and D_{EE} used in (3). Chorus was assumed to have a Gaussian distribution of wave power spectral density as a function of frequency with a peak at $0.35f_{ce}$, bandwidth of $0.15f_{ce}$, and lower and upper frequency cut-offs at $0.125f_{ce}$ and $0.575f_{ce}$. The distribution of wave normal angles was Gaussian in $X = \tan(\psi)$, peaked along the magnetic field direction ($X_m = 0$) with an angular spread of $X_w = \tan(30°)$, and assumed independent of MLT.

Landau ($n = 0$) up to and including $n = \pm 5$ cyclotron harmonic resonances were included, with integration over wave normal angles between $X = 0$ and 1, and bounce averaged over latitudes of $-15° < \lambda_m < 15°$. The diffusion rates were calculated as a matrix with a constant wave intensity $B_{wave}^2 = 10^4$ pT^2 (and later scaled according to observations, see below) for f_{pe}/f_{ce} of 1.5, 2.5, 5.0, 7.5 and 10, electron energies of 10, 30, 100, 300, 1000 and 3000 keV, and $L = 2.5$, 3.5, 4.5, 5.5 and 6.5 with a resolution of less than 1° equatorial pitch angle. Outside these values, diffusion coefficients were set to zero. The cross diffusion coefficient $\langle D_{\alpha E} \rangle$ was not included.

Plate 6 shows the diffusion coefficients as a function of energy and pitch angle for $L = 6.5$ for whistler mode chorus waves in units of s^{-1} for a fixed wave amplitude of 100 pT. For $f_{pe}/f_{ce} = 1.5$ pitch angle diffusion (top left) peaks near 100 keV and extends into the loss cone. Diffusion into the loss cone is important from about 10 keV up to several hundred keV. However, as f_{pe}/f_{ce} is increased up to 10, scattering into the loss cone is restricted to lower energies below about 20 keV (top right), indicating that the waves are generated by lower energy electrons for larger f_{pe}/f_{ce}. For low $f_{pe}/f_{ce} \approx 1.5$ energy diffusion (bottom left) peaks near 30 keV but extends to larger pitch angles >60° up to 1 MeV and beyond. Thus there is a region at high energies and large pitch angles where electrons can be diffused in energy but pitch angle scattering into the loss cone is ineffective. These electrons can be accelerated and trapped in the magnetic field. As f_{pe}/f_{ce} is increased energy diffusion is restricted to large pitch angles and lower energies, and is ineffective above a few hundred keV. Thus wave acceleration is most effective for low f_{pe}/f_{ce}.

To scale the diffusion rates according to magnetic activity, and different MLT characteristics of the waves, a wave model was constructed from the CRRES data. The wave intensities B_{wave}^2 and ratio f_{pe}/f_{ce} measured by CRRES between $-15° < \lambda_m < 15°$ [Meredith et al., 2003a] were determined for $Kp < 2$, $2 \leq Kp \leq 4$ and $Kp \geq 4$ between $L = 1$ to 8, with a resolution of $0.1L$ and 1 hour in MLT. Plate 7 shows the results. As Kp increases (top panels) chorus wave intensity increases and peaks outside the plasmapause from just before midnight through dawn to just after 1200 MLT. At the same time, the ratio f_{pe}/f_{ce} decreases in the same MLT sector providing the right conditions for efficient wave acceleration.

The bounce averaged diffusion rates were scaled and converted to D_{yy} and D_{EE} as a function of E, y, L using interpolation. The diffusion matrix was also interpolated in f_{pe}/f_{ce}. Thus for a given Kp, energy, L shell and pitch angle, the diffusion rates were calculated in each MLT bin according to the values of f_{pe}/f_{ce}, B_{wave}^2, and α, summed over all MLT and then divided by the number of MLT bins. Since electron-chorus interactions are most efficient for low f_{pe}/f_{ce} and high wave intensities, they were only included outside the plasmapause, which is defined by the empirical relation $L_{pp} = 5.6 - 0.46 Kp'$ [Carpenter and Park, 1973], where Kp' is the maximum value of the Kp index during the last 24 hours of simulation.

The outer boundary of the model is set at $L = 8$. The phase space density at the outer boundary is the same as that used by Shprits et al. [2005] and is derived from the average near equatorial flux measured by CRRES at apogee ($J = 8222.6 \exp(-7.068K)$ in cm^{-2} s^{-1} sr^{-1} keV^{-1} where K is kinetic energy in MeV) and kept constant at the outer boundary.

4.1. Global Modeling Results

Plate 8 shows the results of a simulation to represent the passage of a fast solar wind stream. The Kp index is set initially to $Kp = 1.8$ (bottom panel) and the model is set up to find a steady state solution that includes radial diffusion and whistler mode chorus (day 1-2). The phase space density is converted to omnidirectional flux, according to $J = f/p^2$ and integrated over solid angle, for easier interpretation with satellite observations. A radiation belt is formed with a peak in the omnidirectional flux at 1 MeV near $L = 4.5$. After 1 day Kp is stepped up to $Kp = 4$ and kept constant at that level for the rest of the simulation (4 days) to represent a prolonged period of moderate magnetic activity, as shown in Plate 1 during the declining phase of the solar cycle. Whistler mode chorus is switched off, and only radial diffusion is then included. At 1 MeV (top panel) electrons are diffused inwards so that the inner edge moves to about $L = 3$ over a period of about 3 days. The peak flux in the radiation belt increases and moves to slightly lower $L \approx 4.0$. Inward diffusion and flux increase is also found at 400 keV and 100 keV. However, when radial diffusion and chorus are included (Plate 9) the flux at 1 MeV increases significantly for $3.5 < L < 5.5$ once Kp is increased. There is a slight reduction in flux at 400 keV and a major decrease at 100 keV for $L < 6.6$. The decrease at ~100 keV is due to a combination of losses due to pitch angle scattering into the loss cone and energy diffusion to higher energies caused by whistler mode chorus.

Figure 3 shows the phase space density at selected times for four simulations, the initial steady state (which includes radial diffusion and chorus acceleration and loss), and at the end of the simulations for radial diffusion only, chorus only, and both chorus and radial diffusion. For all M, the initial steady state shows that the phase space density is flat for large L and is in equilibrium with the conditions imposed by the outer boundary where radial diffusion is most effective, and falls at low L due to the decreasing efficiency of radial diffusion and the increasing importance of losses. After the initial steady state, the simulation for chorus without radial diffusion shows the development of a large peak near $L = 5.5$ for 3000 MeV/G (top panel) leading to a positive (negative)

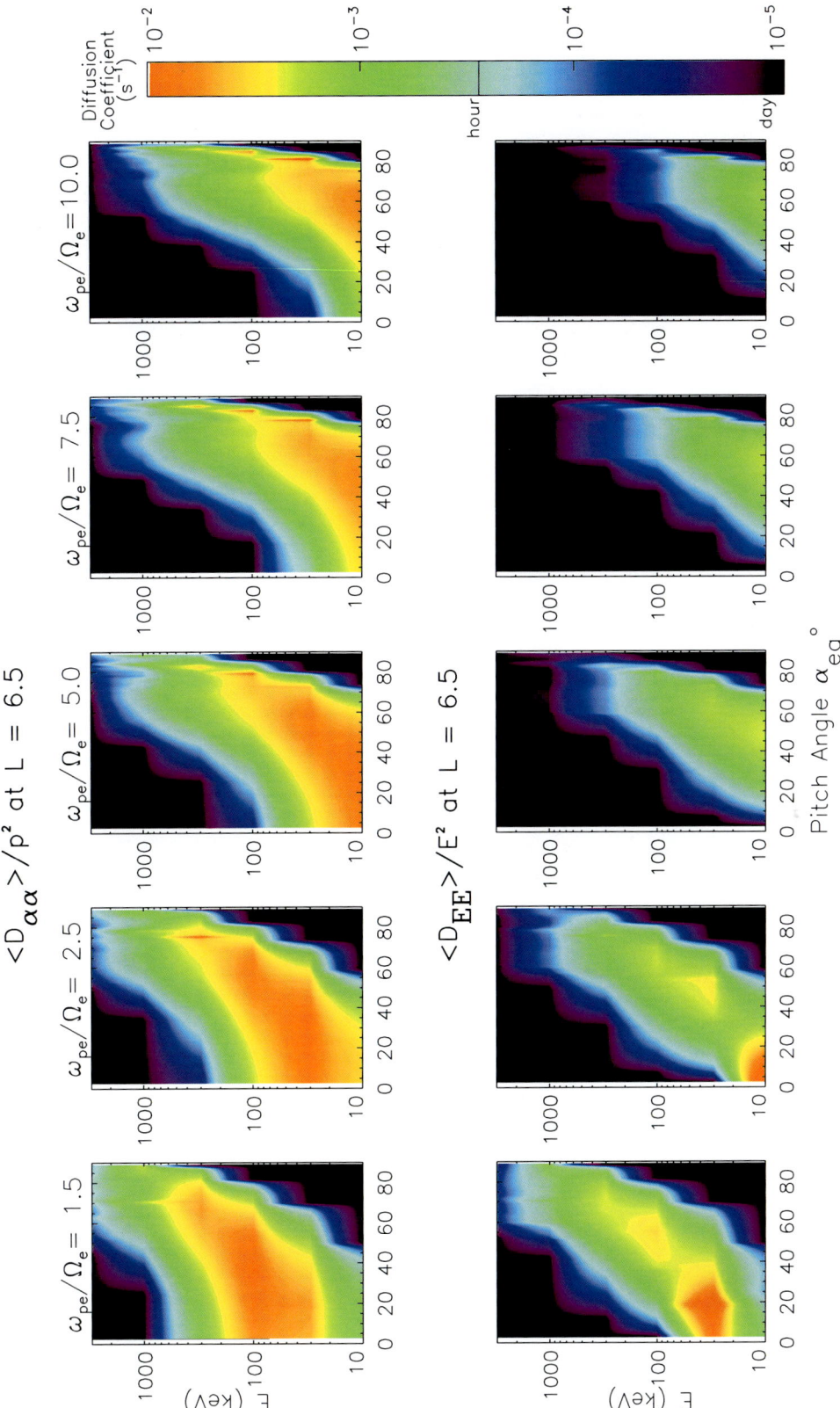

Plate 6. Electron diffusion rates for whistler mode chorus waves at $L = 6.5$. Pitch angle diffusion rates $\langle D_{\alpha\alpha}\rangle/p^2$ (top) and energy diffusion rates D_{EE}/E^2 (bottom) in units of s^{-1} are shown as a function of energy and equatorial pitch angle for different values of f_{pe}/f_{ce} increasing left to right.

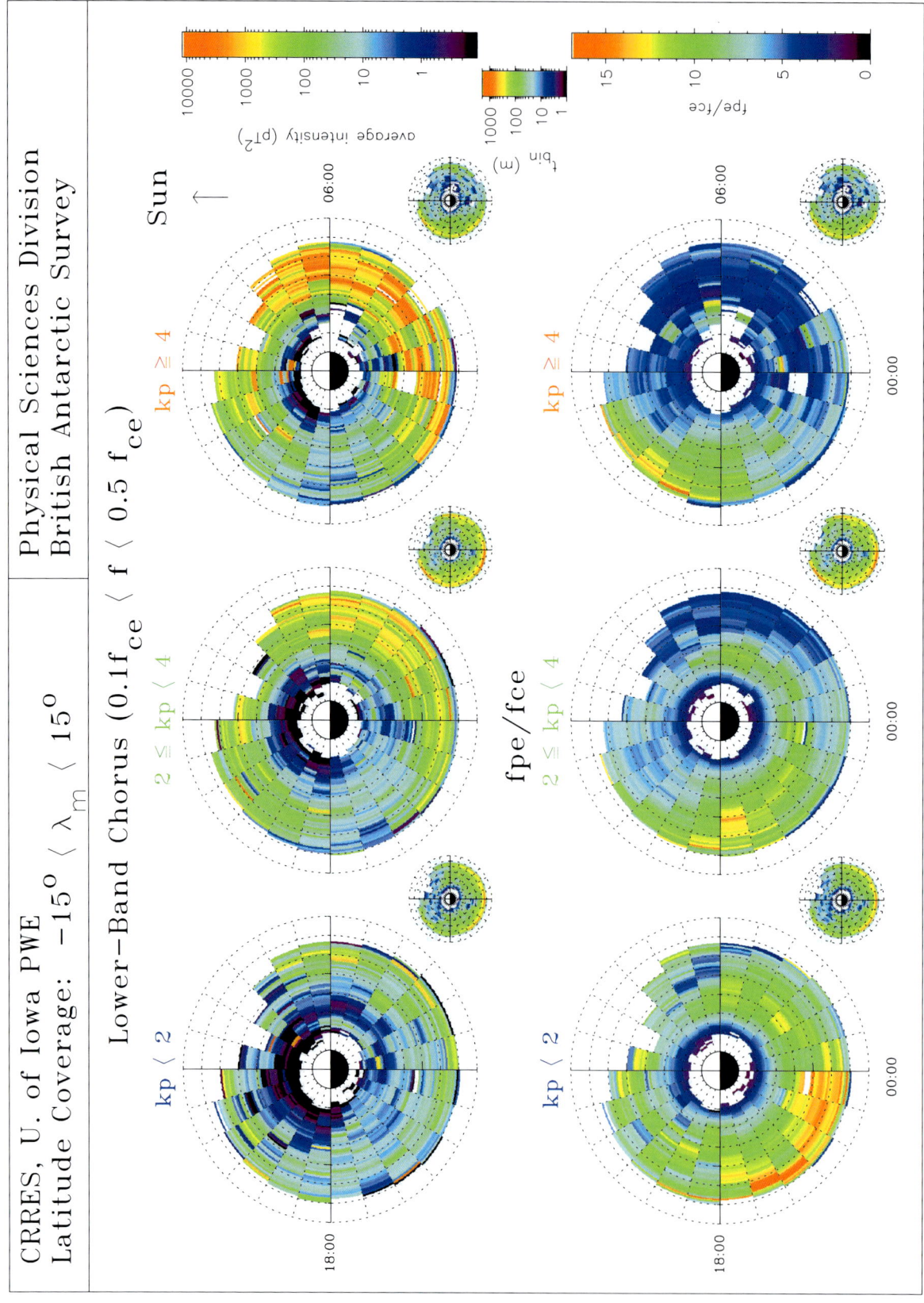

Plate 7. MLT distributions of lower band chorus wave intensities B^2_{wave} in the frequency range $0.1 < f/f_{ce} < 0.5$ (top) and fpe/fce (bottom) for three values of the Kp index increasing left to right.

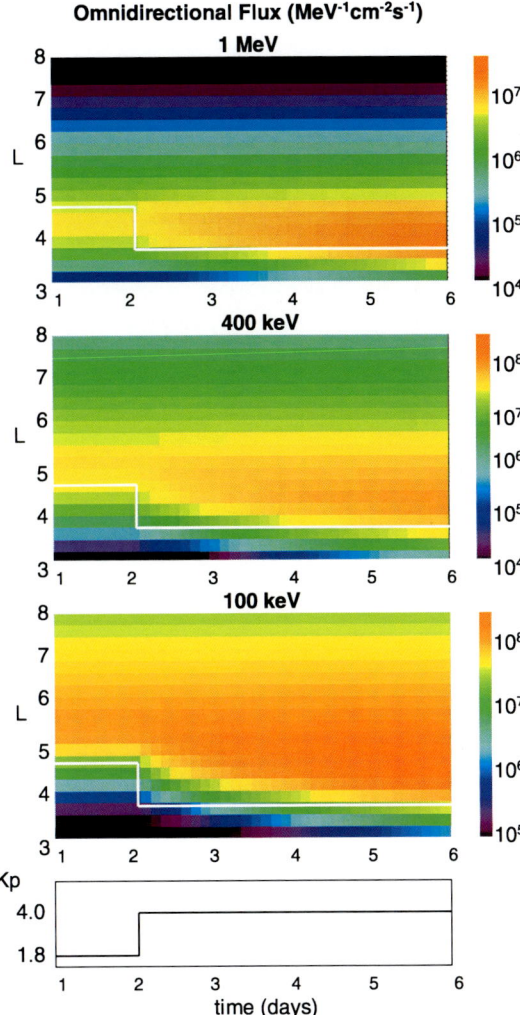

Plate 8. Global simulation results to represent the effects of a fast solar wind stream on the outer radiation belt. From top to bottom are the omnidirectional electron flux at 1 MeV, 400 keV, and 100 keV, and the model Kp index used to drive the model. The results are for radial diffusion without whistler mode chorus waves. The solid white line shows the plasmapause location.

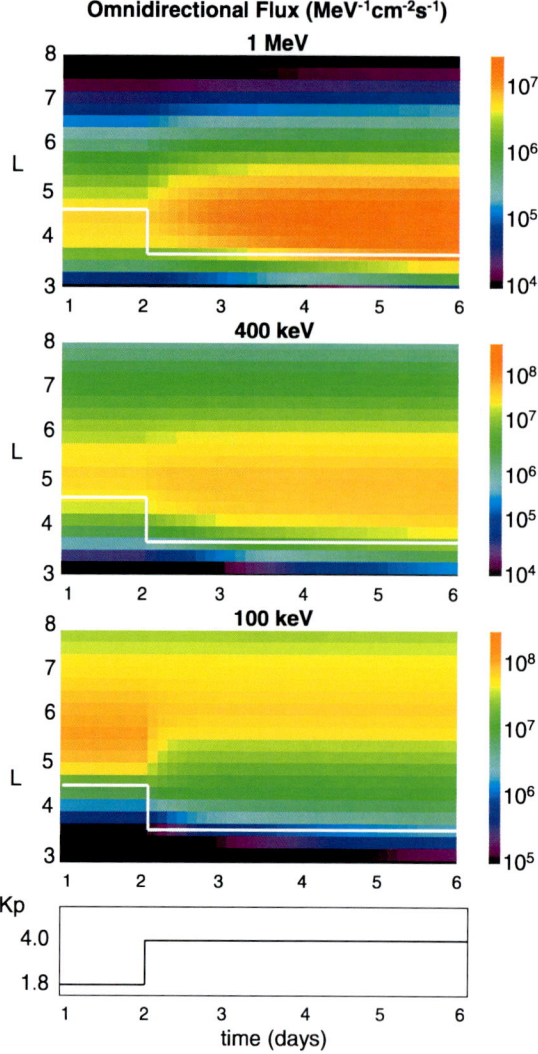

Plate 9. Same as Plate 8 but for radial diffusion and pitch angle and energy diffusion by whistler mode chorus waves outside the plasmapause.

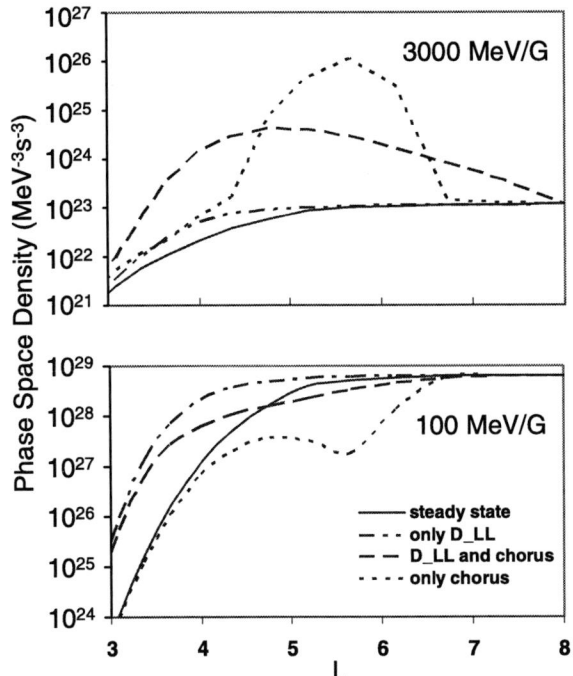

Figure 3. Phase space density for constant first invariant M for (top) 3000 MeV/Gauss and (bottom) 100 MeV/Gauss. The initial steady state is shown by the solid line. Results at the end of the simulation, 4 days after the initial steady state, are for (dotted) chorus without radial diffusion, (dash dot) radial diffusion only, and (dashed) for combined chorus and radial diffusion.

radial gradient inside (outside) $L = 5.5$. When both radial diffusion and chorus are included (long dashed) the peak is reduced and smoothed by radial diffusion both towards smaller and larger L caused by the gradients. In the region near $L = 4$, and $L = 7$, there is a major increase in phase space density, much larger than that obtained by either radial diffusion alone (dot dash line) or chorus alone (dotted line). The results show that although wave acceleration contributes mostly to the region $4.5 < L < 6.5$ the combination of wave acceleration and radial diffusion can increase the electron phase space density by a factor of ~10 over a much wider range of L, approximately $3.5 < L < 7$.

At lower values of M (100 MeV/G) the major effect of chorus is a reduction in phase space density near $L = 5.5$ due to loss via pitch angle scattering, and as a result of diffusion to higher energies (Figure 3, bottom panel). The combination of chorus and radial diffusion smoothes the radial gradient and leads to a positive radial gradient for low L up to about $L = 6.5$. Radial diffusion alone gives a much flatter radial profile from large L down to about $L \approx 4.5$.

The simulation results depend critically on the boundary conditions. Probably the major source of uncertainty is the outer boundary condition for the electron flux. The peak in phase space density for 3000 MeV/G for both chorus and radial diffusion in Figure 3 is a result of this fixed outer boundary. If the boundary were able to 'float' upwards, then we expect the radial profile to be much flatter for $L > 5$ since radial diffusion is very effective at large L.

The simulations presented here do not take into account losses due to EMIC, or chorus microburst precipitation. These waves should contribute to losses of ~MeV electrons. EMIC waves should be very effective inside high density regions, particularly as the plasmapause expands during storm recovery. These waves would reduce the flux increase due to wave acceleration. However, outside the plasmapause where the density is low, losses due to EMIC waves should be small. Since chorus diffusion into the loss cone is small at high energies and low f_{pe}/f_{ce} (Plate 6) it is unlikely that they would suppress the increase in MeV electron flux. However, chorus will precipitate the source electrons (~1-100 keV) and thus a process of continual injection is required to continue driving the waves and accelerate the electrons.

5. CONCLUDING REMARKS

One key difference between electron acceleration during large magnetic storms and acceleration associated with fast solar wind streams, is that in a large magnetic storm the plasmapause can be compressed to very low L so that f_{pe}/f_{ce} becomes small outside the plasmapause, chorus waves are intense, and hence wave acceleration can be very efficient. This is the case for the 2003 Halloween storms [*Baker et al.*, 2004; *Horne et al.*, 2005b]. In a weak or moderate magnetic storm the plasmapause is more likely to extend to larger L, chorus waves are less intense, and therefore wave acceleration should take longer. Precipitation due to chorus will also be important for 1-100 keV electrons. Thus continuous particle injection is required to maintain chorus wave power. This suggests that the key ingredient of weak to moderate storms that occur during the declining phase of the solar cycle is the strong Alfvénic wave activity in the fast solar wind which provides additional inward particle transport inside the magnetosphere to power whistler mode chorus.

The results suggest the following scenario for electron acceleration associated with fast solar wind streams. A southward component of the magnetic field in a CIR causes magnetic reconnection and results in a weak to moderate magnetic storm. Alfvénic wave activity in the fast solar wind following the CIR drives additional reconnection and inward particle transport that results in an unusually long recovery phase which may last several days. This provides a continuous source of free energy to drive whistler mode chorus waves outside the plasmapause from the nightside, though dawn to the dayside. The growth of chorus waves causes pitch angle scattering and loss at low energies, which is continually replenished by inward particle transport caused by the fast solar wind stream. Chorus accelerates a fraction of the low energy electrons to ~MeV energies, principally in the region $4.5 < L < 6.5$, which are then transported to both smaller and larger L by radial diffusion to populate the entire outer radiation belt at high energies.

As noted in the introduction, the radiation belts are a hazard for manned spacecraft and for satellites on orbit. The ability to specify the electron flux in the radiation belts, particularly for orbits where there may be little or no data available, and to predict variations, is an important application for satellite operators, satellite designers and space insurance. Several models are being constructed with various degrees of success [e.g., *Li et al.*, 2001, 2005]. The results here demonstrate that wave acceleration and loss are of major importance, and must be included in physics based models.

Acknowledgments. This research was funded in part by the UK Natural Environment Research Council, and NASA grant NAG5-11922. We thank one of the referees for very helpful comments.

REFERENCES

Abel, B., and R.M. Thorne, Electron scattering loss in Earth's inner magnetosphere 1. Dominant physical processes, *J. Geophys. Res.*, 103, 2385, 1998a.

Abel, B., and R.M. Thorne, Electron scattering and loss in Earth's inner magnetosphere 2. Sensitivity to model parameters, *J. Geophys. Res.*, 103, 2397, 1998b.

Albert, J.M., Nonlinear interaction of outer zone electrons with VLF waves, *J. Geophys. Res.*, 29, doi:10.1029/2001GL013941, 2002.

Albert, J.M., Evaluation of quasi-linear diffusion coefficients for EMIC waves in a multispecies plasma *J. Geophys. Res.*, 108, doi:10.1029/2002JA009792, 2003.

Anderson, B.J., R.E. Erlandson, and L.J. Zanetti, A statistical study of Pc 1-2 magnetic pulsations in the equatorial magnetosphere, 1, Equatorial occurrence distributions, *J. Geophys. Res.*, 97, 3075, 1992a.

Anderson, B.J., R.E. Erlandson, and L.J. Zanetti, A statistical study of Pc 1-2 magnetic pulsations in the equatorial magnetosphere, 2, Wave properties, *J. Geophys. Res.*, 97, 3089, 1992b.

Baker, D.N., J.B. Blake, R.W. Klebesadel, and P.R. Higbie, Highly relativistic electrons in the Earth's outer magnetosphere 1. Lifetimes and temporal history 1979-1984, *J. Geophys. Res.*, 91, 4265-4276, 1986.

Baker, D.N., R.L. McPherron, T.E. Cayton, and R.W. Klebesadel, Linear prediction filter analysis of relativistic electron properties at 6.6RE, *J. Geophys. Res.*, 95, 15,133-15,140, 1990.

Baker, D.N., X. Li, N. Turner, J.H. Allen, L.F. Bargatze, J.B. Blake, R.B. Sheldon, H.E. Spence, R.D. Belian, G.D. Reeves, S.G. Kanekal, B. Klecker, R.P. Lepping, K.W. Olgilvie, R.A. Mewaldt, T. Onsager, H.J. Singer, and G. Rostoker, Recurrent geomagnetic storms and relativistic electron enhancements in the outer magnetosphere: ISTP coordinated measurements, *J. Geophys. Res.*, 102, 14,141-14,148, 1997.

Baker, D.N., J.H. Allen, S.G. Kanekal, and G.D. Reeves, Disturbed space environment may have been related to pager satellite failure, EOS transactions, AGU, 79, 477, 1998.

Baker D.N., S.G. Kanekal, T.I. Pulkkinen, and J.B. Blake, Equinoctial and solstial averages of magnetospheric relativistic electrons: A strong semi-annual modulation, *J. Geophys. Res.*, 26, 3193-3196, 1999.

Baker, D., Satellite anomalies due to space storms, in *Space Storms and Space Weather Hazards*, I.A. Daglis ed. ch. 10, p251-284, Kluwer, Dordrecht, The Netherlands, 2001.

Baker, D.N., S.G. Kanekal, X. Li, S.P. Monk, J. Goldstein, and J.L. Burch, An extreme distortion of the Van Allen belt arising from the 'Hallowe'en' solar storm in 2003, *Nature*, 432, 878-881, 2004.

Beutier, T., and D. Boscher, A three-dimensional analysis of the electron radiation belt by the *Salammbô* code, *J. Geophys. Res.*, 100, 14,853, 1995.

Blake, J.B., D.N. Baker, N. Turner, K.W. Ogilvie, and R.P. Lepping, Correlation of changes in the outer-zone relativistic-electron population with upstream solar wind and magnetic field measurements, *J. Geophys. Res.*, 24, 927-929, 1997.

Brautigam, D.H. and J.M. Albert, Radial diffusion analysis of outer radiation belt electrons during the October 9, 1990, magnetic storm *J. Geophys. Res.*, 105, 291-309, 2000.

Brautigam, D.H., G.P. Ginet, J.M. Albert, J.R. Wygant, D.E. Rowland, A. Ling, and J. Bass, CRRES electric field power spectra and radial diffusion coefficients *J. Geophys. Res.*, 110, A02214, doi:10.1029/2004JA010612291, 2005.

Bühler, P., and L. Desorgher, Relativistic electron enhancements, magnetic sotmrs, and substorm activity, *J. Atmos. Solar-Terr. Phys.*, 64, 593-599, 2002.

Cahill, L.J. Jr. and J.R. Winckler, Periodic magnetopause oscillations observed with the GOES satellites on March 24, 1991, *J. Geophys. Res.*, 97, 8293, 1992.

Carpenter, D.L., and C.G. Park, On what ionosphere workers should know about the plasmapause-plasmasphere, *Rev. Geophys.*, 11, 133, 1973.

Carpenter, D.L., and R.R. Anderson, An ISEE/whistler model of equatorial electron density in the magnetosphere, *J. Geophys. Res.*, 97, 1097, 1992.

Cliver, E.W., Y. Kamide, and A.G. Ling, Mountains versus valleys: Semiannual variation of geomagnetic activity, *J. Geophys. Res.*, 105, 2413, 2000.

Cornwall, J.M., Diffusion processes influenced by conjugate-point wave phenomena, *Radio Sci.*, 3, 740, 1968.

Elkington, S.R., M.K. Hudson, and A.A. Chan, Acceleration of relativistic electrons via drift resonant interactions with torroidal-mode Pc-5 ULF oscillations, *J. Geophys. Res.*, 26, 3273-3276, 1999.

Elkington, S.R., M.K. Hudson, and A.A. Chan, Resonant acceleration and diffusion of outer zone of electrons in a asymmetric geomagnetic field, *J. Geophys. Res.*, 108(A3), 1116, doi:10.1029/2001JA009202, 2003.

Falthammar, C-G., Effects of time dependent electric fields on geomagnetically trapped radiation, *J. Geophys. Res.*, 70, 2503-2516, 1965.

Fraser, B.J., H.J. Singer, W.J. Hughes, J.R. Wygant, R.R. Anderson, and Y.d. Hu, CRRES Poynting vector observations of electromagnetic ion cyclotron waves near the plasmapause, *J. Geophys. Res.*, 101, 15,331, 1996.

Fraser, B.J., and T.S. Nguyen, Is the plasmapause a preferred source region of electromagnetic ion cyclotron waves in the magnetosphere?, *J. Atmos. Solar-Terr. Phys.*, 63, 1225, 2001.

Friedel, R.H.W., G.D. Reeves, and T. Obara, Relativistic electron dynamics in the inner magnetosphere–a review, *J. Atmos. Solar-Terr. Phys.*, 64, 265, 2002.

Glauert, S.A., and R.B. Horne, Calculation of pitch angle and energy diffusion coefficients with the PADIE code, *J. Geophys. Res.*, 110, A04206, doi:10.1029/2004JA010851, 2005.

Gonzalez, W.D., B.T. Tsurutani, A.L.C. Gonzalez, E.J. Smith, F. Tang, and S.-I. Akasofu, Solar wind-magnetosphere coupling during intense magnetic storms (1978-1979), *J. Geophys. Res.*, 94, 8835, 1989.

Green, J.C., and M.G. Kivelson, Relativistic electrons in the outer radiation belt: Differentiating between acceleration mechanisms, *J. Geophys. Res.*, 109, A03213, doi:10.1029/2003JA010153, 2004.

Greenstadt, E.W., J.V. Olsen, P.D. Loewen, H.J. Singer, and C.T. Russell, Correlation of Pc 3, 4, and 5 activity with solar wind speed, *J. Geophys. Res.*, 84, 6694, 1979.

Harvey, K.L., and F. Recely, Polar coronal holes during cycles 22 and 23, *Solar Phys.*, 211, 31-52, 2002.

Helliwell, R.A., A theory of discreet emissions from the magnetosphere. *J. Geophys. Res.*, 72, 4773-4790, 1967.

Horne, R.B., The contribution of wave particle interactions to electron loss and acceleration in the Earth's radiation belts during geomagnetic storms, in *Review of Radio Science 1999-2002*, edited by W.R. Stone, Chapter 33, p801-828, John Wiley, Bognor Regis, 2002.

Horne, R.B., and R.M. Thorne, On the preferred source location for the convective amplification of ion cyclotron waves, *J. Geophys. Res.*, 98, 9233, 1993.

Horne, R.B., and R.M. Thorne, Potential waves for relativistic electron scattering and stochastic acceleration during magnetic storms, *J. Geophys. Res.*, 25, 3011, 1998.

Horne, R.B., and R.M. Thorne, Relativistic electron acceleration and precipitation during resonant interactions with whistler-mode chorus, *J. Geophys. Res.*, 30, doi:10.1029/2003GL016973, 2003.

Horne, R.B., N.P. Meredith, R.M. Thorne, D. Heynderickx, R.H.A. Iles, and R.R. Anderson, Evolution of energetic electron pitch angle distributions during storm time electron acceleration to MeV energies, *J. Geophys. Res.*, 108, doi:10.1029/2002JA009468, 2003a.

Horne, R.B., S.A. Glauert, and R.M., Thorne, Resonant diffusion of radiation belt electrons by whistler-mode chorus, *J. Geophys. Res.*, 30, doi:10.1029/2003GL016963, 2003b.

Horne, R.B., R.M. Thorne, S.A. Glauert, J.M. Albert, N.P. Meredith, and R.R. Anderson, Timescale for radiation belt electron acceleration by whistler mode chorus waves, *J. Geophys. Res.*, 110, A03225, doi:10.1029/2004JA010811, 2005a.

Horne, R.B., R.M. Thorne, Y.Y. Shprits, N.P. Meredith, S.A. Glauert, A.J. Smith, S.G. Kanekal, D.N. Baker, M.J. Engebretson, J.L. Posch, M. Spasojevic, U.S. Inan, J.S. Pickett, and P.M.E. Decreau, Wave acceleration of electrons in the Van Allen radiation belts, *Nature*, 437, doi:10.1038/nature03939, 2005b.

Hudson, M.K., S.R. Elkington, J.G. Lyon, V.A. Machenko, I. Roth, M. Temerin, J.B. Blake, M.S. Gussenhoven, and J.R. Wygant, Simulation of radiation belt formation during sudden storm commencements, *J. Geophys. Res.*, 102, 14,087, 1997.

Hudson, M.K., S.R. Elkington, J.G. Lyon, C.C. Goodrich, and T.J. Rosenberg, Simulation of radiation belt dynamics driven by solar wind variations, in *Sun-Earth Plasma Connections*, edited by J.L. burch, R.L. Carovillano, and S.K. Antiochos, vol. 109, p171, AGU, Washington D.C., 1999.

Iles, R.H.A., A.N. Fazakerley, A.D. Johnstone, N.P. Meredith, and P. Bühler, The relativistic electron response in the outer radiation belt during magnetic storms, *Ann. Geophys.*, 20, 957-965, 2002.

Iles, R.H.A., N.P. Meredith, A.N. Fazakerley, and R.B. Horne, Phase space density analysis of the outer radiation belt energetic electron dynamics *J. Geophys. Res.*, 111, A03204, doi:10.1029/2005JA011206, 2006.

Junginger, H. and W. Baumjohann, Dayside long period magnetospheric pulsations: Solar wind dependence, *J. Geophys. Res.*, 93, 877, 1988.

Kamide, Y., W. Baumjohann, I.A. Daglis, W.D. Gonzalez, M. Grande, J.A. Joselyn, R.L. McPherron, J.L. Phillips, E.G.D. Reeves, G. Rostoker, A.S. Sharma, H.J. Singer, B.T. Tsurutani, and V.M. Vasyliunas, Current understanding of magnetic storms: Storm-substorm relationships, *J. Geophys. Res.*, 103, 17,705-17,728, 1998.

Kennel, C.F., and F. Engelmann, Velocity space diffusion from weak plasma turbulence in a magnetic fields, *Phys. Fluids*, 9, 2377, 1966.

Kennel, C.F., and H.E. Petschek, Limit on stably trapped particle fluxes, *J. Geophys. Res.*, 71, 1-28, 1966.

Kim, H.-J., and A.A. Chan, Fully adiabatic changes in storm time relativistic electron fluxes, *J. Geophys. Res.*, 102, 11,107, 1997.

Kokubun, S., K.N. Erickson, T.A. Fritz, and R.L. McPherron, Local time asymmetry of Pc 4-5 pulsations and associated particle modulations at synchronous orbit, *J. Geophys. Res.*, 94, 6607, 1989.

Kozyra, J.U., T.E. Cravens, A.F. Nagy, and E.G. Fontheim, Effects of energetic heavy ions on electromagnetic ion cyclotron wave generation in the plasmapause region, *J. Geophys. Res.*, 89, 2217, 1984.

Kozyra, J.U., V.K. Jordanova, R.B. Horne, and R.M. Thorne, Modelling of the contribution of electromagnetic ion cyclotron (EMIC) waves to storm-time ring current erosion, in *Geophys. Monogr. Ser.*, vol. 98, edited by B.T. Tsurutani et al., p187, AGU, Washington, D.C., 1997.

Krieger, A.S., A.F. Timothy, and E.C. Roelof, A coronal hole and its identification as the source of a high velocity solar wind stream, *Sol. Phys.*, 29, 505-525, 1973.

Li, X., D.N. Baker, M. Temerin, D. Larson, R.P. Lin, G.D. Reeves, M. Looper, S.G. Kanekal, and R.A. Mewaldt, Are energetic electrons in the solar wind the source of the outer radiation belt? *J. Geophys. Res.*, 24, 923-926, 1997.

Li, X., and M.A. Temerin, The electron radiation belt, *Space Sci. Rev.*, 95, 569, 2001.

Li, X., M.Temerin, D.N. Baker, G.D. Reeves, and D. Larson, Quantitative prediction of radiation belt electrons at geostationary orbit based on solar wind measurements, *J. Geophys. Res.*, 28, 1887-1890, 2001.

Li, X., D.N. Baker, M. Temerin, G. Reeves, R. Friedel, and C. Shen, Energetic electrons, 50 keV to 6 MeV, at geosynchronous orbit: Their responses to solar wind variations, *Space Weather*, 3, S04001, doi:10.1029/2004SW000105, 2005.

Li, X., The Role of Radial Transport in Accelerating Radiation Belt Electrons, in *Recurrent Magnetic Storms: Corotating Solar Wind Streams*, edited by B.T. Tsurutani, R.L. McPherron, W.D. Gonzalez, G. Lu, J.H.A. Sobral, and N. Gopalswamy, AGU, Washington, D.C., 2006.

Loewe, C.A., and G.W. Prölss, Classification and mean behaviour of magnetic storms, *J. Geophys. Res.*, 102, 14,209, 1997.

Lorentzen, K.R., M.P. McCarthy, G.K. Parks, J.E. Foat, R.M. Millan, D.M. Smith, R.P. Lin, and J.P. Treilhou, Precipitation of relativistic electrons by interaction with electromagnetic ion cyclotron waves, *J. Geophys. Res.*, 105, 5381, 2000.

Lorentzen, K.R., J.B. Blake, U.S. Inan, and J. Bortnik, Observations of relativistic electron microbursts in association with VLF chorus, *J. Geophys. Res.*, 106, 6017, 2001.

Lyons, L.R., R.M. Thorne and C.F. Kennel, Pitch-angle diffusion of radiation belt electrons within the plasmasphere, *J. Geophys. Res.*, 77, 3455, 1972.

Lyons, L.R., and D.J. Williams, Quantitative aspects of magnetospheric physics, D. Reidel, Norwell, Mass, 1984.

Lyons, L.R., D.-Y. Lee, R.M. Thorne, R.B. Horne, and A.J. Smith, Solar-wind magnetosphere coupling leading to relativistic electron energization during high-speed streams, *J. Geophys. Res.*, 110, A11202, doi:10.1029/2005JA011254, 2005.

Lysak, R.L., and D. Lee, Response of the dipole magnetosphere to pressure pulses, *J. Geophys. Res.*, 19, 937, 1992.

Mann, I.R., T.P. O'Brien, and D. Milling, Correlations between ULF wave power, solar wind speed, and relativistic electron flux in the magnetosphere: solar cycle dependence, *J. Atmos. Solar-Terr. Phys.*, 66, 187-198, 2004.

Mathie, R.A., and I.R. Mann, A correlation between extended intervals of ULF wave power and storm-time geosynchronous relativistic electron flux enhancements, *J. Geophys. Res.*, 27, 3261, 2000.

McPherron, R.L., and J. Weygand, The solar wind and geomagnetic activity as a function of time relative to corotating interaction regions, in *Recurrent Magnetic Storms: Corotating Solar Wind Streams*, edited by B.T. Tsurutani, R.L. McPherron, W.D. Gonzalez, G. Lu, J.H.A. Sobral, and N. Gopalswamy, AGU, Washington, D.C., 2006.

Meredith, N.P., A.D. Johnstone, R.B. Horne, and R.R. Anderson, The temporal evolution of injected electron distributions in the inner magnetosphere, *J. Geophys. Res.*, 105, 12,907, 2000.

Meredith, N.P., R.B. Horne, and R.R. Anderson, Substorm dependence of chorus amplitudes: Implications for the acceleration of electrons to relativistic energies, *J. Geophys. Res.*, 106, 13,165, 2001.

Meredith, N.P., R.B. Horne, R.H. A. Iles, R.M. Thorne, R.R. Anderson, and D. Heynderickx, Outer zone relativistic electron acceleration associated with substorm enhanced whistler mode chorus, *J. Geophys. Res.*, 107, 10.1029/2001JA900146, 2002.

Meredith, N.P., R.B. Horne, R.M. Thorne, and R.R. Anderson, Favored regions for chorus-driven electron acceleration to relativistic energies in the Earth's outer radiation belt, *J. Geophys. Res.*, 30, 1871, doi:10.1029/2003GL017698, 2003a.

Meredith, N.P., M. Cain, R.B. Horne, R.M. Thorne, D. Summers, and R.R. Anderson, Evidence for chorus driven electron acceleration to relativistic energies from a survey of geomagnetically disturbed periods, *J. Geophys. Res.*, 108(A6), 1248, doi:10.1029/ 2002JA009764, 2003b.

Meredith, N.P., R.M. Thorne, R.B. Horne, D. Summers, B.J. Fraser and R.R. Anderson, Statistical analysis of relativistic electron energies for cyclotron resonance with EMIC waves observed on CRRES, *J. Geophys. Res.*, 108, doi:10.1029/2002JA009700, 2003c.

Meredith, N.P., R.B. Horne, R.M. Thorne, D. Summers, and R.R. Anderson, Substorm dependence of plasmaspheric hiss, *J. Geophys. Res.*, 109, A06209, doi:1029/2004JA010387, 2004.

Meredith, N.P., R.B. Horne, S.A. Glauert, R.M. Thorne, D. Summers, J.M. Albert, and R.R. Anderson, Energetic outer zone electron loss timescales during low geomagnetic activity, *J. Geophys. Res.*, 111, A05212, doi: 10.1029/2005JA011516, 2006.

Milan, R.M., R.P. Lin, and D.M. Smith, K.R. Lorentzen, and M.P. McCarthy, X-ray observations of MeV electron precipitation with balloon-borne germanium spectrometer, *J. Geophys. Res.*, 29, 2194, 10.1029/2002GL015922, 2002.

Miura, A., A Kelvin-Helmholtz instability at the magnetospheric bounary: Dependence on the magnetosheath sonic Mach number, *J. Geophys. Res.*, 97, 10665, 1992.

Miyoshi, Y., A. Morioka, T. Obara, H. Misawa, T. Nagai, and Y. Kasahara, Rebuilding process of the outer radiation belt during the November 3, 1993, magnetic storm: NOAA and EXOS-D observations, *J. Geophys. Res.*, 108, 10.1029/2001JA007542, 2003.

Miyoshi, Y.S., V.K. Jordanova, A. Morioka, and D.S. Evans, Solar cycle variations of the electron radiation belts: Observations and radial diffusion simulation, *Space Weather*, 2, S10S02, doi:10.1029/2004SW000070, 2004.

O'Brien, T.P., R.L. McPherron, D. Sornette, G.D. Reeves, R. Friedel, and H.J. Singer, Which magnetic storms produce relativistic electrons at geosynchronous orbit? *J. Geophys. Res.*, 106, 15,533, 2001.

O'Brien T.P., M.D. Looper, and J.B. Blake, Quantification of relativistic electron microbursts losses during the GEM storms, *J. Geophys. Res.*, 31, L04802, doi:10.1029/2003GL018621, 2004.

Parks, G.K., Spatial characteristics of auroral zone x-ray microbursts, *J. Geophys. Res.*, 72, 215, 1967.

Paulikas, G.A., and J.B. Blake, Effects of the solar wind on magnetospheric dynmaics: Energetic electrons at the synchronous orbit, in *Quantitative modelling of Magnetospheric processes, Geophys. Mongr. Ser.*, vol. 21, edited by W.P. Olsen, p180-202, AGU, Washington D.C., 1979.

Perreault, P.D. and S.-I. Akasofu, A study of geomagnetic storms, *Geophys. J. R. Astron. Soc.*, 54, 547, 1978.

Phillips, J.L., S.J. Bame, W.C. Feldman, B.E. Goldstein, J.T. Gosling, C.M. Hammond, D.J. McComas, M. Neugebauer, E.E. Scime, and S.T. Suess, Ulysses solar wind plasma observations at high southerly latitudes, *Science*, 268, 1030, 1995.

Posch, J.L., M.J. Engebretson, V.A. Pilipenko, W.J. Hughes, C.T. Russell, and L.J. Lanzerotti, Characterizing the long-period ULF response to magnetic storms, *J. Geophys. Res.*, 108, (A1), 1029, doi:10.1029/ 2002JA009386, 2003.

Pytte, T., R.L. McPherron, E.W. Hones Jr., and H.I. West, Multiple-satellite studies of magnetospheric substorms: distinction between polar magnetic substorm and convection driven negative bay, *J. Geophys. Res.*, 83, 663, 1978.

Randall, C.E., V.L. Harvey, G.L. Maney, Y. Orsolini, M. Codrescu, C. Sioris, S. Brohede, C.S. Haley, L.L. Gordley, J.M. Zawodny, and J.M. Russell III, Stratospheric effects of energetic particle precipitation in 2003-2004, *J. Geophys. Res.*, 32, L05802, doi:10.1029/2004GL022003, 2005.

Reeves, G.D., K.L. McAdams, R.H.W. Friedel, and T.P. O'Brien, Acceleration and loss of relativistic electrons during geomagnetic storms, *J. Geophys. Res.*, 30, doi:10.1029/2002GL016513, 2003.

Rhodes, E.J. Jr., and E.J. Smith, Evidence for a large scale gradient in the solar wind velocity, *J. Geophys. Res.*, 81, 2123-2134, 1976a.

Rhodes, E.J. Jr., and E.J. Smith, Further evidence of a heliographic latitude gradient in the solar wind velocity, *J. Geophys. Res.*, 81, 5833-5840, 1976b.

Russell, C.T., and R.L. McPherron, Semianual variation of geomagnetic activity, *J. Geophys. Res.*, 78, 92, 1973.

Schulz, M., and A. Eviatar, Diffusion of equatorial particles in the outer radiation zone, *J. Geophys. Res.*, 74, 2182, 1969.

Schulz, M., and L. Lanzerotti, Particle diffusion in the radiation belts, Springer, New York, 1974.

Schulz, M., The Magnetosphere, Geomagnetically Trapped Radiation, in *Geomagnetism*, vol. 4, edited by J.A. Jacobs, p202-256, Academic Press, 1991.

Selesnick, R.S., and J.B. Blake, Relativistic electron drift shell splitting, *J. Geophys. Res.*, 107, 10.1029/2001JA009179, 2002.

Sergeev, V., and W. Lennartsson, Plasma sheet at $X \approx 20$ Re during steady magnetospheric convection, *Planet. and Space Sci.*, 36, 353-370, 1988.

Sheeley, N.R. Jr., J.W. Harvey, and W.C. Feldman, Coronal holes, solar wind streams and recurrent geomagnetic disturbances 1973-1976, *Solar Phys.*, 49, 271, 1976.

Shprits, Y.Y., R.M. Thorne, G.D. Reeves, and R. Friedel, Radial diffusion modeling with empirical lifetimes: comparison with CRRES observations, *Ann. Geophys.*, 23, 1467-1471, 2005.

Smith, A.J., R.B. Horne, and N.P. Meredith, Ground observations of chorus following geomagnetic storms, *J. Geophys. Res.*, 109, A02205, doi:10.1029/2003JA010204, 2004a.

Smith, A.J., N.P. Meredith, and T.P. O'Brien, Differences is ground-observed chorus in geomagnetic storms with and without enhanced relativistic electron fluxes, *J. Geophys. Res.*, 109, A11204, doi:10.1029/ 2004JA010491, 2004b.

Smith, E.J., A.M.A. Frandsen, B.T. Tsurutani, R.M. Thorne, and K.W. Chan, Plasmaspheric hiss intensity variations during magnetic storms, *J. Geophys. Res.*, 79, 2507, 1974.

Smith, E.J. and J.H. Wolfe, Obsevations of interaction regions and corotating shocks between one and five AU: Pioneers 10 and 11, *J. Geophys. Res.*, 3, 137, 1976.

Southwood, D.J., Magnetopause Kelvin-Helmholtz instability, in *Proceedings of the Magnetospheric Boundary Layers Conference, Eur. Space Agency Spec. Publ., ESA SP* 148, 357, 1979.

Summers, D., R.M. Thorne, and F. Xiao, Relativistic theory of wave-particle resonant diffusion with application to electron acceleration in the magnetosphere, *J. Geophys. Res.*, 103, 20,487, 1998.

Summers, D., C. Ma, N.P. Meredith, R.B. Horne, R.M. Thorne, D. Heynderickx, and R.R. Anderson, Model of the energization of outer-zone electrons by whistler-mode chorus during the October 9, 1990 geomagnetic storm, *J. Geophys. Res.*, 29, 10.1029/2002GL016039, 2002.

Summers, D., and R.M. Thorne, Relativistic electron pitch-angle scattering by electromagnetic ion cyclotron waves during geomagnetic storms, *J. Geophys. Res.*, 108, doi:10.1029/2002JA009489, 2003.

Summers, D., C. Ma, N.P. Meredith, R.B. Horne, R.M. Thorne, and R.R. Anderson, Modeling outer-zone relativistic electron response to whistler mode chorus activity during substorms, *J. Atmos. Solar-Terr. Phys.*, 66, 133-146, 2004.

Thorne, R.M., Energetic radiation belt electron precipitation: a natural depletion mechanism for stratospheric ozone, *Science*, 287, 1977.

Thorne, R.M., R.B. Horne, S.A. Glauert, N.P. Meredith, Y.Y. Shprits, D, Summers, and R.R. Anderson, The influence of wave-particle interactions on relativistic electron dynamics during storms, in *Inner Magnetosphere Interactions: New Perspectives from Imaging, Geophys. Monogr. Ser.*, vol. 159, edited by James L. Burch, Michael Schulz, and Harlan Spence, p101–112, AGU, Washington, D. C., 2005a.

Thorne, R.M., T.P. O'Brien, Y.Y. Shprits, D. Summers, and R.B. Horne, Timescale for MeV electron microbursts loss during storms, *J. Geophys. Res.*, 110, A09202, doi:10.1029/2004JA010882, 2005b.

Tsurutani, B.T., and E.J. Smith, Postmidnight chorus: A substorm phenomenon, *J. Geophys. Res.*, 79, 118, 1974.

Tsurutani, B.T., and E.J. Smith, Two types of magnetospheric ELF chorus and their substorm dependencies, *J. Geophys. Res.*, 82, 5112, 1977.

Tsurutani, B.T., and W.D. Gonzalez, The cause of high-intensity long-duration continuous AE activity (HILDCAAS): Interplanetary Alfven wave trains, *Planet and Space Sci.*, 35, 405, 1987,

Tsurutani, B.T., W.D. Gonzalez, F. Tang, Y.T. Lee, and M. Okada, Reply to L.J. Lanzerotti: Solar wind ram pressure corrections and an estimation of the efficiency of viscous interaction, *J. Geophys. Res.* 19, 1993-1994, 1992.

Tsurutani, B.T., N. Gopalswamy, R.L. McPherron, W.D. Gonzalez, G. Lu, and F.L. Guarnieri, Magnetic storms caused by corotating solar wind streams, in *Recurrent Magnetic Storms: Corotating Solar Wind Streams*, edited by B.T. Tsurutani, R.L. McPherron, W.D. Gonzalez, G. Lu, J.H.A. Sobral, and N. Gopalswamy, AGU, Washington, D.C., 2006.

Varotsou, A., D. Boscher, S. Bourdarie, R.B. Horne, S.A. Glauert, and N.P. Meredith, (2005), Simulation of the outer radiation belt electrons near geosynchronous orbit including both radial diffusion and resonant interaction with whistler-mode chorus waves, *J. Geophys. Res.*, 32, L19106, doi:10.1029/2005GL023282, 2005.

Vassiliadis, D., A.J. Klimas, S.G. Kanekal, D.N. Baker, and R.S. Weigel, Long-term-average, solar cycle, and seasonal reponse of magnetospheric energetic electrons to the solar wind speed, *J. Geophys. Res.*, (A11), 107, 1383, doi:10.1029/2001JA000506, 2002.

Webb, D.F., and J.H. Allen, Spacecraft and ground anomalies related to the October-November 2003 solar activity, *Space Weath.*, 2, doi:10.1029/2004SW000075, 2004.

West, H.I., Jr., R.M. Buck, and R.J. Walton, Electron pitch angle distributions throughout the magnetosphere as observed by OGO 5, *J. Geophys. Res.*, 78, 1064, 1973.

Winckler, J.F., P.D. Bhavsar, and K.A. Anderson, A study of the precipitation of energetic electrons from the geomagnetic field during magnetic storms, *J. Geophys. Res.*, 67, 3717, 1962.

Wrenn, G.L., Conclusive evidence for internal dielectric charging anomalies on geosynchronous communications spacecraft, *J. Spacecraft and Rockets*, 32, 514, 1995.

Wrenn, G.L., D.J. Rodgers, and K.A. Ryden, A solar cycle of spacecraft anomalies due to internal charging, *Ann Geophys.*, 20, 953-956, 2002.

R.R. Anderson, Department of Physics and Astronomy, University of Iowa, Iowa City, IA 52242-1479, USA. (roger-r-anderson@uiowa.edu)

S.A. Glauert, British Antarctic Survey, Natural Environment Research Council, Madingley Road, Cambridge CB3 0ET, England. (sagl@bas.ac.uk)

R.B. Horne, British Antarctic Survey, Natural Environment Research Council, Madingley Road, Cambridge CB3 0ET, England. (r.horne@bas.ac.uk)

N.P. Meredith, British Antarctic Survey, Natural Environment Research Council, Madingley Road, Cambridge CB3 0ET, England. (nmer@bas.ac.uk)

Y.Y. Shprits, Department of Atmospheric and Oceanic Sciences, University of California, Los Angeles, 405 Hilgard Avenue, Los Angeles, CA 90095-1565, USA. (yshprits@atmos.ucla.edu)

R.M. Thorne, Department of Atmospheric and Oceanic Sciences, University of California, Los Angeles, 405 Hilgard Avenue, Los Angeles, CA 90095-1565, USA. (rmt@atmos.ucla.edu)

A. Varotsou, ONERA/Department of Space Environment, BP 4025 - 2, avenue Edouard-Belin, 31055 Toulouse CEDEX 4, France. (Athina.Varotsou@onecert.fr)

D. Boscher, ONERA/Department of Space Environment, BP 4025 - 2, avenue Edouard-Belin, 31055 Toulouse CEDEX 4, France. (Daniel.Boscher@onecert.fr)

Magnetospheric Energetics During HILDCAAs

W.D. Gonzalez, F.L. Guarnieri, A.L. Clua-Gonzalez, E. Echer, and M.V. Alves

Instituto Nacional de Pesquisas Espaciais (INPE), S.J. dos Campos, SP, Brazil

T. Ogino

Solar Terrestrial Environment Lab, Nagoya University, Toyokawa-Nagoya, Japan

B.T. Tsurutani

Jet Propulsion Laboratory, California Institute of Technology, Pasadena, California, USA

Some aspects of High-Intensity, Long-Duration, Continuous AE activity (HILDCAA for short) events are discussed in the context of storm and substorm energy flux and dynamics. The issue of energy transferred from the solar wind to the magnetosphere during HILDCAAs is of particular interest. We make estimates of the latter under the assumption of a summation of short magnetopause reconnection events associated with the quasiperiodic behavior of the IMF Bz component (part of interplanetary Alfven wave trains). A global magnetospheric MHD simulation is used to illustrate the response of the near-Earth plasma sheet to the changing behavior of IMF Bz during idealized HILDCAA events. In conclusion, it is argued that HILDCAAs are a new type of geomagnetic activity, substantially different than either substorms or storms. The relationship between HILDCAAs and substorms from both the interplanetary driver perspective and magnetospheric energy are discussed.

1. INTRODUCTION

High-intensity, long-duration, continuous AE activity (HILDCAA) events were defined by *Tsurutani and Gonzalez* [1987] (hereafter called TG87) as intervals where (1) AE peak values exceed 1000 nT, (2) the duration is greater than 2 days, and (3) the AE values never drops below 200 nT for more than 2h at a time. A further requirement was that (4) HILDCAAs must be separate from magnetic storm main phases.

It should be noted that the original TG87 criteria set for HILDCAA events was a strict one. A sort of extreme criteria were imposed to illustrate the geophysical phenomena. Clearly, the same physical process may occur when one or more of the four criteria are not strictly followed. It should also be noted that the definition of HILDCAA events contains the term "AE activity", and not necessarily substorm or auroral activity.

This was because the initial study was made using mainly the AE index to characterize the enhanced high latitude activity observed during HILDCAAs. Recently, *Tsurutani et al.* [2004] and *Guarnieri* [2005] have shown that substorms do not frequently occur in association with the HILDCAA-AE enhancements and that HILDCAAs represent a different category of auroral activity than substorms. *Tsurutani et al.* [2004] suggest that the AE increases may be related more to convection increases than to substorm expansion phases.

Tsurutani et al. [1990] used IMP-8 solar wind plasma and magnetic field data and the AE indices to show that there is a fairly high correlation between AE and the southward component of the IMF (Bs, where Bs is the negative IMF Bz values in GSM coordinates). The Bs variability during HILDCAAs was initially observed by TG87 as being associated with large-amplitude interplanetary Alfvén waves. The IMP-8 study of *Tsurutani et al.* [1990] also showed that the scale size of such Alfvén waves was smaller than the distance between the Earth and the inner Lagrangian L1 point, since similar correlations between AE and Bz were found to be much smaller or nonexistent when solar wind data collected at the L1 region was used.

The large correlations found by *Tsurutani et al.* [1990] between AE and Bs demonstrate that HILDCAAs are a consequence of solar wind energy transfer to the magnetosphere caused by magnetic reconnection between the southward component of the interplanetary Alfvén waves and magnetospheric magnetic fields. The reconnection intervals are expected to be shorter than during storms or even substorms, due to the oscillatory nature of the Alfvén waves, with scale lengths that are expected to be smaller than the distance between the Earth and the L1 region.

Thus, in this article we try to show some main differences between energy transfer from the solar wind to the magnetosphere during HILDCAAs and contrast it with that obtained during substorms and storms. Further, we also show some results of the inner magnetospheric response to IMF-Bz and solar wind parameters that typically exist near the Earth during HILDCAA events, obtained using a global MHD simulation model.

2. RESULTS

Tsurutani et al. [2006] and *Guarnieri* [2005] have shown that HILDCAA events are associated with high speed streams emanating from the sun. Figure 1 gives an example of this association for the time interval of July 20-30, 1998, as measured by the ACE satellite at L1. At the top, this figure gives the solar wind speed, showing the high speed stream that started by the end of July 22. At the bottom, the AE index shows the HILDCAA event that accompanied the stream, until about the end of July 25. The other panels in this figure give the solar wind density, pressure, IMF components and magnitude, the Dst index and the AL and AU indices. The three components of the IMF, and particularly the Bz component, show the presence of interplanetary Alfvén waves accompanying the stream and the HILDCAA event.

Gonzalez et al. [1994] have initially discussed the nature of HILDCAAs as representing a different category of geomagnetic activity, when compared to substorms and storms. In their Figure 11, these authors illustrated the association of HILDCAAs to a quasi-periodic Bz structure related to interplanetary Alfvén waves and compared it to the Bs structures that are typically related to substorms and storms.

Figure 2 shows ACE data for the interval January 21-30, 2000 in the same format as Figure 1. In this Figure we illustrate the presence of an intense magnetic storm with peak Dst value of about −100 nT (near the end of January 22), a typical isolated substorm with peak AE value of about 600 nT (from January 26, 20:45 UT until January 27, 03:12 UT) and a HILDCAA event starting late on January 27 and lasting until the end of January 30.

The intense storm of Figure 2 is associated with a CME, as observed in the magnetic cloud structure seen in the IMF and Bz rotation. Bs follows the criteria given by *Gonzalez and Tsurutani* [1987] for intense storms, namely Bz < −10 nT amplitude and duration >3 hr. The typical substorm selected in this Figure is associated with a Bs structure of about −5 nT amplitude and about 2 hr duration. The HILDCAA event is associated with the high speed stream, as seen in the solar wind speed data, and with the interplanetary Alfvén wave train that accompanied the stream, as seen in the IMF and solar wind speed data.

Thus, storms, substorms and HILDCAAs have different interplanetary origins, as seen mainly in their related Bz structures. If magnetic reconnection is the main mechanism of energy transfer between the solar wind and the magnetosphere, from the relative amplitudes and durations of the Bz structures, one could expect that reconnection should occur with different intensity and duration for each of the three categories of geomagnetic activity.

Plate 1 illustrates the difference in the intensity of polar and auroral particle precipitation during an intense storm, a typical substorm and a HILDCAA event, as given by a few minutes averages of ultraviolet images taken by the UVI (Ultraviolet Imager) instrument of the POLAR satellite [*Guarnieri*, 2005]. The intensity difference between the magenta and green colors is larger by a factor of 10, whereas for the yellow color it is in between. For the intense storm of July 15, 2000, the auroral oval is represented by intense precipitation, as expected, almost for the full oval. For the typical substorm of January 27, 2000 (discussed with Figure 2), the intense precipitation is highly localized to the midnight sector. On the other hand, for the HILDCAA event of July 24, 1998 (taken from Figure 1), the full auroral oval shows moderate precipitation. This is different from an isolated substorm and is more similar to a precipitation during a storm but with lesser intensity.

Because a HILDCAA event lasts for several days, the integrated precipitation energy can have values as large as that during a very intense storm [*Guarnieri*, 2005].

In Plate 1 the typical Bz and AE signatures, discussed above, are shown for the three geomagnetic active events. We note that during HILDCAAs, sometimes Bz can have an average value near zero, and at times the average can have a net negative value. It has been observed that, for the latter case, there is a tendency to find more substorms developing during the HILDCAA event than for the former case, probably because the net negative Bz implies more effective energy transfer to the magnetosphere via reconnection.

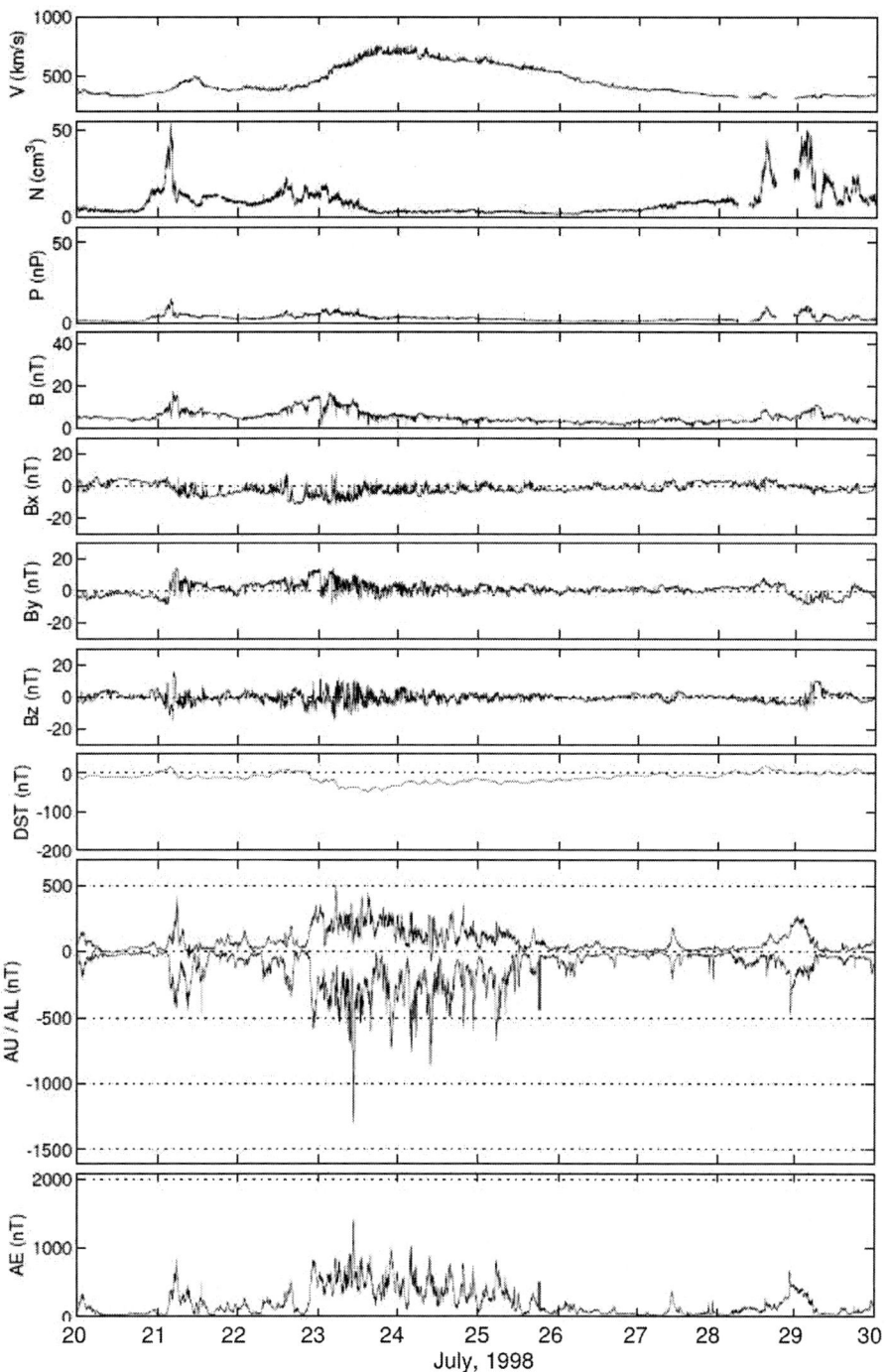

Figure 1. Example of association of a HILDCAA event with a high speed stream for the interval of July 20-30, 1998, as observed by the ACE satellite. See text for description of the data.

2.1. Energy Transfer During HILDCAAs

Guarnieri [2005], through a wavelet analysis, has shown the existence of mainly three quasi-periodicities in the Bz component of the interplanetary Alfvenic fluctuations associated with HILDCAA events. Those correspond to periodicities near 100 min, 40 min and 10 to 15 min. The first two periodicities seem to be associated with incursions in the

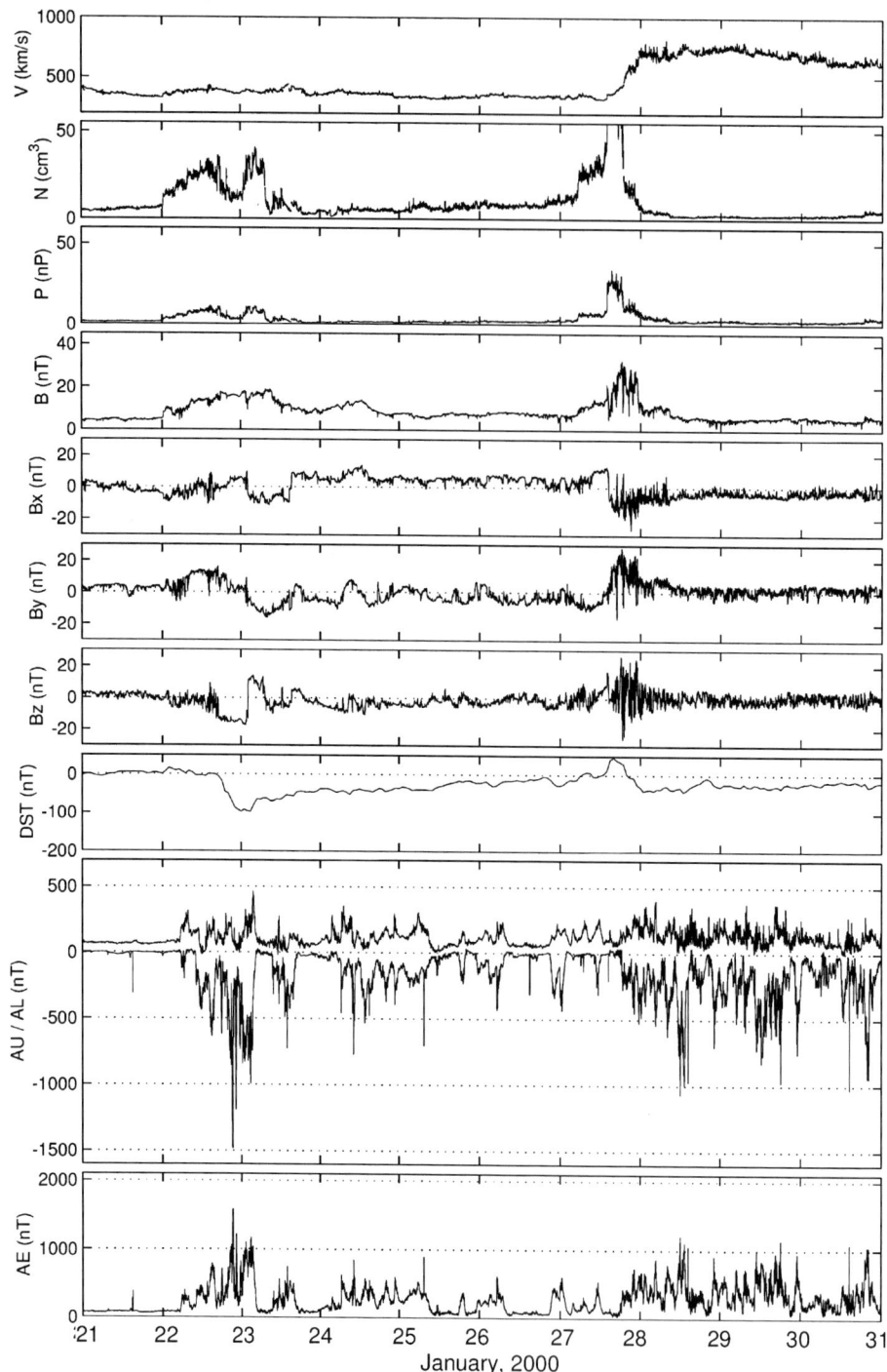

Figure 2. Example of an intense storm, a typical substorm and a HILDCAA event during the interval of January 21-30, 2000, as observed by the ACE satellite. See text for details.

negative Bz domain of the higher frequency fluctuations of 10 to 15 min, such as that illustrated at the bottom of Plate 1. AE tends to respond with broader amplitude increases to the first two periodicities (like in packages), whereas it tends to show only more pronounced peaks in response to the shorter periodicities. Thus, the main correlation between AE and Bz seems to be associated mainly with the longer periodicities. This is illustrated in Figure 3, in which the

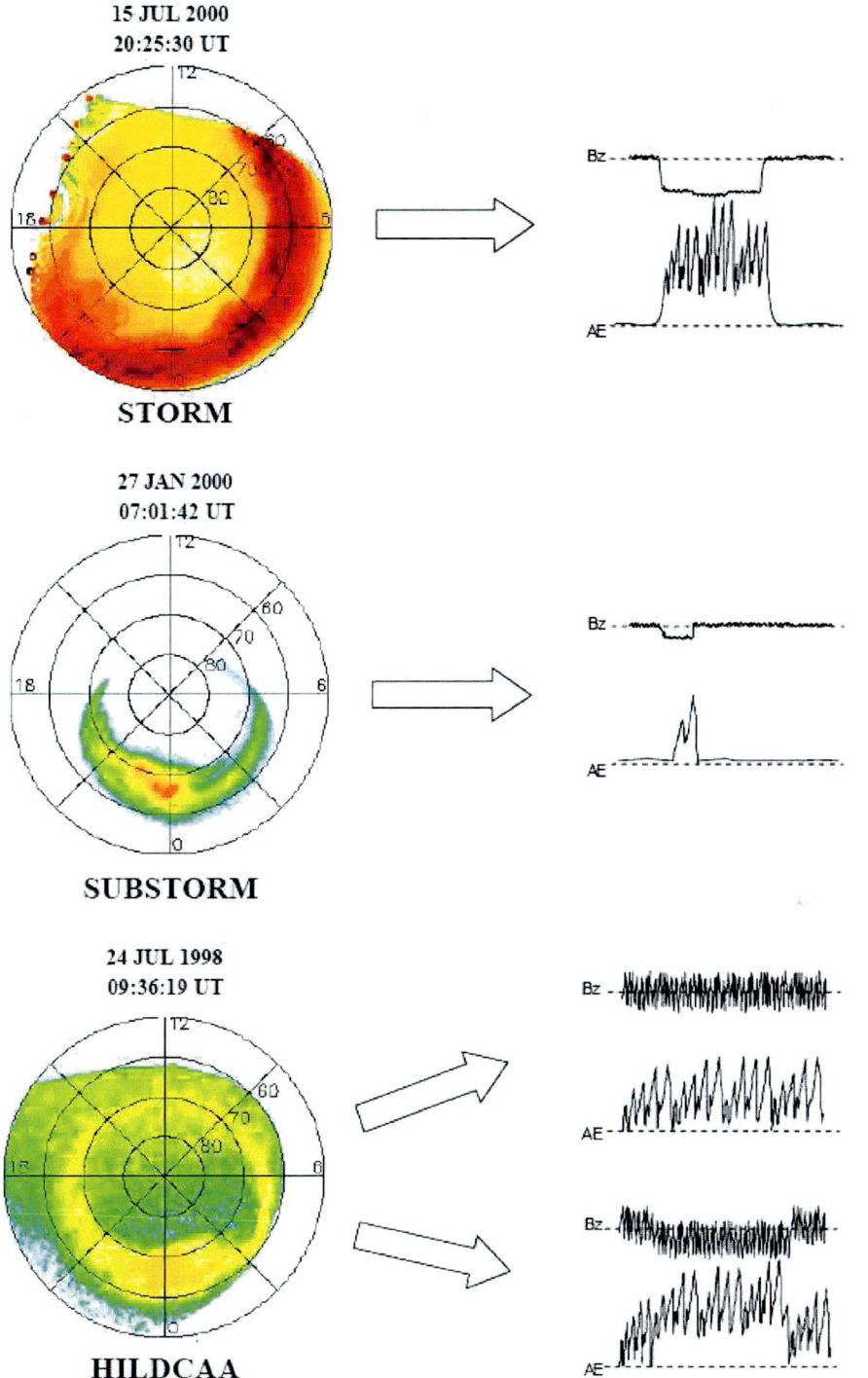

Plate 1. Intensities of polar and auroral particle precipitation, as observed by the POLAR satellite, for an intense storm, a typical substorm and a HILDCAA event, together with their Bz and AE signatures. See text for details.

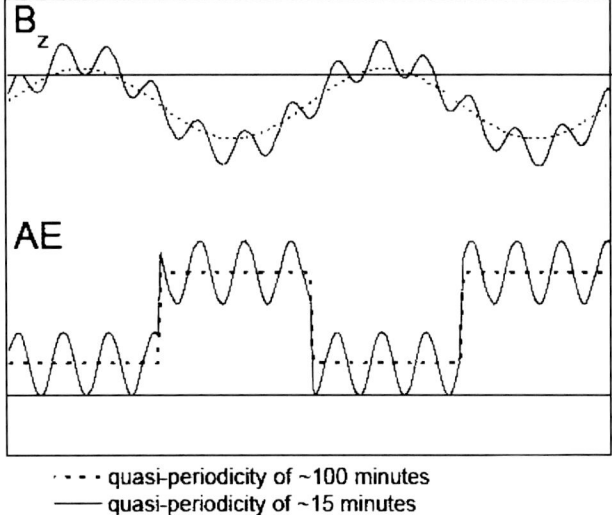

Figure 3. Typical Bz periodicities during HILDCAA events and the corresponding modulation of the AE response.

longer Bz periodicities (modulation with incursions in the negative domain) tend to elevate the broad AE amplitude response. Some of these periodicities are probably associated with previous studies of the filter effect between the solar wind and the response of the magnetosphere [e.g., *Iyemori et al.*, 1979; *Bargatze et al.*, 1985].

As mentioned above, the broad Bz incursion in the negative Bz domain probably represents a time interval in which magnetopause reconnection operates on a time scale similar to that of the longer periodicity (modulation).

On the other hand, the shorter periodicity, say of 10 min, represents in general a north-south Bz variability when the average Bz field of the Alfvén wave is around zero. This situation can be visualized as a short scale switch between intervals of magnetopause reconnection on and off, with a duration of the order of that expected during flux transfer events [*Russell and Elphic*, 1979].

In order to estimate the reconnection energy that goes into the magnetosphere during such reconnection events of short duration (10 min.), we could use the reconnection power as given by the *Perreault-Akasofu* [1978] parameter:

$$\varepsilon = v\, B^2\, F(\Theta)\, l_0^2$$

where v is the solar wind speed, B is the IMF amplitude, $F(\Theta)$ is a geometric factor related to the inclination of the reconnection line and l_0 is a constant radius of the magnetopause (= 7 Earth radii). For our estimate we make $B^2 F(\Theta) = B_z^2$.

Using typical solar wind values for v = 700 km/s, $B_z \approx -5$ nT, we get $\varepsilon \cong 3.5 \times 10^{18}$ ergs/s. If we multiply this power by 10 minutes (duration of a reconnection event), we get an energy of $\approx 1.8 \times 10^{21}$ ergs transferred to the magnetosphere.

For a typical energy transferred during a substorm, during similar wind conditions lasting during about 2 hours, we get a value of about $\approx 1.8 \times 10^{22}$ ergs.

Thus, we need about 10 HILDCAA reconnection events, of 10 minutes duration, to reach a typical substorm energy. This may imply that substorms could be triggered during HILDCAA events that have consecutive negative B_z fields with amplitudes of about −5 nT during about three hours or so. This condition is more easily obtained when the B_z fluctuation is immersed in the negative B_z domain, as discussed in the previous section. During this latter situation, a substorm could be obtained in a shorter HILDCAA interval. Also for substorms of shorter duration and intensity one may need less number of HILDCAA reconnection events. For instance, for a substorm of about 40 minutes to 1 hour duration, one may need only about 5 HILDCAA reconnection events.

This difference in energy transfer intensity for periods when the fluctuating Bz field has an average near zero and when the fluctuation is immersed in the negative Bz domain (net negative average in Bz) can also explain the related poor or larger intensification of the ring current, as seen in the Dst index, for each of these cases, respectively. Examples of such a ring current response can be seen in the HILDCAA intervals of Figures 1 and 2.

For an intense storm with peak Dst = −100 nT, and an associated Bs of −10 nT lasting three hours, the energy input to the magnetosphere is about an order of magnitude larger than for a typical substorm. For more intense storms, the corresponding energies are also larger.

The reconnection events could also be related to contractions and expansions of the magnetosphere, associated with the alternating more negative and more positive values of the fluctuating Bz field. Such contractions and expansions could trigger hydromagnetic waves, such as PC5 waves [e.g., *Tsurutani et al.*, 2006 and references given therein], which could additionally energize the precipitating particles during HILDCAAs, globally observed at high latitudes, as illustrated in Plate 1.

2.2. MHD Simulation of HILDCAAs

In order to have a simple idea on how the inner magnetosphere responds during HILDCAA events, we have used the global MHD simulation model described by *Ogino* [1986]. The code solves the MHD and Maxwell's equations as an initial value problem by using the modified two step Lax-Wendroff scheme in 3 dimensions. A quarter box model with north-south and dawn-dusk symmetry has been used.

We assume a simple periodic square wave for the Bz structure of the Alfvénic fluctuation associated with HILDCAAs.

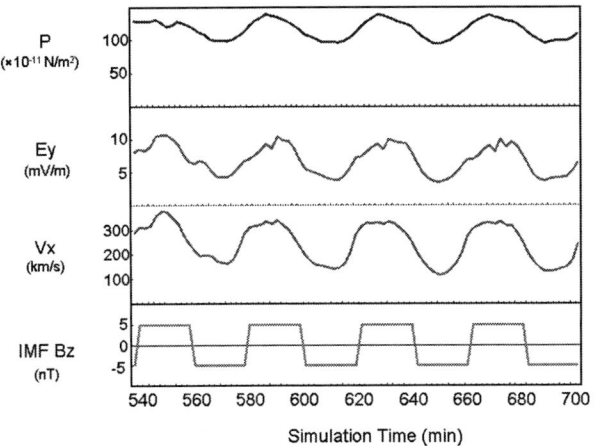

Figure 4. Time variation of pressure P, convection electric field Ey and convection speed Vx in the near Earth plasma sheet as a function of the periodic IMF Bz driver. See text for details.

This periodic square wave has an amplitude of −5 nT alternated with 5 nT, with a period of 40 minutes (the intermediate periodicity mentioned in the previous Section). The solar wind speed is taken as a typical value of 600 km/s during high speed streams, with a density of 5/cc and a solar wind pressure of 2.8 N/m^2 [*Ogino*, 1986].

The grid number of the simulation is (nx,ny,nz) = (300,100,100) and the grid size is dx = dy = dz = 0.3Re and the time step is 0.3s.

The simulation region is −60.3 Re < x < 30.3 Re, −0.15 < y < 30.1 Re and −0.15 < Re < z < 30.1 Re.

Figure 4 gives the time variation of the pressure P, the convection electric field Ey and the speed Vx in the plasma sheet depending on the variation of IMF Bz. The values mean the maximum on the sun-earth line for −15 Re < x < 0 Re in the plasma sheet. The region corresponds to the earth side of the NENL (Near-Earth Neutral line).

One can see that the response of P, Ey and Vx in the near earth plasma sheet is delayed by about 15 minutes from IMF Bz. For example, Vx begins to increase 15 minutes after IMF Bz turns southward (−5 nT), and Vx begins to decrease 15 minutes after IMF Bz turns northward (5 nT). As a result, the maximum in Vx appears at the center of the northward IMF period, whereas the minimum appears at the center of the southward IMF period.

3. DISCUSSION

In terms of event duration, one can say that substorms only last for about two hours, whereas storms last for a day or more (including the main and recovery phases). On the other hand, HILDCAAs can last for several days up to a week or even more.

In terms of energy involved, as illustrated in Plate 1, storms and substorms tend to accumulate more energy transmitted from the solar wind to the magnetosphere in their related duration intervals, substorms being more localized in space in the inner magnetosphere, whereas storms can include larger regions of deposited energy, both in the ring current as in the auroral region. HILDCAAs, on the other hand, tend to involve a large area of deposited energy in the auroral region, although with a less intense regime than, specially, in storms. However, if one integrates the energy of many HILDCAA days, one can end up with more total energy than in storms, except for very large storms, as discussed by *Guarnieri* [2005]. This represents an important issue about practical space weather results of HILDCAAs, such as in the consideration of particle energization at the outer radiation belt, where the well known "killer electrons" are located and can damage satellites orbiting at that region [*Tsurutani et al.*, 2006].

The reason for a quasi-global precipitation behavior shown in Plate 1 for HILDCAAs is not clear at the moment. However, among other issues, probably the persistence of HILDCAAs for several days could give ample time for the participation of several longitudinal sectors in particle precipitation.

The fact that the oscillating Bz field of the Alfvén waves related with HILDCAAs tend to be more geoeffective for the AE response, when the average Bz amplitude is in the negative Bz domain, was seen to be related to more effective reconnection events, leading to substorm reconnection energy levels probably in a shorter time than during HILDCAA events in which the average Bz fluctuation amplitude is less negative or near zero. This could also explain why *Guarnieri* [2005] has observed the occurrence of substorms during HILDCAAs with the former Bz fluctuation behavior more frequently than during HILDCAAs with the latter Bz variability.

From the reconnection power estimated with the ε expression, using typical solar wind values during HILDCAA events, and assuming a main periodicity of 10 minutes for the Bz field, we obtain an integrated energy of ≈1.8 × 10^{21} ergs that become transferred to the magnetosphere during a 10 minutes interval. When we compare this energy with typical substorms energies that run from about 5 × 10^{21} ergs to 10^{22} ergs, depending on the duration of the Bz field, we can say that about 3 to 6 HILDCAA reconnection events, of 10 minutes duration, are necessary to reach a threshold of a substorm energy. However, even a shorter number of HILDCAA reconnection events may be necessary to trigger a substorm when the Bz field varies totally in the negative Bz domain.

The results found in the global MHD simulation, with a periodic square wave for IMF Bz driving reconnection at the magnetopause, seem to indicate that the near-Earth plasma sheet is strongly modulated, both in pressure changes as in the convection speed and electric field. This modulation at the tested period of 40 minutes, illustrated in Figure 4, could signify an effective magnetospheric response to the Bz variability, in which one can also note the presence of a lag of about 15 minutes. This lag time is probably related to the response time of the magnetosphere to the driven system. This is in close agreement with the 20 minutes delay of the driven system reported by *Bargatze et al.* [1985].

Similar simulations carried out, and not shown, for longer periods, such as 90 minutes, do also show a similar but less intense modulation in the amplitude of the plasma sheet parameters. If Bz oscillates but stays in the negative domain, e.g., between -5 nT and 0 nT, the modulation in the plasma parameters near-Earth plasma sheet is much less pronounced, probably related to the less drastic variability in the changing Bz configuration of this latter case.

The reason why the magnetosphere tends to respond more effectively to the 40 minutes Bz oscillation, shown in Figure 4, is not certain, but it may have to do with the idea that this time interval could represent a typical response time of the magnetosphere to solar wind energy transfer and, in such a way, periodic changes of Bz with that period could lead into a sort of resonant and effective magnetospheric response. Thus, we suggest that this idea be tested in future similar simulations by varying the Bz oscillation period in increments of, say, 10 minutes.

Another interesting question raised by the simulation results presented in Figure 4 is about the relationship, if any, of consecutive convection enhancements to the dynamics and formation of substorms.

With respect to the main issues of this paper, namely energy transfer during HILDCAA events and inner tail response to the periodic Bz variations, we would like to suggest more detailed studies about the magnetopause and tail reconnection processes during HILDCAAs, which could help to elucidate not only the process of magnetospheric energization during HILDCAAs but also during substorms. Perhaps, in general, the magnetospheric response to the Bz variability during HILDCAAs is a more basic and simpler topic to study, with results that could be very helpful for the understanding of that response during more complex situations, such as during substorms and storms.

Acknowledgments. The authors W.D.G and A.L.C. de G. would like to acknowledge the CNPq (Brazil) fellowship (bolsa de produtividade de pesquisa). F.L.G. would like to thank CAPES and FAPESP for the fellowships. The author B.T.T. was partially supported by the Jet Propulsion Laboratory, California Institute of Technology under contract with NASA. The MHD simulations were performed at the Computer Center of the Solar Terrestrial Environment Lab and of the Institute of Plasma Physics of Nagoya University, Japan.

REFERENCES

Akasofu, S.-I., Energy coupling between the solar wind and the magnetosphere, *Space Sci. Rev.*, 28, 121, 1981.

Bargatze, L.F., D.N. Baker, R.L. McPherron, and E.W. Hones Jr., Magnetospheric impulse response for many levels of geomagnetic activity, *J. Geophys. Res.*, 90, 6387, 1985.

Gonzalez, W.D., J.A. Joselyn, Y. Kamide, H.W. Kroehl, G. Rostoker, B.T. Tsurutani and V.M. Vasyliunas, What is a geomagnetic storm?, *J. Geophys. Res.*, 99, 5771, 1994.

Guarnieri, F.L., Study of the solar and interplanetary origin of long-duration and continuous auroral activity events, PhD thesis, INPE – S.J. dos Campos, SP, Brazil, February, 2005.

Iyemori, T., H. Maeda, and T. Kamei, Impulse response of geomagnetic indices to interplanetary magnetic field, *J. Geomag. Geoelectr.*, 6, 577, 1979.

Ogino, T., A three dimensional MHD simulation of the interaction of the solar wind with the Earth's magnetosphere: The generation of field aligned currents, *J. Geophys. Res.*, 91, 6791, 1986.

Perreault P. and S.I. Akasofu, A study of geomagnetic storms, *Geophys. J. Roy. Astron. Soc.*, 54, 547, 1978.

Russell, C.T. and R.C. Elphic, ISEE-3 observations of flux transfer events at the dayside magnetopause, *Geophys. Res. Lett.*, 6, 33, 1979.

Tsurutani, B.T. and W.D. Gonzalez, The cause of High-Intensity, Long-Duration, Continuous AE activity (HILDCAAs): interplanetary Alfvén wave trains. *Planetary and Space Science*, 35, 405, 1987.

Tsurutani, B.T., T. Gould, B.E. Goldstein, W.D. Gonzalez, and M. Sugiura, Interplanetary Alfvén waves and auroral (substorm) activity: IMP-8, *J. Geophys. Res.*, 95, 2241, 1990.

Tsurutani, B.T., W.D. Gonzalez, F. Guarnieri, Y. Kamide, X. Zhou and J.K. Arballo, Are high-intensity, long-duration, continuous AE activity (HILDCAA) events substorm expansion events?, *J. Atm. Sol. -Terr. Phys.*, 66, 167, 2004.

Tsurutani, B.T., W.D. Gonzalez, F.L. Guarnieri, N. Gopalswamy, M. Grande, Y. Kamide, Y. Kasahara, G. Lu, I. Mann, R. McPherron, and V.M. Vasyliunas, Corotating solar wind streams and recurrent geomagnetic activity: A Review, *J. Geophys. res.*, in press, 2006.

Energetic Neutral Atom Observations During Recurrent Magnetic Storms

J.-M. Jahn and H.A. Elliott

Space Science Department, Southwest Research Institute, San Antonio, Texas, USA

Recurrent storms are triggered by Earth's encounters with corotating interaction regions (CIR) in the solar wind, they occur most frequently in the declining phase of the solar cycle. CIRs are regions of strong magnetic field accompanied by high pressure. They form in the interplanetary medium by the interaction of low- and high-speed solar wind streams co-rotating with the Sun. Typically, CIRs are not fully developed at 1 AU. A forward shock is commonly lacking, and a reverse shock is absent in the majority of events. Compared to interplanetary coronal mass ejection (ICME) driven events this results in "weak" storms, with minimum *Dst* excursions that can fall short of even −50 nT. On the other hand, the recovery of *Dst* to background levels during high-speed stream (HSS) intervals following a CIR can take many days, much longer than for storms caused by ICMEs. Recovery is accompanied by prolonged intervals of Auroral Electroject index (AE) activity, indicating sustained levels of substorm and plasma sheet activity during *Dst* recovery. In this paper we present IMAGE energetic neutral atom (ENA) observations of recent CIR/high-speed stream intervals, contrasting those observations with recent findings of ENA imaging during ICME-induced storms. We discuss the presence and composition (H^+ and O^+) of the ring current during recurrent storms, characterize the magnetospheric activity during the prolonged recovery of *Dst*, and describe the overall magnetospheric convection associated with the solar wind Alfvén wave activity associated with those events.

1. INTRODUCTION

Geomagnetic storms are caused by the interaction of Earth's magnetosphere with relatively dense, high-velocity solar wind streams that carry extended periods of southward IMF with them. During solar maximum conditions, storms are predominantly caused by interplanetary coronal mass ejections (ICME), whereas during the declining phase of a solar cycle, geomagnetic storm activity is more frequently induced by the interaction with corotating interaction regions (CIR) and associated high speed stream (henceforth abbreviated HSS) regions of the solar wind. In either case, solar wind magnetic field topology and dynamic pressure throughout the event are key ingredients to the geoeffectiveness of an ensuing storm. A large body of literature exists examining the magnetospheric storm response to this solar wind driving. Research seems to emphasize ICME driven storms, probably owing to the fact that these storms show a more varied magnetospheric response. However, the total energy input into the magnetosphere is about the same for either storm type [*Turner and Mitchell*, 2006]. Until recently, the experimental study of geomagnetic storms of any type utilized ground-based observations, geomagnetic indices, and *in situ* observations from various spacecraft. Truly global magnetospheric observations were really only available from a variety of MHD models of the magnetosphere. After early reports of the detection of energetic neutral atoms [*Hovestadt and*

Scholer, 1976; *Roelof et al.*, 1985] and first imaging results of the ring current [*Roelof*, 1987; *Henderson et al.*, 1997], the IMAGE mission [*Burch*, 2000] became a magnetospheric mission exclusively dedicated to the remote sensing of magnetospheric processes. In this paper we will present selected ENA observations from IMAGE during a number of CIR/HSS interactions with Earth's magnetosphere.

1.1. Energetic Neutral Atoms

Magnetospheric energetic neutral atoms (ENA) are created by charge exchange during collisions between energetic ions and cold neutral atoms of Earth's geocorona. Since neutral atoms travel freely across magnetic field lines, ENA measurements can be used to remotely sense magnetospheric particle populations. Although the basic process of charge exchange itself is rather straightforward, several factors determine the exact nature of magnetospheric ENA "emissions". The charge exchange cross section controls the process on the atomic level. The geocoronal neutral density and the local plasma density determine the amount of ion-neutral charge exchange reactions as a function of distance and direction from Earth. The local plasma pitch angle distribution exacts control over the directionality of ENA emissions. The most important facts to remember are that (1) the probability for charge exchange shrinks rapidly as a function of distance, (2) ENA emissions are highly *directional* due to plasma pitch angle distributions, and (3) ENA emissions are volume emissions, i.e., remote sensing of ENAs amounts to line-of-sight integrations along a specified look direction. From any observation point, the directions from which particle emissions arrive are well-known, yet the distances from the observer to the source regions—and thus the source locations themselves—are a priori *not* known. Unfortunately, source locations can also not be deduced easily. To remedy this situation, mathematical inversion techniques and forward modeling approaches have been used to recover the source population distribution of plasma [e.g., *Perez et al.*, 2001; *DeMajistre et al.*, 2004; *Roelof and Skinner*, 2000]. At present, ENA inversions are still labor- and computation-intensive and do not easily lend themselves to automated analysis of long-duration measurements of even a few days. Analyzing raw ENA data instead allows, within limits, both a more direct analysis of highly time-variable ENA measurements, and an easier implementation of long-term and statistical studies [e.g., *Ohtani et al.*, 2006].

1.2. Plasma Sheet

Two of the most interesting regions in the inner magnetosphere to observe with ENA imaging during CIR/HSS induced storms are Earth's ring current and the near-Earth to mid-tail (<30 R_E) plasma sheet. A large portion of plasma in the inner magnetosphere will have passed through Earth's plasma sheet, regardless of whether it originated from the ionosphere or the solar wind. Rather than being only a passive way station, the plasma sheet is an important participant and, at times, controller of inner magnetospheric dynamics. It is the dominant source of ring current plasma. Plasma sheet densities have been shown to quantitatively explain the storm-time variation of the ring current [*Jordanova et al.*, 1998, 1999b,a; *Kozyra et al.*, 1998, 2002; *Ebihara et al.*, 1998; *Ebihara and Ejiri*, 2000; *Liemohn et al.*, 2001]. A reasonable correlation between plasma sheet density at geosynchronous orbit and ring current strength exists [*Thomsen et al.*, 1998]. Due to the shielding electric field, the plasma sheet control of the total energy of the ring current appears to be nonlinear, following $\sim \sqrt{N_{ps}}$ [*Ebihara et al.*, 2005]. The composition of the plasma sheet directly affects the ring current composition, as discussed below.

ENA imaging of the plasma sheet proves difficult, although it has been shown that plasma sheet emissions can be monitored both in the near-Earth and more distant plasma sheet region [*Denton et al.*, 2005a; *McComas et al.*, 2002; *Brandt et al.*, 2002b]. Due to quickly diminishing fluxes, plasma sheet ENA imaging beyond geosynchronous orbit is much more difficult than ring current ENA imaging.

1.3. Ring Current

The ring current itself is a toroidal current centered in the equatorial plane, flowing around Earth at altitudes of ~10,000 – 60,000 km. It consists of energetic particles that participate in a collective near-circular motion around Earth as a results of gradient and curvature drift in the magnetic field. The main current carrier of the ring current are ions from energies of ~20.0 keV up to several hundred keV. While the quiet-time ring current is dominated by H^+ ions, the abundance of O^+ ions can significantly increase during storm times and, especially during solar maximum conditions, at times surpass the H^+ abundance. Other ion species are usually only minor contributors. Ring current ions are supplied by the solar wind and (in the case of O^+ exclusively) by the terrestrial ionosphere.

The ring current is most dynamic and strongest during geomagnetic storms, when it can decrease the Earth's surface magnetic field by several hundred nano-Teslas. The changes in ring current composition, strength, dynamics and morphology are the result of a complicated interaction between many effects, including: solar wind plasma entry into the magnetosphere, ionospheric outflow, plasma sheet density, substorm injection activity, magnetospheric convection and associated electric fields, and particle loss through charge exchange, precipitation, and the dayside magnetopause.

Owing to its capability to alter Earth's surface magnetic field during storm times, the terrestrial ring current has been studied since the early days of space research, and many advances have been made both experimentally and in modeling studies [e.g., *Daglis et al.*, 1999, and references therein].

1.4. Ring Current Sources and Composition

Although *Chappell et al.* [1987] estimated that the ionospheric particle source is sufficient to supply the entire magnetospheric plasma content, the relative contributions of ionospheric and solar wind plasma to the plasma sheet population is still a controversial topic. Several studies have used trajectory models to examine the ionospheric contribution to the plasma sheet and ring current [*Moore et al.*, 1999; *Delcourt et al.*, 1988; *Sauvaud and Delcourt*, 1987; *Delcourt et al.*, 1994, 1989]. *Ashour-Abdalla et al.* [1997] showed that the ionosphere can at times be a significant contributor to the distribution of particles in the near-Earth magnetotail, though often the solar wind is the dominant source [*Ashour-Abdalla et al.*, 1999]. *Borovsky et al.* [1997, 1998] showed that the plasma sheet density is well correlated with the solar wind density. They concluded that the plasma sheet is likely to be of solar wind origin. *Lennartsson* [1992] found the plasma sheet to be more like the solar wind during northward IMF than during southward IMF. They assumed solar wind particles entering the plasma sheet along the flanks of the lower latitude boundary layer (LLBL) during northward IMF. Recent multi-fluid modeling of the magnetosphere by *Winglee* [2003] reaffirmed that the ionosphere might be more important than the solar wind for the plasma sheet for southward IMF conditions, and vice versa for northward IMF conditions.

1.5. Imaging the Ring Current During ICME Storms

The launch of the IMAGE spacecraft has brought a wealth of new data and new insight to the study of the ring current. Combined with ENA image inversion techniques [*Perez et al.*, 2001; *DeMajistre et al.*, 2004] and other modeling approaches [*Fok et al.*, 2003], the global morpology and development of the ring current is now much more accessible experimentally. The transition from asymmetric to symmetric ring current has been observed across a wide range of energies [*Brandt et al.*, 2001; *Pollock et al.*, 2001]. *Brandt et al.* [2002a] observed the rapid response of the ring current to IMF changes. Storm-time electric fields can have significant effects even on high energy ring current particles, moving the peak of the energetic proton flux away from midnight towards dawn [*Brandt et al.*, 2002; *Ebihara and Fok*, 2004]. ENA observations during a strong storm, combined with *in situ* and ground-based measurements, showed that under special circumstances the tail currents rather than the ring current can be the dominant contributor to Dst [*Skoug et al.*, 2003]. The storm-substorm relationship as well as individual isolated substorms have been investigated using ENAs [*Brandt et al.*, 2003; *Reeves et al.*, 2003; *Pollock et al.*, 2003; *Jahn et al.*, 2006]. Of particular interest are also sawtooth events, which globally affect the magnetosphere and ring current in a substorm-like fashion [*Huang et al.*, 2003; *Henderson et al.*, 2006]. Also, for the first time it has become possible to image the ring current oxygen content during storms and sawtooth events, showing the important role of substorms in the delivery of oxygen [*Mitchell et al.*, 2003, 2005; *Ohtani et al.*, 2005; *Ohtani et al.*, 2006; *Henderson et al.*, 2006].

1.6. The IMAGE Mission

Energetic neutral atom images from the IMAGE spacecraft [*Burch*, 2000, and references therein] allow us to globally quantify the ring current strength and near-Earth plasma sheet during long continuous time periods. The medium energy neutral atom imager MENA [*Pollock et al.*, 2000] and high energy neutral atom imager HENA [*Mitchell et al.*, 2000] take data continuously for more than 11 hours out of the IMAGE orbital period of 14 hours (both imagers are not operated inside the radiation belts). The cadence of all IMAGE imaging data is two minutes. IMAGE is in an elliptical polar orbit with 90° inclination, an apogee altitude of 7 R_E and a perigee at 1,000 km. The line of apsides of IMAGE precesses 45°/year, it was over the pole in 2001 and in the equatorial plane in 2003. IMAGE significantly expands the spatial coverage of magnetospheric plasmas compared to *in situ* measurements, and allows for direct comparison with ring current simulation results on a global scale.

2. RING CURRENT OBSERVATIONS

In this paper, we will limit our discussion of the geomagnetic response to CIR/HSS events as seen in ENA images to three areas: (1) existence and shape of the ring current, (2) composition of the ring current, and (3) plasma sheet related measurements. Since ENA observations of CIR/HSS induced storms are still new, we will further concentrate our discussion on *characterizing* the observation and *comparing* them to already published (and thus more widely known) observations during ICME-induced storms.

We have selected two time intervals in 2000 and in 2004 to illustrate the magnetospheric response to CIR/HSS episodes. The 2004 time interval was chosen for a series of clean and recurring high-speed stream intervals. The 2000 time interval was selected since it provides longer, better quality ENA observations at low energies (<10 keV) than 2004, which is necessary to address magnetospheric convection and plasma sheet effects.

2.1. 2004 CIR Observations

To study the ring current response observed with ENAs, we examine a group of four consecutive recurrent storms in the spring of 2004. These storms were caused by two coronal holes, located roughly on opposite sides of the sun. This geometry resulted not only in a higher frequency of high-speed stream encounters (twice every 27 days), it also provided the opportunity to observe high-speed stream of opposite polarity in short succession. An overview over the whole time period is given in Plate 1.

11 February 2004: This first event in the series has inward (i.e., toward the Sun) polarity. What sets this particular event apart from all other events discussed in this paper is its leading edge. For 5 hours before the CIR the IMF points southward in the region of increased density. This magnetic field signature causes a fairly rapid intensification of Dst down to −100 nT, as well as a spike in Kp. Then, before the arrival of the high speed stream, the IMF turns northward. Dst recovers to −50 nT at virtually the same rate as it was previously dropping. After the arrival of high-speed solar wind plasma, Dst recovery continues much more slowly, in a fashion typical for high-speed stream induced storms.

The rise in solar wind speed coincided with a significant increase in Alfvén wave activity typical for recurrent storms. Plate 1 shows fluctuations of IMF B_z for several days after the storm onset, with a 18-hour period of relative quiet around February 17. Once significant Alfvén wave activity has started in the solar wind, Dst recovers slowly over the following seven days at an, on average, near constant rate.

29 February 2004: The second event in this sequence stems from the second coronal hole mentioned above. It shows outward (i.e., away from the Sun) polarity. We detect a clear pressure pulse at the leading edge of the event, followed by a rise in solar wind speed and the onset of Alfvén wave activity. As is more usual for CIR/HSS events, there is no extended period of southward IMF preceding the event. The Alfvén wave activity is sustained for 8+ days, with slowly declining amplitude. In contrast to the previous and the following event, the drop in Dst is very moderate, never venturing below −30 nT. In terms of geoeffectiveness this storm would be considered much weaker, despite the fact that solar wind speed, dynamic pressure history, and Alfvén wave activity are comparable between the first three storms of the study period. Elevated Kp values still indicate that there is a significant amount of disturbance present in the magnetospheric system.

09 March 2004: On this day we see the recurrence of the inward polarity high-speed stream first encountered early February. Obviously, this high-speed stream has evolved over one solar rotation. Close inspection reveals that the onset of the event is still marked by an extended IMF B_z southward region, although B_z has become both less negative and is repeatedly interrupted by short northward excursions. The pressure pulse has become more distinct (no intermittent drop in dynamic pressure) and is lasting longer (over one day). The level of Alfvén wave activity is slightly reduced compared to the previous encounter, however, there is no sign anymore of an intermittent drop in activity.

The differences in solar wind parameters between both interactions with this particular HSS bear out in the behavior of the ring current. At the start of the March 09 event, Dst rapidly drops to about −70 nT and then starts a recovery to quiet time levels over the coming 8 days. In contrast to its previous occurrence, there is no "overshoot" at the onset of the event, likely because of the lack of pronounced long-duration southward IMF conditions. Secondly, Dst is considerably more variable during the recovery. Multiple variations with amplitudes of up to a few 10's of nT peak-to-peak make for a much more dynamic Dst timeline than for the February 11 event.

This CIR/HSS returns a third time on 07 April 2004, but it is preceded by an ICME two days earlier and is therefore not included in this discussion.

27 March 2004: This is the last event in this series, again with outward polarity of the solar wind. It shows the highest solar wind speed of all cases presented, with peak values of 950 km/s, 150 km/s higher than in the previous storms. At the same time, the Dst response is weak (minimum Dst reaches −50 nT), though stronger than during the first occurrence of this high-speed stream region.

2.2. ENA Observations of the Ring Current

Ring current ENA observations corresponding to the 2004 interval have been summarized in Plate 2. Red crosses indicate hydrogen and blue squares oxygen ENA particle fluxes. For each data point, we show the total ENA flux surrounding Earth for 10-60 keV hydrogen and 29-222 keV oxygen, respectively, averaged over 10 minutes. In order to remove any orbital effects, we only show data as long as the IMAGE spacecraft was located within 7.5° of Earth's dipole axis. Clusters of data points around each passage of Earth's dipole axis will therefore be 14 hours apart. The close proximity to the dipole axis was mainly chosen since during early 2004 the apogee of the IMAGE spacecraft was still less then 45° latitude above the equatorial SM plane (apogee was located in the southern hemisphere in 2004). During large portions of the orbit, ENAs were viewed "side-on" rather than "from the top", potentially exaggerating orbital viewing effects. Longer gaps in ENA data between individual CIR/HSS events are not indicative of satellite data gaps but rather reflect our choice to limit the data presentation to the more active time periods. Despite imposing these limits on

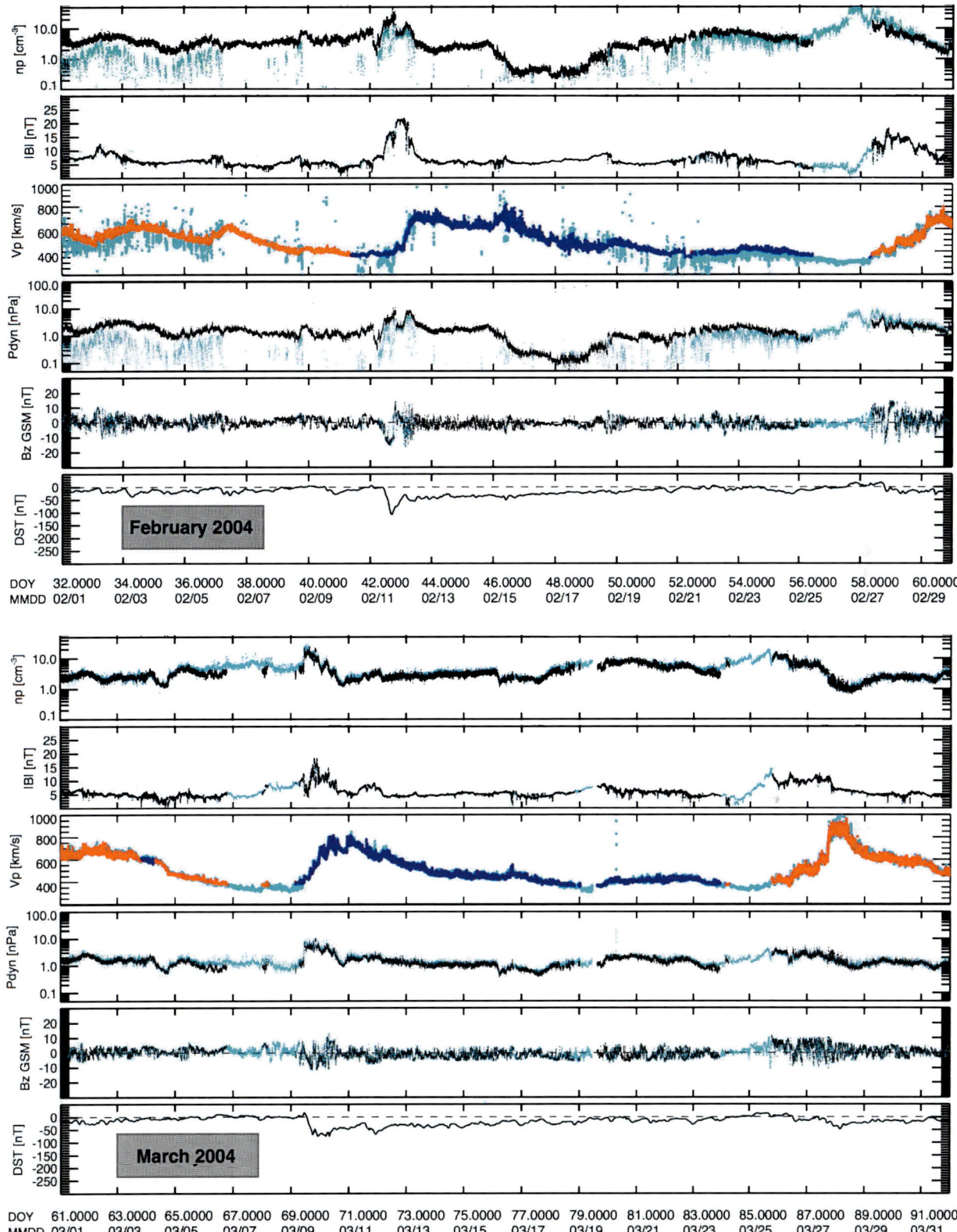

Plate 1. Overview over solar wind data for four consecutive high-speed stream intervals in early 2004 from Wind (pale blue) and ACE (black). Plotted are (from the top): the solar wind density, the total magnetic field, the solar wind speed, with color indicating outward (red) and inward (dark blue) magnetic field polarity based on ACE measurements; the solar wind dynamic pressure, the B_z component of the interplanetary magnetic field, and the Dst index. Note the alternating polarity of high-speed stream events during the two months presented.

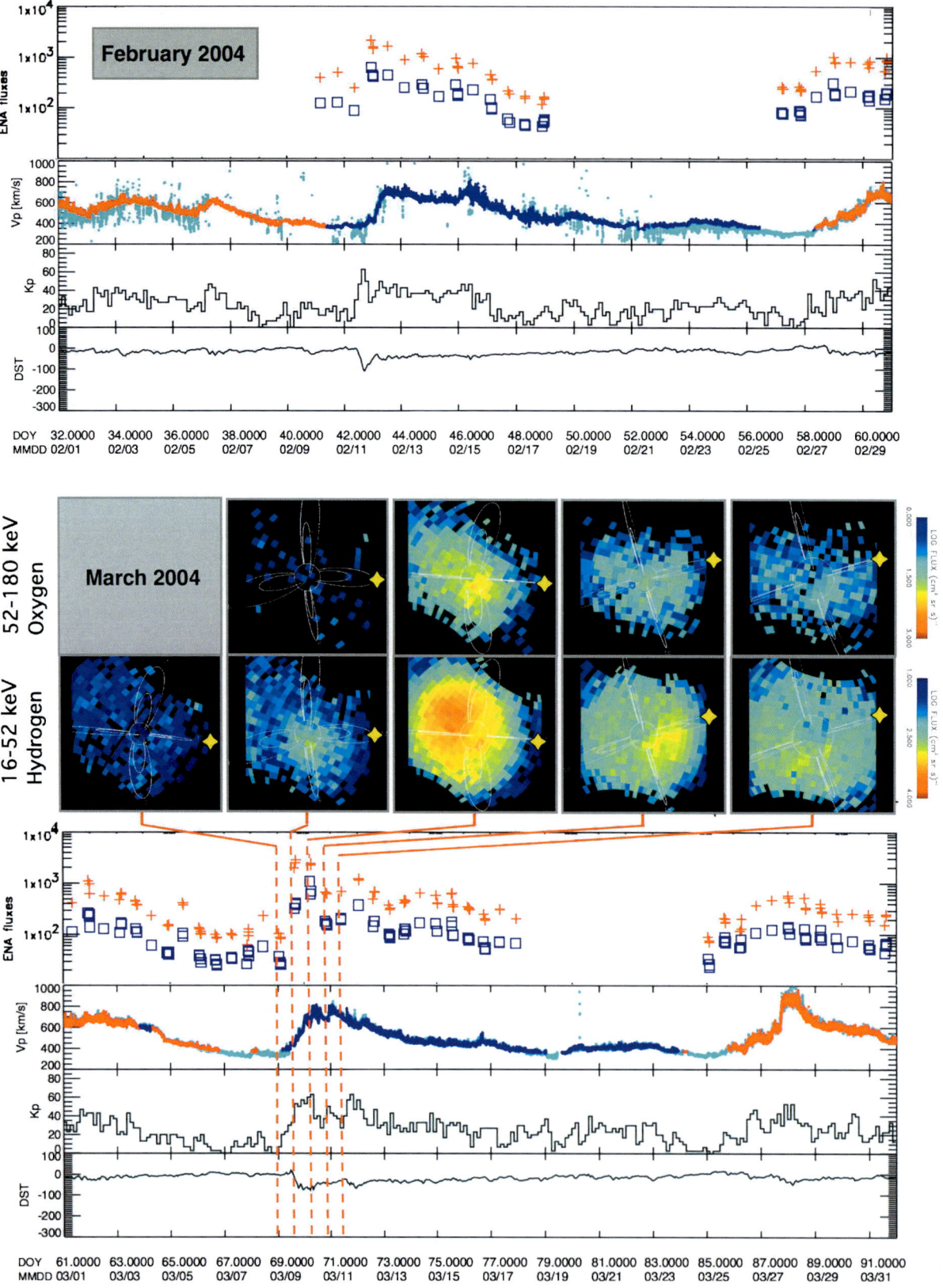

Plate 2. Overview over integrated ENA oxygen (blue squares) and hydrogen (red crosses) ENA ring current emissions from the HENA instrument during Early 2004 for times when IMAGE was located within 7.5° of Earth's dipole axis. Panels below that show the solar wind speed, with polarity encoded as before, *Dst*, and *Kp*. For March 2004 we also show four and five ENA ENA images for 52-180 keV oxygen (top row) and 16-52 keV hydrogen (bottom row) as viewed form the southern hemisphere. In all images, the Sun (indicated by a yellow star) is to the right, and dipole field lines are plotted for $L = 4$ and $L = 8$.

the data, we can easily follow the time development of CRIR/HSS events. High-speed stream last multiple days, for a long-term characterization of ring current behavior the chosen data coverage is sufficient.

From *Dst* recordings during recurrent storms it has already been established that the ring current responds to CIR/HSS events [e.g., *Tsurutani et al.*, 1995]. Based on *Dst* measurements during the 2004 events, we can expect to find a response of global ENA emissions to the geomagnetic activity. To lowest order, Plate 2 shows that observed ENA emissions follow variations in *Dst*. The highest levels of ENA emission coincide with *Dst* minima (please note that our sampling misses the minimum of −100 nT on 11 February 2004). Secondary variations of *Dst*, like the peak on 11–12 March 2004, are also visible as separate ENA emission peaks. This should come as no surprise, as it has already been established that charge exchange can, depending on storm phase, be the dominant loss mechanism for ring current particles [*Jorgensen et al.*, 1997; *Ebihara et al.*, 1999]. Subsequently, *Dst* and integrated ENA emissions in the ring current energy range should generally track each other well. Exceptions to this rule exist, either because other loss mechanisms become dominant [e.g., *Jorgensen et al.*, 2001], or because tail currents contribute significantly to *Dst* [*Skoug et al.*, 2003; *Ohtani et al.*, 2006]. But neither exception should occur during CIR/HSS events.

ENA emissions of the ring current also track a secondary effect which can be observed in this sequence of high-speed stream events. Owing to the Russell-McPherron effect, high-speed streams with inward polarity can show a stronger *Dst* response during spring equinox than high-speed streams with outward polarity for otherwise comparable solar wind conditions [*Crooker*, 2000] (during fall the situation is reversed). This effect can readily be seen in the first events in both February and March of 2004, which are stronger both in *Dst* and ENA emissions than their counterparts of opposite polarity later in those two months. This happens even though the intervals of outward polarity appear to have stronger Alfvén wave activity associated with them.

Another interesting observation on the ring current proper concerns the location of the peak of the ENA flux during the height of geomagnetic activity. Plate 2 shows five ENA image snapshots during the 09 March 2004 CIR/HSS event for 15-60 keV hydrogen (bottom row) and four concurrent ENA image snapshots for 52-180 keV oxygen. In all images the Sun (indicated by a yellow star) is located at the right hand side. Field lines at $L = 4$ and $L = 8$ as well as Earth's geometric outline are shown. IMAGE was located in the southern hemisphere, so the dawn side of the magnetosphere is at the top, the dusk side at the bottom of each picture. Midnight is to the left. At 0240 UT on 10 March 2004 (center column), there was a short period of strong solar wind driving of the magnetosphere with $v_{sw} \approx 650$ km/s, $B_z \approx -5$ nT, and $B_y \approx -5$ nT. These conditions are similar to event studies for ICME driven storms by *Brandt et al.* [2002], which showed that under those solar wind conditions the peak of the main phase storm time ring current can be moved past midnight towards the dawside—though no strong correlation with the chosen solar wind properties was demonstrated. The hydrogen ENA image for 0240 UT in Plate 2 clearly shows the peak of ENA emissions to lie in the dawn sector. Note that the viewing for this time interval from approximately (−1.5, 0.0, −4.0) R_E in *GSM* coordinates obscures the true local time location of the peak somewhat. Nonetheless, this demonstrates that the location of the ring current peak can also vary during CIR/HSS storm time periods, and studying it may indeed expand our abilities to determine the causes of this ring current morphology shift along the lines suggested by *Ebihara et al.* [2005]. Please note that the *following* ENA snapshot at 1700 UT the same day shows a peak ENA emission near noon. Considering that this image was taken when IMF $B_z \approx 0$ nT after an hour of northward IMF conditions, this would not be the peak of the ring current under strong driving, but more likely a gradient curvature drifting particle population.

2.3. The Role of Oxygen

During the recent solar maximum the study of ring current composition has received renewed interest. Oxygen ions can make a significant, and sometimes even dominant, contribution to the storm-time ring current during strong ICME induced storms. Oxygen ions can play a major role especially during main phase and early recovery [e.g., *Daglis*, 1997; *Daglis et al.*, 1999; *Pulkkinen et al.*, 2001; *Greenspan and Hamilton*, 2002]. Differences in charge exchange rates between different species contribute to the more rapid decay of oxygen during the storm recovery phase [*Fu et al.*, 2001; *Ohtani et al.*, 2005]. The most effective agents in delivering oxygen into the ring current appear to be substorms, but they need to be supported by enhanced ionospheric oxygen outflow to deliver ions into the plasma sheet first [*Daglis and Axford*, 1996; *Daglis et al.*, 1999; *Moore et al.*, 1999; *Nosé et al.*, 2000; *Korth et al.*, 2002]. The role of substorms can be seen especially well during sawtooth events. This type of storm is characterized by extended strong driving of the magnetosphere by a steady southward IMF B_z, and by efficient delivery of oxygen into the ring current through recurring substorms [*Mitchell et al.*, 2003; *Henderson et al.*, 2006]. Compared to ICME storms, sawtooth events are more moderate storms ($Dst \approx -150$ nT). Both event types are nonetheless stronger (in terms of *Dst* effects) than CIR/HSS induced storm periods. If any, a comparatively weak oxygen signature in ENA observations may be expected for the latter.

Since CIR/HSS induced storms tend to be associated with the declining phase of the solar cycle, any observation will be affected by the solar cycle variation of ionospheric oxygen outflow. This dependence of ring current oxygen content on the solar cycle has been reported from in situ ring current observations [*Pulkkinen et al.*, 2001]. The solar cycle dependence of outflow is, however, not a simple question of outflow density, but rather a matter of changed particle acceleration processes [e.g., *Yau et al.*, 1985; *Peterson et al.*, 2001].

Can we expect oxygen to be observed during CIR/HSS events? Simple estimates exist to estimate the amount of oxygen outflow and plasma sheet oxygen content based on solar wind input parameters. While not proof in itself, these estimates give an indication of the amount of oxygen available for delivery into the ring current by appropriate processes (convection, substorms). They can be compared to other events where oxygen ENAs have been observed. Plate 3 shows estimates of oxygen outflow density based on a linear relationship between ionospheric oxygen outflow and solar wind dynamic pressure found by *Elliott et al.* [2001] as well as an estimate of plasma sheet density mainly based on solar wind kinetic energy flux [*Lennartsson*, 1989, 1995]. Outflow density estimates are indicated by red circles, plasma sheet density estimates by purple triangles. The plot shows estimated outflow and plasma sheet oxygen densities for the peak solar wind dynamic pressure during the four events described above. In addition, we show the estimates for (1) 07 April 2004, the third time an inward polarity HSS first observed on 11 February 2004 interacts with Earth's magnetosphere, (2) 15 February 2004, a time period inside the 11 February 2004 CIR/HSS event when standard ICME indicators in the solar wind indicated the presence of CME conditions, (3) 21 October 2001, an ICME driven event for which strong ENA observations were reported [*Mitchell et al.*, 2003], and (4) 28 October 2001, another ICME with strong oxygen ENA emissions.

The 21 October event marks possibly the highest oxygen density in the plasma sheet ever measured by TIMAS [*Peterson, private communication*]. The peak oxygen plasma sheet density above 900 eV in the near-midnight plasma sheet around $L = 10$ was about 0.105 cm^{-3}, the peak total (H$^+$ + O$^+$) plasma sheet density above 900 eV about 0.235 cm^{-3}, compared to a 0.22 cm^{-3} oxygen density estimated according to *Lennartsson* [1989] and *Lennartsson* [1995]. For 28 October 2001, the peak oxygen plasma sheet density above 900 eV was 0.078 cm^{-3}, the total plasma sheet density 0.155 cm^{-3} [*Trattner, private communication*], this time compared to an estimate of 0.12 cm^{-3} for oxygen. Although the actual measurement is up to a factor of two smaller than the (simple) estimator of oxygen plasma sheet density, these numbers indicate that plasma sheet estimates plotted in Plate 3 can guide our expectations of oxygen ENA emissions—provided, oxygen can be energized and transported into the ring current.

All three CIR/HSS intervals with inward solar wind polarity (11 February, 09 March and 07 April 2004) have higher estimates of plasma sheet densities than the 28 October 2001 event. Since that ICME event had significant oxygen ENA emissions, albeit reduced compared to the 21 October 2001 ICME, we can reasonably expect that the source oxygen density is strong enough to make oxygen injections into the ring current possible. In principle, we should be able to observe oxygen ENAs during these events.

Oxygen ENA fluxes observed in February/March are plotted as blue squares in Plate 2. Oxygen measurements were treated the same way as hydrogen measurements, i.e., we limited data to within 7.5° of the dipole axis. Several observations can be made. First, we can clearly see that oxygen ENA fluxes are always weaker than hydrogen fluxes. Second, to lowest order, hydrogen and oxygen ENA fluxes track each other very well most of the time. Oxygen fluxes peak when *Dst* peaks, and secondary peaks occur concurrent with increased geomagnetic activities. Exceptions exist, for example during the CIR/HSS in late March 2003, when hydrogen emissions rise more then oxygen emissions. This supports the notion that tracking between both species is not believed to be an instrumental effect, i.e., there is no cross-talk between the hydrogen and oxygen signal suspected in HENA [*Ohtani et al.*, 2006]. Third, oxygen as well as hydrogen peak fluxes are clearly weaker during outward polarity CIR/HSS events.

The perhaps most interesting observation during all of these events is that the oxygen fluxes generally track the hydrogen fluxes. Oxygen should decay much quicker than hydrogen. It has a larger charge exchange rate, and its larger gyroradius makes loss to the dayside magnetopause more likely. One possible explanation for this seeming discrepancy could be the repeated delivery of fresh oxygen by substorm particle injections [e.g., *Ohtani et al.*, 2005]. Sawtooth event observations in particular highlight the difference between hydrogen and oxygen ENA emissions. Hydrogen emissions are rather steady from one sawtooth injection to the next, while oxygen emissions are highly time variable, showing strong injection signatures and rapid decay [*Mitchell et al.*, 2003; *Henderson et al.*, 2006]. Even though the cadence of ENA observations is chosen to be rather long (approx. 14 hours), the hydrogen and oxygen fluxes track each other for days throughout each of the four events presented. This suggests that oxygen plasma is available for charge exchange for extended amounts of time, possibly delivered repeatedly through substorm activity during the high-speed stream interval.

3. ENHANCED CONVECTION

Next to the density peak at the leading edge of a high-speed stream region, the other geoeffective component of CIR/HSS events is the presence of large-scale Alfvén waves

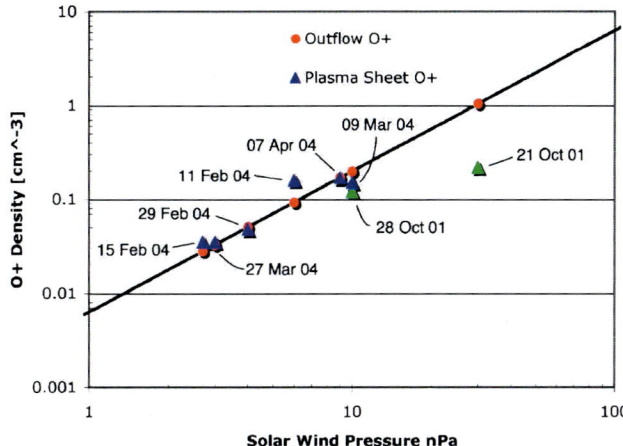

Plate 3. Estimates of oxygen ionospheric outflow density (circles) and plasma sheet density (triangles) as a function of solar wind input. Shown are maximum estimates for the four CIR/HSS events, additional two data points during the 2004 study interval, and two data points for oxygen ENA-rich ICME events in 2001 (green triangles).

in the solar wind. While the average IMF B_z component is near zero for high-speed stream intervals, these large-scale variations can produce repeated periods of sufficiently long southward IMF B_z, repeatedly providing significant energy input into the magnetosphere during several consecutive days. The most prominent result of this energy input is the build-up of relativistic electron fluxes in Earth's radiation belts. Energetic neutral atom observations are, of course, blind to this build-up. They should, however, be sensitive to the increased magnetospheric convection which goes hand-in-hand with southward IMF conditions. Changes in convection affect the plasma sheet, most notably the location of the inner edge of the plasma sheet. As convection increases, the electron plasma sheet has been shown to penetrate deeper into the inner magnetosphere [*Thomsen et al.*, 2002]. There is a good correlation between Kp and bulk plasma properties at geosynchronous orbit [*Korth et al.*, 1999; *Denton et al.*, 2005b], showing that the ion plasma sheet also responds to changes in magnetospheric convection. If ENAs are observed at plasma sheet energies of a few keV, it can be seen that ENA emission enhancements occur during southward IMF conditions and resulting extended substorm growth phases [*Jahn et al.*, 2006]. Those particular emissions do not mimic the injection of energetic particles regularly seen at geosynchronous altitudes during substorm onset, but respond rather slowly on timescales of 10's of minutes [*Pollock et al.*, 2003; *Jahn et al.*, 2006]. This should be expected based on *in situ* observations at geosynchronous altitudes for those lower energy particles [*Birn et al.*, 1997].

To study possible ENA observations of enhanced convection, we selected the summer of 2000. This was a geomagnetically very active period with three significant ICME triggered storms ($Dst_{min} \approx -200$ nT). While 2000 (i.e., solar maximum) is clearly not a optimum period for the study of high-speed stream events, several moderate high-speed stream intervals were present. More importantly, due to the IMAGE orbit with apogee high above the equatorial plane we have better long-term data coverage during times of high Alfvén wave activity than in 2004. We want to concentrate on two high-speed stream time intervals: 28 August through 02 September, and 25 September through 29 September 2000. These events are weak compared to the events in 2004, and they are dwarfed by the ICME induced storms that bracket them. Nonetheless, they are significant events when it comes to enhanced convection caused by solar wind Alfvén wave activity. Plate 4 summarizes the observations for this time period, in the same format as before. The ENA data in the bottom panel of Plate 4 represent MENA observations [*Pollock et al.*, 2000] of below 30 keV integrated night-side ENA rates roughly covering radial distances of geosynchronous orbit $\pm 3 R_E$. Due to the geocoronal density profile, these emissions stem predominantly from inside and including geosynchronous orbit. Data were acquired by integrating ENA rates from all nightside local times. A $\cos^2(\vartheta)$ correction (ϑ is the polar angle of the spacecraft position vector with respect to the z-axis) for the orbital motion of the spacecraft was applied, although those corrections do not necessarily improve data quality [*Ohtani et al.*, 2006]. Data were then binned into 3-hour time bins to mimic the cadence of Kp data. We also required observations to be from at least 5 R_E geocentric distance from Earth.

The main feature of these ENA observations is the excellent correlation with geomagnetic activity. To the casual observer, there appears to be a good correlation with Dst, similar to earlier reports of ENA-ring current correlations. However, closer inspection yields that the log of the medium energy ENA rates from the near-Earth tail correlates very well with Kp. In fact, the ENA-Kp correlation even holds during times when higher levels of Kp are not accompanied by drops in Dst (e.g., 24 August 2000). The correlation is good at a variety of geomagnetic activity levels and time scales. This suggests that the near-Earth tail ENA measurements are indeed sensitive to changes in Earth's plasma sheet, as was already used by *Denton et al.* [2005a] and *McComas et al.* [2002]. Together with the fact that Kp is a good measure of magnetospheric convection controlled by IMF B_z [*Thomsen*, 2004], this indicates that this type of ENA measurement can be utilized to determine the strength of magnetospheric convection directly from ENA nightside and tail imaging measurements.

4. SUMMARY AND CONCLUSIONS

The study of high-speed streams and corotating interaction regions has so far received little to no attention in the energetic neutral atom community. This may have several reasons. First, geomagnetic storms induced by recurring high-speed streams are considered "weak" storms with a less geoeffective solar wind—despite the fact that they build up significant relativistic electron populations of interest to Space Weather researchers and satellite operators. Second, since recurring storms result in a weaker ring current response, they ultimately also show a weaker ENA response from the ring current. The composition of the ring current is significantly less likely to be swayed towards heavier ions. This is driven by the characteristics of the solar wind during those events, but could also be attributed to the fact that recurrent geomagnetic storms occur mainly in the declining phase of the solar cycle. Escape of heavier ionospheric ions from the ionosphere into the magnetosphere is reduced, in turn reducing the amount of ionospheric oxygen in the ring current. The study of the role of oxygen in ring current dynamics is not best done during those times.

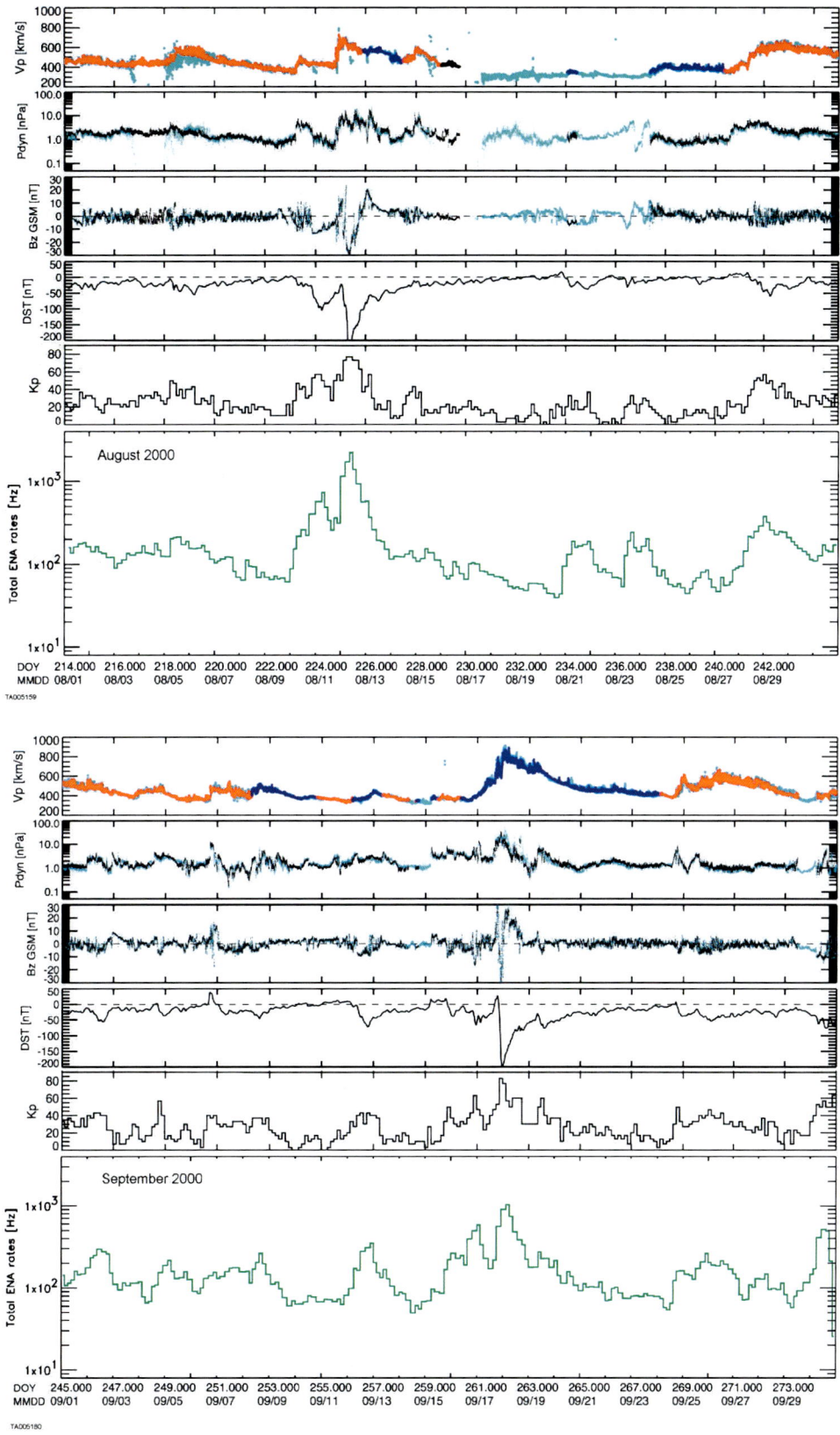

Plate 4. Overview over the summer 2000 enhanced convection signature in medium energy ENAs. From the top are shown (same format as before): solar wind speed with polarity encoded in color, dynamic pressure, IMF B_z, Dst, and Kp. The bottom panel shows integrated night-side medium energy ENAs below 30 keV around geosynchronous orbit, binned into Kp-like 3-hour time bins.

Observations of CIR/high-speed stream intervals with ENA images from the IMAGE spacecraft confirm observations made by other (usually *in situ*) methods, on a more global level. As could be expected from measurements of *Dst* during those events, a (hydrogen) ring current is present. ENA emissions on average stay elevated for several days during a high-speed stream event, indicating that ring current decays very slowly and is replenished repeatedly. Oxygen is present in the ring current, but emissions are again much weaker than during prominent ICME and sawtooth events. This agrees with the expectation that less oxygen plasma is available in the plasma sheet and the ionospheric outflow, though it is not clear how much the delivery of oxygen into the ring current (via substorms and convection) can be expected to be different from ICME and sawtooth events.

Aside from some expected observations during high-speed stream events, the study of recurrent storms even from the narrow perspective of energetic neutral atom imaging holds promise to some important new insight. High-speed streams have the potential of providing a significant interplanetary $E_y^{(sw)} = v_{sw} * B_z$ electric field component, giving rise to enhanced magnetospheric convection. Granted, the $E_y^{(sw)}$ electric field is not applied continuously as is the case during steady magentospheric convection (SMC) or during sawtooth events. It may, on average, not even be as strong. But the continuously high levels of AE and *Kp*, as well as the long duration of enhanced ENA observations demonstrate that the magnetosphere receives significant long-term energy input. The effects of that energy input *can* be followed with ENA imaging at energies where the particles dynamics are not magnetically but electrically dominated. There is a good correlation between *Kp* and medium energy ENA fluxes, indicating that at least in principle ENA fluxes could serve as a more direct, quantitative measure of magnetospheric convection and energy input into the magnetosphere.

In addition to monitoring convection directly using medium energy ENAs stemming from plasma sheet populations, the effect of $E_y^{(sw)}$ on the *position* of the ring current peak may be investigated as long as the ring current is sufficiently strong to be measured well. This provides an additional avenue to studying the ring current topology, covering different conditions than encountered during a strong ICME induced storm. This may help unravel the causes of the postmidnight storm-time enhancement of the ring current.

Acknowledgments. We are indebted to Andreas Isaksson for his help in preparing HENA data for analysis. We would like to thank the HENA team for supporting our data analysis efforts. We thank the ACE and Wind missions for providing access to solar wind measurements. This work was supported by the NASA IMAGE mission under NAS5-96020 and by NASA grant NAG5-11481 (JMJ) as well as by NASA grant NNG04-GR03G (HAE).

REFERENCES

Ashour-Abdalla, M., M. El-Alaoui, V. Peroomian, R.J. Walker, J. Raeder, L.A. Frank, and W.R. Paterson, Source distributions of substorm ions observed in the near-Earth magnetotail, *Geophys. Res. Let.*, 26, 955-958, 1999.

Ashour-Abdalla, M., et al., Ion sources and acceleration mechanisms inferred from local distribution functions, *Geophys. Res. Let.*, 24, 955-958, 1997.

Birn, J., M.F. Thomsen, J.E. Borovsky, G.D. Reeves, D.J. McComas, and R.D. Belian, Characteristic plasma properties during dispersionless substorm injections at geosynchronous orbit, *J. Geophys. Res.*, 102, 2309-2324, 1997.

Borovsky, J.E., M.F. Thomsen, and D.J. McComas, The superdense plasma sheet: Plasmaspheric origin, solar wind origin, or ionospheric origin?, *J. Geophys. Res.*, 102, 22089-22106, 1997.

Borovsky, J.E., M.F. Thomsen, R.C. Elphic, T.E. Cayton, and D.J. McComas, The transport of plasma sheet material from the distant tail to geosynchronous orbit, *J. Geophys. Res.*, 103, 20297-20332, 1998.

Brandt, P.C., D.G. Mitchell, E.C. Roelof, and J.L. Burch, Bastille Day storm: Global response of the terrestrial ring current, *Solar Physics*, 204, 377-386, 2001.

Brandt, P.C., D.G. Mitchell, Y. Ebihara, B.R. Sandel, E.C. Roelof, J.L. Burch, and R. Demajistre, Global IMAGE/HENA observations of the ring current: Examples of rapid response to IMF and ring current-plasmasphere interaction, *J. Geophys. Res.*, 107, 12-1, 2002a.

Brandt, P.C., R. Demajistre, E.C. Roelof, S. Ohtani, D.G. Mitchell, and S. Mende, IMAGE/high-energy energetic neutral atom: Global energetic neutral atom imaging of the plasma sheet and ring current during substorms, *J. Geophys. Res.*, 107, 21-1, 2002b.

Brandt, P.C., S. Ohtani, D.G. Mitchell, M.-C. Fok, E.C. Roelof, and R. Demajistre, Global ENA observations of the storm mainphase ring current: Implications for skewed electric fields in the inner magnetosphere, *Geophys. Res. Let.*, 29, doi:10.1029/2002GL015160, 2002.

Brandt, P.C., D.G. Mitchell, S. Ohtani, R. Demajistre, E.C. Roelof, J.-M. Jahn, C. Pollock, and G. Reeves, *Storm-Substorm relationships during the 4 October, 2000 Storm. IMAGE Global ENA Imaging Results*, pp. 103-118, Disturbances in Geospace, Geophysical Monograph Series, vol. 142, 2003.

Burch, J.L., IMAGE mission overview, *Space Sci. Rev.*, 91, 1-14, 2000.

Chappell, C.R., T.E. Moore, and J.H. Waite, The ionosphere as a fully adequate source of plasma for the earth's magnetosphere, *J. Geophys. Res.*, 92, 5896-5910, 1987.

Crooker, N.U., Solar and heliospheric geoeffective disturbances, *jatp*, 62, 1071-1085, 2000.

Daglis, I.A., *The Role of Magnetosphere-Ionosphere Coupling in Magnetic Storm Dynamics*, p107, Magnetic Storms, Geophysical Monograph Series, vol. 98, 1997.

Daglis, I.A., and W.I. Axford, Fast ionospheric response to enhanced activity in geospace: Ion feeding of the inner magnetotail, *J. Geophys. Res.*, 101, 5047-5066, 1996.

Daglis, I.A., R.M. Thorne, W. Baumjohann, and S. Orsini, The terrestrial ring current: Origin, formation, and decay, rg, 37, 407-438, 1999.

Delcourt, D.C., J.L. Horwitz, and K.R. Swinney, Influence of the interplanetary magnetic field orientation on polar cap ion trajectories - Energy gain and drift effects, *J. Geophys. Res.*, 93, 7565-7570, 1988.

Delcourt, D.C., T.E. Moore, J.H. Waite, and C.R. Chappell, Polar wind ion bands after neutral sheet acceleration, *J. Geophys. Res.*, 94, 3773-3778, 1989.

Delcourt, D.C., T.E. Moore, and C.R. Chappell, Contribution of low-energy ionospheric protons to the plasma sheet, *J. Geophys. Res.*, 99, 5681-5689, 1994.

DeMajistre, R., E.C. Roelof, P. C:son Brandt, and D.G. Mitchell, Retrieval of global magnetospheric ion distributions from high-energy neutral atom measurements made by the IMAGE/HENA instrument, *J. Geophys. Res.*, 109, 4214, 2004.

Denton, M.H., V.K. Jordanova, M.G. Henderson, R.M. Skoug, M.F. Thomsen, C.J. Pollock, S. Zaharia, and H.O. Funsten, Storm-time plasma signatures observed by IMAGE/MENA and comparison with a global physics-based model, *Geophys. Res. Let.*, 32, 17102, 2005a.

Denton, M.H., M.F. Thomsen, H. Korth, S. Lynch, J.C. Zhang, and M.W. Liemohn, Bulk plasma properties at geosynchronous orbit, *J. Geophys. Res.*, 110, 7223, 2005b.

Ebihara, Y., and M. Ejiri, Simulation study on fundamental properties of the storm-time ring current, *J. Geophys. Res.*, 105, 15843-15860, 2000.

Ebihara, Y., and M.-C. Fok, Postmidnight storm-time enhancement of tens-of-keV proton flux, *J. Geophys. Res.*, 109, 12209, 2004.

Ebihara, Y., M. Ejiri, and H. Miyaoka, Simulation on ring current formation: A case study of a storm on February 13, 1972, *Proceedings of the NIPR Symposium on Upper Atmosphere Physics, No. 12. December, 1998. Papers presented at the 21st Symposium on Coordinated Observations of the Ionosphere and the Magnetosphere in the Polar Regions held 24-25 July, 1997 at NIPR, Tokyo, Edited by Hisao Yamagishi, National Institute of Polar Research, Tokyo.*, p1, 12, 1, 1998.

Ebihara, Y., S. Barabash, and M. Ejiri, On the global production rates of energetic neutral atoms (ENAs) and their association with the Dst index, *Geophys. Res. Let.*, 26, 2929-2932, 1999.

Ebihara, Y., M.-C. Fok, R.A. Wolf, M.F. Thomsen, and T.E. Moore, Nonlinear impact of plasma sheet density on the storm-time ring current, *J. Geophys. Res.*, 110, 2208, 2005.

Elliott, H.A., R.H. Comfort, P.D. Craven, M.O. Chandler, and T.E. Moore, Solar wind influence on the oxygen content of ion outflow in the high-altitude polar cap during solar minimum conditions, *J. Geophys. Res.*, 106, 6067-6084, 2001.

Fok, M.-C., et al., Global ENA Image Simulations, *Space Sci. Rev.*, 109, 77-103, 2003.

Fu, S.Y., Q.-G. Zong, B. Wilken, and Z.Y. Pu, Temporal and Spatial Variation of the Ion Composition in the Ring Current, *ssr*, 95, 539-554, 2001.

Greenspan, M.E., and D.C. Hamilton, Relative contributions of H^+ and O^+ to the ring current energy near magnetic storm maximum, *J. Geophys. Res.*, 107, 3-1, 2002.

Henderson, M.G., G.D. Reeves, H.E. Spence, R.B. Sheldon, A.M. Jorgensen, J.B. Blake, and J.F. Fennell, First energetic neutral atom images from Polar, *Geophys. Res. Let.*, 24, 1167-1170, 1997.

Henderson, M.G., G.D. Reeves, R. Skoug, M.F. Thomsen, M.H. Denton, S.B. Mende, T.J. Immel, P.C. Brandt, and H.J. Singer, Magnetospheric and auroral activity during the 18 april 2002 sawtooth event, *J. Geophys. Res.*, 111, 2006.

Hovestadt, D., and M. Scholer, Radiation belt-produced energetic hydrogen in interplanetary space, *J. Geophys. Res.*, 81, 5039-5042, 1976.

Huang, C.-S., J.C. Foster, G.D. Reeves, H.U. Frey, C.J. Pollock, and J.-M. Jahn, Periodic magnetospheric substorms: Multiple space-based and ground-based instrumental observations, *J. Geophys. Res.*, 108, doi:10.1029/2003JA009992, 2003.

Jahn, J.-M., C.J. Pollock, T.J. Immel, and S.B. Mende, Spatial correlation of precipitating and trapped protons associated with an isolated substorm, *Geophys. Res. Let.*, 33, doi:10.1029/2006GL025917, 2006.

Jordanova, V.K., C.J. Farrugia, J.M. Quinn, R.B. Torbert, J.E. Borovsky, R.B. Sheldon, and W.K. Peterson, Simulation of off-equatorial ring current ion spectra measured by Polar for a moderate storm at solar minimum, *J. Geophys. Res.*, 104, 429-436, 1999a.

Jordanova, V.K., R.B. Torbert, R.M. Thorne, H.L. Collin, J.L. Roeder, and J.C. Foster, Ring current activity during the early $B_z < 0$ phase of the January 1997 magnetic cloud, *J. Geophys. Res.*, 104, 24895-24914, 1999b.

Jordanova, V.K., et al., October 1995 magnetic cloud and accompanying storm activity: Ring current evolution, *J. Geophys. Res.*, 103, 79-92, 1998.

Jorgensen, A.M., H.E. Spence, M.G. Henderson, G.D. Reeves, M. Sugiura, and T. Kamei, Global Energetic Neutral Atom (ENA) measurements and their association with the Dst index, *Geophys. Res. Let.*, 24, 3173-3176, 1997.

Jorgensen, A.M., M.G. Henderson, E.C. Roelof, G.D. Reeves, and H.E. Spence, Charge exchange contribution to the decay of the ring current, measured by energetic neutral atoms (ENAs), *J. Geophys. Res.*, 106, 1931-1938, 2001.

Korth, A., R.H.W. Friedel, F. Frutos-Alfaro, C.G. Mouikis, and Q. Zong, Ion composition of substorms during storm-time and non-storm-time periods, *jatp*, 561-566, 2002.

Korth, H., M.F. Thomsen, J.E. Borovsky, and D.J. McComas, Plasma sheet access to geosynchronous orbit, *J. Geophys. Res.*, 104, 25047-25062, 1999.

Kozyra, J.U., V.K. Jordanova, J.E. Borovsky, M.F. Thomsen, D.J. Knipp, D.S. Evans, D.J. McComas, and T.E. Cayton, Effects of a high-density plasma sheet on ring current development during the November 2-6, 1993, magnetic storm, *J. Geophys. Res.*, 103, 26285-26306, 1998.

Kozyra, J.U., M.W. Liemohn, C.R. Clauer, A.J. Ridley, M.F. Thomsen, J.E. Borovsky, J.L. Roeder, V.K. Jordanova, and W.D. Gonzalez, Multistep Dst development and ring current composition changes during the 4-6 June 1991 magnetic storm, *J. Geophys. Res.*, 107, 33-1, 2002.

Lennartsson, O.W., Statistical investigation of IMF B_z effects on energetic (0.1- to 16-keV) magnetospheric O^+ ions, *J. Geophys. Res.*, 100, 23621-23636, 1995.

Lennartsson, W., Energetic (0.1- to 16-keV/e) magnetospheric ion composition at different levels of solar F10.7, *J. Geophys. Res.*, 94, 3600-3610, 1989.

Lennartsson, W., A scenario for solar wind penetration of earth's magnetic tail based on ion composition data from the ISEE 1 spacecraft, *J. Geophys. Res.*, 97, 19221, 1992.

Liemohn, M.W., J.U. Kozyra, M.F. Thomsen, J.L. Roeder, G. Lu, J.E. Borovsky, and T.E. Cayton, Dominant role of the asymmetric ring current in producing the stormtime Dst^*, *J. Geophys. Res.*, 106, 10883-10904, 2001.

McComas, D.J., P. Valek, J.L. Burch, C.J. Pollock, R.M. Skoug, and M.F. Thomsen, Filling and emptying of the plasma sheet: Remote observations with 1-70 keV energetic neutral atoms, *Geophys. Res. Let.*, 29, 36-1, 2002.

Mitchell, D.G., P. C:Son Brandt, E.C. Roelof, D.C. Hamilton, K.C. Retterer, and S. Mende, Global imaging of O^+ from IMAGE/HENA, *ssr*, 109, 63-75, 2003.

Mitchell, D.G., P.C. Brandt, and S.B. Mende, Oxygen in the ring current during major storms, *Advances in Space Research*, 36, 1758-1761, 2005.

Mitchell, D.G., et al., High energy neutral atom (HENA) imager for the IMAGE mission, *ssr*, 91, 67-112, 2000.

Moore, T.E., et al., Ionospheric mass ejection in response to a CME, *Geophys. Res. Let.*, 26, 2339-2342, 1999.

Nosé, M., A.T.Y. Lui, S. Ohtani, B.H. Mauk, R.W. McEntire, D.J. Williams, T. Mukai, and K. Yumoto, Acceleration of oxygen ions of ionospheric origin in the near-Earth magnetotail during substorms, *J. Geophys. Res.*, 105, 7669-7678, 2000.

Ohtani, S., P.C. Brandt, D.G. Mitchell, H. Singer, M. Nosé, G.D. Reeves, and S.B. Mende, Storm-substorm relationship: Variations of the hydrogen and oxygen energetic neutral atom intensities during storm-time substorms, *J. Geophys. Res.*, 110, 7219, 2005.

Ohtani, S., P.C. Brandt, H.J. Singer, D.G. Mitchell, and E.C. Roelof, Statistical characteristics of hydrogen and oxygen ENA emissions from the storm-time ring current, *J. Geophys. Res.*, 111, doi:10.1029/2005JA011201, 2006.

Perez, J.D., G. Kozlowski, P.C. Brandt, D.G. Mitchell, J.-M. Jahn, C.J. Pollock, and Z.Z. Zhang, Initial ion equatorial pitch angle distributions from medium and high energy neutral atom images obtained by IMAGE, *Geophys. Res. Let.*, 28, 1155-1158, 2001.

Peterson, W.K., H.L. Collin, A.W. Yau, and O.W. Lennartsson, Polar/Toroidal Imaging Mass-Angle Spectrograph observations of suprathermal ion outflow during solar minimum conditions, *J. Geophys. Res.*, 106, 6059-6066, 2001.

Pollock, C.J., et al., Medium energy neutral atom (MENA) imager for the IMAGE mission, *ssr*, 91, 113-154, 2000.

Pollock, C.J., et al., First medium energy neutral atom (MENA) images of Earth's magnetosphere during substorm and storm-time, *Geophys. Res. Let.*, 28, 1147-1150, 2001.

Pollock, C.J., et al., The Role and Contributions of Energetic Neutral Atom (ENA) Imaging in Magnetospheric Substorm Research, *ssr*, 109, 155-182, 2003.

Pulkkinen, T.I., et al., Ring current ion composition during solar minimum and rising solar activity: Polar/CAMMICE/MICS results, *J. Geophys. Res.*, 106, 19131-19148, 2001.

Reeves, G.D., et al., *IMAGE, POLAR and Geosynchronous Observations of Substorm and Ring Current Ion Injection*, Disturbances in Geospace, Geophysical Monograph Series, pp. 91-102, vol. 142, 2003.

Roelof, E.C., Energetic neutral atom image of a storm-time ring current, *Geophys. Res. Let.*, 14, 652-655, 1987.

Roelof, E.C., and A.J. Skinner, Extraction of ion distributions from magnetospheric ENA and EUV images, *ssr*, 91, 437-459, 2000.

Roelof, E.C., D.G. Mitchell, and D.J. Williams, Energetic neutral atoms (E approximately 50 keV) from the ring current - IMP 7/8 and ISEE 1, *J. Geophys. Res.*, 90, 10991, 1985.

Sauvaud, J.A., and D. Delcourt, A numerical study of suprathermal ionospheric ion trajectories in three-dimensional electric and magnetic field models, *J. Geophys. Res.*, 92, 5873-5884, 1987.

Skoug, R.M., et al., Tail-dominated storm main phase: 31 March 2001, *J. Geophys. Res.*, 108, 23-1, 2003.

Thomsen, M.F., Why Kp is such a good measure of magnetospheric convection, *Space Weather*, 2, 11004, 2004.

Thomsen, M.F., J.E. Borovsky, D.J. McComas, and M.R. Collier, Variability of the ring current source population, *Geophys. Res. Let.*, 25, 3481-3484, 1998.

Thomsen, M.F., H. Korth, and R.C. Elphic, Upper cutoff energy of the electron plasma sheet as a measure of magnetospheric convection strength, *J. Geophys. Res.*, 107, 25-1, 2002.

Tsurutani, B.T., W.D. Gonzalez, A.L.C. Gonzalez, F. Tang, J.K. Arballo, and M. Okada, Interplanetary origin of geomagnetic activity in the declining phase of the solar cycle, *J. Geophys. Res.*, 100, 21717-21734, 1995.

Turner, N.E., and E.J. Mitchell, *Energetics of Magnetic Storms Associated with High Speed Streams: a Study of Geoeffectiveness, this issue*, Corotating Solar wind Streams and Recurring Geomagnetic Activity, Geophysical Monograph Series, 2006.

Winglee, R.M., Circulation of ionospheric and solar wind particle populations during extended southward interplanetary magnetic field, *J. Geophys. Res.*, 108, doi:10.1029/2002JA009819, 2003.

Yau, A.W., P.H. Beckwith, W.K. Peterson, and E.G. Shelley, Long-term (solar cycle) and seasonal variations of upflowing ionospheric ion events at DE 1 altitudes, *J. Geophys. Res.*, 90, 6395-6407, 1985.

J.-M. Jahn and H.A. Elliott, Space Science Department, Southwest Research Institute, 6220 Culebra Rd., San Antonio, Tx 78238

Global Auroral Response to Interplanetary Media With Emphasis on Solar Wind Dynamic Pressure Enhancements

Kan Liou

The Johns Hopkins University Applied Physics Laboratory, Laurel, Maryland, USA

This paper reviews some important progress made on large-scale auroral dynamics under a variety of interplanetary conditions with emphasis on solar wind dynamic pressure increases. In the last decade, the availability of high time and spatial resolution far-ultraviolet auroral images from Polar has allowed such work to be performed statistically. Like geomagnetic disturbances, the southward component of interplanetary magnetic fields (IMFs) and the solar wind velocity are the major contributors to the hemispheric auroral power through magnetic merging/ reconnection and solar wind viscous processes, respectively. Global auroral imaging from Polar has also demonstrated the importance of sudden increases in the solar wind dynamic pressure. When the magnetosphere is compressed, auroral transients (<10–20 min) are produced first on the dayside and then on the nightside of the oval. The dayside auroral transient is associated mainly with enhancements in both energy and energy flux of precipitating electrons from the plasma sheet. On the nightside the compression-induced aurora is more intense, especially during southward IMF, and often leads to widening of the oval and the closure of the polar cap. Observations seem to suggest an increased efficiency in the solar wind-magnetosphere coupling and an enhancement in the convection-related DP 2 current system by shocks, but whether or not the magnetospheric compression can enhance dayside merging/nightside reconnection and/or viscous processes remains uncertain. To evaluate the importance of magnetospheric compression on large-scale aurora, 27 co-rotating interaction region (CIR) events, which are an important source of high solar wind plasma density in the solar wind, are studied. A superposed epoch analysis of the solar wind plasma data and auroral power during the 27 CIR events indicates a modest increase (80%) in the global auroral power across the CIRs. Global auroral power is also more intense in a high-speed stream than in a low-speed stream. When these results and a high recurrent rate are combined, the CIR and its high-speed downstream region are an important interplanetary event that can exert a profound impact on the auroral ionosphere.

1. INTRODUCTION

The aurora is the result of the bombardment of the upper atmosphere by magnetically guided energetic particles, which appear most frequently in a continuous circular band known as the "auroral oval" in the high-latitude polar region [*Feldstein*, 1960]. Statistically speaking, the occurrence of aurora is clustered in the premidnight sector [e.g., *Newell et al.*, 1996; *Liou et al.*, 1997], the postnoon sector between 1400 and 1600 magnetic local time (MLT) [*Cogger et al.*, 1977], and the prenoon sector [*Meng and Lundin*, 1986] (see Figure 1). Many popular and basic issues of magnetospheric physics that are linked directly or indirectly to these regions demonstrate the importance of these three optically and energetically intense auroral sectors. For example, the premidnight sector is the most aurorally intense region because auroral substorms take place predominantly there [*Akasofu*, 1964; *Craven and Frank*, 1991; *Liou et al.*, 2001]. The second most intense region of aurora is at ~1500 MLT, where spatially periodic auroral "hot" spots (i.e., 1500 MLT auroral bright spots [*Liou et al.*, 1999]) are frequently observed [*Lui et al.*, 1989; *Vo and Murphree*, 1995; *Liou et al.*, 1997]. These spots are associated with the dayside maximum of the region-1 upward field-aligned currents (FACs) [*Iijima and Potemra*, 1976]. The third most intense region is located in the morning sector at ~0900 MLT and coincides with the mantle or "region 0" currents out of the ionosphere [*Newell et al.*, 1996].

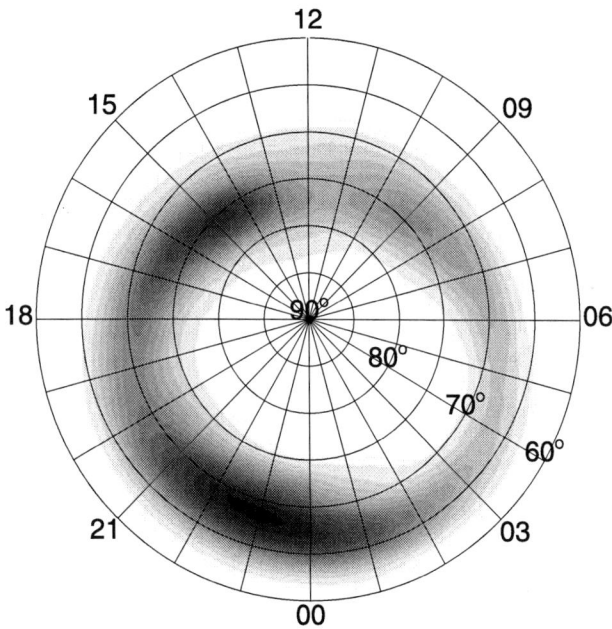

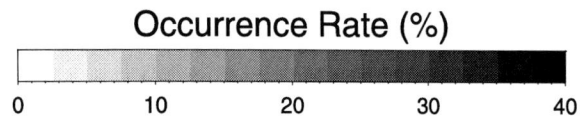

Figure 1. Occurrence rate of northern hemispheric aurora with an energy flux threshold of 2 ergs/cm^2-s in the typical geomagnetic latitude-local time format. The map is derived from ~17,000 Polar ultraviolet imager (UVI) images in the N_2 Lyman-Birge-Hopfield (LBH) long band (160–180 nm).

Such a global auroral morphology was first demonstrated by long-time averages of the global images acquired from the Polar ultraviolet imager (UVI) [*Torr et al.*, 1995] during the solar minimum year (1 April – 28 July 1996) [*Liou et al.*, 1997]. Auroras in each of the three regions have distinct characteristics from each other and respond to the solar wind plasma and interplanetary magnetic field (IMF) conditions in different ways. Different processes may be responsible for auroras occurring in different regions. Current knowledge about the auroras in these three regions is still limited to their morphological characteristics. The first half of this paper will review previous work pertaining to the response of the large-scale auroras in these three regions to the solar wind plasma and IMF. We will then focus on the solar wind dynamic pressure effect on the auroral dynamics. At the end we will discuss the response of global auroral power to co-rotating interaction regions (CIRs).

2. AURORAL RESPONSE TO SOLAR WIND

The aurora is to some extent an optical manifestation of the response of the magnetosphere-ionosphere system to the interplanetary media. Studying auroral characteristics under different solar wind plasma and IMF conditions is a common practice in auroral research. The general goal is to understand the driving processes of the magnetosphere and how energy is fed into the magnetosphere from the solar wind. Physical processes responsible for generating auroral arcs have been thoroughly reviewed by *Borovsky* [1993] and will not be elaborated here.

2.1. Interplanetary Magnetic Field

The north-south component of the IMF plays a major role in the solar wind-magnetosphere coupling that drives the large-scale convection [*Dungey*, 1961]. The kinetic energy of the solar wind is channeled through the interconnected Earth's magnetic field lines and the IMF in the form of an electric field imposed across the magnetosphere and is subsequently released partially into the Earth's upper atmosphere in the form of Joule and particle heating and partially into the solar wind as a plasmoid by reconnection in the nightside magnetotail. This unique process drives the global-scale convection and is believed to be responsible for major geomagnetic disturbances. As geomagnetic activity increases, the size of the auroral oval increases [*Feldstein and Starkov*, 1967] and the occurrence of auroral arcs also increases [*Danielsen*, 1980].

The work of *Lassen and Danielsen* [1978] provides the first solid evidence of a close relationship between the auroral distribution and IMF orientation. Using an extensive database acquired by the Greenland all-sky camera network, they found that the occurrence rate of arcs is higher when IMF B_z is southward and becomes lower after IMF B_z turns northward. They also found that the size of the auroral oval contracts for northward IMF and expands for southward IMF, which is further supported by *in situ* particle measurements from DMSP satellites [*Holzworth and Meng*, 1984]. Because of the spatial limit in ground-based auroral imaging, a large number of observations are required in order to establish such a statistical result. These results were confirmed by auroral images from the high-latitude DE-1 satellite on an instantaneous basis [*Craven and Frank*, 1991].

The y-component of IMF is thought to affect the dayside auroral morphology most and is often attributed to the dawn-dusk asymmetry of auroral distribution. Early case studies have shown that transient auroral events moving eastward or westward in the midday sector are controlled by the polarity of IMF B_y [*Sandholt et al.*, 1986]. A statistical study performed later by *Karlson et al.* [1996] indicated that an asymmetric occurrence distribution of the prenoon-postnoon aurora is associated with the IMF B_y polarity during negative IMF B_z. Postnoon (1200–1400 MLT) auroral events are more frequent than prenoon (1000–1200 MLT) auroral events for negative IMF B_y, while the trend is reversed for positive IMF B_y.

The far-ultraviolet (FUV) auroral imager on-board the Viking satellite was the first true two-dimensional high-time and high-spatial resolution camera. It provided more detailed information about the temporal and spatial responses of aurora to the solar wind variations. Based on Viking FUV images, *Murphree and Elphinston* [1988] reported that during positive IMF B_z and away sector configurations, localized regions of auroral emission on the dayside could propagate either duskward or dawnward depending on the sign of IMF B_y. This pattern is usually linked to the response of the convection pattern to the y-component of IMF when dayside merging is at work. Studies of Viking FUV images have generally shown that the occurrence of afternoon arcs depends strongly on the existence of a negative IMF B_y component [*Murphree et al.*, 1981; *Elphinstone et al.*, 1990; *Vo and Murphree*, 1995; *Karlson et al.*, 1996] but not on IMF B_z [*Vo and Murphree*, 1995]. Because the duskside convection cell is more crescent shaped for IMF $B_y < 0$ than for IMF $B_y > 0$ [*Ruohoniemi and Greenwald*, 1996], stronger upward FACs and therefore more intense electron precipitation are generally expected in the afternoon sector. Partial "penetration" of IMF B_y into the dayside magnetosphere [e.g., *Cowley et al.*, 1991; *Newell et al.*, 1995; *Wing et al.*, 1995] has also been proposed to explain the preferred negative IMF B_y for the afternoon aurora [*Liou et al.*, 1998]. A non-zero curl of IMF B_y in the z direction associated with a gradual decrease in the earthward penetration of IMF B_y may induce an interhemispheric FAC. This current flows into the Northern Hemisphere for IMF $B_y > 0$ and into the Southern Hemisphere for IMF $B_y < 0$. Such a current system would introduce a hemispherical asymmetry in the dayside aurora. Further studies are needed to justify this line of thought.

However, *Liou et al.* [1998] correlated instantaneous northern hemispheric auroral power, inferred from ~17,000 Polar UVI images, with concurrent solar wind plasma and IMF parameters and found that postnoon (1300–1800 MLT) auroral enhancements are mainly associated with a large transverse component of the solar wind electric field, $(V \times B)_\perp$. An increase in the total auroral power occurs not only as the IMF B_y becomes more negative but also as the IMF B_y becomes more positive. The preferred postnoon merging for IMF $B_y > 0$ may be the main cause [*Crooker*, 1979]. Again, an implied hemispherical asymmetry in the dayside aurora is expected but has yet to be confirmed.

Auroral power deposited in the premidnight sector (1900–0100 MLT) has, not unexpectedly, a maximum response to a negative IMF B_z component when the IMF is lagged (shifted forward in time) by 60 min (see Figure 2) and is approximately linearly proportional to the magnitude of negative IMF B_z with a slope of 2 GW/nT [*Liou et al.*, 1998]. A higher nightside auroral power was often observed for IMF $B_y < 0$ than for IMF $B_y > 0$ [*Liou et al.*, 1998; *Liou et al.*, 1999; *Shue et al.*, 2002]. The negative IMF B_y effect may be caused by partial penetration of IMF B_y into the nightside, closed magnetic field [*Stenbaek-Nielsen and Otto*, 1997]. The 60-min lag time for a maximum auroral response suggests a mechanism of loading-unloading solar wind energy responsible for the nightside aurora. The strong IMF B_z effect supports the general concept that auroral activities are highly correlated with geomagnetic activities.

2.2. Solar Wind Velocity

In addition to the IMF, the solar wind velocity is also an important parameter in the transport of mass, momentum, and energy of solar wind to the magnetosphere through "viscous-like interaction" [*Axford and Hines*, 1961] and Kelvin-Helmholtz instability [e.g., *Hasegawa et al.*, 2004, and references therein]. A new type of interaction between the solar wind and the Earth's magnetosphere in association with the freestream turbulence effect is also being introduced [*Borovsky and Steinberg*, this volume, 2006, and references therein].

Although the solar wind electric field is a linear function of both the solar wind velocity and the z-component of the IMF, the velocity effect is usually smaller than the IMF effect. This is probably because the solar wind velocity effect is folded into the solar wind dawn-dusk electric field and the solar wind dynamic pressure. The lack of dynamic range in the solar wind velocity may also be a probable cause.

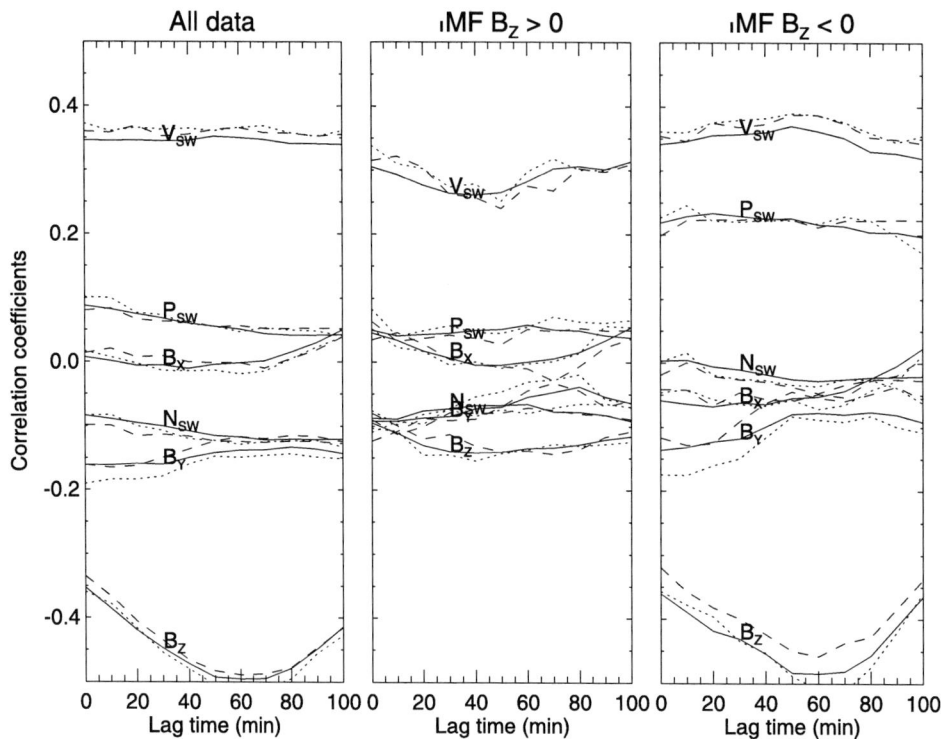

Figure 2. Lagged cross-correlation analysis for six solar wind plasma and IMF parameters acquired from Wind and nightside auroral power, inferred from Polar UVI LBH long band images integrated between 60° and 75° MLAT and 2000 and 0100 MLT, for the (right) entire data set, (middle) $B_z > 0$, and (left) $B_z < 0$. In each panel the solid lines represent the whole data set whereas the dotted lines and dashed lines represent two randomly divided data subsets (from *Liou et al.* [1998]).

Nonetheless, an approximately linear relationship is also found between the solar wind speed and the nightside auroral power [*Liou et al.*, 1998]. The effect of solar wind speed on the auroral power always occurs regardless of the sign of IMF B_z, and it plays the dominant role at times during northward IMF conditions (see Figure 2, middle panel).

The solar wind velocity probably plays a key role in the production of dayside aurora and auroral transients. *Vo and Murphree* [1995] investigated dayside auroras during 68 Viking passes and found that auroral bright spots occur most frequently in the afternoon sector and during high-speed solar wind (>500 km/s). The spatially periodic appearance of the bright spots has been attributed to the Kelvin-Helmholtz instability acting on the low-latitude boundary layer (LLBL) [*Lui et al.*, 1989; *Potemra et al.*, 1990; *Vo and Murphree*, 1995]. The LLBL is probably not the main source region. A later study of the plasma source of 65 postnoon bright events using DMSP particle data showed that the majority (2/3) of events were associated with the plasma sheet and the rest of events were associated with precipitation from various magnetospheric plasma boundary regions [*Liou et al.*, 1999].

2.3. Solar Wind Dynamic Pressure

The solar wind dynamic pressure does not play a major role in the coupling of solar wind parameters compared to other factors such as the z-component of the IMF and the solar wind velocity. Similarly, the dynamic pressure effect on auroral morphology is generally small except during southward IMF (see Figure 2) and/or during extreme conditions such as interplanetary shocks and large pressure pulses. The significance of sudden changes in the solar wind ram pressure on the dynamics of the magnetosphere-ionosphere system was not appreciated until the ISTP era. Intensive studies of the 6–11 January 1997, coronal mass ejection (CME)/magnetic cloud (MC) event demonstrated global auroral enhancements associated with the crossing of interplanetary shocks and pressure pulses [e.g., *Spann et al.*, 1998; *Zhou and Tsurutani*, 1999].

2.3.1. Interplanetary shocks. *Craven et al.* [1986] were the first to present events of global auroral intensification after the magnetospheric compression during both southward and northward conditions. The intensity of the auroral luminosities at dawn, dusk, and midnight seen by DE-1 showed significant increases (2–4 times) after the shock crossings. Compression-induced auroras are transient and are resolved later with higher-speed imagers such as those on board the Polar satellites. Using Polar UVI images, Zhou and Tsurutani [1999] identified a new type of auroral transient that is characterized by a sudden brightening of the aurora on either side or both sides of noon at the time of shock arrival at the magnetosphere. Plate 1 shows one of the events they studied, which occurred on 10 January 1997. Before the shock arrived at the magnetopause, the aurora was very faint due to northward IMF (Plate 1a). Upon arrival of the shock (Plate 1b), dayside auroras brightened (bottom left), and then the brightened region expanded and moved anti-sunward along the dawn and dusk flanks of the oval with a speed roughly corresponding to the speed of the shock in the solar wind [Zhou and Tsurutani, 1999]. The initial brightening was very localized but it immediately propagated/extended to a much wider local time and could sometimes reach the midnight sector, occupying most of the oval in ~20 min [Zhou and Tsurutani, 1999].

The bottom right panels of Plates 1(a–d) show the ratio of Lyman-Birge-Hopfield (LBH) long (160–180 nm) to LBH short (140–160 nm) auroral intensity. The LBH long to short emission ratio is proportional to the average energy of precipitating electrons. Because the UVI LBH short pass band resides in the peak Schumann-Runger O_2 absorption band and the LBH long pass band does not, LBH short photons produced by higher-energy particles that penetrate deeper experience more absorption than LBH long photons. Such an idea was first propagated by Strickland et al. [1983], and the application of this technique to UVI images was developed by, e.g., Germany et al. [1998].

An interesting finding is that there is no significant change in the energy of precipitating electrons except for those at the low-latitude part of the auroral transient after the shock crossing. This result is to some extent consistent with the FAST and DMSP particle data reported by Zhou et al. [2003], in which the typical sub-keV structured electrons at higher latitudes also show a mild increase. Because of the sensitivity of the UVI and because of the large binning of the image pixels, the small increase in the energy of the sub-keV electrons may not be distinguishable by UVI. However, one needs to bear in mind that in situ measurements cannot distinguish between spatial and temporal effects.

Magnetospheric compression can also produce auroras from particle sources deep in the inner magnetosphere. Liou et al. [2002] reported the existence of midday sub-auroral patches (MSPs), which are characterized by a sudden brightening of auroral patches in the midday sub-auroral zone upon the arrival of an interplanetary shock. MSPs have a low occurrence rate (12.5%) and a short lifetime of an order of ~56 min and are confined in local noon from 1000 to 1400 UT. Based on a fortuitous event with concurrent DMSP particle observations, MSPs were associated with dayside extension of central plasma sheet (CPS) particles (i.e., diffuse aurora).

The effect of magnetospheric compression is not limited to the dayside. Compression of the magnetosphere by shocks may enhance auroral activity in the night sector [Zhou and Tsurutani, 2001; Liou et al., 2003; Boudouridis et al., 2003], widen the auroral zone width, and reduce the polar cap size [Boudouridis et al., 2003; 2004].

2.3.2. Pressure pulses. Solar wind dynamic pressure pulses represent another type of solar wind plasma extreme and are often associated with magnetic cavities. These pulses occur more frequently than shocks and are often observed during interplanetary events. Here we present the 8 November 2000 event to demonstrate their geoeffectiveness. According to observations from the Wind spacecraft, which was in the solar wind ~70–80 R_E upstream from the Earth, a magnetic cloud (from 2200 UT, 6 November to 1800 UT, 7 November) preceded the events by several hours. Inside the cloud the magnetic field was initially large and southward (~–10 nT). The magnetic field turned smoothly northward into the second half of the cloud and stayed large (~15 nT) for more than 18 hours. After the passage of the cloud, Wind observed a number of "pressure balanced" magnetic cavities because they were accompanied by large density/pressure pulses (see Figures 3a and 3b). Inside the cavities, the z-component of 1-min averages of the magnetic fields sharply decreased but remained positive for most of the time. The large negative IMF B_z component in the magnetic cloud sheath caused a large magnetic storm ($Dst < \sim$–160 nT), which subsided several hours before the event time period (02–12 UT, 8 November 2000). The planetary K_P index was 2 for the 6-hour period prior to the start of the event and rose to 5– in 3 hours after the arrival of the pulses/cavities on Earth. These solar wind features also produced two large positive deflections in Dst. The positive increase in Dst at times of high solar wind dynamic pressures is well known to be caused by the increase in the magnetopause currents and the inward (earthward) compression of the magnetopause.

The global aspect of auroral dynamics in response to these pressure pulses is demonstrated with global auroral power (GAP) in Figure 3 and a local-time auroral keogram (LAK) in Figure 3d [Meng and Liou, 2002]. These pressure pulse-induced auroras are characterized by dayside enhancements and subsequent nightside extensions of auroras, similar to those associated with interplanetary shocks. Figure 3d shows a clear one-to-one correspondence between the pressure pulses and the GAP enhancements. The correlation coefficient between GAP and the solar wind dynamic pressure is 0.85,

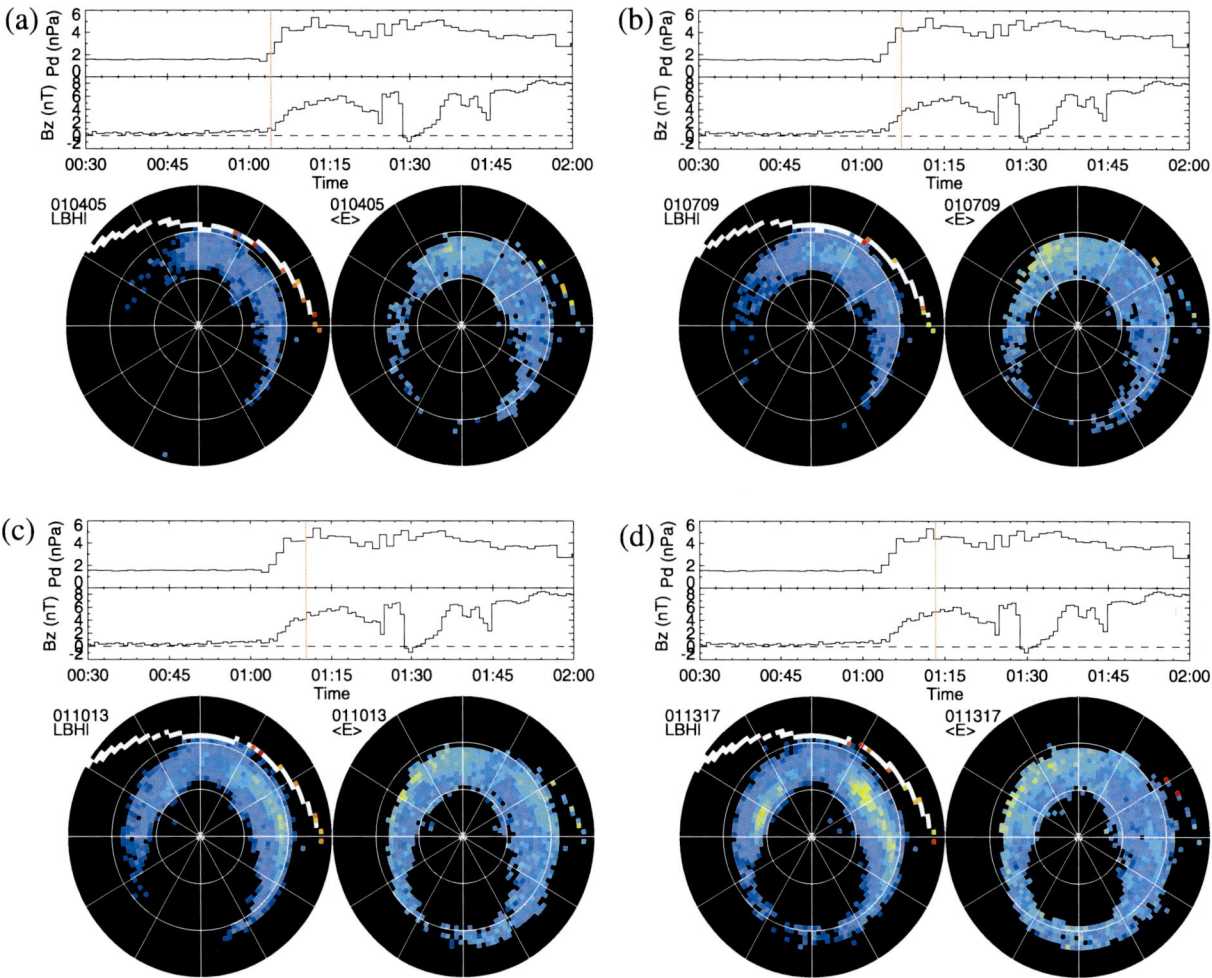

Plate 1. Panels from (a) to (d) are four snapshots of (top line-plot panels) solar wind dynamic pressure and IMF B_z, (bottom left) color-coded energy flux and (bottom right) average energy maps of precipitating electrons inferred from Polar UVI images for the 10 January 1997 shock event. The red vertical line in each line plot indicates the time at which the UVI image was acquired and the white contour in each map demarcates the field of view of UVI.

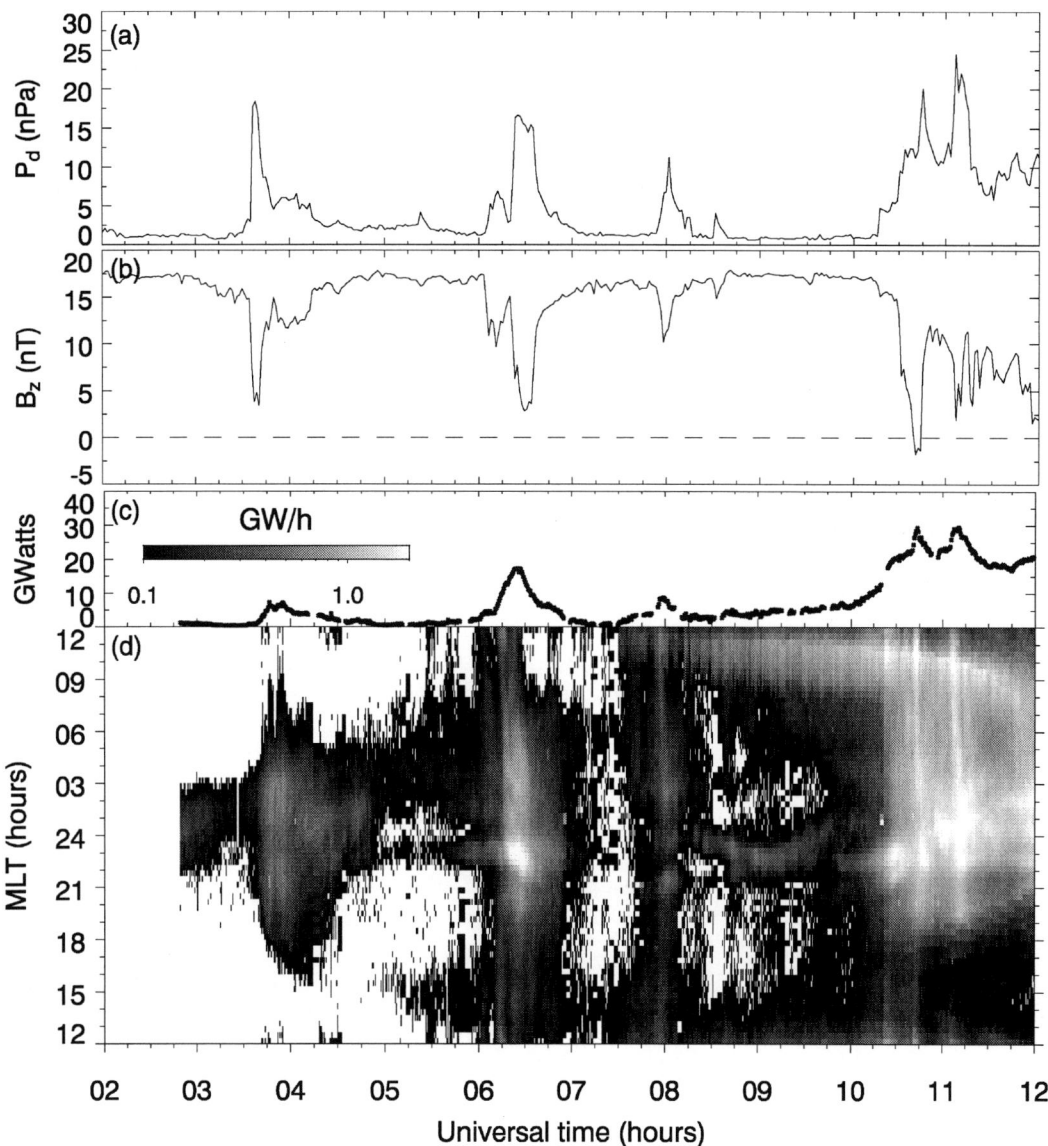

Figure 3. Stack plots from top to bottom are (a) solar wind dynamic pressure, (b) IMF B_z component from Wind, (c) global auroral power integrated from 60° to 90° MLAT, and (d) and magnetic local time keogram (LAK) on 8 November 2000 from 0200 to 1200 UT. The auroral power is inferred from Polar UVI auroral images in the LBH long band.

a somewhat surprising result. Such a good correlation strongly suggests that the intensification of auroras must be associated with the compression of the magnetosphere.

Solar wind pressure pulses can also produce auroral emissions in the subauroral region. *Zhang et al.* [2002] reported auroral emissions detached from the dayside oval in the dusk sector in association with sudden enhancements of solar wind dynamic pressure. They concluded, using the IMAGE FUV SI-12 imager [*Mende et al.*, 2000], which is sensitive to the Doppler-red shifted hydrogen Lyman α line, that the enhancement of the dayside detached aurora is caused by proton precipitation. DMSP particle data were used by these workers and indicated that both electrons and protons (probably other ions) appeared in the field lines of the sub-auroral transients. *Hubert et al.* [2003] also studied the 8 November 2002 pressure pulses events and reached the same conclusion.

2.3.3. Ramp pressure. In addition to the two types of solar wind dynamic pressure variations discussed in the previous sections, there is a third type in which the pressure gradually increases or decreases. The auroral response to this type of dynamic pressure change has been investigated recently by

Zhou and Tsurutani [2004]. Specifically, they examined dayside auroral responses to nine selected gradual, intense solar wind ram pressure (GISWRP) events and found that there is a very good linear relationship between the auroral luminosity at the dawn or dusk and the solar wind dynamic pressure. They proposed that the Kelvin-Helmholtz instability is a probable mechanism because the increased solar wind dynamic pressure can lower the threshold for this instability to occur.

The solar wind "ramp pressure" effect is still not conclusive, however, and it will take more elaborated work to make a quantitative assessment of the pressure effect on the dawn and dusk auroras possible. In the work of *Zhou and Tsurutani* [2004], contributions from the solar wind electric field and solar wind velocity were not considered. The statistical analysis of Polar UVI images by *Liou et al.* [1998] has suggested that auroral power in the postnoon sector (1300–1800 MLT) is controlled mainly by the transverse component of the solar wind electric field (see Section 2.1). In order to quantify the pressure effect, one has to be able to isolate the effect of the solar wind electric field from the data. Furthermore, the possible use of uncorrected auroral images has made their results doubtful. Whether or not the auroral images used in their study were calibrated was not mentioned. The two auroral image figures shown in their papers suggest that the images were not fully calibrated. Finally, the work of *Zhou and Tsurutani* [2004] correlated photon flux at a single point inside the oval with solar wind dynamic pressure; it is also important to study the pressure effect on the other parts of the oval.

2.4. Mechanisms for Dayside Auroral Transients

The work of *Borovsky* [1993] is a good resource on general auroral mechanisms. In connection with the magnetospheric compression, *Tsurutani et al.* [2001] and *Zhou et al.* [2003] have discussed a number of possible mechanisms responsible for the compression aurora. Most of the mechanisms they discussed, such as those associated with FAC intensifications, viscous interactions, and Alfvén wave generation, are associated with processes occurring at the magnetopause boundary layer and can only address the weaker, poleward part of the dayside auroral intensification. As indicated by the auroral images and FAST and DMSP data [*Zhou et al.*, 2003], the major part of dayside auroral brightening occurs at lower latitudes and is associated with enhancements (both energy and energy flux) of central plasma sheet particle precipitation. In this review, we will focus on possible mechanisms responsible for the most intense aurora only.

The occurrence of auroral transients upon the arrival of pressure pulses is interpreted by *Spann et al.* [1998] as a result of the lowering of the mirror points of trapped electrons to altitudes below 100 km due to the highly compressed magnetosphere by shocks. Another possible and often mentioned mechanism is pitch- angle scattering of particles by waves. Diffuse auroras are generally believed to result from pitch-angle diffusion by electrostatic electron cyclotron harmonic waves [e.g., *Kennel et al.*, 1970; *Lyons*, 1974] and whistler mode waves [e.g., *Johnstone et al.*, 1993]. When the magnetosphere is compressed by shocks, perpendicular heating owing to conservation of the first adiabatic invariant can increase the temperature anisotropy of the particles and trigger loss-cone instability; pre-existing, trapped CPS particles can be scattered into the loss cone by wave-particle interactions [*Zhou and Tsurutani*, 1999]. The enhancements of plasma waves during magnetospheric compression have been reported by *Anderson and Hamilton* [1993] as EMIC waves, by *Lauben et al.* [1998] as very low frequency chorus emission in the magnetosphere, and recently by *Zhou et al.* [2003] as broadband electromagnetic waves in the ionosphere.

However, the dayside magnetospheric configuration is greatly distorted by shock impacts. Distortion of the field configuration is expected to be most pronounced in the equatorial magnetosphere. *Liou et al.* [2002] proposed that such an uneven increase in the magnetic field strength by compression would reduce the mirror ratio. Moreover, the new particle distribution may be unstable and may result in loss-cone instabilities that pitch angle scatter particles into the loss cone (perhaps within a few minutes), producing auroral emissions in the dayside oval and the sub-auroral regions. This simple mechanism requires no pitch-angle diffusion and exists only during changes in the solar wind ram pressure, and is probably the dominant process for enhanced diffuse aurora in the equatorward part of the oval. Observations from FAST seem to suggest that enhanced diffuse precipitation is not associated with plasma waves and FACs in the equatorward part of the oval [*Zhou et al.*, 2003]. This consideration is complicated by the fact that ELF (10–1500 Hz) chorus, which may be responsible for the diffuse aurora (~0.1–10 keV) [e.g., *Inan et al.*, 1992], can occur between plasmapause and magnetopause in all local times and predominantly in the dayside equatorial regions [*Tsurutani and Smith*, 1977]. The relative role the two processes play in the compression-triggered diffuse aurora is still an open question.

Many case studies have demonstrated enhancements of auroras, particularly in the day sector, by a shock impact. Whether the enhancements are associated with the shock front or the long-duration, high solar wind pressure downstream of the shock is not known. The quick response of the auroral power to the solar wind pressure pulses shown in Plate 1 and Figure 3 indicates an immediate response of the high-latitude ionosphere to the change in solar wind dynamic pressure. Furthermore, the high resemblance in the shape of auroral power with the solar wind pressure pulses suggests that reconfiguration of the magnetosphere takes place within a time scale of ~10–20 min (the width of the pressure pulses), which is comparable to the response time of the convection to an IMF change [*Ridley et al.*, 1998; *Ruohoniemi and Greenwald*, 1998]. The rise-time/fall-time of solar wind pressure for interplanetary shocks/pulses is probably one of the key parameters in regulating the state of the magnetosphere. If the transition time is long, a slow response of the magnetosphere is expected and the state-to-state transition may be quasi-stationary without producing significant perturbations. On the other hand, if the transition is short, large perturbations are expected. Systematically studying the auroral response and response time scale to shocks/pulses of various sizes should provide valuable information about the solar wind-magnetosphere coupling process.

While the enhancement of the solar wind dynamic pressure is an important source of magnetic activity during enhanced convection, the exact consequence on the electrodynamics of the ionosphere from pressure enhancements has yet to be determined. An outstanding question in this new emerging field is whether solar wind dynamic pressure changes affect magnetic merging/reconnection. It is reasonable to believe that compression of the magnetosphere can enhance the cross-tail currents, as a requirement of pressure balance, and hence the tail reconnection. This question was to some extent addressed recently by Boudouridis and his coworkers [*Boudouridis et al.*, 2003; 2004]. Using DMSP particle and Polar UVI image data in several events, they demonstrated an increase in the latitudinal width of the oval and its associated polar cap closure. The increase is more pronounced in the night than in the day sector and more significant during southward than during northward IMF.

These studies not only provide further evidence about the direct control of the solar wind dynamic pressure on the global aurora but also suggest the enhancement of nightside reconnection by compression. However, a question still remains about how the solar wind pressure exactly affects the nightside reconnection. In an attempt to address this question *Boudouridis et al.* [2005] demonstrated that solar wind dynamic pressure increases can improve the overall coupling efficiency between the solar wind and the magnetosphere. Because of the complexity of the solar wind structures and because of the small number of events being studied, more work needs to be done to quantify the geoeffectiveness of solar wind pressure discontinuities.

2.5. Substorms Triggered by Shocks

Triggering of substorm expansion phase onset by discontinuities in the solar wind remains one of the outstanding issues in magnetospheric physics (see a recent review by *Akasofu* [2005]). In addition to the well-known substorm triggering by northward IMF turnings, interplanetary shocks have long been considered a substorm trigger. It was first noted by *Heppner* [1955], before the concept of substorm was invented, that sudden storm commencements (SSCs), caused by shock impingement, are often followed immediately by a "negative magnetic bay" [*Akasofu and Meng*, 1967]. Later, a number of extensive studies showed strong evidence of magnetic substorm triggers by SSCs or sudden impulses (SIs) produced by solar wind plasma discontinuities [*Schieldge and Siscoe*, 1970; *Kawasaki et al.*, 1971; *Burch*, 1972; *Kokubun et al.*, 1977], making the compression trigger mechanism plausible. However, these studies also indicated that not every SSC/SI was followed by a substorm, indicating that magnetospheric compression is not a sufficient condition for substorm onset triggering. From a theoretical point of view, as the high-pressure solar wind in the shock downstream moves across the Earth's magnetosphere and tailward, the lobe magnetic fields become more stretched and the magnetic pressure in the tail increases, thus enhancing the cross-tail currents and causing the plasma sheet to thin. If the magnetic field becomes very stretched such that ion motion becomes non-adiabatic, magnetic reconnection will be initiated and a substorm can be triggered if the rate of reconnection grows explosively [*Coroniti*, 1985]. It is also possible that the cross-tail current enhancement may disrupt due to plasma current instabilities and ultimately lead to the formation of a substorm current wedge [e.g., *Lui et al.*, 1991, and references therein].

The substorm compression triggering mechanism was recently tested using auroral breakups identified from the Polar UVI images [*Liou et al.*, 2003]. It was found, surprisingly, that most of the "negative magnetic bays" that occur concurrently with SSCs/SIs were not associated with auroral breakups. A non-substorm bay event is shown in Plate 2. The shock-induced aurora in the night sector is characterized by enhancements of aurora on the poleward edge of the oval, probably associated with polar boundary intensification [*Lyons et al.*, 1999]. In a total of 43 interplanetary shock events they studied, two (four) auroral breakups were identified to occur within 10 (20) min of SSCs/SIs. Therefore, negative excursions in the *H*-component of Earth's ground magnetic field in polar latitudes are not always associated with auroral substorms; they can be induced by magnetospheric compression (hence named "compression bays" by *Liou et al.* [2003]).

In contrast to the result of *Liou et al.* [2003], *Zhou and Tsurutani* [2001] demonstrated using images and/or *AE/AL* that substorms can be triggered within 10 min of shock arrival. They concluded, based primarily on *AL*, that a preconditioned magnetosphere, e.g., by a southward IMF component for about 1.5 hours, is a necessary condition for a substorm to be triggered by a shock. Although a southward

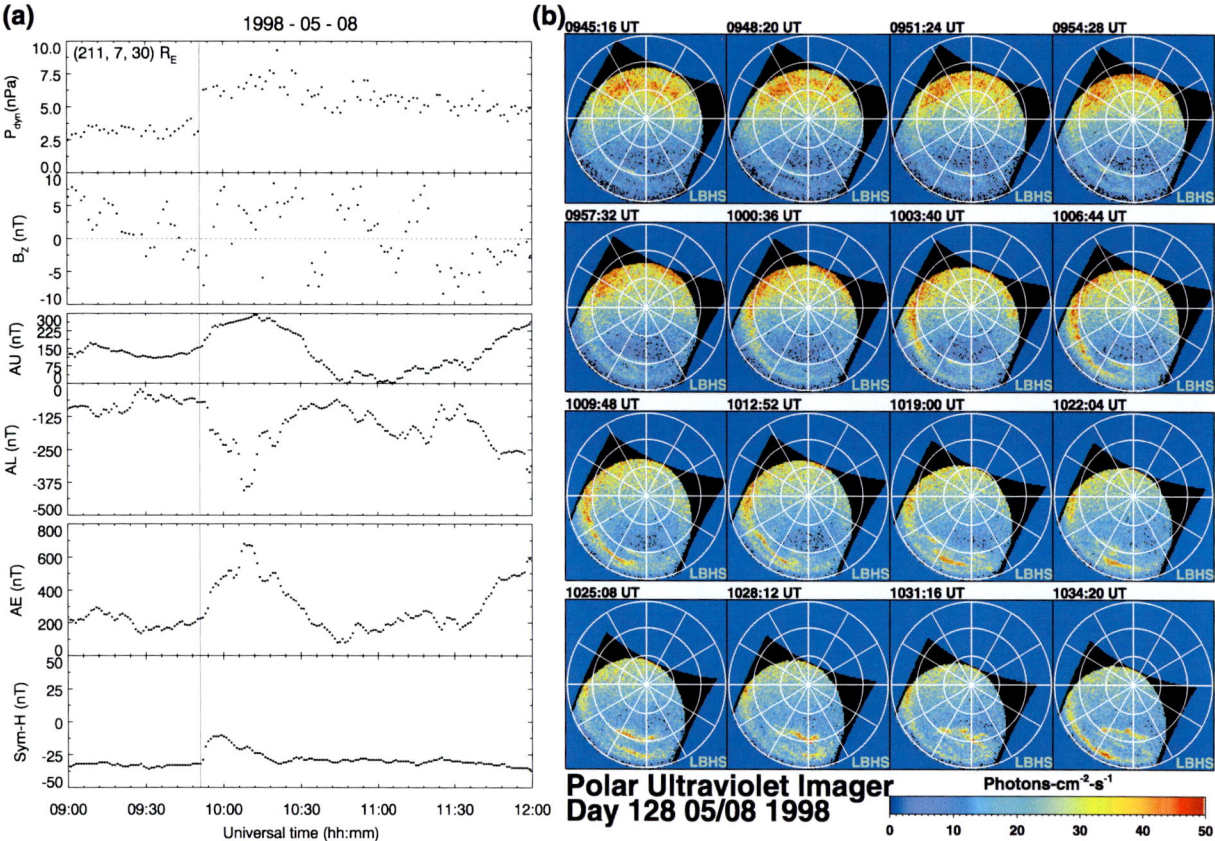

Plate 2. (a) Panels from top to bottom are the time-shifted solar wind pressure, IMF B_z component, auroral electrojet *AU*, *AL*, and *AE* indices, and the *Sym-H* index. The light vertical line indicates the onset time of the SSC (0951 UT). (b) A sequence of Polar UVI auroral images in the LBH short band (~140–160 nm) during an interplanetary shock impact on the magnetosphere (from *Liou et al.* [2003]).

IMF component existing prior to the shock is a preferred condition that leads to the formation of a magnetic bay, auroral intensification, and the closure of the nightside polar cap [*Boudouridis et al.*, 2003; 2004], a question exists about why it requires 90 min for the precondition. It is well established that there is a 40- to 60-min delay response of *AE* to the southward IMF B_z component [*Arnoldy*, 1971; *Meng et al.*, 1973]. If a southward IMF is important, it need not persist for a period as long as that required for a non-triggered onset.

More evidence exists that can differentiate a compression bay from a typical substorm bay. *Liou et al.* [2004] employed a superposed epoch analysis of the auroral electrojet *AU/AL* and the polar cap PC indices. They found that the auroral electrojet enhancements in response to magnetospheric compression is associated with an increase in AU and PC, indicating that compression bays are associated with the DP 2 current system (convection) rather than the DP 1 current system (substorm expansion). They also demonstrated that the compression bay correlates better with current than previous solar wind conditions, suggesting that the compression bay is associated with a direct-driven process. Furthermore, there is a difference in the auroral energy spectrum produced by substorms and by pressure pulses. *Chua et al.* [2001] have compared the average energy of precipitating electrons associated with a substorm-induced and a pressure pulse-induced aurora using Polar UVI images. They found that the average energy for the pressure-induced aurora is less than that for the substorm-induced aurora. They also found that there is no signature of auroral surges associated with a substorm current wedge for the pressure pulse event. The case study of *Boudouridis et al.* [2003] also concluded (quoted from their paper): "a solar wind pressure pulse does not trigger a substorm, but instead causes a global effect on the magnetosphere-ionosphere system that constitutes an entirely separate class of disturbance."

The response of auroral luminosities and cross-polar cap potential drops to magnetospheric compression has not been systematically studied. Although the quick response of auroral power to pressure pulses shown in Figure 3 strongly suggests a direct driven process, a conclusive result can only be made with statistical studies. To understand the magnetospheric compression process, establishing the electrodynamic relationship among the three parameters is essential.

3. GLOBAL AURORAL RESPONSE ACROSS CO-ROTATING INTERACTION REGIONS

In addition to interplanetary shocks, which are driven mostly by CMEs/MCs, co-rotating interaction regions, which are driven by fast solar wind streaming off corona holes, are rich with density enhancements and provide a convenient target for studying geoeffectiveness of dynamic pressure effects. Some recurrent, long-duration geomagnetic storms have been associated with the high-speed stream behind the CIRs [*Joselyn*, 1995]. The high-speed flows downstream of CIRs are associated with a large fluctuating IMF B_z component. The increase in the IMF fluctuation is believed to be associated with large-amplitude Alfvén waves [*Tsurutani et al.*, 1996, and references therein]. There are many studies on the CIRs and their effect on the geomagnetic disturbances (see *Tsurutani et al.* [this volume, 2005] for a recent review); however, there are very few studies of the response of auroral ionosphere to CIRs. In this section we will quantify the global auroral response to 27 CIRs and their neighboring low- and high-speed streams.

The CIR events were provided by Ian Richardson (1996–1998) and Robert McPherron (1994–1996) through the conference email communications. The stream-stream interaction region is determined with the criteria of McPherron [this volume, *McPherron and Weygand*, 2006]: (1) a large, rapid increase from ~300 to 600 km/s in solar wind velocity, (2) a large, slow increase in density followed by a rapid drop in density, and (3) a large increase in magnetic field strength. Based on these criteria, 27 CIR events with concurrent Polar UVI images are selected for further analysis. Note that some of the CIRs that are accompanied by CMEs/MCs are ignored. These events are listed in Table 1.

The solar wind proton density, proton bulk speed, proton thermal speed, IMF magnitude, and IMF z-component for the first event are shown in the bottom five panels of Figure 4. A sharp increase in the solar wind proton temperature and a sharp decrease in the solar wind proton density are used to determine the trailing edge of the CIR. In general, this is the

Table 1. List of 27 CIR events

#	Date	UT Wind	UT Earth	#	Date	UT Wind	UT Earth
1	961209	2338	2340	2	970111	0256	0315
3	970126	0920	1022	4	970205	2346	0049
5	970208	1327	1407	6	970217	1800	1901
7	970221	0408	0510	8	970226	2341	0118
9	970606	0954	1143	10	970622	0625	0653
11	970627	1108	1129	12	970731	0308	0308
13	970809	1213	1227	14	970828	0724	0756
15	970927	1335	1350	16	971101	0947	1001
17	971118	0222	0300	18	971130	0821	0914
19	971204	1857	2002	20	980208	1313	1438
21	980310	1200	1256	22	980321	0513	0610
23	980404	1627	1732	24	980417	0055	0157
25	980529	0516	0607	26	980606	1250	1337
27	980619	0007	0047				

UT referred to the passage time of the CIR trailing edge.

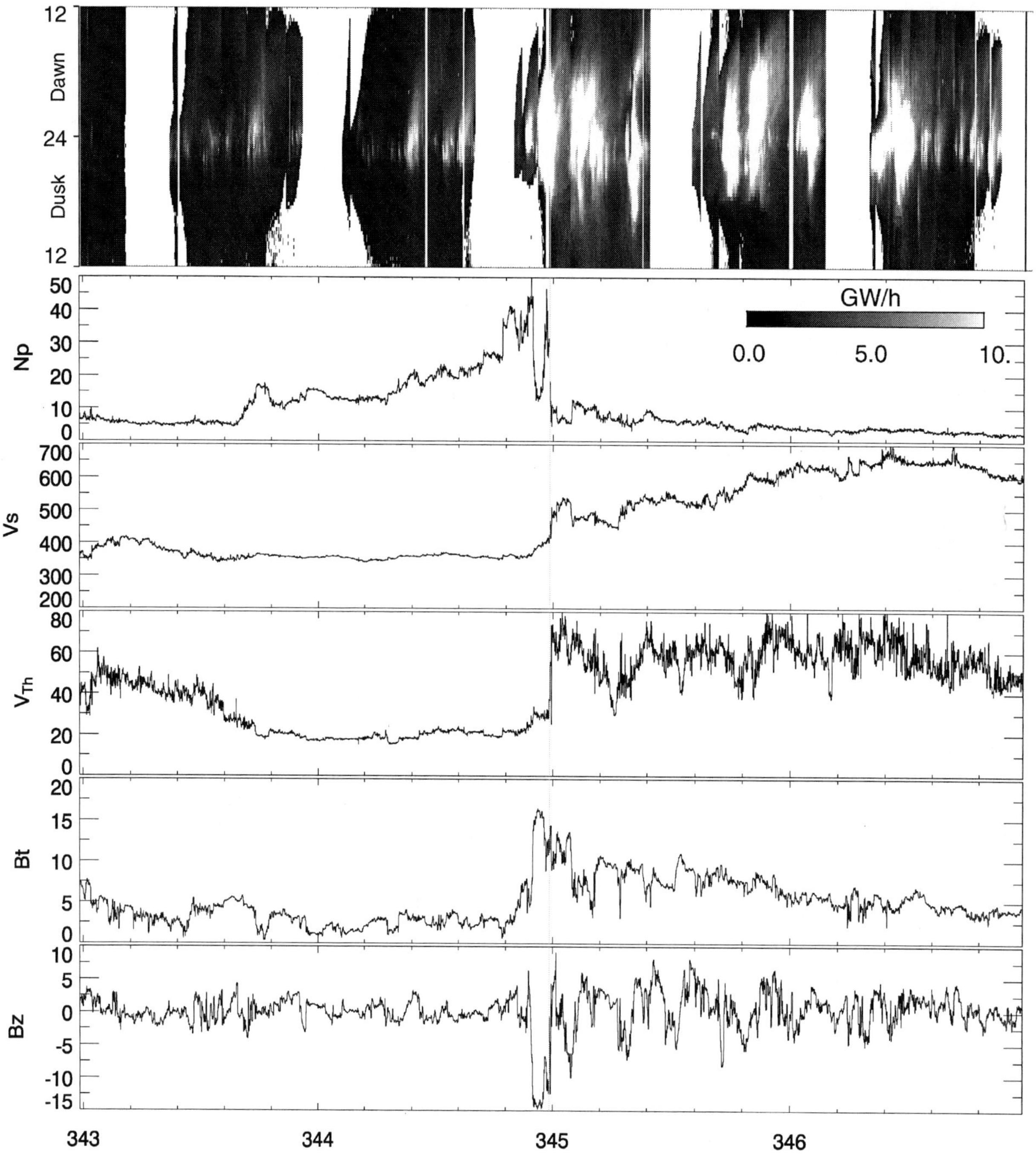

Figure 4. Panels from top to bottom show the magnetic LAK from Polar UVI, solar wind proton density, bulk speed, thermal speed, and the total and the z-component of the IMF from Wind. The vertical line at the center of each panel defines the trailing edge of the CIR. The white areas in LAK indicate image data gaps. The horizontal axis indicates day of the year of 1996.

boundary that separates the low-speed solar wind from the high-speed solar wind. To demonstrate the global auroral response to the CIR events, LAK) is plotted in the first panel of Figure 4. The keogram has a unit of auroral power per unit hour sectors and is derived from auroral images in the LBH band acquired by the Polar UVI imager. Note that auroral power is a physical quantity and is better suited for gauging the state of the magnetosphere than magnetic indices [*Meng and Liou*, 2002]. Significant data gaps (white areas) can be seen in LAK. This is because the northern oval is outside the field-of-view of the UVI. Nonetheless, a large difference in the auroral power across the CIR is discernible. The auroral power is clearly dominant in the night sector and is more intense during high-speed than during low-speed solar wind stream.

Selected IMF and solar wind properties measured by Wind for the 27 CIR periods (48 hours before and after the stream interfaces) are plotted in Figures 5a–5c. The superposed epoch zero time is chosen as the time of a rapid density drop at the trailing edge of the stream interface after propagating to the ionosphere. Figure 5b shows a clear transition from a low-speed solar wind to a high-speed solar wind at the CIRs.

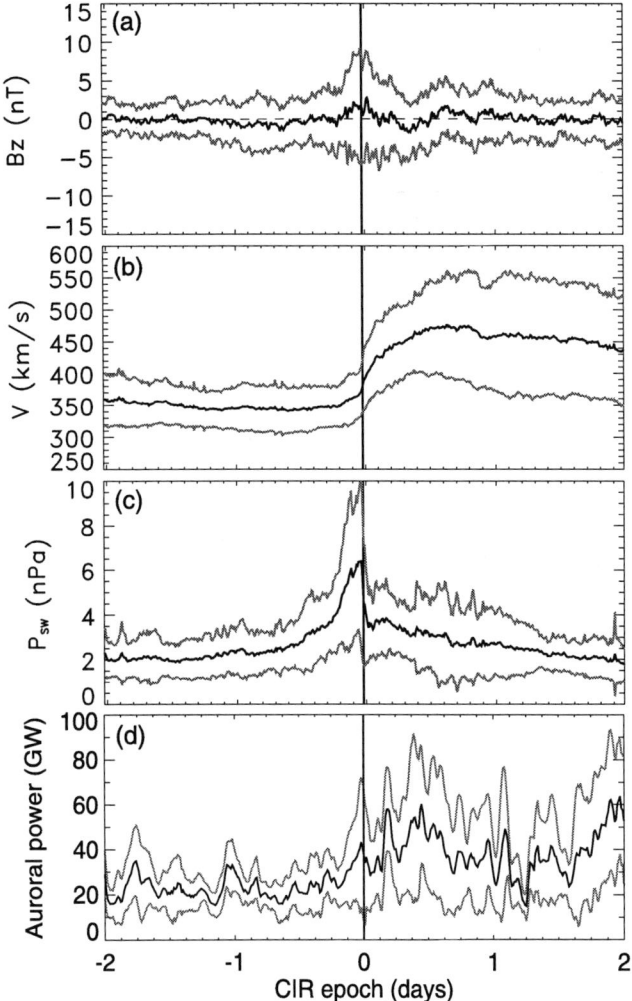

Figure 5. Superposed epoch analysis of (a) the interplanetary B_z component, (b) solar wind speed, (c) dynamic pressure, and (d) northern hemispheric auroral power for 27 co-rotating interaction region (CIR) events. The zero epoch is the arrival time of the CIR trailing boundary. In each panel black lines indicate average values of each parameter and gray lines indicate one standard deviation from the mean for each data bin (5-min bins for solar wind parameters and 7.5-min bins for auroral power).

The start of the speed increase roughly corresponds to the sharp density drop. The z-component of IMF shown in Figure 5a is generally small in both low- and high-speed streams. The northern hemispheric auroral power for the 27 CIR periods is plotted in Figure 5d. Note that the auroral power is derived only from auroral images with full coverage of the entire oval. A clear difference in the auroral power between the slow and fast stream regions is evident: auroral power is larger in the high-speed stream region and smaller in the low-speed stream region. The increase is modest (~80%). The larger (smaller) auroral power in the high-speed stream region is due to larger (smaller) amplitude fluctuating IMF B_z, although the average IMFs in both regions are comparable and close to zero.

High-intensity, long-duration, continuous *AE* activity (HILDCAA) has been reported to occur during high-speed solar wind streams [e.g., *Tsurutani and Gonzalez*, 1987]. It is proposed that prompt penetration of interplanetary electric fields associated with the large-amplitude Alfvénic waves through magnetic merging causes such a large geomagnetic activity [*Tsurutani et al.*, 2004, and references therein]. The auroral result presented here indicates that enhanced ionospheric conductivity is also an important contributor to HILDCAA in addition to the electric field.

A close examination of the auroral power indicates that the start of the power increase does not line up with the CIRs. A gradual increase in the auroral power starts several hours before the passage of the CIR trailing edge, which corresponds to the region of a rapid increase in the solar wind dynamic pressure. During this time the average of IMF B_z was positive and solar wind velocity remained small; therefore, the increase in auroral power is most likely due to the pressure increase within the CIR.

4. SUMMARY

Magnetospheric compression by solar wind dynamic pressure enhancements is an important mechanism of geomagnetic disturbances in addition to the southward IMF-initiated magnetic merging/reconnection and viscous interaction in the sheared flow on the flanks of the magnetopause. This paper reviews some important progress made recently on global auroral responses to solar wind parameters, with emphasis on the compression of the magnetosphere by solar wind dynamic pressure enhancements. The general characteristics of compression-induced auroras include the following: (1) a dayside auroral transient (<~10–20 min) appears that often shows day-to-night extension/propagation, (2) auroras are in general much weaker on the dayside than on the nightside, (3) global auroral power is well correlated with the solar wind pressure pulses when reconnection shuts off, (4) the most intense dayside auroral transients are associated with the enhancement in both energy and energy flux of dayside extension of the CPS precipitation, and (5) compression effects are most dramatic for southward IMF, often resulting in a significant auroral intensification, widening of the auroral oval, and a closure of the polar cap.

There is renewed interest in the geoeffectiveness of co-rotating interaction regions. A superposed epoch analysis of global auroral power in 27 CIR events indicates the importance of the compression mechanism. The global auroral power is generally larger in the high-speed stream than in the low-speed stream regions and shows a modest increase (~80%) across the CIRs. This is consistent with the general belief that the magnetosphere should respond the same to the solar wind and IMF of different origins because the physics underlying the solar wind-magnetosphere coupling is the same.

While there is enough evidence to show the effect of enhancements of the solar wind dynamic pressure on geomagnetic activity, especially when the IMF is directed southward, the physical process associated with magnetospheric compression and the extent to which magnetospheric compression can influence high-latitude ionospheric electrodynamics are still not clear. The physics mechanisms responsible for the compression-enhanced aurora is still an open question. It is also not clear whether magnetospheric compression increases the efficiency of the well-known magnetic field merging and/or viscous interactions or it is a fundamentally different type of coupling mechanism from these processes. These questions are central to the understanding of the magnetospheric compression effects and require further studies.

Acknowledgments. We acknowledge M. Torr, who built the Polar UVI, and G. Parks, the current principal investigator. Wind MFI and SWE data were courtesy of R. Lepping and K. Ogilvie, respectively. This material is based upon work supported by the National Science Foundation grant no. ATM-0318606 and the National Aeronautics and Space Administration under grant no. NNG05GB72G issued through the Polar Mission. The author also acknowledges the many constructive comments from the three reviewers.

REFERENCES

Akasofu, S.-I. The development of the auroral substorm, *Planet. Space Sci*, 12, 273-282, 1964.
Akasofu, S.-I., Several "controversial" issues on substorms, *Space Sci. Rev.*, 120(1-2), 27-65, 2005.
Akasofu, S.-I., and C.-I. Meng, Intense negative bays inside the auroral zone I. The evening sector, *J. Atmos. Terr. Phys.*, 29, 965, 1967.
Anderson, B.J., and D.C. Hamilton, Electromagnetic ion-cyclotron waves stimulated by modest magnetospheric compressions, *J. Geophys. Res.*, 98, 11,369-11,382, 1993.
Arnoldy, R.L., Signature in the interplanetary medium for substorms, *J. Geophys. Res.*, 76, 5189, 1971.
Axford, W.I., and C.O. Hines, A unifying theory of high-latitude geophysical phenomena and geomagnetic storms, *Can. J. Phys.*, 39, 1433, 1961.
Borovsky, J.E., Auroral arc thicknesses as predicted by various theories, *J. Geophys. Res.*, 98, 6101-6138, 1993.
Borovsky, J.E., and J.T. Steinberg, The freestream turbulence effect in solar-wind/magnetosphere coupling: Analysis through the solar cycle and for various types of solar wind, in *this volume*.
Boudouridis, A., E. Zesta, L.R. Lyons, P.C. Anderson, and D. Lummerzheim, Effect of solar wind pressure pulses on the size and strength of the auroral oval, *J. Geophys. Res.*, 108, 8012, doi:10.1029/2002JA009373, 2003.
Boudouridis, A., E. Zesta, L.R. Lyons, P.C. Anderson, and D. Lummerzheim, Magnetospheric reconnection driven by solar wind pressure fronts, *Ann. Geophys.*, 22, 1367-1378, 2004.
Boudouridis, A., E. Zesta, L.R. Lyons, P.C. Anderson, and D. Lummerzheim, Enhanced solar wind geoeffectiveness after a sudden increase in dynamic pressure during southward IMF orientation, *J. Geophys. Res.*, 110, A05,214, doi:10.1029/2004JA010704, 2005.
Burch, J.L., Preconditions for the triggering of polar magnetic substorms by storm sudden commencements, *J. Geophys. Res.*, 77, 6529, 1972.
Chua, D., G. Parks, M. Brittnacher, W. Peria, G. Germany, J. Spann, and C. Carlson, Energy characteristics of auroral electron precipitation: A comparison of substorms and pressure pulse related auroral activity, *J. Geophys. Res.*, 106(A4), 5945-5956, 2001.
Cogger, L.L., J.S. Murphree, S. Ismail, and C.D. Anger, Characteristics of dayside 5577 Å aurora, *Geophys. Res. Lett.*, 4, 413-416, 1977.
Coroniti, F.V., Explosive tail reconnection: The growth and expansion phases of magnetospheric substorms, *J. Geophys. Res.*, 90, 7427, 1985.
Cowley, S.W.H., J.P. Morelli, and M. Lockwood, Dependence of convective flows and particle precipitation in the high-latitude dayside ionosphere on the x and y components of the interplanetary magnetic field, *J. Geophys. Res.*, 96, 5557, 1991.
Craven, J.D., and L.A. Frank, Diagnosis of auroral dynamics using global auroral imaging with emphasis on large-scale evolution, in *Auroral Physics*, edited by C.-I Meng, M. J. Rycroft, and L. A. Frank, p. 273, Cambridge Univ. Press, New York, 1991.
Craven, J.D., L.A. Frank, C.T. Russell, E.J. Smith, and R.P. Lepping, Global auroral responses to magnetospheric compressions by shocks in the solar wind -Two case studies, in *Solar Wind Magnetosphere Coupling,* edited by Y. Kamide and J. A. Slavin, Terra Scientific, Tokyo, pp. 367-380, 1986.
Crooker, N.U., Dayside merging and cusp geometry, *J. Geophys. Res.*, 84, 951, 1979.
Danielsen, C., The dependence of auroral activity upon K_P as determined by the use of an extensive database, *Geophysical Papers R-60*, Danish Meteorological Institute, Copenhagen, 1980.
Dungey, J.W., Interplanetary magnetic field and the auroral zones, *Phys. Rev. Lett.*, 6, 47, 1961.
Elphinstone, R.D., K. Jankowska, J.S. Murphree,, and L.L. Cogger, The configuration of the auroral distribution for interplanetary magnetic field B_z northward, 1, IMF B_x and B_y dependencies as observed by the Viking satellite, *J. Geophys. Res.*, 95, 5791, 1990.
Feldstein, Y.I., Geographical distribution of aurorae and auroral arcs azimuths, in *Investigations of Aurorae*, edited by B.A. Bagarjatsky, pp. 61-78, Publ. House Acad. of Sci., Russia, 1960.
Feldstein, Y.I., and G.V. Starkov, Dynamics of auroral belt and polar geomagnetic disturbances, *Planet. Space Sci.*, 15, 209, 1967.
Germany, G.A., J.F. Spann, G.K. Parks, M.J. Brittnacher, R. Elsen, L. Chen, D. Lummerzheim, and M.H. Rees, Auroral observations from the Polar Ultraviolet Imager (UVI), in *Geospace Mass and Energy Flow: Results From the International Solar-Terrestrial Physics Program, Geophys. Monogr. Ser.*, vol. 104, edited by J. Horwitz, D. Gallagher, and W. Peterson, pp. 149-160, AGU, Washington, D. C., 1998.
Hasegawa, H., M. Fujimoto, T.-D. Phan, H. Réme, A. Balogh, M.W. Dunlop, C. Hashimoto, and R. TanDokoro, Transport of solar wind into earth's magnetosphere through rolled-up Kelvin-Helmholtz vortices, *Nature*, 430, 755-758, 2004.
Heppner, J.P., Note on the occurrence of world-wide SSC's during the onset of negative bays at College, *J. Geophys. Res.*, 60, 29-32, 1955.
Holzworth, R.H., and C.-I. Meng, Auroral boundary variations and the interplanetary magnetic field, *Planet. Space Sci.*, 32, 25, 1984.
Hubert, B., J.C. Gérard, S.A. Fuselier, and S.B. Mende, Observation of dayside subauroral proton flashes with the IMAGE-FUV imagers, *Geophys. Res. Lett.*, 30, 45-48, 2003.

Iijima, T., and T.A. Potemra, The amplitude distribution of field-aligned currents at northern high latitudes, *J. Geophys. Res.*, 81, 2165-2174, 1976.

Inan, U.S., Y.T. Chiu, and G.T. Davidson, Whistler-mode chorus and morningside aurorae, *Geophys. Res. Lett.*, 19, 653-656, 1992.

Johnstone, A.D., D.M. Walton, R. Liu, and D.A. Hardy, Pitch angle diffusion of low-energy electrons by whistler mode waves, *J. Geophys. Res.*, 98, 5959-5967, 1993.

Joselyn, J.A., Geomagnetic activity forecasting: The state of the art, *Space Sci. Rev.*, 33(3), 383-401, 1995.

Karlson, K.A., M. Oieroset, J. Moen, and P.E. Sandholt, A statistical study of flux transfer event signatures in the dayside aurora: The IMF B_y-related prenoon-postnoon asymmetry, *J. Geophys. Res.*, 101, 59, 1996.

Kawasaki, K., S.-I. Akasofu, F. Yasuhara, and C.-I. Meng, Storm sudden commencements and polar magnetic substorms, *J. Geophys. Res.*, 76, 6781-6789, 1971.

Kennel, C.F., F.L. Scaft, R.W. Fredricks, J.H. McGhee, and F.V. Coroniti, VLF electric field observations in the inner magnetosphere, *J. Geophys. Res.*, 75, 6136, 1970.

Kokubun, S., R.L. McPherron, and C.T. Russell, Triggering of sub-storms by solar wind discontinuities, *J. Geophys. Res.*, 82, 74-86, 1977.

Lassen, K., and C. Danielsen, Quiet time pattern of auroral arcs for different directions of the interplanetary magnetic field in the Y-Z plane, *J. Geophys. Res.*, 83, 5277, 1978.

Lauben, D.S., U.S. Inan, T.F. Bell, D.L. Kirchner, G.B. Hospodarsky, and J. S. Pickett, VLF chorus emissions observed by POLAR during the January 10, 1997 magnetic cloud, *Geophys. Res. Lett.*, 25, 2995, 1998.

Liou, K., P.T. Newell, C.-I. Meng, A.T.Y. Lui, M. Brittnacher, and G. Parks, Synoptic auroral distribution: A survey using Polar ultraviolet imagery, *J. Geophys. Res.*, 102, 27,197-27,205, 1997.

Liou, K., P.T. Newell, C.-I. Meng, A.T.Y. Lui, M. Brittnacher, and G. Parks, Characteristics of the solar wind controlled auroral emissions, *J. Geophys. Res.*, 103, 17,543, 1998.

Liou, K., C.-I. Meng, A.T.Y. Lui, P.T. Newell, M. Brittnacher, G. Parks, G.D. Reeves, R.R. Anderson, and K. Yumoto, On relative timing in substorm onset signatures, *J. Geophys. Res.*, 104, 22,807-22,817, 1999.

Liou, K., C.-C. Wu, R.P. Lepping, P.T. Newell, and C. Meng, Midday subauroral patches (MSPs) associated with interplanetary shocks, *Geophys. Res. Lett.*, 29, 1771, 2002.

Liou, K., P.T. Newell, C.-I. Meng, C.-C. Wu, and R.P. Lepping, Investigation of external triggering of substorms with Polar ultraviolet imager observations, *J. Geophys. Res.*, 108(A10), 1364, doi:10.1029/2003JA009984, 2003.

Liou, K., P.T. Newell, C.-I. Meng, C.-C. Wu, and R.P. Lepping, On the relationship between shock-induced polar magnetic bays and solar wind parameters, *J. Geophys. Res.*, 109(A06), 306, doi:10.1029/2004JA010400, 2004.

Lui, A.T.Y., D. Venkatesan, and J.S. Murphree, Auroral bright spots on the dayside oval, *J. Geophys. Res.*, 94, 5515, 1989.

Lui, A.T.Y., C.-L. Chang, A. Mankofsky, H.-K. Wong, and D. Winske, A cross-field current instability for substorm expansions, *J. Geophys. Res.*, 96, 11,389-11,401, 1991.

Lyons, L.R., Electron diffusion driven by magnetospheric electrostatic waves, *J. Geophys. Res.*, 79, 575, 1974.

Lyons, L.R., G.T. Blanchard, J.C. Samson, T.Yamamoto, T. Mukai, A. Nishida, and S. Kokobun, Association between GEOTAIL plasma flows and auroral poleward boundary intensifications observed by CANOPUS photometers, *J. Geophys. Res.*, 104, 4485-4500, 1999.

McPherron, R.L., and J. Weygand, The solar wind and geomagnetic activity as a function of time relative to corotating interaction regions, in *Chapman Conference on Recurrent Magnetic Storms: Corotating Solar Wind Streams*, edited by B.T. Tsurutani, W. Gonzales, G. Lu, and R.L. McPherron, pp. submitted, AGU, Tropical Manaus Eco-Resort and Convention Center, Manaus, Brazil, 2005.

Mende, S.B., et al., Far ultraviolet imaging from the IMAGE spacecraft: 3. Spectral imaging of Lyman-α and O I 135.6 nm, *Space Sci. Rev.*, 91, 287-318, 2000.

Meng, C.-I., and K. Liou, Global auroral power as an index for geospace disturbances, *Geophys. Res. Lett.*, 29(12), 1600, doi:10.1029/2001GL013902, 2002.

Meng, C.-I., and R. Lundin, Auroral morphology of the midday oval, *J. Geophys. Res.*, 91, 1572, 1986.

Meng, C.-I., B. Tsurutani, K. Kawasaki, and S.-I. Akasofu, Cross-correlation analysis of the *AE* index and the interplanetary magnetic field B_z component, *J. Geophys. Res.*, 78, 617, 1973.

Murphree, J.S., and R.D. Elphinstone, Correlative studies using the Viking imagery, *Adv. Space Res.*, 8, 9, 1988.

Murphree, J.S., L.L. Cogger, and C.D. Anger, Characteristic of the instantaneous auroral oval in the 1200-1800 MLT sector, *J. Geophys. Res.*, 86, 7657, 1981.

Newell, P.T., D.G. Sibeck, and C.-I. Meng, Penetration of the interplanetary field B_y and magnetosheath plasma into the magnetosphere: Implications for the predominant magnetopause merging site, *J. Geophys. Res.*, 100, 235, 1995.

Newell, P.T., C.-I. Meng, and K.M. Lyons, Discrete aurorae are sup-pressed in sunlight, Nature, 381, 766, 1996.

Potemra, T.A., H. Vo, D. Venkatesan, L.L. Cogger, R.E. Erlandson, L.J. Zanetti, P.F. Bythrow, and B.J. Anderson, Periodic auroral forms and geomagnetic field oscillations in the 1400 MLT region, *J. Geophys. Res.*, 95, 5835, 1990.

Ridley, A.J., G. Lu, C.R. Clauer, and V.O. Papitashvili, A statistical study of the ionospheric convection response to changing inter-planetary magnetic field conditions using the assimilative mapping of iono-spheric electrodynamics technique, *J. Geophys. Res.*, 103, 4023-4039, 1998.

Ruohoniemi, J.M., and R.A. Greenwald, Statistical patterns of high-latitude convection obtained from Goose Bay HF radar observations, *J. Geophys. Res.*, 101, 21,743-21,763, 1996.

Ruohoniemi, J.M., and R.A. Greenwald, The response of high-latitude convection to a sudden southward IMF turning, *Geophys. Res. Lett.*, 25, 2913, 1998.

Sandholt, P.T., C.S. Deehr, A. Egeland, B. Lybekk, and R. Viereck, Signatures in the dayside aurora of plasma transfer from the magnetosheath, *J. Geophys. Res.*, 91, 10,063, 1986.

Schieldge, J.P., and G.L. Siscoe, A correlation of the occurrence of simultaneous sudden magnetospheric compressions and geomagnetic bay onsets with selected geophysical indices, *J. Atmos. Terr. Phys.*, 32, 1819-1830, 1970.

Shue, J.-H., P.T. Newell, K. Liou, C.-I. Meng, Y. Kamide, and R.P. Lepping, Two-component auroras, *Geophys. Res. Lett.*, 29(10), 1379, doi:10.1029/2002GL014657, 2002.

Spann, J.F., M. Brittnacher, R. Elsen, G.A. Germany, and G.K. Parks, Initial response and complex polar cap structures of the aurora in response to the January 10, 1997, magnetic cloud, *Geophys. Res. Lett.*, 25, 2577-2580, 1998.

Stenbaek-Nielsen, H.C., and A. Otto, Conjugate auroras and the inter-planetary magnetic field, *J. Geophys. Res.*, 102, 2223, 1997.

Strickland, D.J., J.R. Jasperse, and J.A. Whallen, Dependence of auroral FUV emissions on the incidence electron spectrum and neutral atmosphere, *J. Geophys. Res.*, 88, 8051, 1983.

Torr, M.R., et al., A far ultraviolet imager for the international solar-terrestrial physics mission, *Space Sci. Rev.*, 71, 329-383, 1995.

Tsurutani, B.T., and W.D. Gonzalez, The cause of high-intensity long-duration continuous *AE* activity (HILDCAAs): Interplanetary Alfvén wave trains., *Planet. Space Sci.*, 35, 405, 1987.

Tsurutani, B.T., and E.J. Smith, Two types of magnetospheric ELF chorus and their substorm dependences, *J. Geophys. Res.*, 82, 5112-5128, 1977.

Tsurutani, B.T., C.M. Ho, and J.K.A. and, Interplanetary discontinuities and Alfvén waves at high heliographic latitudes: Ulysses, *J. Geophys. Res.*, 101, 11,027-11,038, 1996.

Tsurutani, B.T., W.D. Gonzalez, F. Guarnieri, Y. Kamide, X. Zhou, and J.K. Arballo, Are high-intensity long-duration continuous *AE* activity (HILDCAA) events substorm expansion events?, *J. Atmos. Solar Terr. Phys.*, 66, 167-176, 2004.

Tsurutani, B.T., et al., Auroral zone dayside precipitation during magnetic storm initial phases, *J. Atmos. Terr. Phys.*, 63, 513-522, 2001.

Tsurutani, B.T., N. Gopalswamy, R.L. McPherron, W.D. Gonzalez, G. Lu, and F.L. Guarnieri, Magnetic storms caused by corotating solar wind streams, in *this volume*.

Vo, H.B., and J.S. Murphree, A study of dayside auroral bright spots seen by the Viking auroral imager, *J. Geophys. Res.*, 100, 3649, 1995.

Wing, S.P., P.T. Newell, D.G. Sibeck, and K.B. Baker, A large statistical study of the penetration of interplanetary magnetic field y component into the magnetosphere, *J. Geophys. Res.*, 22, 2083, 1995.

Zhang, Y., L.J. Paxton, and T.J. Immel, Sudden enhancement of solar wind dynamic pressure and dayside detached aurora, *Eos Trans. AGU*, pp. 82(47), SM41B-0814, 2001.

Zhang, Y., L.J. Paxton, T.J. Immel, H.U. Frey, and S.B. Mende, Sudden solar wind dynamic pressure enhancements and dayside detached auroras: IMAGE and DMSP observations, *J. Geophys. Res.*, 108, doi:10.1029/2002JA009355, 2002.

Zhou, X., and B.T. Tsurutani, Rapid intensification and propagation of the dayside aurora: Large scale interplanetary pressure pulses (fast shocks), *Geophys. Res. Lett.*, 26, 1097-1100, 1999.

Zhou, X., and B.T. Tsurutani, Interplanetary shock triggering of night-side geomagnetic activity: Substorms, pseudobreakups, and quiescent events, *J. Geophys. Res.*, 106, 18,957-18,967, 2001.

Zhou, X., and B.T. Tsurutani, Dawn and dusk auroras caused by gradual, intense solar wind ram pressure events, *J. Atmos. Solar Terr. Phys.*, 66, 153-160, 2004.

Zhou, X., R.J. Strangeway, P.C. Anderson, D.G. Sibeck, B.T. Tsurutani, G. Haerendel, H.U. Frey, and J.K. Arballo, Shock aurora: FAST and DMSP observations, *J. Geophys. Res.*, 108, 8019, doi:10.1029/2002JA00970, 2003.

Kan Liou, The Johns Hopkins University Applied Physics Laboratory, Laurel, MD 20723, USA (kan.liou@jhuapl.edu).

IMF B_y and the Spatio-Temporal Structure of the Dayside Aurora

P.E. Sandholt[1], C.J. Farrugia[2], E.J. Lund[2], and W.F. Denig[3]

With dayside auroral observations we may probe the spatial and temporal structure of solar wind-magnetosphere coupling during intervals when the IMF has a substantial B_y component. This work captures some of the magnetosphere-ionosphere features expected to be present when the Earth is embedded in corotating solar wind streams. Specifically, we examine auroral observations made at 76° magnetic latitude to distinguish spatial from temporal magnetosphere features. To this end, we examine how the aurora changes as the station rotates with the Earth under a stationary magnetosphere structure during intervals of quasi-steady IMF orientation. Characterisitic features of the dayside aurora as a function of IMF orientation may be grouped as follows. (I) Strongly northward IMF: elongated cusp arc within 75-80°, associated with sunward convection in the polar cap. (II) Strongly southward IMF: elongated, narrow, and stable cusp arc within 70-75°, and an ionospheric convection symmetric about noon. (III) Southwest(east) IMF with $|B_y/B_z| \geq 1$: (i) near-simultaneous auroral brightenings on field lines convecting noonward in the pre- and postnoon sectors. These are separated by a stable, but strongly attenuated, cusp aurora near noon. The polar convection is asymmetric; (ii) pre- and postnoon brightenings followed by noonward expansions accompanied by poleward moving auroral forms; (iii) strong auroral forms at the polar cap boundary on old, open field lines associated with a newly-discovered flow channel. Deflections in local ground magnetometers are seen whose cause is this flow channel. The phenomena under (iii) are presumably related to dynamo action in the high-latitude boundary layer downstream of the cusp.

INTRODUCTION

Conceptual models of magnetosphere-ionosphere coupling (M-I coupling) processes tend to emphasize those regions of the magnetosphere which have been most explored *in situ* by spacecraft. For example, coupling processes at the low latitude boundary layer south of the cusp feature more prominently than those downstream of the cusp. Further examples are *in situ* spacecraft observations documenting the different modes of magnetopause magnetic reconnection, such as quasi-steady reconnection [*Sonnerup et al.*, 1981; *Sonnerup*, 1984; *Paschmann et al.*, 1979, 1986; *Phan et al.*, 1996] and flux transfer events [*Russell and Elphic*, 1978; *Paschmann et al.*, 1982; *Farrugia et al.*, 1988] which are observed at various magnetopause locations [*Haerendel et al.*, 1978; *Rijnbeek et al.*, 1984; *Thompson et al.*, 2004]. The dependence of the magnetopause reconnection process on conditions external to the magnetosphere (boundary conditions), primarily the north-south (B_z) component of the interplanetary magnetic field (IMF), has been much in the focus of the discussion [*Cowley*, 1982, 1984]. The corresponding signatures in particle precipitation/ aurora of continuous reconnection [*Shelley et al.*, 1976; *Reiff*

---

[1] Department of Physics, University of Oslo, Oslo, Norway
[2] Space Science Center, University of New Hampshire, Durham
[3] Air Force Research Laboratory, Hanscom AFB

Recurrent Magnetic Storms: Corotating Solar Wind Streams
Geophysical Monograph Series 167
Copyright 2006 by the American Geophysical Union.
10.1029/167GM18

et al., 1977; *Trattner et al.*, 1999; *Frey et al.*, 2003; *Zhang et al.*, 2004] and flux transfer events [*Sandholt et al.*, 1986; *Newell and Meng*, 1991; *Escoubet et al.*, 1992; *Lockwood and Smith*, 1992; *Smith*, 1994; *Farrugia et al.*, 1998, 2004a] are now well-documented. Thus, the interesting question at present is not so much whether these reconnection modes are realized in nature, but to integrate observations acquired at various places to form a global picture of where on the magnetopause these reconnection modes occur and how these locations are controlled by external conditions.

An important aim of present and future magnetospheric studies is to merge the different fragments of the spatio-temporal structure into a unified picture of magnetospheric behaviour under electromagnetic stresses imposed from outside. A feasible approach to this goal, which we shall apply in this paper, is to combine the different essential data sets (e.g., plasma convection, particle precipitation/aurora, field-aligned currents) from different critical locations of the magnetosphere-ionosphere system corresponding to specific solar wind/IMF conditions.

An important aspect of the reconnecting magnetosphere is the spatial structure introduced by the east-west (B_y) component of the interplanetary magnetic field (IMF) [*Jørgensen et al.*, 1972; *Cowley*, 1981b; *Gosling et al.*, 1990]. Its most evident effect is to introduce various asymmetries about noon. These asymmetries extend to polar cap convection and magnetic deflections (the latter being often referred to as the Svalgaard-Mansurov effect) [*Svalgaard*, 1973; *Mozer et al.*, 1974; *Heppner and Maynard*, 1987]. Corresponding asymmetries are also observed in auroral precipitation [*Farrugia et al.*, 2004a; *Newell et al.*, 2004; *Sandholt et al.*, 2004b] and field-aligned currents [*Potemra and Saflekos*, 1979; *Friis-Christensen*, 1984; *Taguchi et al.*, 1993; *Watanabe et al.*, 1996; *Farrugia et al.*, 2003]. In the traditional development of the subject, Maxwell stresses on field lines newly opened at the equatorward side of the cusp have been considered the only agent responsible for some of these asymmetries (see e.g., *Lockwood et al.* [1990] on polar cap convection).

In this work we shall show that this view is incomplete. We do this by combining *in situ* magnetospheric plasma observations at two altitudes from spacecraft Cluster and Polar with particle precipitation, ion drift and field-aligned currents from low-altitude spacecraft FAST and DMSP, as well as complementary data sets on ionospheric convection, aurora, and magnetic deflections obtained from ground-based facilities. We shall show that consideration of magnetosphere ionosphere (m-i) coupling at the high latitudes downstream of the cusp is essential to complete this description. Coupling processes at high latitudes have hitherto been largely neglected, not least because observations from this region have up till recently been rather sparse. We shall limit ourselves to the IMF orientations given by the clock angle range 90-135°. (Parameter θ is the polar angle in the Y-Z plane of the GSM coordinate system). By focussing on this range of IMF orientations, we shall be able to study dawn - dusk asymmetries of the open magnetosphere introduced by IMF B_y. A different category of dawn-dusk asymmetry in auroral activity has been reported by *Akasofu and Tsurutani* [1984]. We note that a B_y component is generally present due to the Parker spiral orientation of the IMF in corotating solar wind streams [*Rich and Hairston*, 1994; *Burlaga*, 1995; *Vennerstrøm*, 2001].

We demonstrate that the high-latitude boundary layer [*Siscoe et al.*, 1991] (called hereafter HBL) is an additional source of vigorous momentum transfer from the solar wind to the ionosphere, at the dusk (dawn) side of the polar cap in the northern hemisphere during IMF $B_y < 0$ ($B_y > 0$) conditions. We propose that this aspect of M-I coupling contributes significantly to the asymmetric polar cap convection extending beyond the dawn-dusk terminator and the associated Svalgaard-Mansurov effect in the ground magnetic deflections at high latitudes.

The present study is a continuation of an initial investigation of M-I coupling at this locale based on data acquired during a Cluster-FAST-SuperDARN Søndrestrømfjord conjunction on Jan. 21, 2001 [*Farrugia et al.*, 2004b]. That study is here supplemented by additional ground-based observations of the auroral activity and by spacecraft data. Thus the additional data sets complement the pilot study in various respects: (i) plasma convection: extended documentation of the B_y-related flow channel in the polar cap, (ii) field-aligned currents (FACs), and (iii) auroral precipitation. On (i), we provide an extended documentation of the B_y-related flow channel in the polar cap. On (ii), additional information is given of the FACs responsible for the magnetosphere-ionosphere coupling at the HBL. On (iii), the two-dimensional auroral activity in the prenoon ($B_y > 0$) and postnoon ($B_y < 0$) sectors is carefully investigated.

The auroral signature referred to under (iii) is the high-latitude component of the activity which is commonly referred to as poleward moving auroral forms (PMAFs) [*Sandholt et al.*, 1986, 1990; *Fasel*, 1995]. This auroral activity, with its characteristic evolution in space and time, is endemic to the IMF orientation we selected ($B_z < 0$; $|B_y/B_z| > 1$) and consists of a sequence of brightening events in the pre- and postnoon sectors, well away from noon, each of which is followed by noonward expansions and forms advancing poleward. In recent case studies we have demonstrated how the sequential activation of dayside auroral forms during magnetopause reconnection events [*Sandholt et al.*, 2004a, b; *Farrugia et al.*, 2004a] can be related to the different stages in the evolution of open field lines in the dayside magnetosphere. In line with our stated intentions here, we shall concentrate on the later, higher latitude (>75° magnetic latitude) phase of these auroral events. During IMF $B_y < 0$ conditions these are relatively weak, recurrent 630.0 nm forms expanding antisunward in the region of downward-directed, cusp field-aligned current. This cusp current in this case is coupled to the aforementioned flow channel in the postnoon polar cap. The relevant aurora during B_y positive conditions appears in the prenoon sector of the polar cap boundary as a more intense form, strongly manifest also in the green line emission. The local auroral forms in the pre- and postnoon sectors are placed in the larger context of dayside particle precipitation regimes and plasma convection (i.e. items (i) and (ii) above).

The correspondence between the aurora and the cusp FAC system (the cusp currents of *Taguchi et al.* [1993]), for positive IMF B_y conditions, has been reported by *Farrugia et al.* [2003] and *Sandholt et al.* [2004b]. Magnetic field lines at these latitudes are often referred to as old open field lines, reflecting the fact that "time since reconnection" [*Smith*, 1994] is $>\sim 5$ min. Our distinction between two different stages of M-I coupling along open field lines ("newly open" versus "old open") is consistent with the two-branch open field line model of solar wind-magnetosphere coupling proposed by *Vasyliunas* [1995] on theoretical grounds.

In other work *in situ* magnetic field and plasma observations in the plasma mantle have been obtained by Cluster during the passage of open magnetic flux tubes [*Thompson et al.*, 2004]. These events were called high-latitude flux transfer events. They consist of reversals in the local GSM magnetic field B_z component, recurring at 8 min intervals, and interpreted as disturbances caused by spatially confined currents flowing perpendicular to the magnetic field. The field-aligned current aspect of such events, as well as ionospheric signatures, such as pulsed ionospheric flows and mantle type ion precipitation, have been documented in the recent single case study by *Farrugia et al.* [2004b]. A further documentation will be given in the present paper.

This paper is organized as follows: An initial description of the dayside auroral morphology for conditions characterized by IMF $B_z < 0$ ($|B_y/B_z| > 1$) is followed by the presentation of three case studies. In these we combine and inter-compare observations acquired at different altitudes within the M-I system to obtain a comprehensive view of the spatio-temporal structure of solar wind-magnetosphere reconnective coupling. In the discussion section we focus on the spatio-temporal structure of the dayside aurora in their progression from lower to higher latitudes. The corresponding FAC structure and plasma convection pattern (rotational convection reversal and flow channel at its poleward boundary) may be discussed in relation to theoretical predictions and MHD modelling efforts on dynamo processes in the coupled system of low-latitude (reconnection layer) and high-latitude boundary layers of the dayside magnetosphere [*Siscoe et al.*, 1991, 2000; *Sonnerup and Siebert*, 2003; *Tanaka*, 2003]. A summary is given at the end.

GENERAL REMARKS ON DAYSIDE AURORAL ACTIVITY

A schematic illustration of the association between auroral forms/activities and plasma convection for positive and negative IMF B_y conditions is shown in Figure 1. Motion of auroral forms are indicated by the light arrows. The schematical

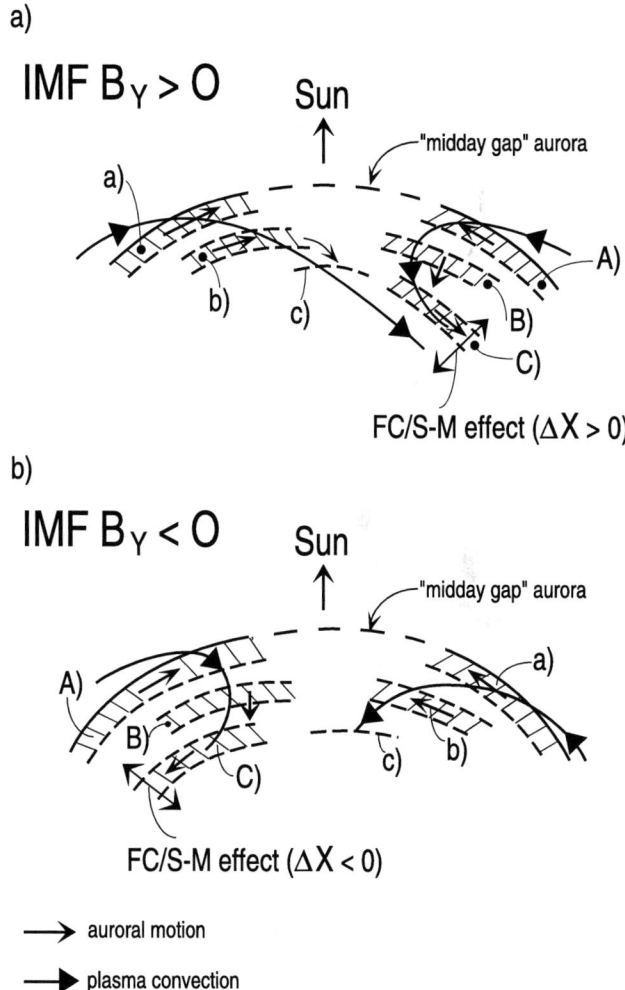

Figure 1. Schematic illustration of the association between dayside auroral forms/activities and plasma convection corresponding to positive (top) and negative IMF B_y conditions ($B_z < 0$). Aurorae and their motions are indicated by dashed lines and lighter arrows. Plasma flow channels (FC) and the associated Svalgaard-Mansurov (S-M) effect in the X component of the ground magnetic deflection (ΔX) are marked by double-arrowed lines. Magnetic noon is at the top and the postnoon side is to the left.

convection patterns are extracted from the statistical study of *Weimer* [1995]. We distinguish between auroral forms associated with the dusk and dawn convection cells, which are strongly asymmetric (round versus crescent-shaped) during the IMF B_y-dominated conditions. Auroral forms marked A - B - C and a - b - c represent a spatio-temporal evolution of the individual auroral brightening events (sequential activations/ motions) associated with the two convection cells. While forms A-B and a-b represent the early stages of the auroral events in the pre- and postnoon sectors (newly open field lines), forms C-c represents the later stage of event evolution, which occur on old open field lines. A characteristic vorticity in the A-B-C auroral activity is evident, with counter-clockwise rotation (looking down onto the polar ionosphere) in the $B_y > 0$/prenoon case, and clockwise in the $B_y < 0$/postnoon case.

The evolution from stage A/a to C/c is the activity often referred to as poleward moving auroral forms (PMAFs) [*Sandholt et al.*, 1986]. This auroral activity appears in strong contrast to the much weaker and more stable emission in the midday sector which is often referred to as the "midday gap aurora". The latter aurora is characterized by the weak (<1 kR) green line emission at 557.7 nm. Particularly when IMF $B_y > 0$, the midday cusp aurora is strongly attenuated compared to its intensity on either side of noon [*Sandholt et al.*, 2004b]. This occurs when the IMF B_y - regulated cusp current is directed into the ionosphere and the mantle current at the poleward edge of the cusp is directed out of the ionosphere [*Lee et al.*, 1985].

DATA DESCRIPTION

Overview of Observations for IMF $B_y < 0$

Plate 1 shows the observation geometry in magnetic latitude versus magnetic local time (MLT) coordinates for three B_y negative cases: (i) Nov. 30, 1997, (ii) Jan. 21, 2001, and (iii) December 7, 2000. In this paper we shall emphasize the Jan. 21, 2001 and Nov. 30, 1997 data. The figure shows (i) spacecraft trajectories or their magnetic footprint, on which are marked various Universal Times needed in the description; (ii) FACs (i.e., moving from higher to lower latitudes, the cusp currents C1 and C2 and the traditional Regions 1 and 2 currents) and their directions; (iii) auroral forms labeled A, B, C, D, and arcs detected from spacecraft data.

For Jan 21, 2001 we shall discuss: (I) Spacecraft FAST data at an altitude of about 1000 km: FACs and particle precipitation along the track indicated in Plate 1 (~1300 MLT meridian), (II) Spacecraft Cluster (The magnetic footprint of spacecraft Cluster (12 R_E), located at 1300 MLT/76° magnetic latitude, is marked by blue star): magnetospheric plasma and magnetic field observations, and (III) Spacecraft DMSP F13 data (black arrowed line): particle precipitations, ionospheric ion drift, and magnetometer data revealing FAC structure, (IV) Ground based observations of ionospheric ion drift from SuperDARN radars and the Søndrestrømfjord incoherent scatter radar in Greenland (the latter data not shown).

The latitudinal location of a channel of strong eastward (antisunward) convection in the dusk side polar cap, as obtained from DMSP spacecraft and the Søndrestrøm radar, is marked by red bars (along the FAST track for Jan. 21, 2001). The location of discrete arcs, as inferred from the spacecraft data, are indicated by a bar along the track, on the equatorward side of the convection channel (red sector).

Auroral observations on Nov. 30, 1997 (113 0UT) are represented by the fields of view of the all-sky camera (circle) and meridian scanning photometers (double-arrowed meridional line along 1500 MLT meridian) in Ny Ålesund. Four latitudinally separate auroral forms are marked A, B, C, and D in the figure. Additional DMSP F13 tracks under very similar IMF conditions as Jan. 21, 2001 are shown by green (Nov. 30, 1997) and blue (Dec. 7, 2000) arrowed lines.

Case 1: Nov. 30, 1997 (Postnoon): DMSP F13-Ground Conjunction

The IMF data for this day are obtained from spacecraft Wind located at (196, –3, 27) R_E.

We shall concentrate on the interval from 0900-1100 UT (Figure 2), an interval following the onset of a high-speed stream at ~08:30 UT (data not shown). A sharp southward turning was recorded at 1002 UT. Between 1002 UT and 1030 UT the field pointed southwest (average $B_z = -5$; $B_y = -5$ nT; $\theta \sim 135°$). Later it rotated toward a more due west orientation. B_y remained at –5 nT, but B_z became slightly positive after 1040 UT. B_x remained positive throughout. The delay time for the IMF seen at Wind to affect the cusp ionosphere may be calculated by using the average convection speed of the solar wind (420 km/s). The sharp southward turning at 1002 is estimated to affect the cusp ionosphere at 1056 UT, after a propagation delay of 54 min. A strong auroral brightening was recorded at the station in Ny Ålesund (1330 MLT) at 1100 UT. Thus, the aurora is expected to be controlled by the southwest directed IMF from appr. 1100 to 1140 UT. After that the IMF affecting the Earth is directed almost due west.

Plate 2 shows meridian scanning photometer observations of the red line aurora from Ny Ålesund on Nov. 30, 1997 during a representative interval of the IMF conditions prevailing after the southward turning of the IMF recorded by spacecraft Wind at 1002 UT. Line-of-sight intensities at 630.0 nm are plotted as a function of zenith angle and time. The photometer field-of-view along the 1500 MLT meridian is indicated in Plate 1.

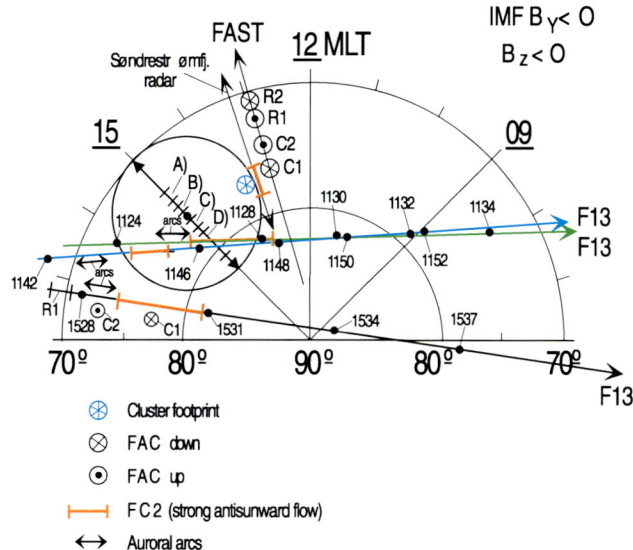

Plate 1. Overview of observations in the dayside magnetosphere-ionosphere system as obtained from the IMF B_y negative cases reported below. The emphasis is placed on (i) aurora, (ii) plasma convection, (iii) particle precipitation and (iv) FAC structure. The reference frame is magnetic local time (MLT)/magnetic latitude (MLAT). See text for details.

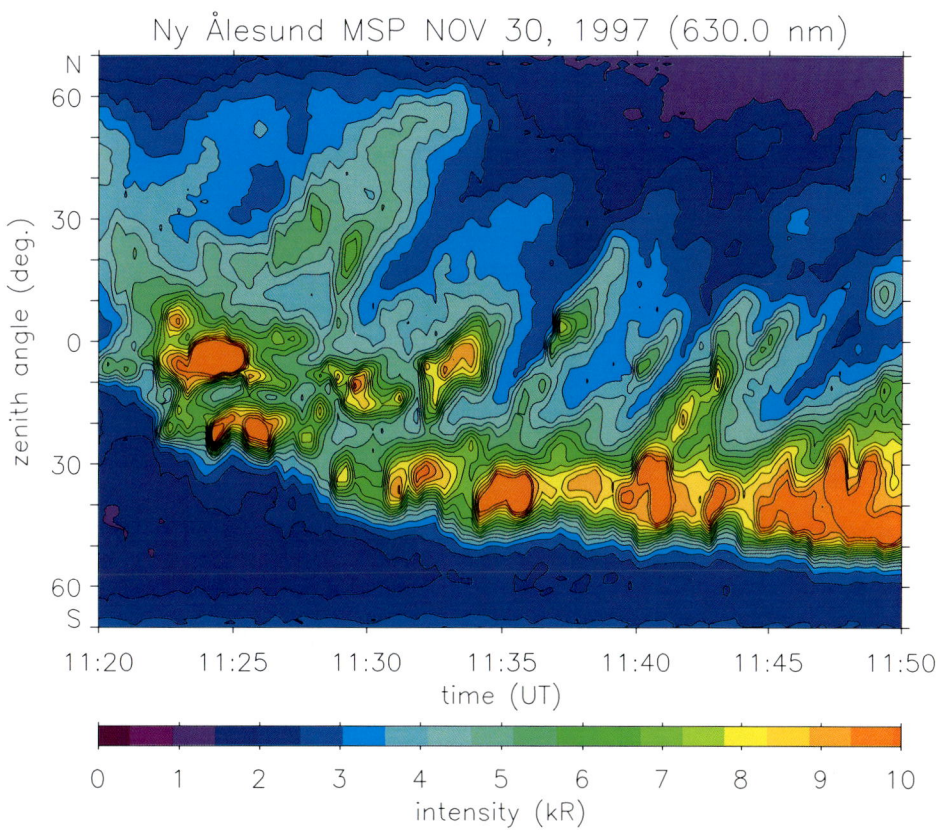

Plate 2. Meridian scanning photometer (MSP) observations of the 630.0 nm aurora from Ny Ålesund (line of sight auroral intensity versus zenith angle and time) during the interval 1120-1150 UT on Nov. 30, 1997. North is up. The auroral intensity is color-coded according to the scale at the bottom.

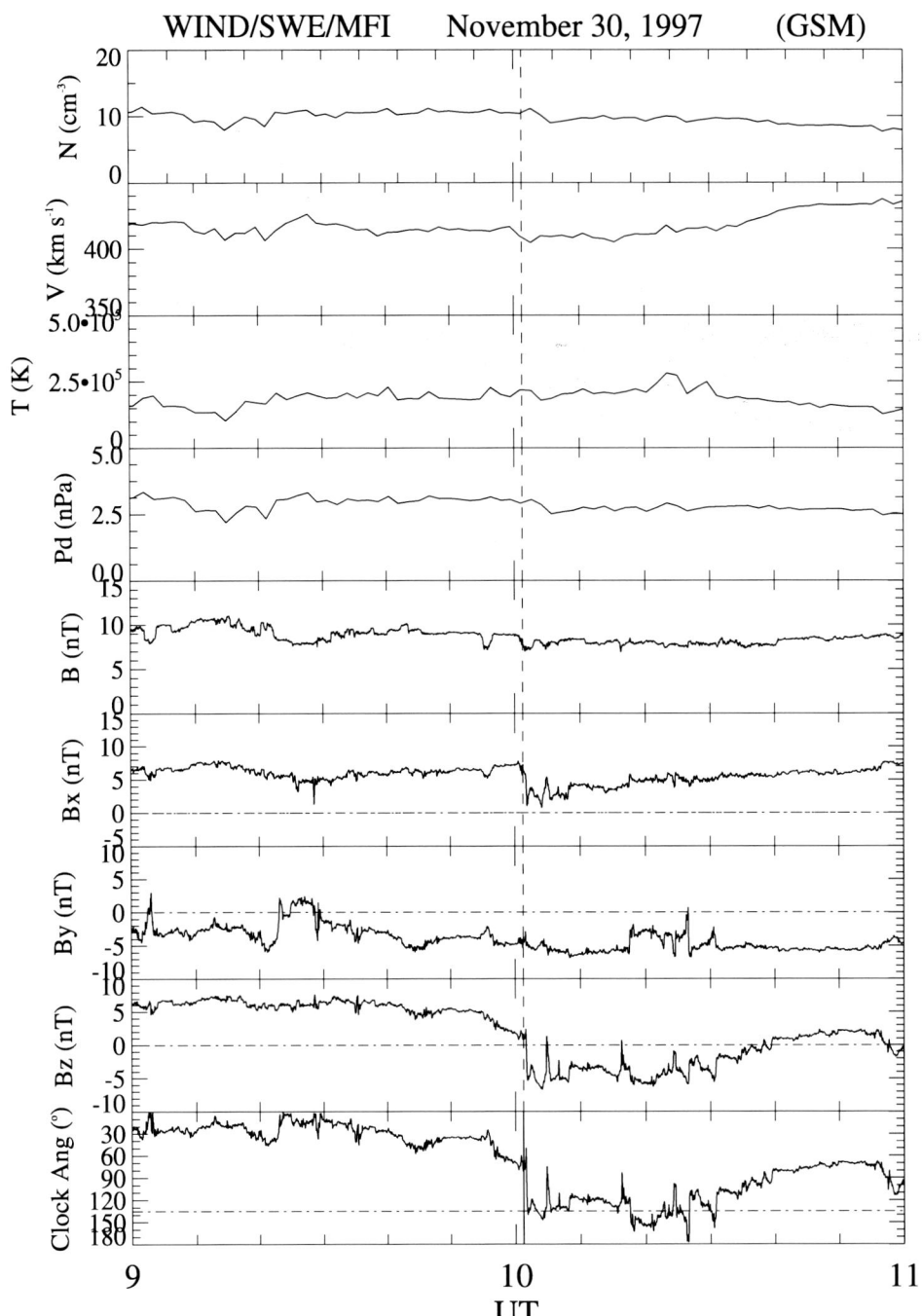

Figure 2. Solar wind plasma and IMF data from spacecraft Wind. Panels from top to bottom shows (i) solar wind density, (ii) velocity, (iii) temperature, (iv) dynamic pressure, (v) total magnetic field, (vi) the field components B_x, B_y, B_z in GSM coordinates, and (vii) the IMF clock angle.

The auroral activity during the period 1122-1150 UT consists of a sequence of brightening events. Each event consists of a 1-2 min-long intensification near the equatorward boundary of the pre-existing luminosity, which is followed by brightenings located at successively higher latitudes. The brightening forms corresponding to each event are observed within the photometer field of view for a period of 5-10 min. Seven major events occurred within 1122-1150 UT, implying a mean recurrence time of 4 min. The 3-4 latitudinally separate forms that are successively activated in each event are those marked A, B, C, and D in Plate 1. Within 1130-1140 UT we can identify 5-6 brightenings of the equatorward boundary (red spots in lower section of the plot). Three of these are followed by poleward moving auroral forms all of which reach latitudes poleward of zenith. This activity consists of the successive brightening of latitudinally separate forms. Four latitudinally separate forms are seen during the intervals 1120-1130 and 1140-1150 UT. The northernmost form is very weak. The two-dimensional aspects of this activity are illustrated below.

By the all-sky camera images in Plate 3 we document the two-dimensional evolution of the auroral events depicted in the scanning photometer plot in Plate 2. We shall describe the activity observed during the interval 1130-1140 UT. Initially we focus on the 1132 UT image (center, top panel). The brightening of the southernmost form (A) at this time appears as a red spot in the lower part of the photometer plot in Plate 2. This brightening form intersected the local meridian at ~30-35° south of zenith. A second form, which we label form B in Plate 1, is centered at ~5° south of zenith. A third, weaker form (C) is located at ~40-50° north of zenith. The direction of motion of the luminosity structures representing forms A, B, and C are westward (noonward), poleward, and eastward (antisunward), respectively. The latter motion appears when comparing the three images in the upper panel (1130, 1132, and 1134 UT) in video mode. Thus, the auroral brightening events are characterized by a clockwise vorticity (looking down). This is most clearly seen when the image sequence is inspected in the video mode.

Plate 4 shows particle precipitation, ionospheric ion drift and magnetic deflection components obtained from spacecraft DMSP F13 during the dusk - dawn pass on Nov. 30, 1997 (that which is marked by a green line in Plate 1). We shall focus on the following features of particle precipitation, ionospheric ion drift, and field-aligned currents (not shown): (i) Boundary plasma sheet (BPS) precipitation, Region 1 FAC, and sunward convection in the latitude range appr. 74.5-76°. (ii) a strong keV electron arc in the vicinity of the convection reversal, poleward of the Region 1 current, within 76-77° magnetic latitudes, and (iii) a channel of strong antisunward convection within the regime of mantle/polar rain precipitation, at 77-82° magnetic latitudes.

Case 2: Jan. 21, 2001 (Postnoon): Cluster-Fast-Søndrestrøm Conjunction

In this subsection we discuss M-I coupling in the postnoon sector during stable IMF B_y negative conditions by relating data sets obtained at different altitudes during a fortuitous conjunction of the magnetic footprints in the ionosphere of spacecraft Cluster (altitude: ~12 R_E) and spacecraft Fast (altitude: ~1000 km). The conjunction, which occurred in the close vicinity of the incoherent scatter radar in Søndrestrømfjord, Greenland, is illustrated in Plate 1 (see also *Farrugia et al.* [2004b]). These data are supplemented by a plot of the instantaneous large-scale plasma convection pattern obtained by the SuperDARN network of ground-based radars and particle precipitation data acquired during a dusk-dawn pass of spacecraft DMSP F13.

IMF Data

The upper section of Figure 3 shows IMF observations from the ACE/MAG instrument for the time interval 1430-1630 UT. ACE was located at (242.5, –1.7, 22.4) R_E (GSE coordinates). The ACE - Cluster propagation delay has been estimated to be 84 min. (see *Farrugia et al.* [2004b]). The ACE measurements have been shifted forward by this amount. After the field perturbation at ~1500 UT the IMF is a steady, sunward-tilted ($B_x > 0$) field with strong westward ($B_y = -9$ nT) and southward ($B_z = -8$ nT) components corresponding to $\theta \sim 130°$. We note that the DMSP satellite pass considered in this paper (Plate 1) occurred within this interval. The lower section of Figure 3 shows IMF observations from spacecraft Wind on Dec. 3, 1997, which is our B_y positive case, reported in a later section.

Plate 5 shows DMSP F13 data for the interval 1519-1546 UT. The satellite track is marked in Plate 1. We shall focus on the following features: (i) a zone of BPS type precipitation/sunward convection in the regime of Region 1 FAC on both sides of the polar cap, (ii) the presence of two distinct polar arcs (keV electrons) at the convection reversal, in the regime of cusp (C2) FAC (directed out of the ionosphere), poleward of the Region 1 current at dusk (17 MLT), (iii) a channel of strong antisunward convection and weak cusp (C1) FAC (directed into ionosphere) in the latitude range 74-85° magnetic latitude, characterized by very weak electron precipitation typical of the polar rain.

Cluster and Fast Data

Plate 6 shows field and flow data from Cluster 1, situated in the postnoon sector (15 MLT) HBL, downstream of the cusp ($B_x > 0$) at an altitude of 12 R_E and an invariant latitude of 79.5°. The magnetic footprint of Cluster at 1545 UT is given in Plate 1. Plotted from top to bottom are the density, pairwise flow and field components in GSM coordinates, and the total field and flow. The bottom panel shows the variation in the east-west magnetic field component at FAST around the time of its magnetic conjugacy with Cluster 1. During

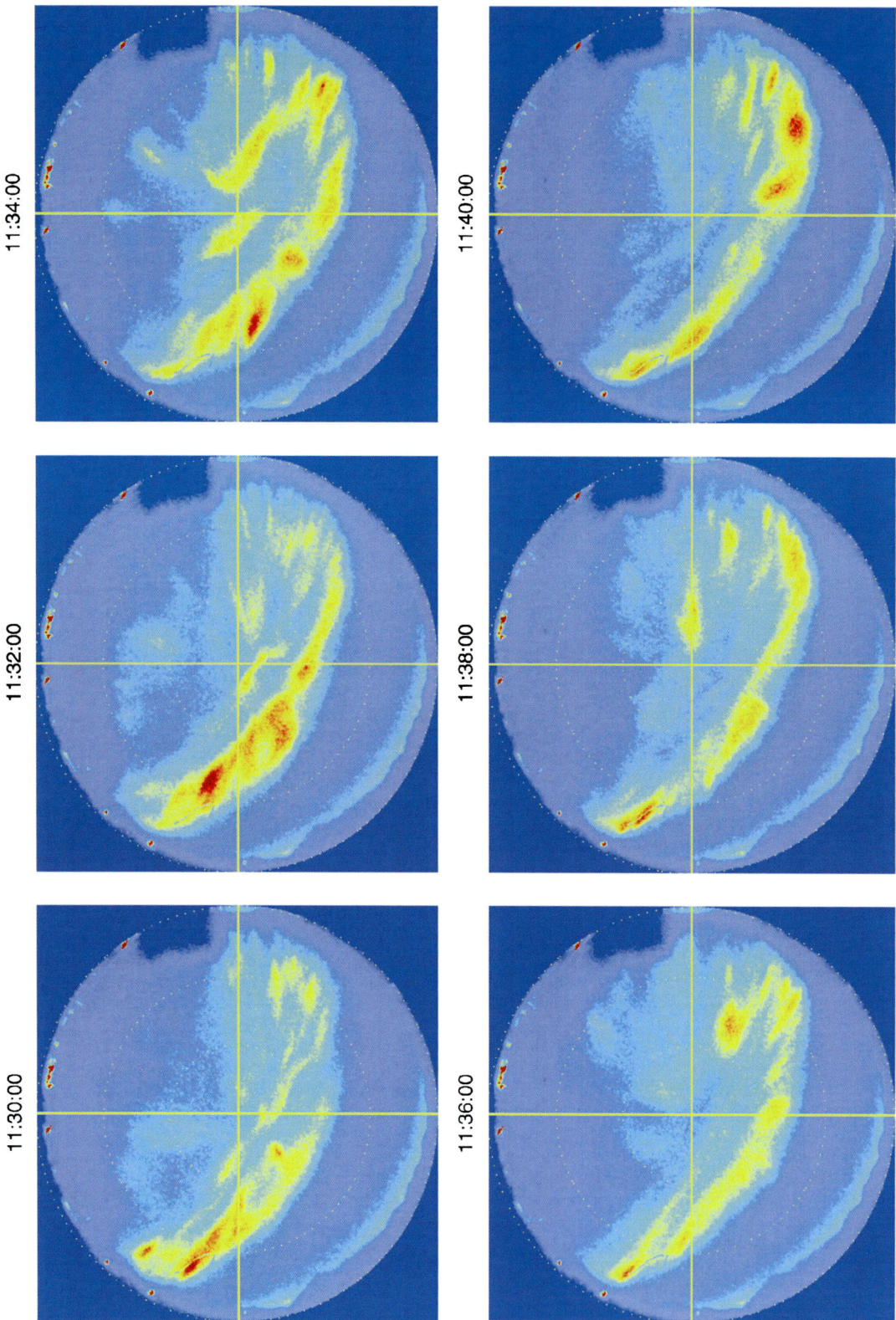

Plate 3. Sequence of all sky camera (ASC) images at 630.0 nm obtained from Ny Ålesund for the interval 1130-1140 UT. The central vertical yellow line in each frame marks the photometer meridian at ~1500 MLT. North is up and west to the left. Zenith angles 30, 60 and 90° are marked by concentric circles.

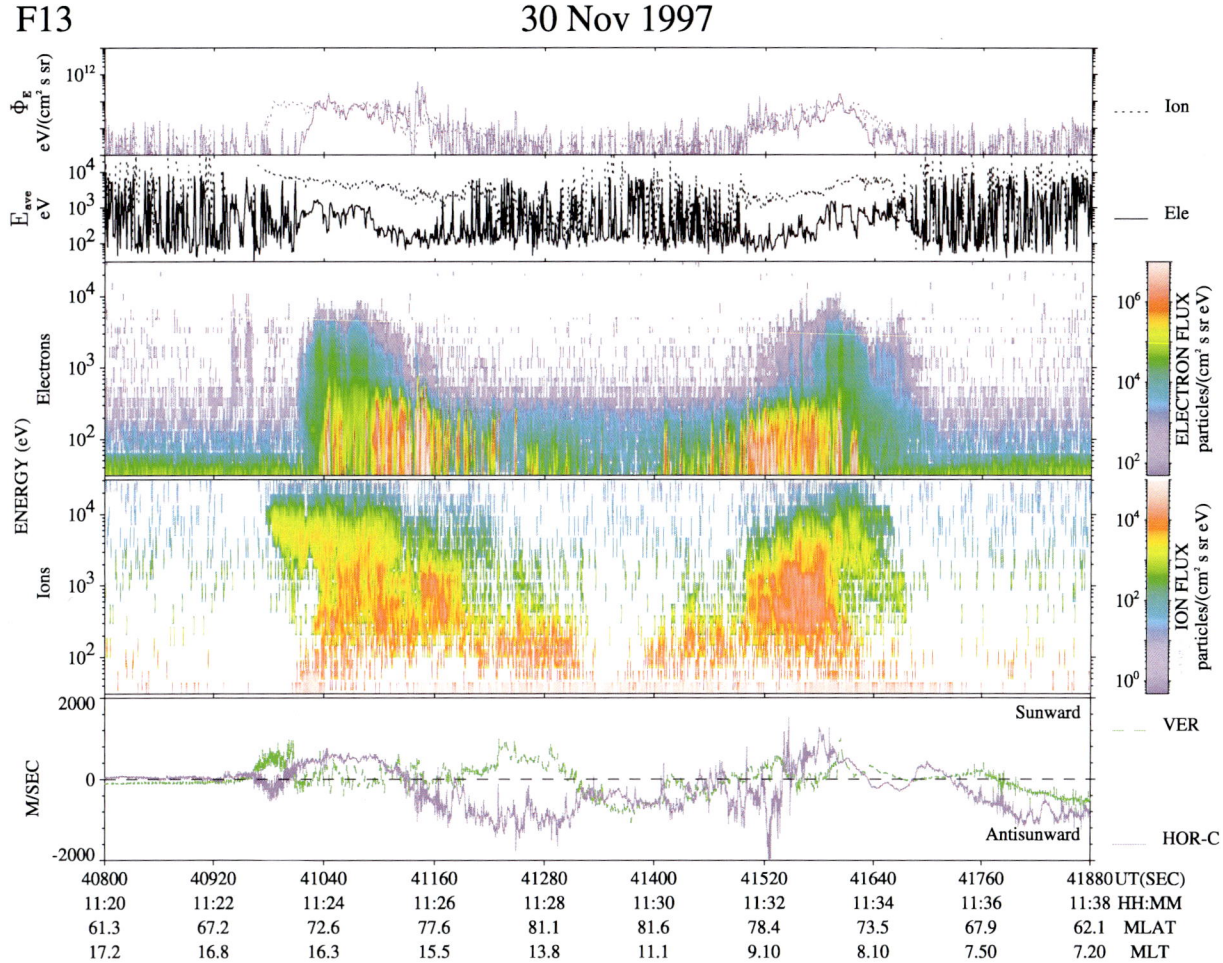

Plate 4. Particle precipitation, ionospheric ion drift, and magnetic deflections obtained from spacecraft DMSP F13 during the interval 1120-1138 UT on Nov. 30, 1997. Panels from top to bottom shows: (i) electron and ion (dotted line) differential energy fluxes, (ii) electron and ion (dotted line) average energy, (iii) color-coded electron differential flux versus energy within the range 30 eV-30 keV, (iv) color-coded ion differential flux, (v) vertical (green) and horizontal (cross - track; violet) ion drift.

222 IMF B_Y AND SPATIO-TEMPORAL STRUCTURE

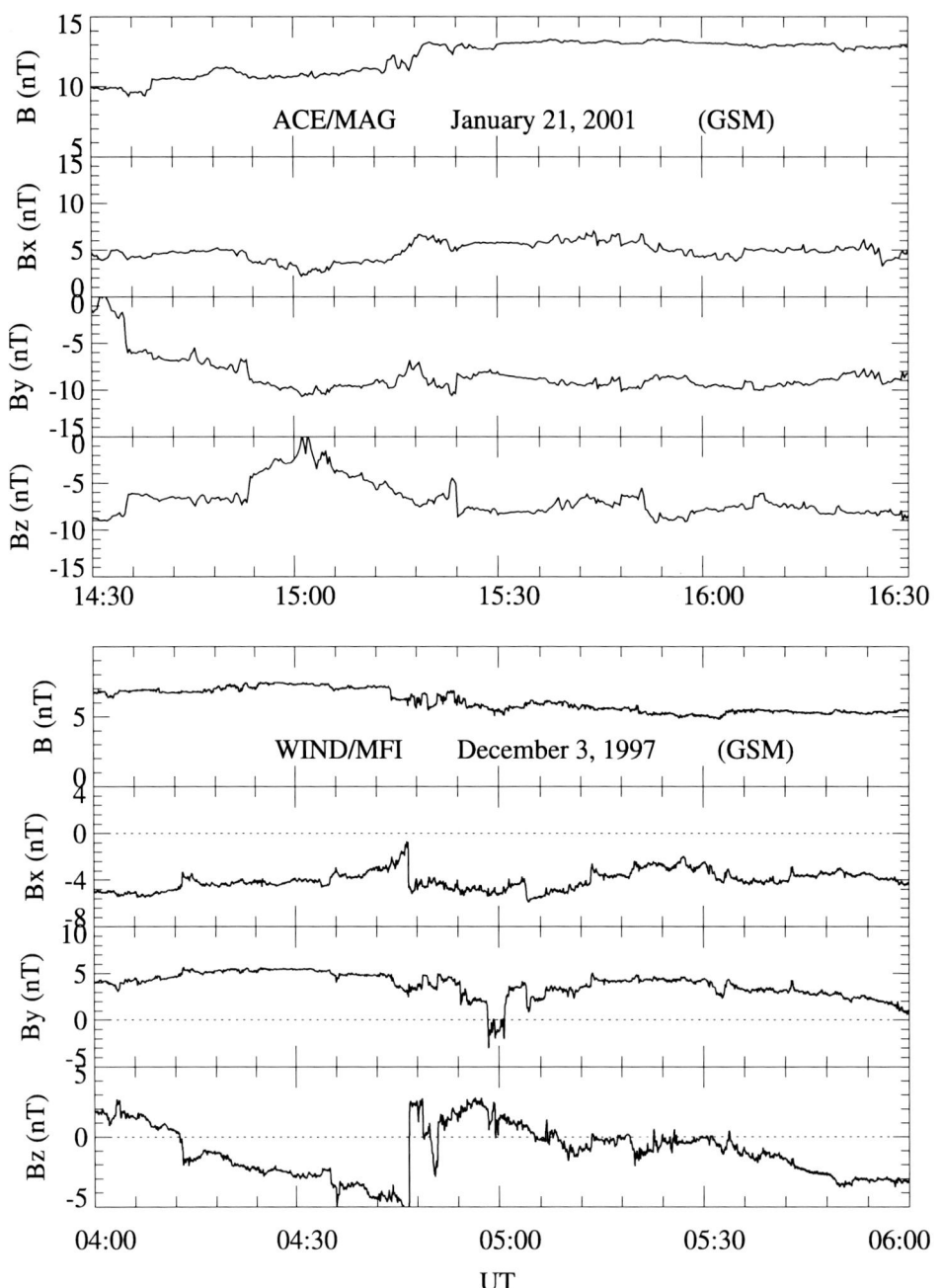

Figure 3. Upper section: The total magnetic field and field components B_x, B_y, and B_z recorded by spacecraft ACE during the interval 1430-1630 UT on Jan. 21, 2001. 84 min has been added to the ACE UT for comparison with ground data. Lower section: The total magnetic field and field components B_x, B_y, and B_z recorded by spacecraft WIND during the interval 0400-0600 UT on Dec. 3, 1997. In this case a 70 min propagation delay has been added for comparison with ground data.

the interval shown (1543-1546 UT), FAST is moving equatorward along the 1300 MLT meridian, traversing the MLAT range 79-68° (see Plate 1). Different FAC regimes were recorded along the FAST orbit, as marked in the figure. The cusp currents C1 and C2 are inward and outward directed currents in the regions of mantle and LLBL-type precipitations, respectively, while R1 (out) and R2 (in) are the traditional Region 1 and Region 2 currents.

About eight flow bursts ($\Delta v > 0$) may be seen exceeding the background value of 140 km/s. Each flow bursts is

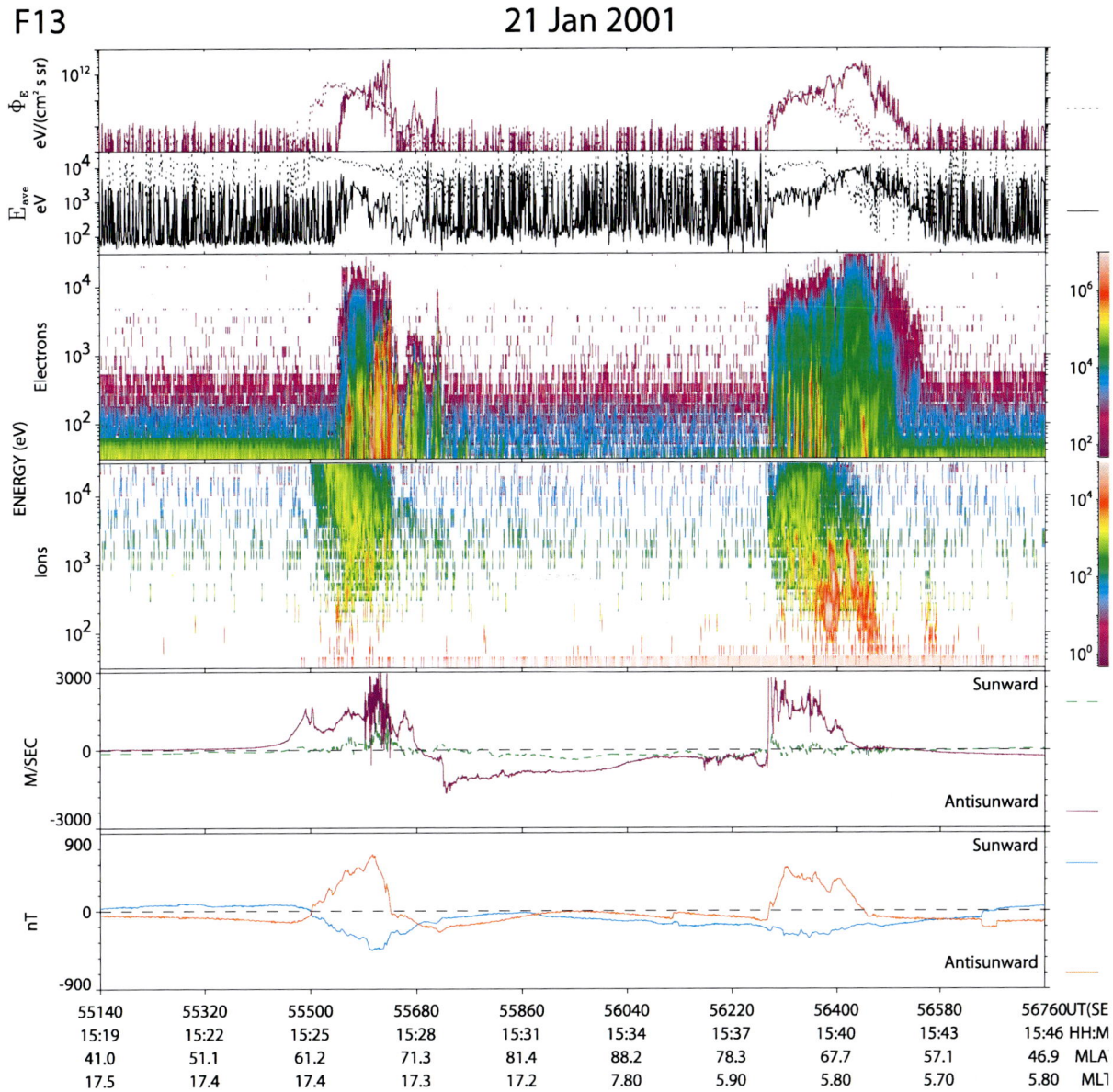

Plate 5. DMSP F13 data for the interval 1519-1546 UT. Panels from top to bottom shows (i) differential energy fluxes of electrons and ions (dotted), (ii) color-coded electron and (iii) ion spectrograms, (iv) ion drifts perpendicular to the satellite track (blue), and (v) horizontal magnetic deflections transverse to (B_z) and along (B_y) the track.

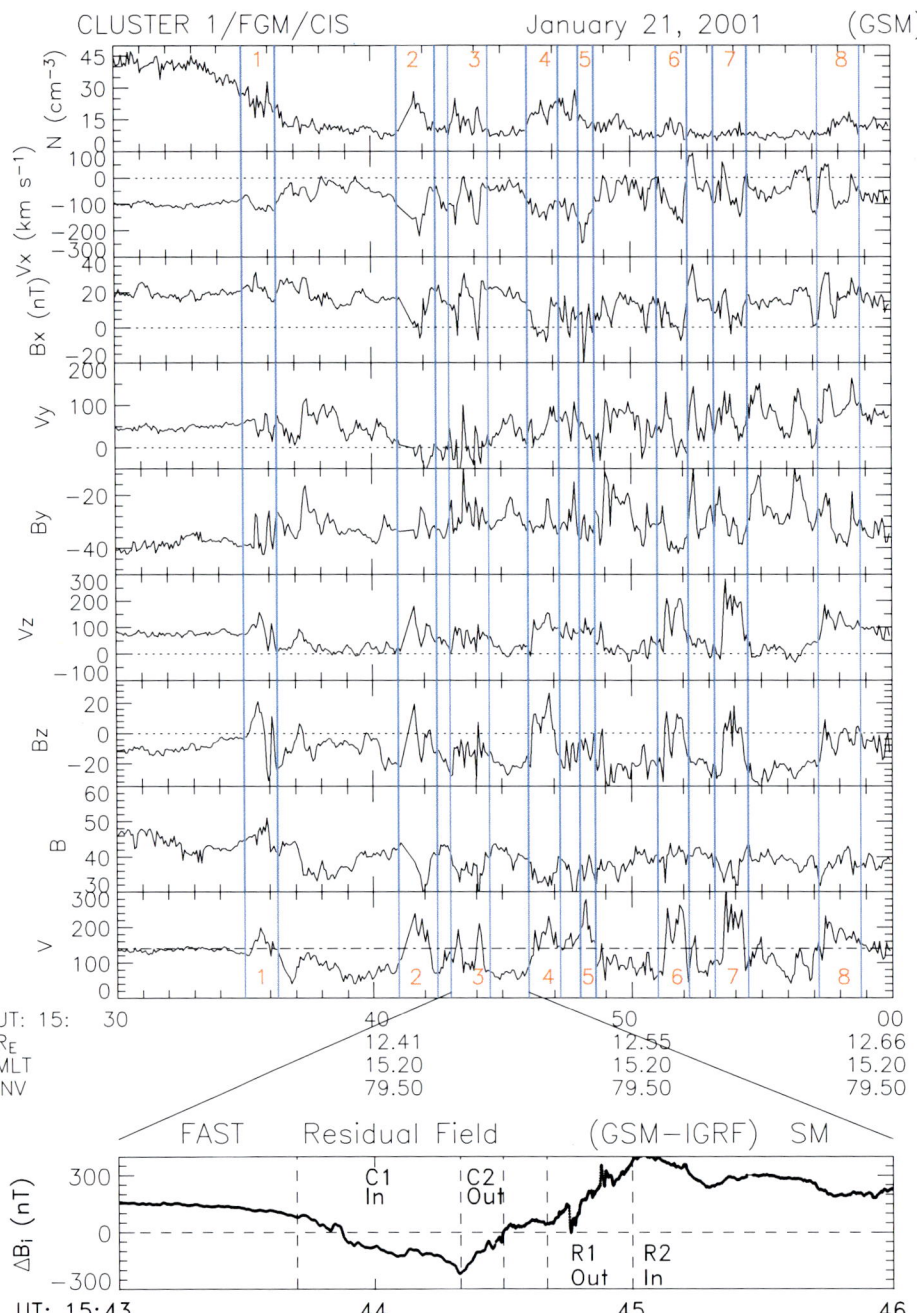

Plate 6. Field and flow data from Cluster 1, situated in the HBL, downstream of the cusp ($B_x > 0$) at invariant latitude of 79.5° (see footprint in Figure 2). The flow bursts (bottom panel) are accompanied by perturbations in the both the field and flow components, which satisfy the Walen relation. The bottom panel shows the deflection in the east-west magnetic field component at FAST around the time of its magnetic conjugacy with Cluster 1 (see Plate 1). After *Farrugia et al.* [2004b].

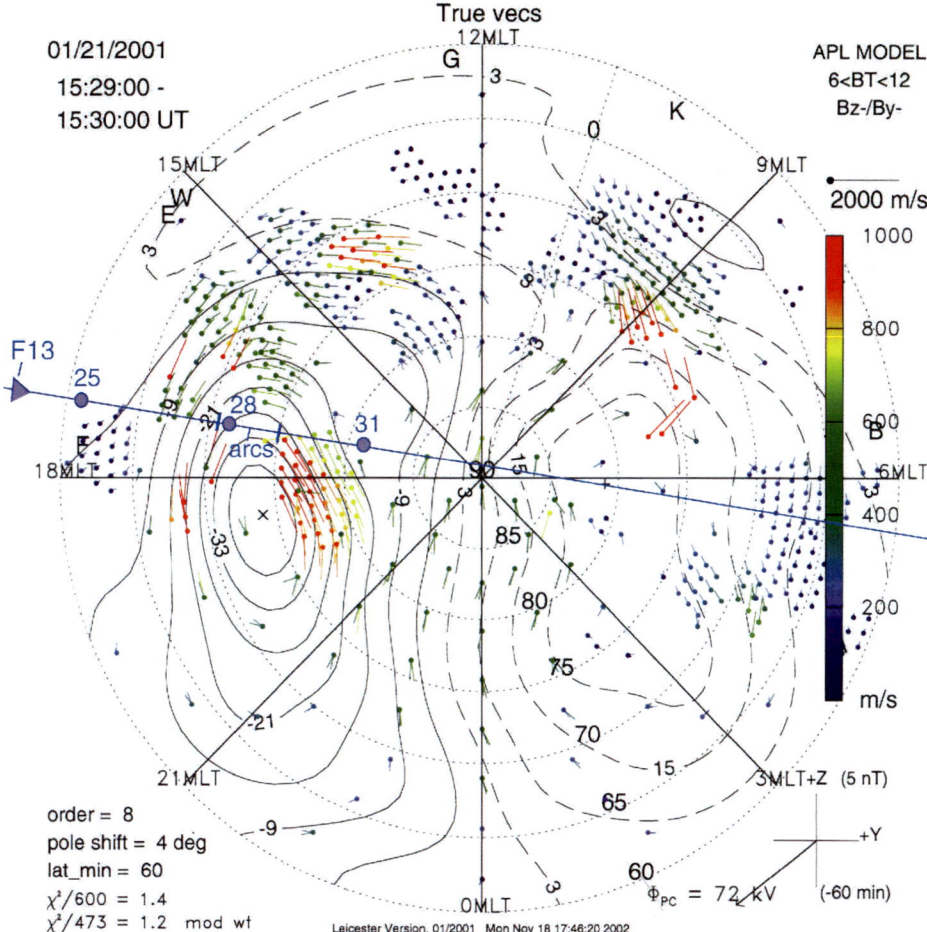

Plate 7. Spatial plot of plasma convection vectors and streamlines obtained by ground based radars of the SuperDARN network. The coordinate system is magnetic local time and magnetic latitude (60-90°). Noon is up and dusk to the left. Convection velocities are color-coded according to the scale at right. The track of the DMSP F13 spacecraft has been drawn in.

accompanied by deflections in the velocity and magnetic field components, satisfying the Walen relation. These disturbances are particularly strong in the V_x (negative) and V_z (positive) components, indicating a flow predominantly deflected tailward and northward. Note the general tendency for the density to increase within the events. This indicates a connection to the dense magnetosheath plasma. The characteristic sequence of B_z reversals associated with the events is similar to the high-latitude flux transfer events reported by *Thompson et al.* [2004] and mentioned in the introduction.

FAST particle data [*Farrugia et al.*, 2004b] shows that the various FAC regimes are associated with characteristic particle precipitation signatures: (i) The cusp current C1 with mantle-type precipitation, (ii) the cusp current C2 with LLBL-type, (iii) R1 with boundary plasma sheet (BPS), and (iv) R2 with central plasma sheet (CPS) precipitations. Polar rain precipitation is seen poleward of the C1 current. The cusp currents C1-C2 comprise the region of stepped cusp precipitation, i.e., ions whose low-energy cutoff increases with decreasing latitudes. This type of ion dispersion has been interpreted in terms of pulsed reconnection [*Lockwood and Smith*, 1992].

SuperDARN Plasma Convection Data

Plate 7 shows a spatial plot of plasma convection ($\vec{E} \times \vec{B}$ - drift) at 1530 UT obtained by the ground-based radars of the SuperDARN network [*Greenwald et al.*, 1995]. The radar data (ion drift vectors) are supplemented by a theoretical model (APL) of the electrostatic potential to obtain maps of plasma convection streamlines over the polar cap. Plasma flow vectors are obtained by using a beam-swinging algorithm, as explained by *Villain et al.* [1987] and *Ruohoniemi et al.* [1989].

The track of spacecraft DMSP F13 is indicated by blue arrowed line, with min after 1500 UT marked. We shall focus on the pass across the dusk sector during the interval 1525-1531 UT. The crossing of discrete auroral arcs at 1528-1529 UT (see Plate 1) is marked in the convection figure. We notice the fast (red) noonward directed convection at either side of noon (centered at 0900 and 1400 MLT) and the channel of fast, antisunward convection at dusk, within 75-80° magnetic latitude. The latter convection channel is located immediately poleward of the discrete arcs recorded by spacecraft DMSP F13 (see Plate 5).

Case 3: Dec. 3, 1997 (Prenoon): Polar-Ground Conjunction

Figure 4 shows a schematic summary of the auroral configuration on Dec. 3, 1997, as inferred from the ground observations in Ny Ålesund during intervals centered at 0550 UT (A), 0740 UT (B), 0950 UT (C), and 1050 UT (D). These

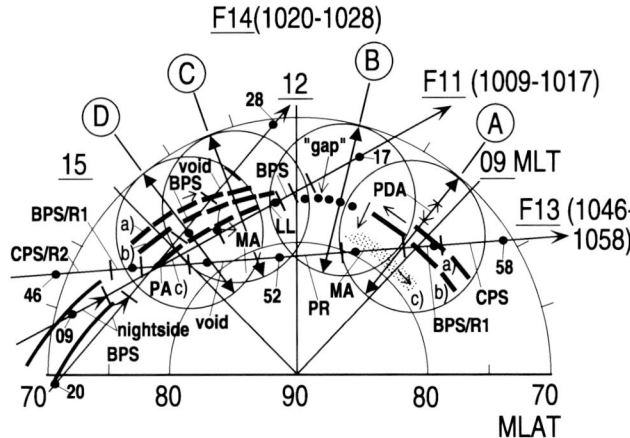

Figure 4. Schematic illustration of dayside auroral configuration applicable to the interval 0550-1100 UT on December 3, 1997. The coordinate system is MLAT/MLT. The fields-of-view at 630.0 nm of the MSP and all-sky camera (ASC) in Ny Ålesund for 0550 UT (A), 0740 UT (B), 0950 UT (C), and 1050 UT (D) are marked by double-arrowed meridional lines and circle, respectively. The tracks of DMSP spacecraft F11, F13, and F14 during the intervals 1009-1017 (F11), 1020-1028 UT (F14), and 1046-1058 (F13), respectively, are indicated. The types of precipitation regimes encountered are marked along the tracks. Multiple, fragmented BPS arcs in the pre- and postnoon sectors are marked a, b, as well as directions of auroral motion (arrows). A deep minimum of auroral intensity in the ~1100-1200 MLT sector is marked with solid dots and labeled "gap".

intervals were selected in order to demonstrate typical features of the aurora in the pre-noon, midday, and postnoon sectors during the prevailing IMF $B_y > 0$ conditions on this day.

The phenomenon of "midday gap aurora" located in the ~1100-1200 MLT sector is characterized by a deep minimum in auroral intensity/activity when compared with the pre- and postnoon sector activities [*Dandekar and Pike*, 1978; *Sandholt et al.*, 2004b]. The multiple, fragmented BPS arcs in the prenoon (~0800-1000 MLT) and postnoon (~1230-1600 MLT) sectors, characterized by noonward motions (see arrows), are shown (labels a and b), as well as the "midday gap aurora" (solid dots; "gap") in the ~1100-1200 MLT sector. Particle precipitation and FAC regimes are marked by bars along the tracks of DMSP spacecraft F11, F13, and F14 during the intervals 1009-1017 UT (F11), 1020-1028 UT (F14), and 1046-1058 UT (F13). The prenoon sector perspective reveals, in addition to the BPS arcs located in the vicinity of 75° MLAT, the presence of a pulsed, diffuse aurora (labeled PDA) to the south, within the regime of CPS precipitation, and a pulsing mantle (labeled MA) type aurora

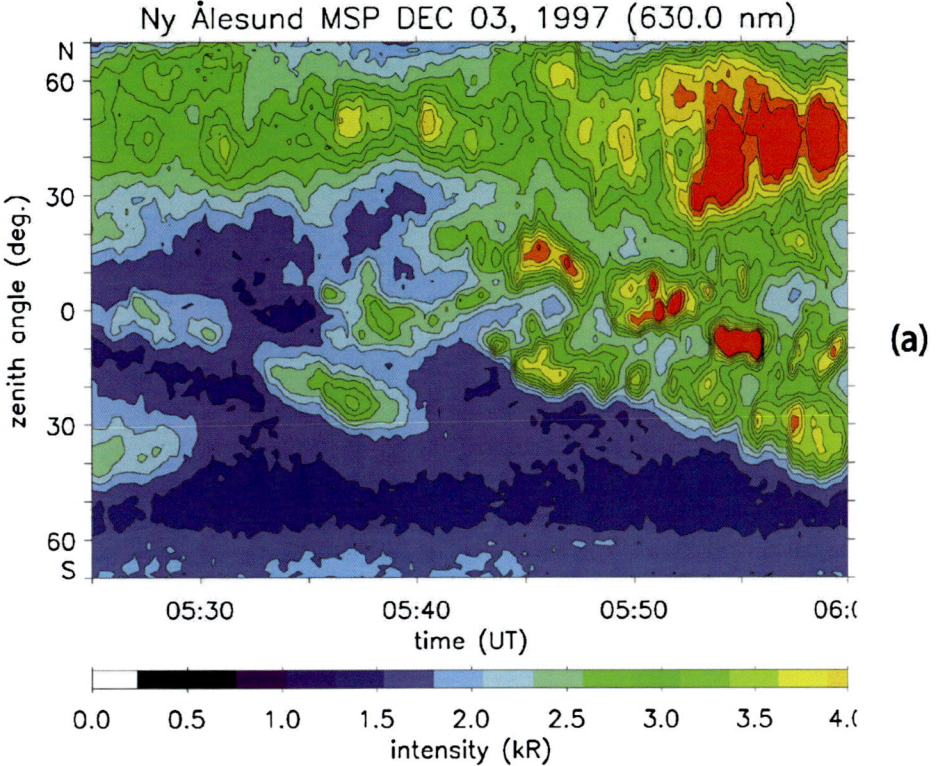

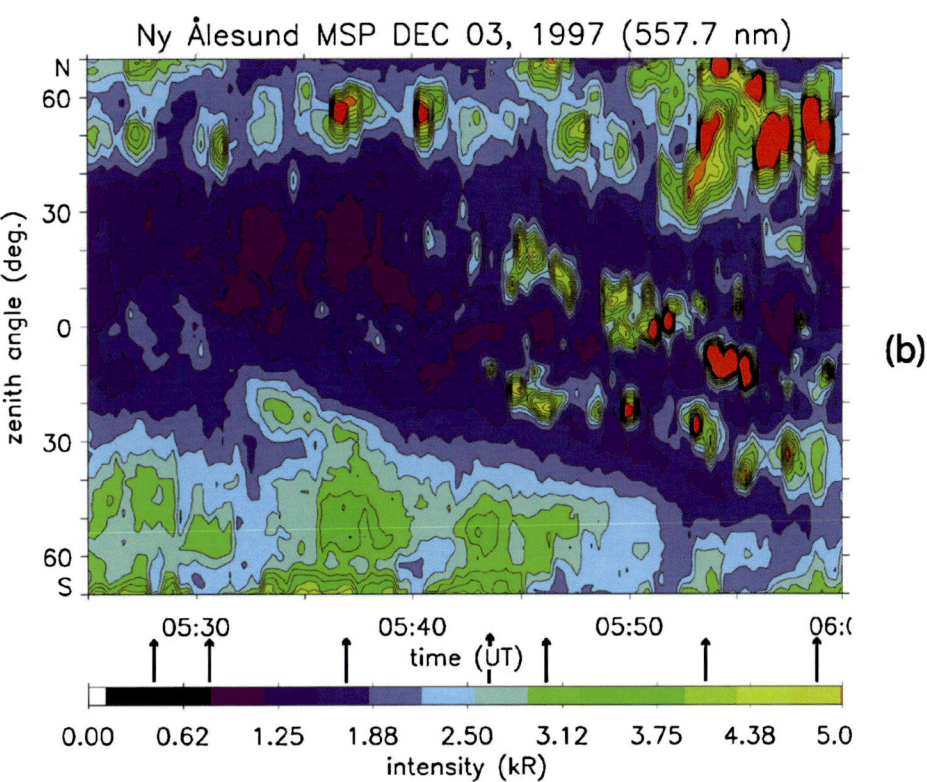

Plate 8. MSP observations from Ny Ålesund during the interval 0525-0600 UT. The red (a) and green (b) line emissions at 630.0 nm and 557.7 nm, respectively. Line-of-sight intensities are plotted as a function of zenith angle and time, color-coded according to the scales at the bottom of each panel. North is up. After *Farrugia et al.* [2003].

in the north (form c). The latter aurora will be described in detail below.

Plate 8 shows meridian scanning photometer observations obtained at Ny Åleasund during the interval 0525-0600 UT on Dec. 3, 1997. The IMF conditions during this interval are given in the lower section of Figure 3. Auroral forms located within the latitude range ~70-82° MLAT are captured by the scanning photometer. The fields-of-view of the meridian scanning photometers and all-sky cameras at this time are marked A in Figure 4. We distinguish between the following latitudinally separate categories of auroral forms: (i) a pulsating patch of diffuse aurora at the southern boundary of the field of view (~73-75° MLAT) appearing in the green line emission (marked by arrows in bottom panel of Plate 8), (ii) two discrete arcs near zenith (~75-77° magnetic latitude), which are intense in both emission lines, and (iii) polar arcs near the poleward boundary of the field-of-view (~78-80° MLAT), appearing in the form of a sequence of brightenings. The magnetospheric plasma sources and FAC regimes corresponding to these auroral forms were monitored by spacecraft Polar, as documented below. The equatorward migration of the discrete arcs (Plate 8) during the interval 0540-0600 UT corresponds to the southward rotation of the IMF at this time (see Figure 3).

Plate 9 shows ion, electron, and magnetic field data obtained from spacecraft Polar at an altitude of 6 R_E during the interval 0400-0600 UT on December 3, 1997. The ionospheric footprint of the spacecraft migrated equatorward, following the 0900 MLT meridian closely, from 81-73° magnetic latitude during the interval 0400-0600 UT, thereby traversing the latitude range of the three types of auroral forms shown in Plate 8. Details of the Polar-ground conjunction is given in *Farrugia et al.* [2003]. The following observations are of interest: (i) the magnetic field data shown in the two bottom panels of the figure indicates the presence of 4 latitudinally separate FAC regimes which we label C1 (up), C2 (down), (cusp currents) and R1 (down), and R2 (up) (traditional Regions 1 and 2 currents), (ii) the corresponding regimes of magnetospheric plasmas (ions and electrons) shown in the color-coded spectrograms reveal the following particle populations in the different FAC regimes: CPS-type in R2, BPS-type in R1, mantle-type in C1 and C2. The BPS-type plasma in R1 consists of a mixture of magnetosheath and magnetospheric particles. The plasma of magnetosheath origin in the C1 FAC regime consists of a sequence of plasma injections. The diamagnetic effect is marked by arrows in the bottom panel. Further details of the Polar data are reported in *Farrugia et al.* [2003]. Plasma and FAC regimes C1, R1, and R2 are all clearly manifested in the aurora. The northernmost aurora, located well north of the R1 current and the corresponding BPS arcs, in regime C1, is interpreted as a type of polar cap arc (see *Kullen et al.* [2002]). The sequence of plasma injections in regime C1 recorded by Polar at an altitude of 6 R_E corresponds to the auroral brightenings recorded by the ground auroral observations.

DISCUSSION

The spatio-temporal structure of the magnetosphere can be studied with different approaches. One approach is through statistics. Large statistical studies of the magnetosphere based on data from satellites in polar orbit have revealed important features of the IMF B_y-regulated dawn-dusk asymmetry in particle precipitation and its relation to plasma convection [*Newell et al.*, 2004]. A summary of their results for the IMF orientation investigated in this study is included in their Figure 1. The approach adopted in the present study, i.e., that of combining ground and satellite observations in representative case studies, may add important aspects of the spatio-temporal structure of the precipitation and plasma convection patterns. Both approaches have their merits. We are particularly interested in the evolution of particle precipitation and convection during reconnection transients during a southward-directed IMF having also a substantial east/west component and relatively stable solar wind plasma conditions. Thus, our event category is different from the aurora excited by interplanetary shocks as reported by *Zhou and Tsurutani* [1999] and *Zhou et al.* [2003]. In this study we focus on the dawn (dusk) side convection cell during positive (negative) IMF B_y ($B_z < 0$) conditions.

The statistical study of *Newell et al.* [2004] shows, for example, that the following precipitation regimes are successively traversed along the convection streamlines in the 1300-1630 MLT (the B_y negative case) and the 0800-1000 MLT (the B_y positive case) sectors: (i) boundary plasma sheet (BPS), (ii) low-latitude boundary layer (LLBL), and (iii) mantle. We document the corresponding latitudinal structure in the aurora during reconnection transients. Thus, we distinguish between latitudinally separate auroral forms which are successively activated during the evolution of the individual events. We pointed out that the spatial structure consisting of discrete steps in the latitudinal precipitation profile, which we shall attribute to the layered (cellular) structure of the corresponding magnetospheric boundary layer [*Sonnerup and Siebert*, 2003], is more clearly resolved in the ground-based auroral observations than in the spacecraft data. Thus, our auroral data reveal the true spatio-temporal structure of particle precipitation along the convection streamlines traversing the convection reversal and the open-closed field line boundary.

The highest-latitude component of the precipitation feature, consisting of a series of plasma injection events in the regime of mantle precipitation, corresponding to the auroral brightening sequence in an oval-aligned polar arc, is documented

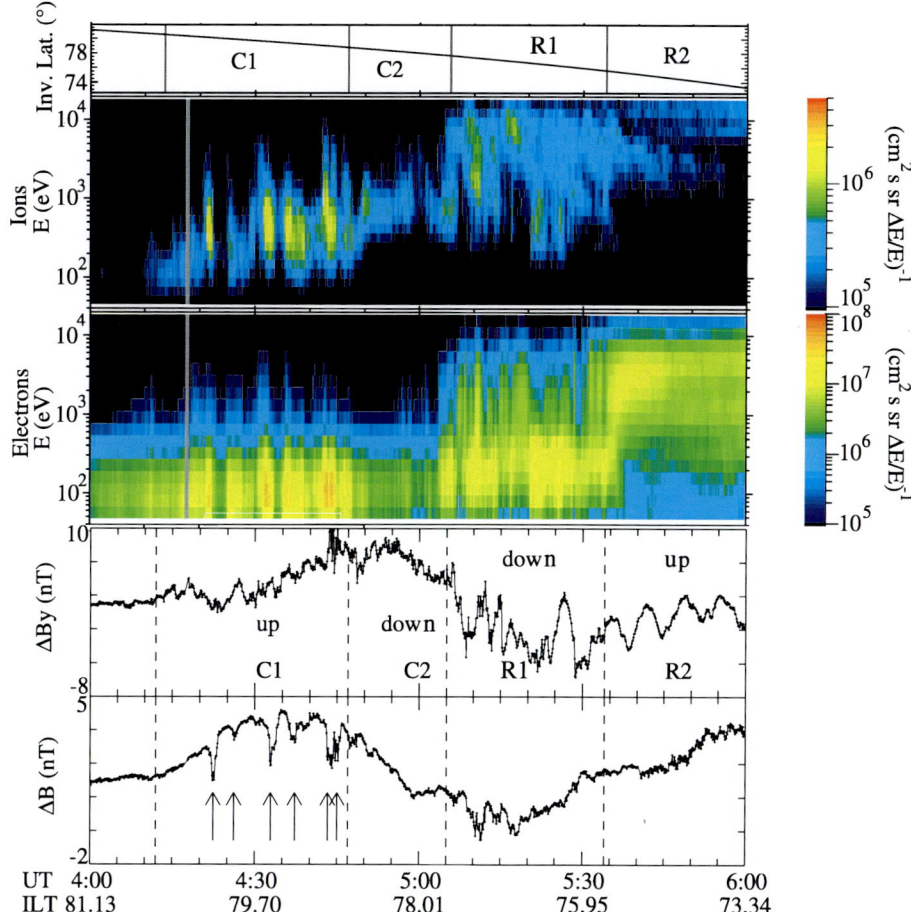

Plate 9. Polar plasma and magnetic field observations during the interval 0400-0600 UT. The two first panels shows Polar Hydra electron and proton data, respectively: differential energy fluxes color-coded according to the respective color bars on the right. Panels 3 and 4 show residual magnetic field component DBy and the total field less the IGRF - 1995 reference field. The extent of the large-scale currents C1, C2, R1, and R2, identified from local measurements of fields and plasma, are deliminated by vertical guidelines. Adapted from *Farrugia et al.* [2003].

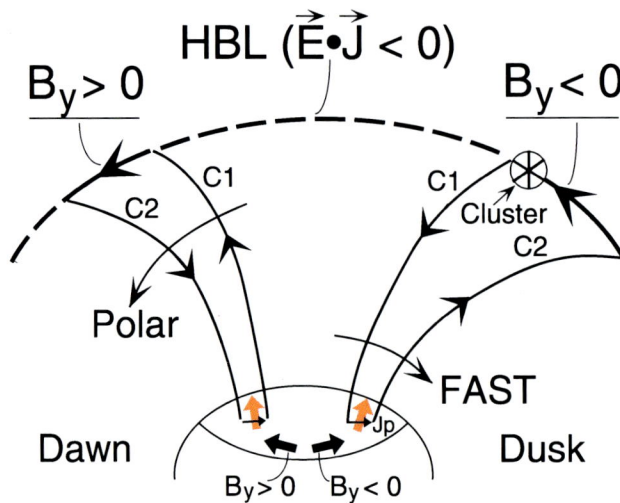

Plate 10. Schematic illustration of IMF B_y-dependent high-latitude field-aligned current system located poleward of the Region 1 (R1) current. Plasma convection channels (FC2) at the dawn (dusk) polar cap boundary, associated with the ionospheric current closure, is marked by red arrow. Locations of spacecraft Cluster and FAST on Jan. 21, 2001, and Polar on Dec. 3, 1997 are indicated.

in the mid-altitude (6 R_E) plasma observations from spacecraft Polar in our case 3 (Plate 9). The corresponding system of cusp FACs (i.e., what we called the C1-C2 current system) transmits momentum to the underlying ionosphere, where it gives rise to a channel of strong antisunward flow on the dawn (dusk) side of the polar cap, according as IMF B_y is positive (negative). We thus conclude that this high-latitude flow, which is often referred to as pulsed ionospheric flows (PIFs; see *Provan et al.* [2002]), is the result of substantial momentum transfer along field lines which have reconnected some time in the past and have their "feet" on the downstream side of the cusp. This flow gives rise to a Svalgaard-Mansurov effect (ground magnetic deflection related to IMF B_y, as decribed by *Svalgaard* [1973]) which is an addition to the effect of the east-west directed flows and attendant Hall currents in the cusp region resulting from magnetic tension forces ($\vec{j} \times \vec{B}$) acting on newly reconnected field lines. Therefore we find it appropriate to distinguish between two types of flow channel, representing different phases of evolution of open field lines, the one located on newly open field lines, and the other on older open field lines. These flow channels are referred to as FC1 and FC2, respectively, in *Sandholt et al.* [2004b]. Both convection channels are marked (as FC1 and FC2) in the schematic illustration given below.

Plate 10 shows a schematic illustration of IMF B_y-related C1-C2 cusp FACs at the dawn ($B_y > 0$)/dusk ($B_y < 0$) side of the polar cap and their relation to the low-altitude convection channel, FC2, and a generator region at high altitudes. The origin of the FACs and the FC2 flow channel is the deflected magnetosheath flow and the associated asymmetric solar wind-magnetosphere coupling introduced by IMF B_y, as discussed theoretically by *Cowley* [1981a] and modelled by *Siscoe et al.* [2000]. Concerning the physics of the magnetospheric boundary layers at low and high latitudes the reader is referred to the theoretical work by *Siscoe et al.* [1991] and *Sonnerup and Siebert* [2003]. An essential element of this study is the documentation of the morphology/dynamics of the auroral forms associated with the cusp C1/C2 FACs.

We note that a polarization effect at the ionospheric level [*Marklund*, 1984] may contribute to the relatively strong electric field responsible for FC2 since the ionospheric conductance in the polar cap ionosphere is low (see Plate 5). This effect is similar to the mechanism giving rise to the strong E-field/ion drift associated with the ionospheric closure of the Region 2 FAC appearing at sub-auroral latitudes during storm intervals, the so-called SAID/SAPS [*Foster and Vo*, 2002].

Before we summarize our results, a note is in order about how our work relates to the magnetospheric response to corotating interaction regions. We have done only a zeroth order approximation of the magnetospheric response, emphasizing the effects of the large IMF B_y on auroras, ionospheric convection, precipitation, FACs. To do this, we use observations made in three case studies so that the results are more representative of the magnetospheric response. However, the short-period fluctuations in IMF B_z in Alfven waves on the fast stream, and the sharp density decrease (implying also a large dynamic pressure decrease) typical of corotating interaction regions (see *Tsurutani et al., Magnetic storms caused by corotating solar wind streams, This Publication.*) have not been addressed here. Yet, we believe our work forms an essential pre-requisite to understand the further ramifications introduced by these important features. As the reader can judge, even considering just IMF B_y leads to a rich complex of phenomena, some only recently established. This forms the groundwork on which further elaborations, taking into account the Alfven waves and the density profile in CIRs, can be superposed.

Clearly, essential for our task was to ascertain that the interplanetary monitors whose data we used were indeed reading what affects the Earth later. These monitors were near the L1 libration point upstream of Earth. Many investigations have shown a lack of correlation of IMF and solar wind features at such distances, though the dependence on the X (Sun-Earth direction) is not as stringent as that on Y. What we have done is to compare the ACE and WIND readings with a monitor in Earth orbit, Geotail. We find the agreement was satisfactory when Geotail was sampling the near-Earth solar wind. The comparison for Jan 21, 2001 may be seen in Figure 3 of *Farrugia et al.* [2004b]. This lends confidence that we have indeed presented representative features of M-I coupling affected by the large IMF B_y. Further work will need to be done to address the complex of reactions when the finer features of CIRs are added. This work is underway.

SUMMARY

The present study may be summarized by the following points:

(i) We have reported essential elements of the spatio-temporal structure of M-I coupling during events of transient magnetopause reconnection combining ground-based and satellite observations, often taken in magnetic conjugacy, and at different magnetospheric altitudes.

(ii) The auroral manifestation appears strongly in the pre- and postnoon sectors as PMAF activity spanning the convection reversal boundary. These PMAFs consist of the successive activations of latitudinally separate auroral forms corresponding to boundary plasma sheet (BPS), low-latitude boundary layer (LLBL), and mantle-type precipitation regimes.

(iii) The high-latitude component/phase of the events is accompanied by a series of plasma injections in the regime of mantle precipitation, and the setting-up of an FAC system, separate from, and situated poleward of, the R1 current. The FAC systen lies in the prenoon sector of the polar cap for IMF $B_y > 0$ and in the postnoon for $B_y < 0$.

(iv) Current closure in the polar cap ionosphere, giving rise to a specific channel of strong (1-2 km/s) antisunward flow (pulsed ionospheric flows; PIFs) and a polar cap convection asymmetry extending to the dawn-dusk terminator.

(v) The ground magnetic signature of the convection channel represents a Svalgaard-Mansurov effect extending to the dawn-dusk terminator.

(vi) By contrast, the standard view on the Svalgaard-Mansurov effect involves only a tension force acting on newly open field lines, causing B_y-regulated east-west convection in the cusp region.

(vii) Relationship with the so-called "high-latitude flux transfer events" (local B_z reversals and plasma flow transients) in the high-latitude boundary layer (HBL), on old open field lines on the downstream side of the cusp, is documented with observations from spacecraft Cluster. This is a high-latitude manifestation of transient reconnection events.

(viii) A consequence of this study is that effective momentum transfer from the solar wind during reconnection events is not limited to the early phase of the events at lower latitudes, but may continue to higher latitudes.

(ix) The solar wind-magnetosphere coupling mode under study is present during B_y-dominated IMF orientation: $B_z < 0$; $|B_y/B_z| > 1$; clock angle range ~90-135°.

(x) It is therefore expected to occur during intervals of corotating solar wind streams, characterized by a Parker spiral orientation of the IMF. On this we have, however, given only an approximate picture, which needs to be elaborated in further work.

Acknowledgments. The Wind data are courtesy of R.P. Lepping and K.W. Ogilvie. We thank J. Wild for providing the SuperDARN plot used in this study. We thank Bjørn Lybekk and Espen Trondsen for technical assistance during the auroral observation campaigns in Ny Ålesund, and with the presentation of the optical data. The auroral observation program on Svalbard is supported by the Norwegian Research Council, the Norwegian Polar Research Institute and AFOSR Task 2311AS. This work is supported by NASA Guest Investigator grant NNG05GG25G, NASA Wind grant NAG5-11803, NASA LWS grant NAG5-12189.

REFERENCES

Akasofu, S.-I., and B. Tsurutani, Unusual auroral features observed on January 10-11, 1983 and their possible relationships to the interplanetary magnetic field, *Geophys. Res. Lett.*, 11, 1086, 1984.

Burlaga, L.F., *Interplanetary Magnetohydrodynamics*, Oxford University Press, 1995.

Cowley, S.W.H., Magnetospheric and ionospheric flow and the interplanetary magnetic field, in *The Physical Basic of the Ionosphere in the Solar-Terrestrial System*, no. 295 in Conference Proceedings, pp. 4-1 — 4-12, AGARD (Advisory Group for Aerospace Research & Development), NATO, Neuilly sur Seine, France, 1981a.

Cowley, S.W.H., Magnetospheric asymmetries associated with the Y-component of the IMF, *Planet. Space Sci.*, 29, 79, 1981b.

Cowley, S.W.H., The causes of convection in the Earth's magnetosphere: A review of developments during the IMS, *Rev. Geophys.*, 20, 531, 1982.

Cowley, S.W.H., Solar wind control of magnetospheric convection, in *Achievements of the International Magnetospheric Study (IMS)*, no. SP-217 in ESA Special Publ., pp. 483-494, ESA (European Space Agency), Noordwijk, The Netherlands, 1984.

Dandekar, B.S., and C.P. Pike, The midday discrete auroral gap, *J. Geophys. Res.*, 83, 4227, 1978.

Escoubet, C.P., M.F. Smith, S.F. Fung, P.C. Anderson, R.A. Hoffman, E.M. Basinska, and J.M. Bosqued, Staircase ion signature in the polar cusp: A case study, *Geophys. Res. Lett.*, 19, 1735, 1992.

Farrugia, C.J., P.E. Sandholt, W.F. Denig, and R.B. Torbert, Observation of a correspondence between poleward-moving auroral forms and stepped cusp ion precipitation, *J. Geophys. Res.*, 103, 9309, 1998.

Farrugia, C.J., P.E. Sandholt, N.C. Maynard, R.B. Torbert, and D.M. Ober, Temporal variations in a four-sheet field-aligned current system and associated aurorae as observed during a Polar-ground magnetic conjunction in the midmorning sector, *J. Geophys. Res.*, 108(A6), 1230, doi:10.1029/2002JA009619, 2003.

Farrugia, C.J., P.E. Sandholt, R.B. Torbert, and N. Østgaard, Temporal and spatial aspects of the cusp inferred from local and global ground- and space-based observations in a case study, *J. Geophys. Res.*, 109, A04209, doi:10.1029/2003JA010121, 2004a.

Farrugia, C.J. *et al.*, A multi-instrument study of flux transfer event structure, *J. Geophys. Res.*, 93, 14,465, 1988.

Farrugia, C.J. *et al.*, Pulsed flows at the high-altitude cusp poleward boundary, and associated ionospheric convection and particle signatures, during a CLUSTER - FAST - SuperDARN - Søndrestrøm conjunction under a southwest IMF, *Ann. Geophys.*, 22, 2891-2905, 2004b.

Fasel, G., Dayside poleward moving auroral forms: A statistical study, *J. Geophys. Res.*, 100, 11,891, 1995.

Foster, J.C., and H.B. Vo, Average characteristics and activity dependence of the subauroral polarization stream, *J. Geophys. Res.*, 107(A12), 1475, doi:10.1029/2002JA009409, 2002.

Frey, H.U., T.D. Phan, S.A. Fuselier, and S.B. Mende, Continuous magnetic reconnection at earth's magnetopause, *Nature*, 426, 533, 2003.

Friis-Christensen, E., Polar cap current systems, in *Magnetospheric Currents, Geophysical Monograph*, vol. 28, edited by T.A. Potemra, pp. 86-95, AGU, Washington, D.C., 1984.

Gosling, J.T., M.F. Thomsen, S.J. Bame, R.C. Elphic, and C.T. Russell, Plasma flow reversals at the dayside magnetopause and the origin of asymmetric polar cap convection, *J. Geophys. Res.*, 95, 8073, 1990.

Greenwald, R.A., *et al.*, DARN/SUPERDARN, *Space Sci. Rev.*, 71, 761, 1995.

Haerendel, G., G. Paschmann, N. Sckopke, H. Rosenbauer, and P.C. Hedgecock, The frontside boundary layer and the problem of reconnection, *J. Geophys. Res.*, 83, 3195, 1978.

Heppner, J., and N.C. Maynard, Empirical high-latitude electric field models, *J. Geophys. Res.*, 92, 4467-4489, 1987.

Jørgensen, T.S., E. Friis-Christensen, and J. Wilhjelm, Interplanetary magnetic field direction and high-latitude ionospheric currents, *J. Geophys. Res.*, 77, 1976, 1972.

Kullen, A., M. Brittnacher, J.A. Cumnock, and L.G. Blomberg, Solar wind dependence of the occurrence and motion of polar auroral arcs: A statistical study, *J. Geophys. Res.*, 107(A11), 1362, doi:10.1029/2002JA009245, 2002.

Lee, L.C., J.R. Kan, and S.-I. Akasofu, On the origin of the cusp field-aligned currents, *J. Geophys.*, 57, 217, 1985.

Lockwood, M., and M.F. Smith, The variation of reconnection rate at the magnetopause and cusp ion precipitation, *J. Geophys. Res.*, 97, 14841, 1992.

Lockwood, M., S.W.H. Cowley, and M.P. Freeman, The excitation of plasma convection in the high-latitude ionosphere, *J. Geophys. Res.*, 95, 7061, 1990.

Marklund, G., Auroral arc classification scheme based on the observed arc-associated electric field pattern, *Planet. Space Sci.*, 32, 193, 1984.

Mozer, F.S., W.D. Gonzalez, F. Bogott, M.C. Kelley, and S. Schutz, High latitude electric fields and the three-dimensional interaction between the interplanetary and terrestrial magnetic fields, *J. Geophys. Res.*, 79, 56, 1974.

Newell, P.T., and C.-I. Meng, Ion acceleration at the equatorward edge of the cusp: Low-altitude observations of patchy merging, *Geophys. Res. Lett.*, 18, 1829, 1991.

Newell, P.T., J.M. Ruohoniemi, and C.-I. Meng, Maps of precipitation by source region, binned by IMF, with inertial convection streamlines, *J. Geophys. Res.*, 109, A10206, doi:10.1029/2004JA10499, 2004.

Paschmann, G., G. Haerendel, I. Papamastorakis, N. Sckopke, S.J. Bame, and J.T. Gosling, Plasma and magnetic field characteristics of magnetic flux transfer events, *J. Geophys. Res.*, 87, 2159, 1982.

Paschmann, G., I. Papamastorakis, W. Baumjohann, N. Sckopke, C.W. Carlson, B.U.O. Sonnerup, and H. Luhr, The magnetopause for large magnetic shear: AMPTE/IRM observations, *J. Geophys. Res.*, 91, 11099, 1986.

Paschmann, G., et al., Plasma acceleration at the Earth's magnetopause: Evidence for reconnection, *Nature*, 282, 243, 1979.

Phan, T., G. Paschmann, and B.U.O. Sonnerup, Low-latitude dayside magnetopause and boundary layer for high magnetic shear 2. occurrence of magnetic reconnection, *J. Geophys. Res.*, 101, 7817, 1996.

Potemra, T.A., and N.A. Saflekos, Birkeland currents and the interplanetary magnetic field, in *Magnetospheric Boundary Layers*, edited by B. Battrick, ESA SP-148, pp. 193-198, Noordwijk, Netherlands, 1979.

Provan, G., S.E. Milan, M. Lester, T.K. Yeoman, and H. Khan, Simultaneous observations of the ionospheric footprint of flux transfer events and dispersed ion signatures, *Ann. Geophys.*, 20, 281, 2002.

Reiff, P.H., T.W. Hill, and J.L. Burch, Solar wind plasma injections at the dayside magnetospheric cusp, *J. Geophys. Res.*, 82, 479, 1977.

Rich, F.J., and M. Hairston, Large-scale convection patterns observed by dmsp, *J. Geophys. Res.*, 99, 3827, 1994.

Rijnbeek, R.P., S.W.H. Cowley, D.J. Southwood, and C.T. Russell, A survey of dayside flux transfer events as observed by ISEE 1 and 2 magnetometers, *J. Geophys. Res.*, 89, 786, 1984.

Ruohoniemi, J.M., R.A. Greenwald, K.B. Baker, J.P. Villain, C. Hanuise, and J. Kelly, Mapping high-latitude plasma convection with coherent HF radars, *J. Geophys. Res.*, 94, 13,463, 1989.

Russell, C.T., and R.C. Elphic, Initial ISEE magnetometer results: Magnetopause observations, *Space Sci. Rev.*, 22, 681, 1978.

Sandholt, P.E., C.S. Deehr, A. Egeland, B. Lybekk, R. Viereck, and G.J. Romick, Signatures in the dayside aurora of plasma transfer from the magnetosheath, *J. Geophys. Res.*, 91, 10,063, 1986.

Sandholt, P.E., M. Lockwood, T. Oguti, S.W.H. Cowley, K.S.C. Freeman, B. Lybekk, A. Egeland, and D.M. Willis, Midday auroral breakup events and related energy and momentum transfer from the magnetosheath, *J. Geophys. Res.*, 95, 1039, 1990.

Sandholt, P.E., C.J. Farrugia, and W.F. Denig, Dayside aurora and the role of IMF $|B_y/B_z|$: detailed morphology and response to magnetopause reconnection, *Ann. Geophys.*, 22, 613, 2004a.

Sandholt, P.E., C.J. Farrugia, and W.F. Denig, Detailed dayside auroral morphology as a function of local time for southeast IMF orientation: implications for solar wind - magnetosphere coupling, *Ann. Geophys.*, 22, 3537, 2004b.

Shelley, E.G., R.D. Sharp, and R.G. Johnson, He and H flux measurements in the dayside cusp: Estimates of convection electric field, *J. Geophys. Res.*, 81, 2363, 1976.

Siscoe, G.L., W. Lotko, and B.U.O. Sonnerup, A high-latitude, low-latitude boundary layer model of the convection current system, *J. Geophys. Res.*, 96, 3487, 1991.

Siscoe, G.L., G.M. Erickson, B.U.O. Sonnerup, N.C. Maynard, K.D. Siebert, D.R. Weimer, and W.W. White, Deflected magnetosheath flow at the high-latitude magnetopause, *J. Geophys. Res.*, 105, 12,851, 2000.

Smith, M.F., Transient dayside reconnection and its effect on the ionosphere, in *Physical Signatures of Magnetospheric Boundary Layer Processes, NATO ASI Series*, vol. 425, edited by J.A. Holtet and A. Egeland, pp. 275-289, Kluwer Academic Publishers, Dordrecht, Holland, 1994.

Sonnerup, B.U.O., Magnetic field line reconnection at the magnetopause: An overview, in *Magnetic reconnection in Space and Laboratory Plasmas, Geophysical Monograph*, vol. 30, edited by E.W. Hones, pp. 92-103, AGU, Washington, D.C., 1984.

Sonnerup, B.U.O., and K.D. Siebert, Theory of the low-latitude boundary layer and its coupling to the ionosphere: A tutorial review, in *Earth's Low-Latitude Boundary Layer, Geophysical Monograph*, vol. 133, edited by P.T. Newell and T. Onsager, pp. 13-32, American Geophysical Union, Washington, D.C., 2003.

Sonnerup, B.U.O., G. Paschmann, I. Papamastorakis, N. Sckopke, G. Haerendel, S.J. Bame, J.R. Asbridge, J.T. Gosling, and C.T. Russell, Evidence for magnetic reconnection at the Earth's magnetopause, *J. Geophys. Res.*, 86, 10049, 1981.

Svalgaard, L., Polar cap magnetic variations and their relationship with the interplanetary magnetic sector structure, *J. Geophys. Res.*, 78, 2064, 1973.

Taguchi, S., M. Sugiura, J.D. Winningham, and J. Slavin, Characterization of the IMF By-dependent field-aligned currents in the cleft region based on DE 2 observations, *J. Geophys. Res.*, 98, 1393, 1993.

Tanaka, T., Formation of magnetospheric plasma population regimes coupled with the dynamo process in the convection system, *J. Geophys. Res.*, 108(A8), 1315, doi:10.1029/2002JA009668, 2003.

Thompson, S.M., M.G. Kivelson, K.K. Khurana, A. Balogh, H. Reme, A.N. Fazakerley, and L.M. Kistler, Cluster observations of quasi-periodic impulsive signatures in the dayside northern lobe: High-latitude flux transfer events?, *J. Geophys. Res.*, 109, A02213, doi:10.1029/2003JA010138, 2004.

Trattner, K.J., S.A. Fuselier, W.K. Peterson, J.-A. Sauvaud, H. Stenuit, N. Dubouloz, and R.A. Kovrazhkin, On spatial and temporal structures in the cusp, *J. Geophys. Res.*, 104, 28,411, 1999.

Vasyliunas, V.M., Multiple-branch model of the open magnetopause, *Geophys. Res. Lett.*, 22, 1145, 1995.

Vennerstrøm, S., Interplanetary sources of magnetic storms: A statistical study, *J. Geophys. Res.*, 106, 29,175, 2001.

Villain, J.P., R.A. Greenwald, K.B. Baker, and J.M. Ruohoniemi, HF radar observations of E region plasma irregularities produced by oblique plasma streaming, *J. Geophys. Res.*, 92, 12327, 1987.

Watanabe, M., T. Iijima, and F.J. Rich, Synthesis models of dayside field-aligned currents for strong interplanetary magnetic field B_y, *J. Geophys. Res.*, 101, 13,303, 1996.

Weimer, D.R., Models of high-latitude electric potentials derived with a least error fit of spherical harmonic coeffcients, *J. Geophys. Res.*, 100, 19595, 1995.

Zhang, Y., C.-I. Meng, L.J. Paxton, D. Morrison, B. Wolven, H. Kil, P. Newell, and S. Wing, Far-ultraviolet signature of polar cusp during southward IMF observed by TIMED/Global Ultraviolet Imager and DMSP, *J. Geophys. Res.*, 110, A01218, doi:1029/2004JA010707, 2006.

Zhou, X., and T. Tsurutani, Rapid intensification and propagation of the dayside aurora: Large scale interplanetary pressure pulses (fast shocks), *Geophys. Res. Lett.*, 26, 1097, 1999.

Zhou, X.-Y., R.J. Strangeway, P.C. Anderson, D.G. Sibeck, B.T. Tsurutani, G. Haerendel, H.U. Frey, and J.K. Arballo, Shock aurora: FAST and DMSP observations, *J. Geophys. Res.*, 108(A4), 8019, doi:10.1029/2002JA009701, 2003.

W.F. Denig, Air Force Research Laboratory, Hanscom AFB, Mass.

C.J. Farrugia, Space Science Center, University of New Hampshire, Durham, NH 03824.

P.E. Sandholt, Department of Physics, University of Oslo, P. O. Box 1048, Blindern, N-0316, Norway. (p.e.sandholt@fys.uio.no)

The Nature of Auroras During High-Intensity Long-Duration Continuous AE Activity (HILDCAA) Events: 1998 to 2001

F.L. Guarnieri

Instituto Nacional de Pesquisas Espaciais (INPE), São José dos Campos, Sao Paulo, Brazil

The nature of auroras during High-Intensity Long-Duration Continuous AE Activity (HILDCAA) events occurred from 1998 to 2001 is studied using auroral images in Lyman-Birge-Hopfield Long (LBHL - 160 to 180 nm) wavelengths taken by the UVI instrument onboard the POLAR satellite. It is found that during HILDCAA events, auroras are occurring not only in the midnight sector, but at all local times along the auroral oval. During some events, auroras are noted even over the polar cap as well. The typical auroral intensities observed during HILDCAA events are in the range from 30 to 60 photon cm^{-2} s^{-1}. Using a method to evaluate relative photon fluxes based on POLAR images, the photon fluxes during HILDCAAs were qualitatively compared with those during magnetic storm main and recovery phases. It is found that, for the auroral zone midnight sector, the relative photon fluxes during the main phases of storms are much higher than those during HILDCAA intervals, as expected. However, photon fluxes observed during HILDCAA events are generally higher than that during the recovery phase of storms. Thus, if integrated over long periods of time (a year), the energy input from particle precipitation into the ionosphere during the declining phase/ minimum phase of the solar cycle, when HILDCAAs/corotating solar wind streams dominate, could be greater than during solar maximum, when interplanetary coronal mass ejections (ICMEs) predominate. This is in agreement with several parallel studies based on the AE and Dst indices that indicate similar conclusions.

1. INTRODUCTION

In 1987 *Tsurutani* and *Gonzalez* observed that some magnetic storms presented exceptionally long "recovery" phases. The Dst index did not recover to pre-storm values for days or even weeks. Realizing that normal particle loss processes could not take this long, they studied these events and found that geomagnetic activity, albeit at a lower level, was continuing to take place during these "recoveries". This phenomenon was called High-Intensity Long-Duration Continuous AE Activity or HILDCAA events. The Dst indices during HILDCAAs are typically weak, with values Dst > -50 nT, which imply weak ring current enhancements. At the time of that study, satellite ultraviolet images were not available, and the AE indices [*Davis and Sugiura*, 1966] were the main parameters used to define the HILDCAA events. The four conditions for HILDCAA identification are: (1) "high-intensity" – the AE index peak must be higher than 1000 nT during the event; (2) "long-duration" – the high and continuous AE activity must last for at least two days; (3) "continuous" AE activity – the AE index value must not fall below 200 nT for periods longer than 2 hours at a time; and (4) the event must occur outside the main phase of a geomagnetic storm.

The occurrence of HILDCAAs in the Earth's magnetosphere is related to the presence of large-amplitude magnetic field fluctuations in the interplanetary medium [*Tsurutani and Gonzalez*, 1987; *Tsurutani et al.*, 1990]. These fluctuations were studied using cross-correlation analysis between magnetic field and velocity components [in the manner of *Belcher and Davis*, 1971] and found to be interplanetary Alfvén waves propagating outward from the Sun.

Considering that the interplanetary structures responsible for HILDCAAs are different than those for storms and substorms, one might ask the question, "are the auroral forms also distinct for HILDCAA auroras?". Auroras during magnetic storms have been studied widely [*Craven et al.*, 1986; *Spann et al.*, 1998; *Zhou and Tsurutani*, 1999; *Brittnacher et al.*, 2000; *Zhou and Tsurutani*, 2004]. In sharp contrast, HILDCAA auroras have never been extensively explored, and thus their auroral forms and intensities are poorly known.

Originally it was not clear whether HILDCAA events were a form of continuous substorms, or plasma sheet oscillations, or some other form of auroral phenomena. Recently *Tsurutani et al.* [2004], *Guarnieri et al.* [2004], and *Guarnieri* [2005] showed that HILDCAA events were different from substorms. These studies found that substorms did occur within HILDCAA intervals, but they were not the dominant causes for the AE increases which define HILDCAAs.

The goal of this paper is to report the results of a study of HILDCAA events that occurred between 1998 and 2001. POLAR ultraviolet images were employed to examine the relative auroral intensities and the morphological auroral features associated with this new type of geomagnetic activity. The results of a comparison between HILDCAAs and magnetic storms intensities will be shown. The spatial and temporal differences in the auroral characteristics of HILDCAAs and storms will be discussed.

2. DATA SETS/METHODOLOGY

HILDCAA events were identified based on the AE and Dst indices. The AE indices, with one-minute temporal resolution, were obtained from the World Data Center for Geomagnetism – Kyoto (http://swdcwww.kugi.kyoto-u.ac.jp). The Dst index was used in order to avoid time intervals during main phases of geomagnetic storms (one of the HILDCAA requirements). All selected events have high AE activity (peak AE > 1000 nT) sustained for more than two days. This activity never drops below 200 nT for more than two hours at a time. In this way, all the "strict" HILDCAA criteria established by *Tsurutani and Gonzalez* [1987] were followed.

Images from the UVI (Ultraviolet Imager) instrument onboard the POLAR satellite [*Torr et al.*, 1995] were used to analyze auroral spatial and temporal distributions during HILDCAA events. The UVI images were taken in the Lyman-Birge-Hopfield (LBH) range, in the long bandwidth mode, from ~160 to ~180 nm. These images have a cadence between 3 and 6 minutes.

The survey images are false color figures showing the north pole (for the time interval analyzed) and the colors indicate the photon counts in photon cm^{-2} s^{-1}. The pixels in the image do not correspond directly to the UVI CCD pixels, since all these images were already pre-processed by the UVI team. All images used in the current analysis were taken when the satellite was at a geocentric distance larger than 8 Re and with the UVI instrument pointing towards the polar region. The "corrected" view from the top of the pole for each image was used in this study.

The POLAR/UVI images for the identified events were employed to generate movies showing the temporal evolution and the most typical auroral features during HILDCAAs. The images were also grouped into panels to illustrate these features. In addition, in order to increase the separation in the color scale for some of the images, an auxiliary procedure of changing the original colors was applied. It was done by multiplying the value of each "pixel" in the image by a statistically determined factor, and then replotting the figure. This method is only an artifact to make some relatively dim features more apparent.

For a comparison of the auroral photon fluxes observed during HILDCAA auroras with fluxes observed during storm auroras, a method based on the UVI survey images was employed. Due to the POLAR orbit, there are very few images showing the entire auroral oval. In order to have a considerable number of images for statistical purposes, a limited portion of the auroral region was selected. The region chosen is the midnight sector, between magnetic latitudes 50° and 80° MLAT, and between 21 and 03 LT. The high latitude region, between 80° and 90° MLAT was excluded because it frequently is dominated by dayglow (for analysis involving absolute values, the dayglow must be removed by a formal method, see *Lummerzheim et al.*, 1997). These survey images are intended only for qualitative analyses, and cannot be used to obtain absolute values.

For each image, the pixels were grouped and summed accordingly to their color, and then multiplied by the corresponding value in photon cm^{-2} s^{-1}. All the resulting values of each binned color were summed, resulting in a value in photon pixel cm^{-2} s^{-1} for the photon flux in the selected region (midnight sector, between 50° and 80° MLAT and between 21 and 03 LT). The factor "pixel cm^{-2}" indicates the area represented by each pixel of the image (not CCD pixel). This "area" should be almost constant, since the images were taken far from Earth, and all the images were already corrected for a polar view.

The next step consisted of taking the value obtained for a single image and integrating it in time until the next available image ($\Delta t = \sim$ 3 to 6 minutes), which results in a relative flux measured in photon pixel cm^{-2}. This procedure was adopted for periods of continuous images of the selected region (hereafter referred as Continuous Interval - C.I.).

Since this method utilizes a limited spatial region and also requires integration over time between images, it is suitable only for phenomena which are spread in all local times along the auroral oval, and which evolve slowly in time. These requirements exclude substorm events, since they may occur in a limited spatial region (which may be outside the selected region), and occur during very short time intervals (which may change significantly between images, mainly when the cadence is longer). HILDCAAs and storms can be analyzed by this technique since they are broad in local time extension and last for days. The scope of this study will be limited to the comparison of HILDCAA auroras photon fluxes with those of magnetic storm main and recovery phases. Subsequent work on substorms will take place in future studies.

3. RESULTS AND DISCUSSION

Using the methodology described above, 14 HILDCAA events which occurred between years 1998 and 2001 were identified. These events are listed in Table 1. The columns from left-to-right are the HILDCAA events: year, event number within the year, the month and day of the start time of the event, the UT time at the start of the event, the event end date and time (columns 5 and 6), and finally the total duration of the HILDCAA event in minutes.

All these events follow strictly the HILDCAA criteria, having durations of more than 2 days. All events identified were used in the following analyses (POLAR UVI images permitting).

3.1. HILDCAA Auroras

The auroras during the 14 events listed in Table 1 were analyzed using the constructed movies and the individual panels showing sequences of images. One example of a HILDCAA aurora is given in Plate 1a, which shows a sequence of POLAR/UVI images during the HILDCAA event identified as event 2_1998.

These images are taken during the interval from 13:45:50 UT to 16:13:02 UT, on July 23, 1998. The cadence was ~3 minutes between images. The main feature of these images is that the aurora is widely distributed in local time along the auroral oval. Aurora is present along the entire oval that is within the view of the POLAR UVI instrument. The aurora may extend also to local noon (unfortunately the instrument field of view limits this determination). The typical photon flux during this aurora is around 50 photon cm^{-2} s^{-1}. There is also dayglow present near the dayside. The intensity of the latter is, at most, 30 photon cm^{-2} s^{-1}. There are some intensifications occurring sporadically. One of them starts at 14:34:54 UT and lasts until 14:50:14 UT. This feature is located between 60° and 70° MLAT in the local time sector between 00 and 03 LT. The maximum intensity is reached between 14:37:58 and 14:41:02 UT, with a peak photon flux of ~70 photon cm^{-2} s^{-1}. Another intensification occurs from 15:11:42 UT until 15:17:50 UT. It is located between 60° and 70° MLAT, in the sector from 21 LT to 00 LT. The photon flux in this region reaches values higher than 130 photon cm^{-2} s^{-1}. The first event may be identified as a substorm expansion phase using the *Akasofu* [1964] definition [see *Tsurutani et al.*, 2004 for further comments]. The latter event may be a feature that has propagated from lower latitudes into the view of POLAR. Further comment cannot be made on this event at this time.

Expansions of the auroras over the polar cap may occur for short time intervals during some events. An example of this can be found in Plate 1a. This aurora expansion begins at ~15:54:38 UT. Some of these images are magnified in Plate 1b. The two images were taken at 15:54:38 UT and 15:57:42 UT. These images were submitted to the procedure of rescaling the colors, discussed previously in the Methodology section. In the sector from 09 LT to 18 LT, the auroras that were located between 65° and 80° MLAT start to expand towards the pole, occupying the region between 80° and 90° MLAT.

Another typical feature of HILDCAA auroral activity is the presence of dayside auroras. Auroras in local daytime hours are somewhat difficult to analyze because of the presence of dayglow. Dayglow can be a strong "background", making auroral features more difficult to identify. The color rescaling method was used to make the regions of the aurora more evident. Plate 2 shows the results of this procedure for an image taken at 09:38 UT on July 24, 1998 during HILDCAA event 2_1998. The midnight sector is the brightest, with photon fluxes higher than 50 photon cm^{-2} s^{-1}. On the dayside there is clearly the presence of a continuous auroral oval. There is no break in it. At local noon, the auroral latitude is 70° MLAT, in agreement with the statistical results of *Feldstein and Starkov* [1967] for the auroral oval location. The dayside intensities range from ~30 to 50 photon cm^{-2} s^{-1}. The superposed dayglow contributes a flux of ~20 photon cm^{-2} s^{-1}.

The above results are typical. HILDCAA auroras are found to be located continuously along the *Feldstein and Starkov* auroral oval. There are typically no breaks in the distribution. The dayside portion has intensities of ~30 to 60 photon cm^{-2} s^{-1}. HILDCAA auroras persist for several days to even as long as weeks.

Table 1. Selected HILDCAA events for years from 1998 to 2001. The columns are, from left-to-right, the year, event number within the year, the month and day of the start time of the event, the UT time at the start of the event, the event end date and time, and the total duration of the HILDCAA event in minutes

Year	Event	Start		End		Duration (min)
		Date (mm/dd)	Time (hh:mm)	Date (mm/dd)	Time (hh:mm)	
1998	1	04/24	18:03	04/27	06:05	3603
	2	07/22	21:09	07/25	12:25	3797
1999	1	04/29	11:20	05/03	11:16	5757
	2	08/17	22:52	08/20	12:00	3669
	3	08/31	15:32	09/02	20:30	3179
	4	10/10	20:00	10/14	17:38	5619
	5	10/23	13:21	10/25	20:57	3337
	6	11/07	17:00	11/10	04:47	3588
	7	12/03	10:00	12/06	00:15	3736
2000	1	01/27	18:10	01/31	03:15	4866
	2	02/05	16:01	02/08	05:33	3693
	3	02/24	0:03	02/27	22:10	5648
	4	05/24	10:00	05/26	18:07	3368
2001	1	05/11	14:04	05/14	10:51	4128

3.2. Auroral Spatial Distribution and Time Scales: Comparison of HILDCAAs, Substorms, and Storms

A comparison of the relative auroral intensities during HILDCAAs and geomagnetic storms can be found in Plate 3. For completeness, an image taken during an auroral substorm has also been included. The image chosen for the storm event was taken on July 15, 2000 at 20:25:30 UT, during the main phase of the "Bastille Day" storm. This is a very strong storm that reached a peak negative Dst value of −301 nT. The chosen substorm event occurred at 07:01:42 UT of January 27, 2000 (unfortunately the instrument field of view does not include most of the dayside for this event). The HILDCAA aurora image was taken at 09:36:19 UT of July 24, 1998 during event 2_1998. Other magnetic storm, substorm and HILDCAA auroral images were analyzed. These events were selected as representative of typical cases.

Magnetic storms usually have the most intense auroral photon fluxes, with intensities ranging from ~150 to ~600 photon cm^{-2} s^{-1}. The most intense fluxes are generally located near local midnight with a range of approximately ±3 hours. During magnetic storms, auroras can be noted at all local times. For extremely intense storms such as this event, auroras are located at middle latitudes [*McIlwain*, 1974].

Substorm auroral intensities can frequently reach peak values as high as ~100 to ~150 photon cm^{-2} s^{-1}. Cases where they were as intense as the Bastille Day storm were not found. Substorm auroras are confined to a small region, usually located in the midnight sector or close by (see *Akasofu*, 1964). A two or three hours local time span would be typical. POLAR UV substorm auroras last from 15 min to several hours, consistent with the *Akasofu* [1964] visible aurora scenario.

HILDCAAs, on the other hand, are characterized by much lower auroral fluxes, from ~30 to ~60 photon cm^{-2} s^{-1}. Some short duration intensifications may reach ~100 photon cm^{-2} s^{-1}, but this is not a typical case. As mentioned previously HILDCAA auroras cover the entire auroral oval, giving them a much greater longitudinal span than for substorm auroras.

For the 4 years studied, it is found that typically magnetic storm auroras are most intense, followed by substorm auroras and then by HILDCAA auroras.

For auroral event durations, magnetic storms last from a few hours to a few days (if considered storm main phases only, without HILDCAA occurrences in their recovery phases; see *Gonzalez et al.*, 1994). Substorm timescales are tens of minutes to up to a few hours [*Akasofu*, 1964]. In contrast, HILDCAA auroras can last from several days to weeks or even a month. This is consistent with the *Tsurutani et al.* [2006, this issue] results, the latter based on AE indices and other geomagnetic data.

In summary, magnetic storm auroras are extremely intense but sporadic and short duration events. In contrast HILDCAA auroras are low-intensity events that are relatively constant, global, and can last for weeks.

3.3. Auroral Photon Flux: Comparison Between Magnetic Storms and HILDCAAs

In this section, auroral photon fluxes (relative measurements) in a limited area on the nightside are compared between HILDCAAs and magnetic storms. All HILDCAA

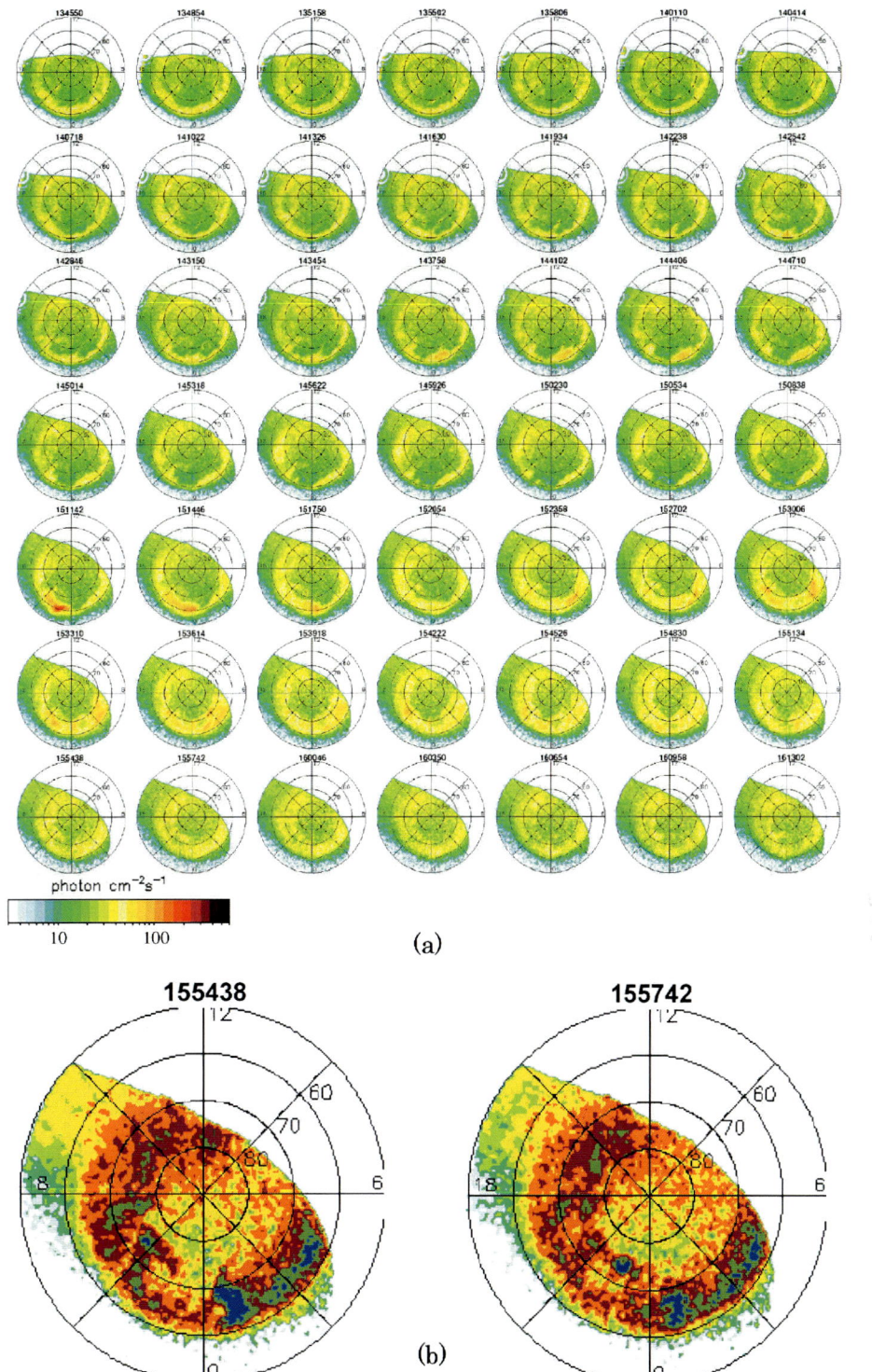

Plate 1. (a) A sequence of POLAR/UVI images for July 23, 1998, from 13:45:50 to 16:13:02 UT, during HILDCAA event 2_1998. (b) Magnified view of two images with the new color scale showing the auroral forms distributed in almost all local times and over the pole.

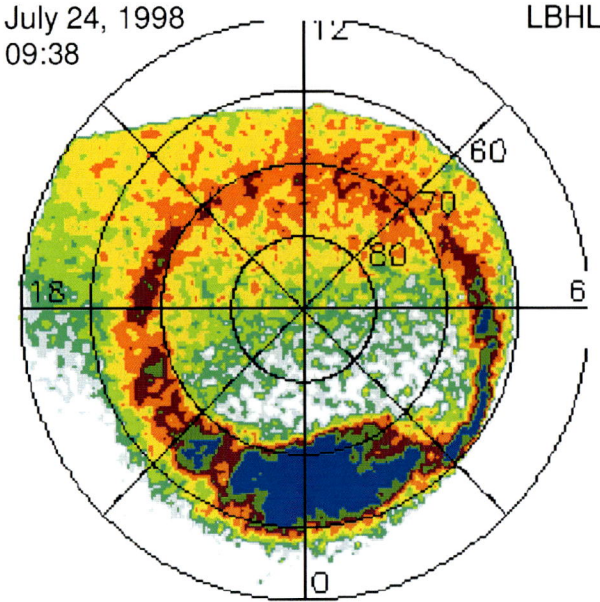

Plate 2. A POLAR image of aurora with the increased color difference. A dayside aurora is well identifiable despite the presence of dayglow.

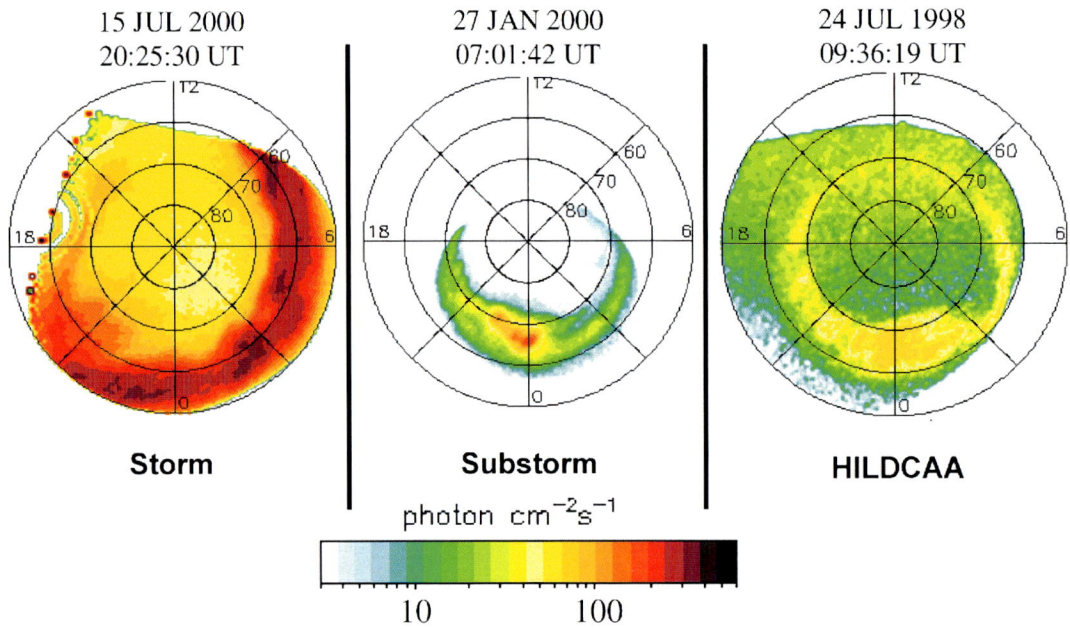

Plate 3. Differences observed in the auroral shapes and distribution for storms, substorms and HILDCAAs. The three types of activity are also different in the time scales and event duration.

events identified along the 4 year of study were used when the UVI images covering the selected area (located between 21 and 03 LT and between 50° and 80° MLAT) were available. The procedure to calculate the relative photon flux has been already described in the Methodology section.

The results of the analyses of the HILDCAA auroras are given in Table 2. From left-to-right, the 6 columns are: the event name, day of year for the event onset, the number of continuous intervals, the month of the event, the total length of the event in seconds and the average photon flux of the event. In the above, some HILDCAAs were divided due to breaks in viewing, related to the satellite orbit. Therefore, if there is only one continuous viewing interval in one day, it is listed as "1". If there is a break in the viewing, and there are two viewing intervals within the same day, then this is listed as a "2", and so on. The results obtained for HILDCAA events are shown in the 6th column in Table 2. The averaged photon flux in the area analyzed during each HILDCAA event is in the range from 6.1×10^4 to 2.3×10^5 photon pixel cm^{-2} s^{-1}.

Two magnetic storms without HILDCAA occurrences in their recovery phases were chosen to compare with the average HILDCAA case. The first is a very intense storm occurred on September 22-25, 1999. This storm has a negative peak Dst of −173 nT. The second storm chosen occurred on July 15-19, 2000, the Bastille Day event discussed previously. It had a peak Dst of −301 nT.

The photon fluxes for the storm events are indicated in Table 3. The first column indicates the event, and the middle column gives the day of year analyzed. Indicated in parentheses is the number of continuous intervals analyzed. The right hand column gives the photon fluxes in photon pixel cm^{-2} s^{-1}. The bold designations are for intervals that are closest to the storm main phases and the non-bold numbers are for storm recovery phases.

During the main phases of the two storms the auroral photon fluxes were higher than any HILDCAA interval studied. The values are 3.64×10^5 for the September 1999 storm and 7.97×10^5 for the July 2000 storm. This result is not surprising, considering the "explosive" nature of magnetic storms. The actual differences may be much larger, since very intense storms (as in the case of the Bastille Day event) have auroral displays at midlatitudes, regions outside of our sampling region. These limited local time and limited latitudinal (midnight sector auroral) average results are consistent with the results shown in Plate 3.

During the recovery phases of magnetic storms, the auroral intensities are much lower than during the main phase. For the September 1999 storm, the photon fluxes in the studied region are in the range from 5.0×10^4 to 1.5×10^5 photon pixel cm^{-2} s^{-1}. For the Bastille Day storm, auroral values are in the range from 5.9×10^4 to 1.8×10^5 photon pixel cm^{-2} s^{-1}. It should be noted that storm recovery phases may not be

Table 2. Average photon flux in the selected region located in the midnight sector, between magnetic latitudes 50° and 80° MLAT, and between 21 and 03 LT. The columns are, from left-to-right, the event number, the day of year, the continuous interval inside each day, the month of occurrence, the event duration in seconds, and the photon flux for each continuous interval of images. A total of 53 continuous intervals during HILDCAA events were analyzed

Event	Day of Year	C.I.*	Month	Length (s)	Photon Flux**
1_1998	114	1	Apr	5704	8.86×10^4
1_1998	115	1	Apr	6624	1.08×10^5
1_1998	116	1	Apr	10488	1.08×10^5
1_1998	116	2	Apr	8832	1.27×10^5
1_1998	117	1	Apr	8483	1.08×10^5
2_1998	203	1	Jul	4784	1.63×10^5
2_1998	204	1	Jul	4196	1.03×10^5
2_1998	204	2	Jul	8832	2.29×10^5
2_1998	204	3	Jul	11224	1.53×10^5
2_1998	205	1	Jul	3680	9.72×10^4
2_1998	205	2	Jul	7157	1.17×10^5
2_1998	205	3	Jul	2411	1.83×10^5
2_1998	206	1	Jul	7323	7.73×10^4
2_1998	206	2	Jul	2429	1.02×10^5
1_1999	119	1	Apr	9568	1.11×10^5
1_1999	120	1	Apr	5814	6.83×10^4
2_1999	230	1	Aug	4416	1.04×10^5
2_1999	230	2	Aug	11298	1.13×10^5
2_1999	230	3	Aug	13358	1.11×10^5
2_1999	231	1	Aug	3938	1.25×10^5
2_1999	231	2	Aug	2723	1.48×10^5
2_1999	232	1	Aug	6072	1.22×10^5
3_1999	243	1	Aug	5060	1.26×10^5
3_1999	244	1	Aug	7213	1.19×10^5
3_1999	244	2	Aug	3864	1.24×10^5
5_1999	298	1	Oct	14590	1.47×10^5
6_1999	312	1	Nov	12034	1.28×10^5
6_1999	312	2	Nov	4894	9.47×10^4
6_1999	313	1	Nov	8721	1.45×10^5
6_1999	313	2	Nov	5225	1.71×10^5
6_1999	314	1	Nov	12254	6.85×10^4
7_1999	337	1	Dec	1472	8.06×10^4
7_1999	338	1	Dec	21822	6.71×10^4
7_1999	338	2	Dec	24546	7.42×10^4
7_1999	339	1	Dec	3422	8.31×10^4
7_1999	339	2	Dec	14867	9.08×10^4
1_2000	027	1	Jan	5907	1.33×10^5
1_2000	028	1	Jan	6955	8.32×10^4
1_2000	029	1	Jan	9421	1.33×10^5
1_2000	030	1	Jan	23460	6.39×10^4
1_2000	030	2	Jan	18658	8.76×10^4

(Continued)

Table 2. Continued

Event	Day of Year	C.I.*	Month	Length (s)	Photon Flux**
2_2000	036	1	Feb	19983	1.45×10^5
2_2000	037	1	Feb	4416	8.70×10^4
2_2000	038	1	Feb	12144	6.22×10^4
2_2000	039	1	Feb	16781	6.52×10^4
3_2000	055	1	Feb	15088	9.40×10^4
3_2000	056	1	Feb	7011	6.16×10^4
3_2000	057	1	Feb	23294	6.98×10^4
3_2000	057	2	Feb	18216	6.84×10^4
3_2000	058	1	Feb	1766	9.34×10^4
3_2000	058	2	Feb	16560	6.13×10^4
4_2000	146	1	May	11224	2.07×10^5
1_2001	132	1	May	3441	2.19×10^5

* C.I. = Continuous Interval
** in photon pixel cm^{-2} s^{-1} (average)

devoid of substorms or other geomagnetic activity. So, the photon fluxes did not decrease monotonically with time during the recovery phase, as can be seen by the values in Table 3. The photon flux increases occurring in the recovery phase are much lower than those observed during the storm main phase.

It is noted that HILDCAA auroral intensities are slightly higher than those of magnetic storm recovery phase. There is some overlap from case to case, but in general, the HILDCAA intensities are higher.

Table 3. Photon fluxes calculated for each interval of continuous images during magnetic storms. The columns indicate the storm event, the day of year with the continuous interval in parenthesis, and the photon flux calculated for each continuous interval of images

Event	Day of Year	Photon flux*
September, 1999	265(1)	3.64×10^5
Peak Dst = –173 nT	266(1)	1.46×10^5
	266(2)	5.08×10^4
	267(1)	5.01×10^4
	268(1)	6.26×10^4
	268(2)	6.91×10^4
July, 2000	197(1)	7.97×10^5
Peak Dst = –301 nT	198(1)	1.83×10^5
(Bastille Day Storm)	199(1)	5.86×10^4
	200(1)	7.20×10^4
	200(2)	1.47×10^5
	201(1)	1.01×10^5

* in photon pixel cm^{-2} s^{-1}

Considering that the dayglow has a seasonal variability, and since the dayglow was not removed from these images for this analysis, a comparison of HILDCAA and storm events occurring close to the same month of the year was included to improve the accuracy of the results. The September 1999 storm event was compared with HILDCAA events occurred in August and October. The mean values and standard deviation (s.d.) for HILDCAA events 2_1999 (August), 3_1999 (August), and 5_1999 (October) are: 1.21×10^5 (s.d. 1.55×10^4), 1.23×10^5 (s.d. 3.61×10^3), and 1.47×10^5 (single value) photon pixel cm^{-2} s^{-1}, respectively. The values for these HILDCAAs events are higher or close to the values observed for the recovery phase of the storm, which are in the range from 5.01×10^4 to 1.46×10^5 photon pixel cm^{-2} s^{-1} (see Table 3 for the storm results). For HILDCAA events 2_1999 and 3_1999, the values are only smaller than the first continuous interval after the main phase. The event 5_1999 has photon fluxes even higher than this first interval.

For the July 2000 storm, the comparison is with HILDCAA event 2_1998, which also occurred in July. The mean photon flux for this HILDCAA event is 1.36×10^5 (s.d. 4.93×10^4) photon pixel cm^{-2} s^{-1}. This value is higher than the fluxes of three continuous intervals in the storm recovery phase (the values for the storm recovery phase are in the range from 5.86×10^4 to 1.83×10^5) photon pixel cm^{-2} s^{-1}. The remaining storm intervals provide values of photon fluxes that are only a little higher than the HILDCAA values.

So, even considering the dayglow variability along the year, the results show that HILDCAA photon fluxes can be comparable or even higher than the recovery phase of some very intense storms.

4. CONCLUSIONS

This paper describes the nature of the auroral intensities observed during HILDCAA events from 1998 to 2001. HILDCAA auroras have moderate intensities, with typical emission values of ~30 to 60 photon cm^{-2} s^{-1}. These HILDCAA emissions are spread along the whole auroral oval covering all local times. The emissions can last from days to weeks. For some of the events, during short time intervals, the aurora also covers the polar cap as well. It is concluded that HILDCAA auroras are different from storm and substorm auroras, and as such represent a new form of energy deposition into the polar upper atmosphere.

The comparison of relative photon fluxes in the midnight sector of the auroral zone showed that HILDCAA auroral intensities are higher than magnetic storm recovery phase auroras, but are far less intense than main phase aurora. Since storm recovery phases last for only a few days, one can expect that integrated fluxes during HILDCAA events that can last several days or weeks, may reach large amounts. The preliminary results given here are in general agreement that at times more energy may be transferred from the solar wind to the magnetosphere/ionosphere during the declining phase/minimum phase of the solar cycle. A more comprehensive, quantitative study of this is needed to address the above issue. The author is presently in progress with this work.

Further studies can use this same technique but with high resolution images instead of the survey images, and a dayglow removal method, in order to estimate absolute values of auroral photon fluxes and auroral emissions during HILDCAAs.

Acknowledgments. The author would like to thank CAPES (Brazil), project 2547/02-3, for the PhD fellowship which allowed a long stay at JPL/NASA where the most part of this research was conducted, and also to FAPESP (Brazil) project 04/14784-4 for the fellowship during the time this paper was written. The author would also like to acknowledge the POLAR/UVI team for providing the data used in this work, as well as the World Data Center for Geomagnetism – Kyoto for the geomagnetic indices.

REFERENCES

Akasofu, S.-I. The development of the auroral substorm, *Planet. Space Sci.*, 12, 273, 1964.

Belcher, J.W., and L. Davis, Jr., Large-amplitude Alfvén waves in the interplanetary medium, 2, *J. Geophys. Res.*, 76, 3534-3563, 1971.

Brittnacher, M., M. Wilber, M. Fillingim, D. Chua, G. Parks, J. Spann, and G. Germany, Global auroral response to a solar wind pressure pulse, *Adv. in Space Res.*, 25, 1377, 2000.

Craven, J.D., L.A. Frank, C.T. Russell, E.E. Smith, and R.P. Lepping, Global auroral responses to magnetospheric compressions by shocks in the solar wind: two case studies, in *Solar Wind-Magnetosphere Coupling*, edited by Kamide, Y., and J.A. Slavin, pp. 367-380, Terra Scientific, Tokyo, 1986.

Davis, N.T., and M. Sugiura, Auroral electrojet activity index AE and its universal time variations, *Journal of Geophys. Res.*, 71, 785, 1966.

Feldstein, Y.I. and G.V. Starkov, Dynamics of auroral belt and polar geomagnetic disturbances, *Planet. Space Sci.*, 15, 209, 1967.

Gonzalez, W.D., J.A. Joselyn, Y. Kamide, H.W. Kroehl, G. Rostoker, B.T. Tsurutani, and V.M. Vasyliunas, What is a geomagnetic storm?, *J. Geophys. Res.*, 99, 5771, 1994.

Guarnieri, F.L., B.T. Tsurutani, W.D. Gonzalez, Y. Kamide, and X.-Y. Zhou, Intense, continuous auroral activity related to high speed streams with interplanetary Alfvén wave trains. *Finnish Meteorological Institute - Special Issue*, 2004.

Guarnieri, F.L., *Study of the solar and interplanetary origin of long-duration and continuous auroral activity events*, PhD thesis, INPE – S.J. dos Campos, SP, Brazil, February, 2005.

Lummerzheim, D., M. Brittnacher, D. Evans, G.A. Germany, G.K. Parks, M.H. Rees, and J.F. Spann, High time resolution study of the hemispheric power carried by energetic electrons into the ionosphere during the May 19/20, 1996 auroral activity, *Geoph. Res. Letters*, 24, 8, 987-990, doi: 10.1029/96GL03828, 1997.

McIlwain, C.E., Substorm injection boundaries, in *Magnetospheric Physics*, edited by McCormac, p143, D. Reidel, Norwell, Mass., 1974.

Spann, J.F., M. Brittnacher, R. Elsen, G.A. Germany, and G.K. Parks, Initial response and complex polar cap structures of the aurora in response to the January 10, 1997 magnetic cloud, *Geoph. Res. Lett.*, 25, 2577, 1998.

Torr, M.R., D.G. Torr, M. Zukic, R.B. Johnson, J. Ajello, P. Banks, K. Clark, K. Cole, C. Keffer, G.K. Parks, B.T. Tsurutani, and J. Spann, A far ultraviolet imager for the international solar-terrestrial physics mission. *Space Science Reviews*, 71, 329, 1995.

Tsurutani, B.T., Gonzalez, W.D., The cause of high-intensity long-duration continuous AE activity (HILDCAAs): Interplanetary Alfvén wave trains, *Planet. Space Sci.*, 35, 405, 1987.

Tsurutani, B.T., T. Gould, B.E. Goldstein, W.D. Gonzalez, and M. Sugiura, Interplanetary Alfvén waves and auroral (substorm) activity: IMP-8, *J. Geophys. Res.*, 95, A3, 2241-2252, 1990.

Tsurutani, B.T., W.D. Gonzalez, F.L. Guarnieri, Y. Kamide, X.-Y. Zhou, and J.K. Arballo, Are high-intensity long-duration continuous AE activity (HILDCAA) events substorm expansion events? *J. Atmosph. Sol. Terr. Phys.*, 66, 167, 2004.

Tsurutani, B.T., W.D. Gonzalez, F.L. Guarnieri, N. Gopalswamy, M. Grande, Y. Kamide, Y. Kasahara, G. Lu, I. Mann, R. McPherron, and V.M. Vasyliunas, Corotating solar wind streams and recurrent geomagnetic activity: A Review, *J. Geophys. Res.*, 111, A07S01, doi:10.1029/2005JA011273, 2006.

Tsurutani, B.T., N. Gopalswamy, R.L. McPherron, W.D. Gonzalez, G. Lu, and F.L. Guarnieri, Magnetic storms caused by corotating solar wind streams, this volume.

Zhou, X.-Y., and B.T. Tsurutani, Rapid intensification and propagation of the dayside aurora: Large scale interplanetary pressure pulses (fast shocks), *Geophys. Res. Lett.*, 26, 1097, 1999.

Zhou X.-Y., and B.T. Tsurutani, Dawn and dusk auroras caused by gradual, intense solar wind ram pressure events, *J. Atmosph. Sol. Terr. Phys.*, 66, 153-160, 2004.

Dayside Ionospheric (GPS) Response to Corotating Solar Wind Streams

B.T. Tsurutani[1,3], A.J. Mannucci,[1] B.A. Iijima,[1] A. Komjathy[1], A. Saito[2], T. Tsuda[3], O.P. Verkhoglyadova[4,3], W.D. Gonzalez[5], and F.L. Guarnieri[5]

The dayside ionospheric modifications associated with the three phases of corotating stream-generated magnetic storms (initial, main and recovery) are investigated. The high density heliospheric current sheet plasmasheet (HCSPS) pressure pulse impingements onto the magnetosphere, the collision of high magnetic field intensity corotating interaction regions (CIRs) onto the magnetosphere, and high speed stream Alfvénic interval ionospheric effects are all investigated using both GPS ground and satellite receiver data. The above results are compared with results associated with magnetic storms caused by interplanetary coronal mass ejections (ICMEs). The main effect found is that southward GSM Bz magnetic fields within CIRs cause an uplift of the dayside ionosphere with the consequence of enhanced ionospheric total electron content (TEC). This limited dayside effect is believed to be due to the variable magnetic field z-components within CIRs. Global dayside and nightside auroral zone ionospheric TEC enhancements are noted associated with the high solar wind Alfvénic intervals. Interpretation of our findings will be given, plus directions for future research in this new and emerging area of space weather.

INTRODUCTION

The ionosphere and its variations can be studied by a powerful new technique: the use of global positioning system (GPS) satellite-based and ground-based receiver data. The receivers track the dual frequency ~1.2 GHz and ~1.5 GHz GPS signals. Using phase delay and ranging information, ground-based receivers are used to measure the total electron content (TEC) along the radio path with ~30s temporal resolution. Using ~100 ground stations, a global map of the ionosphere is obtained. GPS receivers onboard low-altitude Earth-orbiting satellites can be used to identify ionospheric uplifts/downdrafts, as well to obtain rapid latitudinal cuts (latitudinal profiles) through the dayside or nightside ionosphere. This review article will focus on ionospheric variations that occur during corotating interaction region (CIR)-generated magnetic storms and the following high-speed solar wind streams. The results will be compared/contrasted to those that occur during interplanetary coronal mass ejection (ICME)-generated magnetic storms. The former interplanetary phenomena occur during the solar cycle declining phase while the latter phenomena are dominant during the solar maximum phase. The three individual storm phases: initial, main and recovery will be discussed separately. Significant differences are noted between the two types of magnetic storm (declining phase and solar maximum) events.

---

[1]Jet Propulsion Laboratory, Pasadena, California, USA
[2]Kyoto University, Kyoto, Japan
[3]RISH, Kyoto University, Uji, 6110-0011, Japan
[4]IGPP, University of California, Riverside, California, USA
[5]Brazil National Space Research Institute (INPE), Sao Jose dos Campos, SP, Sao Paulo, Brazil

Recurrent Magnetic Storms: Corotating Solar Wind Streams
Geophysical Monograph Series 167
Copyright 2006 by the American Geophysical Union.
10.1029/167GM20

BACKGROUND

There are 28 GPS satellites in polar orbit around the Earth, flying with an altitude of ~20,200 km. The satellites continuously transmit the dual frequency signals of ~1.2 and ~1.5 GHz. The receipt of these ranging signals are used to determine the total electron content (TEC) along the radio path between the satellite and the receiver. A schematic is shown in Figure 1. For ground-based receivers, the TEC of the slant path through the entire ionosphere is measured. For satellites, GPS can be used to measure the (slant path) amount of the ionosphere that is present above the satellite. Other techniques such as ocean altimeters (e.g., TOPEX), can also be used to determine ionospheric TEC as well.

The "slant path" measurements are corrected ("verticalized") by assuming a model ionosphere. Such models are discussed in *Mannucci et al.* [1998; 1999]. Since multiple GPS satellite signals can be tracked simultaneously by each receiver, whether on the ground or in space, the verticalized data from several different slant paths can be intercompared for accuracy. If the verticalized results from a variety of satellites give the same or nearly the same value, then one can have confidence that the assumptions in the model are reasonably valid. However if the verticalized values from a variety of paths through the ionosphere give highly variable values, this is an indication that either the model may be inaccurate at that instant or that the ionosphere may be highly irregular due to geomagnetic disturbances. Both types of cases will be noted in this paper.

Plate 1 shows the CHAMP orbit on November 5-6, 2001. CHAMP was at an altitude of ~430 km and was in a 7 am – 7 pm local time polar orbital trajectory. The SAC-C and TOPEX satellite orbits and local times are shown as well.

Figure 2 shows a strong interplanetary electric field event that is correlated with dayside ionospheric effects on 5 to 6 November, 2001. The top 6 panels of the figure include the interplanetary plasma and magnetic field parameters taken from the ACE satellite. ACE was positioned in orbit around the L1 libration point, ~0.01 AU upstream of the Earth. The bottom two panels are the geomagnetic AE and D_{ST} indices. The data in the top 6 panels have been shifted by ~34 mins to remove the solar wind convection time (associated with the measured solar wind velocity of ~690 km/sec).

An interplanetary shock compresses the southward interplanetary magnetic field of a slow magnetic cloud at ~0154 UT day 310. The shock is indicated by a solid vertical line labeled "S". The time of the enhanced southward IMF (sixth panel from the top) is correlated with the Dst decrease

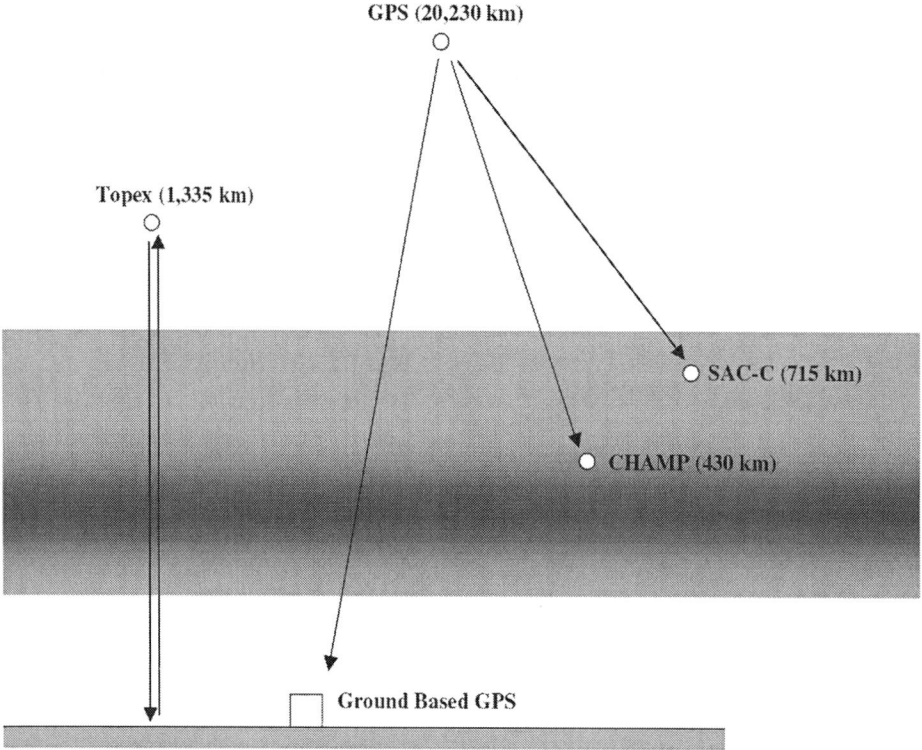

Figure 1. A schematic of transmission from a GPS satellite to ground and low altitude satellite receivers. The shaded area respresents the ionosphere. The TOPEX-Poseidon satellite altimeter signal transmission is also illustrated.

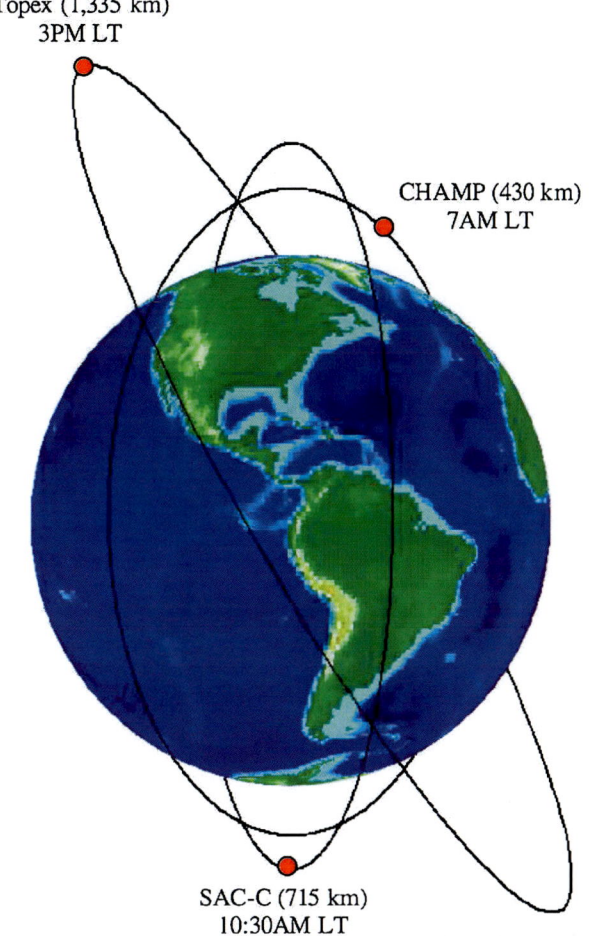

Plate 1. CHAMP, SAC-C and TOPEX satellite orbits on 5-6 November, 2001.

248 DAYSIDE IONOSPHERIC (GPS) RESPONSE TO COROTATING SOLAR WIND STREAMS

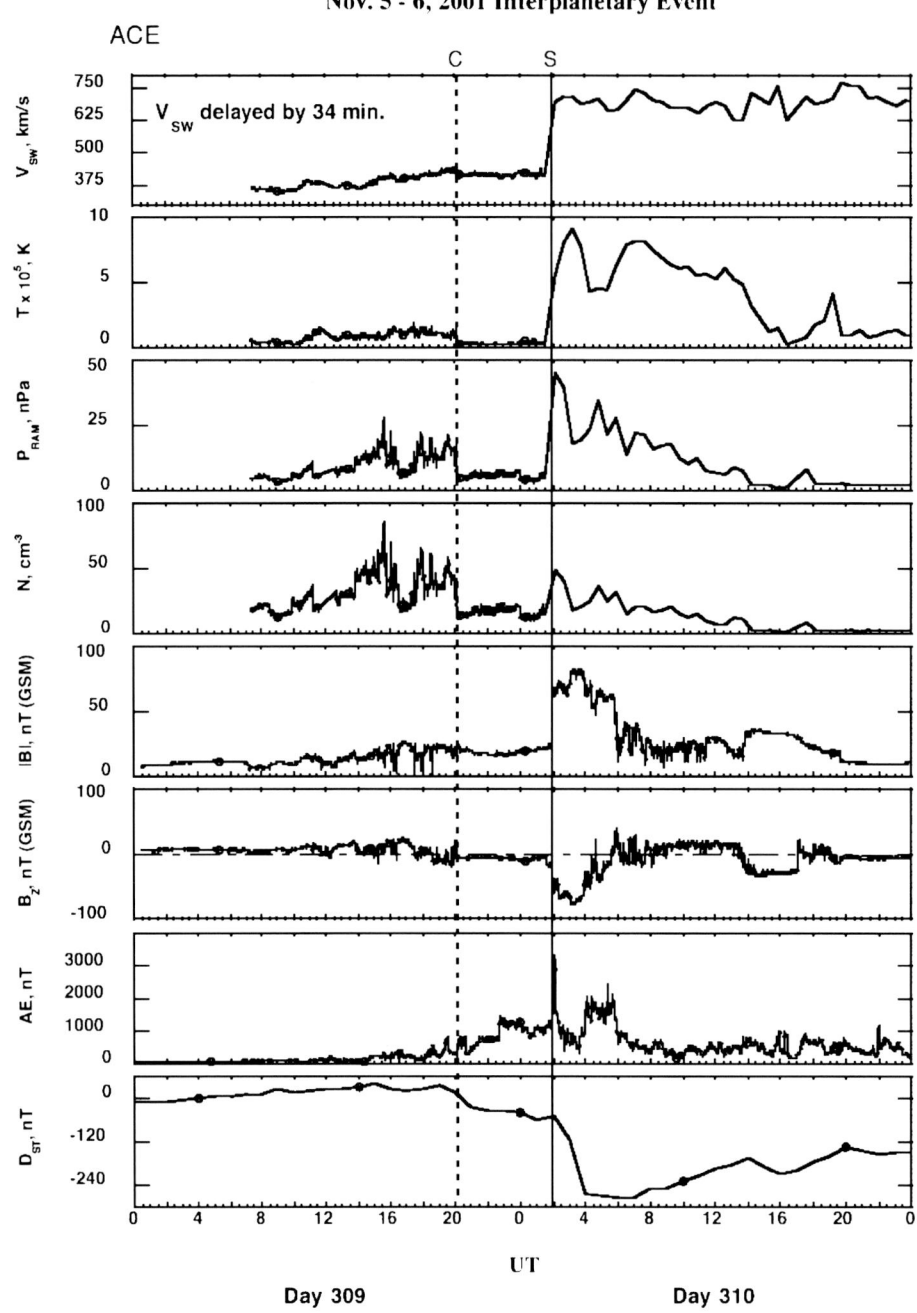

Figure 2. The interplanetary event of 5-6 November 2001 taken by the ACE spacecraft. The resultant geomagnetic indices are given in the bottom two panels.

(bottom panel). The Dst decrease is the magnetic storm "main phase". It is believed that magnetic reconnection [*Dungey*, 1961] between the interplanetary magnetic field and the magnetopause magnetic field is the cause of energy transfer from the solar wind to the magnetosphere, and thus the storm main phase [*Tsurutani et al.*, 1988; *Gonzalez et al.*, 1994].

The effects of the interplanetary electric field on the dayside ionosphere are shown in Figure 3. The top panel contains the CHAMP TEC data for 4 November 2001 and the bottom panel that for CHAMP on 6 November 2001. The event of interest is on 6 November. The 4 November TEC data are shown for "baseline/background" purposes. CHAMP retraces

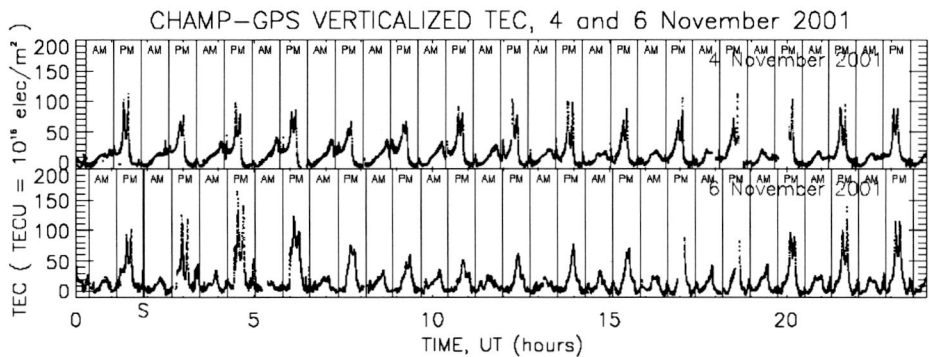

Figure 3. The CHAMP verticalized TEC data for 6 November 2001 (bottom panel). Nov 4 is shown as the top panel as a "baseline" day. The shock passage at ~154 UT (the interplanetary propagation delay time has been added) 6 November is indicated.

its orbit over (essentially) the same geographic features every 48 hrs, so the data 2 days earlier than the event of interest can be used for comparative purposes. The estimated arrival time of the time-delayed interplanetary shock is indicated by a vertical line labelled "S" in the bottom panel. The ~7 am and ~7 pm passes are indicated by "am" and "pm" labels.

It can be noted that after the shock passage on 6 November, the TEC values in the ~7 pm passes are elevated for the 3 passes from ~0300 through ~0620 UT. It is believed that promptly penetrating dawn-to-dusk oriented electric fields caused this TEC enhancement. The TEC values at altitudes above CHAMP become depressed after that time from ~0920 UT to ~1220 UT. The latter effect is believed to be caused by the disturbance dynamo [*Tsurutani et al.*, 2004], but will not be discussed further here. We direct the reader to the above reference and *Blanc and Richmond* [1980], *Scherliess and Fejer* [1997], *Richmond and Lu* [2000], *Lu et al.* [2001] and *Fuller-Rowell et al.* [2002] for further discussion on this topic.

Plate 2 shows that this TEC enhancement detected by CHAMP at ~7 pm on 6 November (Figure 3) is a global dayside effect. Plate 2 shows the TEC data from ~100 ground-based GPS receivers. All values are verticalized TEC values plotted at the approximate intersection point of the GPS raypath with an ionospheric "shell" at 450 km altitude. The top panel is for 4 November from 0409 to 0456 UT and is used as a "baseline" value. The TEC for the entire globe is shown. In this panel, the top and bottom of the figure are the north and south poles, and the center of the graph is the sub-solar equator. It is noted that the region where TEC exceeds 100 TEC units (orange-to-red color) extends from the equator to ~±20° MLAT and from ~900 to ~1900 local time. The CHAMP pass is superposed on the top panel. The color scale is in the same units as that for the ground-based data.

The bottom panel, Plate 2b, shows the TEC ionospheric values from 0414 to 0500 UT on 6 November, ~2 hrs after the impingement of the interplanetary dawn-to-dusk electric field onto the magnetosphere. The ionospheric response to the imposed electric field can be noted by the difference between this panel and the top panel of the Figure. The bottom panel shows that the enhanced TEC region (>100 TECU) extends from the equator to ~±50° MLAT, a much broader region of enhanced TEC than on 4 November. The CHAMP, SAC-C and TOPEX satellite TEC data are also shown to the same scale.

Figure 4 illustrates two "Halloween" 2003 interplanetary CME (ICME) events and their resultant magnetic storms. The ACE solar wind parameters are given in the top five panels and the geomagnetic Dst index at the bottom. The two ICME events are identified by the two fast solar wind streams detected on 29 and 30 September, 2003. The southward component of the IMF Bz that is part of the ICMEs cause the magnetic storm main phases (Dst decreases) *Mannucci et al.* [2005]. The ICMEs were ejected from the solar coronal at approximately the same time as the solar flares which occurred on 28 and 29 September (see *Tsurutani et al.* [2005] for a discussion of flare effects on the ionosphere).

Plate 3 illustrates the response of the ionosphere above CHAMP (now at ~400 km altitude) for 3 dayside passes on October 30, 2003. CHAMP was located at ~1300 local time on that day. The first dayside pass shown occurs prior to the impingement of the enhanced solar wind electric field onto the magnetosphere. This data is labeled 1840 UT on the right of the curve and 1900 UT on the left. There are two peak TEC values at ~±10° MLAT which are called the "equatorial" or "Appleton" anomalies. They are caused by the post-dawn electric field convective uplift of the equatorial ionosphere, followed by a gradual diffusion down the Earth's dipolar magnetic field lines due to gravitation to the latitudes noted in the plot [*Kelley*, 1995]. The TEC intensities above CHAMP are ~75 TECU at the anomalies. The passes after the impingement of the interplanetary electric fields are also

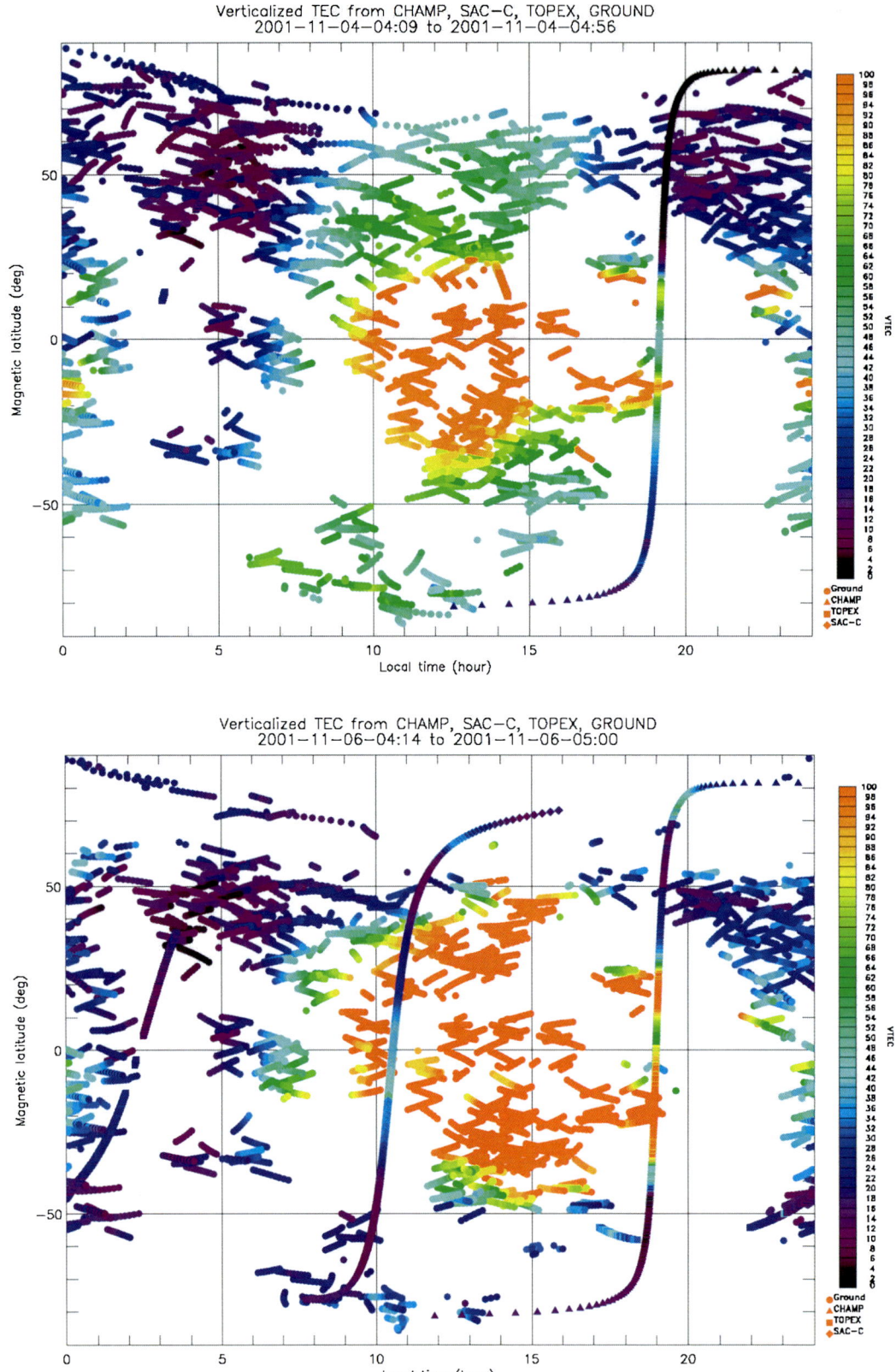

Plate 2. The total ionospheric verticalized TEC from ~100 ground stations and the TEC above various satellites for 0414 to 0500 UT 6 November 2001 (bottom panel) and for 0409 to 0456 UT 4 November 2001 (top panel).

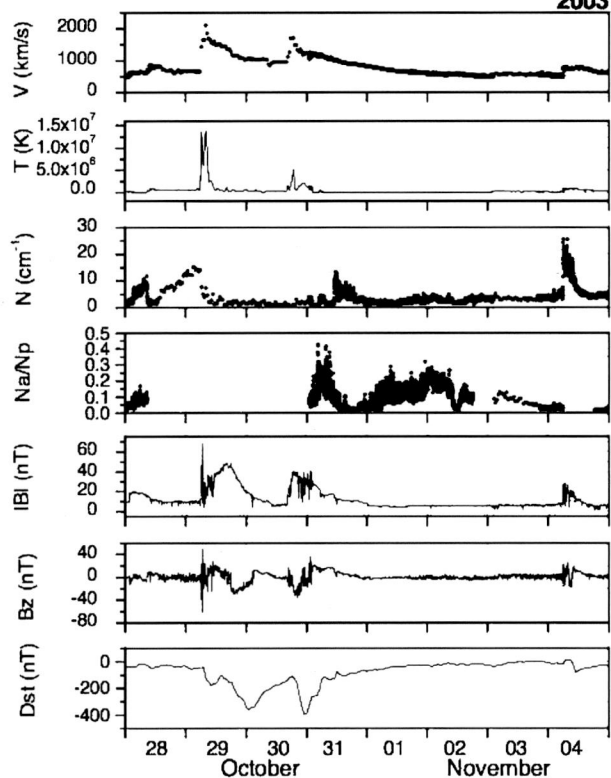

Figure 4. The ACE solar wind parameters for the 29 and 30 September 2003 "Halloween" events.

shown in the Figure, the first ~one hour and 15 min (2032 UT) after the imposition of the electric field event onto the magnetosphere, and the second ~2 hours 45 min afterward (2204 UT). The ionospheric changes that take place are dramatic. For the first pass, the TEC above CHAMP reaches a peak value >200 TECU, with the anomalies located at ~±20° MLAT. In the last pass, the peak values are >300 TECU and are located at ~±30° MLAT.

We attempt to explain the dayside ionospheric observations shown in Figure 3 and Plates 2 and 3 through the sequence of figures: 5, 6 and 7. Figure 5 shows the interplanetary dawn-to-dusk electric field that is caused by the solar wind flow of the embedded southward interplanetary magnetic electric field past the magnetosphere,

$$\vec{E} = -(\vec{V} \times \vec{B}).$$

Magnetic reconnection between the interplanetary magnetic fields and the magnetopause magnetic fields [*Dungey*, 1961], lead to magnetospheric electric fields with the same directionality but diminished in magnitude [*Gonzalez et al.*, 1989, 1994; *Wygant et al.*, 1998; see also *Tsurutani et al.*,

2004]. These magnetospheric electric fields are responsible for the sunward convection of the plasmasheet and for the formation of the storm-time partial ring current in the magnetosphere proper.

If the magnetospheric electric field can "penetrate" to the equatorial ionosphere prior to the formation of magnetospheric-ionospheric "shielding currents" [*Vasyliunas*, 1982; *Fejer and Scherliess*, 1995; *Sobral et al.*, 1997; *Abdu et al.*, 2003; *Kelley et al.*, 2003], or if the interplanetary electric field can promptly penetrate from the polar region ionosphere to the equatorial ionosphere [through the Earth-ionosphere wave guide: *Kikuchi and Araki*, 1979], the effects will be an upward $\vec{E} \times \vec{B}$ convection of the equatorial ionosphere at local noon and a downward $\vec{E} \times \vec{B}$ convection of the equatorial ionosphere at midnight. This is indicated in the Figure. In our present scenario, it does not matter by what path/mechanism the electric fields arrive to the equatorial ionosphere. The only requirement is that the electric fields arrive at the equatorial regions relatively promptly. Clearly, knowledge of the actual path that the electric fields take is extremely important, and deserves considerable further attention. Unfortunately, this is beyond the scope of the present paper.

The schematic in Figure 5 gives one possible scenario for the "uplift" of the dayside equatorial ionosphere noted in Figure 3. Figure 6 indicates how this uplift can lead to higher ionospheric TEC values. The top panel shows the normal dayside (quiet time) ionospheric height versus log density profile. Solar photoionization produces electron-ion pairs and recombination causes destruction of unbound electrons. At equilibrium, the two processes are in balance. The peak density occurs at ~300 km altitude. With the uplift of the dayside equatorial ionosphere (due to dawn-to-dusk or "eastward" electric fields, as viewed from the northern hemisphere), the plasma is brought to heights where the recombination process is much slower (the ion-electron recombination rates are strongly height dependent). The recombination timescales are seconds at ~100 km, ~hrs at ~300 km, and days at ~600 km altitude. Thus the plasma is much longer-lived at these higher altitudes. The ionospheric uplift also diminishes ionospheric plasma densities at lower altitudes. Solar EUV/UV photoionization will create a new ionosphere in these depleted regions, leading to an overall increase in ionospheric TEC. This is shown in the bottom panel of Figure 6. The creation of enhanced dayside ionospheric plasma/ electron column densities (Figure 3 and Plate 3) are explained by this scenario.

Although the general uplift and enhancement of dayside ionospheric TEC column densities are described above in a general manner, it should be noted by the reader that considerable modeling effort is needed. The slow uplift and replacement of ionospheric low altitude plasma will be a continuous process and not a simplified two-step mechanism as shown in

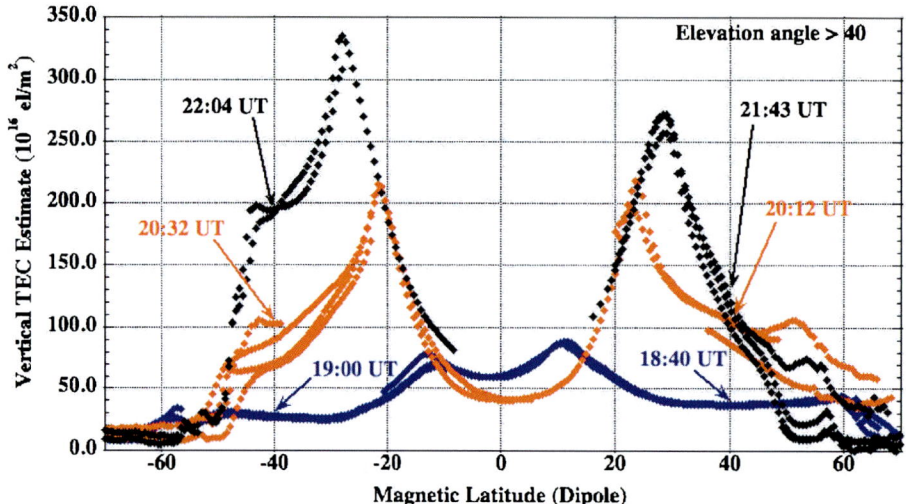

Plate 3. Verticalized TEC above the CHAMP satellite (400 km altitude) before and after the October 30, 2003 interplanetary electric field event.

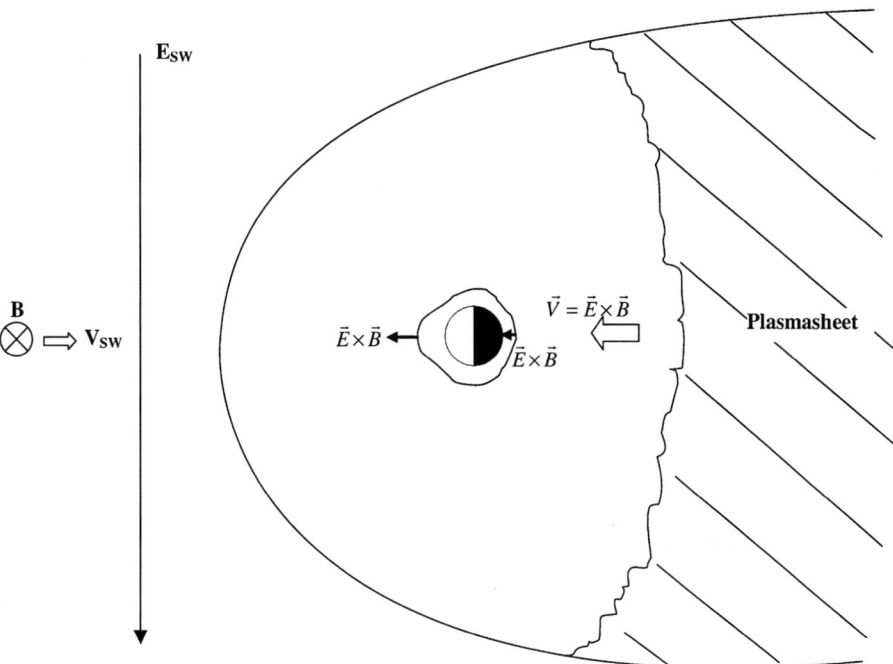

Figure 5. The interplanetary dawn-to-dusk electric field, cross-tail magnetospheric storm convection electric fields and noon and midnight equatorial ionospheric electric fields.

Figure 6 (this was mainly for illustrative purposes). The process strongly depends on the magnitude of the electric field, the time dependence of the electric field, the ambient magnetic field directionality, the magnetic latitude and the local time at Earth. It is highly probable that the ionosphere never reaches an equilibrium state throughout the uplift process. Modeling should be able to indicate the electron density height versus time profiles as a function of eastward electric field magnitudes (as further functions of time and magnetic latitudes). Conversely, these modeling results can be used to estimate expected observational values and put constraints on electric field penetration mechanisms. The latter will be helpful in explaining the overall solar wind-ionospheric coupling process.

The top panel of Figure 7 shows the ordinary fountain effect. The bottom panel shows the "dayside super fountain effect". The physical mechanism is the same, the near-equatorial ionospheric uplift by eastward electric fields and downward flow along the Earth's magnetic field lines. However for the dayside superfountain effect, the ionosphere is lifted to very high heights (as previously shown by CHAMP data), by much stronger and long lasting electric fields. The uplifted plasma flows down the Earth's magnetic field lines to much higher latitudes. This is the explanation for the broader latitudinal range of the dayside plasma and the middle latitude anomaly locations shown in Plates 2 and 3. Again, computer modeling will help understand further ionospheric effects in the GPS data. This will be briefly addressed later in this paper. The reader is directed to *Tanaka* [1981] and *Basu et al.* [2001] for related observations.

*High-Speed Solar Wind Streams
and Geomagnetic Activity*

An example of an interplanetary heliospheric current sheet (HCS) plasma sheet [*Winterhalter et al.*, 1994], a corotating interaction region [CIR: *Smith and Wolf*, 1976; *Balogh et al.*, 1999], and a high-speed stream proper, are shown in Figure 8. The panels from top to bottom are: the solar wind speed, proton density, temperature, the three components of the interplanetary magnetic field (in GSM coordinates) and field magnitude, and the AE and Dst indices. The top panel shows the high-speed stream (proper) on the right side of the vertical dashed line, and the slow speed stream at the far left. The corotating interaction region (CIR) is formed by the interaction of the fast stream with the slow stream, and is identified by the compressed magnetic fields (next to bottom panel) from the beginning of 25 January to the abrupt end (at the vertical dashed line) at ~2000 UT 25 January. The abrupt magnetic field magnitude decrease, velocity increase (top panel) and density decease (second from top panel) at the vertical dashed line indicate that this event is a fast reverse

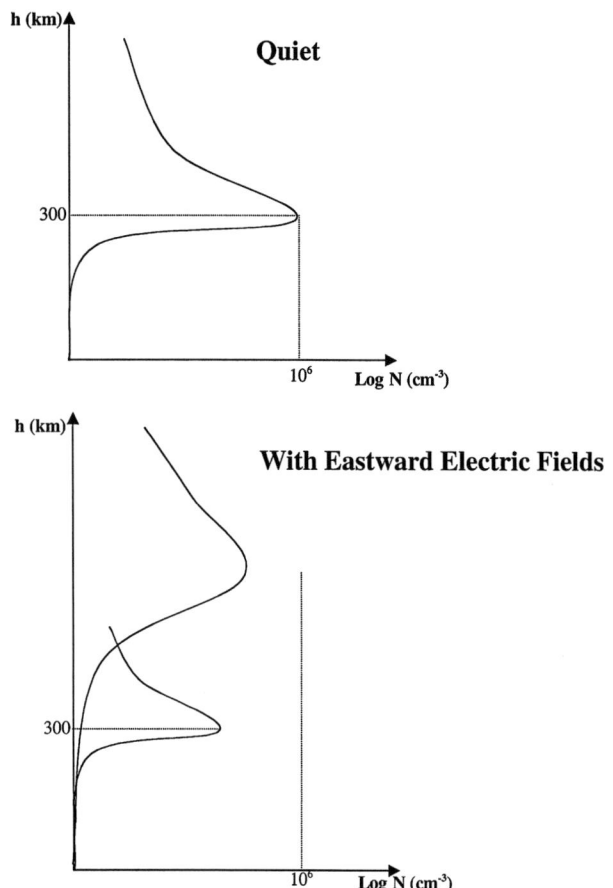

Figure 6. Consequences of dayside ionospheric uplift.

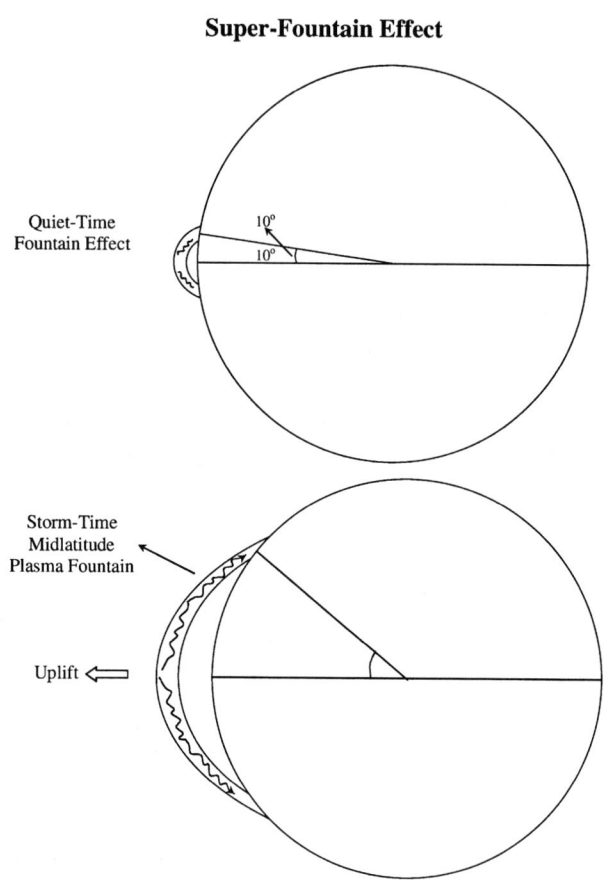

Figure 7. The dayside super fountain effect.

shock. The CIR occurs in between the fast stream and the slow stream and is composed of heated and accelerated slow stream plasma at the leading antisolar portion and heated and decelerated fast stream plasma at the trailing portion. The two portions of the CIR are typically separated by an interface discontinuity [see discussion in *Pizzo*, 1985].

The heliosphere sector boundary or heliospheric current sheet [HCS; *Smith et al.*, 1978] is where the IMF B_X and B_Y components reverse sign. For positive sector (outward from the sun) magnetic fields, $B_X < 0$ and $B_Y > 0$ (in GSM coordinates) for Parker spiral magnetic field lines. The magnetic fields during January 24 are therefore in a positive sector. The negative sector magnetic fields ($B_X > 0$ and $B_Y < 0$) are generally present in the CIR and high speed stream proper. Thus the heliospheric current sheet crossing is somewhere on 24 January (possibly multiple crossings). Near the heliospheric current sheet is a slow speed ($V_{SW} < 400$ km/s), high density "plasmasheet" (HCSPS). The high ion densities can be noted from ~1800 UT 24 January to ~0100 UT 25 January.

The solar wind features that cause the different magnetic storm phases (initial, main and recovery) are: 1) the high density HCS plasmasheet, due to its high ram pressure causes the storm initial phase, 2) the southward IMF B_Z intervals within the CIR causes the storm main phases (through magnetic reconnection), and 3) the high speed streams and Alfvénic B_Z fluctuations are associated with the extended recovery phases of the storms. For the main phases of magnetic storms, the CIR B_Z fields are highly irregular (intervals of southward B_Z are sporadic and short-lived), unlike the smoothly varying fields of ICME sheaths and/or magnetic clouds. Thus CIR-related magnetic storms are weak ($D_{ST} > -50$ nT) to moderate (-100 nT $\leq D_{ST} \leq -50$ nT) in intensity. The continuous and elevated AE intensities (not shown) in the high speed streams proper are due to the southward components of the IMF B_Z fluctuations associated with large amplitude Alfvén waves present in the streams. Presumably short duration southward IMF B_Z events lead to sporadic reconnection, and consequential plasma injections into the nightside magnetosphere and concomitant auroras very much like during high-intensity, long-duration, continuous AE activity (HILDCAA) events [for a discussion of the latter, see *Tsurutani et al.*, 2006a,b].

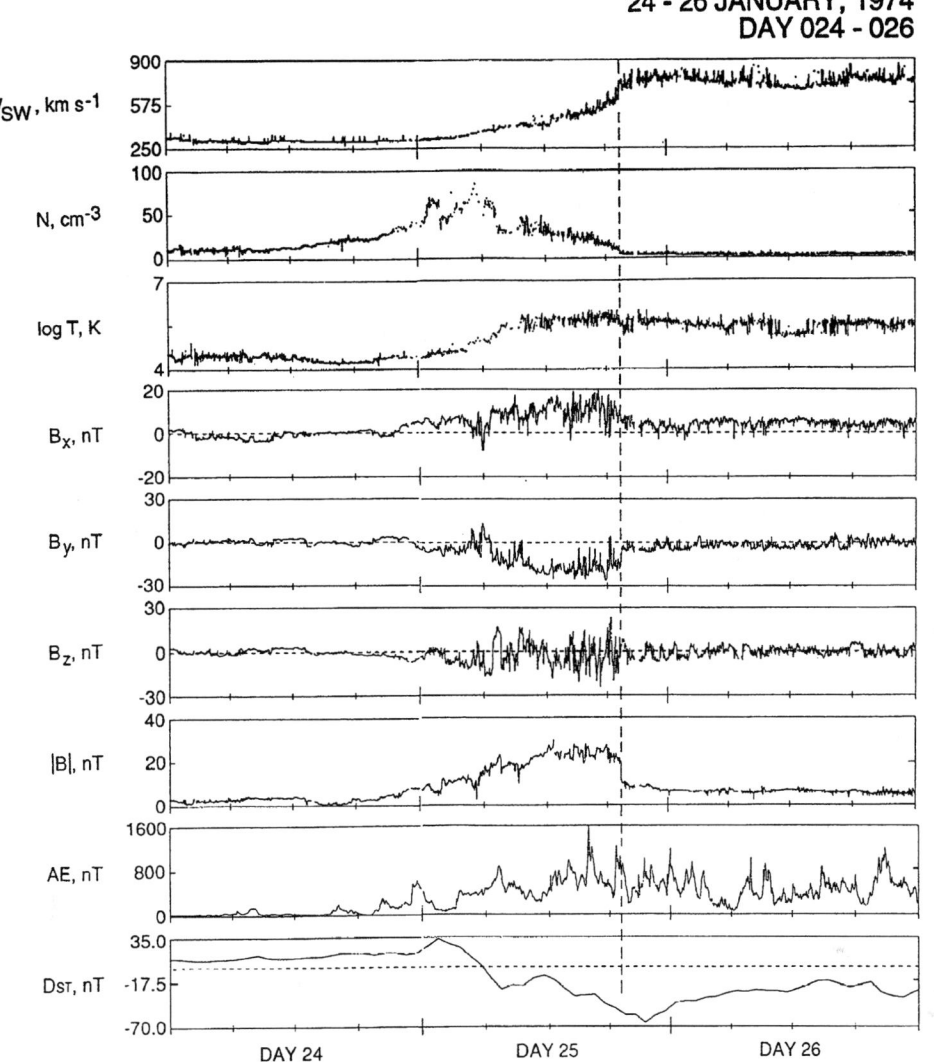

Figure 8. An example of a high speed stream and slow speed stream interaction. The various features of the slow stream, stream-stream interaction and high speed stream proper cause the storm initial phases, main phases and recovery phases.

The bottom panel in Figure 9 shows a summary of the CIR-generated magnetic storm of Figure 8. Because CIRs typically do not have fast forward shocks at 1 AU distances from the sun, there are no storm sudden commencements (SSCs). The initial phase is caused by high-density, low-speed streams. In contrast, ICME initial phases (top panel) are generated by density compressions across fast-forward shocks and thus SSCs/sudden impulses (SIs) are typically present. The CIR-generated storm main phases are associated with magnetic reconnection due to southward interplanetary magnetic fields within the CIRs. The sporadic IMF B_Z events lead to irregularly shaped, weak-to-moderate intensity main phases. In contrast, the ICME storm main-phases are generated by either sheath and/or magnetic cloud southward magnetic fields which are often smooth, regular and of long-duration. The magnetic storm intensities can reach $D_{ST} = -500$ nT or even more negative values. The ICME-produced storms can vary in intensity from weak to as intense as "great" (Dst < −250 nT) or even "extreme" (Dst < −1,000 nT).

The CIR/high speed stream magnetic storm "recovery phases" can last days to even as long as weeks. Actually, the term "recovery", in this case, is somewhat of a misnomer. It has been shown that the southward components of the IMF B_Z fluctuations (Alfvén waves) within high speed streams proper cause small sporadic injections of plasma and energy into the outer portions of the magnetosphere (presumably through magnetic reconnection), preventing the magnetosphere from fully "recovering". In the case of typical ICME-generated

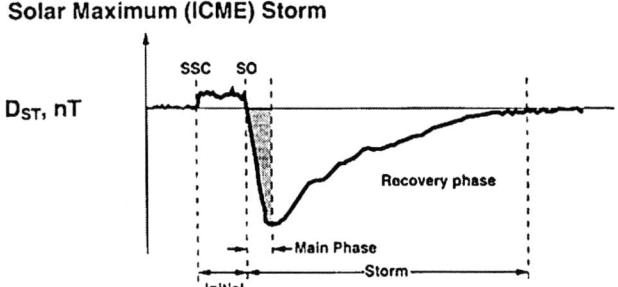

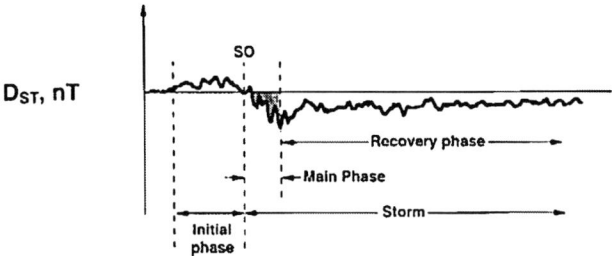

Figure 9. A schematic of a CIR generated magnetic storm. The various interplanetary features shown in Figure 8 are the causes of the magnetic storm phases.

storms, the decay of the D_{ST} index from its maximum negative excursion back to a baseline value is due to the physical process of energetic ring current ion losses via wave-particle interactions, charge-exchange with thermal atoms, Coulomb collisions and convection. Of course, similar loss processes are taking place during the CIR-generated storm "recovery phases", but sporadic energy injection is also occurring leading to a stable, long-lasting "recovery".

Further discussion of the similarity and differences between the initial, main and recovery phases of magnetic storms induced by ICMEs and CIRs can be found in the introductory chapter of this book [*Tsurutani et al.*, 2006a] and in *Tsurutani et al.*, 2006b.

Dayside Ionospheric TEC Effects due to Corotating Solar Wind Streams

To compare/contrast the dayside ionospheric TEC effects during corotating solar wind (slow) and fast streams, we have selected 4 streams to examine in detail (out of an initial selection of ~15 events). All of the interplanetary data were taken by the ACE spacecraft instrumentation. As indicated previously in Figures 8 and 9, the magnetic storm main and recovery phases are due to the CIRs and high speed streams proper. However it has also been shown that the upstream slow streams also play a role, that of causing the storm initial phases. Thus when we refer to "corotating streams", we mean the slow speed stream just upstream of the CIR, the CIR and the fast stream proper. The 4 "corotating stream" events chosen are: June 5-12, 1998 (days 156 to 162), January 2 to 6, 2003, January 17 to 23, 2003 and December 7 and 19, 2003 (days 341 to 352). Each of the four interplanetary events has been selected based on their specific features. The interplanetary events were selected without knowledge of the ionospheric responses. Thus, it is hoped that the results of this sample will prove to be representative of the broader range of events.

January 16 to 22, 2003 Event

Figure 10 shows the January 16-22, 2003 high speed stream and portions of the precursor low speed stream. From top down are: the solar wind speed, proton density, temperature, magnetic field magnitude, and B_X, B_Y, and B_Z (in GSM coordinates). The plasma beta (plasma thermal pressure divided by magnetic pressure) is shown beneath the B_Z panel.

The bottom two panels are the IMF B_Z shown in higher (1 min ave) resolution, and the IMF Bz with one hr boxcar averaging, respectively. The point of the latter panels will be discussed later.

Many of the relationships between interplanetary features and magnetic storm phases (shown in Figure 8 and 9) can be noted here as well. There is a sharp plasma density (~40 cm^{-3}) spike at ~1200 UT January 17. Note that this density is greater than the typical values behind ICME shocks. The interplanetary HCS is noted by the sharp change in B_X from +2 nT to –4 nT and B_Y from –5 nT to +8 nT immediately after the density spike. The CIR extends from ~00 UT January 18 to ~0600 UT January 19 (here we use the heightened densities as a secondary guide). The IMF B_Z fluctuation amplitudes are largest within the CIR. However the hourly B_Z data (bottom panel) shows that the B_Z value is on average positive throughout the CIR.

The high speed stream proper lasts from ~0600 January 19 until the apparent end at ~0000 UT January 22. The IMF B_Z fluctuations (Alfvén waves) have large amplitudes from the stream onset until ~00 UT January 21 where the amplitudes decrease abruptly.

Ram Pressure Pulse Effects

On January 10, 1997, instruments onboard the WIND spacecraft detected a fast-forward interplanetary shock (not shown to conserve space). This shock preceded a well-studied ICME/magnetic cloud. Plate 4 illustrates the northern hemispheric auroral response to the shock as viewed in the POLAR UV imaging experiment. The images are of the Lyman-Birge-Hopfield (UV) band and are taken at a ~1 min 13 s cadence (only every ninth image is show here to conserve space).

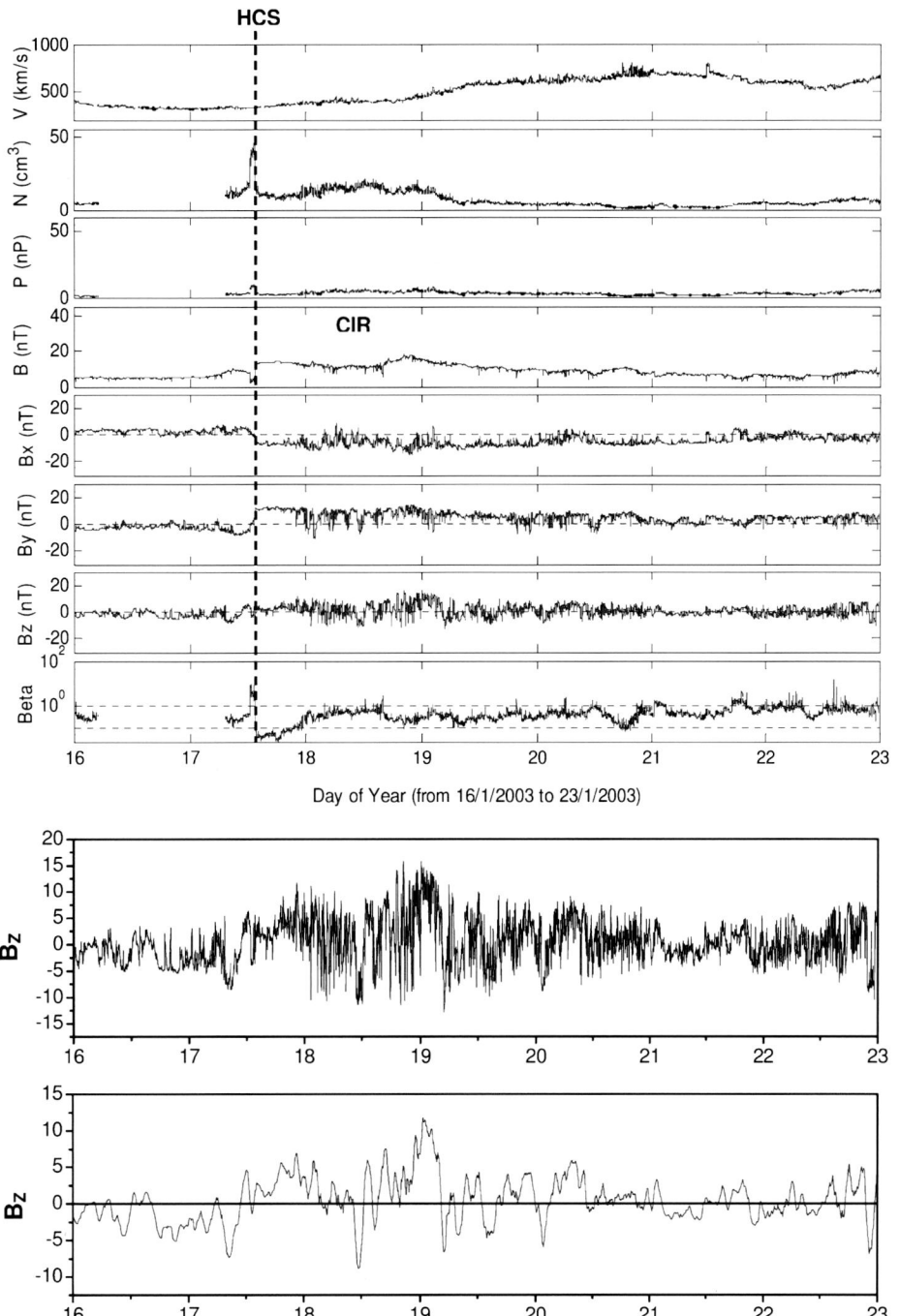

Figure 10. The January 16-22, 2003 high speed stream, upstream slow speed stream. The interplanetary features causing geomagnetic activity will be discussed.

The shock arrived at the magnetosphere between 0058:11 and 0059:25 UT (the high resolution images are not shown to conserve space). Note that the auroral brightening occurs first on the dayside from ~0600 to ~1300 local time and then later extends to both earlier and later local times. This is particularly clear in panels d) through f). It is clear that the auroral brightenings are present first on the dayside (it should be noted that the normal to the shock wave was not in the GSM/GSE negative x– direction, but obliquely, as shown here), and then extended past dawn and to dusk. The auroral

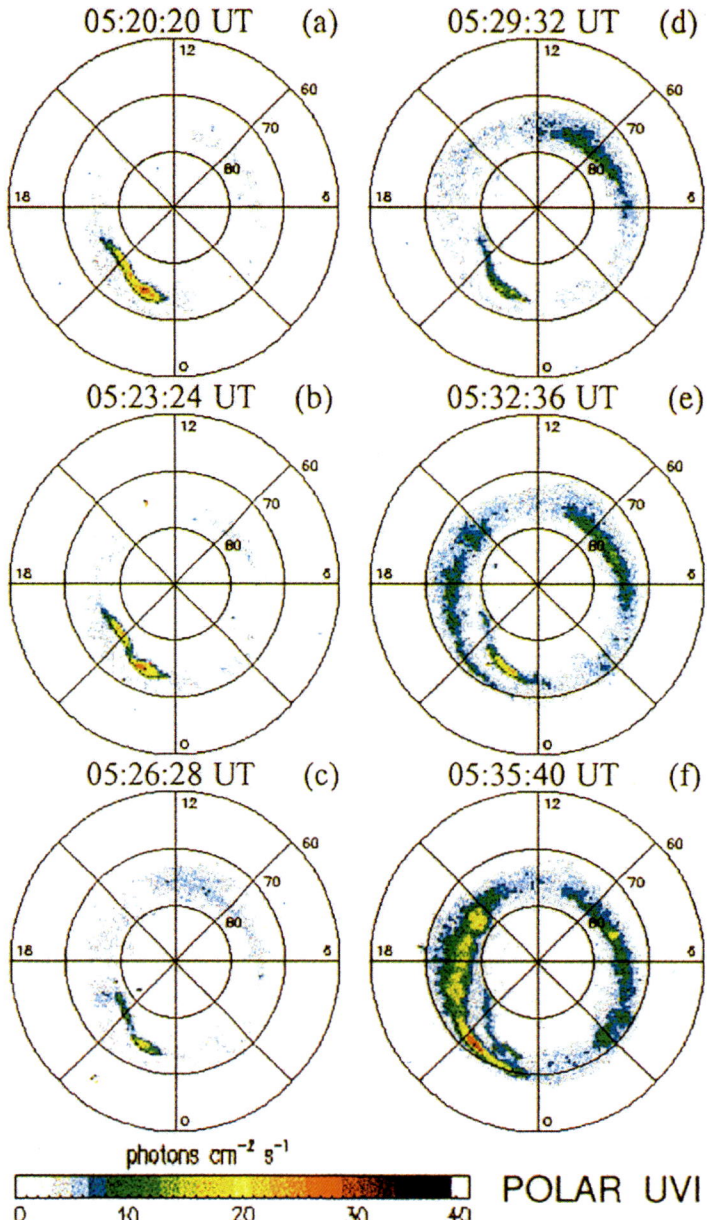

Plate 4. The auroral response to an interplanetary shock on 10 January 1997.

"propagation" speed was mapped out to the equatorial plane of the magnetosphere, and it was found that this calculated value of speed was consistent with the measured solar wind speed (taken from WIND measurements). Potential scenarios (plasma betatron acceleration and kinetic Alfvén waves) for the cause of the auroras by local magnetospheric compression (by the shock pressure pulse propagation) have been presented by *Zhou and Tsurutani* [1999] and *Tsurutani et al.* [2001a], respectively.

It should be noted that the impingement of interplanetary shocks onto the magnetosphere can trigger magnetospheric substorms on the nightside with intense auroral displays. This occurs only when the magnetosphere has been "preconditioned" with southward interplanetary Bz fields [*Zhou and Tsurutani*, 2001; *Tsurutani and Zhou*, 2003].

Because the January 17 HCSPS ram pressure pulse (Figure 10) increase would be much larger than the shock event pressure pulse (the density increases across shocks is typically less than 4.0), it is almost certain that the HCSPS pressure pulse would cause significant dayside auroras. Particle (betatron) acceleration in the magnetosphere and losses into the polar ionosphere (via the loss cone instability) will create not only auroras via precipitating electron excitation processes, but enhanced electron densities due to the energetic electron losses in the upper atmosphere (the dominant energy loss mechanism is atomic and molecular ionization). The length of time such ionization effects will last will depend on the particle kinetic energies and altitudes of the particle deposition.

There is another solar wind/magnetospheric energy transfer mechanism, that of a "viscous interaction". It has been empirically noted that slow ram pressure increases (over time) can cause gradual auroral intensifications [*Zhou et al.*, 2003; *Zhou and Tsurutani*, 2004]. Although the exact physical mechanism has yet to be identified, several observational examples have been published. Thus in addition to a fast pressure pulse aurora mechanism (betatron acceleration of pre-existing magnetospheric plasma and/or the creation and damping of Alfvénic waves), a slowly developing process is also present.

Plate 5 shows a slow solar wind ram pressure buildup during 3-1/2 hours of March 13, 1997. The panels, from top to bottom, are: the solar wind ram pressure (N mV_{SW}^2), the proton density, the solar wind speed, and the interplanetary magnetic field GSM components. The images at the top are the POLAR UV images shown at a reduced cadence of ~15 min. It is clear that the auroral intensities at ~1800 local time (far left side of images) increases over the 3 hrs where there is increasing solar wind ram pressure.

It can be reported here that no obvious ground- or satellite-based dayside TEC effects of hemispheric scale were noted during or after the solar wind ram pressure pulse. There were no obvious TEC effects for ram pressure pulses during the other events studied as well.

For this CIR event there were no obvious GPS TEC effects during the CIR-storm main phase, and no effects during the high speed stream-Alfvénic interval as well. It should be noted, however, that the IMF B_Z during the CIR was generally positive throughout the event. This point will be discussed later in the paper.

January 1 to 6, 2003 Event

We next look at the 1 to 6 January event, Figure 11. The format is the same as for Figure 10. Here again there is a high density HCSPS (~1800 to 2400 UT January 2), with a HCS at ~0000 UT 3 January (demarked by a vertical dashed line). A CIR extends from ~0000 UT January 3 to ~0200 UT January 4. The IMF B_Z fluctuations are largest within the CIR. One important difference of this event from that in Figure 10 is that the hourly average IMF B_Z is more southward in the latter half of the CIR (~1200 UT January 3 to ~0200 UT January 4). The high speed stream proper extends only to ~00 UT 5 January. The stream is characterized by Alfvénic fluctuations.

Plate 6a and b illustrates that there is no obvious ground- or satellite-based hemispheric-scale ionospheric TEC effect caused by the HCSPS pressure pulse. The format is the same as in Plate 2. The top panel is a baseline taken two days prior to the event, ~2347 UT December 31, 2002 to ~0034 UT January 1, 2003, and the bottom panel the event time of interest, ~2343 UT January 2, 2003 to ~0029 UT January 3, 2003. There is essentially no indication of significant changes between the two figures.

Plate 7a and b show the global TEC changes during the CIR passage on January 3. The top panel for 1953 to 2040 UT January 1, 2003 are the baseline values, and the time of interest is shown in the bottom panel for 1948 to 2035 UT January 3, 2003. The format is the same as in Plate 6, except the peak scale (red) is now lowered to 70 TEC units (from 100 TECU). One remarkable feature of this "baseline" is the low TEC values. This is most likely due to the low solar irradiation at Earth due to the presence of a solar coronal hole facing it. In the opposite sense, high TEC backgrounds were noted when solar active regions faced the Earth during the Halloween 2003 events [*Tsurutani et al.*, 2006c,d]. From the latter works, it was noted that the dayside TEC value dropped precipitously when the Halloween flare/CME active region EUV source went over the solar limb on November 4, 2003.

In Plate 7 there is a substantial TEC change between the baseline interval and the interval during the CIR passage. In the top panel, the intense TEC region (in red) occurs primarily near 10 LT within ±20° of the magnetic equator. However the bottom panel indicates that the enhanced dayside ionospheric TEC extends over a much broader local time region

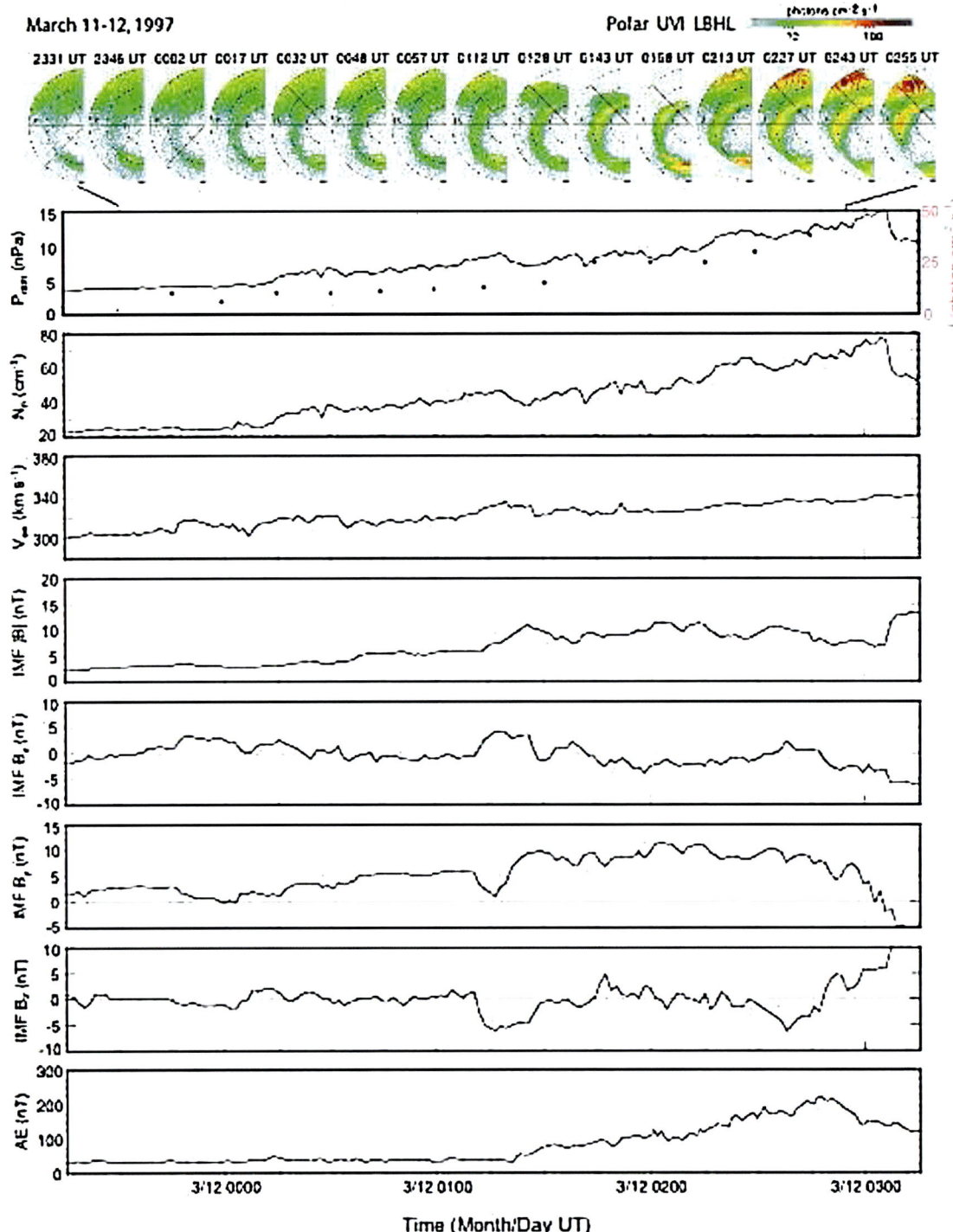

Plate 5. A slow solar wind ram pressure event and consequential auroras on March 13, 1997.

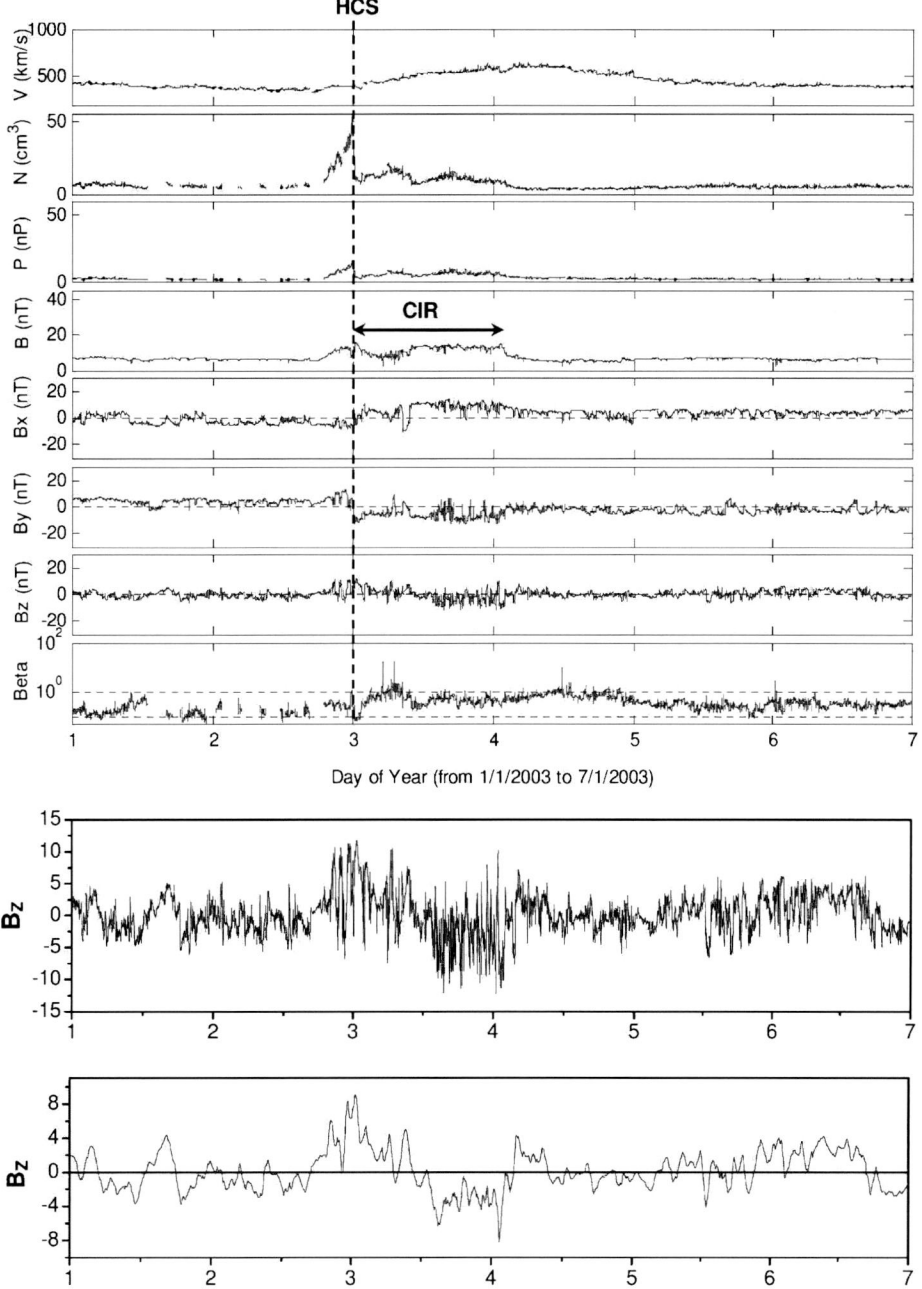

Figure 11. The interplanetary data for a 1 to 6 January 2003 event. The format is the same as that for Figure 13.

from 10 to 17 LT during the CIR passage. The latitudinal extent is similar, but may extend to slightly lower southern latitudes (~−30°). This TEC enhancement is different than what was noted for the ionospheric response to the interplanetary dawn-to-dusk ICME electric fields noted in Plate 2. The ICME electric fields caused a large latitudinal extension (up to ~±50°) over a broad longitude range. The primary difference between the two cases is that the sources of interplanetary electric fields are different. In Figure 2 the source was southward ICME magnetic cloud magnetic fields. Here it is time-averaged (but fluctuating) southward CIR magnetic fields. It is possible that the fluctuating IMF Bsouth fields lead to small equatorial ionospheric uplifts with enhanced photoionization at lower altitudes, leading to the TEC enhancements. Thus the "dayside superfountain effect" is occurring, but it is limited in scope.

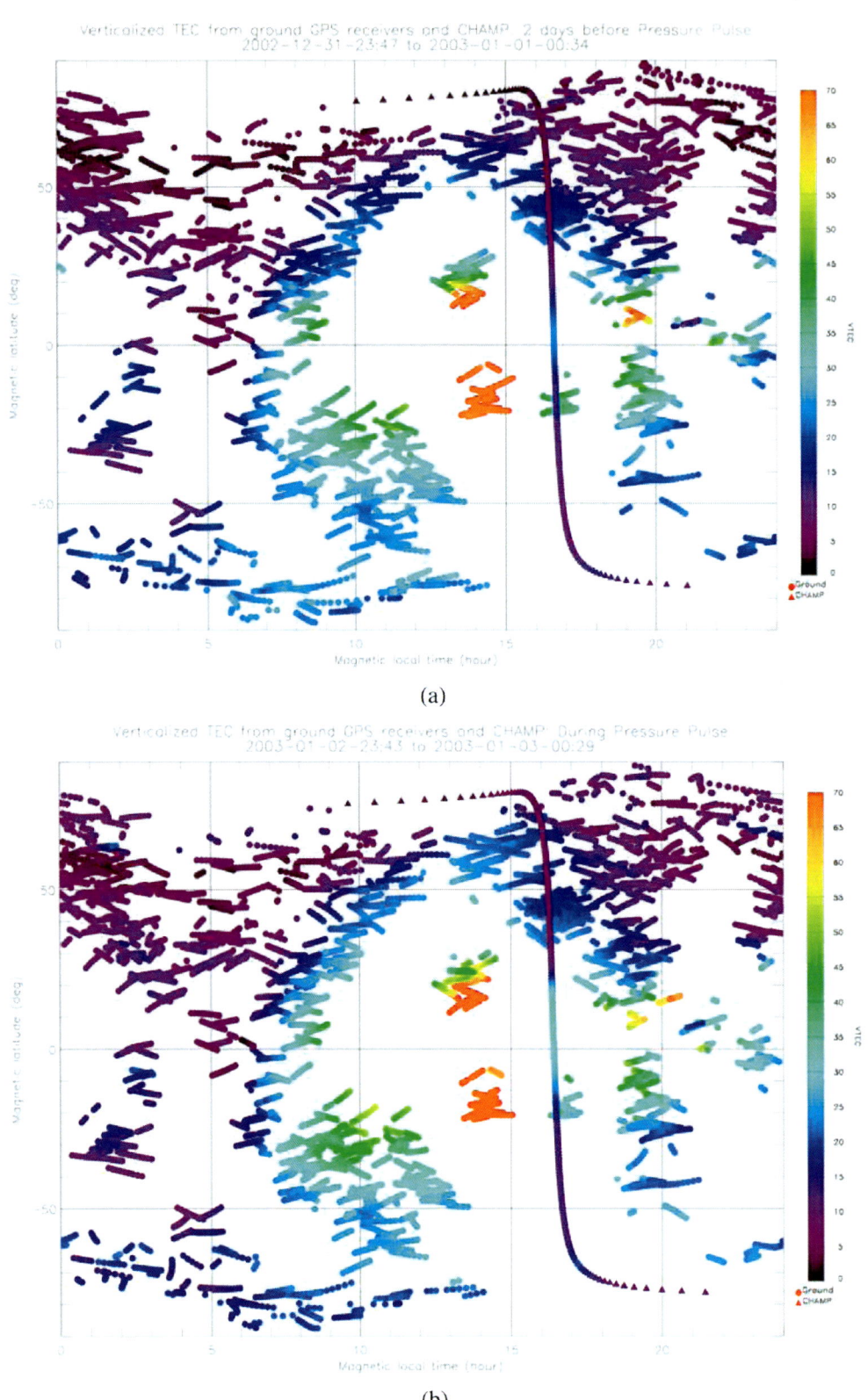

Plate 6. The verticalized TEC data for the HCSPS pressure pulse noted in Figure 16. The top panel is the baseline data and the bottom panel is the interval of interest: ~2343 UT Janueary 2 to ~0029 UT January 3, 2003.

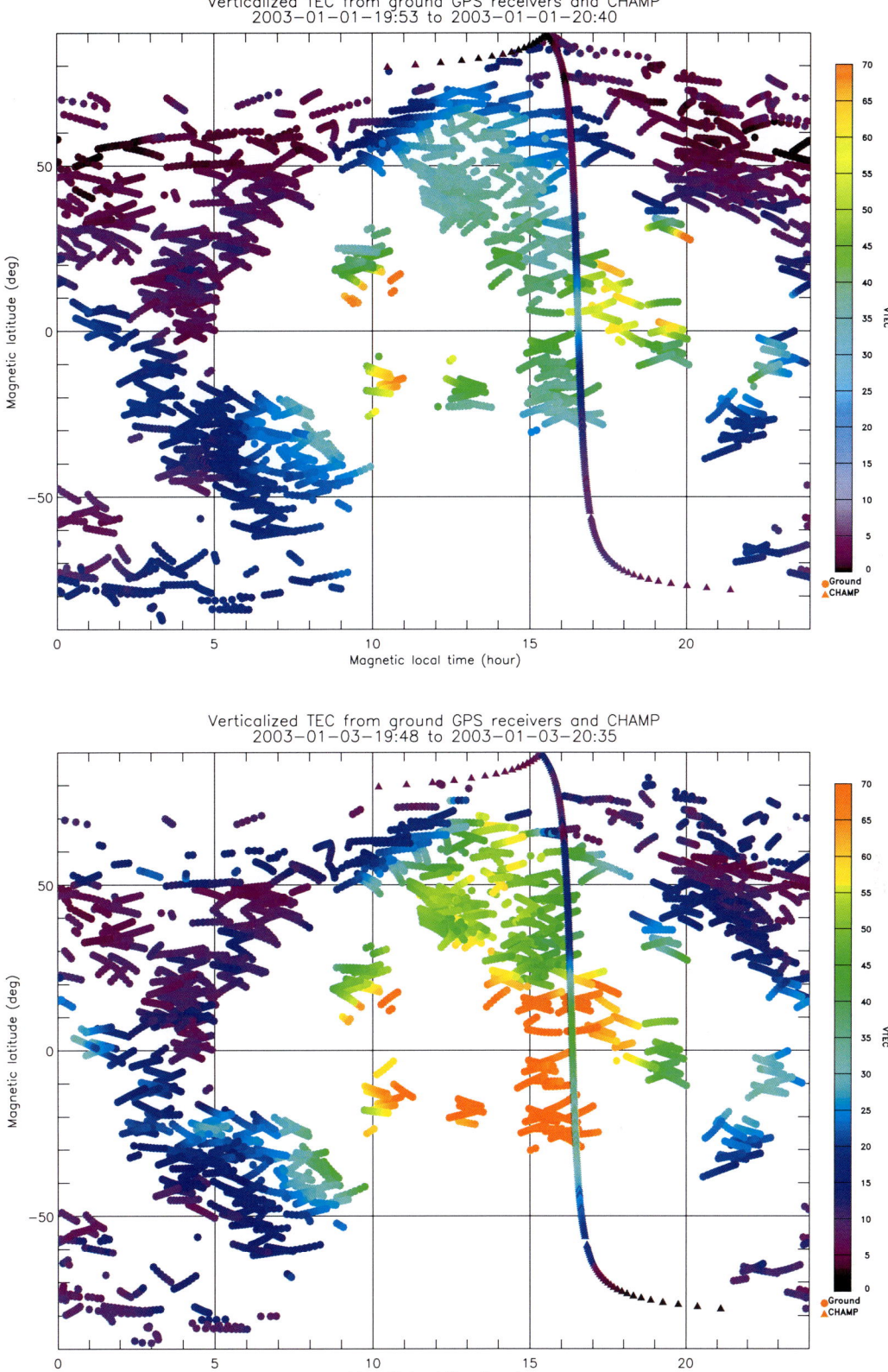

Plate 7. Global TEC changes during the CIR of January 3, 2003. The top panel is baseline data and the event of interest is shown on the bottom: 0115 to 0200 UT.

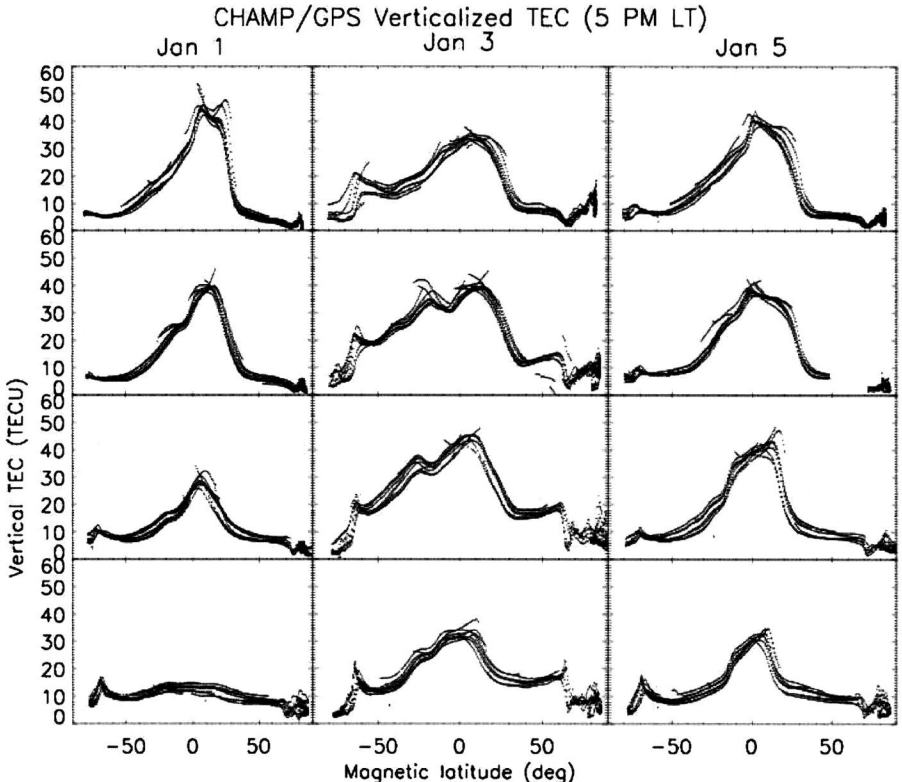

Figure 12. The verticalized TEC data above the CHAMP satellite. Four different Universal Times during January 1, 3 and 5 are shown.

Figure 12 shows the CHAMP verticalized TEC values as a function of magnetic latitude and time. CHAMP passed through the dayside ionosphere at a nearly constant ~5 pm local time. Three days (January 1, 3 and 5) and 4 universal times (~1706 UT, ~1839 UT, ~2012 UT, and ~2145 UT) are shown. The days are shown in vertical columns, with January 1 on the left, January 3 in the center, and January 5 at the right. The rows correspond to ~1706 UT, ~1839 UT, ~2012 UT and ~2145 UT (the times given are the plot center times within a 5 min accuracy). For each of the 9 panels, the verticalized TEC values (in TECUs) are given as the ordinate and the horizontal axis extends from the south magnetic pole to the north magnetic pole.

The beginning of January 1 corresponds to an interval of slow solar wind with low plasma densities. The high density HCSPS hits the magnetosphere during the latter-part of the next day (from ~2000 UT onward). The CIR impinges on the magnetosphere on January 3. Within the CIR, the IMF B_Z has both northward and southward excursions, but is mostly northward during the first half of the day and mostly southward during the latter half of the day, as mentioned previously. The stream peak speed is attained at ~0300 UT January 4, and decreases monotonically thereafter. The magnetic field is ~7 nT and there are Alfvénic B_Z fluctuations throughout the interval.

Let us first compare January 1 ~1706 UT and ~1839 UT TEC data against those on January 5. The peak TEC values on the different days are comparable, ~40-45 TECU. However the width of the enhanced ionization region is slightly broader on January 5. On January 1 at ~1706 and ~1839 UT, the 20 TECU levels were extended from ~−20° MLAT to ~+28° MLAT. On January 5 the 20 TECU levels were at ~−25° and ~+28° MLAT at ~1706 UT and ~−23° and ~+29° MLAT at ~1839 UT. It is the southern hemisphere border that is slightly more poleward. The broader latitudinal extent is again noted at ~2012 UT. On January 1, the 20 TECU levels are at ~−8° MLAT and ~+18° MLAT, considerably smaller than at ~1706 and ~1839 UT. The broadness of this region on January 5 is ~−23° MLAT and ~+22° MLAT, only slightly smaller than at ~1706 UT and ~1839 UT.

The most significant dayside ionospheric change occurs on January 3, during the time interval when the CIR hits the magnetosphere. To illustrate the differences, we will compare the January 3 ~1706, ~1839, ~2012 and ~2145 UT panels with the same ones on January 1. As mentioned previously, CHAMP was over the same geographical features during the two days, eliminating potential orographic differences.

January 3 CIR Effects

The most striking feature of the CHAMP ionospheric TEC data is the high TEC values at all magnetic latitudes from the equator to the auroral zones (~−60° MLAT and +60° MLAT). Take as an example the changes at −50° MLAT. At ~1839 and ~2012 UT, the ionosphere above CHAMP is ~20 TECU. On January 1 and 5, the values at the same latitude are 5 and 7 TECU, respectively.

Another striking feature of the January 3 data is the extremely elevated TEC values at ~−63° MLAT. The peak values were as high as ~23 TECU at ~1839, ~2012 and ~2145 UT, compared with a January 1 background level of 6 to 12 TECU. Seven different GPS satellites were tracked simultaneously. The TEC values along the different paths to the GPS satellites all indicated a local peak. This peak intensity corresponds to the southern auroral zone.

At 1706 UT, the peak intensity at the equator of ~34 TECU on January 3 versus ~42 TECU (on January 1) is noted. This may be due to an overall northward IMF average. The latter would correspond to a dusk-to-dawn directed electric field on the dayside ionosphere, which would cause a downward convection of the ionospheric plasma. Recombination of electron-ion pairs could lead to the observed decrease. The latitudinal locations of 20 TECU levels are about comparable: ~−23° and ~+28° MLAT for January 3 and ~−20° and ~+28° MLAT for January 1.

~1839 and ~2012 UT January 3

At ~1839 UT on January 3, besides a ~40 TECU peak at ~+10° MLAT (a similar peak is present on January 1), there is an additional peak of ~36 TECU at ~−19° MLAT. At ~2012 UT, this peak moves to ~25° MLAT.

At ~2012 UT, there are now peaks in both auroral zones. At ~1706 UT there is a ~20 TECU peak at ~−63° MLAT and no peak (~5 TECU) at ~+60° MLAT. At ~2012 UT these two peaks remain at approximately the same locations while the northern hemispheric intensity increased to ~18 TECU.

This strong ionospheric effect on January 3 can also be noted in the ground-based data. This is illustrated in Plate 8. In this Plate, only TEC data from ~5 pm (±15 min) local time stations were used. The TEC data has been "verticalized". The data is limited to GPS satellites well above the horizon (elevation angles >35°).

The plot shows the unusual ionosphere disturbances on January 3. There is a sudden ionospheric TEC increase on this day which is not present on any of the other days. The time interval corresponds to a few hours after the CIR southward IMF B_z fields hit the magnetosphere. This is the most noteworthy item of this 5-day interval during the HCSPS, CIR and high-speed stream impingement onto the magnetosphere.

The GPS ground receiver data from an auroral zone station, Kiruna, Sweden is shown in Plate 9a and b. Panel a shows a "baseline" day, January 2, and panel b shows the CIR day, January 3. The time is given in UT hours. Local noon is at ~0100 UT, while local midnight is at ~1300 UT. What is particularly interesting about this multiple satellite track data are the much higher TEC values on 3 January 2003. There is also a much greater spread in ionospheric TEC values from one satellite track to another. This is present not only at midnight where auroral precipitation dominates during substorms, but also at all local times, including daytime. It should be noted that *Guarnieri* [2006], from the analysis of POLAR UV image data, found that during high speed streams/HILDCAAs, there was significant auroral precipitation over the whole auroral zone. The precipitation was weaker than during substorms, but was still significant. Our ionospheric TEC results are also consistent with precipitation occurring at all local times.

SUMMARY

HCSPS Pressure Pulses and Storm Initial Phases

Although dayside and nightside auroras might be expected during HCSPS impacts (impulsive increases in solar wind ram pressure) onto the magnetosphere, no obvious TEC signatures were noted in the GPS data. The GPS technique is probably too insensitive to detect/identify such levels of precipitation.

CIRs and Storm Main Phases

Significant dayside ionospheric TEC increases are noted at middle latitudes and at auroral latitudes (60°–65° MLAT) during intervals when the average CIR interplanetary B_z magnetic fields were southward (January 3, 2003). This was shown in Figures 2, 3 and Plate 2. A possible mechanism for this TEC enhancement is the same one as described for dayside TEC enhancements during ICME-related magnetic storms: prompt penetration of interplanetary electric fields to the dayside equatorial ionosphere plus $\vec{E} \times \vec{B}$ upward convection of the ionospheric plasma. On the dayside, the upward convection to higher altitudes where the electron-ion recombination time scales increase, plus a continuation of solar photoionization at lower altitudes, leads to an overall TEC enhancement (Plate 3 and Figure 6). Gravity and pressure-driven diffusion of the high altitude plasma down the Earth's

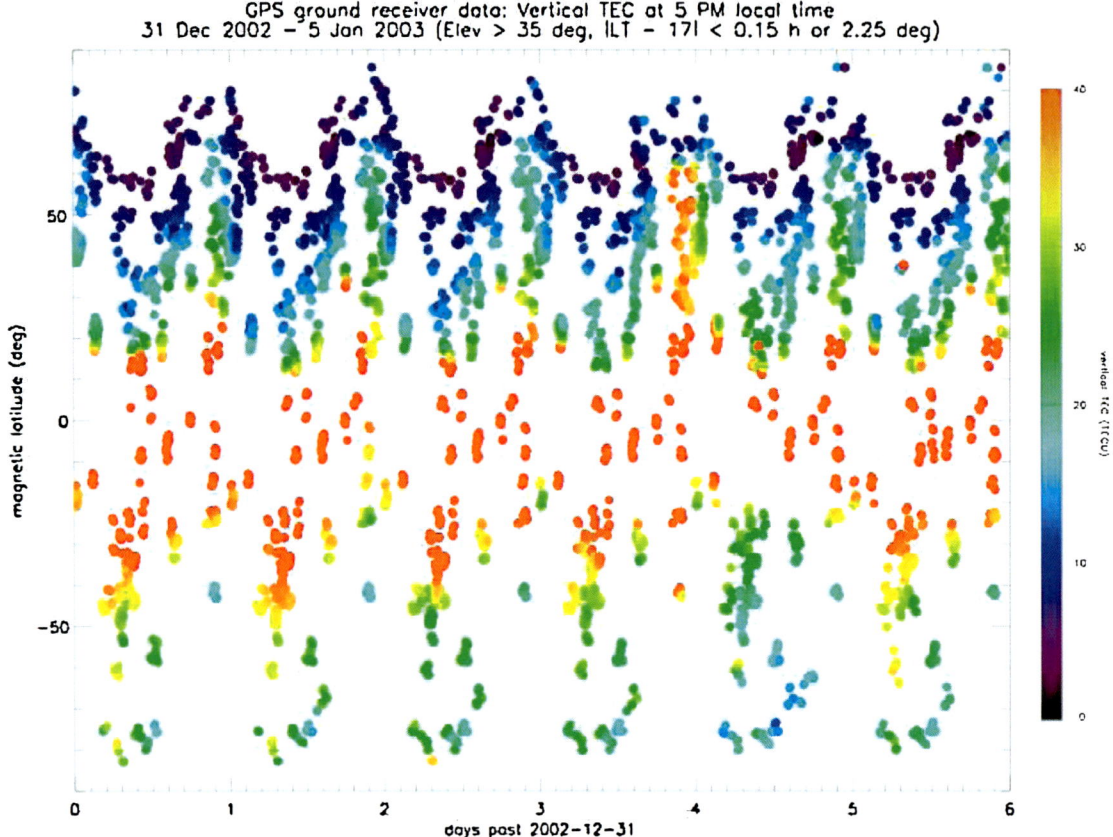

Plate 8. Ground based TEC data at 5 pm local time on January 3, 2003.

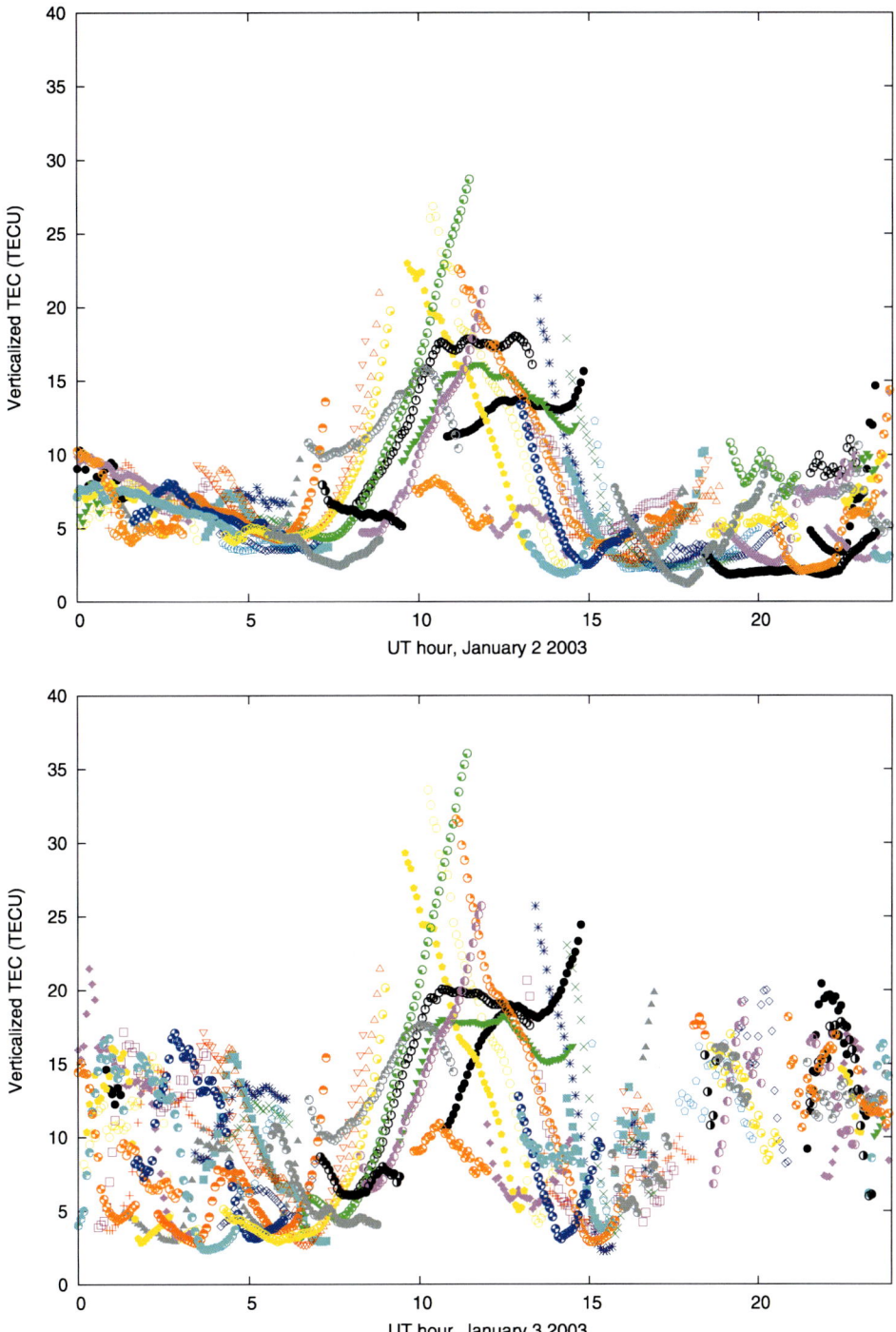

Plate 9. Verticalized TEC data from Kiruna Sweden on January 2 (baseline data) in top panel, and during the CIR event on January 3 (bottom panel).

magnetic lines of force, leads to the displacement of this plasma to middle latitudes (Figure 7). This latter process has been called the "dayside super-fountain effect". It is plausible that this same process occurs both during CIR-generated magnetic storms and ICME-generated magnetic storms. Another process that increases TEC during storms is enhanced equatorial winds due to high-latitude Joule heating [Buonosanto et al., 1999; Crowley et al., 2006]. However whether this mechanism can affect the dayside equatorial region is questionable [see discussion in Tsurutani et al., 2005, 2006c,d]. We note that the CHAMP observations (Figure 12) indicate that the TEC increases occur primarily at near-equatorial latitudes, consistent with the dayside superfountain effect.

The third part of the CIR-storm findings, the auroral zone TEC enhancements at all local times, appears to be a CIR-related effect alone. Previous ICME studies did not find similar high latitude effects. We propose an explanation for this. CIR-related magnetic storms are caused by short duration southward IMF B_Z fields (parts of the Alfvén waves). These fields cause short duration sporadic magnetic reconnection on the dayside magnetopause, launching local magnetospheric kinetic Alfvén waves [Tsurutani et al., 2001b]. Because the orientation of the interplanetary magnetic fields are quite random, the reconnection patches and the kinetic Alfvén waves will be randomly located over the magnetopause. The aurora will be global and not necessarily centered at midnight as is the case for substorms and storms. Clearly this is a hypothesis that can explain our observations. However considerable more research is necessary to determine whether this hypothesis is correct or not.

The auroral latitude ionospheric TEC enhancements may be due to electron precipitation ionization losses. If the CIR-storm auroras are created by the precipitation of typical ~1–10 keV electrons, then the precipitating electron stopping heights would be 110 to 90 km above Earth (satellites such as CHAMP at 400 km altitude would not detect such effects). The recombination rates at these altitudes will be rapid, and the effects would not be long lasting. However the disturbance dynamo will lift the freshly created plasma to higher altitudes where their lifetimes will be considerably longer. Fresh precipitation of electrons at the lower altitudes will lead to increased TEC with time.

The above idea could be tested by using auroral zone ground-based TEC plus ionosonde data from the same locations during CIR magnetic storms. One would expect to note a gradual increase in TEC and a higher height distribution with increasing time.

The High Speed Steam (Proper) Ionospheric Effects

Very little ionospheric equatorial or middle latitude ionospheric effects were noted in the GPS ground and satellite data during high speed streams. It is indeed possible that some effects are present, but if so, they are small and of second-order in magnitude (but not necessarily second-order in importance) compared with the large effects noted in ICME and CIR southward IMF events.

From magnetospheric particle observations [see Soraas et al., 2004; 2005] and from the interplanetary CIR observations, one would not expect to have equatorial or middle latitude ionospheric effects. As previously mentioned, the IMF B_Z is highly fluctuating within CIRs, and one would not expect large, deep magnetospheric convection from such events. Particles are injected only to L > 4 during HILDCAA events [Soraas et al., 2004; 2005].

The high activity auroral zone ionosphere was not examined in this present effort to determine if ionospheric TEC enhancements were present (as during CIRs) or not. However, it was noted from UV auroral images [Guarnieri, 2006] that there is low-level particle precipitation throughout all local times in the auroral zone during such intervals. Whether auroral zone ground station GPS receivers are sensitive enough to detect this effect or not is left for a future study.

CONCLUSION AND FUTURE WORK

Dayside Equatorial and Middle Latitudinal Effects

We have noted dayside equatorial and middle latitude ionospheric TEC enhancements where the CIR has a long term average net southward interplanetary magnetic field orientation. When the CIR IMF was northward-oriented, there were no major large-scale ionospheric effects.

It is probable that these ionospheric effects are due to a "dayside super fountain" mechanism, where the equatorial ionospheric uplift is caused by promptly penetrating interplanetary dawn-to-dusk electric fields plus transport to middle latitudes.

This hypothesis can be easily tested by ionosonde observations during southward IMF Bz events. It would be interesting to determine how effective this ionospheric uplift is. As one example, does the E region ionosphere also get uplifted to high altitudes (perhaps only during long duration events such as during ICMEs)? If so, then does this uplifted ionosphere get "inverted" as the plasma is transported to middle latitudes? The heavier E-region ions will then be placed on top before further mixing can occur. Thus at middle latitudes, the heavy ions may have two different locations, one at regular E-region heights and a second F-region layer at higher altitudes.

In the study, it was noted that the TEC baselines were particularly low (peak equatorial values of ~70 TECU) during high speed stream intervals. These would be intervals when solar coronal holes face the Earth. In comparison, TEC baselines much greater than ~100 TECU were observed during

A Statistical Study of Ionospheric Irregularities Observed With a GPS Network in Japan

Y. Otsuka, T. Aramaki, and T. Ogawa

Solar-Terrestrial Environment Laboratory, Nagoya University, Toyokawa, Japan

A. Saito

Graduate School of Science, Kyoto University, Kyoto, Japan

The Geographical Survey Institute of Japan has installed a network of about 1000 dual-frequency GPS receivers in Japan with their mutual distance of about 25 km. Phases and pseudo ranges of dual-frequency GPS signals are recorded every 30 s. The dense distribution of the GPS receivers allows us to reveal two-dimensional structures of the ionospheric plasma density irregularities with scale sizes of the order of several kilometers. We analyzed TEC data obtained from the GPS network of Japan in 2000. It was found that the irregularity characteristics over Japan depended on latitude. The results are as follows: (1) at the northern part, the irregularities appeared only during geomagnetic storms. (2) At the middle part, they had the most frequent occurrence in summer nighttime and were usually accompanied by Medium-Scale Traveling Ionospheric Disturbances (MSTID). (3) At the southern part, they were associated with the equatorial plasma bubbles and their occurrences were highest in the equinoctial nighttime.

1. INTRODUCTION

Global Positioning System (GPS) has been used to measure total electron content (TEC) along ray path. GPS phase fluctuations obtained from TEC variations were examined to study the ionospheric irregularities at equatorial region [*Aarrons et al.*, 1996] and high latitudes [*Aarons*, 1997]. *Pi et al.* [1997] presented global maps of ionospheric irregularities measured at more than 165 GPS stations over the world to demonstrate the capability of the worldwide GPS network monitoring ionospheric irregularities continuously on global scale. *Beach and Kintner* [1999] investigated the relationship between GPS amplitude scintillations and TEC variations for the same line of sight at equatorial region. Gigahertz scintillations of satellite radio signals observed at equatorial region is related to the ionospheric irregularities that occurred within the plasma bubbles [e.g., *Basu and Basu*, 1981]. At mid-latitude region, gigahertz scintillations are very rare phenomena. However, they were observed during severe geomagnetic storms [*Ogawa and Kumagai*, 1985] and during a event of the ionospheric trough moving equatorward over the northern U.S. with a storm time-enhanced density [*Ledvina et al.*, 2002]. In Japan, using a radio wave from a geostationary satellite, *Fujita et al.* [1978] presented that 1.7 GHz scintillation activity was enhanced at night in June. They also showed that total electron content variations were associated with the scintillations. *Sinno and Kan* [1978] presented some results of amplitude scintillations of 136 MHz radio waves transmitted from a geostationary satellite, and showed that the scintillations were related to spread *F* and sporadic *E*. *Kersley et al.* [1980] also showed that VHF scintillations were associated with nighttime enhancement of total electron content, using 137 MHz transmission from a geostationary satellite. They suggested that the irregularities causing the scintillations could arise from penetration of atmospheric gravity waves to the ionosphere.

Recurrent Magnetic Storms: Corotating Solar Wind Streams
Geophysical Monograph Series 167
Copyright 2006 by the American Geophysical Union.
10.1029/167GM21

In the present study, we have analyzed TEC data obtained from the GPS network of Japan in 2000 to reveal the characteristics of the ionospheric irregularities over Japan.

2. DATA AND METHOD

The Geographical Survey Institute of Japan has installed a network of more than 1000 dual-frequency (1.57542 GHz and 1.22760 GHz) GPS receivers in Japan to monitor crustal deformation. The mean distance between receivers is approximately 25 km. Phases and pseudo ranges of dual-frequency GPS signals are recorded at 30-second interval. Phase difference between the dual-frequency GPS signals tracks precisely change of total electron content (TEC) along ray path. Thus TEC fluctuations are called as phase fluctuations. GPS phase fluctuations are examined using the rate of change of TEC (ROT) and the rate of TEC index (ROTI) [e.g., *Aarons et al.*, 1996; *Pi et al.*, 1997]. ROT is defined as differentiation of TEC at 30-second interval and converted to the unit of TECU/min, where 1 TECU is 10^{16} electrons/m^2. To study the phase fluctuation statistically, ROTI is defined as the root mean square deviation of ROT within 5 minutes. In this study, ROTI is computed for each 5-minute interval. Since GPS satellites move in a circular orbit at 20,200 km altitude with a period of 12 hours in sidereal coordinate, projection of the GPS satellite to the ionospheric altitude moves at approximately 80 m/s at zenith. Thus, ROT and ROTI is a measurement of the irregularities with scale size of the order of several kilometers.

3. STATISTICAL RESULTS

Figure 1 shows locations where occurrence rate of the ionospheric irregularities were investigated in this study. The six regions were selected to investigate latitudinal variation of the ionospheric irregularities. We defined an irregularity event as the event in which ROTI exceeds 0.1 TECU/min. In this study, we excluded the data with elevation angle less than 40°. ROTI was averaged over one hour and within the area enclosed by solid lines in Figure 1.

Plate 1 shows local time and seasonal variations of the occurrence rate of the event, for which ROTI exceeds 0.1 TECU/min, versus local time and season in 2000. The occurrence rate shows strong dependences on local time, season, and latitude. At the northern parts of Japan, regions 1 and 2 ([44°N, 142°E; geomagnetic latitude 35°N] and [38°N, 140°E; geomagnetic latitude 29°N]), the occurrence rate has two peaks in summer and autumn night. At middle parts, regions 2-5 (29-38°N; geomagnetic latitude 15-29°N), the occurrence rate is highest in the summer nighttime. It increases with increasing latitude. At southern pats, regions 5 and 6 ([29°N, 130°E; geomagnetic latitude 19°N] and [25°N, 128°E; geomagnetic latitude 15°N], respectively), the occurrence

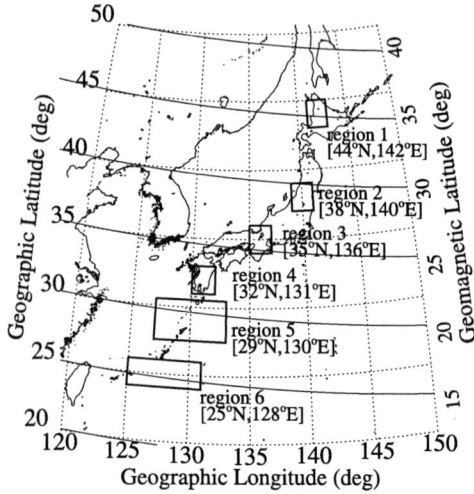

Figure 1. Map showing locations where occurrence rate of the ionospheric irregularities are investigated in this study. Geomagnetic latitudes are shown by solid curves. ROTI is averaged within the area enclosed by solid lines.

rate is highest in equinoctial nighttime. Especially, this feature can be seen clearly at region 6. At dawn (06-07 LT), the occurrence rate of the GPS phase fluctuations is high, especially at region 2 during vernal equinox. This enhancement seems to be caused by rapid increases of TEC in the morning due to the plasma production by the solar radiation.

Plate 2 is same as Plate 1, but the data only during geomagnetically quiet conditions ($K_p \leq 2$) are used. High occurrence rate in equinoctial nighttime, which can be seen in Plate 1, disappears in Plate 2. This indicates that the ionospheric irregularities in equinoctial nighttime is related to the geomagnetic disturbed condition.

4. A CASE STUDY OF THE GEOMAGNETIC STORM EVENT

As shown in the previous section, GPS phase fluctuations occur at the northern parts of Japan in the geomagnetic disturbed conditions. A case study of GPS phase fluctuations during a geomagnetic storm is presented in this section. A geomagnetic storm occurred on February 12 with SSCs at 0853 JST. *Dst* reached a minimum of −133 nT at 2100 JST on February 12.

Plate 3 shows time sequence of two-dimensional maps of the GPS phase fluctuation index ROTI for the time period between 2030 JST on February 12, 2000 and 0430 JST on the subsequent day. The irregularities are seen in two regions, middle and northern parts of Japan. Between 2030 and 2230 JST, during development phase of the geomagnetic storm,

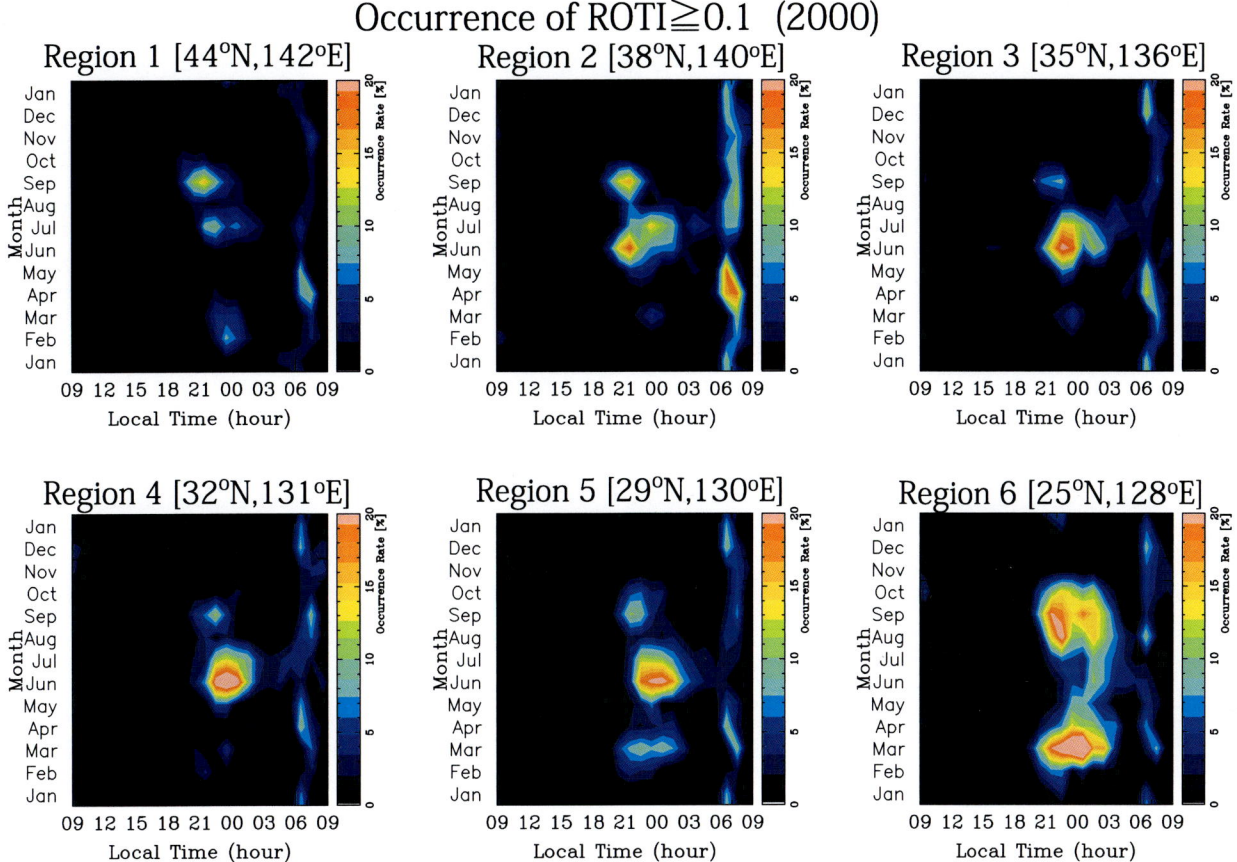

Plate 1. Local time and seasonal variations of occurrence rate of the ionospheric irregularities (ROTI ≥ 0.1 TECU/min) at six regions shown in Figure 1. GPS data obtained from GEONET in 2000 are used.

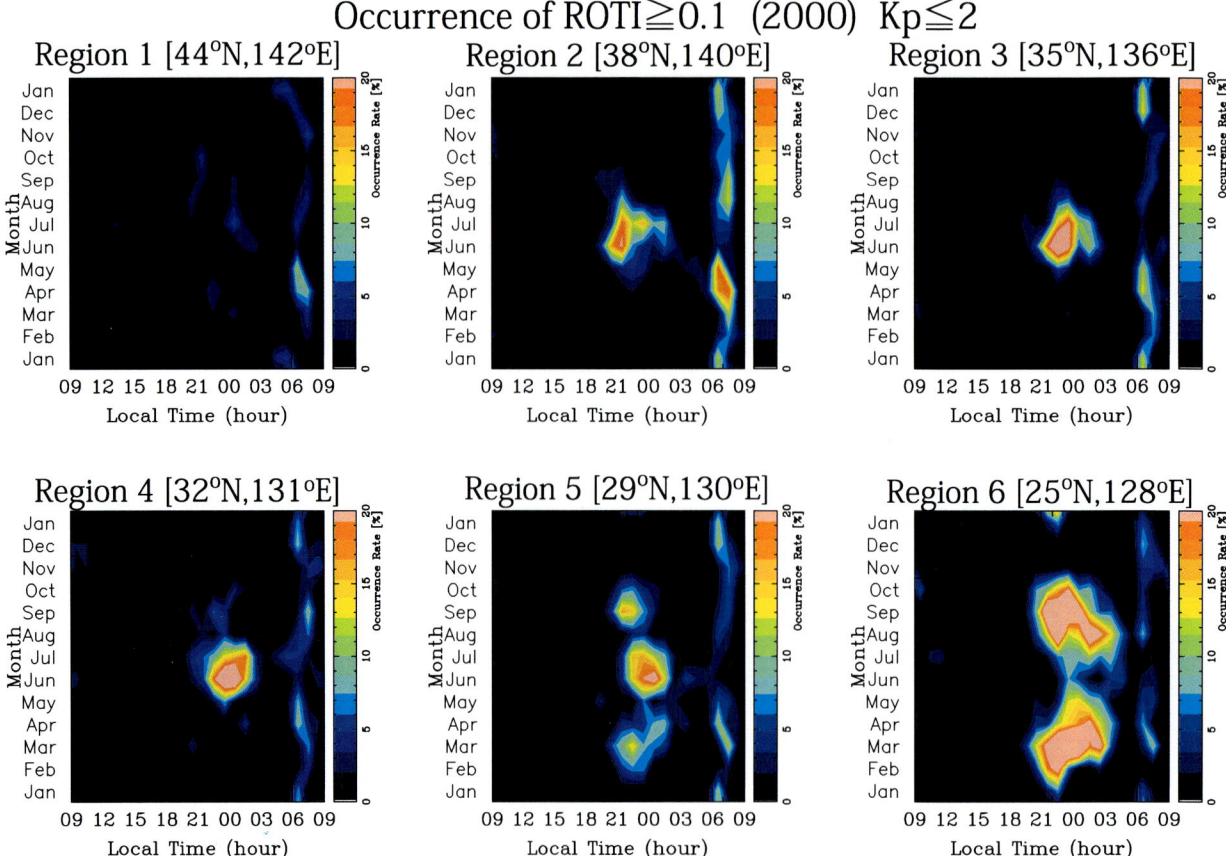

Plate 2. Same as Plate 1 but the data only during geomagnetically quiet condition ($K_p \leq 2$) are used.

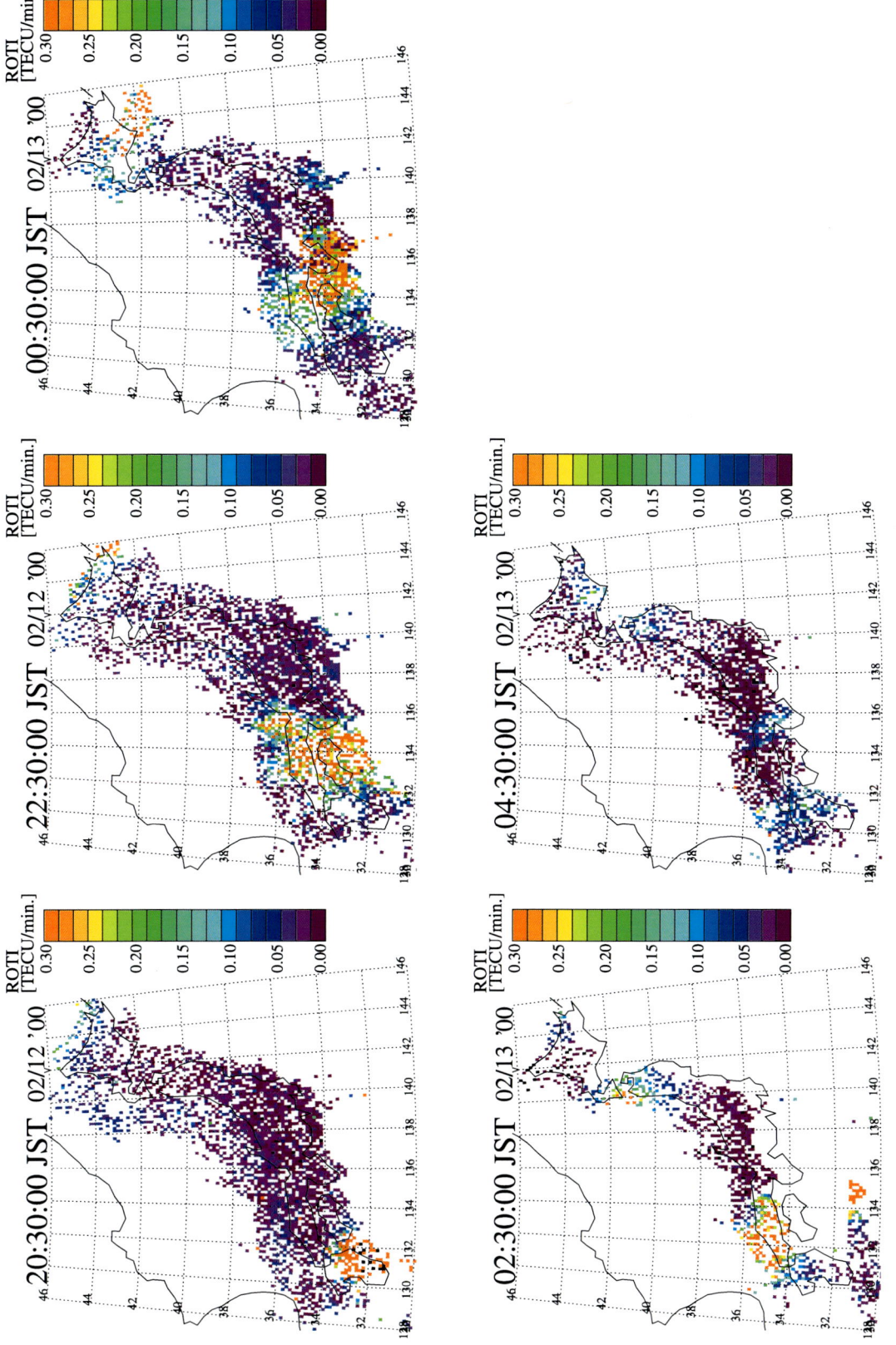

Plate 3. Time sequence of two-dimensional maps of GPS phase fluctuation index ROTI for the time period between 2030 JST on February 12 and 0430 JST on the subsequent day in the interval of 2 hours.

the ionospheric irregularities appeared at the northern edge of Japan, and propagated to the southwest direction at about 70 m/s. The irregularity region stretched from the northwest to the southeast and their width was approximately 100 km. The other irregularities appeared in the middle of Japan at 2030 JST, became stronger during 2030-2300 JST, and disappeared at 0430 JST on the subsequent day. The irregularity region which was stretching along meridional direction with sharp boundary moved eastward until midnight and then moved westward.

We compared the spatial distribution of the GPS phase fluctuation to that of Traveling Ionospheric Disturbance (TID). Plate 4 shows comparison of two-dimensional maps between the GPS phase fluctuations and the perturbation component of TEC at 23:30 JST on February 12, 2000. TEC data shown in Plate 4 (b) have been filtered to pass only perturbation component with the period between 5 and 60 minutes for each satellite-receiver pair. This band-pass width is chosen to track TEC perturbations caused by TID. The TEC perturbations with the period shorter than 5 minutes were cut off in this study to distinguish the TEC perturbations caused by the TID from the GPS phase fluctuations. Comparison of Plate 4 (a) and (b) indicates that the structure of TEC perturbations caused by TID is coincident with that of the GPS phase fluctuation index ROTI. *Sahai et al.* [2001] presented all-sky images of OI 630-nm airglow emission observed at northern and middle part of Japan, Rikubetsu (43.5°N, 143.8°E) and Shigaraki (34.8°N, 136.1°E), respectively, during the present event. They showed that TEC enhancement at the northern part of Japan coincided with 630-nm airglow enhancement, and that TEC perturbations at the middle part of Japan coincided with airglow depletions. These results suggest that the spatial gradient of the plasma density plays an important role to generate the ionospheric irregularities.

5. COMPARISON WITH MEDIUM-SCALE TRAVELING IONOSPHERIC DISTURBANCES

As shown in Plate 1, occurrence rate of the ionospheric irregularities at the middle part of Japan is highest in summer nighttime. Medium-Scale Traveling Ionospheric Disturbances (MSTID) are also frequently observed as TEC perturbations derived with GEONET during summer nighttime [e.g., *Saito et al.*, 1998]. A typical example of the MSTID is shown in Plate 5. The large-scale TEC variations whose time scale was longer than one hour were subtracted to derive the perturbation component of TEC. The derivation method of these two-dimensional TEC perturbation maps was described by *Saito et al.* [1998]. Clear wavy structure of the TEC perturbations can be seen in Plate 5. The structure stretches from the northwest to the southeast and their wavelengths are from 200 km to 500 km. They propagated to the southwest direction at about 100 m/s.

Plate 6 shows local time and seasonal variations of the MSTID activity, which is defined as $\delta I/\bar{I}$, where δI is the standard deviation of the perturbed TEC within one hour, and $\bar{I}$ is the one-hour average of absolute TEC. The MSTID activity obtained from different satellites were averaged. The background TEC is estimated by using the method of *Otsuka et al.* [2002]. Nighttime MSTID activity shows semiannual variation with a major peak in summer. From comparison of Plates 1 and 6, it is found that the MSTID activity coincides with occurrence of the ionospheric irregularities with scale size of the order of several kilometers. Therefore, we expect that the irregularities during nighttime in summer could be caused by the plasma density gradients with spatial scale-size of several hundred kilometers due to MSTID.

6. PHASE FLUCTUATIONS AND LOSS OF LOCK IN GPS RECEIVER

Plate 1 shows that the occurrence rate of the ionospheric irregularities at southern parts of Japan is highest in two equinoxes. This seasonal variation is similar to that of the equatorial plasma bubbles which are depletions in the equatorial F region plasma. Magnetic flux tubes with low-density plasma at the bottomside of the equatorial F region rise to higher altitudes to become elongated plasma depleted flux tubes. Within the plasma density depletions caused by the plasma bubbles, plasma density irregularities with various spatial scale-sizes were observed by VHF/UHF/L-band radars [e.g., *Woodman and LaHoz*, 1976; *Tsunoda*, 1980, 1983], and satellite beacons [e.g., *Basu and Basu*, 1981]. As described in Section 1, ROTI indicates existence of irregularities with scale size of the order of several kilometers.

A radio signal passing through the plasma density irregularities fluctuates in amplitude and phase since the irregularities act as diffraction gratings. This phenomenon is known as scintillation. Tracking performance of GPS receiver is degraded in the presence of the ionospheric scintillations. Rapid phase variations cause a Doppler shift in the GPS signal, resulting in a loss of phase lock [*Leik*, 1995]. *Kintner et al.* [2001] demonstrated some cases in which amplitude scintillations were responsible for the loss of lock. Scale-size of the irregularities which cause scintillations corresponds to the first-Fresnel dimension of the radio wave. The Fresnel dimension is given by $\sqrt{2\lambda z}$, where λ is the radio wavelength and z is the altitude of the ionosphere [e.g., *Basu et al.*, 1999]. For the case of GPS scintillation, the Fresnel dimension is about 400 m because the GPS L1 frequency is 1.57542 GHz and the ionospheric altitude is about 400 km. Thus, statistics of the losses of lock indicates the presence of the ionospheric irregularities with scale size of 400 m.

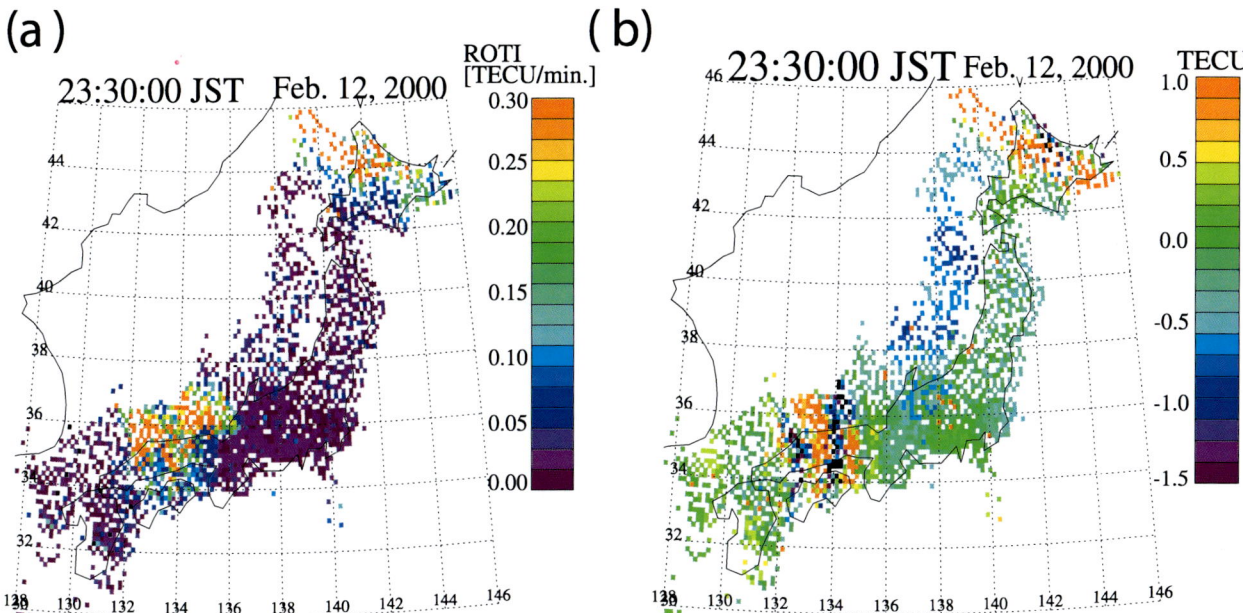

Plate 4. Two-dimensional maps of (a) GPS phase fluctuation index ROTI and (b) perturbation component of TEC with the period of 5-60 minutes at 23:30 JST on February 12, 2000.

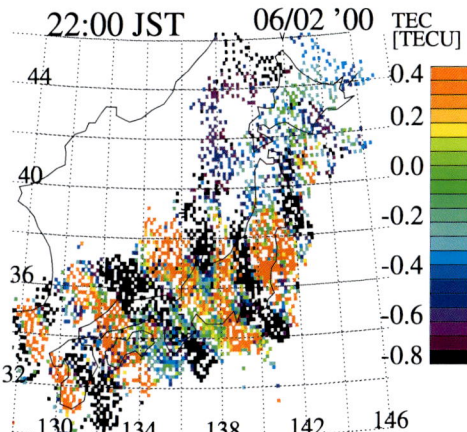

Plate 5. Example of medium-scale traveling ionospheric disturbance (MSTID) over Japan.

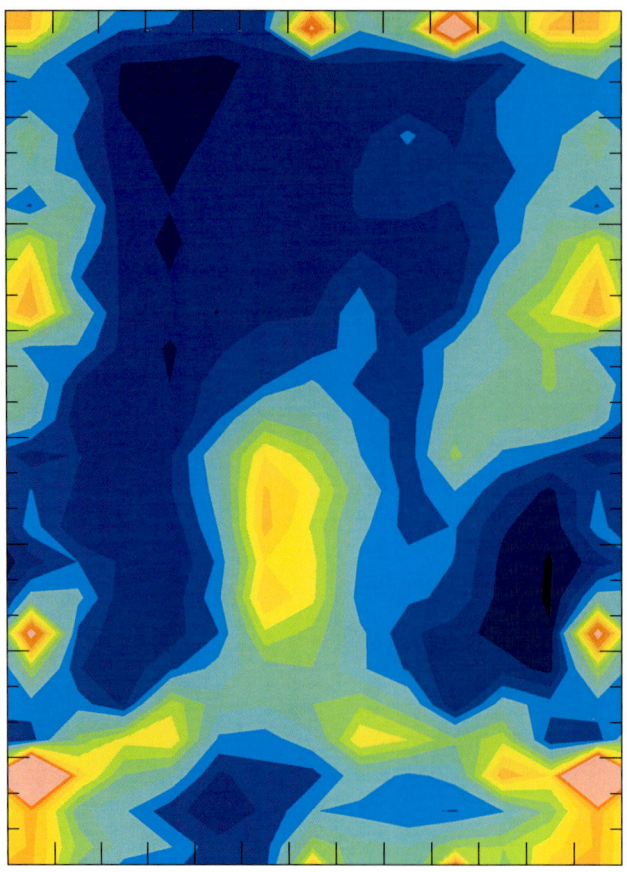

Plate 6. Local time and seasonal variations of MSTID activity at 35°N, 136°E in 2000. The activity is defined as ratio of the TEC standard deviation to the background value of TEC.

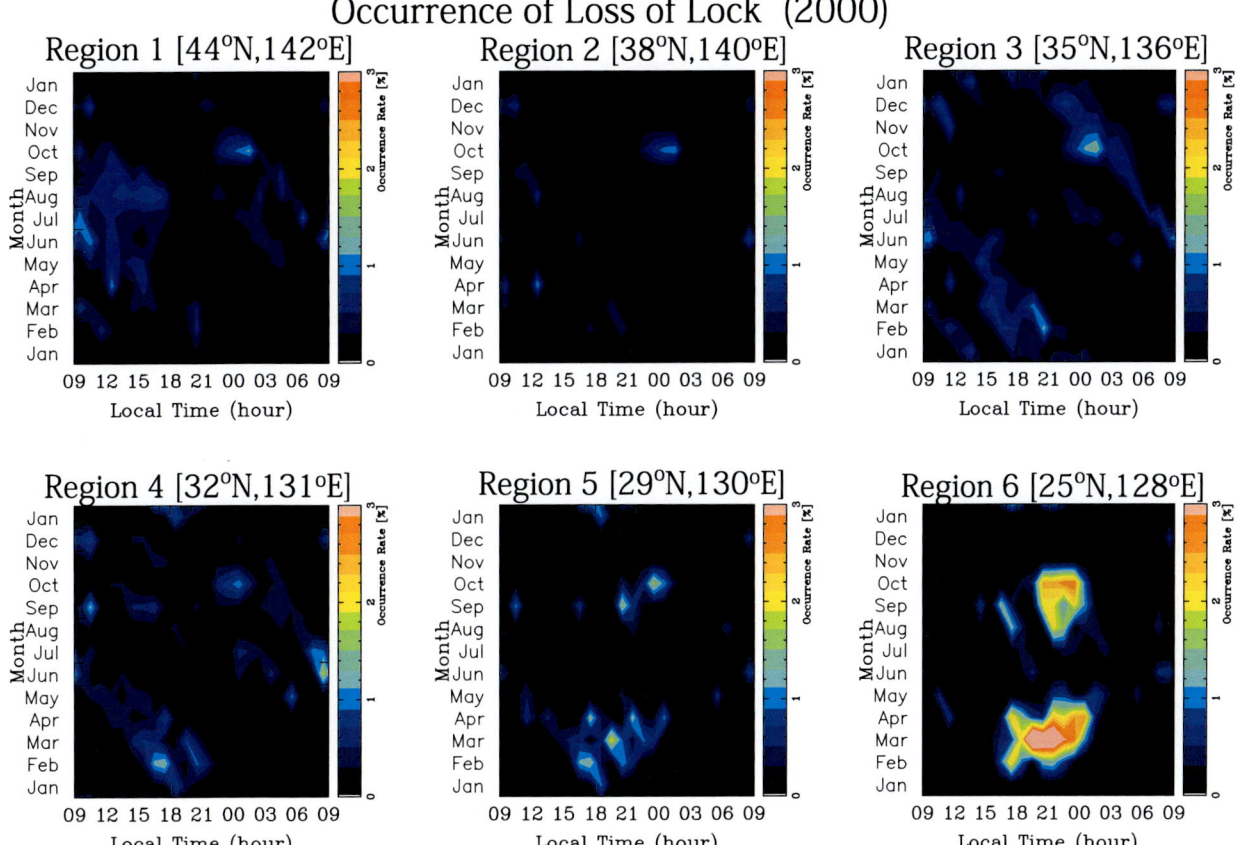

Plate 7. Occurrence rate of losses of lock in GPS receiver versus local time and season at the six regions shown in Figure 1. GPS data obtained from GEONET in 2000 are used.

Using GPS data in 2000, we have studied local time and seasonal variations of the occurrence rate of losses of lock in GPS receiver (Plate 7). The losses of lock were not observed except for Okinawa (regions 5 and 6), and at Okinawa the occurrence was highest in the nighttime of equinoxes. This seasonal variation is consistent with that of the ionospheric irregularities obtained from GPS phase fluctuations shown in Plate 1. Therefore, we expect that the loss of lock in the GPS receiver is caused by the irregularities associated with the equatorial plasma bubbles. The GPS phase fluctuations (ROTI) were observed between 2000 and 0300 LT (Plates 1 and 2), while the losses of lock were observed between 2000 and 0100 LT. Since plasma bubbles are usually generated near sunset, this result indicates that 400-meter scale irregularities that cause GPS scintillations decay faster than large scale (ROTI) irregularities. This feature is consistent with that reported by *Basu et al.* [1999], who compared ROTI and scintillation index S_4 obtained at an equatorial region. Previous observations showed that scintillations of VHF beacon radio waves were observed during the whole night [*Valladares et al.*, 1996]. 630-nm airglow intensity depletions caused by plasma bubbles were observed with all-sky airglow imager until sunrise [e.g., *Pimenta et al.*, 2003; *Martinis et al.*, 2003]. These results indicate that large-scale structures, such as plasma bubbles which have more than several ten-kilometer scale-size, are maintained long after small scale irregularities disappear [*Basu et al.*, 1978]. *Basu et al.* [1978] have shown that during generation phase of the equatorial irregularities in the evening hours, meter-scale irregularities that cause VHF radar backscatter echoes coexist with kilometer-scale irregularities that cause scintillations of VHF and L-band radio waves, whereas in the later phase, the meter-scale irregularities decay but the large-scale ones continue to exist. The decay of the plasma structures is probably caused by plasma diffusion in the direction perpendicular to the geomagnetic field lines [*Kelley*, 1989]. Time constant of the diffusion is given as $(k^2 D)^{-1}$, where k is the wavenumber of the plasma structure, D is the perpendicular diffusion coefficient. Since D in the F region is approximately 1 m/s^2, the time constant of the diffusion for 1-km scale structures is approximately 8 hours. Therefore, plasma structure with several kilometer or lager scale-size can be maintained throughout the night. However, the small-scale plasma structures decay earlier than large-scale structures and disappear during nighttime. Consequently, disappearance of GPS loss of lock after 0100 LT could be attributed to the decay of the 400-meter scale irregularities due to the plasma diffusion.

7. SUMMARY AND CONCLUSIONS

We have analyzed TEC data obtained from the GPS network of Japan in 2000. It was found that the irregularity characteristics over Japan depended on latitude. The results are as follows: 1) at the northern part, the irregularities appeared only during geomagnetic storms. 2) At the middle part, they had the most frequent occurrence in the summer night and were usually accompanied by medium-scale traveling ionospheric disturbances (MSTID). 3) At the southern part, they were associated with equatorial plasma bubbles and their occurrences were highest in the equinoctial nighttime.

Acknowledgments. GPS data of GEONET were provided by the Geographical Survey Institute of Japan.

REFERENCES

Aarons, J., Global positioning system phase fluctuations at auroral latitudes, *J. Geophys. Res.*, 102, 17219-17231, 1997.

Aarons, J., M. Mendillo, R. Yantosca, and E. Kudeki, GPS phase fluctuations in the equatorial region during the MISETA 1994 campaign, *J. Geophys. Res.*, 101, 26851-26862, 1996.

Basu, S., and S. Basu, Equatorial scintillations - a review, *J. Atmos. Terr. Phys.*, 43, 473-489, 1981.

Basu, S., Su. Basu, J. Aarons, J.P. McClure, and M.D. Cousins, On the coexistence of kilometer- and meter-scale irregularities in the nighttime equatorial F region, *J. Geophys. Res.*, 83, 4219-4226, 1978.

Basu, S., K.M. Groves, J.M. Quinn, and P. Dherty, A comparison of TEC fluctuations and scintillations at Ascension Island, *J. Atmos. Solar-Terr. Phys.*, 61, 1219-1226, 1999.

Beach, T.L., and P.M. Kintner, Simultaneous Global Positioning System observations of equatorial scintillations and total electron content fluctuations, *J. Geophys. Res.*, 104, 22553-22565, 1999.

Fujita, M., K. Sinno, and T. Ogawa, Frequency dependence of ionospheric scintillations and its application to spectral estimation of electron density irregularities, *J. Atmos. Terr. Phys.*, 44, 13-18, 1982.

Kelley, M.C., The Earth's ionosphere, Plasma Physics and Electrodynamics, *Academic*, San Diego, Calif, 1989.

Kersley, L., J. Aarons, and J.A. Klobuchar, Nighttime enhancements in total electron content near Arecibo and their association with VHF scintillation, *J. Geophys. Res.*, 85, 4214-4222, 1980.

Kintner P.M., H. Kill, T.L. Beach, and E.R. de Paula, Fading timescales associated with GPS signals and potential consequences, *Radio Sci.*, 36, 731-743, 2001.

Ledvina, B.M., J.J. Makela, and P.M. Kintner, First observation of intense GPS L1 amplitude scintillations at midlatitude, *Geophys. Res. Lett.*, 29, doi:10.1029/2002GL014770, 2002.

Leik, A., GPS Satellite Surveying, second edition, John Wiley and Sons, U.S.A, 1995.

Martinis, C., J.V. Eccles, J. Baumgardner, J. Manzano, and M. Mendillo, Latitude dependence of zonal plasma drifts obtained from dual-site airglow observations, *J. Geophys. Res.*, 108, 1129, doi:10.1029/2002JA009462, 2003.

Ogawa, T., and H. Kumagai, Deep depletions of total electron content associated with severe mid-latitude gigahertz scintillations during geomagnetic storms, *J. Geophys. Res.*, 90, 6652-6656, 1985.

Otsuka, Y., T. Ogawa, A. Saito, T. Tsugawa, S. Fukao, and S. Miyazaki, A new technique for mapping of total electron content using GPS network in Japan, *Earth Planets Space*, 54, 63-70, 2002.

Pi, X., A.J. Mannucci, U.J. Lindqwister, and C.M. Ho, Monitoring of global ionospheric irregularities using the worldwide GPS network, *Geophsy. Res. Lett.*, 24, 2283-2286, 1997.

Pimenta, A.A., P.R. Fagundes, Y. Sahai, J.A. Bittencourt, and J.R. Abalde, Equatorial F-region plasma depletion drifts: latitudinal and seasonal variations, *Ann. Geophysicae*, 21, 2315-2322, 2003.

Sahai, Y., K. Shiokawa, Y. Otsuka, C. Ihara, T. Ogawa, K. Igarashi, S. Miyazaki, and A. Saito, Imaging observations of midlatitude ionospheric disturbances during the geomagnetic storm of February 12, 2000, *J. Geophys. Res.*, 106, 24,481-24,492.

Saito, A., S. Fukao, and S. Miyazaki, High resolution mapping of TEC perturbations with the GSI GPS network over Japan, *Geophys. Res. Lett.*, 25, 3079-3082, 1998.

Sinno, K. and M. Kan, Mid-latitude ionospheric scintillations of VHF radio signals associated with peculiar fluctuations of Faraday rotation, *J. Atmos. Terr. Phys.*, 40, 503-506, 1978.

Tsunoda, R.T., Backscatter measurements of 11-cm equatorial spread-F irregularities, *Geophys. Res. Lett.*, 7, 848-850, 1980.

Tsunoda, R.T., On the generation and growth of equatorial backscatter plumes 2. Structuring of the west walls of upwellings, *J. Geophys. Res.*, 88, 4869-4874, 1983.

Valladares, C.E., R. Sheehan, S. Basu, H. Kuenzler, and J. Espinoza, The multi-instrumented studies of equatorial thermosphere aeronomy scintillation system: Climatology of zonal drifts, *J. Geophys. Res.*, 27, 26,839-26,850, 1996.

Woodman, R.F. and C. LaHoz, Radar observations of *F*-region equatorial irregularities, *J. Geophys. Res.*, 81, 5447-5466, 1976.

T. Aramaki, T. Ogawa, and Y. Otsuka, Solar-Terrestrial Environment Laboratory, Nagoya University, Honohara 3-13, Toyokawa 442–8507, Japan. (otsuka@stelab.nagoya-u.ac.jp)

A. Saito, Graduate School of Science, Kyoto University, Kitashirakawa-Oiwakecho Sakyo-ku, Kyoto 606–8502, Japan. (saitoua@kugi.kyoto-u.ac.jp)

Magnetic Storm Associated Disturbance Dynamo Effects in the Low and Equatorial Latitude Ionosphere

M.A. Abdu, J.R. de Souza, J.H.A. Sobral, and I.S. Batista

Instituto Nacional de Pesquisas Espaciais, Caixa Postal 515, São José dos Campos, SP, Brazil

During geomagnetic storms, magnetospheric energy injection at high latitudes leads to global scale disturbances in ionospheric electric fields and thermospheric winds. Over middle and low latitudes the disturbance dynamo (DD) electric field sets in within a few hours from the high latitude energy deposition. As a result, the major low latitude ionospheric phenomena, such as the ionization anomaly, the electrojet and the plasma bubbles irregularity processes, can be inhibited during times of their normal developments, while enhancing them at other times. Depending upon the strength and duration of the magnetospheric disturbances, the disturbance dynamo effects could last from several hours to a few days. An empirical model has been developed which successfully describes the diurnal pattern of the DD electric field. The basic theory has been successfully used in simulation models to predict the typical diurnal behavior of this electric field. However, the precise relationship of the intensity and duration of the effects at low latitudes to that of the causative storm seems poorly known, for validation purpose in case studies. This paper reviews our present state of knowledge on the disturbance dynamo effects in the low/equatorial latitude ionosphere, and discusses some new results from data obtained in the South American longitudinal sector. The DD electric field characteristics in response to magnetic disturbances arising from the CIRs (co-rotating interaction regions) and co-rotating streams, that are of significantly longer duration than those arising from the CMEs (Coronal mass ejections), are also discussed in some detail.

1. INTRODUCTION

The global ionosphere-thermosphere (I-T) system undergoes drastic modifications as a result of the interactive processes set off by the impulsive impact of the high speed solar wind plasma cloud, having southward interplanetary magnetic field B_z, with the earth's magnetosphere marking the onset of a magnetic storm/substorm event. The initial shock is followed by energy input into the high latitude I-T system and in a subsequent phase by the transfer of momentum and energy towards lower latitudes of the earth. During the magnetospheric compression from the initial impact, transient electric fields due to sudden impulse (SI)/sudden storm commencement (SSC) of seconds to a few minutes duration are observed around the globe. The dawn-dusk polar cap electric field that is established during the growth/development phase of a storm/substorm event, that often follows the initial impulse, promptly penetrates to equatorial latitudes until partially shielded by the Region-2 field aligned current that develops in the inner magnetosphere [*Nishida et al.*, 1966; *Vasyliunas*, 1970; *Gonzales et al.*, 1979; *Kelley et al.*, 1979, *Fejer et al.*, 1979; *Fejer and Scherliess*, 1995; *Kikuchi et al.*, 2000; *Richmond et al.*, 2003].

Recurrent Magnetic Storms: Corotating Solar Wind Streams
Geophysical Monograph Series 167
Copyright 2006 by the American Geophysical Union.
10.1029/167GM22

This undershielding electric field has eastward (westward) polarity on the day (night) side of the earth, but the polarity reverses under the overshielding condition that remains at the end of a substorm growth marked by IMF B_z turning northward. The magnetospheric energy input into the auroral region that causes heating and expansion of the high latitude I-T system, and the acceleration of neutrals through drag force from rapid ion convection under strong electric fields, are subsequently responsible for setting off disturbance winds as part of global scale disturbances in the thermospheric general circulation, and equatorward propagating large scale gravity waves that in turn lead to the generation of longer lasting electric fields by wind dynamo, (known as disturbance dynamo electric fields). The DD electric fields occur over the middle- and low-latitude regions after some time delay with respect to the initial prompt penetration electric fields. As we will be discussing later the polarity of this electric field is generally opposite to that of the quiettime wind dynamo electric field driven by solar heating and to that of the under-shielding electric field. These disturbance electric fields and the associated disturbance wind system are responsible for important changes in the dynamical and electrodynamical processes of the low latitude regions whereby all the major phenomena of the equatorial ionosphere-thermosphere system suffer important modifications. In this paper we intend to focus mainly on the disturbance dynamo electric fields and associated disturbance winds in terms of their generation mechanism, temporal and spatial distribution and their roles in the major phenomenology of the equatorial/low latitude ionosphere-thermosphere system. Drastic modifications of the low latitude ionospheric F region plasma can occur due to prompt penetrating as well as disturbance dynamo electric fields in the form of large increases or decreases in the height of the F layer [e.g., *Abdu et al.*, 1995; *Sobral et al.*, 1997] in the height integrated plasma content (TEC-Total Electron Content) of the dayside ionosphere [e.g., *Tsurutani et al.*, 2004; *Mannucci et al.*, 2005], severe TEC decreases/depletion in dusk sector ionosphere [*Batista et al.*, 1991; *Abdu*, 1997; *Greenspan et al.*, 1991; *Basu et al.*, 2001], or anomalous plasma fountain leading to plasma structuring by instability processes whereby field aligned irregularities in a wide spectrum of scale sizes, widely known as spread F/plasma bubble irregularities, could dominate the nighttime ionosphere [*Abdu et al.*, 2003a]. Such changes can have significant impact on a variety of space application systems, including, for example, the wide-spread use of the trans-ionospheric signals from GPS satellites. Thus the scientific interest in the problem as well as possible impacts on space application systems poses important questions as to the physical mechanisms of the relevant coupling processes and our modeling capability for predicting their occurrences. It is important to improve our understanding of the cause-effect relationship from the following view points: (a) the physical mechanisms, in terms of the dynamical and electrodynamical interactions, by which the disturbances originating from high latitudes extend/expand to equatorial latitudes, and the seasonal, solar activity and longitude dependences of the related processes, and (b) the contributions of these disturbances to a wide range of variabilities, including on short term and day-to-day scales widely observed in the major phenomenology of the low latitude/equatorial regions.

Before we discuss the processes under disturbed conditions it is useful to present a brief overview of the coupling processes operating under quiet conditions in the low latitude ionosphere-thermosphere system, of which we do have a reasonably good understanding. [See for example *Rishbeth*, 1972, *Abdu*, 1999; *Sastri et al.*, 2003]. Briefly, these coupling processes involve the generation of electric fields by the wind dynamo driven by the solar heating of the lower atmosphere and the thermosphere, and the resulting plasma transport and fountain effects that determine the equatorial ionosphere phenomenology at large and small scales. Atmospheric forcing due to upward propagating tidal waves, often modulated by planetary- and gravity-waves, establishes the wind field of the upper layers, whose interaction with the magnetized conducting ionosphere at E layer heights produces, through U × B forcing, the dynamo electric field that establishes the ionospheric current system, whose intensifications under the unique effects of the horizontal magnetic field lines lead to the intense equatorial electrojet current that flows in a narrow latitudinal belt of ~±3° centered around the dip equator [e.g., *Forbes*, 1981]. The east-west component of the E layer dynamo electric field mapped from off-equatorial latitudes on to higher levels in the equatorial F layer by the highly conducting geomagnetic field lines causes vertical transport of ionospheric plasma. The dayside eastward electric field causes E × B plasma uplift which constitutes the plasma fountain responsible for the generation of the Equatorial Ionization Anomaly (EIA), also known as the Appleton Anomaly. The plasma uplifted by the fountain effect to higher altitudes above the equator is displaced to off-equatorial latitudes by the force of diffusion and gravity, leading to the characterization of the EIA as a region of low density plasma centered on the dip equator that is flanked by two low latitude belts of enhanced densities on either sides at approximately ±15-18° in latitude. At the sunset, the F layer dynamo becomes the dominant source of electric field [*Rishbeth*, 1971]. An enhancement in the zonal (eastward) electric field, known as prereversal electric field enhancement (PRE), just before its post-sunset reversal to westward, occurs as a result of the U × B action of the zonal thermospheric wind (eastward in the evening hours) in the presence of a longitude/local time gradient in the E layer Pedersen conductivity that is present across the sunset terminator.

The evening eastward thermospheric zonal wind produces a vertical/downward electric field in the F region that can be represented as:

$$E_Z = U_y \times B_0[\Sigma_F/(\Sigma_F + \Sigma_E)] \quad (1)$$

where U_y is the thermospheric zonal wind, B_0 is the geomagnetic field intensity, and Σ_E and Σ_F are the integrated conductivities, respectively, of the E- and F-regions [*Abdu et al.*, 2003b; 2006]. Due to the faster post-sunset decay of Σ_E, as compared to Σ_F, E_z tends to increase towards the nightside, and the application of curl-free condition to such an electric field could lead to the enhanced zonal electric field (PRE), as originally proposed by *Rishbeth* [1971]. The PRE drives an enhanced plasma fountain which causes a resurgence of the EIA during post-sunset hours. With the rapid vertical uplift, as part of the enhanced fountain, the bottom-side gradient region of the F-layer turns unstable to density perturbation leading to the generation of equatorial spread F/plasma bubbles irregularities through generalized Rayleigh-Taylor instability mechanism. The sunset electrodynamics processes, of which the PRE, the resurgent EIA and the plasma bubble irregularities form integral parts, constitute perhaps the most interesting and important research topic of current interest in this field. The electric fields, in general, control a variety of other phenomena as well of the Equatorial Ionosphere-Thermosphere System (EITS) such as sporadic E layers [*Abdu et al.*, 2003b] and the F_3 layer [*Balan et al.*, 1997] which we will not be discussing here. While the electric fields, produced by the wind dynamo, are the immediate/basic driving forces for the major phenomenology during day and night, as discussed above, the wind system that produced this electric field in the first place, acts also in a more direct way to control the intensity of the phenomena. The variability in the wind system thus represents a major source of the variabilities at short-term and day-to-day scales widely observed of the different EITS phenomena [see for example, *Abdu*, 1997; *Abdu et al.*, 2006].

During magnetic storms the magnetospheric energy input over high latitudes causes heating and upwelling of the I-T system which result in global scale disturbances in thermospheric circulation and wind system whereby disturbance wind dynamo electric fields dominate the electrodynamical and related processes over middle- and low-latitudes. As a result, the developments of the major low latitude ionospheric phenomena, such as the ionization anomaly, the electrojet, and the plasma bubbles irregularity events, etc., are inhibited during the times of their normal occurrence, while enhancing their development at other times. Depending upon the strength of a magnetospheric storm, the disturbance dynamo effects could last from several hours to a few days after the development and the main phase of a storm. We discuss in this paper the important characteristics of the disturbance dynamo electric fields, pointing out their consequences to the electrodynamic processes and phenomenology of the low latitude region based on the existing theoretical simulation and empirical modeling results, and observational data, some of which concern new results from the South American longitude sector. We will be discussing the following topics: DD electric field manifestations in the different EITS phenomena, generation mechanism, local time dependence, duration/persistence of the effects, intensity as a function of solar activity phase and high latitude energy input, and possible longitude dependence. As regards the temporal and spatial distributions of the DD electric field, the two important questions that need answers are: (a) does the delay time varies with the local time sectors for a given high latitude energy input? and (b) if yes, how does it vary with the difference between the longitude of the energy input and that of the observation point?. Although there are no observational or model results so far that can fully clarify these questions the results and discussion to be presented here will hopefully provide a basis for forming some useful ideas on possible answers to these questions. Finally a summary of the outstanding features and explanations already known and some additional problems to be focussed on in future investigations will be presented.

2. MAJOR DD ELECTRIC FIELD CHARACTERISTICS AND EFFECTS ON EQUATORIAL PHENOMENOLOGY

A theory with model simulation for the generation of disturbance dynamo electric field was proposed by *Blanc and Richmond* [1980] and the first identification of the distinguishing features of this electric field from that of the prompt penetration electric field in the Jicamarca radar vertical drift results was reported by *Fejer et al.* [1983]. Since then several studies have been conducted on the different characteristics of this electric field including its development and evolution, local time and storm time dependences, and latitudinal and longitudinal variations [*Sastri*, 1988; *Abdu et al.*, 1997; *Scherliess and Fejer*, 1997, 1999; *Richmond and Lu*, 2000; *Fejer and Emmert*, 2003; *Richmond et al.*, 2003]. The equatorial and low latitude signatures of the prompt penetration (PP) or the disturbance dynamo (DD) electric fields can be rather easily identified in responses to relatively shorter duration and isolated storm events. In the case of extended storm events the low latitude electric fields observed after a time longer than the typical propagation time (3-8 hrs) for the auroral disturbance to reach lower latitudes, will have, on an average, weaker intensity, as the PP and DD electric fields have in general opposite polarities [see, for example,

Abdu et al., 1997; *Fejer* 2002, *Richmond et al.*, 2003]. The salient features of the two electric fields can be easily distinguished in the vertical plasma drift velocity variations observed by the Jicamarca radar during the storm disturbance of August 8-10, 1972 as presented in Figure 1 [*Fejer et al.*, 1983]. The rapid variations in the AE index (upper panel) produce corresponding rapid fluctuations in the vertical drift (lower panel), the upward or downward polarities (eastward or westward PP electric fields) depending upon whether the AE index indicates an intensification or weakening of a storm event. In contrast, the slow variation of the vertical drift (increasing upward drift) on the next night (August 9-10, starting at ~22 LT) that is unaccompanied by any AE activity represents a typical case of the DD electric field which is an eastward electric field during these night hours of its observation. A combination of both types of electric field is present during 07-10 LT on August 09 and their separation into the two components is a challenging task. The DD electric field has westward polarity during the day as can be seen in the response of equatorial electrojet (EEJ) current intensity to the March 22, 1979 storm shown in Figure 2 [*Mazaudier and Venkateswaran*, 1990] which presents the magnetic field H-component variations over three longitudinally separated equatorial observatories, Trivandrum, Addis-Ababa, and Huancayo. For this event the storm energy input at auroral latitude lasted for ~6 hours starting at ~11 UT on March 22. The decrease in the H-component amplitude shown by the hatched areas represents a decrease of the EEJ intensity (observable only during the day hours), first over Trivandrum and continued on westward till Huancayo, the total duration of the DD electric field in this case being at least 16 hours. The precise beginning and ending of the DD electric field is not available from these plots, however.

The EEJ current intensity appears to be a very sensitive indicator of the equatorial disturbance electric fields. This can be further demonstrated from the response of the EEJ instability generation process determined by electric field threshold limits modulated by superimposed disturbance electric fields. Figure 3 shows the EEJ 3-m irregularities dynamics presented in the form of the spectrogram of the Doppler velocities (upper panel) as observed by a 50 MHz coherent back-scatter radar operated at Sao Luis, Brazil, and local magnetogram H-component and AE/AL indices (lower panels), during the recovery phase of an extended storm that started on August 26 1998. We note large variabilities in the irregularity strength and velocity spectrum throughout the day. As explained by *Abdu et al* [2003c], the intermittent nature of the radar echoes suggests the presence of a strong DD westward electric field which inhibited the EEJ and its instability process for most of the daytime so that the transient intensifications/fluctuations in the AE index was responsible for corresponding PP eastward electric field transients that was intense enough to cause 3-m irregularity development. The PP electric field transients also caused associated EEJ current intensifications as indicated in local magnetic field H-component variations (middle panel). Here we note a demonstration of the interplay of the PP and DD electric fields in the case of an extended storm event.

An interesting case of equatorial F layer height response to DD electric field which gives a better idea of the evolution of the disturbance electric fields starting from the storm onset, including the approximate beginning and end times of the DD electric field, is presented in Plate 1. Plotted in this figure are the F layer heights at specific plasma frequencies (at interval of 1 MHz staring from 3 MHz) and the hmF2 values (pink curve), during the recent November 09-11, 2004, storm, as obtained from the Digisondes operated at the two equatorial sites, Sao Luis (2.33 S, 44.2 W, dip angle: ~ –.2°) and Fortaleza (3.9 S, 38.45 W, dip angle: –9°) in Brazil. For a comparison with a reference day we have plotted also the hmF2 values (blue curve) for November 15 which was a quiet day. (For the sake of clarity the quiet day variations in the different plasma frequencies are not shown in the figure). The increase in the hmF2 on the quiet day around 18-19 LT (21-22 UT) is due to the prereversal vertical drift (zonal electric field) enhancement that occurs near sunset (as was mentioned before). The SYM-H and ASY-H that are representative of the Dst and AE variations respectively [*Iyemore et al.*, 1993] shown in the top panel indicate the onset of a storm/substorm disturbance at 19 UT of November 09. During the auroral intensification and the Dst decrease that

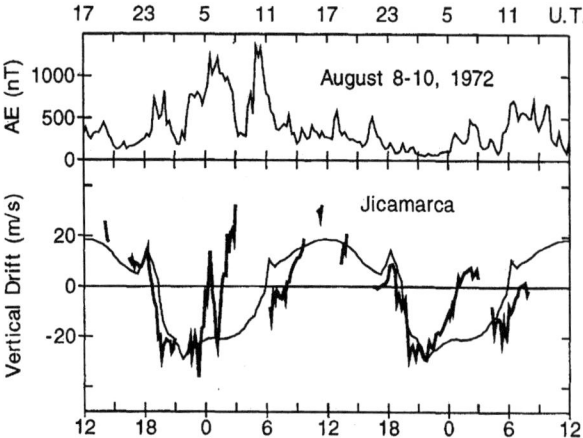

Figure 1. Auroral Activity (AE) index variation during August 8-10, 1972 (top panel) and the corresponding vertical drift variation observed by Jicamarca radar [*Fejer et al.*, 1983]. The rapid PP electric field variations follow the AE activity on the night of 8-9 August, whereas the slowly varying DD electric field on the night of 9-10 occurred after quieting of the AE activity.

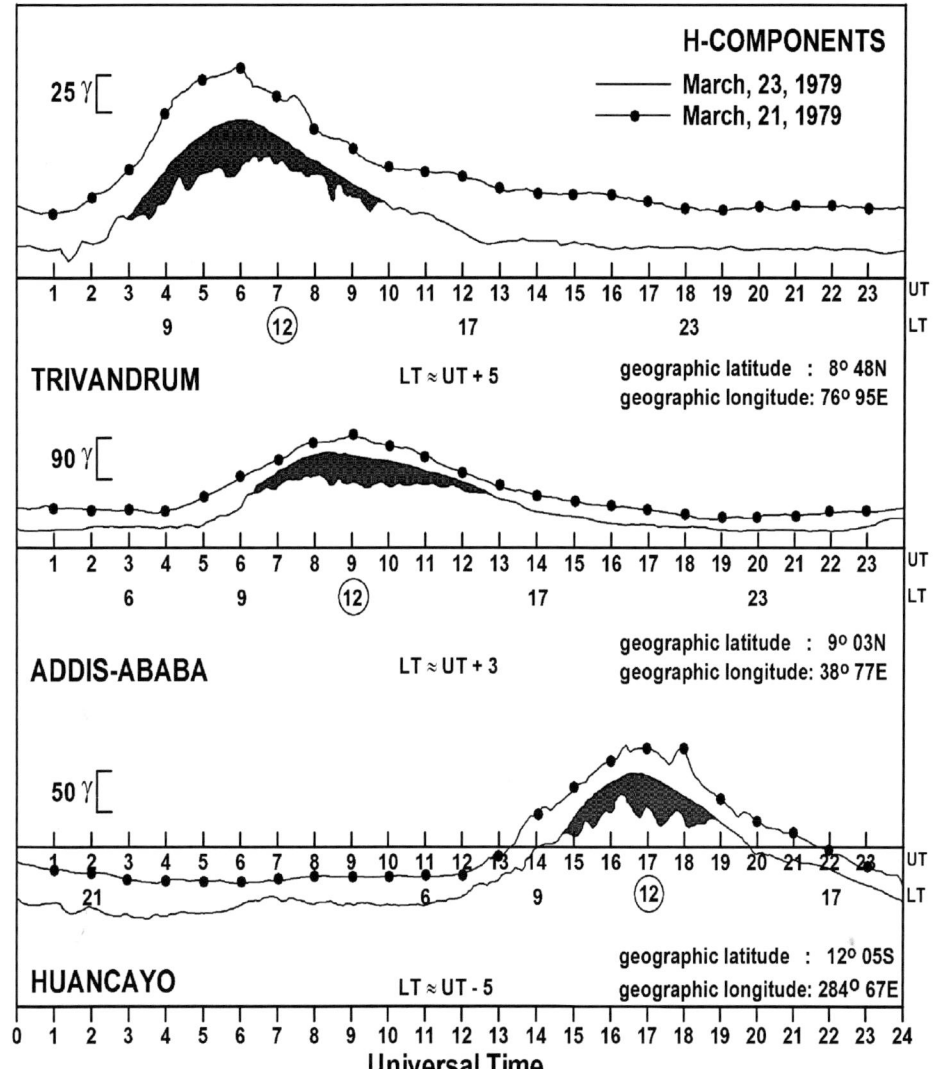

Figure 2. Magnetic field H-component variations during the March 22-23, 1979 storm over Trivandrum, Addis Ababa, and Huancayo. The H-component variation on March 21, a quiet day, is shown by dotted solid line, and that of March 23 is shown by a solid line. The solid line that describes a smooth H variation is obtained by subtracting a difference in Dst of 36 nT between the two days. The hatched area is identified as the disturbance of ionospheric origin, produced by a westward DD electric field [*Mazaudier and Venkateswaran*, 1990].

followed (and ended at ~2100UT) a PP electric field of eastward polarity (under-shielded electric field) caused a significant uplift of the F layer that was followed by a downdraft of the layer coincident with the decrease in the ASY-H and the end of the Dst decrease at ~21 UT. This seems to arise from an over-shielding westward electric field. A DD westward polarity electric field seems to have set in just before 20 LT/23 UT (~4 hours after the storm onset) as can be inferred from the lower than normal F layer heights both at Sao Luis (SL) and Fortaleza (Fz). The DD electric field polarity subsequently reversed to eastward at 22 LT (01 UT) remaining so till ~02 LT(05 UT) when an increase of AE activity seems to be responsible for a PP electric field which seems to have caused the reversal to westward of the net electric field. The combined influence of the PP and DD electric fields is evident till ~15 LT (18 UT) of this day (Nov.10) when the AE and Dst have nearly recovered to their normal/low values, and then the DD electric field of westward polarity dominates the F layer height variation till 22 LT (01 UT) as it was the case on the previous day. The DD electric

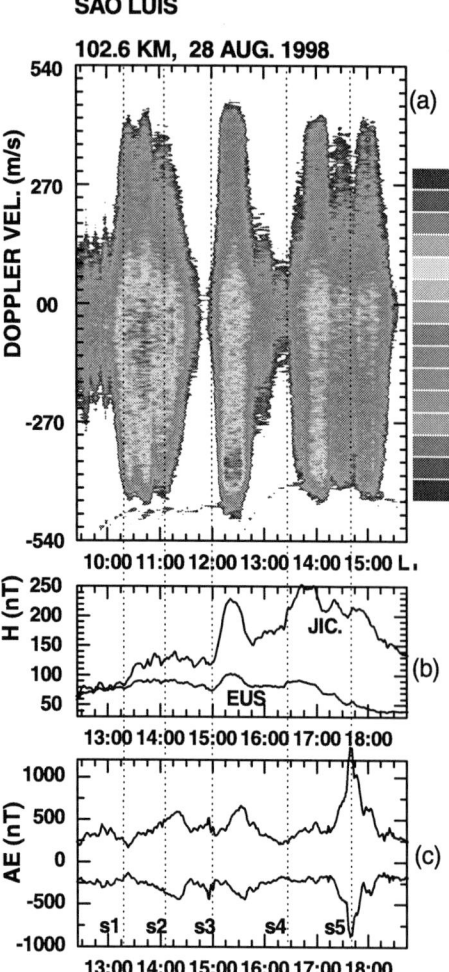

Figure 3. Doppler spectrogram, 3-dimensional plot, corresponding to 102.6 km from the EEJ 3-m irregularity echoes on August 28, 1998 over Sao Luis (top panel); the magnetogram H–component variation over Jicamarca -JIC and Eusebio -EUS (middle panel) and the auroral activity indices, AE and AL (bottom panel). The vertical lines identified as s1, s2 etc. are guide lines to show the association between the specific fluctuation phases of the magnetic field variations and the EEJ 3-m irregularity spectral developments.

field then turns eastward, but with weak intensity (including a brief duration of westward polarity) till around 09 LT after which a weak westward DD electric field seems to persist through the evening of November 11. This description of the event sequence helps us conclude that the DD electric field polarity is westward during most of the day time and evening hours, it has an eastward polarity at least near midnight and post midnight hours, and the PP electric field can influence the identification of the DD electric field characteristics in the case of extended storm/substorm activities as evident in this and previous examples [see also, *Abdu et al.*, 1997].

3. DD-ELECTRIC FIELD MECHANISM, THEORY AND MODELING

Magnetospheric energy input at high latitude drives equatorward directed disturbance thermospheric winds arising from the Joule heating and collisional interaction of the neutrals with rapidly convecting ions, under strong high latitude electric fields. Because of the inertia of the neutral air a few hours are required to set up the disturbance wind system, and once set up they can persist several hours [*Richmond et al.*, 2003]. A theory for the generation of dynamo electric field by the disturbance winds from auroral region Joule heating was first proposed by *Blanc and Richmond* [1980] who modeled the local time and latitude distribution of the disturbance dynamo electric field that could explain many of the features of this electric field observed over middle and low latitudes. A schematic of the mechanism of the DD electric field generation as proposed by them is presented in Plate 2 [see, also *Mazaudier and Venkateswaran*, 1990]. With the transport of the angular momentum, the disturbance winds originating from the auroral heating, initially directed equatorward, acquires westward velocity with respect to the earth under the coriolis effect. The westward disturbance wind over midlatitude produces an equatorward Pedersen current (J_P) that tends to charge the low latitude ionosphere positive until the current flow is stopped by the resulting poleward electric field (E_P). The effect of this poleward electric field which is perpendicular to the upward (downward) directed magnetic field lines in the southern (northern) hemisphere midlatitudes, is to produce a westward plasma drift and an eastward Hall current (J_H). The interruption of this Hall current at the terminators results in two current loops and establishment of a dusk-dawn electric field. This electric field has a polarity opposite to that of the quiet time wind dynamo electric field and the lower latitude current vortex is opposite in direction to the normal S_q current vortex. The dusk-dawn electric field extends to the low and equatorial latitude through the conducting ionosphere (that is through the current loops). In this way the onset of a DD electric field over equatorial/low latitudes in fact marks the arrival of the disturbance winds to midlatitudes. The continuing disturbance electric field effects over equatorial latitudes can be modified further with the later arrival of the disturbance winds at these latitudes by their participation in the 'local' wind dynamo that generates, for example, the PRE.

The DD electric field pattern calculated by the *Blanc and Richmond* [1980] model (which we will often refer to as BR model) has been found to explain reasonably well the empirical model representation of the DD electric field by *Scherliess and Fejer* [1997] that was built on long period statistical data sets from the Jicamarca radar and from the AE-E satellite. In particular they find excellent agreement during

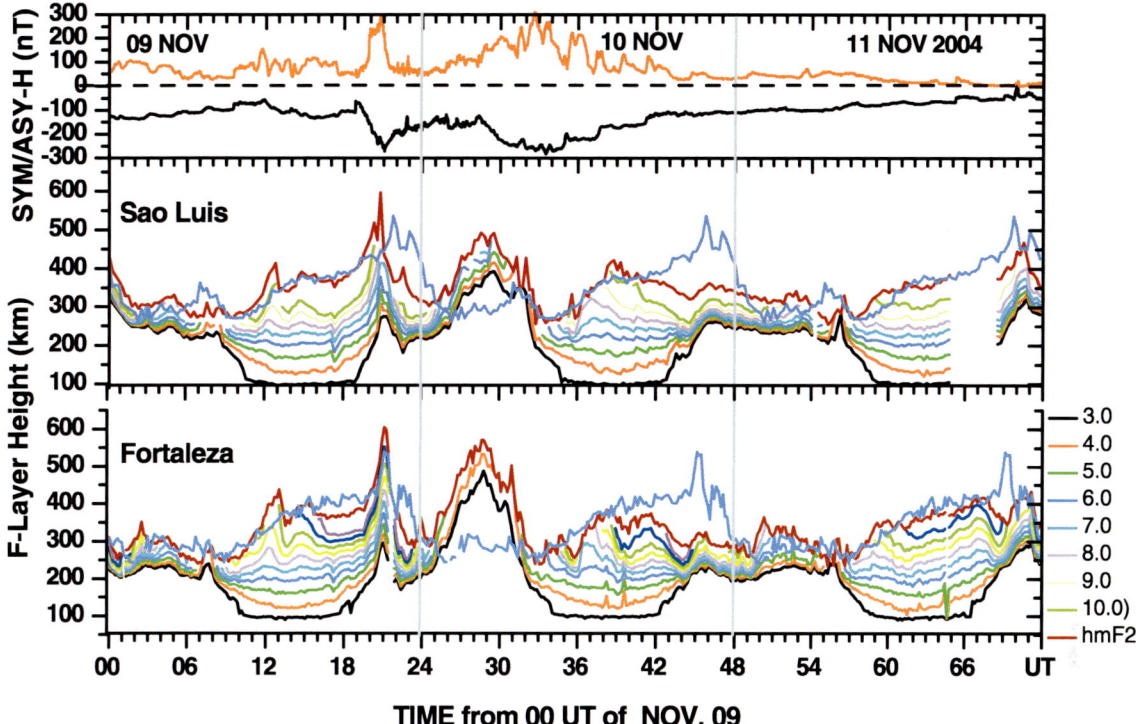

Plate 1. SYM-H and ASY-H variations during the November 09-11, 2004 storm (top panel). Ionospheric F region heights at specific plasma frequencies at interval of 1 MHz starting at 3 MHz as obtained from the Digisonde SAO software over Sao Luis and Fortaleza (bottom panel). Also plotted are the corresponding hmF2 variation (pink curve) and the hmF2 variation for a reference day, Nov. 15, (blue curve). The increase in the hmF2 that occurs in the evening hours seen clearly in the reference curve is due to the equatorial pre-reversal electric field enhancement (PRE).

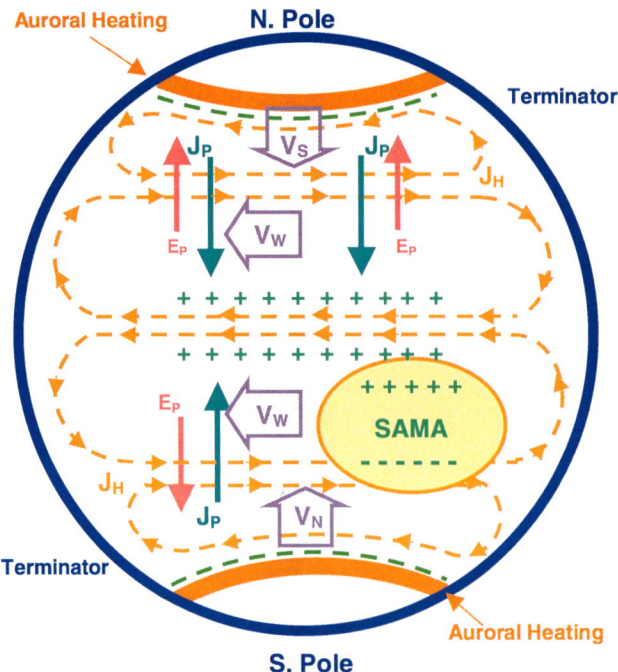

Plate 2. A schematic diagram showing the mechanism for the generation of disturbance dynamo electric field as proposed by *Blanc and Richmond* [1980]. Similar sketch for northern hemisphere has been presented by *Mazaudier and Venkateswaran*, [1990]. The equatorward disturbance winds (indicated as V_N and V_S) acquires a westward velocity (V_W) which produces equatorward directed Pedersen currents (J_P) resulting in poleward directed electric field E_P. The Hall current produced by the $E_P \times B$ action is shown as J_H whose interruption due to the conductivity decay at the solar terminations is the source of the global DD electric field. In the southern hemispheric half we have demarcated an area for the SAMA (South Atlantic Magnetic Anomaly) region where enhanced ionization due to storm induced energetic particle precipitation could modify the ionospheric conductivity.

post-midnight-pre sunrise hours. However important exceptions have been noted in comparative case studies with the BR model, especially in the evening sector [*Abdu et al.*, 1995] that might warrant some explanations beyond what can be accounted for by statistical fluctuations in the data. We shall come back to this point later. The time delay for the DD electric fields observed over equatorial latitudes with respect to the onset time of the high latitude energy input could depend upon such factors as the longitude sector where the bulk of the energy deposition takes place, the intensity of the storm, as well as the local time sectors of the observations. The simulation by *Blanc and Richmond* [1980] does not predict these dependences, but does show that the global (latitude–local time) electrostatic potential pattern and the associated electric field are generated mainly by dayside winds, the nightside wind not generating any substantial dynamo effect on the dayside. Thus depending upon the local time/longitude sector where the auroral heating takes place and the large scale structure in the conductivity distribution of the dynamo region a certain longitudinal dependence in the DD electric field is expected to be present, as has been observed as well [for example, *Abdu et al.*, 1997]. The time delays for the observed DD electric field effects over equatorial latitudes with respect to the storm onset seem to vary from ~5 to 8 hours or more [*Fejer and Scherliess*, 1995; *Abdu et al.*, 1997, *Scherliess and Fejer* 1997]. However, the time delay found in some model simulations for the presence of disturbance winds at low latitudes seems to be much shorter, around two hours [*Fuller-Rowell et al.*, 1996]. Besides the disturbance wind system considered in the BR model there could be other potentially important sources/controlling factors of ionospheric disturbance dynamo, such as the dynamo effect of fast traveling atmospheric disturbances that could reach equatorial latitude within a short time after the onset of high latitude energy deposition [see for example, *Fuller-Rowell et al.*, 1996; *Prolss*, 1995], fossil wind equatorward of the shielding layer (region-2 field aligned current) following the poleward contraction of the equatorward edge of the diffuse aurora after sudden geomagnetic quietening [*Fejer et al.*, 1990; *Spiro et al.*, 1988], and magnetic storm induced composition and conductivity changes. However the relative effectiveness/efficiencies of these factors are yet to be determined. For further details on the influence of the fossil winds, see *Richmond et al.* [2003].

In the southern hemispheric half of the Plate 2 we have considered a possible role of the enhanced ionospheric conductivity of the South Atlantic Magnetic Anomaly (SAMA) that could possibly modify/contribute to the DD electric field effects over this region [see, for example, *Abdu et al.*, 2005]. Enhancement in the ionospheric conductivity that can occur as a result of energetic particle precipitation in the SAMA region during magnetic storms [*Batista and Abdu*, 1977] can significantly modify and enhance the ionospheric conductivity of this region with important consequences for the equatorial electrodynamic processes [*Abdu at al.*, 1998, 2003a; 2005] as we will discuss later.

A more detailed and self-consistent modeling of the wind dynamo electric field effects over the equatorial ionosphere has been presented recently by *Richmond et al.* [2003] based on the Magnetosphere-Thermosphere-Ionosphere–Electrodynamics General Circulation model of *Peymirat et al.* [1998]. [See also, *Peymirat and Fontaine*, 1994; and *Richmond et al.*, 1992]. Their model is capable of simulating the quiet time electric field (mainly due to solar radiation driven wind system) and the disturbance electric fields arising from storm time process involving energy input at high latitude and associated polar cap expansion and contraction phases. The modeling includes the effects from polar cap potential drop build up and decay, particle precipitation and Joule heating by electric fields as well as the contribution from "fossil winds" arising from neutral acceleration by convecting ions (through the ion drag effect) that continues into the over-shielding phase by the region-2 field aligned current accompanied by a contracting polar cap. The effect of the fossil wind is to extend the duration of the overshielding electric field, as modeled by *Spiro et al.* [1988] and *Fejer et al.* [1990] using the Rice Convection Model (RCM). In fact the fossil wind process can be viewed as the first phase of the DD electric field mechanism as noted by *Fejer* [2002]. We will examine some results on the characteristics of the disturbance wind dynamo electric field based on the simulation results presented by *Richmond et al.* [2003] and reproduced in Figure 4a and (b). Figure 4(a) shows the evolution of a storm event in terms of the polar cap steady state, expansion, and contraction phases controlled by the polar cap potential drops across the day-side and night-side gaps as defined by the authors. The potential drop across the day-side gap being maintained larger (smaller) than that of the night-side gap causes expansion (contraction) of the polar cap. Thus the interval A-B represents the storm expansion phase, the interval B-C the main phase, and starting from C (for the case shown by the solid line) the storm recovers. The results of the runs for the times A, C and E shown in Figure 4b corresponds to the simulated equatorial electric field variations at the beginning (pre storm), the end of the main phase and about 3 hours into the post recovery of the storm, respectively. These results correspond to 364 km altitude above the magnetic equator. The solid lines represent the total electric field including (1) ionospheric wind dynamo and (2) penetration of imposed polar cap electric field to equatorial latitudes, and the dotted line shows the contribution only from (2), that is, without wind dynamo effect.

The quiet time electric field pattern is well reproduced by the solid curve at the pre storm time 'A'. The evening

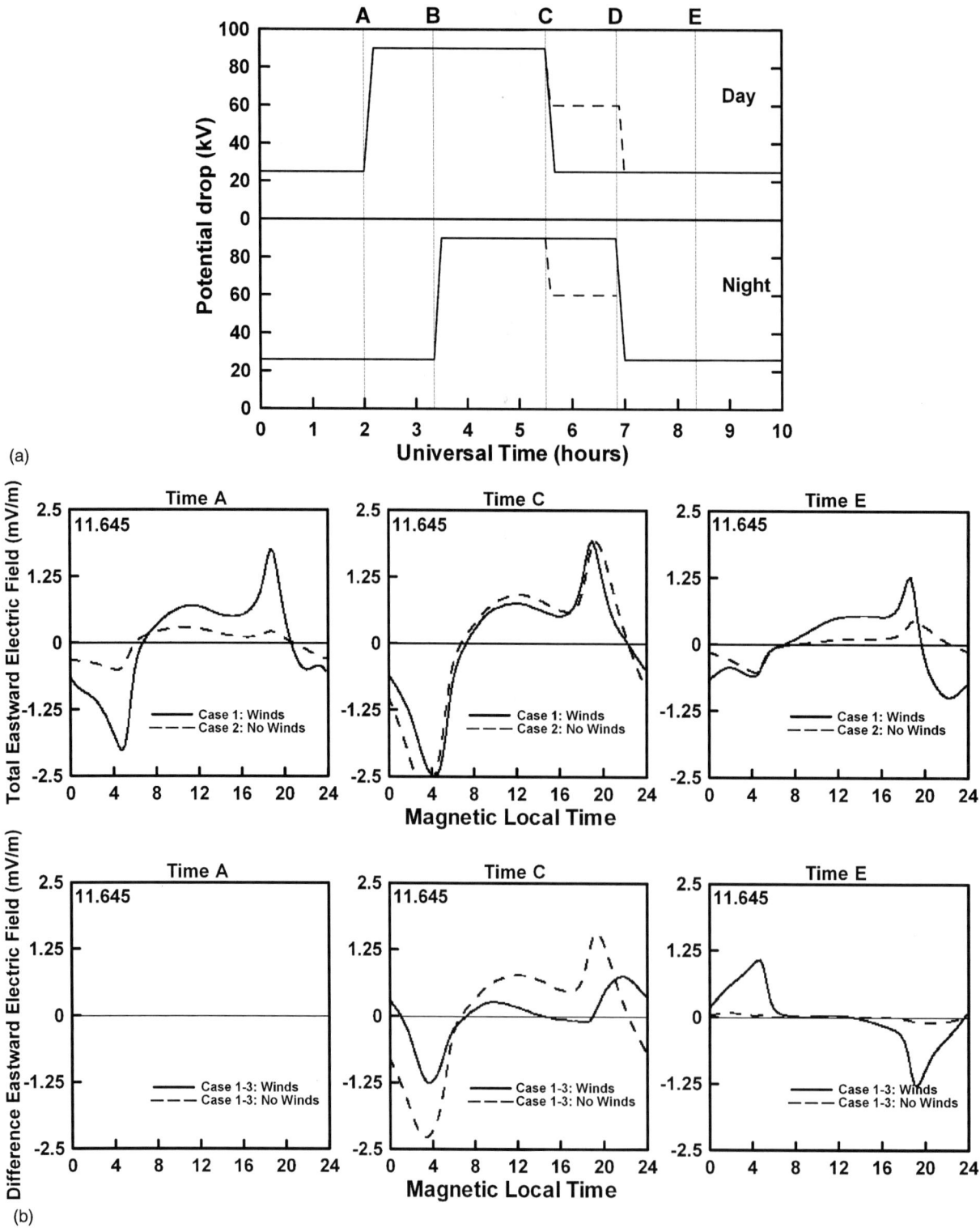

Figure 4. (a) Time evolution of the potential drops (top panel) across the day-side (upper portion) and night-side (lower portion) gaps. The solid line is for polar cap contraction and the dashed line for non-contraction at the storm recovery. The times A, B, C, D and E indicate the different storm phases as explained in the text. (b) The model results of eastward electric fields at 364 km over the dip equator for time A (pre storm) in the top left panel, time C (pre recovery) in the top middle panel, and time E (post recovery) in the top right panel. The lower panels show the difference between the eastward electric fields with respect their pre storm values, for runs with (solid) and without (dashed) winds. [*Richmond et al.*, 2003].

prereversal electric field enhancement may be noted, in particular, in agreement with such basic features brought out also in a previous simulation by the TIEGCM [*Fesen et al.*, 2000]. Under the conditions of quiet time energy input considered in the model there is also a weaker electric field (dashed curve) from equatorward penetration of an imposed polar electric field due to incomplete steady state shielding. Well into the storm, just before the end of the main phase (time 'C') the disturbance wind effect tends to be opposite to that of the normal solar heating so that the total electric field indicated by solid line is dominated by penetrating polar electric field (dashed line). The electric field from disturbance wind opposes also the penetrating polar cap electric field so that the net disturbance electric field at time C (solid line, the lower middle panel of Figure 4b) is considerably weaker than the pre storm electric field for the conditions simulated. Our main interest here is in the post-storm electric field pattern represented by that at the time E. Here the disturbance electric field is almost totally contributed by the disturbance winds. The most notable features of its diurnal pattern are a pronounced westward electric field during the evening and pre-midnight hours and a similarly pronounced but eastward electric field during the post-midnight/pre-sunrise hours. These results appear to present a better local time variation of the DD electric field than did the BR model in some important respects. For example the evening/post sunset westward electric field which is significantly more intense after the storm (at storm time E) in Figure 4b than in the BR model (see Figure 5b to be discussed later) occurring around the local time of the usual prereversal eastward electric field enhancement seems to be capable of explaining the near total inhibition of the PRE observed after the storm on November 10 in Plate 1. (We will discuss this point again later). On the other hand the near absence of any DD electric field/disturbance wind effect during the daytime hours 07-14 LT in Figure 4b appears to be inadequate to explain certain observational results (to be specified later) some of which seem to be better explained for by the BR model. This might be a consequence of the local time sector of the wind disturbance associated with the high latitude energy deposition used in these models. The models, however, do not address some specific questions such as for example the minimum time delay, or the time delay for the maximum effect, for the DD electric field to be observed at a specific location in the equatorial region as a function of the longitude and local time sectors of the initial high latitude energy deposition. The observational results so far available also do not provide answers to such questions.

3.1. Empirical Model Results on the DD Electric Field

An extensive data set on vertical drift velocities covering 20 years (1968-88) of measurements by the Jicamarca radar was analyzed by *Scherliess and Fejer* [1997] to develop an empirical model describing the average characteristics of the equatorial vertical plasma drifts due to the disturbance dynamo, as a function of local time and the time history of the auroral electrojet (AE) activity index used as an indicator of the high latitude energy input. The method by *Fejer and Scherliess* [1995] utilized linear regression analysis, simultaneous multiple parameter fitting and storm-time dependent binning. The results show that there are two basic components in the DD electric field over the equator characterized by (1) short term delay of

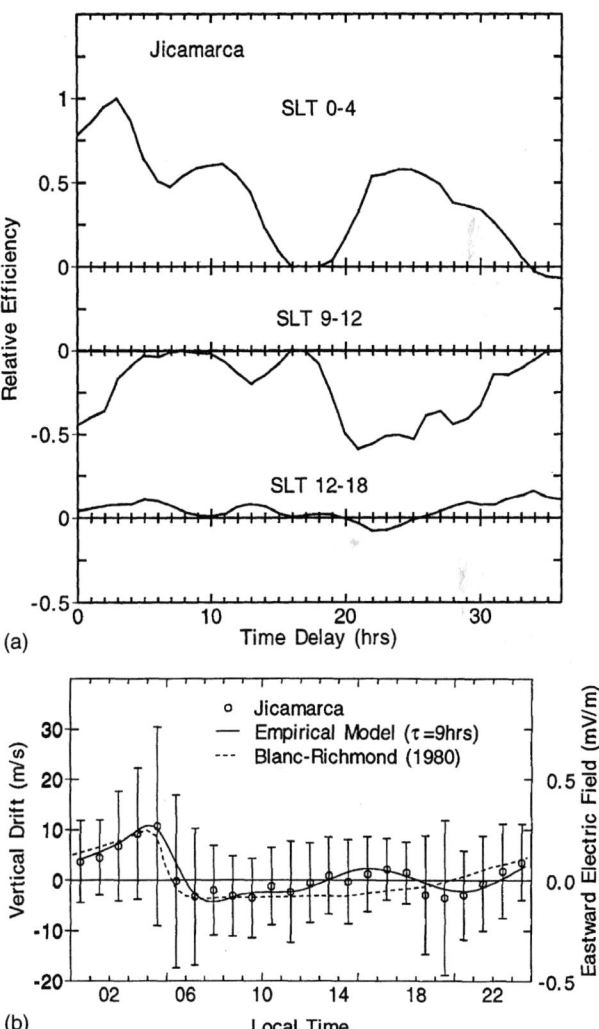

Figure 5. (a) Relative efficiency of the disturbance dynamo process in three local time sectors as function of the time delay between the auroral and equatorial disturbances (*Scherliess and Fejer*, 1997). (b) Comparison of the empirical disturbance dynamo drift pattern with the results from the *Blanc-Richmond* [1980] model, both for an increase in the hemispheric power input corresponding to about 400 nT over quiet time level [*Scherliess and Fejer*, 1997].

1-12 hours and (2) long term delay of 22-28 hours, between the high latitude energy input (AE enhancement) and equatorial velocity perturbations. In general, the short term delay events have westward polarity during most of the daytime and eastward polarity with a post-midnight peak in intensity during the night. The long term delay events have strong eastward electric field during the night and a short duration westward electric field of weak intensity during morning hours, when geomagnetically quiet conditions over several hours are preceded by large or moderate storms (AE > 350 nT). Figure 5a shows the storm time variation (that is, the variation as a function of the delay between the auroral and equatorial disturbances) of the relative efficiency of the disturbance dynamo process for three local time sectors. The largest disturbance dynamo induced upward drift (eastward electric field), which corresponds to large positive efficiencies, occurs with time delays of a few hours in the post- midnight sector. In this sector the component with the longer delay (20-30 hours) also has strong intensity (efficiency). For the morning sector (09-12 LT) the disturbance drift is downward (westward electric field) and has smaller intensity for the short term delay component, and the intensity is larger for the longer delay component. It is possible however that the larger efficiencies obtained at smaller time delays (1-3 hours) in Figure 5a could contain significant contribution from the over-shielding electric fields. In the afternoon sector the disturbance electric field is nearly absent. The result for sunset hours (not included in this figure) presents a weak but downward disturbance drift with a delay time of 2-12 hours. The shorter term delay effects have a time constant of around 6 hours for recovery to quiet time consistent with that of thermospheric winds following magnetic storms [*Fuller-Rowell*, 1996]. While the shorter term delay effects arise from the disturbance winds as modeled by *Blanc and Richmond* [1980], the mechanism of the long-term disturbances is not clearly identified. However, as pointed out by *Scherliess and Fejer* [1997], it is suspected to be caused by thermospheric composition changes that reach the lower latitudes about a day after large storm events [*Fuller-Rowell*, 1996] and by the associated electron density and conductivity modifications that accompany it.

These average characteristics of the DD electric field are in general agreement with the theoretical/model results of *Blanc and Richmond* [1980]. Especially, the agreement is excellent, as explained by *Scherliess and Fejer* [1997], during the post-midnight hours as can be noted in the comparison of the two results for a time delay of 9 hours shown in Figure 5b. The agreement with the model results of *Richmond et al.* [2003] looks also excellent during the post-midnight hours as well as during the post-sunset hours of the PRE. However, during the morning hours (09-12 LT) the BR model appears to explain better the observed average downward drift of *Scherliess and Fejer* than does the model results of *Richmond et al.* [2003]. These different degrees of the agreement during the morning/day sector may be caused by the fact that strong disturbance winds in the BR model are present during the day while in *Richmond et al.*, they seems to be more dominant during the night hours (see their Figure 3 for time E). It has been pointed out by BR that the electrostatic potential generated by dayside disturbance winds is very similar to that generated by the total wind system whereas the nightside winds do not have any substantial dynamo effect on the dayside.

4. DISCUSSION AND COMPARISON OF MODEL RESULTS WITH DATA IN CASE STUDIES

The duration and intensity of the disturbance wind dynamo electric field effects could vary significantly from event to event depending upon season and solar activity even for cases of high latitude energy deposition events that are of comparable intensity. Thus the agreement between the model results with the observational data during specific events could vary significantly from case to case depending upon the seasonal and solar cycle conditions of the observations as we shall discuss below. The empirical model of *Scherliess and Fejer* [1997] closely represents moderate solar flux equinoctial disturbance drift and does not consider a possible dependence of the disturbance dynamo drifts on season and solar cycle. The BR model does not explicitly specify the representative seasonal and solar activity conditions of their simulation, while the model by *Richmond et al.* [2003] was run for equinox and solar maximum conditions. However, the cases of DD electric field events that we will discuss below will be examined in the light of the empirical and theoretical model results presented above in an attempt to identify their characteristics features as arising from the nature of the high latitude energy deposition, possible factors arising from seasonal and solar activity differences related to the observations, and some of the consequence to the equatorial phenomenology.

It is of interest to look for DD electric field effects arising from long duration AE activity such as those usually associated with corotating events and HILDCAAs (High Intensity Long Duration Continuous AE Activity), [*Tsurutani and Gonzalez*, 1987] in contrast to the shorter duration/isolated activities such as those often associated with CME's. Figure 6 shows examples of AE variations during a few of the 15 events used in an analysis whose results will be presented below. In Figure 6 we note the AE activity continuing through the 4-day interval considered in the analysis. The analysis consisted in determining the average ionospheric critical parameters, hmF2 and foF2, corresponding to Kp index values exceeding 3 for local time intervals (Δt) preceding the observation time (t_o) by different time delays. These values are then compared with the corresponding values obtained by

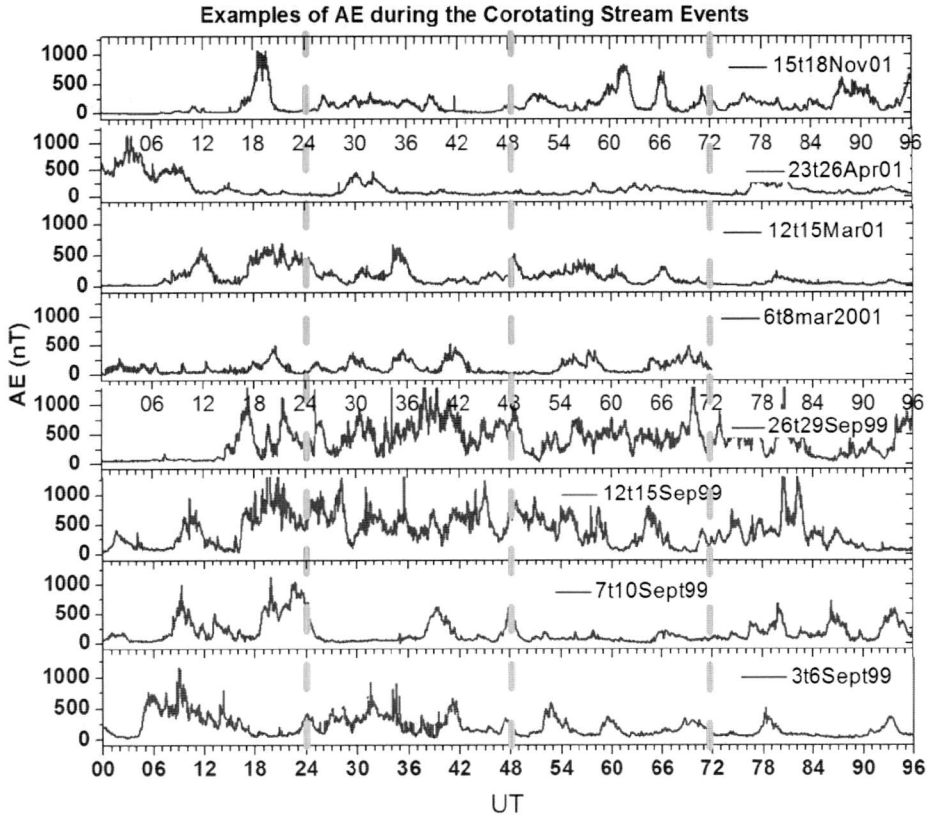

Figure 6. Some examples of extended duration AE activity during corotating solar wind stream events. These cases belong to the set of data analyzed for obtaining the results presented in Plate 3.

considering cases of Kp < 3. The results presented in Plate 3 show the disturbance values (in reddish colors) and quiet time values (in bluish colors) of these parameters, and those of the vertical drift velocity (Vz) calculated as dhF/dt, for Sao Luis (2.33 S, 44.2W, dip angle: –2°) and Cachoeira Paulista (22.6 S, 315 E; dip angle: –28°). The smallest time delay considered was zero and the Δt was 3 hours which means that the observed values correspond to Kp exceeding 3 during the first three hours (0 to –3 hours) preceeding the observation time. These values seem to be influenced by both the PP and DD electric fields. The values corresponding to the next successively delayed intervals (from –3 to –15 hours, –9 to –21 hours, and –18 to –30 hours) should contain significant DD electric field effect. The continuing AE activity has contributed to PP electric field, which is in general of opposite polarity to the DD electric field, [Richmond et al., 2003] so that the net effect obtained from the analysis is a weaker disturbance electric field effect than would have been the case for DD electric field effects occurring under quiet post-storm conditions. The main highlights of the analysis results are the following: Over Sao Luis (near the dip equator) the most perceivable effect is a decrease in the pre-reversal eastward electric field seen both in the post-sunset Vz as well as in the F layer height parameters (hmF2 and h'F). The effect of a PRE inhibition due to DD electric field widely observed in the case of shorter duration or isolated events, in agreement with the model result of Richmond et al. [2003], has suffered a reduction due to the under-shielded PP eastward electric field that must have entered into the averaged values. The post-midnight eastward DD electric field widely observed for appropriate individual events is also suppressed for the same reason. Over Cachoeira Paulista which is located near the equatorial ionization anomaly crest the most perceivable effect seems to be disturbance winds blowing equatorward during most of the night and poleward during afternoon–evening hours, as suggested by the respective increase and decrease of the F layer heights. We may note further that the nighttime equatorial anomaly is somewhat inhibited as judged from the foF2 values over Cachoeira Paulista. These results lead to the conclusion that the low latitude ionosphere disturbance effects, on average tend to get weaker, but continue for several days, under auroral energy

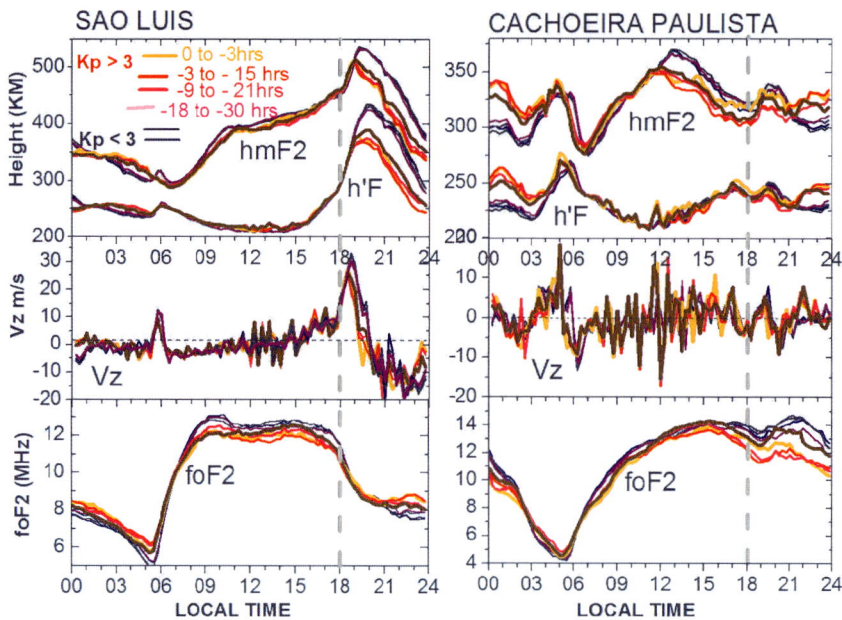

Plate 3. Average responses, as a function of local time, of the F regions critical parameters, hmF2 and h'F (top panel), foF2 (bottom panel) and the vertical drift velocity Vz calculated as dhF/dt (middle panel) during corotating solar wind stream induced magnetospheric disturbances. The results for Sao Luis are on the left and for Cachoeira Paulista on the right. The average values of the parameters were calculated for intervals of Kp values >3 (red) and Kp < 3 (blue) occurring with variable time delay/separation preceding the observation. The Kp intervals and delays considered are: 0 to −3 hrs, −3 to −15 hrs, −9 to −21 hrs, −18 to −30 hrs.

deposition of extended duration produced by the CIRs and corotating solar wind streams interaction with the magnetosphere.

In contrast to the above results the DD electric field that follows CME induced storms of moderate duration and well defined recovery phase such as that presented in Plate 1 has a duration less than a day as can be verified by hmF2 variations over Sao Luis for this case. We will examine additional examples of the equatorial ionospheric response to other CME induced storms for which the recovery occurred after the first day of storm onset. The results for one case study are presented in Plate 4 which shows the AE and Kp variations (top and bottom panels) following the shock event that triggered the November 6, 2001 magnetic storm. (For detailed discussions on the global ionospheric response to this event, see, *Tsurutani et al.*, 2004; and *Maruyama et al.*, 2004). The storm started with an interplanetary-magnetospheric shock event accompanied by an SSC at 0155 UT which was promptly followed by a polar cap dawn-dusk electric field that caused a strong PP electric field of eastward polarity in the dayside equatorial ionosphere (in the Pacific longitude sector) and simultaneously a westward electric field in the midnight sector over Brazil. The shock event occurred when the auroral electrojet was already active (as indicated by the AE indices) for 6-8 hours prior to it. Within ~2 hours, after the shock the disturbance electric field reversed the polarity marking the onset of an intense DD electric field of eastward polarity at ~04 UT, that is ~01 LT in Brazil, as is indicated by the rapid rise of the hmF2 and h'F at this time (pink curve) in Plate 4. It is very likely that the AE activity prior to the SSC is a leading cause of this DD electric field which produced a vertical plasma drift velocity of the order of 80 m/s. (We should point out that the vertical drift velocity/eastward electric field is indicated by the time rate of change of the F layer virtual height, $d(h'F)/dt$, in this case.) The layer formation at lower heights at sunrise seems to have contributed to the abrupt height decrease that occurred at the reversal of the DD electric field to westward that began soon after 05 LT. A continuing westward polarity DD electric field manifestation at F layer heights seems to have been suppressed somewhat by an eastward PP electric field due to the continuing AE activity as indicated by larger than normal values of these heights during the daytime on November 6. However, the higher than normal foF2 values that followed on this day is indicative of EIA inhibition by a westward DD electric field, which is consistent with the global daytime ionospheric total electron content distribution determined from GPS signals received on ground as well as by the CHAMP satellite on this day as reported by *Tsurutani et al.* [2004]. The DD electric field of westward polarity that appears to have dominated the post-sunset hours of November 06 (based on the reduced PRE inferred from the h'F variation) turned to eastward polarity during the post-midnight hours (as to be expected from the theory and model results). The DD electric field with westward polarity seems to have persisted during the daytime and evening hours of the next two consecutive days as suggested by the reduced hmF2 values with the correspondingly enhanced foF2 values over Sao Luis. The energy deposition at auroral latitudes, although with very reduced intensity, that continued into the early hours of November 08 seems to be responsible for the persistence of the DD electric field observed till at least the evening hours of this day (when the data acquisition was interrupted).

Similar results as those presented in Plate 4, but for the more severe storm events of 20-23 November 2003 are shown in Plate 5. With the storm onset at ~0320 UT (0020 LT) and the continuing energy input (indicated by the large AE variations) the first indication of the start of a dominant DD electric field effect is clearly evident at 18 LT when the PRE is almost totally suppressed by a westward electric field. The DD electric field turns eastward just around local midnight as it did in Plate 4 (as well as during the October 2003 storm to be presented in Plate 6). The vertical drift velocity reached the order of 50 m/s. The near total suppression of the PRE by a westward electric field and the large post-midnight layer rise caused by an eastward electric field are in excellent agreement with the DD electric field simulation results of *Richmond et al.* [2003]. The morning reversal to westward electric field occurred exactly at the same time and rate as in Plate 4. The DD westward electric field was present till ~18 LT of November 21 as can be verified by the lower daytime hmF2 values. There is no indication of a DD electric field during the PRE and night hours that followed. Even though a certain degree of mix up with possible PP electric fields from the ongoing AE activity is expected, the nearly total absence of any DD eastward electric field effect during the post-midnight hours, when the AE activity was nearly absent, calls our special attention. The sequence of these disturbance effects strongly suggests that the DD electric field in this case has a shorter duration as compared to that of the November 2001 event (Plate 4) even though the high latitude energy input for this storm is significantly more intense than during the November 2001 event, as can be verified from a comparison of the AE amplitudes for the two cases. We should note however that in the latter case although the DD electric field persisted for longer duration it was dominant mainly during the daytime and evening hours only (as can be observed in the lower than normal hmF2 values on November 7 and 8) with the absence of an expected eastward DD electric field during the post-midnight hours of these days. The possible cause of such a behavior is unknown at present. When we consider that the monthly average solar flux values (F10.7) for November 2001 and November 2003 were, respectively, 208 and 134 respectively, these results would

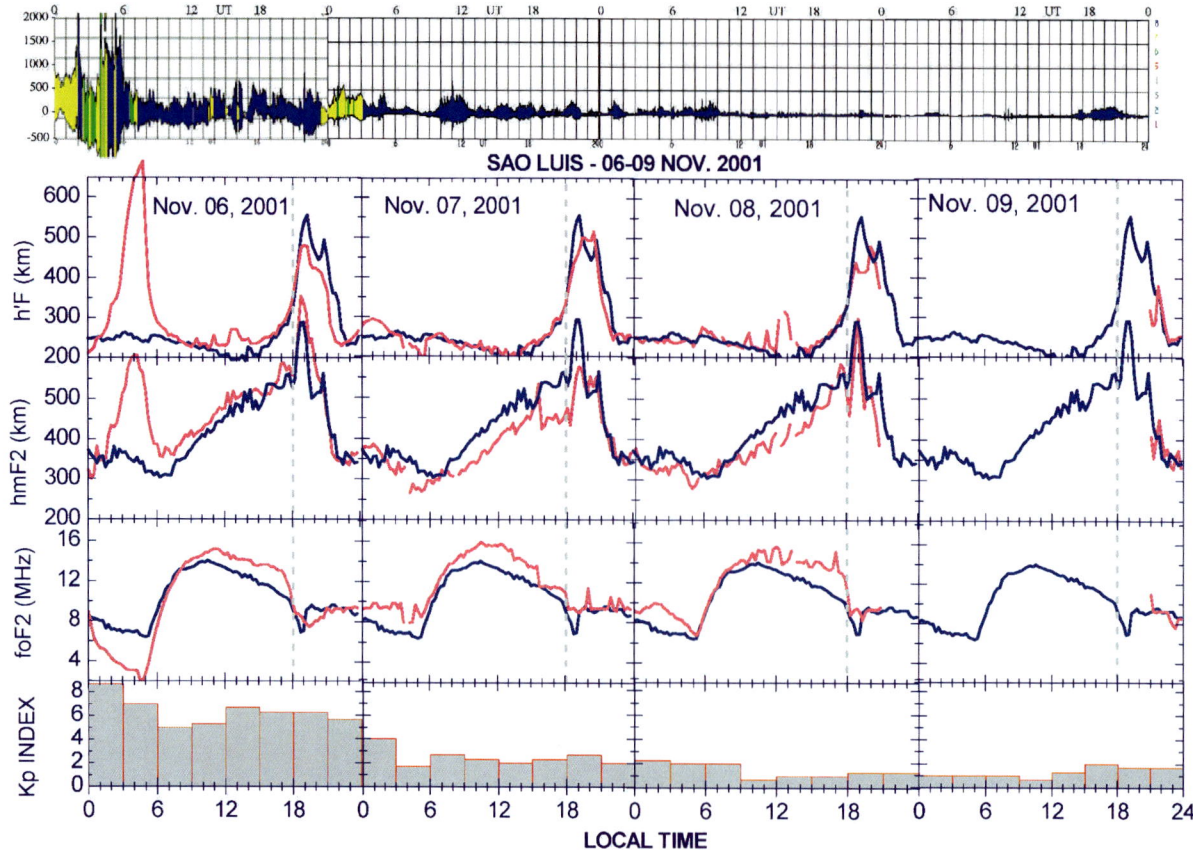

Plate 4. The auroral activity indices, AE and AL, variations during November 06-08, 2001, following the start of a storm at ~02 UT on Nov. 06. (top panel), the Kp variations (bottom panel), and the response of the ionospheric parameters, h'F, hmF2 and foF2 (middle three panels) at Sao Luis. The pink curves represent the disturbed values and the blue curves are the reference day variation patterns.

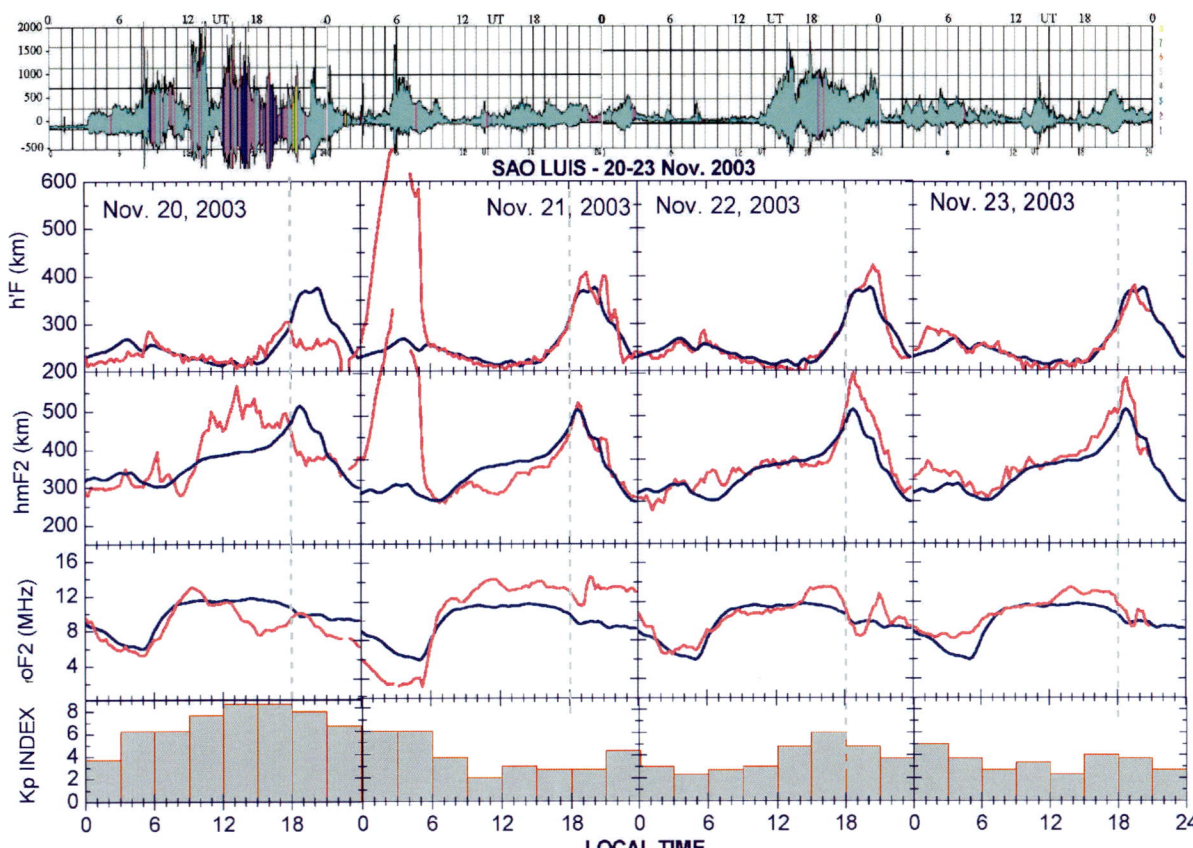

Plate 5. The auroral activity indices, AE and AL, variations during November 20-22, 2003, following the start of an intense storm at ~0320 UT on Nov. 20. (top panel), the Kp variations (bottom panel), and the response of the ionospheric parameters, h'F, hmF2 and foF2 (middle three panels) at Sao Luis.

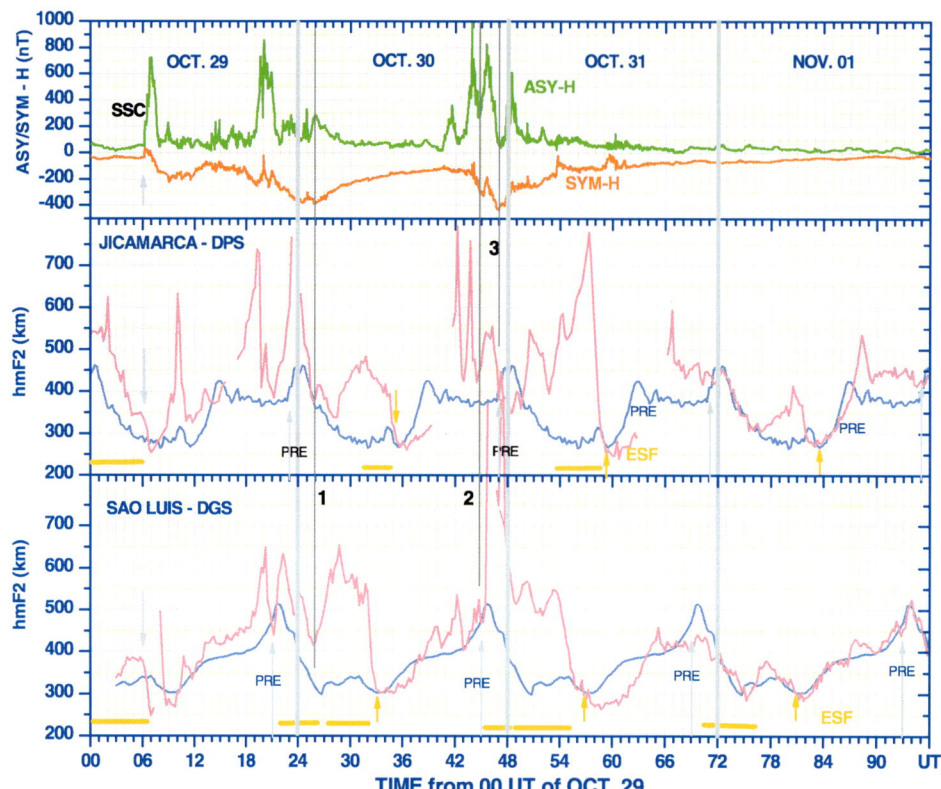

Plate 6. ASY/SYM-H variations (top panel) during October 29-November 01, 2003. The SSC occurred at ~06 UT. hmF2 responses over Jicamarca (middle panel) and Sao Luis (bottom panel). The reference day curves are shown in blue. The quiet time evening height increase associated with the prereversal zonal electric field enhancement occurs near 18 LT indicated by grey vertical lines. The black vertical line 1 marks the onset of the hmF2 increase over Sao Luis due to the eastward disturbance dynamo electric field; The corresponding hmF2 increase over Jicamarca occurs two hours later. The vertical line 2 indicates the large uplift of the F layer during the PRE over Sao Luis, with much smaller hmF2 increase over Jicamarca associated with auroral activity related PP eastward electric field, The vertical line 3 indicates the suppression of the PRE over Jicamarca due DD electric field. 'ESF' refers to Spread-*F* traces in the ionogram whose durations are indicated by horizontal lines/patches.

point to the possible existence of a solar activity dependence for the DD electric field as far as its intensity and duration after the end of the storm are concerned. The storms of higher level of solar activity (as in the case of the November 2001 storm) seem to produce larger intensity and longer duration post-storm DD electric fields than the storms of lower solar activity level (as in the case of the November 2003 storm) even when the latter is significantly more intense than the former.

An explanation for what appears to be a solar cycle effect in the DD electric field may be proposed as follows. According to the model simulation results by *Richmond et al.* [2003] discussed earlier, the electric field generated by the disturbance wind dynamo has in general a polarity that is opposite to that generated by the winds from solar heating of the thermosphere, that is, the quiet time winds. Our previous study [*Pincheira et al.*, 2002], showed that the quiet-time winds over low-to-middle latitudes that are generally poleward during the day, (and equatorward during the night), have larger amplitude during low solar flux as compared to high solar flux years as a result of the increased ion drag effect in the latter case. According to the model simulation by *Blanc and Richmond* [1980], also discussed earlier, the global potential distribution resulting from the disturbance wind dynamo is controlled more by daytime winds, than by nighttime winds. Therefore the daytime poleward wind that has a relatively larger amplitude under low solar flux conditions could oppose the effects from a disturbance equatorward wind that basically drives the DD electric field even when such winds are driven by a higher intensity high latitude energy deposition. This explanation needs to be tested by model calculations, however.

As regards the seasonal dependence of the DD electric field (and its implication on the PRE), *Fejer* [2002] showed that the largest evening downward perturbation during solar maximum occurs in equinox. The westward directed disturbance electric field was considerably weaker in June than during equinox conditions for moderate solar flux conditions. Such a seasonal behavior may be originating from the two distinct causes of the negative effect on the PRE, usually attributed solely to a westward DD electric field at these hours. The possibility that the two source are part of the same disturbance dynamo processes was pointed out by *Abdu et al.* [1995]. The global DD electric field that is westward in the evening equatorial region will promptly contribute to a negative effect on the upward plasma drift normally/under quiet-time produced by an eastward electric field arising from the F layer dynamo as part of the sunset electrodynamics [*Rishbeth*, 1971; *Heelis et al.*, 1974]. It is to be recalled that the wind dynamo mechanism predicts the presence of DD electric fields at equatorial latitudes in time the disturbance winds arrive at middle latitudes. In a subsequent phase we do expect to observe disturbance westward winds hitting the equatorial latitudes. In this case such winds could retard or reverse to westward the solar thermal tide induced eastward wind that is basically responsible for the generation of the quiet time PRE. The reduced eastward (or the disturbance westward) wind could significantly reduce/suppress the development of the PRE leading to a more intense negative effect than could be accounted for by the original global DD electric field alone. Both these contributions seems to be present in the disturbance electric field modeled by *Richmond et al.* [2003] which produces a disturbance evening westward electric field of significantly larger intensity (Figure 4b) than that modeled by *Blanc and Richmond* [1980]. Thus the seasonally dependent negative effect on the evening drift could arise from a seasonal dependence of the equatorial disturbance wind effect as well as the global DD electric field (that is, of the mid-latitude disturbance winds).

One of the important questions related to the DD electric field process concerns the possible longitudinal variation in its intensity during its onset and evolution phases. Thermospheric disturbance circulation cells with restricted longitudinal extension, the local time dependence of the high latitude energy depositions, and a non-uniform ionospheric conductivity distribution may lead to a longitudinally dependent DD electric field distribution. Some evidence for the possible existence of longitudinal differences in the onset time of the DD electric field was presented by *Abdu et al.* [1997] from analysis of ionosonde and HF radar data from simultaneous observations conducted in the Brazilian and Indian longitude sectors. Some new results on the presence of longitudinal effects within the South American longitude sector are presented in Plate 6 which shows the ionospheric response to the intense magnetic storm of Oct 29-31, 2003 as observed by the digisondes at Jicamarca (Peru) and Sao Luis (Brazil). The one minute values of the ASY-H and SYM-H indices are plotted in the top panel. The F region peak height hmF2 for the two locations are plotted in the lower two panels. Starting from a few hours after the storm onset at 0610 UT we may observe an interplay of superimposed PP and DD electric fields associated with the AE (ASY-H) activity of which we will not discuss here. Instead we focus on the dominant DD electric field manifestations evident during the recovery (or weak AE activity) periods such as the post-midnight/pre-sunrise hours of October 30 and 31 and the evening of October 31. Over Sao Luis, the onset of an eastward DD electric field during the night of Oct. 29-30 is clearly indicated by the rapid rise of the hmF2 (vertical line 1) starting at ~23 LT (from a descending trend that preceded it). Under this initially strong eastward electric field followed by a weak westward electric field the F layer height was maintained high until the sunrise related rapid decrease sets in by 05 LT. An exactly similar height variation but displaced forward by

2 hours is present in the Jicamarca data. (This displacement in UT is attributed to the 2 hours difference in local time between Jicamarca and Sao Luis). We may note that the intensity of the eastward electric field and the hence hmF2 values are perceivably larger over Sao Luis than over Jicamarca. The DD field continued to be westward after 06 LT at both locations. The DD electric field on the next day presents similar features although in this case the midnight start of the eastward polarity over Sao Luis is masked by the effect of an unusually large east-west PP electric field that occurred at the time of the PRE and lasted for 2-3 hours interval. During the pre sun-rise hours of this day the DD electric field appears to be more intense over Jicamarca than it was over Sao Luis. It is to be noted that in both cases the layer uplift under the DD eastward electric field caused the development of plasma bubble irregularities over both locations as indicated by 'ESF' (equatorial spread F) in the figure. The plasma bubble irregularity development under the post-midnight eastward DD electric field, as happened in this case, may not always occur since such development conditions will depend also upon the intensity of possible disturbance meridional/transe equatorial winds that could accompany the DD electric field. Cases of bubble development inhibition by such winds during otherwise propitious conditions (offered by a DD eastward electric field) have been presented by *Abdu et al.* [1997]. Coming to the evening sector of October 31 we note that the DD electric field effect in suppressing/reducing the PRE was present over Sao Luis and not over Jicamarca. Thus a considerable degree of longitudinal variation in the DD electric field intensity is found to be present in the South American longitude sector which is examined here.

The factors that contribute to the longitudinal dependence of the DD electric field are not clearly known at present. Longitudinal structures in the disturbance wind field as well as in the ionospheric conductivity distribution should play important roles in the disturbance dynamo mechanism of *Blanc and Richmond* [1980] and *Richmond et al.* [2003]. Little is known about possible longitudinal features in the disturbance wind field that can be related to the observed features of the DD electric field effects. On the other hand there is compelling evidence for the existence of large scale longitudinal (and latitudinal) structures in the form of horizontal gradients in the ionospheric conductivity distribution during magnetic storms in the South American longitude sector due to the South Atlantic Magnetic Anomaly (SAMA). It is well known that the energetic particle precipitation is a regular source of ionization in the SAMA region [*Abdu and Batista*, 1977] and significant enhancement in this ionization occurs during magnetic storms [*Batista and Abdu*, 1977]. The effects of the enhanced conductivity on equatorial electric fields and electrodynamics processes have been the subject of some recent investigations [*Abdu et al.*, 1998, 2003a, 2005, *Lin and Yeh*, 2005]. In Plate 2 we have demarcated in the southern hemisphere a region identified as SAMA wherein conductivity enhancement with large scale horizontal gradients must be occurring during magnetic storms. In this region the poleward electric field produced by the winds of the disturbance dynamo mechanism could get enhanced/modified to cause corresponding changes in the DD electric field. The precise location and the horizontal extension/size of the enhanced conductivity zone within the general geographic area of South America in relation to the central region of the SAMA could vary from one storm to another depending upon the type of the precipitating energetic particles (electrons or ions) and their energy spectra. (The point of the weakest total geomagnetic field intensity that defines the centre of the SAMA is presently located in southern part of Brazil (at ~25°S, 52°W) as a consequence of its well known secular westward drift according the IGRF model). Thus a certain local enhancement of the DD electric field could take place in the general SAMA region whereby different intensities of the DD electric field are generated at Jicamarca and Sao Luis as observed in Plate 6.

5. SUMMARY AND CONCLUSIONS

We have presented a discussion, and some new results, on the different aspects of the disturbance dynamo electric field that dominates the equatorial and low latitude regions following magnetic storms. These electric fields appear at equatorial latitudes within a few hours after the onset of a storm that initiates energy deposition in the high latitude ionosphere-thermosphere system. The energy deposition occurring during the growth and development phases of storms/substorms, causes atmospheric heating and upwelling leading to the equatorward transfer of momentum and energy through disturbance winds. The dynamo electric field generated by the disturbance winds have longer duration than the (under-shielded) electric fields that promptly penetrate to equatorial latitudes during the storm growth/development phases, and they persist longer into the post-recovery of the storm. The basic theory of generation of this electric field seems to be well accepted and offers a good basis for operating models. The simulation results explain reasonably well important features of the diurnal pattern of the DD electric field. Such features also agree well with the empirical drift model built on extensive Jicamarca radar and satellite data sets. The electric field polarities, eastward or westward, oppose those of the normal dynamo electric field produced by solar heating, as well as that of the penetrating (under-shielded) polar electric fields. An attempt was made to evaluate the DD electric field effects associated with storms driven by corotating solar wind stream for which the high latitude energy deposition continues for significantly longer periods than for storms driven by CME's. It was found that generally the disturbance electric fields during the former class of storms are weaker, but longer lasting (as is the associated energy deposition process), than during the latter class of storms. The DD electric fields can modify in a significant way the major phenomenology of the equatorial/low latitude ionosphere in such a way that the developments of the phenomena are inhibited during the times of their normal occurrences, but new effects can occur at other times, of which we have presented and discussed a few examples: the equatorial electrojet and its associated instability processes, mainly daytime phenomena, are inhibited by a westward electric field; the evening prereversal electric field development is inhibited to varying degrees as a result of the complementary effects from a global DD westward electric field and the equatorial disturbance winds that are also westward, in the evening; the equatorial ionization anomaly development gets inhibited by a westward electric field during the day and post-sunset hours, bun it can also get enhanced under the eastward electric field of the post- midnight hours although not discussed in this paper; the post-midnight DD electric field of eastward polarity is often responsible for the generation of spread-F plasma bubble irregularities, although some cases of their possible inhibition by the simultaneous presence of a disturbance meridional wind have also been pointed out. Seasonal variations in the intensity of the evening/post-sunset westward electric field has been identified over Jicamarca. Such dependence at other local times needs to be determined. There are also indications for a solar flux dependence of the DD electric field in the Brazilian longitude sector, the intensity being higher, and duration being longer, for a storm during higher solar flux conditions than for one which occurs during lower solar flux conditions, even when the intensity of the high latitude energy input for the storm occurring during lower solar flux conditions is higher. More case and statistical studies need to be conducted to quantify such dependences. The DD electric field appears to exhibit significant degree of longitude dependence, which, in the South American longitude sector, could be influenced by the storm associated ionospheric conductivity modification caused by particle precipitation in the SAMA region. There is a need for collecting/analyzing more observational data on the DD electric field and improve upon the existing theory, modeling and simulation schemes in order to improve further our understanding of the relevant processes, and especially, those related to possible local time/longitude dependent delay times, the observed seasonal, solar flux and longitude dependences, of the DD electric field.

Acknowledgments. The authors wish to acknowledge the supports from the Fundação de Amparo a Pesquisa do Estado de São Paulo- FAPESP through the project 1999/00437-0, Conselho Nacional de Desenvolvimento Cientifico e Tecnologico- CNPq through grants n° 502804/2004-1. We thank the Jicamarca Radio Observatory for the use of their Digisonde data that we have retrieved from the UML DIDBase. We thank John MacDougall of the University of Western Ontario, London, Canada for useful discussion, and Maria Goreti dos Santos Aquino for assistance in processing the digisonde data and in preparing some of the Figures.

REFERENCES

Abdu, M.A., Major phenomena of the equatorial ionosphere-thermosphere system under disturbed conditions, *J. Atmos. Solar-Terr. Phys.* 59, 1505-1519, 1997.

Abdu, M.A., Coupling and energetics of the equatorial ionosphere-thermosphere system: advances during the STEP period, *J. Atmos. Solar-Terr. Physics*, 61, 153-165, 1999.

Abdu, M.A., and I.S. Batista, Sporadic E layer phenomena in the Brazilian geomagnetic anomaly: Evidence for a regular particle ionization source, *J. Atmos. Terr. Phys.*, 39, 723-731, 1977.

Abdu, M.A., I.S. Batista, G.O. Walker, J.H.A. Sobral, N.B. Trivedi, and E.R. de Paula, Equatorial ionospheric electric fields during magnetospheric disturbances: local time/longitudinal dependencies from recent EITS campaigns, *J. Atmos. Terr. Phys.* 57, 1065-1083, 1995.

Abdu, M.A., J.H. Sastri, J. MacDougall, I.S. Batista, and J.H.A. Sobral, Equatorial disturbance dynamo electric field longitudinal structure and spread F: a case study from GUARA/EITS campaigns, *Geophys. Res. Lett.*, 24, 1707-1710, 1997.

Abdu, M.A., P.T. Jayachandran, J. Mac Dougall, J.F. Cecile, J.H.A. Sobral, Equatorial F-Region Zonal Plasma Irregularity Drifts Under Magnetospheric Disturbances, *Geophys. Res. Lett.*, 25, 4137-4140, 1998.

Abdu, M.A., I.S. Batista, H. Takahashi, J. MacDougall, J.H. Sobral, A.F. Medeiros, and N.B. Trivedi, Magnetospheric Disturbance Induced Equatorial Plasma bubble Development and Dynamics: A Case Study in Brazilian Sector. *J. Geophys. Res.*, 108, (A12), 1449, doi:10.1029/2002JA009721, 2003a.

Abdu, M.A., J. MacDougall, I.S. Batista, J.H.A. Sobral, and P.T. Jaychandran, Equatorial evening prereversal electric field enhancement and sporadic E layer disruption: A manifestation of E and F region coupling, *J. Geophys. Res.*, 108(A6), 1254, doi: 10.1029/2002JA009285, 2003b.

Abdu, M.A., C.M. Denardini, J.H.A. Sobral, I.S. Batista, P. Muralikrishna, K.N. Iyer, O. Veliz, E.R. De Paula, Equatorial electrojet 3-m irregularity dynamics during magnetic disturbances over Brazil: results from the new VHF radar at São Luís, *J. Atmos. Terr. Phys.*, Vol. 65, Nos. 14-15, 1293-1309, 2003, September-October 2003c.

Abdu, M.A., I.S. Batista, A.J. Carrasco, and C.G.M. Brum, South Atlantic magnetic anomaly ionization: a review and a new focus on electrodynamic effects in the equatorial ionosphere, *J. Atmos. Solar-Terrest. Phys.*, 67, 1643-1657, 2005.

Abdu, M.A., T.K. Ramkumar, I.S. Batista, C.G.M. Brum, H. Takahashi, B.W. Reinisch, J.H.A. Sobral, Planetary wave signatures in the equatorial atmosphere-ionosphere system, and Mesosphere- E- and F- region coupling, *J. Atmos. Solar-Terrest. Phys.*, 68, 509-522, 2006.

Balan, N., G.J. Bailey, M.A. Abdu, K.I. Oyama, P.G. Richards, J. MacDougall, and I.S. Bastista, The equatorial plasma fountain and its effects over three locations: evidence for an additional layer, *J. Geophys. Res.*, 102, 2047-2056, 1997.

Basu, S., Su. Basu, K.M. Groves, H.-C. Yeh, S.-Y. Su, F.J. Rich, P.J. Sultan, and M.J. Keskinen, Response of the equatorial ionosphere in the South Atlantic region to the great magnetic storm of July 15, 2000, *Geophys. Res. Lett.*, 18, 3577-3580, 2001.

Batista, I.S., and M.A. Abdu, Magnetic storm associated delayed sporadic E layer enhancement in the Brazilian Geomagnetic Anomaly, *J. Geophys. Res.*, 82, 4777-4783, 1977.

Batista, I.S., E.R. de Paula, M.A. Abdu, and N.B. Trivedi, Ionospheric effects of the March 13, 1989 magnetic storm at low latitudes, *J. Geophys. Res.*, 96, 13,943-13,952, 1991.

Blanc, M., and A.D. Richmond, The ionospheric disturbance dynamo, *J. Geophys. Res.*, 85, 1669-1699, 1980.

Fejer, B.G., Low latitude storm time ionospheric electrodynamics, *J. Atmos. Solar- Terr. Phys.*, 64, 1401-1408, 2002.

Fejer, B.G., C.A. Gonzales, D.T. Farley, M.C. Kelley, and R.F. Woodman, Equatorial electric fields during magnetically disturbed conditions, 1, the effect of the interplanetary magnetic field, *J. Geophys. Res.*, 84, 5797-5802, 1979.

Fejer B.G., M.F. Larsen, and D.T. Farley, Equatorial disturbance dynamo electric fields, *Geophys. Res. Lett.*, 10, 537-540, 1983.

Fejer, B.G., M.C. Kelley, C.D. Senior, O. de La Beaujardiere, J.A. Holt, C.A. Tepley, R.G. Burnside, M.A. Abdu, J.H.A. Sobral, R.F. Woodman, Y. Kamide, and R. Lepping, "Low And Midlatitude Ionospheric Electric Fields During The January 1984 Gismos Campaign, *J. Geophys. Res.*, 85(A3), 2367-2377, 1990.

Fejer, Bela, G., and L. Scherliess, Time dependent response of equatorial ionospheric electric fields to magnetospheric disturbances, *Geophys. Res. Lett.*, 22, 851-854, 1995.

Fejer, B.G., J.T. Emmert, Low-latitude ionospheric disturbance electric field effects during the recovery phase of the 19-21 October 1998 magnetic storm, *J. Geophys. Res.*, 108(A12), 1454, doi:10.1029/2003JA010190, 2003.

Forbes, J.M., The equatorial electrojet, *Rev. Geophys. and Space Physics*, 19(3), 469-504, 1981.

Fesen, C.G., Crowley, R.G. Roble, A.D. Richmond, and B.G. Fejer, Simulation of the pre-reversal enhancement in the low latitude vertical ion drifts, *Geophys. Res. Lett.*, 27, 1851-1854, 2000.

Fuller-Rowell, T.J., M.V. Codrescu, H. Rishbeth, R.J. Moffet, and S. Quegan, On the seasonal response of the thermosphere and ionosphere to geomagnetic storms, *J. Geophys. Res.*, 101, 2343-2353, 1996.

Gonzales, C.A., M.C. Kelley, B.G. Fejer, J.F. Vickrey, and R.F. Woodman, Equatorial electric fields during magnetically disturbed conditions, II, Implications of simultaneous auroral and equatorial measurements, *J. Geophys. Res.*, 84(A10), 5803-5812, 1979.

Greenspan, M.E., C.E. Rasmussen, W.J. Burke, and M.A. Abdu, Equatorial density depletions observed at 840 km during the great storm of March 1989, *J. Geophys. Res.*, 96, 13,931-13,942, 1991.

Heelis, R.A., P.C. Kendall, R.J. Moffet, D.W. Windle, and H. Rishbeth, Electrical coupling of the E- and F- region and its effects on the F-region drifts and winds, *Planet. Space Sci.*, 22, 743-756, 1974.

Iyemore T., Araki, T., Kamei T., and Takeda M, Mid-Latitude Geomagnetic Indices ASY and SYM, Report No.2, Kyoto University, 1993.

Kelley, M.C., B.G. Fejer, and C.A. Gonzales, An explanation for anomalous ionospheric electric fields associated with a northward turning of the interplanetary magnetic field, *Geophys. Res. Lett.*, 6(4), 301-304, 1979.

Kikuchi, T., H. L,hr, K. Schlegel, H. Tachihara, M. Shinohara and T.-I. Kitamura, Penetration of auroral electric fields to the equator during a substorm, *J. Geophys. Res.*, 105, 23251-23261, 2000.

Lin, C.S., and H.C. Yeh, Satellite observations of electric fields in the South Atlantic anomaly region during the July 2000 magnetic storm, *J. Geophys. Res.*, 110, A03305, doi: 10.1029/2003JA010215, 2005.

Mannucci, A.J., B.T. Tsurutani, B.A. Ijima, A. Komjathy, A Saito, W.D. Gonzalez, F.L. Guarnieri, J.U. Kozyra, and R. Skoug, *Geophy. Res. Letts.*, 32, L12S02, doi: 10.1029/2004GL021467, 2005.

Maruyama, T., Ma, G.Y., Nakamura, M., Signature of TEC storm on 6 November 2001 derived from dense receiver network and ionosonde chain over Japan, *J. Geophy. Res.*, 109, (A10) Art No., A10302 Oct 15 2004.

Mazaudier, C.A., and S.V. Venkateswaran, Delayed ionospheric effects of the geomagnetic storms of March 22, 1979 studied by the sixth coordinated data analysis workshop (CDAW-6), *Ann. Geophys.*, 8, 511-518, 1990.

Nishida, A., N. Iwasaki, and T. Nagata, The origin of fluctuations in the equatorial electrojet: A new type of geomagnetic variations, *Ann. Geophys.*, 22, 478-484, 1966.

Peymirat, C., and D. Fontaine, Numerical simulation of magnetospheric convection including the effect of field-aligned currents and electron precipitation, *J. Geophys. Res.*, 99, 11,155-11,176, 1994.

Peymirat, C., A.D. Richmond, B.A. Emery, and R.G. Roble, A magnetosphere-thermosphere-ionosphere electrodynamics general circulation model, *J. Geophys. Res.*, 103, 17,467-17,477, 1998.

Pincheira, X.T., M.A. Abdu, I.S. Batista, and P. Richards, Na investigation of ionospheric response and disturbance winds, during magnetic storms over South American sector, *J. Geophys. Res.*, 107, A11, 1379, doi: 10.1029/2001JA000263, 2002.

Prölss, G.W., Ionospheric F-region storms, in *Handbook of Atmospheric Electrodynamics*, vol. 2, edited by H. Volland, pp. 195-248, CRC Press, Boca Raton, Fla., 1995.

Richmond, A.D., E.C. Ridley, and R.G. Roble, A thermosphere/ionosphere general circulation model with coupled electrodynamics, *Geophys. Res. Lett.*, 19, 601-604, 1992.

Richmond, A.D., and G. Lu, Upper-atmospheric effects of magnetic storms: A brief tutorial, *J. Atmos. Solar- Terr. Phys.*, 62, 1115-1127, 2000.

Richmond, A.D., C. Peymirat, R.G. Roble, Long-lasting disturbances in the equatorial ionospheric electric field simulated with a coupled magnetosphere-ionosphere-thermosphere model, *J. Geophys. Res.*, 108, A3, 1118, doi: 10.1029/2002JA009758, 2003.

Rishbeth, H., Polarization fields produced by winds in the equatorial F region, *Planet. Space Sci.*, 19, 357-369, 1971.

Rishbeth,. H., Thermospheric winds and the F-region: a review, *J. Atmos. Terr. Phys.*, 34, 1-47, 1972.

Sastri, J.H., Equatorial electric fields of ionospheric disturbance dynamo origin, *Ann. Geophys.*, 6, 635-642, 1988.

Sastri, J.H., R. Sridharan, and T.K. Pant, Equatorial ionosphere-thermosphere system during geomagnetic storms, in "Disturbance in Geophysics: The Storm-Substorm Relationship", *Geophyscal Monograph*, AGU 10.1029/142GM16, 2003.

Scherliess, L., and B.G. Fejer, Storm time dependence of equatorial disturbance dynamo zonal electric fields, *J. Geophys. Res.*, 102, 24037-24046, 1997.

Scherliess, L., and B.G. Fejer, Radar and satellite global equatorial F-region vertical drift model, *J. Geophys. Res.*, 104, 6829-6842, 1999.

Spiro, R.W., R.A. Wolf, and B.G. Fejer, Penetration of high latitude electric field effects to low latitudes during SUNDIAL 1984, *Ann. Geophys.*, 6, 39-50, 1988.

Sobral, J.H. A., M.A. Abdu, W.D. Gonzalez, I. Batista, and A.L. Clua de Gonzalez, Low-latitude ionospheric response during intense magnetic storms at solar maximum, *J. Geophys. Res.*, 102, 14,305-14,313, 1997.

Tsurutani, B.T., and W.D. Gonzalez, The cause of high-intensity long-duration continuous AE activity (HILDCAAs): interplanetary Alfvén wave trains, *Planet. Space Sci.*, 35, 405-412, 1987.

Tsurutani, B.T., A. Mannucci, B. Iijima, M.A. Abdu, J.H.A. Sobral, W.D. Gonzalez, F.L. Guarnieri, T. Tsuda, A. Saito, K. Yumoto, B.G. Fejer, T. Fuller Rowell, J.U.O. Kozyra, J.C. Foster, and A. Coster, V.M. Vasyliunas, Global Dayside Ionospheric Uplift And Enhancement Associated With Interplanetary Electric Fields, *J. Geophy. Res.*, 109, A08302, doi: 10.1029/2003JA010342, 2004.

Vasyliunas, V.M., Mathematical models of magnetospheric convection and its coupling to the ionosphere, in *Particles and Fields in the Magnetosphere*, edited by B.M. McCormack, pp. 60-71, D. Reidel, Norwell, Mass., 1970.

M.A. Abdu, J.R. de Souza, J.H.A. Sobral, and I.S. Batista, Instituto Nacional de Pesquisas Espaciais, Ave dos Astronautas 1758, 12245 970 São Jose dos Campos, SP, Brazil. (maabdu@dae.inpe.br)

Selected Upper Atmospheric Storm Effects

Gerd W. Prölss

Institut für Astrophysik und Extraterrestrische Forschung, Universität Bonn, Bonn, Germany

In the polar ionosphere, a prominent increase in the electron temperature is observed beneath the magnetospheric cleft. During geomagnetic storms this temperature peak is displaced towards lower latitudes, sometimes by as much as 18° (and possibly more). This displacement is attributed to the reconfiguration of the magnetosphere during magnetic substorms. The magnitude of this temperature enhancement is only weakly dependent on the level of geomagnetic activity.

At middle latitudes, a significant increase in the electron temperature is observed during geomagnetic storms. This is attributed to a general decrease in the electron density. Thus a close linear correlation exists between both quantities, with the rate of temperature increase possibly depending on height and/or level of solar activity. The decrease in the electron density, in turn, is caused by changes in the neutral gas composition and, specifically, by a decrease in the atomic oxygen to molecular nitrogen density ratio. A simple estimate indicates that the relative changes in both parameters should be of the same magnitude, and this is indeed observed.

The same density decreases encountered by satellites in the topside ionosphere are also observed by ground-based ionosondes. Their measurements indicate that larger negative ionospheric storms are initiated in the night sector of the Earth. Such ground-based measurements also allow us to search for differences in the ionospheric response to an isolated short-duration geomagnetic storm and recurrent high-intensity long-duration continuous auroral activity (HILDCAA). As it turns out, the same kind of ionospheric disturbance effects are observed during both events, with the only difference being that during HILDCAA these perturbations last much longer.

Ionospheric holes are one of the most spectacular disturbance effects observed at equatorial latitudes. These holes are marked by a steep drop in the electron density to very low values. Also their bottom is rather flat and almost without any structure. Different explanations of this phenomenon have been offered, none of which is generally accepted. Evidently, more comprehensive measurements are needed. This also applies to many other poorly understood aspects of upper atmospheric storms.

1. INTRODUCTION

By 'upper atmospheric storms' we mean global perturbations of the outer gas envelope of the Earth above about 100 km altitude. They are caused by an increased dissipation of solar wind energy in the space environment of the Earth.

If only the neutral upper atmosphere is considered, these perturbations are also called 'thermospheric storms'; in the case of the ionized component, the term 'ionospheric storm' is commonly used. Upper atmospheric storms are accompanied by winds and drifts of high velocity, which is in agreement with the narrower meteorological definition. However, they also exhibit large changes in other state parameters like temperature and density, and these are the subject of the present contribution.

The first to explicitly report on magnetic storm associated perturbations of the Kennelly-Heaviside layer (as the ionosphere was called at that time) were *Hafstad and Tuve* [1929]. Thirty years later *Jacchia* [1959] discovered storm-induced changes also in the thermosphere. Since then a large number of papers have been published in this field, and the author's collection comprises almost 1000 articles. In this situation review articles like those by *Abdu et al.* [1991], *Prölss* [1995, 1997], *Fuller-Rowell et al.* [1997], *Buonsanto* [1999], *Mikhailov* [2000], *Richmond and Lu* [2000], *Förster and Jakowski* [2000], *Danilov and Lastovicka* [2001], *Schlegel* [2005], and *Prölss* [2005a] become increasingly important.

Of the many different aspects of upper atmospheric storms, only a few can be discussed in what follows. In our selection we proceed from polar to equatorial latitudes. First, ionospheric heating effects beneath the magnetospheric cleft are investigated (Section 2). Next, perturbations observed at middle latitudes, like changes in the electron temperature and density and in the neutral gas composition, are discussed (Section 3).

In this context we also search for differences in the ionospheric response to isolated short-duration and recurrent long-duration geomagnetic storms (Section 4). Finally, ionospheric holes at equatorial latitudes are described (Section 5). The main results of this study are summarized in Section 6.

2. IONOSPHERIC STORM EFFECTS IN THE CLEFT REGION

The cleft (or cusp) represents an important topological feature of the terrestrial magnetosphere. Here magnetic field lines separate, extending to different parts of the magnetosphere; see Figure 1. Because of this special topology, plasma from the magnetosheath and the magnetospheric boundary layers (low-latitude boundary layer, plasma mantle) have direct access to the dayside polar upper atmosphere. Accordingly, large numbers of low energy electrons are observed to precipitate in this region [e.g., *Burch*, 1968; *Heikkila and Winningham*, 1971; *Newell and Meng*, 1992; and references therein].

This soft electron precipitation is responsible for a significant heating of the electron gas in the upper ionosphere, leading to a prominent peak in the electron temperature; see Figure 2. A sequence of latitudinal profiles of the electron temperature is shown as observed in the dayside polar ionosphere. The sudden increase in the electron temperature at high latitudes is a clear signature of the cleft precipitation. In the following discussion we are interested in how this heating

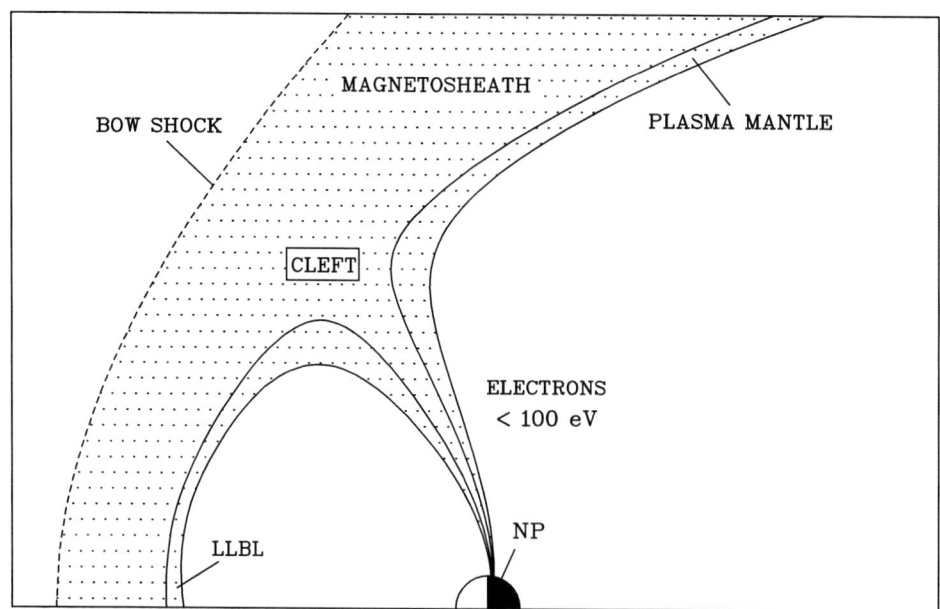

Figure 1. Topology of the magnetosphere in the cleft region and associated plasma populations. The Tsyganenko model T96-1 was used to calculate the field lines for winter solstice conditions in the northern hemisphere. The footpoints of the field lines are located at 70°, 71.5°, 73.5° and 75° magnetic invariant latitude. NP stands for (geographic) northpole, LLBL for low-latitude boundary layer.

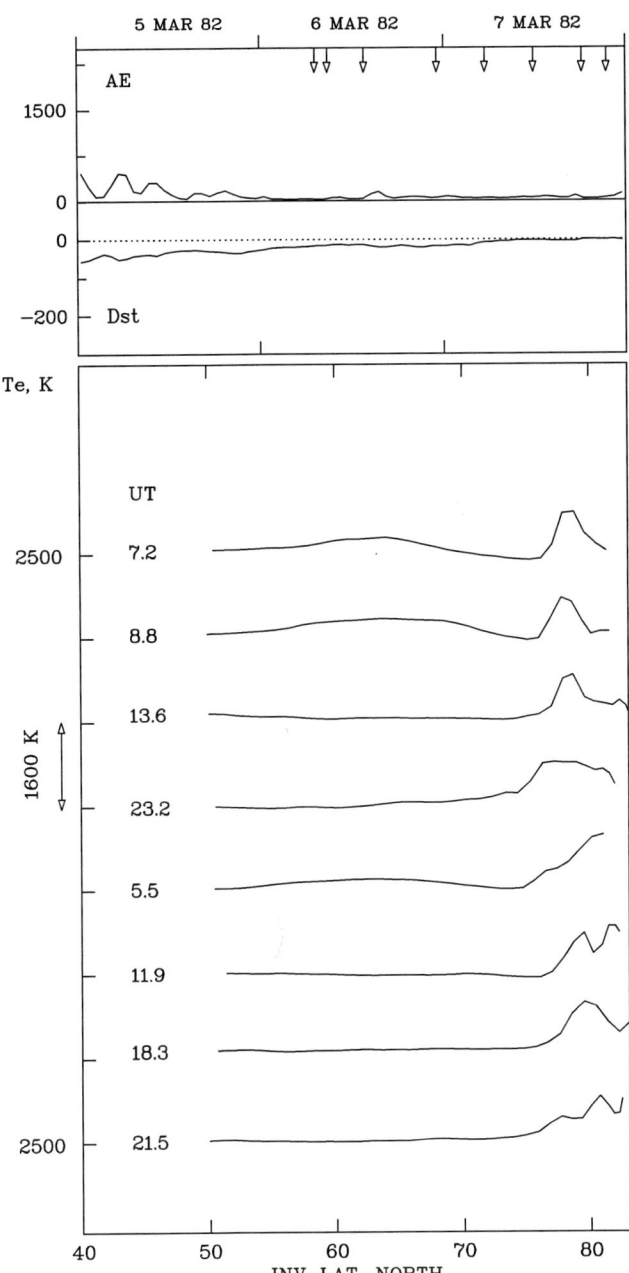

Figure 2. Latitudinal variation of the electron temperature beneath the cleft region during magnetically quiet conditions. In the upper part of the figure, the AE and Dst indices are used to describe the geomagnetic activity during three days in March 1982. During this time eight latitudinal profiles of the electron temperature were obtained by the DE-2 satellite. These are shown in the form of a stack plot in the lower part of the figure. The times of observation are indicated next to each profile, and also by arrows in the upper part of the figure. For each profile a tick mark indicates the 2500 K level, and the distance between the profiles corresponds to 1600 K. Using a constant temperature gradient of 2.5 K/km, all temperature measurements have been adjusted to a common altitude of 380 km.

effect changes during disturbed conditions. The first to investigate this question was *Titheridge* [1976]. He was able to demonstrate that during geomagnetic storms the increase in the electron temperature is displaced towards lower latitudes. However, the temperatures he used were not directly measured but inferred from topside ionograms obtained by the Alouette-1 satellite. Also, the spatial resolution of his data (≥4°) was too low to resolve the latitudinal structure of the heating effect. This prompted the present author to repeat the study of Titheridge, this time, however, using directly measured electron temperatures with better spatial resolution [*Prölss*, 2005b]. A general description of the DE-2 data set used in that study is given in *Hoffman and Schmerling* [1981], *Krehbiel et al.* [1981], *Carignan et al.* [1981], and *Hanson et al.* [1981]. This data set is also used in the present study.

A few days before the data shown in Figure 2 were obtained, a larger geomagnetic storm occurred. Figure 3 illustrates the ionospheric response to this event. The format of data presentation is the same as that used in Figure 2. If the electron temperature profiles shown in both figures are compared, the most obvious storm-induced change is the displacement of the temperature peaks toward lower latitudes. The magnitude of this displacement is of the order of 10 degrees and more.

This is by no means a singular case, and equatorward shifts of similar magnitude were also observed during the storm of September 6, 1982; see Figure 4. This time the data were recorded in the southern hemisphere.

In order to describe the displacement effect in a more systematic manner, the latitudes of the temperature peaks and their associated AE indices were determined for a large number of satellite passes. Only data obtained in the dayside polar region between 9 and 15 h magnetic local time were considered. The latitudes of the temperature peaks were then sorted into 100 nT wide intervals of the AE index. For each of the intervals, the median and the upper and lower quartiles were determined. These are indicated by the dots and bars in Figure 5. Finally, a regression line was fitted to the medians. It is evident that an excellent linear correlation exists between the AE index and the position of the temperature peak.

For completely quiet geomagnetic conditions (AE = 0), the regression line predicts the temperature peak to be located near 79° invariant latitude. This position is indicated by the dotted lines in Figures 3 and 4. On the other hand, the lowest latitude at which a temperature peak was observed during disturbed conditions was 60.8° invariant latitude; see upper panel of Figure 7.

This reference altitude lies within the range of the actual observation heights. The solar local time of observation is approximately 12 h, the magnetic local time varies between 10 and 14 h. Note that the observations were made in the northern hemisphere.

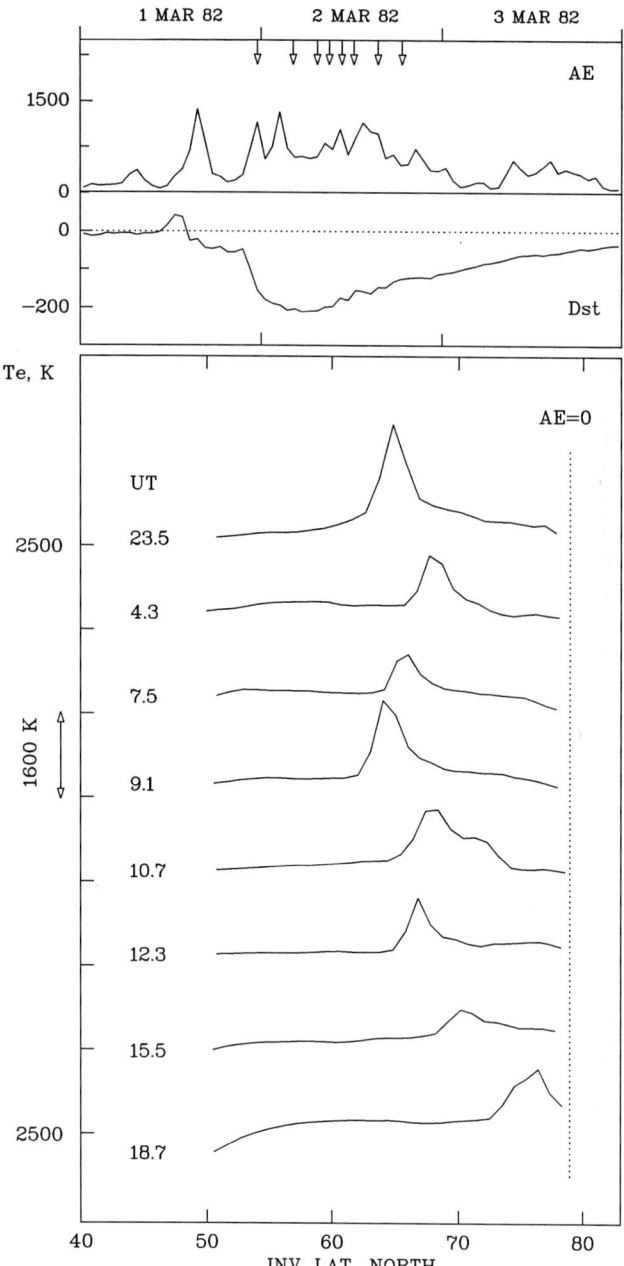

Figure 3. Latitudinal variation of the electron temperature beneath the cleft region during the magnetic storm of March 1/2, 1982. The format of data presentation corresponds to that of Figure 2. A vertical dotted line has been added, indicating the mean position of the temperature peak during very quiet geomagnetic conditions; see also Figure 5.

Returning once more to the data presented in Figures 3 and 4, we note that the cleft-related temperature increases are much larger for the September storm; see the temperature scales indicated by the arrows. This difference is mainly due to the different observation heights, which are approximately 380 km for the March event and 600 km for the September event. As has been shown by *Titheridge* [1976] and *Prölss* [2006], the temperatures increase significantly with increasing altitude, especially within the cleft region.

There is also a tendency for the magnitude of the temperature enhancement to increase with increasing magnetic activity; see Figure 6. This correlation, however, is relatively weak and the scatter of data points substantial. In fact, the largest temperature increase observed occurred during only moderately disturbed conditions; see lower panel of Figure 7.

Even larger temperature peaks are observed if the spatial resolution of the data is increased; see *Fontheim et al.* [1987]. This latter study also demonstrates that the magnitude of the temperature enhancement is primarily controlled by the intensity of the incident electron fluxes in the lowest energy ranges. What, in turn, controls the intensity of these fluxes remains an open question.

3. STORM EFFECTS AT MIDDLE LATITUDES

Upper atmospheric storms are by no means restricted to the polar region. In the following discussion, we are interested in what happens equatorward of the cleft region at middle latitudes. To illustrate these disturbance effects we again use data obtained during the storm event of March 1982. Specifically, we are interested in the satellite pass which took place on March 2 at about 9.1 UT (fourth profile from above in Figure 3). The electron temperatures measured during this pass are shown once more in the uppermost panel of Figure 8, this time, however, over an extended latitudinal range. Also, to make storm-induced changes more discernible, an undisturbed temperature profile has been added. This reference profile was obtained on March 6 at about 8.8 UT; see Figure 2, second profile from above.

Whereas the equatorward shift of the cleft heating has already been discussed, the general increase of the electron temperature, which extends all the way from polar to middle latitudes, is a new disturbance feature. This temperature enhancement is quite substantial and of the same order of magnitude as that observed in the cleft region. How do we explain this heating effect, especially at middle latitudes where no soft electron precipitation is to be expected? The answer to this question is given in the second panel of Figure 8 where the associated electron density measurements are presented. Again the solid line is the stormtime profile, the dashed line the reference profile.

First we note that there is no significant increase in the electron density in the cleft region. This confirms that the cleft heating is produced by *soft* particle precipitation. Next we note the large, storm-induced decrease in the electron density which extends over a wide latitudinal range. Since the onset of this decrease coincides with the rise in the electron temperature, we suspect that both changes are correlated, and this is indeed the case.

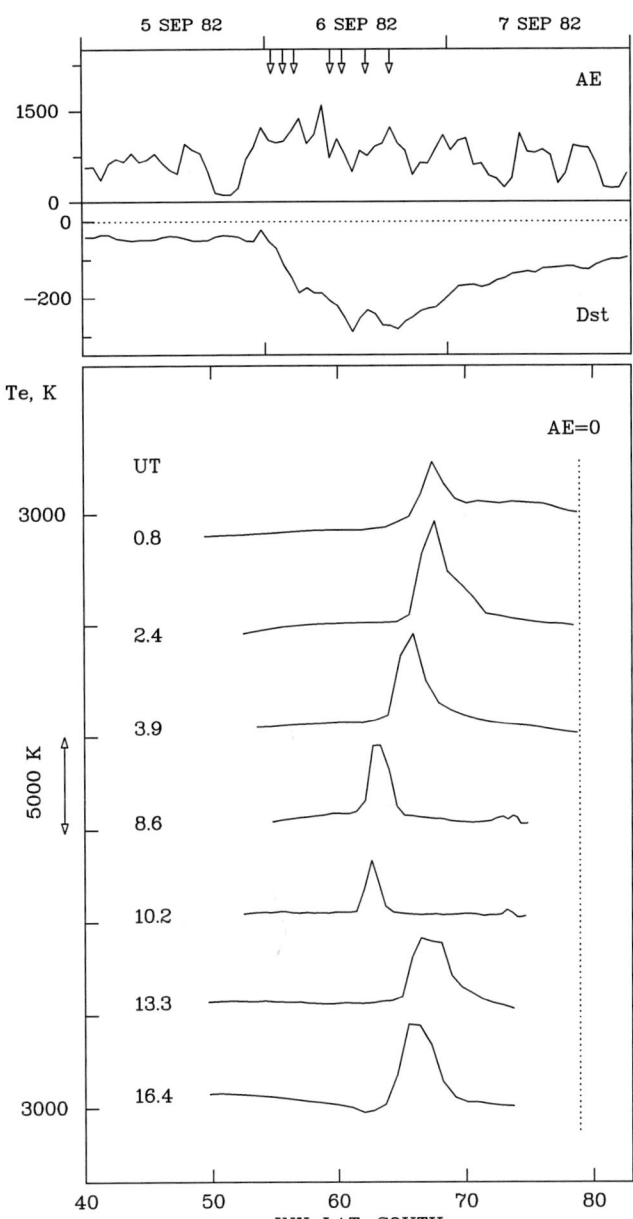

Figure 4. Latitudinal variation of the electron temperature beneath the cleft region during the magnetic storm of September 6, 1982. The general format of data presentation corresponds to that of Figure 3. For each profile, a tick mark indicates the 3000 K level, and the common temperature scale is indicated by an arrow next to the left ordinate. Using a constant temperature gradient of 2.5 K/km, all temperature measurements have been adjusted to a common altitude of 600 km. This reference altitude lies within the range of the actual observation heights. The solar local time of observation is approximately 12 h, the magnetic local time varies between 10 and 12 h. Note that these observations were obtained in the southern hemisphere.

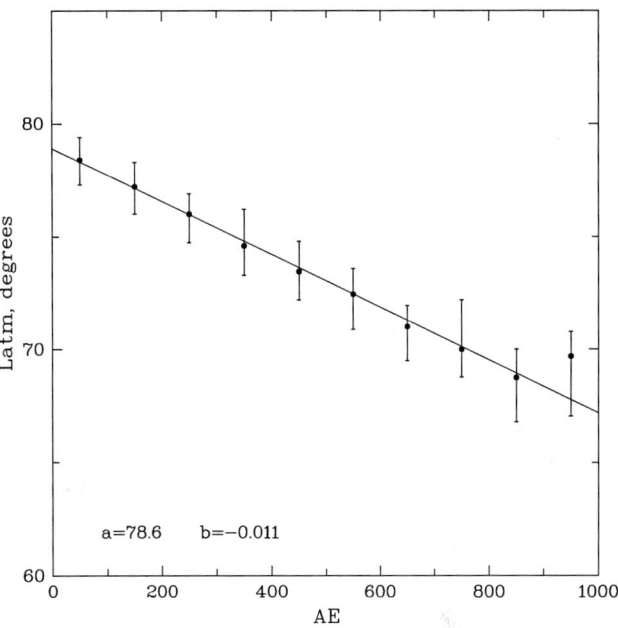

Figure 5. Position of the cleft-associated temperature peak as a function of geomagnetic activity. The ordinate indicates the invariant latitude of the temperature peak, $(Lat)p$, the abscissa the associated AE index. For the AE index, linearly-interpolated, hourly-averaged values were used. Both parameters were determined for a total of 1409 electron temperature profiles. Only data obtained in the 9 to 15 h magnetic local time sector were considered. For each 100 nT wide interval of the AE index, the median and upper and lower quartiles of the latitudes of the temperature peak were determined; these are shown by the dots and bars in the figure. A regression line has been fitted to the medians, and its ordinate intersection a (in K) and slope b (in K/AE) are given in the lower left-hand corner. Note that the mean position of the temperature peak for AE equal zero ($= a$) is indicated by vertical dotted lines in Figures 3 and 4.

In Figure 9 we plotted the electron temperature as a function of the electron density. This plot is based on data obtained during the seven satellite passes shown in Figure 3. Only measurements taken below 60° invariant latitude are considered. The excellent correlation between both quantities is evident.

In an attempt to explain this tight correlation, let us assume that solar radiation deposits a certain amount of heat into the electron gas of the ionosphere. If the density is high, each electron will get only a small portion of this heat, and the electron temperature will be low. If, however, the density decreases, each electron will obtain an increasingly larger share of this energy, and the temperature will rise. As intuitive as this explanation may seem, it would predict an inverse, not a linear relationship between density and temperature. A closer inspection of the associated heat balance equation indicates that complications arise from the complex form of the various heat loss processes. These processes include

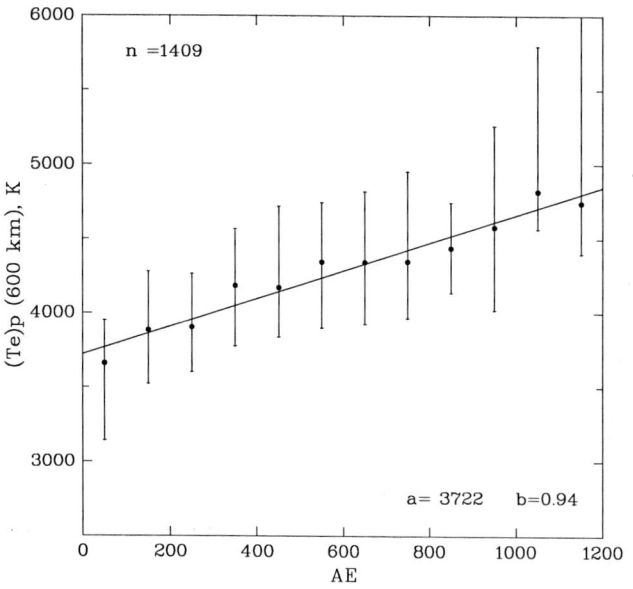

Figure 6. Magnitude of the cleft-related temperature peak as a function of magnetic activity. This figure is based on the same data set used in Figure 5. Also the format of data presentation is the same except that the latitude of the temperature peak is replaced by the magnitude of the temperature peak, $(T_e)p$. A height dependent temperature gradient is used to adjust all temperature values to a common altitude of 600 km [*Prölss*, 2006].

Coulomb collisions with atomic oxygen ions, excitation of fine structure states of neutral oxygen, and vibrational cooling by molecular nitrogen. A quasi-linear relationship between electron temperature and density can be obtained only if all three loss processes are properly taken into account. For a more detailed discussion of this topic see, for example, *Williams and McDonalds* [1987], and references therein.

Given the close correlation between the electron temperature and density, the question remains why the electron density decreases in the first place. The answer is given in the third panel of Figure 8. There we have plotted the atomic oxygen to molecular nitrogen density ratio, O/N_2. As is evident, this density ratio shows a significant decrease in approximately the same latitudinal range as the electron density, indicating that both changes are correlated.

Such a correlation is indeed predicted by theory. In the upper ionosphere, the production of ionization is proportional to the atomic oxygen density, and the loss of ionization increases with increasing molecular nitrogen density. Therefore, any decrease in the O/N_2 density ratio will automatically entail a decrease in the electron density. Moreover, a simple estimate indicates that the *relative* changes in both quantities should be of the same magnitude [*Prölss*, 2004]. This is actually confirmed by the data presented in Figure 10. This figure is based again on measurements obtained at the

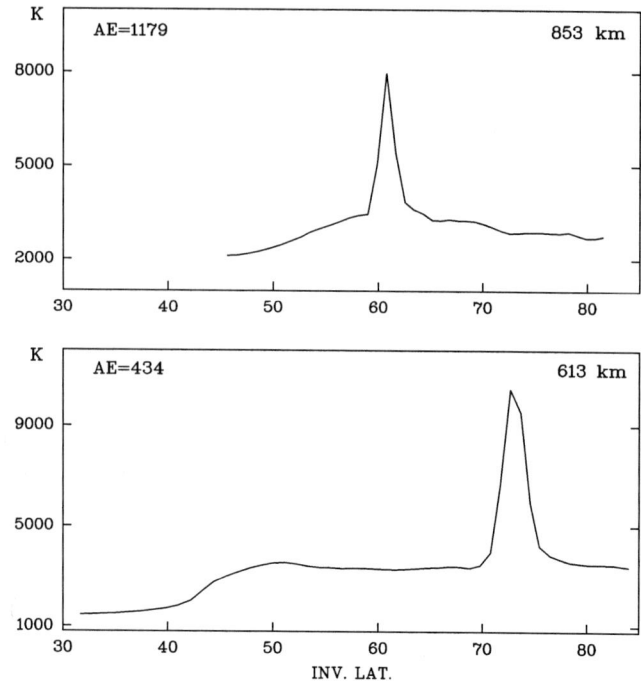

Figure 7. Largest displacement (upper panel) and largest value (lower panel) of the cleft-related temperature peak as identified in the DE-2 data set. The temperature profile shown in the upper panel was obtained on October 14, 1982 at about 2.7 UT, the one shown in the lower panel on September 27, 1982 at about 3.1 UT. In both cases the magnetic local time of observation was close to 10 h.

beginning of March 1982; see Figures 2 and 3. Only data taken below 60° invariant latitude are considered. The good agreement between the data points and the theoretical prediction indicated by the straight line is evident.

Continuing our inquiry, we ask why the O/N_2 density ratio decreases during storms. Answering this question would require a more extensive discussion, which is beyond the scope of this contribution. The interested reader is referred to the pertinent literature in this field [e.g., *Mayr and Volland*, 1972; *Prölss et al.*, 1988; *Burns et al.*, 1991; *Namgaladze et al*, 1996; *Fuller-Rowell et al.*, 1997; *Lu et al.*, 2001; and references therein]. All we want to point out here is that the composition changes observed at middle latitudes are not produced locally. Rather, they originate at higher latitudes and are subsequently transported to middle latitudes. The ion temperature data presented in the lowermost panel of Figure 8 support this claim. Again the solid line describes the storm profile, the dashed line the reference profile.

First we note that there is no correlation between the latitudinal variation of the ion temperature and that of the electron temperature. This is not surprising since the ion temperature is primarily dependent on heating produced by electric currents and *energetic* particle precipitation

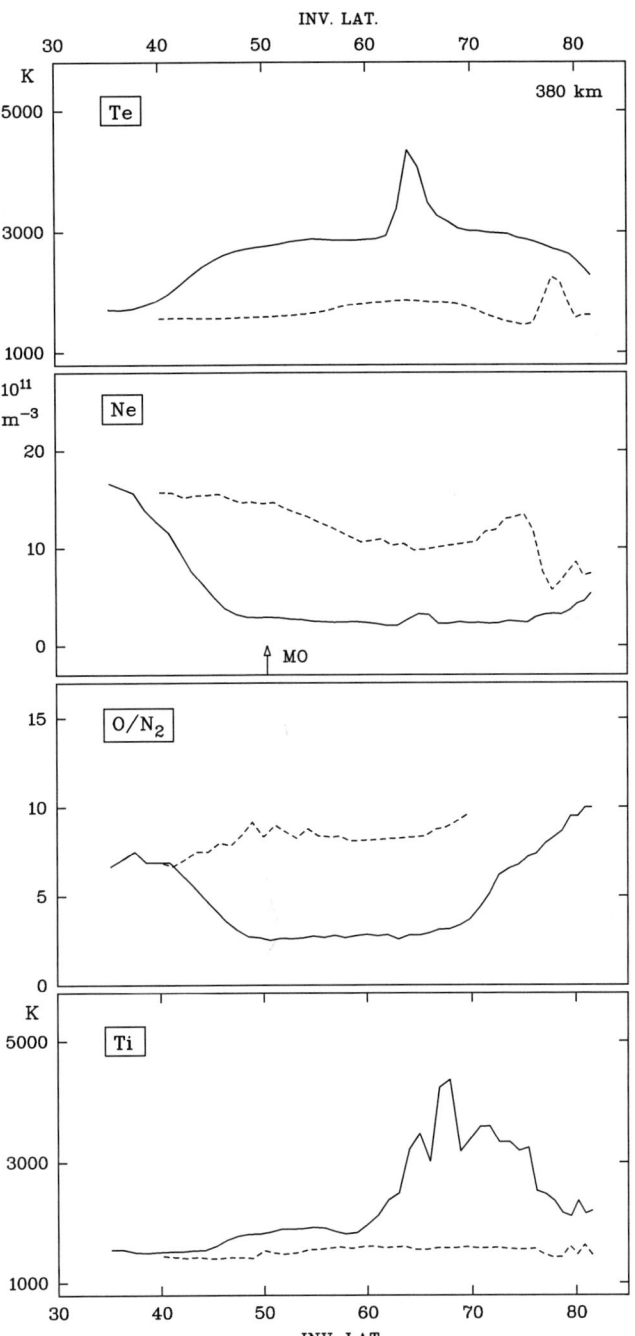

Figure 8. Storm-induced changes in upper atmospheric parameters at high and middle latitudes. The storm data (solid lines) were obtained on March 2, 1982 at about 9.1 UT, the reference data (dashed lines) on March 6, 1982 at about 8.8 UT. The magnetic activity during these two passes is indicated in Figs. 2 and 3. In both cases, the solar and magnetic local times of observation are close to 12 h and 13 h, respectively. Plotted are, from top to bottom, the electron temperature T_e, the electron density N_e, the atomic oxygen to molecular nitrogen density ratio O/N_2, and the ion temperature T_i. The electron temperature and density, and the O/N_2 density ratio have been adjusted to a common altitude of 380 km. For the electron temperature, a constant temperature gradient of 2.5 K/km was assumed; for the other parameters, the measured electron, ion and neutral gas temperatures were used. In the second panel, an arrow indicates the position of the Moscow (MO) ionosonde, which is located close to the subsatellite tracks of the two DE-2 passes considered here.

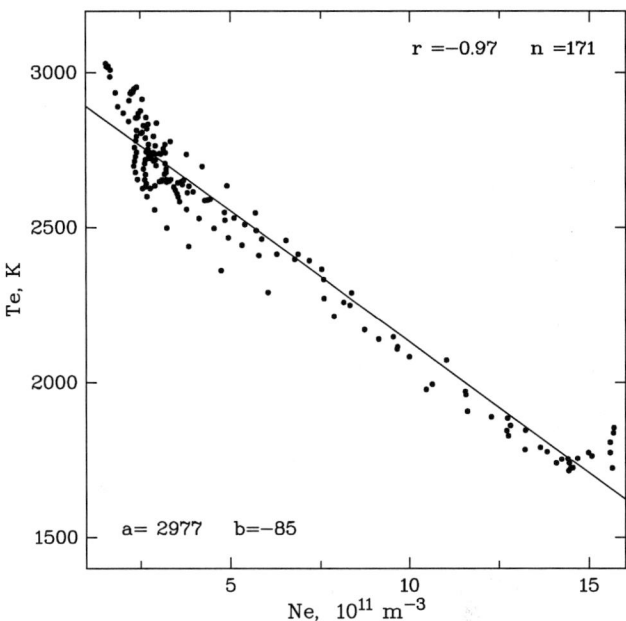

Figure 9. Electron temperature as a function of electron density. The plot is based on data collected during the eight satellite passes shown in Figure 3. Only measurements obtained at middle latitudes (<60° inv. lat.) are considered. Both electron temperature and electron density refer to a common altitude of 380 km. The regression line fitted to the data emphasizes the linear correlation between both quantities. The correlation coefficient r and the number of data points available are given in the upper right-hand corner, the ordinate intersection a and the slope b of the regression line in the lower left-hand corner.

[e.g., *Killeen et al.*, 1984; *Emery et al.*, 1985; *Lu et al.*, 1996]. Now this same energy deposition is also held responsible for the generation of neutral gas composition changes. If the latitudinal profiles of the ion temperature and the O/N_2 density ratio are then compared, it is clear that the composition changes extend way beyond the heating region at high latitudes. Therefore disturbance transport must take place.

In order to find out when this disturbance transport takes place, we turn to ground-based observations. The upper part of Figure 11 describes once more the geomagnetic activity during the March '82 storm, the lower part the ionospheric response to this event. Plotted are the time variations of the maximum electron density in the ionosphere, N_m, as

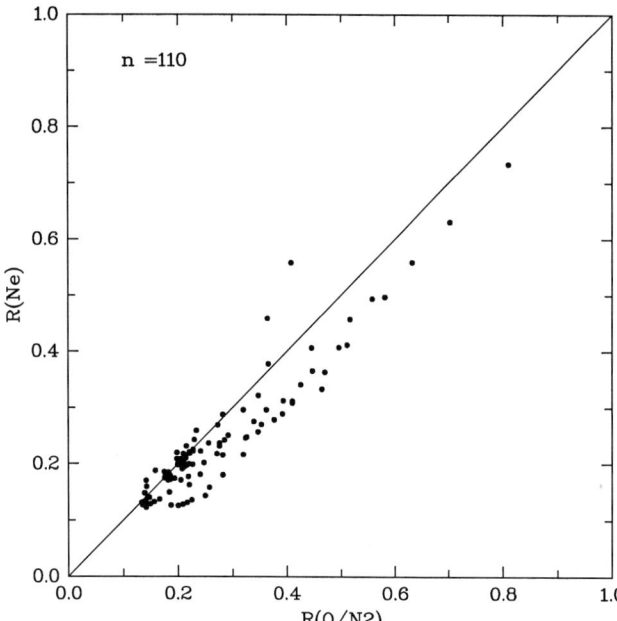

Figure 10. Relative changes in the electron density as a function of relative changes in the atomic oxygen to molecular nitrogen density ratio. The relative changes were calculated using data collected during the satellite passes shown in Figs. 2 and 3. Only measurements obtained at middle latitudes (<60° inv. lat.) are considered. Both electron density and the O/N_2 density ratio refer to a common altitude of 380 km. The straight line represents the prediction of a simple estimate, $R(Ne) \simeq R(O/N_2)$. The number of data points available is given in the upper left-hand corner.

observed at four ionosonde stations. The thin lines without data points describe the undisturbed local time variations of Nm. At sunrise the ionization density increases until in the afternoon a maximum is reached; in the evening and at night the density again decreases. The fatter lines marked by data points indicate how this daily variation changes during storm conditions. Note that the Moscow ionosonde (MO) is located close to the subsatellite tracks of the DE-2 passes, the data of which are presented in Figure 8.

On March 2, a large reduction in the maximum electron density – a so-called negative ionospheric storm – is observed at all four stations. As to be expected, this decrease is of the same order of magnitude as that observed by the DE-2 satellite at this time. However, it is also clear that these negative storm effects begin long before they are eventually observed by this satellite. In fact, they probably set in during the night of March 1 to March 2. This is consistent with current storm models which predict a disturbance transport in the local night sector. Of course transient and more localized disturbance transport may also occur during daytime, as indicated by the perturbations observed on the first storm day [see also *Liou et al.*, 2005].

4. LONG-DURATION GEOMAGNETIC ACTIVITY AND ASSOCIATED IONOSPHERIC PERTURBATIONS

Figure 11 illustrates the typical ionospheric response to an isolated geomagnetic storm of short duration. Smaller and more transient perturbations are observed on the first day; larger and persistent negative storm effects are observed on the second day; and recovery to almost normal conditions is observed on the third day. How does this pattern change if we are not dealing with an isolated storm of short duration but with high-intensity long-duration continuous auroral activity [=HILDCAA; *Tsurutani and Gonzalez*, 1987]? To investigate this question we selected a recurrent HILDCAA event which occurred in spring 1973; see Figure 12.

Five periods of geomagnetic activity are shown. In each panel the upper curve is the AE index, the lower curve the Dst index. Both indices are plotted relative to the same zero line but using different scales. To emphasize periods of larger geomagnetic activity, the area between both curves is shaded in black. Also, each time interval is time-shifted by 28 days, i.e. by approximately one solar rotation period. This way the recurrent nature of the geomagnetic activity becomes more evident. Here we are specifically interested in the magnetic activities observed in February, March and April since these activities – using the stringent criteria laid down by *Tsurutani and Gonzalez* (op. cit.) – qualify as HILDCAA events.

How does the ionosphere respond to this kind of activity? To be consistent with the data set shown in Figure 11, we selected the March event to investigate this question. The upper part of Figure 13 describes once more the geomagnetic activity during this HILDCAA event; in the lower part the maximum electron densities in the ionosphere as observed at the two ionosonde stations Moscow and Tomsk are plotted as a function of time. The format of data presentation is the same as that used in Figure 11.

After some initial perturbations on the first storm day, a negative ionospheric storm develops on the second day. But instead of recovering on the third day, the ionization density stays depressed for another five days. Evidently, during HILDCAA events the same kind of ionospheric perturbations are observed as are recorded during short-duration geomagnetic storms. The only difference is that during HILDCAA events the ionospheric perturbations persist as long as the magnetic activity continues.

5. IONOSPHERIC HOLES AT EQUATORIAL LATITUDES

Ionospheric 'holes' certainly belong to the most spectacular storm phenomena at low latitudes. An example of this kind of perturbation is presented in Figure 14. The upper part shows a subsatellite track of the DMSP-F15 satellite during

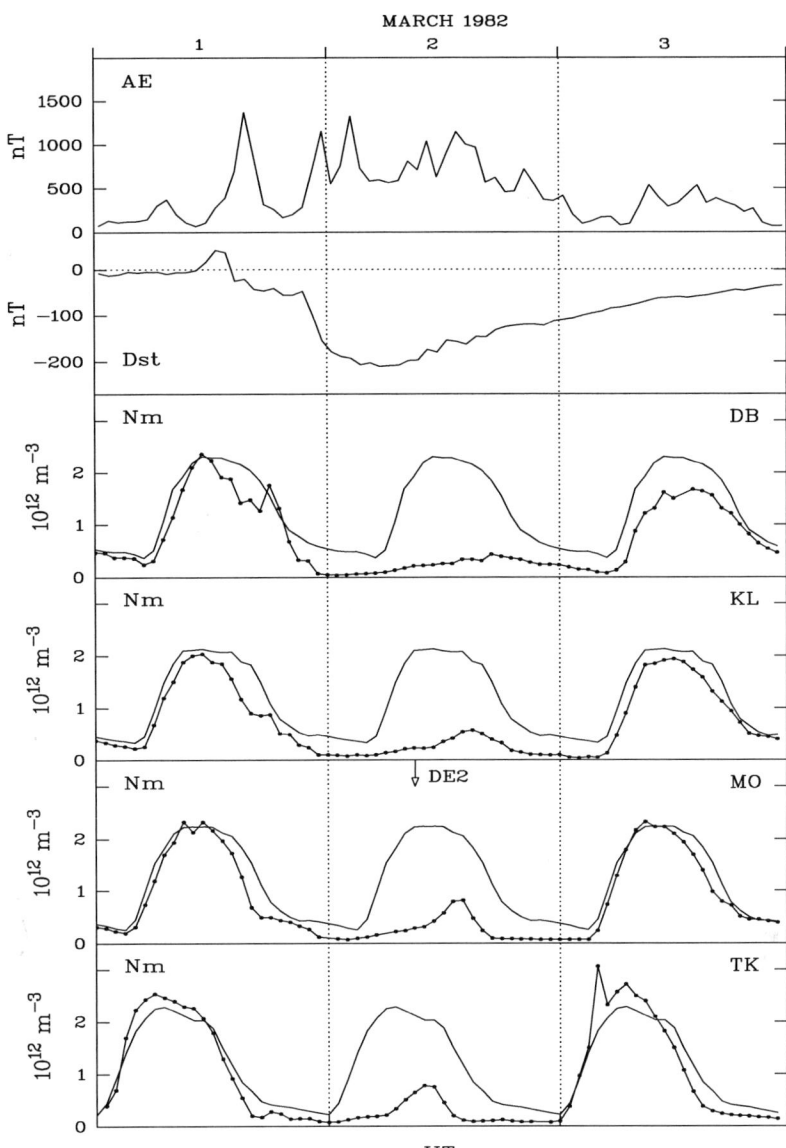

Figure 11. Changes in the maximum electron density of the ionosphere during the geomagnetic storm of March 1/2, 1982. In the upper part of this figure, the AE and Dst indices are used to describe once more the magnetic activity during this event. In the lower part, storm-induced changes in the maximum electron density of the ionosphere, Nm, as observed at the four ionosonde stations Dourbes (DB; 52 °N mag. lat., 5 °E geog. long.; SLT=UT), Kaliningrad (KL; 53 °N mag. lat., 21 °E geog. long.; SLT=UT+1), Moscow (MO; 50 °N mag. lat., 39 °E geog. long.; SLT=UT+2) and Tomsk (TK; 46 °N mag. lat., 85 °E geog. long.; SLT=UT+6) are presented. The storm-time variations are indicated by the lines marked by data points. The mean variations observed on March 6 to 8 serve as quiet-time references (thin lines without data points). An arrow indicates the time when the DE-2 satellite collected data near the Moscow ionosonde station; see also Figure 8.

the storm event of October 29/30, 2003. This was one of the largest solar-terrestrial storms ever observed, with an X17/4B flare on October 28, a solar wind velocity of nearly 2000 km/s and a minimum Dst index of less than −350 nT. In the lower part of this figure, the ion density measured along the satellite trajectory is plotted as a function of geographic latitude. The densities refer to an altitude of 840 km and have been plotted on a logarithmic scale.

The most striking feature of this data set is the large drop in the ion density near the magnetic equator. This ionospheric hole is characterized by rather steep walls on both sides of the density depletion, by a flat and almost structureless bottom,

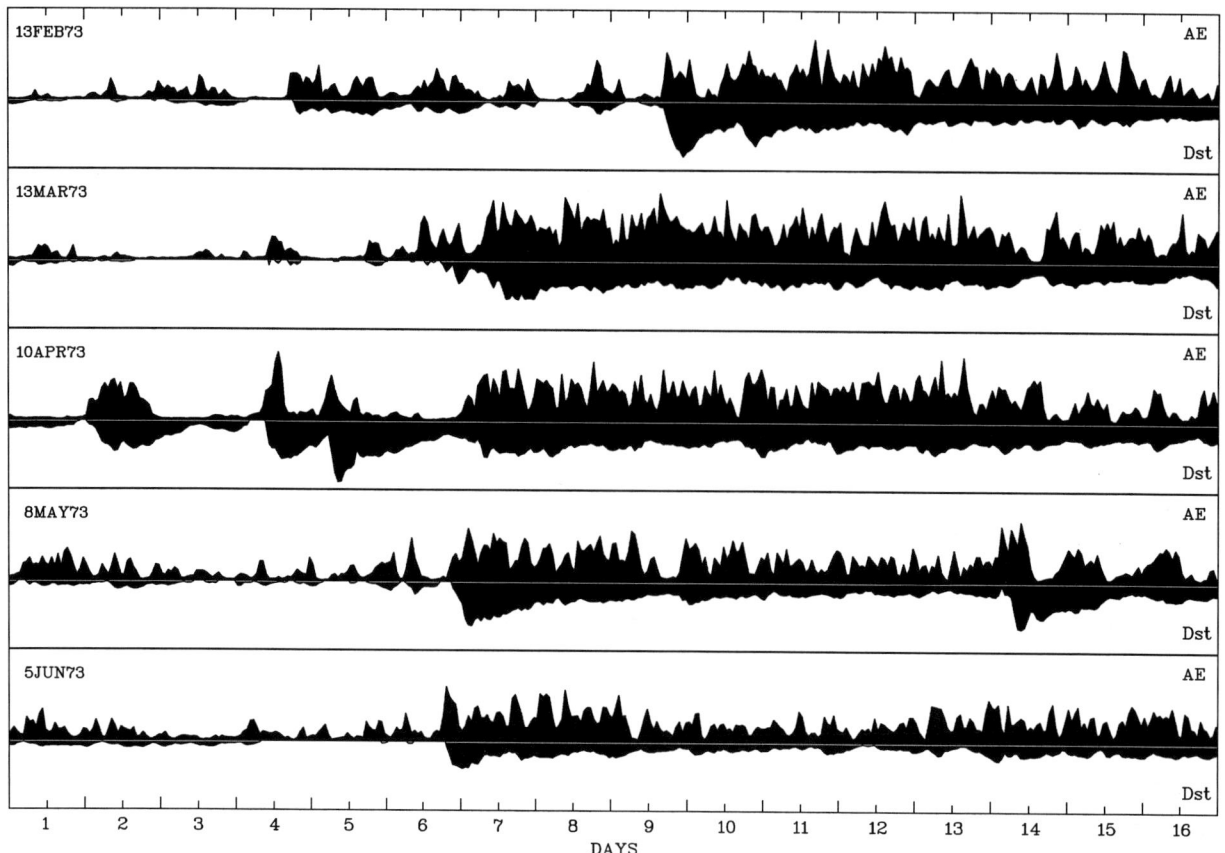

Figure 12. Recurrent geomagnetic activity during spring 1973. Shown are five periods of geomagnetic activity, each time-shifted by 28 days. The upper curve in each panel is the AE index, the lower the Dst index. Both indices use the same zero line but different scales. The distance from the zero line to the upper boundary of each panel corresponds to 1700 nT, the distance to the lower boundary −150 nT. In order to emphasize periods of larger magnetic activity, the areas between both curves are shaded in black. According to the criteria laid down by *Tsurutani and Gonzalez* [1987], the long-duration magnetic activities observed in February, March and April qualify as HILDCAA events.

and by densities which are less than 1/100 of the values recorded outside the hole. Up to today, similar density dropouts have been observed during at least six different geomagnetic storms, all of which were unusually large [e.g., *Greenspan et al.*, 1991; *Basu et al.*, 2001; *Lin et al.*, 2001; *Lee et al.*, 2002; *Su et al.*, 2002; *Vlasov et al.*, 2003; *Lin and Yeh*, 2005; *Kil et al.*, 2005a,b]. Also, all these observations were made after sunset, indicating that we are dealing here with a nighttime phenomenon.

As to the origin of this spectacular disturbance phenomenon, different mechanisms have been proposed. One group of authors suggests that an equatorial super-fountain removes the ionization from the equatorial region [e.g., *Greenspan et al.*, 1991; *Batista et al.*, 1991; *Basu et al.*, 2001]. Others assume that the ionospheric layer rises to above the observation height of the satellites [*Su et al.*, 2002]. Some support for this latter explanation comes from ionosonde measurements which show that the layer height may indeed increase to high altitudes [*Su et al.*, op. cit.; *Sobral et al.*, 1997; *Abdu et al.*, 2005]. Sunward convection of the nighttime plasma represents another possibility [*Vlasov et al.*, 2003]. Finally, *Kil et al.* [2005a,b] propose that ionospheric holes are formed when so-called ionospheric bubbles merge to form a big bubble. Evidently, simultaneous measurements of all the key parameters involved are needed to single out the correct explanation.

6. SUMMARY AND DISCUSSION

Storm effects in the polar ionosphere remain a poorly documented and understood phenomenon. Early attempts to identify such effects were hampered by the strong absorption encountered in this region. Later, satellite studies revealed

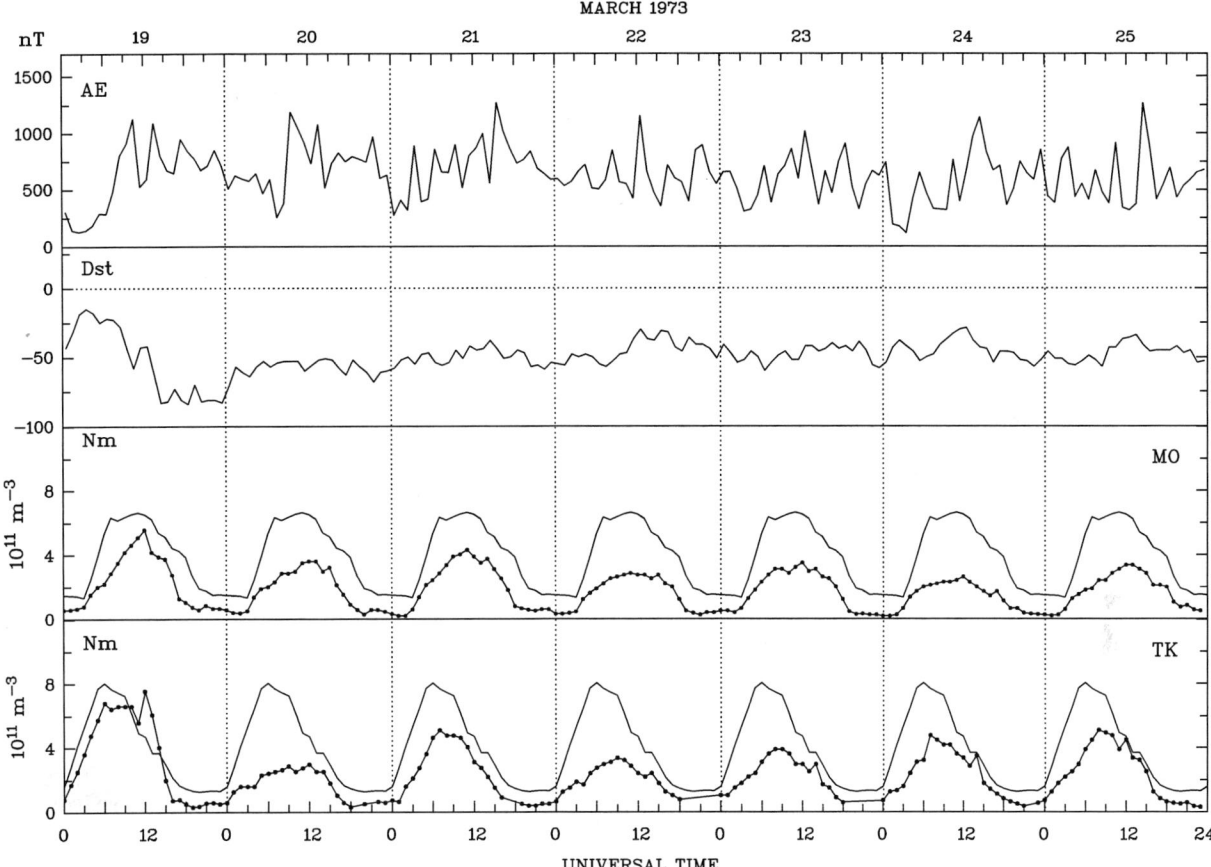

Figure 13. Changes in the maximum electron density of the ionosphere during the HILDCAA event of March 1973. The format of data presentation corresponds to that of Figure 11 except that this time only ionosonde data from Moscow (MO) and Tomsk (TK) are shown. The mean variations observed from March 14 to 17 serve as quiet-time references.

that the polar ionosphere is highly structured, particularly in the absence of sunlight. Typically, a satellite traversing this region will observe a number of peaks and troughs in the electron temperature and density. Since these are not stationary, a determination of a quiet-time pattern which could serve as a reference for storm studies is difficult. In this situation it is more promising to concentrate on the study of specific morphological features.

Prominent among these features is the electron temperature enhancement observed beneath the magnetospheric cleft. This temperature peak is a few degrees wide in latitude and has an average increase of 1200 K above the background temperature at 700 km altitude. Here the storm-time behavior of this phenomenon is investigated. It is found that the temperature peak is significantly displaced towards lower latitudes during such storm conditions. This displacement may be as large as 18 degrees [and possibly more; see also *Meng*, 1982, 1983; *Karpachev and Afonin*, 2004]. It is also found that a close

linear correlation exists between the location of the temperature peak and the AE index. This suggests that, the position of the temperature peak and the associated cleft precipitation are primarily controlled by the magnetospheric reconfiguration during substorms [e.g., *Eather*, 1985; *Stasiewicz*, 1991; *Prölss*, 2006]. On the other hand, it is not clear which of the many magnetospheric and/or interplanetary parameters controls the magnitude of this temperature enhancement.

To improve our knowledge of the disturbed polar ionosphere, similar studies should also be performed for other prominent features such as the temperature enhancement in the night sector of the Earth or the density troughs observed at all local times. Presently, none of these features are described by the International Reference Ionosphere [*Bilitza*, 2001].

At middle latitudes, storm-induced changes in the electron temperature are also a topic which deserves further study. As has been demonstrated, a close linear relationship exists between increases in the electron temperature and decreases

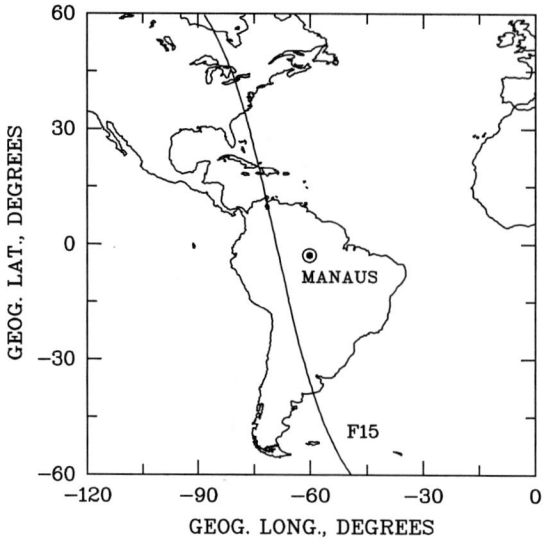
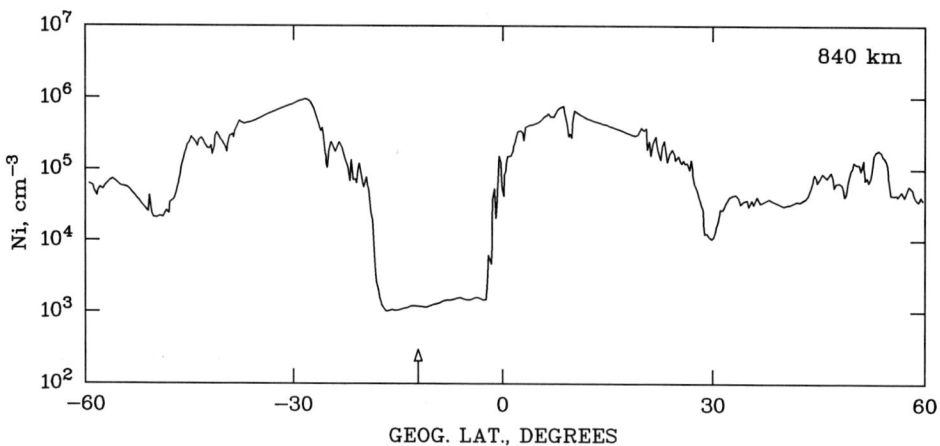

Figure 14. Storm-induced ionospheric hole in the equatorial ionosphere. The upper part of this figure shows a subsatellite track of the DMSP-F15 satellite during the large geomagnetic storm of October 30, 2003. In the lower part, the ion density measured along the associated satellite trajectory is plotted as a function of geographic latitude. Note that this density is plotted on a logarithmic scale. The solar local time of observation is approximately 21:30 h, the observation height 840 km. An arrow indicates the location of the magnetic equator. [After *Kil et al.*, 2005b]

in the electron density. A similar correlation was previously observed during the storm event of February 1973 [*Prölss et al.*, 1975]. At that time, however, the rate of temperature increase was much larger (130 K per decrease in electron density by 10^{11} m^{-3} versus 85 K for the same density decrease in the present case). What causes this difference is not clear, but it may have to do with the different observation heights and the different levels of solar activity.

The same linear correlation between electron density and temperature is also observed during quiet times [e.g., *McDonald and Williams*, 1986; *Breen and Williams*, 1991].

Therefore the physics on which this correlation rests is probably the same for storm and non-storm conditions. This physics, however, turns out to be surprisingly complex considering the simple linear relationship observed.

In contrast to the correlation between electron density and electron temperature, the correlation between electron density and neutral gas composition is well understood. In fact, it is now generally accepted that the large-scale decreases in the ionization density, which have become known as negative ionospheric storms, are caused by decreases in the O/N_2 and O/O_2 density ratios. It is also well established that these composition changes are sufficient to account for the magnitude of the observed perturbations [e.g., *Prölss and Werner*, 2002]. Therefore other mechanisms will play only a secondary role.

Ionospheric perturbations may also be used to search for differences in the response of the upper atmosphere to isolated short-duration and recurrent long-duration geomagnetic storms. As it turns out, the same kinds of storm effects are observed during both events. The only difference is that during periods of high-intensity long-duration continuous auroral activity (HILDCAA) these perturbations last longer. This suggests that for the upper atmosphere it is unimportant where the disturbance energy comes from, and – going a step back – what causes the interplanetary magnetic field to turn southward, as long as sufficient energy is dissipated into the polar regions.

Ionospheric storm effects at low latitudes differ from those observed at middle and high latitudes. This has to do with the low inclination of the geomagnetic field, which supports a number of special phenomena like the equatorial plasma fountain or various plasma instabilities. The most striking disturbance effects observed in this region are so-called ionospheric holes, a term first used by *Eckersley* [1942]. By this we mean regions in the low-latitude ionosphere where the density decreases to very low values, sometimes to less than 1/100 of their original values. Also, these regions are bound by very steep walls and flat bottoms. Different explanations have been offered, none of which is generally accepted. Evidently, additional measurements are needed to specify more accurately the properties of this phenomenon and the conditions under which it occurs.

The situation encountered here is not uncommon in upper atmospheric storm research. Some of the storm effects have been known for decades and have been well-documented in numerous publications. Also, various mechanisms have been proposed to explain these perturbations. However, suitable measurements are missing which would allow us to single out the correct explanation. Therefore progress in this field will critically depend on better and, above all, more comprehensive measurements. These should include, for example, simultaneous measurements of the local electric fields, of the local neutral winds and of the neutral gas composition, to name a few key parameters. In this sense, upper atmospheric storms do represent a real challenge for the experimentalists.

Acknowledgments. The DE-2 data used in this study were kindly provided by the NASA National Space Science Data Center. The ionosonde data were obtained from World Data Center A. I am grateful to all the experimenters who contributed to these data sets. I am also indebted to K. Schrüfer and B. Winkel for their help in preparing this manuscript.

REFERENCES

Abdu, M.A., J.H.A. Sobral, E.R. de Paula, and I.S. Batista, Magnetospheric disturbance effects on the Equatorial Ionization Anomaly (EIA): an overview, *J. Atmos. Terr. Phys.*, 53, 757-771, 1991.

Abdu, M.A., J.R. de Souza, J.H.A. Sobral, and I.S. Batista, Magnetic storm associated disturbance dynamo effects over low and equatorial latitude F-region (this issue), 2006.

Basu, S., K.M. Groves, H.-C. Yeh, S.-Y. Su, F.J. Rich, P.J. Sultan, and M.J. Keskinen, Response of the equatorial ionosphere in the South Atlantic region to the great magnetic storm of July 15, 2000, *Geophys. Res. Lett.*, 28, 3577-3580, 2001.

Batista, I.S., E.R. de Paula, M.A. Abdu, N.B. Trivedi, and M.E. Greenspan, Ionospheric effects of the March 13, 1989, magnetic storm at low and equatorial latitudes, *J. Geophys. Res.*, 96, 13943-13952, 1991.

Bilitza, D., International Reference Ionosphere 2000, *Radio Sci.*, 36, 261-275, 2001.

Breen, A.R., and P.J.S. Williams, The relationship between electron density and electron temperature under quiet conditions in the auroral zone, *Adv. Space Res.*, 11, No. 10, 167-170, 1991.

Buonsanto, M.J., Ionospheric storms – a review, *Space Sci. Rev.*, 88, 563-601, 1999.

Burch, J.L., Low-energy electron fluxes at latitudes above the auroral zone, *J. Geophys. Res.*, 73, 3585-3591, 1968.

Burns, A.G., T.L. Killeen, and R.G. Roble, A theoretical study of thermospheric composition perturbations during an impulsive geomagnetic storm, *J. Geophys. Res.*, 96, 14153-14167, 1991.

Carignan, G.R., B.P. Block, J.C. Maurer, A.E. Hedin, C.A. Reber, and N.W. Spencer, The neutral mass spectrometer on Dynamics Explorer, *Space Sci. Instr.* 5, 429-441, 1981.

Danilov, A.D., and J. Lastovicka, Effects of geomagnetic storms on the ionosphere and atmosphere, *Int. J. Geomag. Aeron.*, 2, 209-224, 2001.

Eather, R.H., Polar cusp dynamics, *J. Geophys. Res.*, 90, 1569-1576, 1985.

Eckersley, T.L., Holes in the ionosphere and magnetic storms, *Nature*, 150, 177, 1942.

Emery, B.A., R.G. Roble, E.C. Ridley, T.L. Killeen, M.H. Rees, J.D. Winningham, G.R. Carignan, P.B. Hays, R.A. Heelis, W.B. Hanson, N.W. Spencer, L.H. Brace, and M. Sugiura, Thermospheric and ionospheric structure of the southern hemisphere polar cap on October 21, 1981, as determined from Dynamics Explorer 2 satellite data, *J. Geophys. Res.*, 90, 6553-6566, 1985.

Förster, M., and N. Jakowski, Geomagnetic storm effects on the topside ionosphere and plasmasphere: A compact tutorial and new results, *Space Sci. Rev.*, 21, 47-87, 2000.

Fontheim, E.G., L.H. Brace, and J.D. Winningham, Properties of low-energy electron precipitation in the cleft during periods of unusually high ambient electron temperatures, *J. Geophys. Res.*, 92, 12267-12273, 1987.

Fuller-Rowell, T.J., M.V. Codrescu, R.G. Roble, and A.D. Richmond, How does the thermosphere and ionosphere react to a geomagnetic storm?, in *Magnetic Storms*, edited by B.T. Tsurutani, W.D. Gonzales, Y. Kamide, and J.K. Arballo, *Geophys. Monogr.* 98, American Geophys. Union, Washington, D.C., 203-225, 1997.

Greenspan, M.E., C.E. Rasmussen, W.J. Burke, and M.A. Abdu, Equatorial density depletions observed at 840 km during the great magnetic storm of March 1989, *J. Geophys. Res.*, 96, 13931-13942, 1991.

Hafstad, L.R., and M.A. Tuve, Further studies of the Kennelly-Heaviside layer by the echo-method, *Proc. Inst. Radio Eng.*, 17, 1513-1522, 1929.

Hanson, W.B., R.A. Heelis, R.A. Power, C.R. Lippincott, D.R. Zuccaro, B.J. Holt, L.H. Harmon and S. Sanatani, The retarding potential analyzer for Dynamics Explorer-B, *Space Sci. Instr.*, 5, 503-510, 1981.

Heikkila, W.J., and J.D. Winningham, Penetration of magnetosheath plasma to low altitudes through the dayside magnetospheric cusps, *J. Geophys. Res.*, 76, 883-891, 1971.

Hoffman, R.A., and E.R. Schmerling, Dynamics Explorer program: An overview, *Space Sci. Instr.*, 5, 345-348, 1981.

Jacchia, L.G., Corpuscular radiation and the acceleration of artificial satellites, *Nature*, 183, 1662-1663, 1959.

Karpachev, A.T., and V.V. Afonin, Variations in the structure of the high-latitude ionosphere during the March 22-23, 1979, storm based on Cosmos-900 and Intercosmos-19 data, *Geomag. Aeron.*, 44, 60-68, 2004.

Kil, H., and L.J. Paxton, Ionospheric disturbance in the low-latitude region during the magnetic storm of July 15, 2000 – Origin of the equatorial plasma density depletions, *J. Geophys. Res.* (submitted), 2005a.

Kil, H., L.J. Paxton, S.-Y. Su, Y. Zhang, and H.C. Yeh, Characteristics of the storm-induced big bubbles (SIBBs), *J. Geophys. Res.* (submitted), 2005b.

Killeen, T.L., P.B. Hays, G.R. Carignan, R.A. Heelis, W.B. Hanson, N.W. Spencer and L.H. Brace, Ion-neutral coupling in the high-latitude F region: Evaluation of ion heating terms from Dynamics Explorer 2, *J. Geophys. Res.*, 89, 7495-7508, 1984.

Krehbiel, J.P., L.H. Brace, R.F. Theis, W.H. Pinkus, and R.B. Kaplan, The Dynamics Explorer Langmuir probe instrument, *Space Sci. Instr.*, 5, 493-502, 1981.

Lee, J.J., K.W. Min, V.P. Kim, V.V. Hegai, K.-I. Oyama, F.J. Rich, and J. Kim, Large density depletions in the nighttime upper ionosphere during the magnetic storm of July 15, 2000, *Geophys. Res. Lett.*, 29, No.3, 2-1, 10.1029/2001GL013991, 2002.

Lin, C.S., and H.-C. Yeh, Satellite observations of electric fields in the South Atlantic anomaly region during the July 2000 magnetic storm, *J. Geophys. Res.*, 110, A03305, doi: 10.1029/2003JA010215, 2005.

Lin, C.S., H.-C. Yeh, and S.-Y. Su, ROCSAT-1 satellite observations of magnetic anomaly density structures during the great magnetic storm of July 15-16, 2000, *Terr. Atmos. Ocean Sci.*, 12, 567-582, 2001.

Liou, K., P.T. Newell, B.J. Anderson, L. Zanetti, and C.-I. Meng, Neutral composition effects on ionospheric storms at middle and low latitudes, *J. Geophys. Res.*, 110, A05309, doi: 10.1029/2004JA010840, 2005.

Lu, G., B.A. Emery, A.S. Rodger, M. Lester, J.R. Taylor, D.S. Evans, J.M. Ruohoniemi, W.F. Denig, O. da la Beaujardiére, R.A. Frahm, J.D. Winningham, and D.L. Chenette, High-latitude ionospheric electrodynamics as determined by the assimilative mapping of ionospheric electrodynamics procedure for the conjunctive SUNDIAL/ATLAS1/GEM period March 28-29, 1992, *J. Geophys. Res.*, 101, 26697-26718, 1996.

Lu, G., A.D. Richmond, R.G. Roble, and B.A. Emery, Coexistence of ionospheric positive and negative storm phases under northern winter conditions: A case study, *J. Geophys. Res.*, 106, 24493-24504, 2001.

Mayr, H.G., and H. Volland, Magnetic storm effects in the neutral composition, *Planet. Space Sci.*, 20, 379-393, 1972.

McDonald, J.N., and P.J.S. Williams, Electron temperature and electron density in the F-region of the ionosphere. I. Observed relationship, *J. Atmos. Terr. Phys.*, 48, 545-557, 1985.

Meng, C.-I., Latitudinal variation of the polar cusp during a geomagnetic storm, *Geophys. Res. Lett.*, 9, 60-63, 1982.

Meng, C.-I., Case studies of the storm time variation of the polar cusp, *J. Geophys. Res.*, 88, 137-149, 1983.

Mikhailov, A.V., Ionospheric F2-layer storms, *Fisica de la Tierra*, 12, 223-262, 2000.

Namgaladze, A.A., A.N. Namgaladze, and M.A. Volkov, Numerical modelling of the thermospheric and ionospheric effects of magnetospheric processes in the cusp region, *Ann. Geophys.*, 14, 1343-1355, 1996.

Newell, P.T., and C.-I. Meng, Mapping the dayside ionosphere to the magnetosphere according to particle precipitation characteristics, *Geophys. Res. Lett.*, 19, 609-612, 1992.

Prölss, G.W., Ionospheric F-region storms, in *Handbook of Atmospheric Electrodynamics*, 2, edited by H.Volland, 195-248, CRC Press / Boca Raton, 1995.

Prölss, G.W., Magnetic storm associated perturbations of the upper atmosphere, in *Magnetic Storms*, edited by B.T. Tsurutani, W.D. Gonzalez, Y.Kamide, and K.Arballo, *Geophys. Monogr.*, 98, 227-241, 1997.

Prölss, G.W., *Physics of the Earth's space environment*, p431, Springer, 2004.

Prölss, G.W., Space weather effects in the upper atmosphere: Low and middle latitudes, in *Space Weather*, edited by H. Fichtner, K. Scherer, U. Mall, and B. Heber, 193-234, Lecture Notes in Physics, Springer, Berlin/Heidelberg, 2005a.

Prölss, G.W., The ionospheric heating beneath the magnetospheric cleft revisited, *Ann. Geophys.*, 23, 827-830, 2005b.

Prölss, G.W., Electron temperature enhancement beneath the magnetospheric cusp, *J. Geophys. Res.* (in press), 2006.

Prölss, G.W., and S. Werner, Vibrationally excited nitrogen and oxygen and the origin of negative ionospheric storms, *J. Geophys. Res.*, 107, 10.1029/2001JA900126, 2002.

Prölss, G.W., U. von Zahn, and W.J. Raitt, Neutral atmospheric composition, plasma density, and electron temperature at F region heights, *J. Geophys. Res.*, 80, 3715-3718, 1975.

Prölss, G.W., M. Roemer, and J.W. Slowey, Dissipation of solar wind energy in the earth's upper atmosphere: The geomagnetic activity effect, CIRA 1986, *Adv. Space Res.*, 8, No. 5, 215-261, 1988.

Richmond, A.D., and G. Lu, Upper-atmospheric effects of magnetic storms: a brief tutorial, *J. Atmos. Solar-Terr. Phys.*, 62, 1115-1127, 2000.

Schlegel, K., Space weather effects in the upper atmosphere: High latitudes, in *Space Weather*, edited by H. Fichtner, K. Scherer, U. Mall, and B. Heber, 215-238, Lecture Notes in Physics, Springer, Berlin/Heidelberg, 2005.

Sobral, J.H.A., M.A. Abdu, W.D. González, B.T. Tsurutani, I.S. Batista, and A.L. Clua de González, Effects of intense storms and substorms on the equatorial ionosphere / thermosphere system in the American sector from ground-based and satellite data, *J. Geophys. Res.*, 102, 14305-14313, 1997.

Stasiewicz, K., Polar cusp topology and position as a function of interplanetary magnetic field and magnetic activity: Comparison of a model with Viking and other observations, *J. Geophys. Res.*, 96, 15789-15800, 1991.

Su, S.-Y., H.C. Yeh, C.K. Chao, and R.A. Heelis, Observation of a large density dropout across the magnetic field at 600 km altitude during the 6-7 April 2000 magnetic storm, *J. Geophys.Res.*, 107, No. A11, 18-1, 10.1029/2001JA007552, 2002.

Titheridge, J.E., Ionospheric heating beneath the magnetospheric cleft, *J. Geophys. Res.*, 81, 3221-3226, 1976.

Tsurutani, B.T., and W.D. Gonzalez, The cause of high-intensity long-duration continuous AE activity (HILDCAAs): Interplanetary Alfvén wave trains, *Planet. Space Sci.*, 35, 405-412, 1987.

Vlasov, M., M.C. Kelley, and H. Kil, Analysis of ground-based and satellite observations of F-region behavior during the great magnetic storm of July 15, 2000, *J. Atmos. Solar-Terr. Phys.*, 65, 1223-1234, 2003.

Williams, P.J.S., and J.N. McDonald, Electron temperature and electron density in the F-region of the ionosphere. II. The role of atomic oxygen and molecular nitrogen, *J. Atmos. Terr. Phys.*, 49, 873-887, 1987.

Response of the Upper/Middle Atmosphere to Coronal Holes and Powerful High-Speed Solar Wind Streams in 2003

J.U. Kozyra[1], G. Crowley[2], B.A. Emery[3], X. Fang[1], G. Maris[4], M.G. Mlynczak[5], R.J. Niciejewski[1], S.E. Palo[6], L.J. Paxton[7], C.E. Randall[6], P.-P. Rong[8], J.M. Russell III[8], W. Skinner[1], S.C. Solomon[3], E.R. Talaat[7], Q. Wu[3], and J.-H. Yee[7]

High-speed solar wind streams originating from large coronal holes reached a maximum in 2003 during the descending phase of solar cycle 23. At the same time, magnetic activity (as indicated by the aa index) reached the highest levels of the last four solar cycles. The rotation of active regions behind the limb and the appearance of coronal holes contribute to a decrease in EUV radiation at Earth prior to the arrival of a high-speed stream. This leads to a cooling of the upper atmosphere and a decrease in the total electron content (TEC). Changes in $\Sigma O/N_2$ are also expected but contributions from local time and seasonal changes as the satellite orbit precesses during a solar rotation will require additional simulations to unravel. These systematic changes are intriguing because auroral disturbances expand further equatorward in latitude and deeper in altitude as F10.7 decreases. Disturbances in nitric oxide (NO), the dominant cooling agent in the upper atmosphere, continue for the duration of the high-speed stream activity and may not fully recover before the next stream hits. These disturbed conditions might actually represent the most common state of the upper atmosphere in years of strong recurrent high-speed streams. Finally, the persistent source of thermospheric NO in the auroral region during high-speed streams, when combined with favorable meteorological conditions in the dark polar middle atmosphere, results in significant enhancements in stratospheric NO_x. These correspond to increases in stratospheric ozone loss, establishing a previously unexplored link between high-speed streams and stratospheric variability.

[1] Atmospheric, Oceanic and Space Sciences Dept., University of Michigan, Ann Arbor, MI
[2] Southwest Research Institute, San Antonio, TX
[3] High Altitude Observatory, National Center for Atmospheric Research, Boulder, CO
[4] Astronomical Institute of the Romanian Academy, Bucharest, Romania
[5] Science Directorate, NASA/Langley Research Center, Hampton, VA
[6] University of Colorado, Boulder, CO
[7] Applied Physics Laboratory, Johns Hopkins Univ., Laurel, MD
[8] Hampton Univ., Hampton, VA

Recurrent Magnetic Storms: Corotating Solar Wind Streams
Geophysical Monograph Series 167
Copyright 2006 by the American Geophysical Union.
10.1029/167GM24

1.0 INTRODUCTION

The recurrent magnetic activity in 2003 provides the best opportunity to date for examining the impacts on the Earth's upper atmosphere and ionosphere of the strong, long-lived energy inputs from corotating interaction regions (CIRs)/high speed streams [c.f., *Tsurutani et al.*, 1999]. The CIRs/high speed streams themselves were exceptional in duration and speed, at times containing unusually strong fluctuating B_z components, and thus should produce unambiguous signatures in the atmosphere. In the past, sparse data sets have only allowed glimpses of this complex interaction but now with the Thermosphere Ionosphere Mesosphere Energetics and Dynamics (TIMED) spacecraft joining the suite of operating spacecraft that constitute

NASA's Sun-Earth Connections Great Observatory and the array of Earth-observing spacecraft currently in orbit, the potential for new discoveries about the nature of the geospace response to solar variability in general, and high-speed streams in particular, is high.

Coronal holes are the sources of high-speed solar wind streams [*Hundhausen*, 1977]. These high-speed streams are responsible for the large-scale structures in the heliosphere imposed by corotating interaction regions [*Maravilla et al.*, 2001]. A corotating interaction region is formed as a high-speed solar wind stream, originating from a low-latitude coronal hole, pushes into and piles up the slower solar wind ahead of it and creates a region of compressed plasma and magnetic fields at its leading edge [c.f., *Pizzo*, 1985; *Crooker and Cliver*, 1994]. Since the slower solar wind originates from, or in the vicinity of, the Sun's streamer belt, coronal mass ejections (CMEs) propagating within the streamer belt can become compressed in the CIR near the Sun or during transit through interplanetary space and contribute to an increase in the intensity of southward IMF. The strong, short-duration southward IMF in the CIR triggers modest magnetic activity that can be intensified even further if a CME has become entrained in the CIR.

The high speed coronal hole wind trailing the CIR contains an elevated level of Alfven wave turbulence which results in rapid fluctuations in the IMF from northward to southward [*Tsurutani et al.*, 1999; *Kessel et al.*, 2004] as the solar wind passes by the Earth. These rapid fluctuations in IMF B_z are not long-lived or intense enough to support the strong convection necessary to drive particles deep into the inner magnetosphere and produce magnetic storms. In fact, observations in the inner magnetosphere during several of these high-speed stream intervals indicated a decrease in energetic ring current particles rather than an enhancement, as one would have expected during magnetic storm activity [*Friedel et al.*, 1997]. The fluctuating fields in the high-speed coronal hole wind are sufficient to trigger recurrent magnetotail current disruptions that produce high-intensity, long-duration, continuous auroral activity (HILDCAA) events [c.f., *Tsurutani et al.*, 1995; *Tsurutani et al.*, 1999; *Gonzalez et al.*, 1999]. HILDCAA events can last more than 10 days and repeat twice per solar rotation. They greatly extend the recovery phase of any magnetic storm produced by the preceding CIR region. This type of high latitude activity can actually produce higher yearly-averaged AE values (a measure of auroral electrojet intensity and thus energy deposition in the auroral atmosphere) in the descending phase of the solar cycle than at solar maximum where transient events produce fewer but stronger magnetic storms. For example, as discussed in *Tsurutani et al.* [1999], the annual average of AE reached 283 nT in 1974 (descending phase) compared to 221 nT in 1979 (solar maximum).

Season is another important factor in the geoeffectiveness of high-speed streams. Magnetic activity is observed to be stronger at the equinoxes than at other times of the year for IMF sectors with toward polarity ($+B_x$) in northern hemisphere spring and with away polarity ($-B_x$) in northern hemisphere autumn. There are a number of theories to explain this effect which rely on the orientation of the Earth's dipole axis with respect to the heliospheric magnetic field topology [*Russell and McPherron*, 1973; *Cliver et al.*, 2000; *O'Brien and McPherron*, 2002] to either maximize the IMF B_z component seen by the Earth or systematically alter the reconnection rate at the dayside magnetopause. This geometric effect significantly enhances magnetic activity due to high-speed streams.

In years of strong high-speed solar wind stream activity, solar wind energy flow into geospace reaches its peak for the solar cycle as CIRs and the trailing high speed streams drive prolonged periodic auroral activity [c.f., *Tsurutani et al.*, 1995]. Because of their spatial extent and velocity, high-speed streams can take up to 10 days or more to pass by the Earth. In contrast, solar eruptive events result in more short-lived (day long) episodes of strong slowly varying southward IMF characteristic of the interplanetary signatures of coronal mass ejections. The result of the interaction of CIRs/high speed streams with geospace is to produce long-term (week to 10 days) perturbations to chemistry, dynamics and energetics of the upper atmosphere. Coupling of these disturbances to the middle atmosphere via chemical species transport takes on a different character than for impulsive solar particle events, as it is modulated by stratospheric meteorology, and builds up on longer time scales [c.f., *Randall et al.*, 2005]. In 2003, unusually large polar coronal holes stretched toward the Sun's equator developing low-latitude extensions that persisted in some form for most of the year. The appearance of low-latitude coronal holes from this mechanism has been seen on other occasions [*Timothy et al.*, 1975; *Bromage et al.*, 2000; *Cranmer*, 2002]. Solar wind speeds from these low-latitude extensions are found to be roughly proportional to their areas [*Nolte et al.*, 1976; *Neugebauer et al.*, 1998]. Because of this relationship to area, the speeds in the largest of these low latitude extensions approach the speeds in the polar coronal hole itself, whereas for smaller coronal hole extensions speeds are much lower than in the polar coronal hole. In 2003, speeds in the high-speed solar wind streams reached 800 km/s. The observed occurrence of these coronal hole extensions as well as isolated coronal holes produced by active regions is forcing a re-evaluation of the simplified picture of recurrent high-speed streams produced by outflows from two polar coronal holes separated by a tilted heliospheric current sheet where outflow speeds reach a minimum [c.f., *Luhmann et al.*, 2002].

In 2003 (descending phase of solar cycle 23), the monthly aa-value reached 37 compared to 25 in 2000 near solar

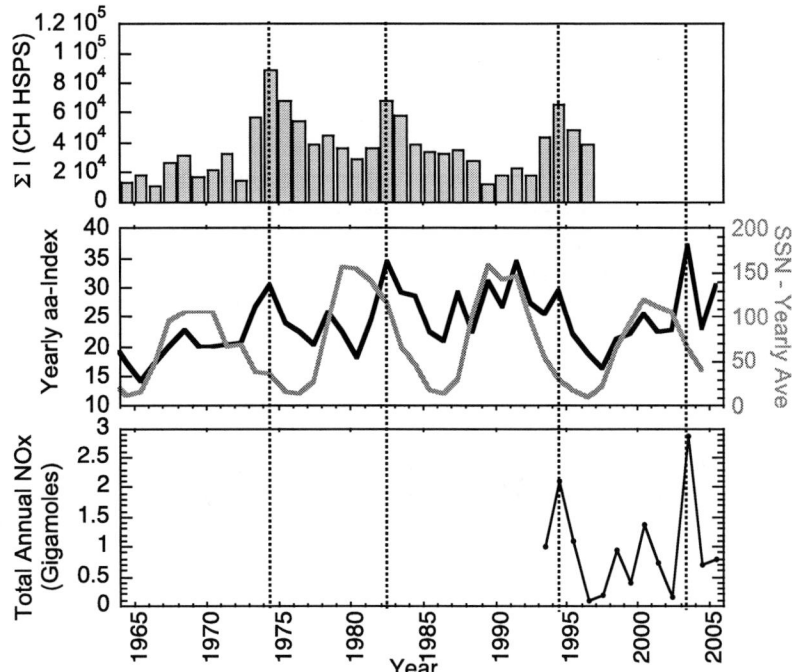

Figure 1. Top Panel: Importance parameter (I) for high-speed solar wind streams from 1964-1996 [*Maris and Maris*, 2005] (histogram). Middle Panel: Sunspot number (grey) and yearly-averaged aa index courtesy of the World Data Center for Geomagnetism in Copenhagen (black) are plotted. Years (dotted vertical lines) with the most powerful high-speed streams (as given by the importance parameter) have some of the strongest peaks in aa index due to recurrent magnetic activity. High-speed streams in 2003 in the descending phase of solar cycle 23 produced some of the strongest magnetic activity of the last 4 solar cycles. Bottom Panel: Total annual NOx (gigamoles) descending into the southern hemisphere stratosphere after production by energetic particles in the mesosphere or thermosphere, as derived from HALOE observations by *Randall et al.* [2006]. Since NOx participates in the destruction of stratospheric ozone, this is a potential pathway linking climate variability to high-speed stream activity.

maximum (see Figure 1). Here the aa-index is taken as a proxy for energy input due to magnetic activity indicating that, in solar cycle 23 as in solar cycle 20 (discussed above), energy input reached a maximum in the descending phase of the solar cycle. For direct comparison purposes, in solar cycle 20, the yearly-averaged monthly aa-value in the descending phase (1974) was 30 and during solar maximum (1979) was 23. One might expect that in years of intense high-speed stream activity the interaction of the streams with geospace would have a significant and observable impact on the ionosphere and upper atmosphere since the solar EUV flux, the coupling to the magnetosphere and the high-latitude energy inputs change. In this paper, some of the dramatic effects observed during the 2003 CIR/high speed stream intervals are examined.

Section 2.1 puts the current solar cycle 23 into historical context with other recent solar cycles. Section 2.2 explores the characteristics of the global energy input into the ionosphere, thermosphere, mesosphere system during 2003. Section 3.0 looks at preconditioning of the neutral atmosphere by variations in solar EUV as large coronal holes with low EUV radiation flux transit the solar disc just prior to the arrival of the high speed stream at Earth. Section 4.0 presents observations of nitric oxide as a thermostat regulating the temperature of the upper atmosphere. Section 5.0 traces NO_x into the stratosphere where it participates in the destruction of ozone. Section 6.0 discusses possible changes in atmospheric dynamics in response to long-lived and recurrent high latitude heating and cooling. Section 7.0 draws these aspects together into one of the first comprehensive views of sun-to-Earth system behavior during high-speed streams.

2.0 HIGH SPEED STREAMS AND RECURRENT MAGNETIC ACTIVITY IN 2003

2.1. Historical Context

Typically high-speed streams maximize in the descending phase of the solar cycle [*Crooker and Cliver*, 1994]. In fact, the presence of a peak in high speed stream activity in the mid to late descending phase has been typical of the last 8 solar cycles (15-22), as evidenced by a strong 13.5 day periodicity in magnetic activity indices at these times [*Mursula and Zieger*, 1998] indicating two streams per solar rotation. This trend continues in the current solar cycle 23. Figure 1 (middle panel) is a plot over the last 4 solar cycles of yearly sunspot number calculated as the average of the daily mean sunspot numbers from the National Geophysical Data Center in Boulder (http://www.ngdc.noaa.gov/stp/SOLAR/ftpsunspotnumber.html) and yearly aa-index calculated as the average of the monthly-mean aa from the World Data Center for Geomagnetism in Copenhagen. The aa-index was developed to investigate the long-term changes in the magnetic activity at Earth [*Mayaud*, 1973]. Data from two antipodal magnetic observatories is combined to cancel out daily and seasonal variations leaving behind planetary-scale magnetic field variations due to magnetic activity. Figure 1 (top panel) is a histogram of the yearly sum of the *Importances* (ΣI) of high speed streams for the 3 solar cycles preceding solar cycle 23 [*Maris and Maris*, 2005], where $I = \Delta V_{max} \times d$, $\Delta V_{max} = (V_{max} - V_0)$, V_{max} is the maximum value of the daily mean speed during the entire d-day interval of the stream, V_0 is the daily mean speed obtained by finding the mean value between the speed immediately preceding and following the stream, and d is the duration of the stream in days.

The importance parameter identifies 1974, 1982-83 and 1994 as years of the strongest high-speed streams. These same years have major peaks in auroral energy input as indicated by the yearly-average of the aa-index. The large peaks in the yearly-averaged aa-index in 1991 and 1989 are due to major magnetic storms driven by coronal mass ejections during the pronounced double-peaked solar maximum in cycle 22 [see Figure 2 in *Richardson et al.*, 2001] and are not related to high speed stream activity. *Richardson et al.* [2001] found peaks in the occurrence frequency of CIR-driven magnetic storms in 1974, 1983, and 1994 consistent with the *Importance Parameter* in Figure 1. Though the importance parameter is not yet available for 2003, there is a clear peak in the aa-index in 2003 during the declining phase of the current solar cycle when CIR-driven activity should be at a maximum repeating the same pattern as the previous 3 solar cycles.

There are some unusual aspects of the current solar cycle 23. If even and odd solar cycles are paired over the last 150 years, starting with solar cycle 10, the odd numbered cycle will have the higher peak sunspot number (see *Cliver et al.*, [1996], and references therein). Magnetic activity due to 27-day recurrent solar wind streams is stronger in the declining phase of even-numbered cycles. Solar cycle 23 breaks this 150-year pattern. High-speed streams in the declining phase of solar cycle 23, rather than producing weaker magnetic activity than solar cycle 22, produced the strongest activity (as indicated by the aa index) of the past 4 solar cycles. Further, solar cycle 23 has a lower peak sunspot number than solar cycle 22 [c.f., *Ahluwalia*, 2003]. The reasons why the current solar cycle breaks the established long-term patterns described above are unknown.

The first active region (AR) of SC 23 appeared on May 1996, just in the minimum phase, very late in comparison with previous cycles. In previous cycles, the first AR of the new cycle arrived earlier appearing in the descending phase of the preceding cycle. The delayed appearance of the first AR in SC23 might provide clues to the reasons for the slow amplification of the new cycle's toroidal magnetic field and the lower activity level. The ARs of SC23 outnumbered those from SC22 after February 1997. Furthermore, the real ascending phase of SC23 only began to develop in September 1997. This delayed appearance is similar to that for SC 20 [*Maris et al.*, 2003].

SC 23 had a series of long duration ARs that were extremely productive in generating high-energy eruptive events (flares, CMEs), which, in turn, created strong geomagnetic effects and notable perturbations in space weather. Those events took place mainly in certain intervals: July 2000, March-April 2001, May-June 2002 and, the most active, Oct.-Nov. 2003. Comparing these very active intervals with the low sunspot numbers in SC 23 it might be suggested that there were some special conditions allowing the accumulation of high energy in individual ARs, rather than lower energy in a larger number of ARs formed during a "normal" cycle such as SC 21 or 22.

The close correspondence between magnetic activity at Earth and the evolution of the large-scale solar magnetic field makes it possible to predict features of the coming solar cycle based solely on magnetic indices at Earth. Even though solar cycle 23 was surprisingly weak, *Ahuluwalia* [2003] was able to predict the peak sunspot number of solar cycle 23 and the rise time to solar maximum based on this technique. The predicted rise time from onset was 43 ± 6 months compared to the actual rise time of 47 months after onset. Predicted peak sunspot number was 131 +33/−20 compared to an actual smooth peak sunspot number of 120.8. This interesting result is thought to be a consequence of the control of the heliospheric structure – and, thus, of the variation in induced interplanetary electric field (E_y) - by the open magnetic flux in coronal hole streams as the solar magnetic field evolves over the solar cycle. The E_y imprints the features of this variability on magnetic activity at Earth. This means that changes in the solar magnetic field will drive changing patterns of magnetic activity at Earth. And conversely, geomagnetic activity indices can give clues about the evolution of the solar magnetic field.

2.2. Energy Input Into Geospace

The dramatic increase in geomagnetic activity during 2003 is clearly seen in the enhanced monthly aa-values in Figure 1.

To further investigate this peak, Plate 1 displays daily-averaged aa-values, which contain information about geomagnetic activity on the time scale of the high-speed streams and eruptive solar events. Each solar rotation appears as a vertical strip of 27 days while time and Bartels Rotation Number increase towards the right along the horizontal axis. In this format, the duration and recurrence of geomagnetic activity, which serves to identify high-speed stream intervals, is very clear. The duration is given by the extent of strong aa in the vertical direction and the number of recurrences by extent in the horizontal direction (across multiple solar rotations). Using this format, the figure provides a strong visual assessment of the relative contributions of superstorms and high-speed streams to the average aa-value. Arrows in the figure identify the contributions from the 2003/10/29-30, 2003/10/30-31 and the 2003/11/20-21 superstorms. Four solar rotations (Bartels rotation #2322-#2325) containing some very powerful high-speed stream activity are indicated by blue shading in the figure. Data analysis efforts in the rest of the paper are largely centered on this shaded interval of high-speed stream activity. Images from SOHO/EIT taken from each of these solar rotations (see Plate 1) show the extent of coronal holes on the solar disc at this time.

Plate 2 explores how the distribution of the aa-index relates to energy flow into the auroral region. On the left is daily aa-index for the last 4 solar cycles. On the right is the intersatellite-adjusted global electron hemispheric power (GHPe) for the last 3 solar cycles. Hemispheric power is an estimate of the power in gigawatts (GW) carried by precipitating electrons into the auroral region and is obtained by scaling statistical patterns using in situ satellite observations of the local electron flux and energy distribution [*Evans*, 1987]. The global intersatellite-adjusted values of hemispheric power are based on comparisons between all NOAA and DMSP satellites [*Emery et al.*, 2006]. Values are adjusted for consistency between satellites and information from both the northern and southern hemispheres is combined making it a global index. The general consistency between regions of high aa-index values and regions of high GHPe is evident in Plate 2 during years of strong high-speed stream activity (1982-3, 1994 and 2003 identified in Figure 1) indicating that auroral activations are an important mechanism for energy input during these intervals. It is also clear that in 2003 the energy input is dominated by long-lived (more than 10 days, indicated by vertical stripes of enhanced energy input in Plate 2), recurrent (over many solar rotations, indicated by horizontal stripes of enhanced energy input in Plate 2) events typical of the geospace response to CIRs/high speed streams rather than the much shorter (only a few days as identified by arrows in Plate 1) non-recurrent energy inputs typical of magnetic storms driven by impulsive solar events.

Given this relationship to high-speed streams, the enhancements in the GHPe-index should be well correlated with daily averaged values of the solar wind velocity. A comparison of daily GHPe (left side) and daily-averaged solar wind velocity (right side) from the OMNI database is given in Plate 3. In the years of strong high-speed stream activity (1982-83, 1994, and 2003), there is a close correspondence between elevated GHPe and high solar wind speeds, as expected.

Plate 4 displays solar, solar wind and geomagnetic activity indices for most of 2003. Each set of stacked plots contains information for three Bartels rotation periods. The top and bottom panels give hourly averages of solar wind speed and proton density, respectively from the SWEPAM instrument on the ACE spacecraft. The fifth panel gives hourly averaged IMF B_z in GSM coordinates from the MAG instrument on the ACE spacecraft. Selected recurrent high-speed streams are highlighted in grey as they evolve throughout the year. One striking aspect of the figure is the amount of time during which CIRs/high speed streams provide the dominant energy input during 2003. The coronal hole that produced the high-speed stream is identified above each highlighted interval along with its designation (if it existed) in the preceding solar rotation [from the coronal hole history maintained by Jan Alvestad as part of the Solar Terrestrial Activity Reports at http://www.dxlc.com/solar/coronal_holes.html]. High values of intersatellite-adjusted global hemispheric power, GHPe (second panel), indicate that the streams and associated corotating interaction regions produced strong responses in geospace. In the interval 12 October – 11 November 2003 (days 285-315), the atmospheric effects of some of the most geoeffective high speed streams of the entire year (as indicated by the GHPe values) are examined in detail in the later sections of this paper. Note that, as discussed earlier, GHPe reaches its largest values for streams in the IMF away (toward) sector in the autumn (spring) during solar cycle 23.

TIMED C10.7 and F10.7 are both plotted in the third panel to indicate the daily variability in solar EUV. F10.7 is the observed radio flux at a wavelength of 10.7 cm reaching the Earth at local noon used as a proxy for the Sun's EUV flux. C10.7 is based on the total ionizing energy flux 0-103 nm measured on a daily-average basis by the Solar EUV Experiment (SEE) instrument on the TIMED satellite (described in *Woods et al.* [2005]). This daily-average flux is then re-scaled into "F10.7 units". Plotted in the same panel in plate 4 is the yearly-averaged sunspot number. Large variations in F10.7 (C10.7), equivalent to those over a whole solar cycle occur, at times, during a single solar rotation (e.g., 12-30 October 2003, days 285-303, during which F10.7 varied from 90 to 280). In most cases, the sunspot number drops as the F10.7 (C10.7) drops indicating a decrease in the number of active regions in addition to the appearance of very large coronal holes as the source of this extreme variation in F10.7

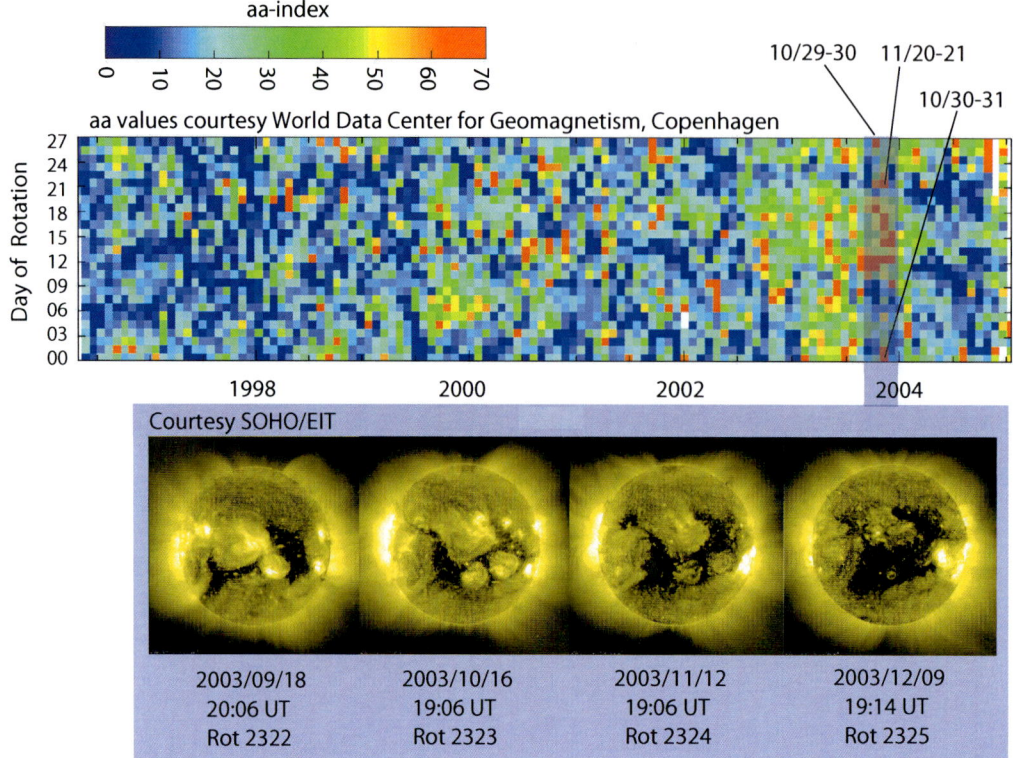

Plate 1. Plot of the daily aa index (a measure of the intensity of magnetic activity) for solar cycle 23 from the World Data Center for Geomagnetism, Copenhagen. Each solar rotation (1-27 days) is displayed as a vertical strip in the plot with UT (and Bartels rotation number) increasing along the horizontal axis. In this format the duration of magnetic activity is indicated by the vertical extent of large values of aa-index in any given solar rotation. The recurrence of magnetic activity appears as elevated aa-index values in the horizontal direction. To place them in context with the recurrent, long-duration high-speed stream activity, aa index values for the 2003/10/29-30, 2003/10/30-31 and the 2003/11/20-21 geomagnetic superstorms are indicated by arrows above. Strong high-speed stream-related magnetic activity recurs over 4 Bartels rotation numbers (highlighted by blue shading). The SOHO EIT images in the shaded box illustrate the large low-latitude coronal holes responsible for this elevated activity.

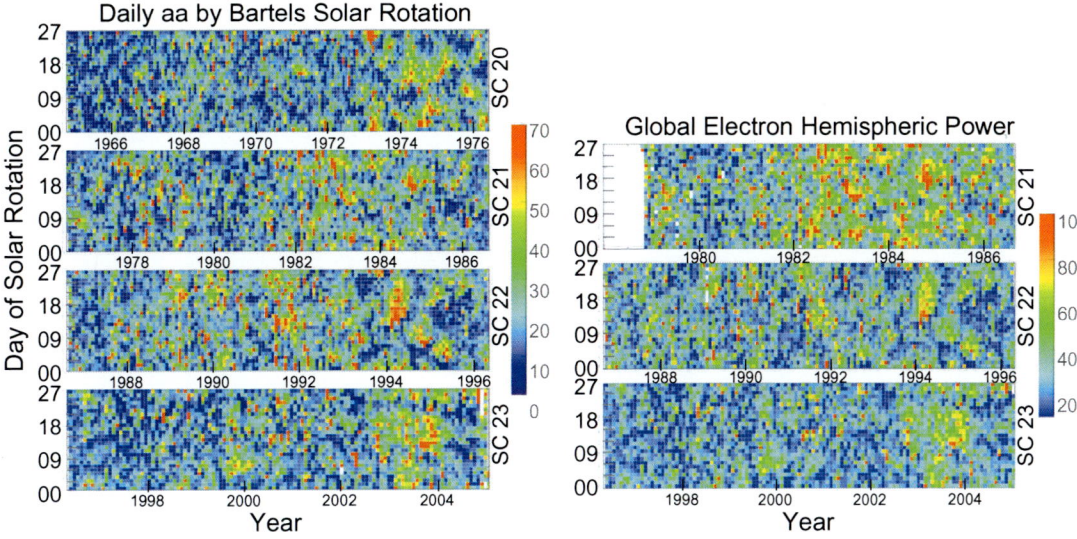

Plate 2. Daily aa values are given on the left for the last 4 solar cycles plotted by Bartels solar rotation number. Daily intersatellite-adjusted global hemispheric power (GW) is plotted back to the start of the data set in November 1978.

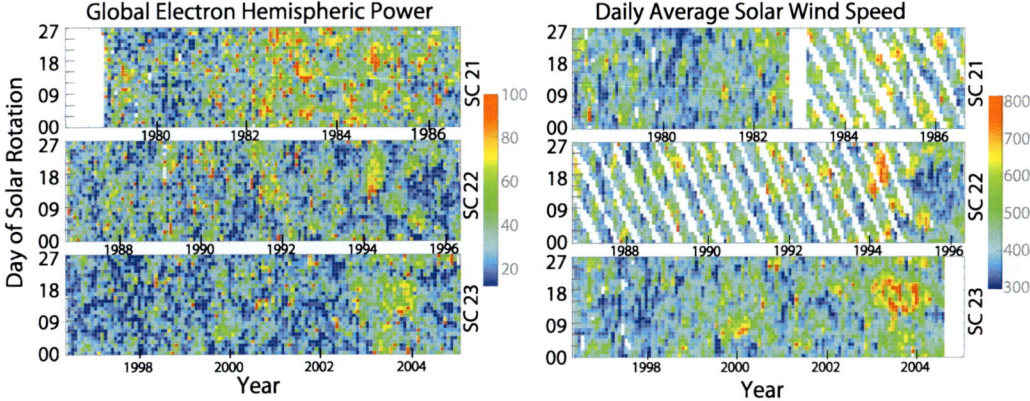

Plate 3. Intersatellite-adjusted global hemispheric power plotted by Bartels solar rotation numbers on the left. Daily-averages of solar wind speed from the OMNI data base plotted in the same format on the right.

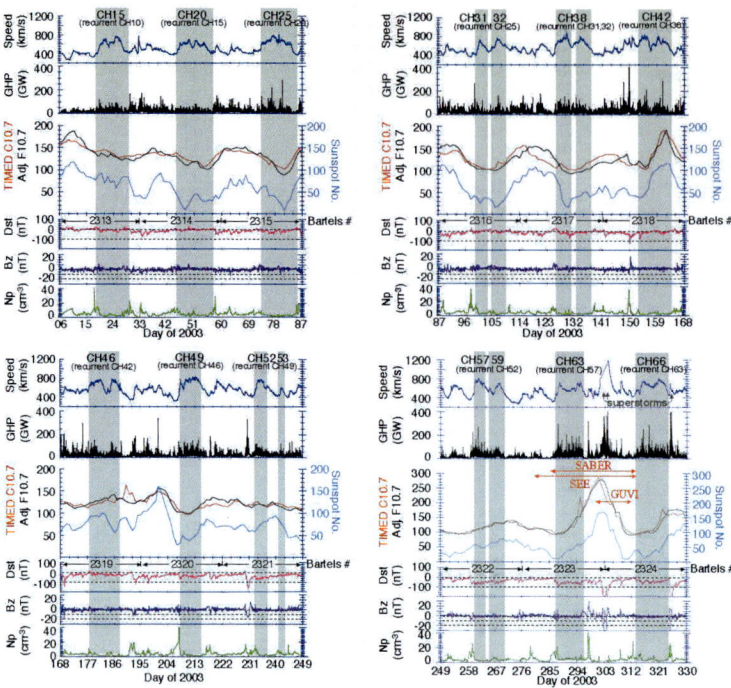

Plate 4. Summary of solar (F10.7, C10.7 and sunspot number), solar wind (speed, density and IMF B_z from ACE) and geophysical parameters (GHPe and Dst) organized by Bartel's Rotation Numbers in 2003. The red lines in panel 3 labeled GUVI, SEE, and SABER highlight the time intervals used to investigate atmospheric responses to the high speed streams in the TIMED data set.

(C10.7). CIRs drive strong activity in the auroral region as indicated by GHPe (panel 2) but they do not trigger large geomagnetic storms (as shown by the Dst-index in panel 4 of Plate 4), which remains greater than –75 nT during the high-speed stream intervals. The minimum Dst generally occurs in association with the CIR at the leading edge of the stream and more modest but long-lived activity follows in association with the high-speed stream itself. The IMF B_z (panel 5 of Plate 4) is moderate and rapidly fluctuating during the high-speed stream intervals. Large, long-lived excursions in IMF B_z are not generally associated with the high speed streams but with interplanetary coronal mass ejections, which drive moderate to strong magnetic storms as indicated by negative excursions in Dst generally exceeding –75 nT (see panel 4 of Plate 4). Finally, solar wind density, given in panel 6 of Plate 4, is enhanced within the corotating interaction region where the high-speed stream is interacting with the slower solar wind ahead of it. It is low within the trailing high-speed stream itself where solar wind speed and density are in general anti-correlated.

3.0 PRECONDITIONING BY DECREASES IN EUV RADIATION

The temperature and density structures that make up the basic large-scale features on the solar disc (i.e., coronal holes, active regions and quiet sun) produce irradiance variations at EUV and X-ray wavelengths [c.f., *Warren et al.*, 2001]. This is because coronal holes are darker and active regions are brighter than the quiet sun at these wavelengths. Prediction of this variation can be made based on a measure of these differences (called emission measure distributions) coupled with estimates of the area covered by each of these features on the solar surface. During solar minimum coronal holes can cover as much as 20% of the solar surface. Exceptional coronal holes in 2003 coupled with active regions on the opposite side of the sun produced variations in EUV during a single solar rotation that were equivalent at times to a full solar-cycle's worth of change. However, in general, there was an ~27-day variation in F10.7 with values ranging between a low of 100-120 and a high of 150-170 giving an ~50% change in F10.7. The decrease in EUV during the passage of these large coronal holes has significant consequences for the atmosphere and ionosphere. Observations from the TIMED satellite of height-integrated or column O/N$_2$ (here labeled ΣO/N$_2$ to distinguish it from [O]/[N$_2$] at a single altitude) and total electron content (TEC) from the CODE database will be examined for systematic variations. Finally modeling of the atmospheric response to the change in EUV input over the interval 7 October – 9 November 2003 (days 280 – 313) will be used to estimate the variations in exospheric temperature. All these features are discussed below.

To investigate the changes in the atmosphere-ionosphere system over a solar rotation, Plate 5 displays observations of ΣO/N$_2$ from the Global Ultraviolet Imager (GUVI) on the TIMED satellite (described in more detail in *Christensen et al.* [2003]) and of TEC from the CODE data base over a 13-day interval within which F10.7 dropped from a high value of 280 to a low value of 90. During this period, the TIMED spacecraft was in the afternoon local time sector moving from 1600 to 1350 LT. Data is displayed for 27 October (day 300) and 8 November (day 312) at the beginning and end of the interval. The magnetic activity on these two dates is relatively low but they bracket an interval of intense magnetic activity triggered by multiple eruptions of an active region. A comparison of these two panels indicates that the atmospheric ΣO/N$_2$ increases and the TEC decreases from the beginning to the end of this interval. Following 8 November (day 312), a high-speed stream encountered the Earth, providing an example wherein the atmosphere was preconditioned by a systematic drop in EUV prior to the arrival of a high-speed stream. This type of background variation in ΣO/N$_2$ and TEC with EUV is seen as a persistent feature of the GUVI and CODE TEC data sets throughout 2003.

Because of the drift in local time of the satellite orbit and the sampling of the TEC data to correspond to the TIMED satellite track, the variations in both TEC and ΣO/N$_2$ contain local time and seasonal contributions in addition to the effects of EUV changes during the solar rotation, which must be separated carefully. For the ΣO/N$_2$ variation, though changes are expected as a result of the solar rotation, it is not clear how the magnitude of these changes compares to contributions due to the local time drift of the satellite orbit and progression of seasons. There are two ways to change the column-integrated composition of the mid- to low-latitude atmosphere: (1) by changing the O or N$_2$ density at the base of the atmospheric column (set at 1×10^{17} cm^{-3} N$_2$ for the GUVI analysis) or (2) by changing the vertical transport across pressure surfaces. Composition changes in the thermosphere are generally thought to be due to the latter process (represented by the divergence term in the mass continuity equation) [*Crowley et al.*, 1989; *Fuller-Rowell and Rees*, 1996; *Burns et al.*, 1991; *Meier et al.*, 2005]. Even though the EUV variation over the interval of 14-30 October 2003 (days 287-303) takes place in roughly half a solar rotation period, the TIMEGCM output indicates there are changes in both the O density at the base of the atmospheric column and in the balance between the EUV-driven and the aurorally driven circulation. There are also indications that seasonal variations in the EUV-driven circulation are not negligible over this interval. Separating all these contributions requires additional simulations.

For the TEC variation, the dominant driver is the change in EUV over the solar rotation. Output from the TIMEGCM

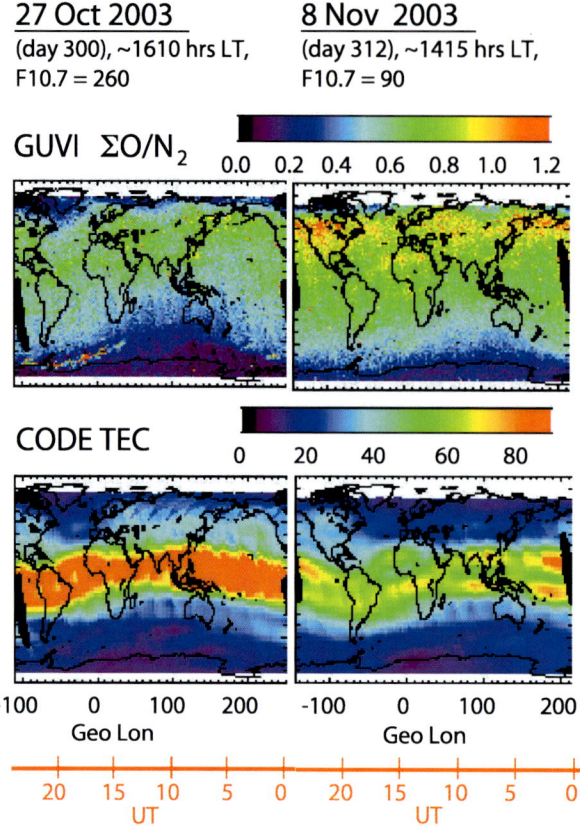

Plate 5. Atmospheric $\Sigma O/N_2$ ratio and TEC from the CODE database is shown for days at the beginning and end of a decrease in F10.7 from 280 to 90 during a 12-day interval from 27 October – 8 November 2003 (days 300-312). Magnetic activity was relatively low on the days shown above but reached extreme levels in the intervening time period (see Plate 4).

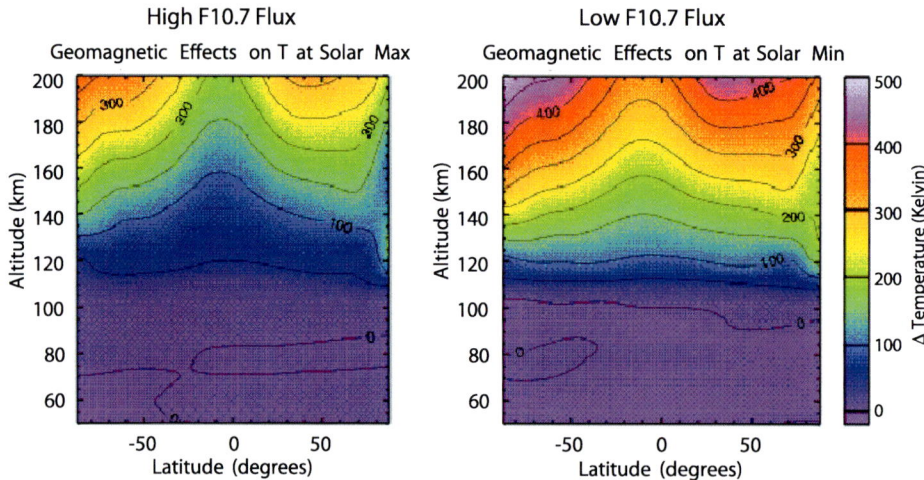

Plate 6. Simulations using the ASPEN-TIMEGCM model of magnetic storms with the same disturbance inputs but under solar maximum, high F10.7 (left) and solar minimum, low F10.7 (right) conditions. The zonal mean temperature difference between the pre-disturbance time (April 14) and the storm (April 19) are shown for each case. Both cases show heating on the order of 100s of degrees Kelvin, but the effects of the storm are predicted to be larger and reach lower altitudes during conditions of low F10.7 characteristic of solar minimum.

[model described in *Roble and Ridley*, 1994; *Crowley et al.*, 1999, 2006] for this interval indicates that the gradient in TEC is negligible over the local time interval covered in Plate 5.

Though it seems intuitive that electron density and thus TEC should vary strongly with EUV flux, the real situation is much more complex due to coupling and feedback within the ionosphere-thermosphere system. A study by *Doherty et al.*, [2000] of TEC variations up to 1100 km altitude showed no systematic variation with daily F10.7. The implication is that other factors overwhelm the F10.7 variation in the TEC below 1100 km. However, CODE TEC observations (shown in Plate 5) use GPS satellites at much higher altitude and thus include a contribution to the TEC from the plasmasphere. It is likely in this case that the variability in TEC is due mostly to this plasmaspheric contribution. *Rich et al.* [2003] propose that the plasmaspheric density should have an F10.7 variation because of the variation with F10.7 in Ni and Te at DMSP altitudes (~880 km) everywhere equatorward of the auroral oval and midlatitude trough in the topside F-layer. When expressed as a percentage of the running 27-day average for a given magnetic latitude (MLAT) and magnetic local time (MLT), the variation is roughly constant for all MLATs and MLTs, increasing with increasing F10.7 during the solar rotation. A plasmaspheric variation with F10.7 has not been experimentally verified. Studies have also failed to show a clear daily variation in Ni and Te at the peak of the F-layer correlated with F10.7 [c.f., *Rich et al.*, 2003]. The source of the observed TEC variation with the solar rotation is another question that will require modeling to resolve.

A numerical experiment was performed using the TIMEGCM model of the thermosphere-ionosphere to simulate the effects of identical magnetic storms under high and low F10.7 conditions using measured magnetic storm inputs on 19 April 2002 and high latitude energy inputs for a reference quiet day on 14 April 2002 prior to the onset of the storm. The zonal mean temperature difference between quiet and storm days for high F10.7 (right) and low F10.7 (left) atmospheric conditions is displayed in Plate 6. The results show that thermospheric composition and temperature changes due to magnetic activity extend to lower latitudes and deeper in altitude in a cool atmosphere typical of low F10.7 conditions compared to a hotter atmosphere typical of high F10.7 conditions. *Burns et al.* [2004] modeled identical magnetic storms in solar minimum (F10.7 = 67, average F10.7 = 72) and solar maximum (F10.7 = 190, F10.7average = 190) using the Thermosphere-Ionosphere Nested Grid (TING) model and arrived at the same conclusion regarding penetration of storm effects. The implication of these studies is that CIRs/high speed streams should produce signatures at systematically lower than expected latitudes and altitudes for extended periods in the declining phase of the solar cycle.

The NCAR Thermosphere-Ionosphere-Electrodynamics General Circulation Model (TIE-GCM) [*Richmond et al.*, 1992] was run from 7 October – 11 November (days 280-315) of 2003, which encompasses the interval in Plate 5. Two different solar EUV inputs were used, one derived from the $F_{10.7}$ variation using the EUVAC proxy model [*Richards et al.*, 1994] and the other using daily TIMED Solar EUV Experiment (SEE) [*Woods et al.*, 2005] observations. The new solar parameterization method of *Solomon and Qian* [2005] was employed to fully account for photoelectron ionization and dissociation effects. This interval includes the last few days before the drop in $F_{10.7}$ to ~90 on 12 October (day 285) (described above) followed by a steady increase in $F_{10.7}$ to ~280 on 29 October (day 302) as the large coronal hole associated with the high speed streams disappeared around the limb and several major active regions rotated into view on the solar disc, and finally to a gradual decrease to $F_{10.7}$ ~90 again as the next coronal hole rotated into view. The TIE-GCM results shown in Plate 7 indicate a decrease in the global mean exospheric temperature to a minimum value of ~900 K in concert with the drop in EUV just prior to the arrival of the high speed stream, and then a gradual rise in exospheric temperature to ~1100 K due both to an increase in EUV and to enhanced auroral activity driven by the CIR/high speed stream interaction with the atmosphere. Finally a steep rise to a double-peaked exospheric temperature maximum of ~1600 K occurs as several intense active regions rotate into view on the solar disc driving $F_{10.7}$ to ~280 and producing two extreme magnetic storms during the intervals 29-30 October (days 302-303) and 30-31 October (days 303-304), respectively, seen in both the Dst-index and GHPe-values. When the strong active regions rotate around the limb and a recurrent coronal hole moves into view on the solar disc, the $F_{10.7}$ drops once again and the exospheric temperature cools to ~900 K just prior to the arrival of another high speed coronal hole stream at Earth on 10 November 2003 (day 314) (see Plate 4).

4.0 NITRIC OXIDE THERMOSTAT

The Sounding of the Atmosphere with Broadband Emission Radiometry (SABER) experiment on TIMED (described in *Russell et al.* [1999]) made global observations of the dramatic increase in the rate of infrared (IR) emission from the thermosphere during geomagnetic storm events in 2002-2004. The response and recovery of the thermosphere to geomagnetic storms has been studied previously [*Maeda et al.*, 1992; *Killeen et al.*, 1997] with the relative importance of radiation and heat conduction assessed by model computations. SABER observed an order of magnitude increase in NO 5.3 micron band emissions within a day of the onset of magnetic storm events. The amount of nitric oxide is

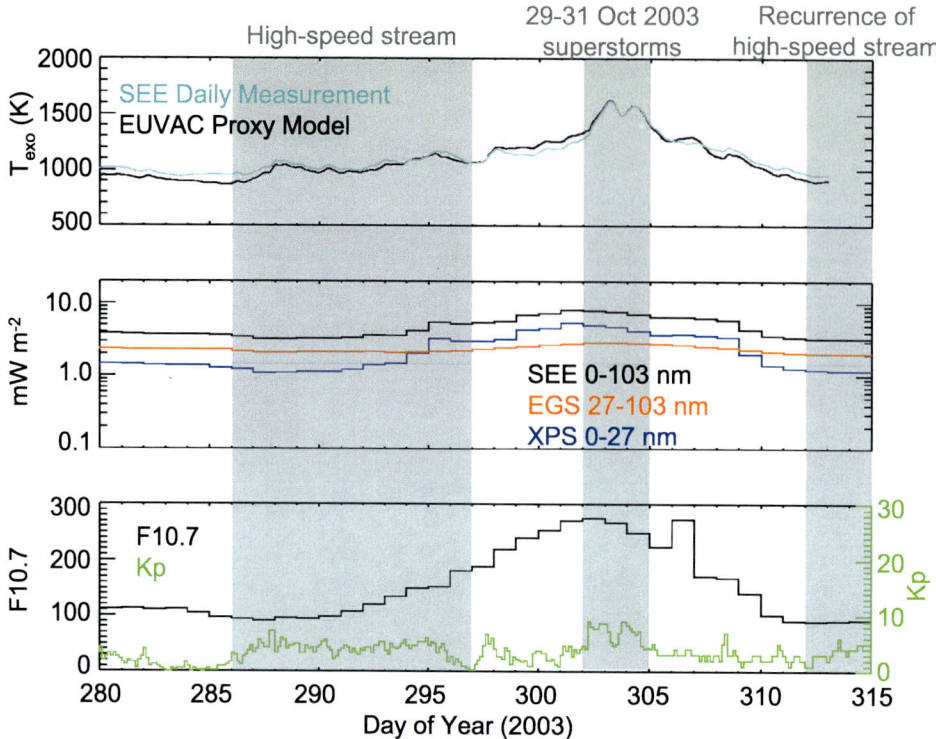

Plate 7. Exospheric temperature calculated by the NCAR TIE-GCM, using TIMED/SEE observations of solar EUV and the EUVAC proxy model as inputs, from 7 October – 11 November 2003 (days 280 to 315). High-speed stream intervals are indicated by grey shading. Top panel: global mean exospheric temperature calculated by the model using the two forms of solar input. Middle panel: SEE daily energy flux measurement integrated over three wavelength intervals, 0-27 nm, corresponding to its X-ray Photodiode System (XPS) channel, 27-103 nm, corresponding to the EUV Grating Spectrometer channel (ionization range only), and their sum. Bottom panel: $F_{10.7}$ and Kp indices during the period. The exospheric temperature reaches a minimum at the time of arrival of a high-speed stream (~13 October, day 286). Throughout the high-speed stream and into the strong magnetic storm activity that follows, the exospheric temperature continues to increase. It returns to low values just prior to the reappearance of the same high-speed stream around 8 November (day 312).

enhanced during solar and magnetic activity due to production by x-ray flare emissions, solar energetic particles and auroral precipitation. The nitric oxide then collides with atomic oxygen. These collisions rapidly convert the oxygen kinetic energy into internal vibrational energy of the NO molecule. This internal energy is either radiated by spontaneous emission from NO, resulting in a cooling of the atmosphere, or it is physically quenched back into the thermal field, resulting in no net change in kinetic temperature. Nitric oxide vibrations may also be excited through impact of particles other than O (such as electrons) but collisions with N_2 and O_2 are not efficient at populating NO vibrational levels.

Mlynczak et al. [2003] found that NO emission acts as a "natural thermostat," allowing the atmosphere to shed storm energy rapidly and recover to pre-storm conditions in as little as three to four days. Emission from the NO molecule is known to be the major radiative cooling mechanism of the thermosphere above about 115 km [*Kockarts*, 1980]. Of the major IR emitters in the thermosphere, NO emission is responsible for 97% of the total observed IR radiation enhancement during the storm periods. SABER observations enable the study of how the energy from the Sun and magnetosphere are converted into chemical potential, thermal, and subsequently IR energy and how promptly the storm or substorm energy is removed and the ionosphere, thermosphere, mesosphere system recovers.

Here SABER observations are applied to the investigation of magnetic activity characteristic of CIR/high speed stream drivers. The energy inputs to geospace from these coronal hole-associated solar wind drivers are quite different from those due to the impulsive solar eruptions. The coronal-hole drivers are in general both more modest and longer-lived (up to 10 days or more compared to the few days time scale associated with solar eruptions such as CMEs). The energy inputs from CIRs/high speed streams are also more confined to high latitudes since they are basically a result of recurrent magnetic substorm activity and are not embedded in major expansions of the auroral oval to low latitudes typical of strong magnetic storms. The NO molecules have longer lifetimes against dissociation by solar photons in the dark polar region. In addition, a given solar rotation may have two high-speed stream intervals each lasting of the order of 10 days (see Plate 4). This indicates that disturbed levels of nitric oxide might actually define the most common state of the atmosphere during years of strong high-speed stream activity such as occurred in 2003.

SABER observations of 5.3 μm volume emission rates (VER) at 125 km altitude in energy per unit volume per unit time over the interval of 12 October – 8 November (day 285-312), in 2003 are displayed in Plate 8. The instrument records about 2000 radiance profiles per day as it continuously scans the limb. These profiles have been combined into a single global view at an approximately fixed local time. All 5.3 μm radiation emitted by NO exits the atmosphere. The white circle poleward of −54 deg in the southern hemisphere indicates the absence of SABER data at these latitudes owing to its view direction during this time period. The interval of 12-24 October 2003 (days 285-297) encompasses CIR/high speed stream activity. During 30 October (day 303) a superstorm occurred and surprisingly the NO VER values did not increase appreciably over the prior-day values. NO volume emission rates, elevated for some 27 days, are beginning to die away on 8 November (day 312) but a new high-speed stream hits on 9 November (day 313) (not shown) creating elevated NO and enhancing 5.3-μm radiative loss again.

NO radiative loss is extremely important for maintaining the temperature structure of the upper atmosphere. The TIMEGCM model was used to simulate identical magnetic storms, one with and one without NO radiative cooling. The results are shown in Plate 9. In the presence of NO cooling, the atmosphere returns to its pre-storm state within a few days. In the absence of NO cooling, the atmosphere is still perturbed 6 days later. Since this is longer than the time between magnetic disturbances, the atmosphere may not return to its quiet state before the next disturbance hits. The implications for the basic state of the thermosphere are significant. How the nearly continuous presence of enhanced NO during years of strong high-speed stream activity effects the atmospheric state is an open question.

5.0 SOURCE OF STRATOSPHERIC NO_X AND OZONE DESTRUCTION

It has long been known that energetic particle precipitation (EPP) produces elevated levels of NO in the thermosphere and mesosphere and that this NO can be transported into the stratosphere in the winter polar region, where NO has a long lifetime against photodissociation [c.f., *Solomon et al.*, 1982]. As pointed out by *Randall et al.* [2005], NO produced in this way acts as a persistent source of NO_x for catalytic ozone destruction; whereas solar particle events are more sporadic and short-lived, producing much smaller amounts of NO_x directly in the stratosphere.

The importance of this downward transport of NO from the winter auroral region was brought home dramatically when unprecedented levels of NO_x were observed in the stratosphere in April 2004 and tentatively attributed to downward transport of NO_x produced in the mesosphere or thermosphere during magnetic superstorms in October and November 2003, five or six months earlier [*Natarajan et al.*, 2004; *Orsolini et al.*, 2005; *Seppala et al.*, 2004]. *Randall et al.* [2005] pointed out the importance of meteorological

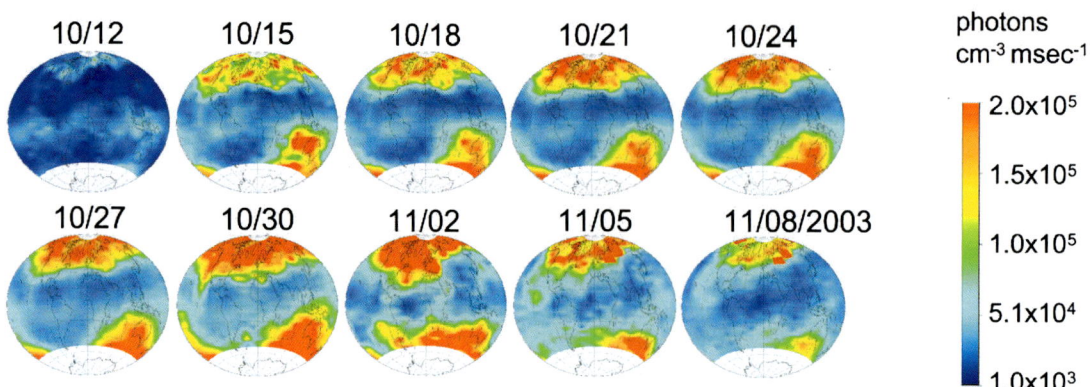

Plate 8. 5.3 micron NO volume emission rate observed by the TIMED SABER instrument. The interval of 12-24 October 2003 (days 285-297) encompasses high-speed stream activity. A superstorm occurred on 30 October (day 303). NO volume emission rate, elevated for some 27 days, is beginning to die away on 8 November (day 312) but a new high-speed stream hits on 9 November (day 313).

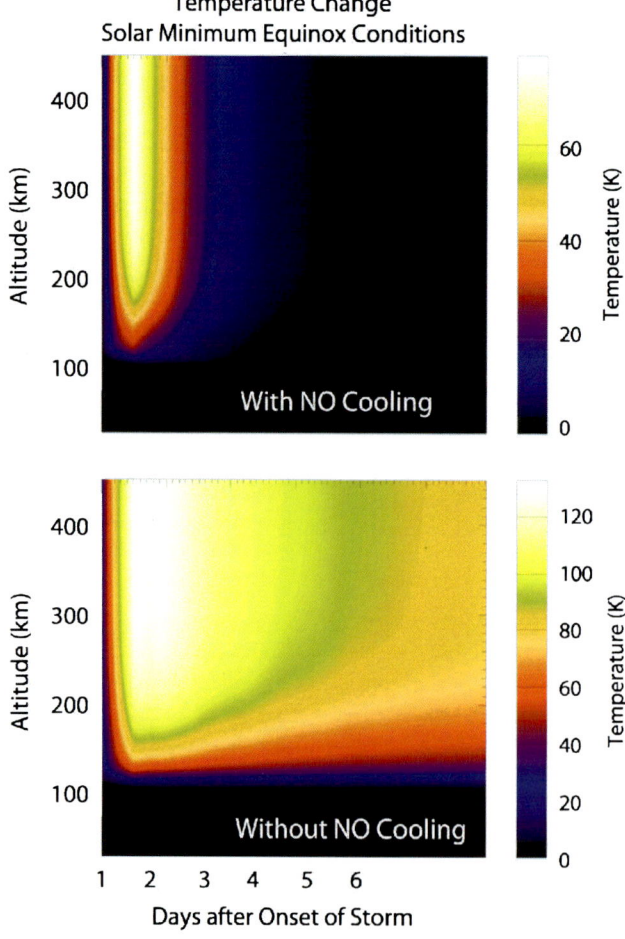

Plate 9. Identical storms in solar minimum equinox conditions, one with and one without NO cooling. With NO cooling the atmosphere returns to pre-storm conditions within a few days. Without NO cooling, the atmosphere is still greatly perturbed more than 6 days later.

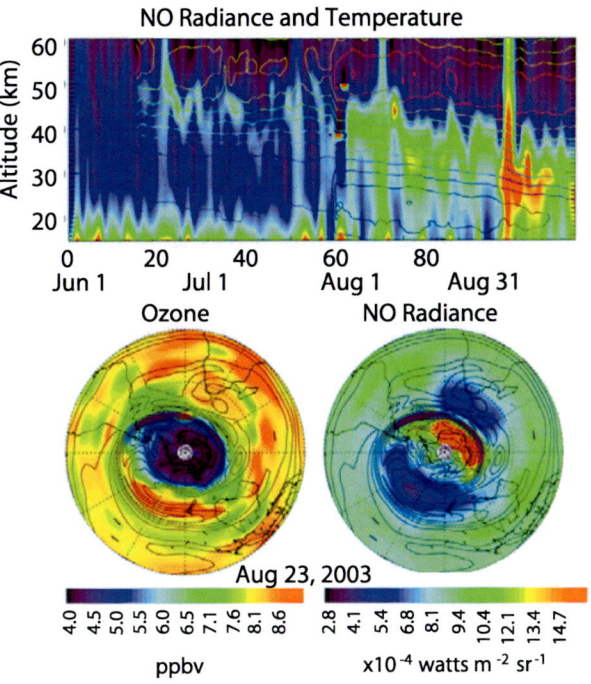

Plate 10. (Top panel) NO radiance (colors, red is high) and temperature (contours) altitude versus time for the period June 1 to August 31, 2003, in the southern hemisphere (45° S – 85° S) derived from TIMED SABER observations. The discontinuity on July 18th occurs because the spacecraft yaws causing SABER to switch from north viewing to south viewing giving a high latitude average covering 45° S to 85° S. Prior to July 18th, the average is over the latitude range 45° S to 55° S. (Bottom panel) NO radiance, O_3 mixing ratio, and potential vorticity contours at 40 km for August 23 from SABER.

conditions to the downward transport. A stable stratospheric polar vortex in the northern hemisphere confined the NO to high latitudes providing an efficient transport path into the stratosphere in early 2004. However, this stable vortex formed late in January 2004 raising the question of whether the source of the April 2004 enhancement might be weaker auroral activity occurring in December 2003 or January 2004 rather than the superstorms in late 2003. This interplay between the meteorology of the stratosphere and the strength and altitude distribution of the aurorally produced NO_x is an interesting refinement to the understanding of chemical coupling between the mesosphere and stratosphere [*Randall et al.*, 1998, 2001; *Siskind et al.*, 1997, 2000].

High-speed stream activity in the descending phase of the solar cycle provides an almost continuous source of energetic aurora particle precipitation and associated NO production. The stratospheric polar vortex in the southern hemisphere is more stable than in the northern hemisphere so the descent of mesospheric NO_x should be very efficient in this region. *Funke et al.* [2005] showed observational evidence of significant NO_x, having been produced by energetic particles in the upper atmosphere, descending in the south polar vortex during 2003. SABER measurements of NO radiance from 45° S to 85° S for the period June 1 to August 31, 2003 (days 152-243) support the *Funke et al.* [2005] results. Plate 10 (upper panel) shows a strong downward shift (nearly 30 km) of the altitude of the NO radiance peak that occurs over the entire season. The discontinuity on July 18 is caused by the yaw of the TIMED satellite but after this date, the complete polar region is included in the sampling. The time before July 18 is included to lengthen the time span and make it easier to see the seasonal trend. NO radiance is strongly dependent on atmospheric temperature and further investigation (not shown) showed that in the 2003 southern hemisphere winter, the stratopause temperature had an extremely weak seasonal trend. Therefore we can conclude that the upward trend of the stratospheric NO radiance magnitude is mostly associated with enhancement of the NO mixing ratio. It therefore can be argued that the tremendous downward shift in the altitude of the NO radiance peak is likely the reflection of downward transport of the NO mixing ratio.

The polar stereographic view of the NO radiance for one day (August 23, 40 km surface) during this time period is shown in Plate 10 (right lower panel). The striking feature is that the high NO radiance is enclosed in the polar vortex region defined by the potential vorticity contours (purple curves). This region also contains low ozone (Plate 10, left lower panel). The persistently strong southern hemisphere polar vortex during June 2003 [*Funke et al.*, 2005] effectively contained the NO_x-rich air and prevented it from being destroyed by sunlight, resulting in its downward transport across the stratopause. Although NO radiance is useful in revealing the stratospheric response to solar EPP events, it must be interpreted carefully because of the strong influence temperature has on the signal. Future work will include studies to determine the feasibility of retrieving stratospheric NO mixing ratio, which is a far easier parameter to interpret when looking for the influence of NO descent from higher altitudes (above 100 km) to the stratosphere.

Using data from the Halogen Occultation Experiment in 1992-2005, *Randall et al.* [2006] calculated the amount of NO_x descending into the southern hemisphere stratosphere after production by energetic particles in the mesosphere or thermosphere. Figure 1 (lower panel) is a plot of the total annual NO_x in gigamoles from that high altitude source. A comparison of panels 1, 2, and 3 in Figure 1 indicates that large amounts of NO_x are transported from the mesosphere into the stratosphere during the years of strong high-speed stream activity (1994 and 2003). There is an interesting and largely unexplored correlation between high-speed stream activity and NO_x abundance in the stratosphere. The variations in NO_x arising from energetic particle precipitation have also been shown to lead to changes in stratospheric ozone distributions [*Randall et al.*, 1998; 2001; 2005]. There is speculation, based on results of a coupled chemistry-climate model that changes in stratospheric ozone and thus the associated absorption of solar radiation in the stratosphere, may result in temperature changes even in the troposphere as a result of disturbances to the dynamical structures in the atmosphere [*Rozanov et al.*, 2005]. If the speculations were confirmed, together with the results described here, they would indicate that high-speed streams are potentially a significant factor in natural forcing of climate changes.

6.0 ATMOSPHERIC DYNAMICS

The Quasi-biennial Oscillation (QBO), a 1.8-2.6 year oscillation of the equatorial mean wind, dominates the low-latitude upper troposphere and stratosphere wind field. It is theorized that the momentum deposition from upward propagating waves generates the QBO and Semi-Annual Oscillation (SAO) in the zonal circulation of the stratosphere and mesosphere, although it is not well understood which waves are responsible and what the involved processes are. These zonal wind oscillations, in turn, are thought to modulate the waves from the lower atmosphere (involved in the formation of the QBO and SAO) as they propagate upwards. SABER has observed a possible modification of the atmospheric tides in the upper mesosphere by the QBO in that the tidal amplitudes are lower in 2003 during the eastward phase of the QBO (phase determined from observations at 30 mb) [*Talaat et al.*, 2005]. However, this picture is complicated by the fact that long-duration, co-rotating, high-speed solar wind streams occurred in 2003. These resulted in prolonged low to mid-level geomagnetic activity, possibly influencing the amplitude of the diurnal tide through related chemical and dynamical changes in the atmosphere. Plate 11 compares the aa-index (bottom panel), a measure of geomagnetic activity with the low-latitude averages of the diurnal tidal amplitudes from the TIMED Doppler Interferometer (TIDI) instrument (described in *Killeen et al.* [1999]) in the altitude range 85 km to 105 km obtained by two different methods (upper and middle panels). The phases of the QBO are marked by red (easterly or westward phase) and black (westerly or eastward phase) bars. In the upper panel, diurnal tidal amplitudes were obtained by differencing ascending and descending node winds at common latitudes near the equator. In the middle panel, the Hough (1,1) fitting of low latitude (within 35 degrees of the equator) winds is used to estimate tidal amplitudes. The two independent techniques provide similar results for the amplitude of the low latitude diurnal tide. The Hough mode fit appears to have more contrast between strong and weak amplitudes and is not as noisy. There is some indication that tidal amplitudes are depressed during the times of maximum high-speed stream inputs.

The daily longitudinally averaged TIDI meridional and zonal winds (shown in Plate 12) were used to obtain more detailed information on the diurnal tides at 90 km from 7 September to 26 November (days 250-330) and search for any link between tidal variations and high-speed solar wind streams in 2003. The upper panel of Plate 12 shows the difference between the ascending and descending nodes of the meridional winds. The ascending and descending nodes differ in local time by ~12 hours. The difference in the nodes is used to eliminate the background and semidiurnal tides in the data and amplify the diurnal tide amplitudes. Times of high-speed stream activity are highlighted in grey and intervals of extreme magnetic storm activity called superstorms were highlighted in red.

The diurnal tide in the meridional winds is usually antisymmetric about the equator and limited within 40° N and 40° S. The meridional winds in the upper panel show that kind of trend. Short periods of enhancements in the diurnal tides are seen after 18 September (day 261) and 11 October (day 284). Around 17 October (day 290), the phase of the diurnal tide in the TIDI data switched (noting the sign change of the meridional winds). That is likely due to the changes in the local time of the TIDI sampling. The enhanced diurnal tide continues until 30 October (day 303) in the northern hemisphere.

The zonal wind data are displayed in the lower panel (Plate 12). The zonal wind diurnal tide is more symmetric about the equator consistent with past observations. Enhancements in the diurnal tide are also seen after 17 September (day 260) and near day 11 October (day 284). After 23 October (day 296), there is an increase of negative winds at the equator, which lasted until 3 November (day 307). There is a transition at 23 October (day 296) from large amplitudes near 20° latitude to large amplitudes near the equator. This may also be due to the local time changes in TIDI sampling.

The data show some associations between changes in the diurnal tides and high-speed stream intervals. Whether such changes in the diurnal tides are related to the solar wind variations is an intriguing question, which needs further study.

With the current dataset spanning 2002-2006, it is impossible to separate high-speed stream effects from the variations in tidal amplitude due to the QBO mechanism. The peak in high-speed stream activity coincides with the westerly phase of the QBO also correlated with depression of tidal amplitudes. Understanding the behavior of the tides is not only crucial to characterizing mesopause variability but also transport in the region. Momentum deposition by the diurnal tide at low latitudes in the lower thermosphere produces indirect circulations that will transport neutral and ionized constituents both vertically and horizontally to higher latitudes. TIMED will only have sampled about two QBO cycles through 2006. This record is insufficient to understand the impact of the QBO on the ionosphere, thermosphere, mesosphere system in a statistically significant fashion and separate QBO effects from those attributable to the high-speed streams.

7.0 CONCLUSIONS

A maximum in high-speed streams produced by low-latitude solar coronal holes and coronal hole extensions occurred in the year 2003 in the declining phase of solar cycle 23. The associated magnetic activity at Earth (indicated by the aa-index) reached a maximum not exceeded in the last 4 solar cycles (20-23). Solar cycle 23 ended a pattern established over the last 150 years of maximum high-speed stream activity in the declining phase of the even-numbered cycle when placed in even-odd pairings starting with solar cycle 10 [*Cliver et al.*, 1996; *Ahluwalia*, 2003].

The recurrent magnetic activity in 2003 provides the best opportunity to date for examining the impacts on the Earth's atmosphere and ionosphere of the strong, long-lived energy inputs from CIRs/high speed streams. The CIRs themselves were exceptional and thus should produce unambiguous signatures in the atmosphere. In the past, sparse data sets have only allowed glimpses of this complex interaction but now with the TIMED spacecraft joining the suite of operating spacecraft that constitute NASA's Sun-Earth Connections Great Observatory and the array of Earth-observing spacecraft currently in orbit around Earth, the potential for new discoveries about the nature of the geospace response to high speed streams is high. The results of this study follow:

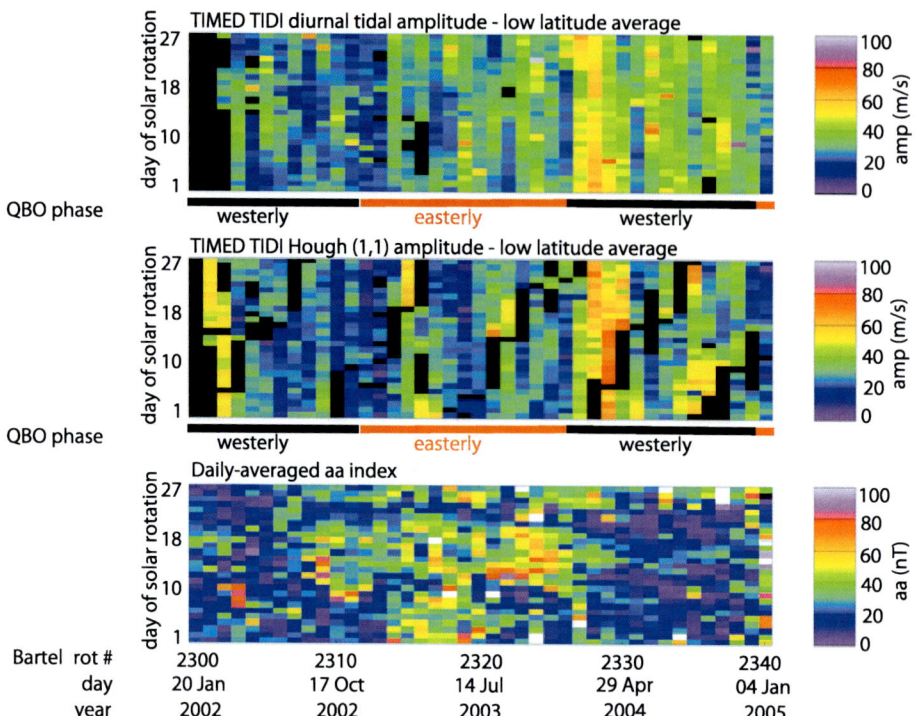

Plate 11. TIDI low-latitude tidal amplitudes determined using two different techniques (described in the text) plotted by Bartels solar rotation numbers (top two panels). Below is a plot of the daily-averaged aa values in the same format for comparison. Phases of the QBO are indicated by black and red bars between panels.

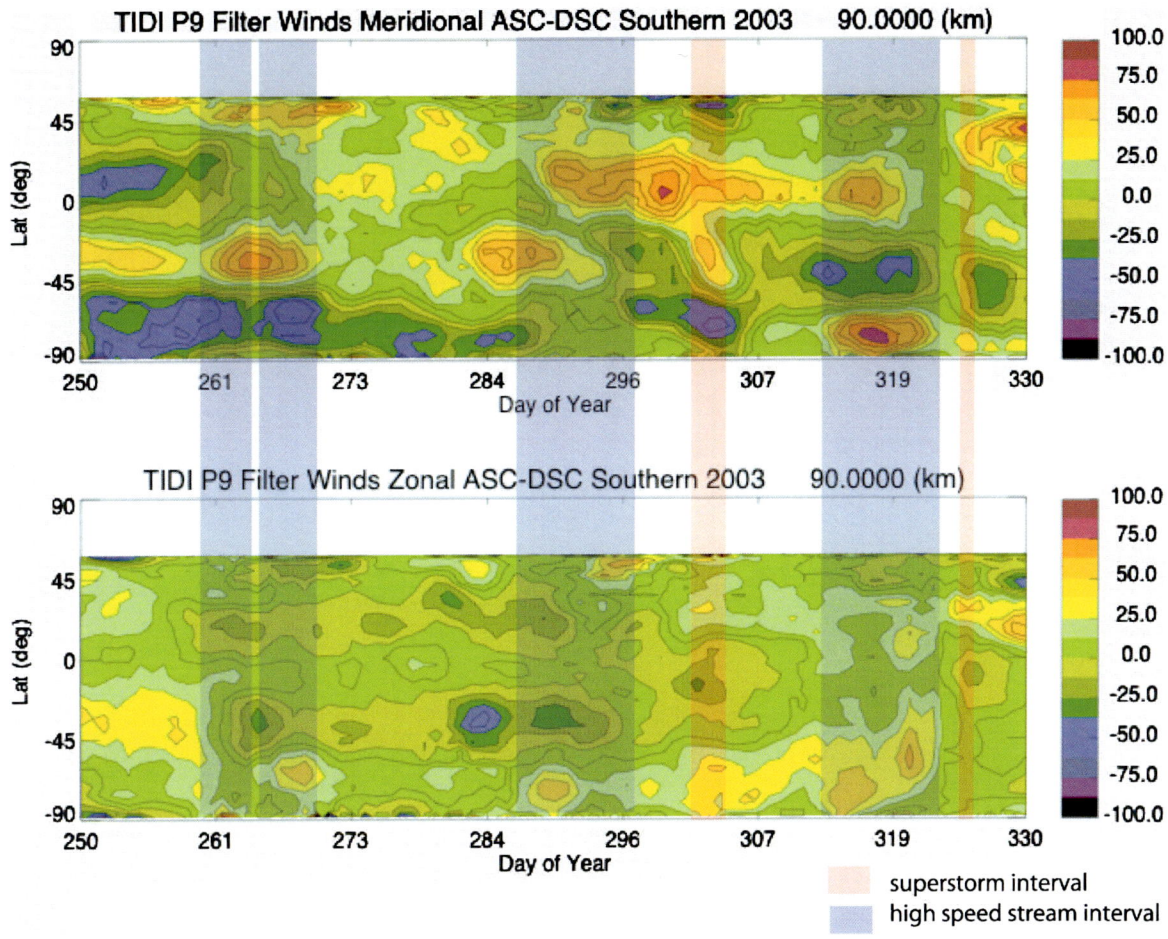

Plate 12. Daily longitudinally averaged TIDI meridional and zonal winds during 7 September – 26 November 2003 (days 250-330). Intervals of recurrent high-speed stream activity are shaded in blue. Intervals of severe magnetic storm activity termed superstorms are shaded in red. Please note that regions of positive wind values are surrounded by whitish contours; regions of negative wind values are not.

1. The correspondence between large values of the *Importance Parameter* (I) for high speed streams (proportional to the duration and the difference in velocity between the stream and the background solar wind) and peaks in the aa-index indicate that these parameters are key to the enhanced geoeffectiveness of the high speed streams in the descending phase of the solar cycle.

2. Prior to the arrival of the high-speed stream and its leading corotating interaction region (CIR) at Earth, the coronal-hole source of the stream (which is darker than the quiet sun in the EUV range) contributes to a large decrease in EUV at Earth. The rotation of active regions behind the solar limb also contributes to the EUV decrease. This cools the upper atmosphere and decreases the total electron content (TEC) seen in the CODE database at mid- and low-latitudes. Though a corresponding variation in $\Sigma O/N_2$ is expected, the magnitude of this variation must be de-convolved from variations due to the local time drift in the TIMED orbit and seasonal variations in the circulation over a 10-day interval. More simulations are needed to resolve this question. This systematic drop in F10.7 before the arrival of a high speed stream is intriguing because the TIMEGCM model indicates that disturbances in composition and temperature due to magnetic activity expand lower in latitude and deeper in altitude for low F10.7 atmospheric conditions.

3. High-speed streams drive recurrent high-latitude auroral activations over ten days or more, with two streams per solar rotation. Nitric oxide is produced by the auroral precipitation. Disturbances in nitric oxide, the dominant IR cooling agent in the upper atmosphere, continue for the duration of the high-speed stream activity and may not fully recover before the next high-speed stream hits. During peaks in high-speed stream activity such as occurred in 2003, disturbed conditions might actually represent the most common state of the upper atmosphere.

4. The persistent source of nitric oxide in the auroral region during high-speed streams, in combination with favorable meteorological conditions in the stratosphere results in strong peaks in energetic-particle-produced (EPP) stratospheric NO_x in years of strong high-speed stream activity. A strong polar vortex provides the favorable meteorological conditions trapping the NO_x at high latitudes where it has a long lifetime against photochemical loss and creating a pathway for efficient transport of NO between the dark polar thermosphere/mesosphere region and the stratosphere. These peaks in NO_x produce corresponding peaks in stratospheric ozone loss suggesting a largely unexplored link between high-speed streams and climate variability.

5. Some evidence exists for disturbed dynamical structures and wave properties in the upper/middle atmosphere during high-speed streams but additional observations and model studies are needed to confirm and then understand these effects.

Acknowledgments. The authors thank the ACE MAG and SWEPAM instrument teams and the ACE Science Center for providing the ACE data. J. Kozyra would like to acknowledge helpful discussion with T. Fuller-Rowell and R. Meier. The authors would like to express appreciation to NASA for funding under grants and contracts as follows: NAG5-5030 (J. Kozyra); NAG5-5049 (W. Skinner, R. Niciejewski, and Q. Wu); NAG5-5334 to NCAR (Q.Wu); NAG5-5335 and NAG5-11410 to NCAR (S. Solomon); NAG5-5028 (S. Palo); NAG5-13613, LWS grant NNG04GN04G, NAG-5001 to the Aerospace Corporation, and NNG0SGE43G to JHUAPL (G. Crowley); and NAS1-97042 (J. Russell III and P.-P. Rong). NCAR is supported by the National Science Foundation. B. Emery would like to acknowledge support under Grant #0208145 from the National Space Weather Program. M. Mlynczak recognizes continued support by the NASA Science Mission Directorate.

REFERENCES

Ahluwalia, H.S., Meandering path to solar activity forecast for cycle 23, *Solar Wind Ten: Proceedings of the Tenth International Solar Wind Conference*, edited by M. Velli, R. Bruno, and F. Malara, American Institute of Physics, 176-179, 2003.

Bromage, B.J.I., D. Alexander, A. Breen, J.R. Clegg, G. DelZanna, C. Deforest, D. Dobrzycka, N. Gopalswamy, B. Thompson and P.K. Browning, Structure of a large low latitude coronal hole, *Solar Phys.*, 193, 181-93, 2000.

Burns, A.G., T.L. Killeen, and R.G. Roble, A simulation of thermospheric composition changes during an impulse storm, *J. Geophys. Res.*, 96, 14153-14167, 1991.

Burns, A.G., T.L. Killeen, W. Wang, R.G. Roble, The solar-cycle-dependent response of the thermosphere to geomagnetic storms, *J. Atmos. Sol. Terr. Phys.*, 66, 1-14, 2004.

Christensen, A.B., et al., Initial observations with the Global Ultraviolet Imager (GUVI) in the NASA TIMED satellite mission, *J. Geophys. Res.*, 108 (A12), 1451, doi:10.1029/2003JA009918, 2003.

Cliver, E.W., V. Boriakoff, and K.H. Bounar, The 22-year cycle of geomagnetic and solar wind activity, *J. Geophys. Res.*, 101, A12, 27091-27109, 1996.

Cliver, E.W., Y. Kamide, and A.G. Ling, Mountains versus valleys: Semiannual variation of geomagnetic activity, *J. Geophys. Res.*, 105 (A2), 2413-2424, 2000.

Cranmer, S.R., Coronal holes and the high speed solar wind, *Space Sci Rev.*, 101, 229-294, 2002.

Crooker, N.U., and E.W. Cliver, Postmodern view of M-regions, *J. Geophys. Res.*, 99, A12, 23383-23390, 1994.

Crowley, G., Emery B.A., Roble R.G., Carlson H.C., and Knipp D.J., Thermospheric dynamics during the equinox transition study I. model simulations for sept 18 and 19, 1984, *J. Geophys. Res.*, vol. 94, 16925, 1989.

Crowley, G., C. Freitas, A. Ridley, D. Winningham, R.G. Roble, and A.D. Richmond, Next Generation Space Weather Specification and Forecasting Model, Proceedings of the Ionospheric Effects Symposium, Alexandria, VA, pp. 34-41, October 1999.

Crowley, G., C. Hackert, R.R. Meier, D.J. Strickland, L.J. Paxton, X. Pi, A. Manucci, A. Christensen, D. Morrison, G. Bust, R.G. Roble, N. Curtis, G. Wene, Global Thermosphere-Ionosphere Response to Onset of November 20, 2003 Magnetic Storm, *J. Geophys. Res.*, submitted April 2006.

Doherty, P.H., J.A. Klobuchar, and J.M. Kunches, Eye on the ionosphere: The correlation between solar 10.7 cm radio flux and ionospheric range delay, GPS Solutions, 3 (4), 75-79, 2000.

Emery, B.A., D.S. Evans, M.S. Greer, K. Kadinsky-Cade, E. Holeman, F.J. Rich, and W. Xu, The low energy auroral electron and ion hemispheric power after NOAA and DMSP intersatellite adjustments, NCAR Scientific and Technical Report, TN-470+STR, 2006. at http://cedarweb.hao.ucar.edu/ instruments/ehp.html).

Evans, D.S., Global statistical patterns of auroral phenomena, Proceedings of the Symposium on Quantitative Modeling of Magnetospheric - Ionospheric Coupling Processes, p325, Kyoto, 1987.

Friedel, R.H.W. and A. Korth, Review of CRRES ring current observations, *Adv. Space Res.*, vol. 20, No. 3, 311-320, 1997.

Fuller-Rowell, T.J., and D. Rees, Numerical simulations of the distribution of atomic oxygen and nitric oxide in the thermosphere and upper mesosphere, *Adv. Space Res.*, vol. 18, No. 9/10, 255-305, 1996.

Funke, B., M. Lopez-Puertas, S. Gil-Lopez, T. von Clarmann, G.P. Stiller, H. Fischer, and S. Kellmann, Downward transport of upper atmospheric NOx into the polar stratosphere and lower mesosphere during the Antarctic 2003 and Arctic 2002/2003 winters, *J. Geophys. Res.*, 110, D24308, doi:10.1029/2005JD006463, 2005.

Gonzalez, W.D., B.T. Tsurutani and A.L. Clúa de Gonzalez, Interplanetary origin of geomagnetic storms, *Space Sci. Rev.*, 88, 529-62, 1999.

Hundhausen, A.J., in *Coronal Holes and High Speed Wind Streams*, edited by J.B. Zirker, Colorado Assoc. Univ. Press, Boulder, 225, 1977.

Kessel, R.L., I.R. Mann, S.F. Fung, D.K. Milling, and N. O'Connell, Correlation of Pc5 wave power inside and outside the magnetosphere during high speed streams, *Ann. Geophys.*, 22, 629-641, 2004.

Killeen, T.L., A.G. Burns, I. Azeem, S. Cochran, and R.G. Roble, A theoretical analysis of the energy budget in the lower thermosphere, *J. Atmos. Sol. Terr. Phys.*, vol. 59, No. 6, 675-689, 1997.

Killeen, Timothy L., Skinner, Wilbert R., Johnson, Roberta M., Edmonson, Charles J., Wu, Qian, Niciejewski, Rick J., Grassl, Heinz J., Gell, David A., Hansen, Peter E., Harvey, Jon D., Kafkalidis, Julie F., TIMED Doppler Interferometer (TIDI), Proceedings of SPIE - The International Society for Optical Engineering, 3756, pp. 289-301, 1999.

Kockarts, G., Nitric oxide cooling in the terrestrial thermosphere, *Geophys. Res. Lett.*, 7, 137-140, 1980.

Luhmann, J.G., Y. Li, C.N. Arge, P.R. Gazis, and R. Ulrich, Solar cycle changes in coronal holes and space weather cycles, *J. Geophys. Res.*, 107, A8, doi:10.1029/2001JA007550, 2002.

Maeda, S., T.J. Fuller-Rowell, and D.S. Evans, Heat budget of the thermosphere and temperature variations during the recovery phase of a geomagnetic storm, *J. Geophys. Res.*, 97 (A10), 14,947-14,957, 1992.

Maravilla, D., A. Lara, J.F. Valdes Galicia, and B. Mendoza, An analysis of polar coronal hole evolution: Relations to other solar phenomena and heliospheric consequences, *Solar Physics*, 203, 27-38, 2001.

Maris, O., and G. Maris, Special features of the high-speed plasma stream cycles, *Adv. Space Res.*, 35, 2129-2140, 2005.

Maris, G., M.D. Popescu, D. Besliu, Solar cycle 23: Forecasts and observations, *Rom. Astron. J.*, 13, 139-142, 2003.

Mayaud, P.N., A hundred year series of geomagnetic data, 1868-1967, indices aa, storm sudden commencements, in IAGA Bull., 33, *Int. Union of Geod. and Geophys.*, Paris, 1973.

Meier, R.R., G. Crowley, D.J. Strickland, A.B. Christensen, L.J. Paxton, D. Morrison, and C.L. Hackert, First look at the 20 November 2003 superstorm with TIMED/GUVI: Comparisons with a thermospheric global circulation model, *J. Geophys. Res.*, vol. 110, A09S41, doi:10.1029/2004JA010990, 2005.

Mlynczak, M.G., et al., The natural thermostat of nitric oxide emission at 5.3 mm in the thermosphere observed during the solar storms of April 2002, *Geophys. Res. Lett.*, 30 (21), 2100, doi:10.1029/2003GL017693, 2003.

Mlynczak, M.G., F. Javier Martin-Torres, G. Crowley, D.P. Kratz, B. Funke, G. Lu, M. Lopez-Puertas, J.M. Russell III, J. Kozyra, C. Mertens, R. Sharma, L. Gordley, R. Picard, J. Winick, and L. Paxton, Energy transport in the thermosphere during the solar storms of April 2002, *J. Geophys. Res.*, 110, A12S25, doi:10.1029/2005JA011141, 2005.

Mursula, K., and B. Zieger, Solar excursion phases during the last 14 solar cycles, *Geophys. Res. Lett.*, 25, 11, 1851-1854, 1998.

Natarajan, M., E.E. Remsberg, L.E. Deaver, and J.M. Russell III, Anomalously high levels of NO_x in the polar upper stratosphere during April, 2004: Photochemical consistency of HALOE observations, *Geophys. Res. Lett.*, 31, L15113, doi:10.1029/2004GL020566, 2004.

Neugebauer, M., R.J. Forsyth, A.B. Galvin, et al., *J. Geophys. Res.*, 103, 14587, 1998.

Nolte, J.T., A.S. Krieger, A.F. Timothy, R.E. Gold, E.C. Roelof, G. Vaiana, A.J. Lazarus, J.D. Sullivan, and P.S. McIntosh, Coronal holes as sources of solar wind, *Solar Phys.*, 46, 303-22, 1976.

O'Brien, T.P., and R.L. McPherron, Seasonal and diurnal variation of Dst dynamics, *J. Geophys. Res.*, 107 (A11), 1341, doi:10.1029/2002JA009435, 2002.

Orsolini, Y., M.L. Santee, G.L. Manney, and C.E. Randall, An upper stratospheric layer of enhanced HNO_3 following exceptional solar flares, *Geophys. Res. Lett.*, 32, L12S01, doi:10.1029/2004GL021588, 2005.

Pizzo, V.J., Interplanetary shocks on the large scale: A retrospective on the last decades' theoretical efforts, in *Collisionless Shocks in the Heliosphere: Reviews of Current Research*, edited by B.T. Tsurutani, and R.G. Stone, 51-68, *Amer. Geophys. Union*, Washington D.C., 1985.

Randall, C.E., D.W. Rusch, R.M. Bevilacqua, K.W. Hoppel, and J.D. Lumpe, Polar Ozone and Aerosol Measurement (POAM) II stratospheric NO_2, 1993-1996, *J. Geophys. Res.*, 103, 28,361-28,371, 1998.

Randall, C.E., D.E. Siskind, and R.M. Bevilacqua, Stratospheric NO_x enhancements in the southern hemisphere vortex in winter/spring of 2000, *Geophys. Res. Lett.*, 28, 2385-2388, 2001.

Randall, C.E., V.L. Harvey, G.L. Manney, Y. Orsolini, M. Codrescu, C. Sioris, S. Brohede, C.S. Haley, L.L. Gordley, J.M. Zawodny, and J.M. Russell III, Stratospheric effects of energetic particle precipitation in 2003-2004, *Geophys. Res. Lett.*, 32, L05802, doi:10.1029/2004GL022003, 2005.

Randall, C.E. et al., Energetic particle precipitation effects on the southern hemisphere stratosphere in 1992-2005, submitted to JGR, 2006.

Rich, F.J., P.J. Sultan, and W.J. Burke, The 27-day variations of plasma densities and temperatures in the topside ionosphere, *J. Geophys. Res.*, 108 (A7), 1297, doi:10.1029/2002JA009731, 2003.

Richards, P.G., J.A. Fennelly, and D.G. Torr, EUVAC: A solar EUV flux model for aeronomic calculations, *J. Geophys. Res.*, 99, 8981, 1994.

Richardson, I.G., E.W. Cliver, and H.V. Cane, Sources of geomagnetic storms for solar minimum and maximum conditions during 1972-2000, *Geophys. Res. Lett.*, 28, 13, 2569-2572, 2001.

Richmond, A.D., E.C. Ridley, and R.G. Roble, A thermosphere/ionosphere general circulation model with coupled electrodynamics, *Geophys. Res. Lett.*, 19, 601, 1992.

Roble, R.G. and Ridley, A., Thermosphere-Ionosphere-Mesosphere-Electro Dynamics General Circulation Model (TIME-GCM): Equinox solar cycle minimum simulations (300-500 km), *Geophys. Res. Lett.*, 22, 417-420, 1994.

Rozanov, E., L. Callis, M. Schlesinger, F. Yang, N. Andronova, and V. Zubov, Atmospheric response to NO_y source due to energetic electron precipitation, *Geophys. Res. Lett.*, 32, L14811, doi:10.1029/1005GL023041, 2005.

Russell, C.T. and R.L. McPherron, Semiannual variation of geomagnetic activity, *J. Geophys. Res.*, 78, 92-108, 1973.

Russell, J.M., M.G. Mlynczak, L.L. Gordley, J. Tansock, and R. Esplin, An overview of the SABER experiment and preliminary calibration results, Proceedings of the SPIE, 44th Annual Meeting, Denver, Colorado, July 18-23, vol. 3756, 277-288, 1999.

Seppala, A., P.T. Verronen, E. Kyrölä, S. Hassinen, L. Backman, A. Hauchecorne, J.L. Bertaux, and D. Fussen, Solar proton events of October–November 2003: Ozone depletion in the Northern Hemisphere polar winter as seen by GOMOS/Envisat, *Geophys. Res. Lett.*, 31, L19107, doi:10.1029/2004GL021042, 2004.

Siskind, D.E., J.T. Bacmeister, M.E. Summers, and J.M. Russell, III, Two-dimensional model calculations of nitric oxide transport in the middle atmosphere and comparison with Halogen Occultation Experiment data, *J. Geophys. Res.*, 102, 3527-3545, 1997.

Siskind, D.E., G.E. Nedoluha, C.E. Randall, M. Fromm, and J.M. Russell, III, An assessment of southern hemisphere stratospheric NO_x enhancements due to transport from the upper atmosphere, *Geophys. Res. Lett.*, 27, 329-332, 2000.

Solomon, S.C., and L. Qian, Solar extreme-ultraviolet irradiance for general circulation models, *J. Geophys. Res.*, 110, A10306, doi:10.1029/2005JA011160, 2005.

Solomon, S., P.J. Crutzen, and R.G. Roble, Photochemical coupling between the thermosphere and the lower atmosphere: 1. Odd nitrogen from 50 to 120 km, *J. Geophys. Res.*, 87, 7206-7220, 1982.

Talaat, E.R., *et al.*, Atmospheric tides from the stratosphere to the lower thermosphere as observed by TIMED. The IAGA Scientific Assembly in Toulouse, 18-29 July 2005, Paper IAGA2005-A-00939, 2005.

Timothy, A.F., A.S. Krieger, and G.S. Vaiana, *Solar Phys.*, 42, 135, 1975.

Tsurutani, B.T., W.D. Gonzalez, A.L.C. Gonzalez, F. Tang, J.K. Arballo, and M. Okada, Interplanetary origin of geomagnetic activity in the declining phase of the solar cycle, *J. Geophys. Res.*, 100, A11, 21717-33, 1995.

Tsurutani, B.T., W.D. Gonzalez, Y. Kamide, C.M. Ho, G.S. Lakhina, J.K. Arballo, R.M. Thorne, J.S. Pickett, and R.A. Howard, The interplanetary causes of magnetic storms, HILDCAAs and viscous interaction, *Phys. Chem Earth* (C), 24, No. 1-3, 93-99, 1999.

Warren, H.P., J.T. Mariska, and J. Lean, A new model of solar EUV irradiance variability. 1. Model formulation, *J. Geophys. Res.*, 106, A8, 15745-15757, 2001.

Woods, T.N., F.G. Eparvier, S.M. Bailey, P.C. Chamberlain, J. Lean, G.J. Rottman, S.C. Solomon, W.K. Tobiska, and D. Woodraska, The Solar EUV Experiment (SEE): Mission overview and first results, *J. Geophys. Res.*, 110, A01312, doi:10.1029/2004JA010765, 2005.